THE JOHN ZINK HAMWORTHY
COMBUSTION HANDBOOK
SECOND EDITION

Volume 2
DESIGN AND OPERATIONS

INDUSTRIAL COMBUSTION SERIES
Series Editors:
Charles E. Baukal, Jr.

The Coen & Hamworthy Combustion Handbook:
Fundamentals for Power, Marine & Industrial Applications
Stephen Londerville and Charles E. Baukal, Jr.

The John Zink Hamworthy Combustion Handbook, Second Edition
Volume 1 — Fundamentals
Volume II — Design and Operations
Volume III — Applications
Charles E. Baukal, Jr.

Industrial Burners Handbook
Charles E. Baukal, Jr.

The John Zink Combustion Handbook
Charles E. Baukal, Jr.

Computational Fluid Dynamics in Industrial Combustion
Charles E. Baukal, Jr., Vladimir Gershtein, and Xianming Jimmy Li

Heat Transfer in Industrial Combustion
Charles E. Baukal, Jr.

Oxygen-Enhanced Combustion
Charles E. Baukal, Jr.

THE JOHN ZINK HAMWORTHY
COMBUSTION HANDBOOK
SECOND EDITION

Volume 2
DESIGN AND OPERATIONS

Edited by
Charles E. Baukal, Jr.

CRC Press
Taylor & Francis Group
Boca Raton London New York

CRC Press is an imprint of the
Taylor & Francis Group, an **informa** business

CRC Press
Taylor & Francis Group
6000 Broken Sound Parkway NW, Suite 300
Boca Raton, FL 33487-2742

© 2013 by Taylor & Francis Group, LLC
CRC Press is an imprint of Taylor & Francis Group, an Informa business

No claim to original U.S. Government works

Printed on Acid-free paper
Version Date: 20130111

Printed and bound in India by Replika Press Pvt. Ltd.

International Standard Book Number-13: 978-1-4398-3964-5 (Hardback)

This book contains information obtained from authentic and highly regarded sources. Reasonable efforts have been made to publish reliable data and information, but the author and publisher cannot assume responsibility for the validity of all materials or the consequences of their use. The authors and publishers have attempted to trace the copyright holders of all material reproduced in this publication and apologize to copyright holders if permission to publish in this form has not been obtained. If any copyright material has not been acknowledged please write and let us know so we may rectify in any future reprint.

Except as permitted under U.S. Copyright Law, no part of this book may be reprinted, reproduced, transmitted, or utilized in any form by any electronic, mechanical, or other means, now known or hereafter invented, including photocopying, microfilming, and recording, or in any information storage or retrieval system, without written permission from the publishers.

For permission to photocopy or use material electronically from this work, please access www.copyright.com (http://www.copyright.com/) or contact the Copyright Clearance Center, Inc. (CCC), 222 Rosewood Drive, Danvers, MA 01923, 978-750-8400. CCC is a not-for-profit organization that provides licenses and registration for a variety of users. For organizations that have been granted a photocopy license by the CCC, a separate system of payment has been arranged.

Trademark Notice: Product or corporate names may be trademarks or registered trademarks, and are used only for identification and explanation without intent to infringe.

Visit the Taylor & Francis Web site at
http://www.taylorandfrancis.com

and the CRC Press Web site at
http://www.crcpress.com

Contents

List of Figures vii
List of Tables xxv
Foreword to the First Edition xxvii
Preface to the First Edition xxix
Preface to the Second Edition xxxi
Acknowledgments xxxiii
Editor xxxv
Contributors xxxvii
Prologue to the First Edition xli

1. **Safety** 1
   Charles E. Baukal, Jr.

2. **Combustion Controls** 45
   Rodney Crockett and Jim Heinlein

3. **Blowers for Combustion Systems** 73
   John Bellovich and Jim Warren

4. **Metallurgy** 95
   Wes Bussman, Garry Mayfield, Jason D. McAdams, Mike Pappe, and Jon Hembree

5. **Refractory for Combustion Systems** 133
   Jim Warren

6. **Burner Design** 151
   Richard T. Waibel, Michael G. Claxton, and Bernd Reese

7. **Combustion Diagnostics** 173
   Wes Bussman, I.-Ping Chung, and Jaime A. Erazo, Jr.

8. **Burner Testing** 191
   Jaime A. Erazo, Jr. and Thomas M. Korb

9. **Flare Testing** 203
   Charles E. Baukal, Jr. and Roger Poe

10. **Thermal Oxidizer Testing** 227
    Bruce C. Johnson and Nathan S. Petersen

11. **Burner Installation and Maintenance** 245
    William Johnson, Mike Pappe, Erwin Platvoet, and Michael G. Claxton

12. **Burner/Heater Operations** 299
    William Johnson, Erwin Platvoet, Mike Pappe, Michael G. Claxton, Richard T. Waibel, and Jason D. McAdams

13. **Burner Troubleshooting** 331
    William Johnson, Erwin Platvoet, Mike Pappe, Michael G. Claxton, and Richard T. Waibel

14. Flare Operations, Maintenance, and Troubleshooting 377
Robert E. Schwartz and Zachary L. Kodesh

15. Thermal Oxidizer Installation and Maintenance 395
Dale Campbell

16. Thermal Oxidizer Operations and Troubleshooting 415
Dale Campbell

Appendix A: Units and Conversions 437

Appendix B: Physical Properties of Materials 447

Appendix C: Properties of Gases and Liquids 463

Appendix D: Properties of Solids 507

Index 513

List of Figures

Figure 1.1	Test heater that has been overpressured	2
Figure 1.2	Fire tetrahedron	5
Figure 1.3	Metal mesh shielding personnel from hot exhaust stack	8
Figure 1.4	Insulated temporary ductwork	8
Figure 1.5	Radiation from a viewport	9
Figure 1.6	Stainless steel fence shielding flow control equipment from thermal radiation during flare testing	9
Figure 1.7	Radiometer for measuring thermal radiation	10
Figure 1.8	Viewport with shutter	10
Figure 1.9	Heat resistant suit	10
Figure 1.10	Quarter wave tube on a propylene vaporizer	11
Figure 1.11	Silencers	12
Figure 1.12	Natural draft burner with no air inlet muffler	12
Figure 1.13	Natural draft burner with an air inlet damper	12
Figure 1.14	Typical muffler for a radiant wall-fired natural draft burner	12
Figure 1.15	Large mufflers on two natural draft burners	13
Figure 1.16	Large mufflers on two radiant wall burners	13
Figure 1.17	Two enclosed flares	14
Figure 1.18	Typical muffler used on a natural draft burner	14
Figure 1.19	Sound pressure versus frequency with and without a muffler	15
Figure 1.20	Cylindrical muffler on a suction pyrometer	15
Figure 1.21	Typical ear plugs	15
Figure 1.22	Typical ear muffs	15
Figure 1.23	Ear muffs designed to be used with hard hats	16
Figure 1.24	Pressurized gas cylinders located outside a building where gases are used in a lab inside the building	17
Figure 1.25	Tube rupture in a fired heater	17
Figure 1.26	Trapped steam in a dead-end that can freeze and cause pipe failure	17
Figure 1.27	CO detector	20
Figure 1.28	Flare stack explosion due to improper purging	21
Figure 1.29	Burner damaged from a flashback	23
Figure 1.30	Premix burner lifting off	23
Figure 1.31	Coen iScan® flame detector	24
Figure 1.32	Typical continuous emission measurement system	25

Figure 1.33	Typical in situ analyzer	25
Figure 1.34	Rubber mat over tripping hazard	28
Figure 1.35	Safety tape around test apparatus	29
Figure 1.36	Furnace camera	29
Figure 1.37	Emergency stop pushbuttons in a control room	29
Figure 1.38	Emergency stop pushbutton next to a sight port on a test furnace	30
Figure 1.39	Liquid fuel containment for diesel storage tanks	30
Figure 1.40	Example of a cabinet used to store flammables	30
Figure 1.41	NO, CO, O_2, and combustible analyzers in a combustion laboratory	31
Figure 1.42	Example of a containment system designed to keep potentially hazardous fluids from leaking onto the ground	31
Figure 1.43	Example of safety signs on the outside of a building	31
Figure 1.44	Fire alarm system in a combustion test facility	32
Figure 1.45	Large portable fire extinguisher	32
Figure 1.46	Emergency pull ring to shut down the entire facility	33
Figure 1.47	Examples of safety and medical kits	33
Figure 1.48	Emergency shower	33
Figure 1.49	Wind sock	34
Figure 1.50	Vapor pressures for light hydrocarbons	36
Figure 1.51	Ethylene oxide plant explosion caused by autoignition	38
Figure 1.52	Photo of a flame arrestor	38
Figure 1.53	Safety documentation feedback flowchart	39
Figure 1.54	Aerial photos of Phillips 66 Incident in Pasadena, TX, on October 23, 1989	40
Figure 2.1	Programmable logic controller	47
Figure 2.2	Touchscreen	48
Figure 2.3	Simplified flow diagram of a standard burner light-off sequence	49
Figure 2.4	Simple analog loop	50
Figure 2.5	Feedforward loop	50
Figure 2.6	Double-block-and-bleed system	51
Figure 2.7	Fail-safe input to PLC	51
Figure 2.8	Shutdown string	52
Figure 2.9	Master fuel trip circuit	53
Figure 2.10	Typical pipe rack	54
Figure 2.11	(a) Large control panel. (b) Small control panel	55
Figure 2.12	Inside the large control panel	55
Figure 2.13	Two different types of pressure switches	56
Figure 2.14	Pneumatic control valve	58

List of Figures

Figure 2.15	Control valve characteristics	58
Figure 2.16	Thermocouple	59
Figure 2.17	Thermowell and thermocouple	59
Figure 2.18	Velocity thermocouple	60
Figure 2.19	Pressure transmitter (left) and pressure gage (right)	60
Figure 2.20	Mechanically linked parallel positioning	62
Figure 2.21	Electronically linked parallel positioning	63
Figure 2.22	A variation of parallel positioning	64
Figure 2.23	Fuel flow rate versus control signal	64
Figure 2.24	Typical butterfly-type valve calculation	65
Figure 2.25	The required shape of the air valve characterizer	65
Figure 2.26	Fully metered control scheme	66
Figure 2.27	Fully metered control scheme with cross limiting	66
Figure 2.28	O_2 trim of airflow rate	67
Figure 2.29	O_2 trim of air SP	68
Figure 2.30	Multiple fuels and O_2 sources	68
Figure 2.31	Controller	69
Figure 2.32	Analog controller with manual reset	69
Figure 2.33	Analog controller with automatic reset	70
Figure 3.1	A centrifugal fan	74
Figure 3.2	Fan wheel designs	74
Figure 3.3	A vane axial fan	75
Figure 3.4	A purge air blower on the side of a combustion chamber	75
Figure 3.5	A multistage high-speed centrifugal blower for a landfill application	75
Figure 3.6	Fan drive arrangements for centrifugal fans AMCA standard 99-2404-03	76
Figure 3.7	An arrangement 4 fan	77
Figure 3.8	Basic centrifugal fan curve	79
Figure 3.9	Basic vane axial fan curve	79
Figure 3.10	Basic centrifugal fan curve with HP	79
Figure 3.11	Forward-tip blade operating curve for 1780 RPM, 70°F, and 0.075 lb/ft^3 density	80
Figure 3.12	Backward-curved blade operating curve for 1780 RPM 70°F and 0.075 lb/ft^3 density	80
Figure 3.13	One primary and one backup fan in the field with ducting	81
Figure 3.14	Six-blade vane axial fan in the field	81
Figure 3.15	Outlet damper effects on fan performance	83
Figure 3.16	Inlet damper effects on fan performance	83
Figure 3.17	Centrifugal fan with inlet and outlet dampers	83
Figure 3.18	Speed change effects on fan performance	84

Figure 3.19	Variable and controlled pitch change effects on fan performance	84
Figure 3.20	Close-up of variable pitch blades on a vane axial fan	84
Figure 3.21	Close-up of a flexible coupling.	85
Figure 3.22	A belt-driven centrifugal blower	86
Figure 3.23	Oil-lubricated bearing with reservoir	86
Figure 3.24	Maintenance of arrangement 8 bearings.	87
Figure 3.25	Fan foundation	87
Figure 3.26	Inlet and outlet expansion joints for vibration isolation of ducting	88
Figure 3.27	Outlet damper fan curve with HP	90
Figure 3.28	Inlet damper fan curve with HP.	91
Figure 3.29	Speed control fan curve with HP.	91
Figure 4.1	Electron microscope image of corroded carbon steel.	97
Figure 4.2	Electron microscope image of chromium oxide layer on SS surface	98
Figure 4.3	Magnified view of SS showing grain boundary where the steel is rich in chromium content.	99
Figure 4.4	Magnified view showing intergranular corrosion.	99
Figure 4.5	Magnified view showing crack propagating across the grain boundary due to stress corrosion cracking	100
Figure 4.6	Magnified view showing removal of the oxide scale from an alloy surface due to thermal cycling	100
Figure 4.7	Magnified view showing the cross section of an alloy suffering from sulfidation attack	101
Figure 4.8	Corrosion of carbon steel process equipment located along the Texas coastline.	102
Figure 4.9	Metal dusting on an inlet tube of a heat exchanger unit	103
Figure 4.10	Charpy impact test apparatus.	104
Figure 4.11	Charpy impact test specimens of carbon steel at −50°F (−10°C)	104
Figure 4.12	Results of Charpy impact tests for high-carbon steel, low-carbon steel, and austenitic SS.	104
Figure 4.13	Normalizing of a forged carbon steel specimen. (a) Magnified view of grain structure before normalizing and (b) after normalizing. The normalizing process involved heating the specimen to 1600°F and then allowing it to cool in still air.	104
Figure 4.14	Process burner schematic.	105
Figure 4.15	Oxide scale formed on the outer surface of process tubes.	109
Figure 4.16	Heater process tubes coated with ceramic	110
Figure 4.17	Rupture process tube after long-term overheating.	111
Figure 4.18	Ruptured process heater tube.	111
Figure 4.19	(a) Process tubes lying on top of a burner. (b) Tubes pulled from the radiant section of the heater	111
Figure 4.20	(a) Failure of a process heater tube. (b) Close-up view of ruptured tube	112
Figure 4.21	Large crack in a process tube used in a SMR.	112
Figure 4.22	External corrosion of a heater process tube	112
Figure 4.23	Various two-phase flow regimes that can occur in process heater tubes	113
Figure 4.24	A vacuum heater tube showing signs of oxidation due to stratified two-phase flow	113

List of Figures

Figure 4.25	Sagging process tube due to high-temperature creep.	113
Figure 4.26	Rupture strength of low-carbon steel and 300-series SS at various temperatures	114
Figure 4.27	Damage to a burner caused by a failure of a support system upstream in the coker reactor	114
Figure 4.28	Damage to a control valve due to sulfidation corrosion	114
Figure 4.29	Damaged oil burner fuel nozzle.	114
Figure 4.30	Orifice spud in service for 10 years.	115
Figure 4.31	Center fuel gas burner tip showing signs of corrosion damage due to high-temperature oxidation.	115
Figure 4.32	(a) Corroded burner tip and (b) new burner tip.	116
Figure 4.33	Photographs of erosion–corrosion damage of a tip from a PSA burner gas.	116
Figure 4.34	Flame retention segments located around the inside perimeter of the flare tip	117
Figure 4.35	Scale and loss of a pilot tip due to sulfidation attack	118
Figure 4.36	(a) External burning on flare tip and (b) pilot that was positioned on the downwind side of the flare tip suffered extreme corrosion damage to the windshield.	119
Figure 4.37	Pilot that was positioned on the downwind side of the flare tip suffered extreme corrosion damage to the windshield.	119
Figure 4.38	(a) Internal burning on an air-assisted flare and (b) failure of an air-assisted flare due to stress corrosion cracking	120
Figure 4.39	(a) Inspection by helicopter of steam leaking from underneath a steam-assisted flare tip and (b) close-up view showing a ruptured lower steam ring.	121
Figure 4.40	Deterioration to the inlet of the lower steam educator tube due to flame impingement for an extended period of time.	121
Figure 4.41	Deterioration of a steam spider due to flame impingement for an extended period of time	122
Figure 4.42	Binary iron–carbon phase diagram.	122
Figure 4.43	Typical TTT diagram for medium-carbon steel.	123
Figure 4.44	Typical TTT diagram for medium-carbon alloy steel AISI 4340.	123
Figure 4.45	Maximum hardness versus % carbon by weight.	124
Figure 4.46	Tensile and bend specimens	124
Figure 4.47	Solidification cracking susceptibility	125
Figure 4.48	Penetrant testing the connection welds of a flare	126
Figure 4.49	Dye penetrant indication example.	126
Figure 4.50	Magnetic particle testing of gas piping using the AC yoke dry method	127
Figure 4.51	(a) Radiographic testing of flare component welds using an iridium 192 radioactive source in a containment vault and (b) radiographic film image of a weld as viewed in a prescribed darkened room	128
Figure 4.52	An ultrasonic, angle-beam apparatus (foreground) with the resulting signal.	129
Figure 4.53	Positive material identification using a handheld analyzer on a process burner riser pipe to confirm the material type	130
Figure 4.54	Metallographic replicas taken from a flare tip during fabrication.	131
Figure 5.1	Example of everyday refractory.	134
Figure 5.2	Several raw materials used in making refractory.	135
Figure 5.3	Drawing representing glue phase.	135

Figure 5.4	Photograph demonstrating glue phase.	135
Figure 5.5	Chemically bonded plastic pieces	136
Figure 5.6	Installation of plastic	136
Figure 5.7	Plastic refractory anchoring system.	136
Figure 5.8	Common firebricks	137
Figure 5.9	Shapes and sizes of standard bricks.	138
Figure 5.10	Example of labor-intensive brick installation.	139
Figure 5.11	Firebricks cut and installed in a circular pattern.	139
Figure 5.12	Use of pogo sticks to ensure brick is properly seated.	139
Figure 5.13	Example of many different types of ceramic refractory.	139
Figure 5.14	Refractory blanket.	139
Figure 5.15	Ceramic refractory anchoring components	140
Figure 5.16	Illustration of anchors penetrating through refractory	140
Figure 5.17	Photograph of installed ceramic refractory.	140
Figure 5.18	Relationship of different refractory physical properties	141
Figure 5.19	Difference between single- and multiple-component lining.	141
Figure 5.20	Several different wire anchors and components.	142
Figure 5.21	Examples of different anchoring	143
Figure 5.22	Typical anchor spacing	143
Figure 5.23	Several different vee-type anchors.	144
Figure 5.24	Example of footed wavy V-anchor installation.	144
Figure 5.25	Anchor welding.	144
Figure 5.26	Extensive use of wavy anchors	144
Figure 5.27	Anchor distance below refractory surface	145
Figure 5.28	Example of steel fibers.	146
Figure 5.29	Example of cold drawn fibers.	146
Figure 5.30	Example of chopped fibers	146
Figure 5.31	Example of curl anchors.	147
Figure 5.32	Example of K-bar anchors.	147
Figure 5.33	Example of S-bar anchors.	147
Figure 5.34	Example of refractory failure	148
Figure 5.35	Prequalifying work.	148
Figure 5.36	Example of vessel having refractory spray installed.	148
Figure 5.37	Example of adequate personnel and equipment.	149
Figure 5.38	Example of poured refractory in a mold	149
Figure 5.39	Cracked refractory	150
Figure 6.1	Graph of sustainable combustion for methane.	153
Figure 6.2	Typical raw gas burner tips	154

List of Figures

Figure 6.3	Typical premix metering orifice spud and air mixer assembly.	154
Figure 6.4	Typical gas fuel capacity curve.	155
Figure 6.5	Internal mix twin fluid atomizer (EA style).	155
Figure 6.6	Internal mix twin fluid atomizer (MEA style).	155
Figure 6.7	Port mix twin fluid atomizer (PM style).	156
Figure 6.8	Typical liquid fuel capacity curve	156
Figure 6.9	Typical throat of a raw gas burner.	157
Figure 6.10	Ledge in the burner tile.	160
Figure 6.11	Flame stabilizer or flame holder	160
Figure 6.12	Swirler.	161
Figure 6.13	Round-shaped flame.	161
Figure 6.14	Flat-shaped flame.	162
Figure 6.15	(a) Regen tile and (b) swirler.	165
Figure 6.16	Combustor schematic.	165
Figure 6.17	View combustor with organ set	166
Figure 6.18	Front view to the organ set of a combustor.	166
Figure 6.19	Axial velocity within combustion chamber (cold air flow model)	167
Figure 6.20	Combustor flame.	167
Figure 6.21	Standard combustor normal—FD combustor short	167
Figure 6.22	Typical conventional raw gas burner.	168
Figure 6.23	Typical premix gas burner.	168
Figure 6.24	Typical round flame combination burner.	169
Figure 6.25	Typical round flame, high-intensity combination burner.	169
Figure 6.26	Typical staged-fuel flat flame burner	169
Figure 6.27	Typical radiant wall burner.	170
Figure 6.28	Downfired burners in a hydrogen reformer furnace	170
Figure 7.1	U-tube manometer	174
Figure 7.2	The Bourdon pressure gauge.	174
Figure 7.3	Internal view of the Bourdon pressure gauge	175
Figure 7.4	Basic components of a Bourdon pressure gauge	175
Figure 7.5	Pressure snubber	176
Figure 7.6	An oil-type deadweight tester.	176
Figure 7.7	Basic components of a deadweight tester	176
Figure 7.8	Orifice plate	177
Figure 7.9	Illustration showing static pressure drop through an orifice metering run	177
Figure 7.10	Common pressure-tap arrangements for orifice metering runs	178
Figure 7.11	Cutaway view of a venturi meter	180

Figure 7.12	Schematic of a venturi meter	180
Figure 7.13	Turbine meter	180
Figure 7.14	NASA satellite image of clouds off the Chilean coast near the Juan Fernandez Islands showing von Karman vortex streets and figure drawn for greater clarity	180
Figure 7.15	Vortex meter	181
Figure 7.16	Principles of magnetic induction	181
Figure 7.17	Time-of-flight ultrasonic flow meter: single-path type	182
Figure 7.18	Time-of-flight ultrasonic flow meter: multipath type	182
Figure 7.19	Thermal mass meter	183
Figure 7.20	Energy balance type thermal mass meter	183
Figure 7.21	Positive displacement meter	183
Figure 7.22	Types of positive displacement meters	183
Figure 7.23	Locations for a Pitot tube traverse in a round or rectangular duct, based on centroids of equal area	184
Figure 7.24	Averaging Pitot tube	185
Figure 7.25	Photograph of an averaging Pitot tube located inside a large duct	185
Figure 7.26	A simple FTIR spectrometer layout	186
Figure 7.27	Photograph of an FTIR system	186
Figure 7.28	Schematic of phase Doppler particle anemometer (PDPA)	187
Figure 7.29	Schematic of oil gun and spray chamber by using PDPA for droplet size measurements	188
Figure 7.30	Typical PDPA droplet size measurements	188
Figure 7.31	PDPA mass accumulation measurements at different mass ratios	188
Figure 7.32	Spray images for liquid planar laser-induced fluorescence (LPLIF) (left) and scattered light (right)	189
Figure 8.1	Aerial view of an industrial combustion testing facility	194
Figure 8.2	Test furnace used primarily for ethylene applications	195
Figure 8.3	Test furnace capable of simulating terrace wall-fired heaters	195
Figure 8.4	Vertical cylindrical furnace for freestanding, upfired burner tests	196
Figure 8.5	Self-contained, portable combustion air heater and blower used for testing forced draft, preheated air burner designs	196
Figure 8.6	Schematic of heat flux probe	197
Figure 8.7	Schematic of heat flux probe mounted in a test furnace	198
Figure 8.8	Schematic of a CO probe mounted in a test furnace	198
Figure 8.9	Test fuel selection process flow diagram	200
Figure 9.1	World-class flare test facility at John Zink Company, LLC, in Tulsa, OK	206
Figure 9.2	An air-assisted flare undergoing testing	206
Figure 9.3	Pressure-assisted flare utilizing the Coanda principle	207
Figure 9.4	Block diagram of flare test facility	208
Figure 9.5	Fuel processing system	208

List of Figures

Figure 9.6	A test in progress as viewed from the control room	209
Figure 9.7	Typical flare test control screen	210
Figure 9.8	Graph of a typical flow control history for manual and automatic operation	211
Figure 9.9	Flare test control center	211
Figure 9.10	(a) Upwind and (b) crosswind views of a flame during a flare test	212
Figure 9.11	Thermogram of a flare flame	212
Figure 9.12	Radiation measurement system (foreground) in use during a flare test	213
Figure 9.13	Isoflux profiles for typical radiation measurement	214
Figure 9.14	Sound measurement system in use during a flare test	214
Figure 9.15	Typical overall sound level as a function of time	215
Figure 9.16	Average sound profile for a given time window	215
Figure 9.17	Flare pilot test stand	216
Figure 9.18	Hydrostatic flare tip test	216
Figure 9.19	Typical ground flare array	217
Figure 9.20	Photo of a row of ground flares, with flame heights compared to those previously measured for single and dual burner tests	218
Figure 9.21	Multiple ground flare burners being tested to determine flame heights and cross lighting distances	218
Figure 9.22	Multiple Indair® flare test to determine minimum operating pressure range and tip spacing for cross lighting	219
Figure 9.23	Air-assisted flare test	220
Figure 9.24	Air flare blower failure test	221
Figure 9.25	Effectiveness of steam in smoke suppression	222
Figure 9.26	Steam-assisted flare test experiencing over-steaming conditions	222
Figure 9.27	Radiation from an offshore flare	223
Figure 9.28	Testing a water-assisted flare	223
Figure 9.29	Radiation reduction by water injection	224
Figure 9.30	Noise reduction by water injection	224
Figure 10.1	Thermal oxidizer test facility	228
Figure 10.2	Schematic of a common configuration of a horizontal thermal oxidizer	229
Figure 10.3	Thermal oxidizer test facility control room	229
Figure 10.4	Cutaway view of a thermal oxidizer test burner capable of both fuel and air staging	231
Figure 10.5	Test facility metering skid	235
Figure 10.6	Commercial test rack with the following components from top to bottom: CO meter, oxygen meter, NOx meter, total hydrocarbons (THC) meter, and gas conditioning system	239
Figure 10.7	Portable combustion gas analyzer	239
Figure 10.8	FTIR gas analyzer	239
Figure 10.9	In situ oxygen analyzer	240

Figure 10.10	Extractive sample probe.	240
Figure 10.11	Orifice plate meter.	241
Figure 10.12	Thermal mass flowmeter	241
Figure 10.13	Ceramic-sheathed thermocouple.	242
Figure 10.14	Handheld radiometer.	242
Figure 10.15	U-tube manometer.	243
Figure 10.16	Inclined manometer.	243
Figure 10.17	Bourdon tube pressure gauge.	243
Figure 11.1	Burner tiles: (a) flat shaped and (b) round shaped	246
Figure 11.2	Burner tile with intricate design features.	246
Figure 11.3	Gas tips with different drill patterns	247
Figure 11.4	Burner mounting sleeve.	248
Figure 11.5	Round burner tile.	249
Figure 11.6	Rectangular or flat flame burner tile.	249
Figure 11.7	Burner tile sections	249
Figure 11.8	Dry "fit" to check dimensions.	249
Figure 11.9	Burner tiles with mortared joints	250
Figure 11.10	Applying mortar to burner tile	250
Figure 11.11	Gas tips protected with masking tape during installation	250
Figure 11.12	Tile with groove-clearing bolt heads.	251
Figure 11.13	Handle burner tile with caution.	251
Figure 11.14	Crane-lifting burner.	252
Figure 11.15	Burner stand and forklift	252
Figure 11.16	Lifting device for burner installation.	253
Figure 11.17	Various lifting techniques.	254
Figure 11.18	Cables and hoist used to mount burners.	255
Figure 11.19	Burner mounted to heater floor.	255
Figure 11.20	Burner designed to be mounted in a common plenum	256
Figure 11.21	P-box-type burner with an integral plenum	256
Figure 11.22	Expansion joint between tile and floor	257
Figure 11.23	Open area around tip for future removal	257
Figure 11.24	Floor is (incorrectly) higher than gas tips	257
Figure 11.25	Gas tips above floor as designed	257
Figure 11.26	Horizontally mounted burner.	257
Figure 11.27	Horizontal burners in service	258
Figure 11.28	Down-fired burners located on top of the heater.	258
Figure 11.29	Down-fired burners in operation	259

List of Figures

Figure 11.30	Horizontally mounted burner.	259
Figure 11.31	Ultralow-NOx radiant wall burners.	260
Figure 11.32	Gas tip/diffuser cone position	260
Figure 11.33	Radiant wall tip location.	261
Figure 11.34	Gas tip location (before mounting in a heater)	261
Figure 11.35	Final burner inspection.	262
Figure 11.36	Dual-blade air dampers.	262
Figure 11.37	Rotary-type air register.	263
Figure 11.38	Jackshaft system to control the combustion air.	263
Figure 11.39	Fuel piping insulated and steam traced	264
Figure 11.40	Burner fuel lines taken from the top of the header to reduce gas tip fouling	264
Figure 11.41	Example pilot conduit boxes.	265
Figure 11.42	Burners with flame scanners.	265
Figure 11.43	Visual inspection of the burner	265
Figure 11.44	Recording operating information.	266
Figure 11.45	Premix and raw gas tips	266
Figure 11.46	Raw gas burner tips.	266
Figure 11.47	Gas tips may look similar, but they are not the same.	267
Figure 11.48	Burner drawings will show number of gas tips and proper orientation.	267
Figure 11.49	Example of gas tip drilling information on burner documentation.	268
Figure 11.50	Overheated gas tip	268
Figure 11.51	Plugged gas tip	268
Figure 11.52	Coke buildup in tip	268
Figure 11.53	Pipe vise for gas tip maintenance.	269
Figure 11.54	Floor-mounted burners in a vertical cylindrical furnace	269
Figure 11.55	Gas riser mounting flange.	270
Figure 11.56	Using steam to blow out the gas riser/tip assembly.	270
Figure 11.57	Thread lubricant.	270
Figure 11.58	Checking ports with correct sized drill bit.	270
Figure 11.59	Using a drill bit to clean a gas port.	271
Figure 11.60	T-handle used to manually clean gas tips.	271
Figure 11.61	Diagram of a HEVD premix burner.	271
Figure 11.62	View of primary air door assembly including fuel orifice	271
Figure 11.63	Single-port orifice spud.	272
Figure 11.64	QD orifice spud	272
Figure 11.65	Plugged QD orifice spud	272
Figure 11.66	JZV premix gas tip	272

Figure 11.67	LPM radiant wall gas tip.	272
Figure 11.68	Burner damaged from flashback.	273
Figure 11.69	Dirty mixer/primary air door	273
Figure 11.70	Spider with severe oxidation from overheating.	273
Figure 11.71	HEVD spider with internal fouling.	274
Figure 11.72	HEVD burner without a primary muffler.	274
Figure 11.73	HEVD burner with muffler installed.	274
Figure 11.74	Primary air muffler removed for inspection.	274
Figure 11.75	Worn gasket and dirty insulation.	275
Figure 11.76	Burner tile in good condition	275
Figure 11.77	Burner tile with large broken pieces.	275
Figure 11.78	Tile with large crack	276
Figure 11.79	Tile crumbling and coming apart.	276
Figure 11.80	Radiant wall burner tile that should be replaced	276
Figure 11.81	Vanadium attack on burner tile.	276
Figure 11.82	Catalyst buildup on regen oil tile.	277
Figure 11.83	Inspection of burner tiles inside a furnace.	277
Figure 11.84	Small cracks can be repaired.	277
Figure 11.85	Tile with large cracks should be replaced.	277
Figure 11.86	Applying a thin layer of mortar.	277
Figure 11.87	Checking gas tip orientation	278
Figure 11.88	Final tile installation.	278
Figure 11.89	Multiple section burner tile	278
Figure 11.90	Diffuser cone used for stabilizing the flame.	278
Figure 11.91	Burner tile ledge used to stabilize the flame.	278
Figure 11.92	Diffuser cones used for flame stability.	279
Figure 11.93	Swirler used to stabilize oil flames.	279
Figure 11.94	Flame deflector ring with stabilizing tabs.	279
Figure 11.95	Diffuser cone in good condition.	279
Figure 11.96	Damaged diffuser cone.	279
Figure 11.97	Diffuser cone location too low.	280
Figure 11.98	Rotating-type air registers	280
Figure 11.99	Rotary air register with E-Z Roll bearings	280
Figure 11.100	Burner damper shown in closed and open position	281
Figure 11.101	Locking air control handle with 18 positions	281
Figure 11.102	Burners with air handles all set at the same position	281
Figure 11.103	Damper linkage for dual-bladed opposed motion design.	281

List of Figures

Figure 11.104	Bearings for smooth damper operation.	282
Figure 11.105	Combination oil and gas LoNOx burner.	282
Figure 11.106	Secondary and primary (regen) tiles.	282
Figure 11.107	Regen tile and one section of secondary tile.	283
Figure 11.108	Regen tile used to stabilize the oil flame.	283
Figure 11.109	Oil and steam spray at the oil tip.	283
Figure 11.110	Regen oil tile with an oil gun in the center	284
Figure 11.111	Typical rotary-type air registers	284
Figure 11.112	Vane-type air register	285
Figure 11.113	Integral plenum box with inlet air damper and muffler.	285
Figure 11.114	Oil gun insert and oil body receiver (with red caps)	286
Figure 11.115	Oil body receiver with copper gaskets for sealing.	286
Figure 11.116	Oil gun bodies	286
Figure 11.117	Oil gun parts.	286
Figure 11.118	EA-/SA-type oil tip.	286
Figure 11.119	Oil tip stamped with "864."	287
Figure 11.120	MEA oil gun parts	287
Figure 11.121	MEA-type oil tip.	287
Figure 11.122	MEA oil gun parts	287
Figure 11.123	HERO oil tip with dual atomizing design	287
Figure 11.124	HERO oil tip, sleeve, and collar.	288
Figure 11.125	HERO oil tips and atomizers.	288
Figure 11.126	HERO oil gun inserts	288
Figure 11.127	Applying high-temperature anti-seize to oil gun threads.	289
Figure 11.128	Atomizer with labyrinth seals and steam ports.	290
Figure 11.129	Checking atomizer location in sleeve	290
Figure 11.130	ST-1S high-stability burner pilot	291
Figure 11.131	Pilot shield glowing in normal operation.	291
Figure 11.132	Pilot parts	292
Figure 11.133	Checking pilot orifice	292
Figure 11.134	Pilot ignition rod assembly.	293
Figure 11.135	Typical relay panel	293
Figure 11.136	Pilot "on" green light illuminated	294
Figure 11.137	Operating pilot with flame rod.	294
Figure 11.138	Pilot with exposed flame rod	294
Figure 11.139	ST-1SE pilot with external igniter and ceramic insulators	295
Figure 11.140	Close-up view of ceramic insulator and ignition rod.	295

Figure 11.141	New-style pilot with no exposed insulators	295
Figure 11.142	New ST-1SE-FR pilot tip	295
Figure 11.143	New ST-1SE-FR pilot with enclosed flame rod	295
Figure 11.144	New pilot with an internal high-energy igniter system	295
Figure 11.145	End view of the internal high-energy exciter	296
Figure 11.146	ST-1SE-FR pilot	296
Figure 11.147	ST-1SE-FR pilots	297
Figure 11.148	Meter showing voltage with power on	297
Figure 11.149	Smart flame scanner mounted on a swivel connector	298
Figure 11.150	Power supply for a smart scanner	298
Figure 12.1	Flames outside the heater	300
Figure 12.2	Flame impingement on process tubes	300
Figure 12.3	Leak in a process tube	301
Figure 12.4	Failure to purge the heater before light-off lead to an explosion	301
Figure 12.5	Heater draft profile	302
Figure 12.6	Draft and O_2 location	302
Figure 12.7	Inclined manometer	303
Figure 12.8	Magnehelic draft gauge	303
Figure 12.9	Draft transmitter	303
Figure 12.10	Arch refractory failed due to positive pressure at the arch	304
Figure 12.11	Excess air indication by oxygen content	304
Figure 12.12	Multiple oxygen analyzer locations	305
Figure 12.13	Large balanced draft cabin heater	305
Figure 12.14	Air leakage around process tube	305
Figure 12.15	Open inspection port allowing tramp air into the heater	306
Figure 12.16	Oxygen probe extending into heater	306
Figure 12.17	Cost of operating with higher excess oxygen levels (natural gas)	307
Figure 12.18	Close-coupled extractive flue gas analyzer	307
Figure 12.19	Flue gas analyzer data panel	308
Figure 12.20	Portable flue gas analyzer	308
Figure 12.21	Fuel pressure gauge	309
Figure 12.22	Maximum and minimum operating pressure	310
Figure 12.23	Fuel pressure guidelines	310
Figure 12.24	Typical burner capacity curves	311
Figure 12.25	Viscosity versus temperature for a range of hydrocarbons	313
Figure 12.26	Velocity thermocouple	314
Figure 12.27	Process tube fouling due to flame impingement	315

List of Figures xxi

Figure 12.28	Process tubes and tube hangers.	315
Figure 12.29	Tube scale due to flame impingement.	315
Figure 12.30	Checking burner gas tip location.	319
Figure 12.31	Typical lower explosion limit (LEL) meter	319
Figure 12.32	Natural-draft furnace adjustments.	323
Figure 12.33	Balanced draft furnace adjustments.	323
Figure 12.34	Uniform flames and refractory color.	325
Figure 12.35	Uneven flame patterns.	326
Figure 12.36	Severe tile damage.	327
Figure 12.37	Open sight door.	327
Figure 12.38	Thermal image of an open explosion door	327
Figure 12.39	Air leaks at sight door and access door.	328
Figure 12.40	Examples of pinched block valves	329
Figure 12.41	Heater casing oxidation	329
Figure 13.1	High-energy portable igniter.	334
Figure 13.2	Portable torch	335
Figure 13.3	Pilot with electric ignition and flame rod	335
Figure 13.4	Pilot with shrouded flame rod.	335
Figure 13.5	Pilot with external flame rod.	336
Figure 13.6	ST-1SE-FR pilot which includes a flame rod	336
Figure 13.7	Test circuit.	336
Figure 13.8	Relay panel	336
Figure 13.9	Flame rod voltage (VAC)/no flame/power on.	337
Figure 13.10	Positive lead/test circuit	337
Figure 13.11	Negative lead/test circuit	337
Figure 13.12	Relay panel/"Flame ON" light energized	337
Figure 13.13	Connecting test probes for flame voltage	338
Figure 13.14	Flame rod voltage (VDC).	338
Figure 13.15	Flame rod current (μA).	338
Figure 13.16	Long flames.	339
Figure 13.17	Fouled gas tip.	340
Figure 13.18	Leaning flames.	341
Figure 13.19	CFD modeling of leaning flames caused by irregular burner spacing.	342
Figure 13.20	Irregular flame pattern	342
Figure 13.21	Nonuniform flame patterns	343
Figure 13.22	Plugged gas tip.	343
Figure 13.23	Eroded gas tip	343

Figure 13.24	Arch brick in burner tile.	343
Figure 13.25	Heater out of air.	344
Figure 13.26	Flames short of air.	345
Figure 13.27	Flame lift-off.	345
Figure 13.28	Flashback inside the burner mixer.	347
Figure 13.29	Flame burning inside gas tip.	348
Figure 13.30	Gas tips glowing due to flashback.	348
Figure 13.31	Flame impingement.	348
Figure 13.32	Tube hangers glowing red.	349
Figure 13.33	Flames short of air and impinging on tubes.	349
Figure 13.34	Tube scale.	349
Figure 13.35	Stages of overheating on process tubes.	350
Figure 13.36	Pinhole leak in tube.	350
Figure 13.37	Tube rupture due to coking.	350
Figure 13.38	Reed walls between burners and tubes.	351
Figure 13.39	Damaged burner tile.	351
Figure 13.40	Flame interaction.	352
Figure 13.41	CFD model shows flames merging as a result of a tight spacing.	352
Figure 13.42	CFD model showing flame interaction.	353
Figure 13.43	CFD model showing flame interaction and recirculation zone.	353
Figure 13.44	Cold firebox operation (floor and walls are darker in color)	354
Figure 13.45	CFD model showing velocity vectors.	354
Figure 13.46	CO versus firebox temperature.	355
Figure 13.47	Plugged fins on convection section tubes.	356
Figure 13.48	Burner capacity curve.	358
Figure 13.49	Identical gas tips with different drillings.	358
Figure 13.50	Fuel pressure gauge.	359
Figure 13.51	Poor mixing due to low fuel pressure.	359
Figure 13.52	Flame lift-off.	360
Figure 13.53	Flame impingement.	360
Figure 13.54	DEEPstar oil burner.	362
Figure 13.55	Combination burner with Regen (oil) tile.	362
Figure 13.56	Oil gun in center of Regen (oil) tile (burner partially assembled).	362
Figure 13.57	Oil spillage on burners.	362
Figure 13.58	Fouled oil guns.	362
Figure 13.59	Flame impingement.	363
Figure 13.60	Coke mound on burner.	363

List of Figures

Figure 13.61	Typical oil gun components	363
Figure 13.62	Vanadium attack on burner tile.	364
Figure 13.63	Catalyst buildup on oil tile.	364
Figure 13.64	Heavy oil system.	365
Figure 13.65	Insulated fuel skid.	366
Figure 13.66	Conventional DBA-style burner.	368
Figure 13.67	Staged air low NOx burner.	369
Figure 13.68	Staged fuel low NOx burner.	369
Figure 13.69	COOLstar™ ultralow NOx burner.	369
Figure 13.70	Air leakage around convection tube.	370
Figure 13.71	Flexible tube seals	370
Figure 13.72	Open inspection port	371
Figure 13.73	Oxygen versus relative humidity%.	372
Figure 13.74	Oxygen versus atmospheric pressure	373
Figure 13.75	Typical fired heater	374
Figure 13.76	Typical CFD model of a duct system.	375
Figure 15.1	Drawing of a typical TO system.	396
Figure 15.2	Typical TO burner.	397
Figure 15.3	Coked-up burner tip.	398
Figure 15.4	Damaged burner tip	398
Figure 15.5	Damaged burner tile	398
Figure 15.6	Damaged burner cone	399
Figure 15.7	Dirty pilot ceramic insulator.	399
Figure 15.8	Typical TO	400
Figure 15.9	Sliding anchor bolt	400
Figure 15.10	Spalled refractory.	401
Figure 15.11	Dirty sight port.	401
Figure 15.12	Example of solids in a TO	401
Figure 15.13	Example of corrosion in the shell	401
Figure 15.14	Typical fire-tube boiler.	402
Figure 15.15	Typical watertube boiler.	402
Figure 15.16	Typical boiler water column and gauge glass.	403
Figure 15.17	Damaged boiler ferrules—before repair	404
Figure 15.18	Boiler ferrules—after repair	404
Figure 15.19	Thermocouples and thermowell.	405
Figure 15.20	Cracked damaged expansion joint.	406
Figure 15.21	Typical quench system.	409

Figure 15.22	Typical heat exchanger.	410
Figure 15.23	Typical liquid seal.	411
Figure 15.24	Typical packed quench absorber tower.	412
Figure 15.25	Typical scrubber.	413
Figure 16.1	Typical gas capacity curve.	419
Figure 16.2	Typical horizontal TO.	420
Figure 16.3	Typical TO refractory.	421
Figure 16.4	Typical watertube boiler.	423
Figure 16.5	Spray gun examples using water with air atomization	429
Figure 16.6	Typical quench system	430
Figure 16.7	Gas to gas exchanger	432
Figure 16.8	Typical liquid seal	433
Figure 16.9	Typical scrubber	434

List of Tables

Table 1.1	Flammability Limits for Common Fuels at Standard Temperature and Pressure	5
Table 1.2	Minimum Ignition Temperatures for Common Fuels at Standard Temperature and Pressure	5
Table 1.3	Laminar Flame Speeds	23
Table 1.4	Flammability and Ignition Characteristics of Liquids and Gases	35
Table 1.5	Ignition Sources of Major Fires	37
Table 1.6	Minimum Ignition Energies	37
Table 1.7	Benefits of a Successful Process Knowledge and Documentation Program	39
Table 2.1	Gas Valve Data	65
Table 2.2	Data for Characterizer	65
Table 3.1	Relative Characteristics of Centrifugal Blowers	75
Table 3.2	Affects of Temperature and Pressure on Volume and Horsepower	77
Table 3.3	Effects of Density on Horsepower	78
Table 3.4	Fan Bearing Vibration Limits	88
Table 3.5	Fan Vibration Diagnostic Clues	88
Table 3.6	Control Options Relative to Design Rate	89
Table 3.7	Potential Controls Cost Savings	89
Table 3.8	Common Fan Problems and Possible Solutions	92
Table 4.1	Classification of Iron–Carbon Alloys	96
Table 4.2	Composition Ranges of a Few Standard Carbon Steel Alloys	97
Table 4.3	Metals Commonly Used in the Process Burner and Flare Industry	99
Table 4.4	Generally Accepted Service Temperatures for Several 300-Series SSs	101
Table 4.5	Estimated Corrosion Rate for Oxidation	107
Table 4.6	Solidification Mode Based on FN	125
Table 5.1	Maximum Use Temperature of Various Grades of Stainless Steel	142
Table 5.2	Standard Spacing for Various Anchors	145
Table 6.1	Theoretical Air Control Opening, Based on No Change in Control Flow Coefficient and 3% Leakage at the Full Closed Position	158
Table 6.2	Theoretical Air Control Opening, Based on No Change in Control Flow Coefficient and 6% Leakage at the Full Closed Position	159
Table 8.1	Example of a Refinery Fuel Gas Composition	199
Table 8.2	Comparison of RFG and Simulated Test Fuel	199
Table 10.1	Incinerability of Several Common Hazardous Air Pollutants	237
Table 10.2	Simulation of Hazardous Vent Stream	237
Table 10.3	Simulator of Exothermic Waste Stream	237

Table 12.1	Some Typical Control Limits	318
Table 12.2	Typical Excess Air Values for Gas Burners	322
Table 12.3	Typical Excess Air Values for Liquid Fuel Firing	323
Table 13.1	Ratio of the Upper and Lower Explosive Limits and Flashback Probability in Premix Burners for Various Fuels	347
Table 13.2	Troubleshooting Gas Burners	361
Table 13.3	Troubleshooting for Oil Burners	367
Table 14.1	Flare Operation	391
Table 14.2	Pilots	393
Table 14.3	Flame Front Generator	393
Table 14.4	Electronic Ignition at or near Pilot Tip	394
Table 14.5	Pilot Verification System	394
Table 16.1	General TO Troubleshooting Guide	422
Table 16.2	U. S. Department of Energy—Federal Energy Management Program, Operations and Maintenance Best Practices—A Guide to Achieving Operational Efficiency	424
Table 16.3	Quench System Troubleshooting	432
Table 16.4	Absorber Scrubber System Troubleshooting	434

Foreword to the First Edition

As we enter the twenty-first century, the importance of energy for industry, transportation, and electricity generation in our daily lives is profound. Combustion of fossil fuels is by far the predominant source of energy today and will likely remain that way for many years to come.

Combustion has played major roles in human civilization, including both practical and mystical ones. Since man discovered how to create fire, we have relied on combustion to perform a variety of tasks. Fire was first used for heating and cooking, and later to manufacture tools and weapons. For all practical purposes, it was not until the onset of the Industrial Revolution in the nineteenth century that man started to harness power from combustion. We have made rapid progress in the application of combustion systems since then, and many industries have come into existence as a direct result of this achievement.

Demands placed on combustion systems change continuously with time and are becoming more stringent. The safety of combustion systems has always been essential, but emphasis on effective heat transfer, temperature uniformity, equipment scale-up, efficiency, controls, and—more recently—environmental emissions and combustion-generated noise has evolved over time. Such demands create tremendous challenges for combustion engineers. These challenges have been successfully met in most applications by combining experience and sound engineering practices with creative and innovative problem-solving.

Understanding combustion requires knowledge of the fundamentals: turbulent mixing, heat transfer, and chemical kinetics. The complex nature of practical combustion systems, combined with the lack of reliable analytical models in the past, encouraged researchers to rely heavily on empirical methods to predict performance and to develop new products. Fortunately, the combustion field has gained considerable scientific knowledge in the last few decades, and such knowledge is now utilized in industry by engineers to evaluate and design combustion systems in a more rigorous manner. This progress is the result of efforts in academia, government laboratories, private labs, and companies like John Zink.

The advent of ever-faster and more powerful computers has had a profound impact on the manner in which engineers model combustion systems. Computational fluid dynamics (CFD) was born from these developments. Combined with validation by experimental techniques, CFD is an essential tool in combustion research, development, analysis, and equipment design.

Today's diagnostic tools and instrumentation—with capabilities unimaginable just a few years ago—allow engineers and scientists to gather detailed information in hostile combustion environments at both microscopic and macroscopic levels. Lasers, spectroscopy, advanced infrared, and ultraviolet camera systems are used to nonintrusively gather quantitative and qualitative information, including combustion temperature, velocity, species concentration, flow visualization, particle size, and loading. Advanced diagnostic systems and instrumentation are being transferred beyond the laboratory to implementation in practical field applications. The information obtained with these systems has considerably advanced our knowledge of combustion equipment and has been an indispensable source of CFD model validation.

Oil refining, chemical processes, and power generation are energy-intensive industries with combustion applications in burners, process heaters, boilers, and cogeneration systems, as well as flares and thermal oxidizers. Combustion for these industries presents unique challenges related to the variety of fuel compositions encountered. Combustion equipment must be flexible to be able to operate in a safe, reliable, efficient, and environmentally responsible manner under a wide array of fuel compositions and conditions.

Combustion is an exciting and intellectually challenging field containing plenty of opportunities to enhance fundamental and practical knowledge that will ultimately lead to the development of new products with improved performance.

This book represents the tireless efforts of many John Zink engineers willing to share their unique knowledge and experience with other combustion engineers, researchers, operators of combustion equipment, and college students. We have tried to include insightful and helpful information on combustion fundamentals, combustion noise, CFD design, experimental techniques, equipment, controls, maintenance, and troubleshooting. We hope our readers will agree that we have done so.

David H. Koch
Executive Vice President
Koch Industries

Preface to the First Edition

Combustion is described as "the rapid oxidation of a fuel resulting in the release of usable heat and production of a visible flame."[1] Combustion is used to generate 90% of the world's power.[2] Regarding the science of combustion, Liñán and Williams wrote the following:

> Although combustion has a long history and great economic and technical importance, its scientific investigation is of relatively recent origin. Combustion science can be defined as the science of exothermic chemical reactions in flows with heat and mass transfer. As such, it involves thermodynamics, chemical kinetics, fluid mechanics, and transport processes. Since the foundations of the second and last of these subjects were not laid until the middle of the nineteenth century, combustion did not emerge as a science until the beginning of the twentieth century.[3]

Chomiak wrote the following: "In spite of their fundamental importance and practical applications, combustion processes are far from being fully understood."[4] In Strahle's opinion, "combustion is a difficult subject, being truly interdisciplinary and requiring the merging of knowledge in several fields."[5] It involves the study of chemistry, kinetics, thermodynamics, electromagnetic radiation, aerodynamics, and fluid mechanics, including multiphase flow and turbulence, heat and mass transfer, and quantum mechanics to name a few. Regarding combustion research,

> The pioneering experiments in combustion research, some 600,000 years ago, were concerned with flame propagation rather than ignition. The initial ignition source was provided by Mother Nature in the form of the electrical discharge plasma of a thunderstorm or as volcanic lava, depending on location. ... Thus, in the beginning, Nature provided an arc-augmented diffusion flame and the first of man's combustion experiments established that the heat of combustion was very much greater than the activation energy—i.e., that quite a small flame on a stick would spontaneously propagate itself into a very large fire, given a sufficient supply of fuel.[6]

In one of the classic books on combustion, Lewis and von Elbe wrote the following:

> Substantial progress has been made in establishing a common understanding of combustion phenomena. However, this process of consolidation of the scientific approach to the subject is not yet complete. Much remains to be done to advance the phenomenological understanding of flame processes so that theoretical correlations and predictions can be made on the basis of secure and realistic models.[7]

Despite the length of time it has been around, despite its importance to man, and despite vast amounts of research, combustion is still far from being completely understood. One of the purposes of this book is to improve that understanding, particularly in industrial combustion applications in the process and power generation industries.

This book is generally organized in two parts. Part I deals with the basic theory of some of the disciplines (combustion, heat transfer, fluid flow, etc.) important for the understanding of any combustion process and consists of Chapters 1 through 13. While these topics have been satisfactorily covered in many combustion textbooks, this book treats them from the context of the process and power generation industries. Part II deals with specific equipment design issues and applications in the process and power generation industries.

References

1. Industrial Heating Equipment Association, *Combustion Technology Manual*, 5th edn. Combustion Division of the Industrial Heating Equipment Association, Arlington, VA, 1994, p. 1.
2. N. Chigier, *Energy, Combustion, and Environment*. McGraw-Hill, New York, 1981, p. ix.
3. A. Liñán and F.A. Williams, *Fundamental Aspects of Combustion*. Oxford University Press, Oxford, U.K., 1993, p. 3.
4. Chomiak, *Combustion: A Study in Theory, Fact and Application*, p. 1.
5. W.C. Strahle, *An Introduction to Combustion*. Gordon & Breach, Langhorne, PA, 1993, p. ix.
6. F.J. Weinberg, The first half-million years of combustion research and today's burning problems, in *Fifteenth Symposium (International) on Combustion*, The Combustion Institute, Pittsburgh, PA, 1974, p. 1.
7. B. Lewis and G. von Elbe, *Combustion, Flames and Explosions of Gases*, 3rd edn. Academic Press, New York, 1987, p. xv.

Preface to the Second Edition

The first edition of *The John Zink Combustion Handbook* was published in 2001. It replaced the previous industry standard book (*Furnace Operations*, 3rd edition, Gulf Publishing, Houston, 1981) written by Dr. Robert Reed, who was the former technical director of the John Zink Company. The first edition of the *Zink Handbook* consisted of 800 oversized pages, was in full color, and was written by 30 authors as compared to *Furnace Operations*, which consisted of 230 pages in black and white and was written by a single author. The first edition of the *Zink Handbook* was a major expansion compared to *Furnace Operations*. The second edition of the *Zink Handbook* is another major expansion compared to the first edition.

The second edition consists of three volumes, collectively about twice as large as the single-volume first edition. Volume I concerns the fundamentals of industrial combustion such as chemistry, fluid flow, and heat transfer. While the basic theory is presented for each topic, the unique treatment compared to standard textbooks is how these topics apply to industrial combustion. Volume II concerns design and operations and includes topics related to equipment used in industrial combustion such as installation, maintenance, and troubleshooting. It also includes an extensive appendix with data relevant to industrial combustion equipment and processes. Volume III concerns applications and covers topics such as process burners, boiler burners, process flares, thermal oxidizers, and vapor control. It shows how the information in volumes I and II is used to design and operate equipment in particular industry applications.

There were several reasons for writing a second edition. The first is the natural improvement in technology with time. For example, the NOx emissions from process burners are lower than ever and continue to decrease with advancements in technology. A second reason for the new edition is to make improvements to the first edition as recommended by readers. One example is to have more property data useful for the design and operation of combustion equipment. A third reason for the new edition is to expand the coverage to include technologies not covered in the first edition such as metallurgy, refractories, blowers, and vapor control equipment.

While these three volumes represent a significant expansion of the first edition, some topics could have been covered in greater detail and some topics have received little if any attention. There is still much to learn on the subject of industrial combustion, which is far more complicated than the average person would ever imagine. This is what makes it such an exciting and dynamic area of technology that has a significant impact on society because it affects nearly every aspect of our lives.

Acknowledgments

The authors would like to collectively thank the John Zink and Coen companies for their help and support during the preparation of these materials. Many colleagues helped with ideas, content, and the preparation of figures and tables. The authors would also like to thank Rick Ketchum, Andrew Walter, Vincent Wong, and Jeffrey Ma for their help in preparing the materials for this book.

Chuck Baukal thanks his wife, Beth, and his daughters, Christine, Caitlyn, and Courtney, for their continued support. He also thanks the good Lord above, without whom this would not have been possible. Wes Bussman thanks his family, Brenda, Sean, and Zach, for their support. He also thanks all of his colleagues at the John Zink Company, LLC (Tulsa, Oklahoma), for their encouragement and for the knowledge they have shared with him throughout his career. Jason McAdams thanks his wife, Heather, and his sons, Ian and Nathan, for their support and for their understanding of the extra time and effort that being in the combustion business sometimes takes. He also thanks all of those who have given their time to help him learn throughout his career. Bill Johnson thanks his father and mother, Harold and Mary Johnson, for giving him a solid foundation and great work ethics. His wife, Judy, deserves a special and loving thank you for allowing him to pursue a career in combustion, which has involved world travel and a lot of time away from home over the years. He also wants to thank his great mentors at John Zink, including Hershel Goodnight, Don Iverson, Charles Summers, Ronnie Williams, Jake Campbell, and Sam Napier. It has always amazed him how a bunch of "cowboys" from Oklahoma built the largest combustion company in the world. A special thanks to Dr. Richard Waibel and Mike Claxton for being his "day-to-day" mentors for many years. Last but not least, thanks to Earl Schnell and Charlene Isley for being supervisors, which allowed him to grow and expand his knowledge and get this knowledge in the best place to benefit the company. Dale Campbell thanks his wife, Jeanne, and his two daughters, Marie and Krista, for their support throughout his career at John Zink. He also thanks the good Lord and the many colleagues at John Zink who have been of great help to him over the years. Mike Pappe thanks his wife, Laresa, for her continued support and acknowledges his coworkers at the John Zink Company. Bruce Johnson especially thanks his wife, Pat, for always encouraging him and helping him with his career. He also thanks his parents, Ruth and Carlyle, for setting an example and for financial support to attend college.

Editor

Charles E. Baukal, Jr., is the director of the John Zink Institute for the John Zink Company, LLC (Tulsa, Oklahoma), where he has been since 1998. He has also been the director of R&D and the director of the R&D Test Center at Zink. He previously worked for 13 years at Air Products and Chemicals, Inc. (Allentown, Pennsylvania) in the areas of oxygen-enhanced combustion and rapid gas quenching in the ferrous and nonferrous metals, minerals, and waste incineration industries. He also worked for Marsden, Inc. (a burner supplier in Pennsauken, New Jersey) for five years in the paper, printing, and textile industries and for Selas Corp. (a burner supplier in Dresher, Pennsylvania) in the metals industry, both of whom make industrial combustion equipment. He has over 30 years of experience in the fields of industrial combustion, pollution control, and heat transfer and has authored more than 100 publications in those areas. Dr. Baukal is an adjunct instructor for Oral Roberts University, the University of Oklahoma, the University of Tulsa, and the University of Utah. He is the author/editor of eight books in the field of industrial combustion, including *Oxygen-Enhanced Combustion* (1998), *Heat Transfer in Industrial Combustion* (2000), *Computational Fluid Dynamics in Industrial Combustion* (2001), *The John Zink Combustion Handbook* (2001), *Industrial Combustion Pollution and Control* (2004), *Handbook of Industrial Burners* (2004), *Heat Transfer from Flame Impingement Normal to a Plane Surface* (2009), and *Industrial Combustion Testing* (2011).

Dr. Baukal has a PhD in mechanical engineering from the University of Pennsylvania (Philadelphia, Pennsylvania), an MBA from the University of Tulsa, is a licensed professional engineer in the state of Pennsylvania, a board-certified environmental engineer (BCEE), and a qualified environmental professional (QEP). He has also served as an expert witness in the field of combustion, has 11 U.S. patents, and is a member of numerous honorary societies and *Who's Who* compilations. Dr. Baukal is a member of the American Society of Mechanical Engineers, the American Society for Engineering Education (ASEE), and the Combustion Institute. He serves on several advisory boards, holds offices in the Combustion Institute and ASEE, and is a reviewer for combustion, heat transfer, environmental, and energy journals.

Contributors

John Bellovich is the manager of the Combustion Rental Group at the John Zink Company, LLC (Tulsa, Oklahoma). He received his BSME from the University of Tulsa and has more than over 20 years of experience in the industrial combustion industry. He has written or co-written three published articles.

Wes Bussman, PhD, is a senior research and development engineer for the John Zink Company, LLC (Tulsa, Oklahoma). He received his PhD in mechanical engineering from the University of Tulsa. He has over 20 years of experience in basic scientific research work, industrial technology research and development, and combustion design engineering. He holds ten patents and has authored several published articles and conference papers. He has also been a contributing author to several combustion-related books. He has taught engineering courses at several universities and is a member of the Kappa Mu Epsilon Mathematical Society and Sigma Xi Research Society.

Dale Campbell, PE, is a senior design engineer in the John Zink Thermal Oxidizer Systems Group. He received his BS in chemical engineering from the University of Tulsa and has worked for the John Zink Company, LLC (Tulsa, Oklahoma), since 1973. His primary responsibility has been the detailed design, equipment application, startup, and project management of waste incinerator systems of varying complexity. He serves as a primary resource for incinerator troubleshooting and design in the thermal oxidizer aftermarket group. He is also a registered professional engineer in the state of Oklahoma.

I.-Ping Chung, PhD, is a senior development engineer in the Technology and Commercial Development Group at the John Zink Company, LLC (Tulsa, Oklahoma). She has worked in the field of industrial combustion and equipment, fluid dynamics, atomization and sprays, spray combustion, and laser diagnosis in combustion and has a PhD in mechanical and aerospace engineering. She has 24 publications and holds 9 patents. She is a registered professional engineer of mechanical engineering in California and Iowa.

Michael G. Claxton is a senior principal engineer in the Burner Process Group of the John Zink Company, LLC (Tulsa, Oklahoma). He received his BS in mechanical engineering from the University of Tulsa and has worked for the John Zink Company, LLC (Tulsa, Oklahoma), in the field of industrial burners and combustion equipment since 1974. He has coauthored a number of papers and presentations covering combustion, combustion equipment, and combustion-generated emissions and is co-holder of several combustion-related patents.

Rodney Crockett is the manager of Controls Engineering for the Flare Systems Group at the John Zink Company, LLC (Tulsa, Oklahoma). He has a BS in electronic engineering technology from DeVry Institute of Technology and has more than 20 years of experience in the industrial combustion industry including flares, incinerators, boiler burners, and vapor control. He has been with John Zink since 2001 serving in several roles including senior controls engineer, lead engineer, and manager of controls for both the boiler burner and flares divisions.

Jaime A. Erazo, Jr., is a process engineer for the Process Burner Group of the John Zink Company, LLC (Tulsa, Oklahoma). He joined John Zink in 2008 after completing his BS in chemical engineering and MS in mechanical engineering at the University of Oklahoma (OU). While at OU, he was a research assistant in the Combustion and Flame Dynamics Laboratory, where his work focused on alternative fuels and spray flame combustion. He has authored several technical papers in the field of combustion.

Jim Heinlein, PE, is a senior controls engineer at the John Zink Company, LLC (Tulsa, Oklahoma). He received his BSEE (honors) from the University of Tulsa and is a registered professional engineer. He is the controls instructor at the John Zink Thermal Oxidizer School and is a member of ISA, Tau Beta Pi, and Eta Kappa Nu. A U.S. Navy veteran, he qualified as chief reactor watch and naval surface warfare expert. He has also worked extensively in the fields of nuclear engineering, computer systems design, and low observables engineering. He has been involved in controls engineering at John Zink since 1991 and has designed several hundred control systems currently operating worldwide.

Jon Hembree is a welding engineer at the John Zink Company, LLC (Tulsa, Oklahoma). He received his BS in mechanical engineering from the University of Tulsa and is currently working on his MS in mechanical engineering. His MS research is about the influence of surface roughness on endurance strength. His primary responsibility is the application of welding processes and procedures.

Bruce C. Johnson, MSc, PE, was the technology development leader for the Thermal Oxidation Systems Group at the John Zink Company, LLC (Tulsa, Oklahoma). He received his MSc in chemical engineering from the University of North Dakota under a U.S. Bureau of Mines research fellowship. He is a registered professional engineer in Oklahoma. He has spent much of his career in research and development and has been employed by the Calgon Corporation, Department of Energy, Combustion Engineering Company, and several thermal oxidation companies. His qualifications include process design, product development, testing, and project management of governmental and corporate research and development groups. He has six patents and has authored numerous technical reports and papers.

William (Bill) Johnson started in the industry in 1969 with Stanley Consultants in Muscatine, Iowa, where he helped design coal-fired steam-generating units up to 650 MW. He moved to Oklahoma in 1976 and joined Williams Brothers Engineering. At Williams Brothers, he was involved in oil and gas transmission and offshore unloading facilities. He then joined John Zink in November 1977 and worked as a projector in the Burner Group.

Zachary L. Kodesh is the technology manager in the Flare Systems Division of the John Zink Company, LLC (Tulsa, Oklahoma). He received his BS and MS in mechanical engineering from Oklahoma State University and has worked for John Zink since 1990. He is a registered professional engineer, a certified functional safety expert, a member of ASME, and is an active participant in the American Petroleum Institute.

Thomas M. Korb, PhD, PE, is the director of research and development at John Zink Co. LLC (Tulsa, Oklahoma), where he is responsible for global research and development, computational fluid dynamics, and testing facilities. Dr. Korb has over 18 years of experience in combustion and thermal sciences. His work has included designing and testing of combustion equipment for the refining and petrochemical industries as well as failure analysis engineering of accidental fires and explosions. He has also worked in the development of both gas turbine and diesel engines. He is a registered professional engineer and a member of Tau Beta Pi National Engineering Honor Society and the American Society of Mechanical Engineers. He is also actively involved in the American Flame Research Committee. He received his PhD in mechanical engineering from Arizona State University (Tempe, Arizona). His research focused on fundamental ignition mechanisms with a particular emphasis on hot surface ignition of hydrocarbon fuels and the impact of hot surface material and surface oxide structure.

Garry Mayfield is the quality director, John Zink Company, LLC (Tulsa, Oklahoma). He is responsible for the maintenance and execution of John Zink–held ASME U and S authorizations, National Board NB and R authorizations, UL listings, ISO 9001 certification, and ASME IX welding and ASME V/SNT TC-1A nondestructive testing programs. He holds AWS CWI and NACE Level III certifications and is lead auditor of ISO 9001 Quality Management System. He previously held certifications in nondestructive testing methods, radiographic, ultrasonic, liquid penetrant, and magnetic particle. He completes ongoing education and training provided by many sources targeting quality control process deployment to provide reliability solutions and ensure conformance of new energy industry equipment.

Jason D. McAdams, PE, is the director of applications engineering for the Process Burner Group of the John Zink Company, LLC (Tulsa, Oklahoma). He has been with John Zink since 1999, serving in several roles including product development, design engineering, project management, technical sales, and management. Prior to this, he worked for more than five years at the Koch Refining Company, where he gained experience with the application of combustion equipment as a project engineer and reliability engineer. He received his BS in mechanical engineering from the University of Tulsa and has authored multiple publications in journals and conference proceedings. McAdams is a registered professional engineer in the state of Oklahoma.

Mike Pappe works for the John Zink Aftermarket Group and has over 30 years of field-related experience in the refining industry. Mike's career began at Kerr-McGee Chemical Corporation in 1973, where he completed an electrical journeyman apprenticeship. He then moved on to Westinghouse, where he was a field instrument and control technician, calibrating temperature controls for the metals industry in Los Angeles. He worked as instrument and control technician at several oil refineries in Los Angeles before moving on to CF Braun in Alhambra, California, where he maintained control systems for the pilot plants at the Heat Transfer Research Institute and the Fractionation Research Institute on the CF Braun campus. In 1990, Mike joined John Zink as a field service technician and currently holds a position of End User Sales and Service in the West Coast of North America.

Nathan S. Petersen PE, currently works as a process engineer at the John Zink Company, LLC (Tulsa, Oklahoma), where he has been since 2005. He has served in various engineering roles in the Process Burner Group, Flare Group, and Thermal Oxidizer Group, consisting of process and mechanical design and equipment testing. He received his BS in chemistry and chemical engineering along with an MS in chemical engineering from the University of Utah (Salt Lake City, Utah). He is also a licensed professional engineer in the state of Oklahoma.

Erwin Platvoet is director of process burner engineering at the John Zink Company, LLC (Tulsa, Oklahoma). He has over 20 years of experience in CFD and heat transfer engineering for various companies in Europe and the United States. He received his MS in chemical engineering from Twente University of Technology in the Netherlands and currently holds six patents.

Roger Poe is a research associate at the John Zink Company, LLC (Tulsa, Oklahoma), where he has worked since 1999. He received his BS in mechanical engineering from Fairmont State University (Fairmont, West Virginia). Prior to this, he served as manager of Callidus Technologies (Tulsa, Oklahoma) Test and Research Center from 1995 to 1999, where he was responsible for the design and testing of specialty burners as well as the development of new burner equipment for the refinery and petrochemical industry. From 1989 to 1995, he managed the facilities and personnel for Penn State University Energy and Fuels Research Center (State College, Pennsylvania). He served as a working manager and researcher for the Donlee Technologies Research and Development Group (York, Pennsylvania) from 1985 to 1989, where he was involved with low NOx boiler burner technologies, as they relate to both liquid and gaseous fuels, coal gasification in pilot scale, fluidized bed reactors, and the development and testing of fluidized bed combustion units. At Penn State University, he conducted further research on fluidized bed combustion, coal gasification, micronized and pulverized coal applications, coal slurry formulation and combustion, as well as low NOx gas and oil development. Most recently, he has focused his work on low NOx process–type burners and large-scale flaring equipment while working at John Zink and Callidus Technologies. He has published more than 24 articles and has been named on over 67 patents relating to burners, flares, and pilots. His research interests include fluid mechanics, combustion, thermodynamics, combustion testing, and manufacturing. During the course of his career, he has jointly worked with the Department of Energy, the Department of Defense, NORAD, the Institute of Gas Technology, the Gas Research Institute, Sandia National Labs, and the Natick Naval Test Labs.

Bernd Reese is leader of process technology at John Zink KEU GmbH, Krefeld, Germany. He has worked in the field of burner design and process technology for over 25 years. From these he spent 15 years in the research department and test center of KEU, focusing on new burner development and the improvement of existing designs. He and his team improved the KEU Denox-Technology. He has coauthored numerous technical articles and papers.

Robert E. Schwartz (retired) PE, is a senior technical specialist at the John Zink Company, LLC (Tulsa, Oklahoma). He has worked in the fields of combustion, flares, pressure relieving systems, fluid flow, and heat transfer for nearly 50 years, which include 42 years with John Zink, where he has provided technical and business leadership in all product areas and has extensive international experience. He has 51 U.S. patents for inventions of apparatus and methods that are in use throughout John Zink. He is also the associate editor of *The John Zink Combustion Handbook*. His areas of technical expertise include development, design, fabrication, and operation of combustion equipment, including flares, incinerators, process burners, boiler burners, and vapor control; reduction of NOx and other emissions from combustion processes; fluid flow and heat transfer in process and combustion equipment; noise elimination and control; vapor emissions control using recovery processes; hazardous waste site remediation; and permission and operation of hazardous waste storage and disposal sites. His professional organizations and awards include member of the American Society of Mechanical Engineers, the American Institute of Chemical Engineers and Sigma Xi, the Scientific Research Society, Registered Professional Engineer in the state of Oklahoma, recipient of the University of Missouri Honor Award for Distinguished Service in Engineering, and election to The University of Tulsa Engineering Hall of Fame. He received his BS and MS in mechanical engineering from the University of Missouri.

Richard T. ("Dick") Waibel (deceased), PhD, was a senior principal engineer in the Burner Process Engineering Group at the John Zink Company, LLC (Tulsa, Oklahoma). He worked in the field of burner design and development and earned a doctorate in fuel science from Pennsylvania State University. He published over 70 technical papers, publications, and presentations. Dr. Waibel was the chairman of the American Flame Research Committee for many years starting in 1995.

Jim Warren is the manager, Mechanical Engineering Group, for the Thermal Oxidizer/Flare Division at the John Zink Company, LLC (Tulsa, Oklahoma). He received his BSME from the University of Tulsa and has over 23 years of experience with John Zink. His area of expertise is refractory and rotating equipment. He holds API-936 certification in refractory installation quality control and is responsible for equipment selection for centrifugal and vaneaxial blowers. He currently serves on the Mechanical Engineering Advisory Board of the University of Tulsa.

Prologue to the First Edition

Fred Koch and John Zink—Pioneers in the Petroleum Industry

The early decades of the twentieth century saw the birth and growth of the petroleum industry in Oklahoma. Drilling derricks sprouted like wildflowers throughout the state, making it among the top oil producers in the nation and Tulsa the "Oil Capital of the World" by the 1920s.

Refining operations accompanied oil production. Many of the early refineries were so small that today they would be called pilot plants. They were often merely topping processes, skimming off natural gasoline and other light fuel products and sending the remainder to larger refineries with more complex processing facilities.

Along with oil, enough natural gas was found to make its gathering and sale a viable business as well. Refineries frequently purchased this natural gas to fuel their boilers and process heaters. At the same time, these refineries vented propane, butane, and other light gaseous hydrocarbons into the atmosphere because their burners could not burn them safely and efficiently. Early burner designs made even natural gas difficult to burn as traditional practice and safety concerns led to the use of large amounts of excess air and flames that nearly filled the fire box. Such poor burning qualities hurt plant profitability.

Among firms engaged in natural gas gathering and sales in the northeastern part of the state was Oklahoma Natural Gas Company (ONG). It was there that John Steele Zink (Figure P.1b), after completing his studies at the University of Oklahoma in 1917, went to work as a chemist. Zink's chemistry and engineering education enabled him to advance to the position of manager of industrial sales. But while the wasteful use of natural gas due to inefficient burners increased those sales, it troubled Zink and awakened his talents first as an innovator and inventor and then as an entrepreneur.

Seeing the problems with existing burners, Zink responded by creating one that needed less excess air and produced a compact, well-defined flame shape. A superior burner for that era, it was technically a premix burner with partial primary air and partial draft-induced secondary air. The use of two airflows led to its trade name, BI-MIX®. The BI-MIX® burner is shown in a drawing from one of Zink's earliest patents (Figure P.2).

ONG showed no interest in selling its improved burners to its customers, so in 1929 Zink resigned and founded Mid-Continent Gas Appliance Co., which he later renamed the John Zink Company.

(a)

(b)

FIGURE P.1
Fred Koch (a) and John Steele Zink (b).

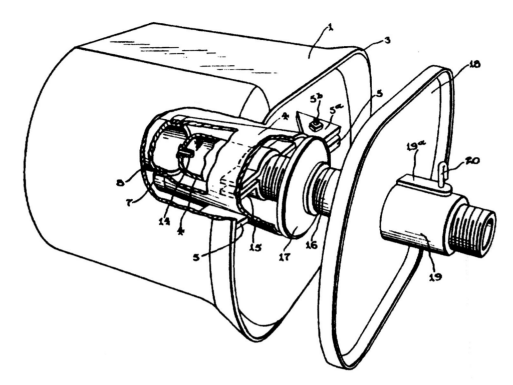

FIGURE P.2
Drawing of BI-MIX® from Zink's patent.

Zink's BI-MIX® burner was the first of many advances in technology made by his company, which to date has seen over 250 U.S. patents awarded to nearly 80 of its employees. He carried out early manufacturing of the burner in the garage of his Tulsa-area home and sold it from the back of his automobile as he traveled the Oklahoma oil fields, generating the money he needed to buy the components required to fabricate the new burners.

The novel burners attracted customers by reducing their fuel costs, producing a more compact flame for more efficient heater operation, burning a wide range of gases, and generally being safer to use. Word of mouth among operators helped spread their use throughout not only Oklahoma but, by the late 1930s, to foreign refineries as well.

Growth of the company required Zink to relocate his family and business to larger facilities on the outskirts of Tulsa. In 1935, he moved into a set of farm buildings on Peoria Avenue, a few miles to the south of the city downtown, a location Zink thought would allow for plenty of future expansion.

As time passed, Zink's company became engaged in making numerous other products, sparked by its founder's beliefs in customer service and solving customer problems. After World War II, Zink was the largest sole proprietorship west of the Mississippi River. Zink's reputation for innovation attracted customers who wanted new burners and, eventually, whole new families of products. For example, customers began asking for reliable pilots and pilot igniters, when atmospheric venting of waste gases and emergency discharges was replaced by combustion in flares in the late 1940s. This in turn was followed by requests for flare burners and finally complete flare systems, marking the start of the flare equipment industry. Similar customer requests for help in dealing with gas and liquid waste streams and hydrocarbon vapor led the Zink Company to become a leading supplier of gas and liquid waste incinerators and also of hydrocarbon vapor recovery and other vapor control products.

Zink's great interest in product development and innovation led to the construction of the company's first furnace for testing burners. This furnace was specially designed to simulate the heat absorption that takes place in a process heater. Zink had the furnace built in the middle of the employee parking lot, a seemingly odd placement. He had good reason for this because he wanted his engineers to pass the test furnace every day as they came and went from work as a reminder of the importance of product development to the company's success.

Zink went beyond encouraging innovation and motivating his own employees. During the late 1940s, Zink and his technical team leader, Robert Reed (who together with Zink developed the first smokeless flare), sensed a need for an industry-wide meeting to discuss technologies and experiences associated with process heating. In 1950, they hosted the first of four annual process heating seminars in Tulsa. Interest in the seminars was high, with the attendance level reaching 300. Attendees of the first process heating seminar asked Zink and Reed to conduct training sessions for their operators and engineers. These training sessions, which combined lectures and practical hands-on burner operation in Zink's small research and development center, were the start of the John Zink Burner School®. The year 2010 marked the 60th anniversary of the original seminar and the 50th year in which the Burner School has been offered. Over the years, other schools were added to provide customer training in the technology and operation of hydrocarbon vapor recovery systems, vapor combustors, and flares.

Included among the 150 industry leaders attending the first seminar was Harry Litwin, former president and part owner of Koch Engineering Co., now part of Koch Industries of Wichita, Kansas. Litwin was a panelist at the closing session. Koch Engineering was established in 1943 to provide engineering services to the oil refining industry. In the early 1950s, it developed an improved design for distillation trays and because of their commercial success the company chose to exit the engineering business. Litwin left Koch at that time and set up his own firm, the Litwin Engineering Co., which grew into a sizeable business.

During the same period that John Zink founded his business, another talented young engineer and industry innovator, Fred C. Koch, was establishing his reputation as an expert in oil processing. The predecessor to Koch Engineering Co. was the Winkler–Koch Engineering Co., jointly owned by Fred Koch with Lewis Winkler, which designed processing units for oil refineries. Fred Koch had developed a unique and very successful thermal cracking process that was sold to many independent refineries throughout the United States, Europe, and the former Soviet Union. One of the first of these processing units was installed in a refinery in Duncan, Oklahoma, in 1928, one year before Zink started his own company.

While the two men were not personally acquainted, Koch's and Zink's companies knew each other well in those early years. Winkler–Koch Engineering was an early customer for Zink burners. The burners were also used in the Wood River refinery in Hartford, Illinois. Winkler–Koch constructed this refinery in 1940 with Fred Koch as a significant part owner and the head of refining operations. Winkler–Koch Engineering, and later Koch Engineering, continued to buy Zink burners for many years.

Fred Koch and two of his sons, Charles and David, were even more successful in growing their family business than were Zink and his family. When the Zink family sold the John Zink Company to Sunbeam Corporation* in 1972, the company's annual revenues were $15 million. By that time Koch Industries, Inc., the parent of Koch Engineering, had revenues of almost $1 billion. Since then, Koch has continued to grow; its revenues in the year 2011 were over $100 billion.

When the John Zink Company was offered for sale in 1989, its long association with Koch made Koch Industries a very interested bidder. Acting through its Chemical Technology Group, Koch Industries quickly formed an acquisition team, headed by David Koch, which succeeded in purchasing the John Zink Company.

Koch's management philosophy and focus on innovation and customer service sparked a new era of revitalization and expansion for the John Zink Company. Koch recognized that the Peoria Avenue research, manufacturing, and office facilities were outdated. The growth of Tulsa after World War II had made Zink's facilities an industrial island in the middle of a residential area. The seven test furnaces on Peoria Avenue at the time of the acquisition, in particular, were cramped, with such inadequate infrastructure and obsolete instrumentation that they could not handle the sophisticated research and development required for modern burners.

A fast-track design and construction effort by Koch resulted in a new office and manufacturing complex in the northeastern sector of Tulsa that was completed at the end of 1991. In addition, a spacious R&D facility adjacent to the new office and manufacturing building replaced the Peoria test facility.

The initial multimillion dollar investment in R&D facilities included an office building housing the R&D staff and support personnel, a burner prototype fabrication shop, and an indoor laboratory building. Additional features included steam boilers, fuel storage and handling, data gathering centers, and measurement instrumentation and data logging for performance parameters from fuel flow to flue gas analysis.

Koch has repeatedly expanded the R&D facility. When the new facility began testing activities in 1992, nine furnaces and a multipurpose flare testing area were in service. Today, there are 14 outdoor test furnaces and 2 indoor research furnaces. Control systems are frequently updated to keep them state of the art.

Zink is now able to monitor burner tests from an elevated customer center that has a broad view of the entire test facility. The customer center includes complete automation of burner testing with live data on control panels and flame shape viewing on color video monitors.

A new flare testing facility (Figure P.3) was constructed in the early 2000s to dramatically expand and improve Zink's capabilities. This project represents the

---
* Sunbeam Corporation was primarily known as an appliance maker. Less well known was Sunbeam's group of industrial specialty companies such as John Zink Company.

FIGURE P.3
Flare testing facility.

company's largest single R&D investment since the original construction of the R&D facility in 1991. The new facilities accommodate the firing of a wide variety of fuel blends (propane, propylene, butane, ethylene, natural gas, hydrogen, and diluents such as nitrogen and carbon dioxide) to reproduce or closely simulate a customer's fuel composition. Multiple cameras provide video images along with the electronic monitoring and recording of a wide range of flare test data, including noise emissions. The facility can test all varieties of flare systems with very large sustained gas flow rates at or near those levels that customers will encounter in the field. Indeed, flow capacity matches or exceeds the smokeless rate of gas flow for virtually all customers' industrial plants, giving the new flare facility a capability unmatched in the world.

These world-class test facilities are staffed with engineers and technicians who combine theoretical training with practical experience. They use the latest design and analytical tools, such as computational fluid dynamics, physical modeling, and a phase Doppler particle analyzer. The team can act quickly to deliver innovative products that work successfully, based on designs that can be exactly verified before the equipment is installed in the field.

Koch's investment in facilities and highly trained technical staff carries on the tradition John Zink began more than 80 years ago: providing our customers today, as he did in his time, with solutions to their combustion needs through better products, applications, information, and service.

Robert E. Schwartz
Tulsa, Oklahoma

1
Safety

Charles E. Baukal, Jr.

CONTENTS

1.1 Introduction 2
    1.1.1 Definitions 2
    1.1.2 Combustion Tetrahedron 4
1.2 Safety Review 4
1.3 Hazards 6
    1.3.1 Excessive Temperature 7
    1.3.2 Thermal Radiation 7
    1.3.3 Noise 9
    1.3.4 High Pressure 10
    1.3.5 Fires 16
        1.3.5.1 Heat Damage 16
        1.3.5.2 Smoke Generation 16
    1.3.6 Explosions 19
        1.3.6.1 Explosions in Tanks and Piping 19
        1.3.6.2 Explosions in Stacks 20
        1.3.6.3 Explosions in Furnaces 21
    1.3.7 Flame Instability 21
    1.3.8 Environmental 22
1.4 Codes and Standards 24
    1.4.1 NFPA Codes and Standards 25
        1.4.1.1 NFPA 86: Standard for Ovens and Furnaces, 2011 Edition 26
        1.4.1.2 NFPA 70: National Electric Code (NEC), Updated Annually 26
        1.4.1.3 NFPA 497: Classification of Flammable Liquids, Gases, or Vapors and of Hazardous (Classified) Locations for Electrical Installations in Chemical Process Areas, 2012 Edition 26
        1.4.1.4 NFPA 54: National Fuel Gas Code, 1999 Edition 26
        1.4.1.5 NFPA 58: Liquefied Petroleum Gas Code, 2011 Edition 26
        1.4.1.6 NFPA 30: Flammable and Combustible Liquids Code, 1996 Edition 26
        1.4.1.7 NFPA 921: Guide for Fire and Explosion Investigations, 2011 Edition 26
    1.4.2 Additional Standards and Guidelines 27
    1.4.3 Industrial Insurance Carriers 27
    1.4.4 Testing Laboratories 27
1.5 Accident Prevention 27
    1.5.1 Ignition Control 27
    1.5.2 General 27
1.6 Accident Mitigation 28
    1.6.1 Design Engineering 31
        1.6.1.1 Flammability Characteristics 32
        1.6.1.2 Ignition Control 33
        1.6.1.3 Fire Extinguishment 36
1.7 Safety Documentation and Operator Training 38
    1.7.1 Design Information 38
    1.7.2 Process Hazard Analysis Reports 39

1.7.3 Standard Operating Procedures 39
1.7.4 Operator Training and Documentation 40
1.8 Recommendations 41
1.9 Sources for Further Information 41
References 42

1.1 Introduction

Industrial combustion can be dangerous for many reasons.[1,2] Fires and explosions are a major concern in industrial combustion processes and account for as much as 95% of the losses in accidents in the process industries.[3] Figure 1.1 shows the bulging walls in a test heater that resulted from an overpressure event caused by a rapid deflagration. The consequences of a fire or an explosion in a chemical or petrochemical plant, for example, can be very severe and very public because of the high volume of flammable liquids and gases handled in those plants.[4,5] An example is the explosion at the Houston Chemical Complex of the Phillips Petroleum Co. in Pasadena, Texas in 1989 where 23 were killed and 300 were injured.[6] Another example is the explosion at the Irving Oil Refinery in St. John, New Brunswick, where one person was killed from an explosion in a hydrocracker unit, apparently due to a tube failure.[7] *Loss Prevention Bulletin* listed all of the major incidents worldwide that occurred from 1960 to 1989 in the hydrocarbon chemical process industries including refineries, petrochemical plants, gas processing plants, and terminals.[8] Some of these involved large property losses and deaths. These types of events have heightened the safety-consciousness of these industries to both prevent such incidents and to effectively handle them if they should occur.[9] The moral, social, economic, environmental, and legal ramifications of an accident make combustion safety a critical element in plant design and operation. Preventing an incident is definitely preferred to protecting people and equipment from the consequences of an incident if it occurs.[10] Burning large quantities of fuel means appropriate precautions must be taken to prevent equipment damage and personnel injury.

There are many factors that can contribute to an accident[11]: human error,[12,13] equipment malfunction, upset conditions, fire or explosion near the apparatus,[14] improper procedures, and severe weather conditions. In a report prepared by the American Petroleum Institute, the following causes were noted for 88 incidents which occurred in refining and chemical unit operations from 1959 to 1978: 28% equipment failures, 28% human error, 13% faulty design, 11% inadequate procedures, 5% insufficient inspection, 2% process upsets, and 13% education.[15] Uehara analyzed the risks to Japan's petrochemical plants in the event of a large earthquake, which has a stronger likelihood in Japan due to the high frequency of seismic activity.[16] There are also many potential dangers caused by fires and explosions: flying shrapnel, pressure waves from a blast, high heat loads from flame radiation,[17–19] and high temperatures. All of these can have severe consequences to both people and equipment and may need to be considered in minimizing the potential impact of an incident. Fry showed how computer models can be used to simulate fires and

FIGURE 1.1
Test heater that has been overpressured. (From Baukal, C., Testing safety, in *Industrial Combustion Testing*, C.E. Baukal, ed., Chapter 2, Fig. 2.1, CRC Press, Boca Raton, FL, 2011.)

explosions in the chemical process industry to help design appropriate measures to prevent these incidents and how to respond if they should occur.[20] Ogle presented a method for analyzing the explosion hazard in an enclosure that is only partially filled with flammable gas.[21] He showed that an explosion pressure at the stoichiometric condition is approximately 50 times greater than the failure pressure of most industrial structures. This obviously can have catastrophic results for industrial combustors.

Kletz[22] lists four circumstances that are frequent causes of accidents or dangerous conditions including performing or preparing for maintenance, making modifications to furnace design, human error, and labeling errors or omissions. When preparing for maintenance, it is important to remove hazards from the maintenance area, isolate the area and/or equipment from operational equipment, and follow maintenance procedures carefully. When modifying a furnace design, even when the modification seems minor, the proposed modification should go through design procedures similar to the procedures used for the original installation of the equipment. Without appropriate design, it is not always possible to anticipate how a change in one piece of equipment will affect other equipment.

Human error is sometimes caused by inattention or poor training but is frequently caused by a deliberate attempt to shortcut a cumbersome procedure or to make an inconvenient piece of equipment more convenient to use. Accidents caused by labeling are frequently the result of out-of-date labeling, incorrect labeling, or no labeling at all resulting in the incorrect operation of equipment.

Safety documentation and operator training provide the backbone of a strong safety program and are absolutely essential to maintain a safe combustion working environment. Safety documentation for combustion-related processes includes design information, process hazard analysis (PHA) reports, standard operating procedures (SOPs), and training documentation. Feedback from each of these documentation elements are linked together as part of a plant's overall process safety program.

This chapter is not intended to be exhaustive, but is designed to highlight some of the more common safety concerns for industrial combustion processes. Additional references are available on combustion safety for the interested reader.[23–29] The National Fire Protection Association (NFPA) is a good source of information including some important standards related to industrial combustion safety (e.g.,[30–33]). Other examples of sources include the European Committee for Standardization (CEN) EN746-1 standard "Industrial Thermoprocessing Equipment—Common Safety Requirements for Industrial Thermoprocessing Equipment," the Canadian Standards Association (CSA) B139-00 standard "Installation Code for Oil Burning Equipment," and the Japanese Standards Association (JSA) JIS B 8415 standard "General Safety Code for Industrial Combustion Furnaces."

A number of good books are available on safety in combustion systems and in the chemical and petrochemical industries.[34–44] Crowl and Louvar have written a textbook designed to teach and apply the fundamentals of chemical process safety.[45] King has written a large book on safety for the process industries, including the chemical and petrochemical industries, with specific emphasis on the U.K. and European standards and regulations.[46] Kletz has written an encyclopedia-format book on safety and loss prevention containing small articles on about 400 different topics.[47] Nolan has written an extensive guide to understanding and mitigating hydrocarbon fires and explosions.[48] Nolan characterized accidents or failures into the following basic areas: ignorance, economic considerations, oversight and negligence, and unusual occurrences. However, it is noted that nearly all incidents are preventable. Nolan listed the following principles as the general philosophy for fire and explosion protection for oil, gas, and related facilities:

1. Prevent the immediate exposure of individuals to fire and explosion hazards.
2. Provide inherently safe facilities.
3. Meet the prescriptive and objective requirements of governmental laws and regulations.
4. Achieve a level of fire and explosion risk that is acceptable to the employees, the general public, the petroleum and related industries, the local and national government, and the company itself.
5. Protect the economic interest of the company for both short- and long-range impacts.
6. Comply with a corporation's policies, standards, and guidelines.
7. Consider the interest of business partners.
8. Achieve a cost-effective and practical approach.
9. Minimize space (and weight if offshore) implications.
10. Respond to operational needs and desires.
11. Protect the reputation of the company.
12. Eliminate or prevent the deliberate opportunities for employee or public-induced damages.

This chapter is designed to be an introduction to safety because a comprehensive discussion of safety is beyond the scope of this book. Chapter 2 provides more detailed information on the use of controls to enhance the safety of combustion systems.

1.1.1 Definitions

Crowl and Louvar,[45] Nolan,[48] and NFPA 86[33] all provide extensive definitions of common combustion safety vocabulary. Some of the most commonly used definitions related to fire and explosion phenomena are provided on the following.

Autoignition—The process through which a flammable liquid's vapors are capable of extracting enough energy from the environment to self-ignite, without the presence of a spark or flame.

Autoignition temperature—The minimum temperature at which a flammable liquid is capable of autoignition.

Auto-oxidation—The process of slow oxidation, resulting in the production of heat energy, sometimes leading to autoignition if the heat energy is not removed from the system.

Burner—A device or group of devices used for the introduction of fuel and oxidizer into a furnace at the required velocities, turbulence, and mixing proportion to support ignition and continuous combustion of the fuel.

Combustion—A chemical process that is the result of the rapid reaction of an oxidizing agent and a combustible material. The combustion reaction releases energy (in the form of heat and light), part of which is used to sustain the combustion reaction.

Combustible—In general, a material that is capable of undergoing the combustion process in the presence of an oxidation agent and a suitable ignition source.

Combustible liquid—A liquid having a flash point (FP) at or above 140°F (60°C) and below 200°F (93°C). A combustible liquid basically becomes a flammable liquid when the ambient temperature is raised above the combustible liquid's FP.

Confined explosion—An explosion occurring within a confined space, such as a building, vessel, or furnace.

Detonation—An explosion that results in a shock wave that moves at a speed less than the speed of sound in the unreacted medium.

Deflagration—An explosion that results in a shock wave that moves at a speed greater than the speed of sound in the unreacted medium.

Explosion—A rapid expansion of gases that results in a rapidly moving shock wave.

Fire—the generic term given to the combustion process.

Flame—A controlled fire produced by a burner.

Flammable—In general, a material that is capable of being easily ignited and burning rapidly.

Flammable liquid—A liquid having a FP below 140°F (60°C) and having a vapor pressure below 40 psia (2000 mm Hg) at 100°F (38°C).

Flare—A device incorporating a large burner, typically on top of a large exhaust stack, used for the burning of combustible exhaust gases vented from an industrial process.

Flash point (FP)—The lowest temperature of a liquid at which it gives off enough vapor to form an ignitable mixture with air immediately over the surface of the liquid.

Ignition—The process of initiating the combustion process through the introduction of energy to a flammable mixture.

Lower flammability limit (LFL)—The minimum concentration of a combustible gas or vapor in air below which combustion will not occur upon contact with an ignition source, sometimes referred to as the lower explosive limit (LEL).

Minimum ignition energy (MIE)—The minimum energy required in order to initiate the combustion process.

Overpressure—The pressure generated by an explosive blast, relative to ambient pressure.

Shock wave—A pressure wave moving through a gas as the result of an explosive blast. The generation of the shock wave occurs so rapidly that the process is primarily adiabatic.

Spontaneous combustion—The combustion process resulting from auto-oxidation and subsequent autoignition of a flammable liquid.

Upper flammability limit (UFL)—The maximum concentration of a combustible gas or vapor in air above which combustion will not occur upon contact with an ignition source. Sometimes referred to as the upper explosive limit (UEL).

Vapor pressure—The pressure exerted by a volatile liquid as determined by the Reid method (ASTM D-323-58). Measured in terms of pounds per square inch (absolute).

1.1.2 Combustion Tetrahedron

Four basic elements must be present for a combustion process: (1) fuel, (2) oxidizer, (3) heat, and (4) a reaction chain. This is usually referred to as the "fire tetrahedron" as shown in Figure 1.2. Fire may be defined as a rapid chemical reaction between a fuel and an oxidant,

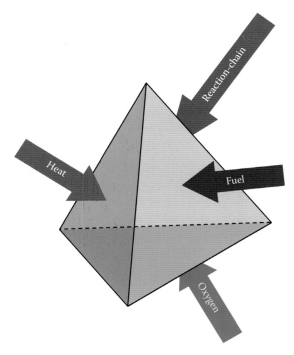

FIGURE 1.2
Fire tetrahedron.

more rapidly and violently.[49] For example, pure oxygen is used to enhance many high-temperature industrial combustion processes.[50] Even metals can burn in pure oxygen. In reality, a fuel and oxidizer can be in the presence of an ignition source without combusting, if the mixture is not within the flammability limits. Table 1.1 shows the flammability limits for a few common fuels.[51] For example, the LFL for methane (CH_4) in air is 5.0% CH_4 by volume with the balance being air. The UFL for CH_4 in air is 15.0% CH_4 by volume. If the mixture contains less than 5.0% or more than 15.0% CH_4 by volume, then the mixture is outside the flammability limits and will not combust at standard temperature and pressure.

The third face of the tetrahedron involves an energy source to both initiate and sustain the combustion reactions. Table 1.2 shows the minimum ignition temperature required to initiate the combustion reaction where there is sufficient heat to both initiate and sustain the reaction. The typical distinction between a flame and a fire is that a flame is controlled and desirable, while a fire is uncontrolled and possibly undesirable. To prevent or extinguish a fire, one or more of the four legs of the combustion tetrahedron must be removed. Fuel may be in the form of a solid, liquid, or gas (see Volume 1, Chapters 3 and 5). Typical solid fuels include coal, wood dust, fibers, metal particles, and plastics. Typical liquid fuels include gasoline, diesel, ether, fuel oil, and pentane. Typical gaseous fuels include natural gas, propane, hydrogen, acetylene, and butane. If the fuel source is removed, then the fire goes out. For example, in a fired heater, the flame goes out when the fuel supply to the burner is shut off.

The second face of the fire tetrahedron is the oxidizer. Oxidizers are also present in solid, liquid, and gaseous forms. Solid oxidizers include metal peroxides and ammonium nitrate. Liquid oxidizers include hydrogen peroxide, nitric acid, and perchloric acid. Gaseous oxidizers include oxygen, fluorine, and chlorine. Oxygen (contained in air) is the oxidizer used in industrial combustion. A fire will be extinguished if the oxidizer is removed. This may be done, for example, by smothering the fire with a blanket or by injecting an inert gas like N_2 or CO_2 in and around the fire to displace the oxidizer. In nearly all cases, the oxidizer is oxygen that is present in normal air at about 21% by volume. Higher concentrations of oxygen can cause a flame to burn

TABLE 1.1

Flammability Limits for Common Fuels at Standard Temperature and Pressure

	Flammability Limits (% Fuel Gas by Volume)			
Fuel	Lower, in Air	Upper, in Air	Lower, in O_2	Upper, in O_2
Butane, C_4H_{10}	1.86	8.41	1.8	49
Carbon monoxide, CO	12.5	74.2	19	94
Ethane, C_2H_6	3.0	12.5	3	66
Hydrogen, H_2	4.0	74.2	4	94
Methane, CH_4	5.0	15.0	5.1	61
Propane, C_3H_8	2.1	10.1	2.3	55
Propylene, C_3H_6	2.4	10.3	2.1	53

Source: Reed, R.J., *North American Combustion Handbook, Vol. I: Combustion, Fuels, Stoichiometry, Heat Transfer, Fluid Flow*, 3rd edn., Table 10.1, North American Manufacturing Company, Cleveland, OH, 1986.

TABLE 1.2

Minimum Ignition Temperatures for Common Fuels at Standard Temperature and Pressure

	Minimum Ignition Temperature			
Fuel	In Air (°F)	In Air (°C)	In O_2 (°F)	In O_2 (°C)
Butane, C_4H_{10}	761	405	541	283
Carbon monoxide, CO	1128	609	1090	588
Ethane, C_2H_6	882	472	—	—
Hydrogen, H_2	1062	572	1040	560
Methane, CH_4	1170	632	1033	556
Propane, C_3H_8	919	493	874	468

Source: Reed, R.J., *North American Combustion Handbook, Vol. I: Combustion, Fuels, Stoichiometry, Heat Transfer, Fluid Flow*, 3rd edn., Table 10.2, North American Manufacturing Company, Cleveland, OH, 1986.

of various gaseous fuels and oxidizers at stoichiometric conditions and standard temperature and pressure. For example, the minimum ignition temperature for a stoichiometric air/CH_4 mixture is 1170°F (632°C). In normal burner operation, the flame is ignited with either a pilot or a spark igniter. A fire or explosion may be initiated if the fuel/oxidizer mixture contacts a hot surface, an unintended spark, or static electricity.[52] A common fire prevention step is to eliminate all ignition sources from an area containing known fuel sources. After the fire has been initiated, energy is still required to sustain the flame. That source is normally the exothermic heat release from the reaction itself that makes most flames self-sustaining. A common method of extinguishing fires is to deluge the area with water, which eventually cools all surfaces below the ignition point. Water is inert and inexpensive and has a very high heat capacity relative to other liquids, which makes it a good extinguishing agent. However, water should not be used, for example, to extinguish a grease fire as the water can actually cause the fire to spread.

The fourth face of the tetrahedron is a reaction chain in order to sustain combustion. If any of the steps in the chemical chain reaction are broken, the flame may be extinguished. This is the principle behind certain types of fire extinguishers, such as dry chemical or halogenated hydrocarbon. These extinguishing agents inactivate the intermediate products of the flame reactions. This reduces the combustion rate of heat evolution, which eventually extinguishes the flame by removing the heat source that makes the flame self-sustaining.

1.2 Safety Review

A safety review is recommended before any type of combustion experiment or modification of an existing process, no matter how routine. The extent of the review will depend on how common the process is and the potential for damage or injury. For processes that are commonly done with relatively little chance for damage or injury, a short informal review may be all that is required. For less common processes with a high potential for damage or injury, a more extensive formal review is recommended that could take a significant amount of time to complete. Some type of checklist should be used to ensure that all appropriate factors have been considered in the review.

The increasing quantities of hazardous materials used at a given plant and the increasing complexity of the plants have made safety analyses both more difficult and more important because of the possibility of catastrophic accidents.[53] There are various types of analyses that are used for a process hazards analysis (PHA) of the design of the equipment and processes, including the effects of human error. Qualitative methods include checklists, What-If, and hazard and operability (HAZOP). Quantitative methods include event trees, fault trees, and failure modes and effect analysis (FMEA). All of these methods require rigorous documentation and implementation to ensure that all potential safety problems and the associated recommendations are addressed.

HAZOP and What-If reviews are two common qualitative methods used to conduct process hazard analyses in the chemical and petrochemical industries.[45,54–57] These methods are used to review equipment and process designs to identify potential hazards[58] to minimize the risk of accidents.[59]

In a What-If review, many questions are asked to determine what the consequences might be in the event of a particular incident. Some of the parameters that are commonly assessed include pressure, temperature, and flow. Questions are then asked as to what happens if that parameter is too high or too low, for example, to see what the consequence would be. If the consequence could cause an accident, then preventative actions can be taken either to prevent that parameter from ever reaching that state or to add a safety system to respond in the event that the parameter does reach that state. For example, if a furnace temperature exceeds a predetermined level indicating some type of problem, then the burners can automatically be shut off to prevent damaging the furnace. The system is designed to respond to prevent the parameter, temperature in that case, from ever reaching a critical level. Another example is causing a relief valve to vent to reduce the pressure, if the pressure becomes too high in a system. In that example, a safety measure is implemented if the variable is too high.

The review team should consist of those with appropriate experience in all of the relevant areas involved in the process. The engineer or scientist responsible for the process should be part of the team, although not necessarily the team leader. At least one experienced technician or operator, preferably one who will be involved in the operation of the equipment, should participate in the review. Many organizations have someone responsible for safety who is normally part of such a review team and may lead the team as appropriate. If a formal review will be conducted, someone trained in the particular review process being used will often lead the process. Some specialists may be needed on the team. For example, if new advanced sensors or controls will be used in the process, then someone familiar with those technologies should be included. Someone involved in the management of the facility may be involved in these safety reviews. Depending on the extent of the review and the potential for damage or injury, it is generally recommended that at least one person on the review team should not be directly involved in the process. This person should be independent of the budgeting and scheduling so they can focus on safety and not be unduly influenced by other factors.

Prior to conducting the review, the team leader should gather all appropriate documentation such as drawings and procedures and make sure they are up to date. Some new documentation may need to be created if it does not already exist. Each component should have a unique label which can be used in the review to identify various scenarios. For example, a fuel supply ball valve might have a label such as BV101. The review can then use this label to refer to conditions where BV101 might be open or closed. The documentation should also include performance specifications for the equipment such as the maximum flow rate, temperature, and pressure. If appropriate, vendor literature may also be needed to determine the limits of operation for a given piece of equipment.

Formal reviews normally require appropriate documentation. For example, Excel spreadsheets or Word templates are available for conducting safety reviews to both ensure that a specific process has been used in the analysis and to document the results of the review. The documentation should include the team members, the dates and times of the review, the functional area of each team member, the purpose of the review, the equipment and processes being analyzed, the key specifications for the equipment (e.g., maximum flow rate), what codes and standards apply to the review if any, and a discussion of past incidents that may be relevant to the what is being reviewed. The review should consider the equipment and materials in the vicinity of what is being reviewed. For example, fuel storage tanks or high-voltage electrical equipment in the vicinity should be considered. All potential deviations (e.g., high flow or no flow when there should be flow) should be analyzed including possible causes for the deviation (e.g., human error or equipment failure), the likelihood (e.g., frequency) the event might occur, the consequences if it did occur, and the safeguards to help mitigate the consequences or prevent the event from occurring. This often means educated estimates as many of the potential events may never have occurred before. If the current equipment or procedures are deemed inadequate, then appropriate recommendations should be made. In some cases, a recommendation might be to further study a particular issue. Each recommendation should be assigned to someone, not necessarily a team member, with a completion date. The recommendations should be tracked to ensure they have been completed on a timely basis. The review team may require that certain recommendations be completed prior to commencing operations, while other recommendations may not need to be completed before operations can begin.

Safety reviews should be scheduled far enough in advance that any recommendations from the reviews can be implemented in time so they do not adversely affect the operations' schedule, especially if any equipment needs to be ordered or any new apparatus needs to be built. Waiting until the last minute for the review may unnecessarily pressure the review team into making hasty decisions or cause them to overlook possible incidents that could occur. This may mean that alternative team members may need to be used if primary team members are unavailable in the time frame of the review.

Another aspect of a safety review that may need to be considered, depending on the nature of the test, is environmental permitting. For example, if a new type of fuel will be used that the facility is not currently permitted to use, then the appropriate approvals will need to be obtained before operations can begin. This is often the case if a fuel contains hazardous components, such as benzene or chlorine-containing compounds. A related scenario is when the exhaust products may contain components that the facility is not currently permitted to emit. For example, if furans or dioxins may be potential effluents, the facility environmental permit may need to be checked to see if these are allowed and if so at what levels. Getting a new permit or a variance of an existing permit can be a lengthy process, so be sure to allow enough time for this. If additional treatment equipment is required to reduce pollution emissions, this can further lengthen the process. For example, obtaining equipment to remove furans or dioxins from the exhaust gases can be both expensive and lengthy, so plan accordingly.

1.3 Hazards

There are many potential hazards that may need to be considered in industrial combustion processes. Some of the common ones are briefly considered next.

1.3.1 Excessive Temperature

Combustion typically involves high temperatures that can cause equipment damage and personnel injuries if not properly handled. For production equipment in operating plants, there is often a maximum surface temperature limit dictated by government regulations. For example, the U.S. Occupational Safety and Health Administration (OSHA) guidelines limit the maximum external temperature of a surface exposed to

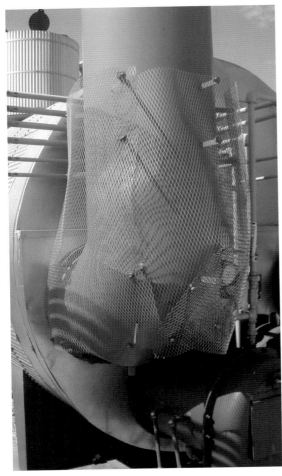

FIGURE 1.3
Metal mesh shielding personnel from hot exhaust stack. (From Baukal, C., Testing safety, in *Industrial Combustion Testing*, C.E. Baukal, ed., Chapter 2, Fig. 2.2, CRC Press, Boca Raton, FL, 2011.)

FIGURE 1.4
Insulated temporary ductwork. (From Baukal, C., Testing safety, in *Industrial Combustion Testing*, C.E. Baukal, ed., Chapter 2, Fig. 2.3, CRC Press, Boca Raton, FL, 2011.)

personnel to 140°F (60°C). If the temperature of equipment in the vicinity of personnel exceeds that temperature, then some type of shielding (see Figure 1.3) is usually required to prevent people from accidentally touching hot surfaces that could burn their skin. An alternative is to add more insulation to the hot surface to reduce the surface temperature (see Figure 1.4). For any experiments that might have high external surface temperatures that could be contacted by personnel, appropriate precautions must be taken to prevent burn injuries.

Equipment could be damaged by excessive temperatures. This includes equipment outside the combustion chamber as well as inside. For example, external equipment such as valves for controlling fuel and air flows may have Teflon® seals that could be damaged by high temperatures. Processes should be analyzed to ensure the equipment can either handle the temperatures that might be encountered, that some type of cooling is provided, or that shields are in place to protect any equipment that could be adversely affected by the heat. Equipment inside the combustion chamber must be capable of handling the temperatures that may be generated. For example, metal furnace walls are normally protected by insulation rated to a suitable temperature. Any uninsulated metal that will be directly exposed to the heat may need to be cooled. The tubes in a process heater are cooled by the hydrocarbon fluids flowing through them. The tubes in a boiler are cooled by water flowing through them. Instrumentation such as metal gas sampling probes is often water-cooled to both protect the metal and to quench the gas sample as part of the measurement process (see Volume 1, Chapter 14).

In some cases, the temperatures may not be known a priori in the case of process conditions that have not been previously tested. While calculations may give an estimate of the expected temperatures, it may be advisable to include some instrumentation, such as thermocouples and infrared detectors, to measure the actual temperatures. If there are any areas that are particularly

Safety

a concern for high temperatures, warnings and alarms could be added to alert operators. Automatic shutdowns could even be added to prevent equipment damage.

1.3.2 Thermal Radiation

Thermal radiation is related to excessive temperature, but no direct contact is necessary for damage or injury to occur which distinguishes it here from the hazard caused by contacting high-temperature surfaces. Thermal radiation is a necessary and important phenomenon in nearly all industrial combustion processes (see Volume 1, Chapter 7).[60] Figure 1.5 shows thermal radiation coming from the viewport of a hot furnace. It is typically the predominant form of heat transfer from the flame to the combustor walls and heat load. Most combustion engineers design systems to optimize the radiation heat transfer to get the proper temperature distribution in the combustor. Some burners are deliberately designed to maximize the radiation from the flame by creating luminous flames.[61]

Excessive external thermal radiation can cause equipment damage and injure personnel. Electronic equipment in particular is susceptible to damage from high-radiant loadings. External radiation can melt plastic parts such as valve seals, dry out lubricated parts such as motors, and make equipment operation more difficult where, for example, operators may need to wear gloves to open or close valves. Figure 1.6 shows a corrugated stainless steel fence designed to shield combustion flow control equipment (not visible in the photograph because it is behind the fence), especially the electronics, from thermal radiation during flare testing (see Chapter 9). An even more serious concern is premature ignition caused by thermal radiation of the air–fuel mixture prior to reaching the burner outlet, where the fuel and oxidant are premixed upstream of the burner outlet.

Personnel injury from excessive thermal radiation should be considered in most industrial combustion applications. Figure 1.7 shows an instrument for measuring thermal radiation that can be used to determine potentially dangerous levels. Heat stress from high thermal radiation loading can cause illnesses ranging from behavioral disorders to heat stroke and even death.[62] Heat stress is the mildest form and is temporary in nature. It could include heat rashes sometimes referred to as "prickly heat" and dehydration. High-humidity ambient conditions can exacerbate the problem for personnel in the vicinity of high thermal radiation conditions. Skin can be damaged from excessive heat. Even

FIGURE 1.5
Radiation from a viewport. (From Baukal, C.E., *Industrial Combustion Pollution and Control*, Fig. 11.13, Marcel Dekker, New York, 2004.)

FIGURE 1.6
Stainless steel fence shielding flow control equipment from thermal radiation during flare testing. (From Baukal, C., Testing safety, in *Industrial Combustion Testing*, C.E. Baukal, ed., Chapter 2, Fig. 2.5, CRC Press, Boca Raton, FL, 2011.)

FIGURE 1.7
Radiometer for measuring thermal radiation.

FIGURE 1.9
Heat resistant suit. (From Baukal, C.E., *Industrial Combustion Pollution and Control*, Fig. 11.16, Marcel Dekker, New York, 2004.)

lower thermal radiation loadings can cause worker fatigue and reduce performance. Older, overweight, or out-of-shape workers are particularly at risk from high thermal radiation loads.

There are two general abatement strategies used to mitigate and control external thermal radiation loading: reducing the source of the radiation or shielding personnel and equipment from the source. Source reduction involves reducing the external radiation source. One way is to insulate the radiation source, for example, a furnace wall, to reduce the external surface temperature and hence the radiation. Another way is to reduce the energy source heating the radiating surface. The easiest way to do that is to reduce the firing rate; however, this will also reduce the production rate, which is normally not desirable. Another way to reduce the radiation source is to cool it, for example, by water cooling the external furnace walls. Viewports should have some type of shutter to minimize the heat escaping through them when they are not in use (see Figure 1.8). Furnace leaks should be repaired, not only to reduce external thermal radiation, but also to improve the thermal efficiency.[63]

The second abatement technique is shielding and cooling which involves shielding personnel and equipment from the external radiation source. This can be done with some type of physical barrier such as a wall (e.g., Figure 1.6). It can be done with screens around the radiation source to prevent personnel from getting too close. Individual pieces of equipment can be shielded with insulation, a reflective surface, or some type of solid material. Personnel can be shielded by wearing appropriate protective clothing designed for high-heat environments (see Figure 1.9).

1.3.3 Noise

Noise is often defined as unwanted sound. There are many possible sources of noise in industrial combustion (see Vol. 1, Chapter 16). Some of these include combustion roar, jet noise for flow through orifices, flow noise for fluids flowing through piping, and equipment noise such as from the combustion air fan. Acoustic resonance can exacerbate the problem by magnifying the noise.

There are several strategies that can be used to reduce combustion noise (see Refs. [64,65] for more details). One strategy is either to move the source of the undesirable sound away from personnel or to move personnel away from the sound. However, this may not be practical for many industrial applications. Another strategy is to put some type of sound barrier between the noise

FIGURE 1.8
Viewport with shutter. (From Baukal, C.E., *Industrial Combustion Pollution and Control*, Fig. 11.15, Marcel Dekker, New York, 2004.)

and personnel. The barrier can be either reflective or absorptive to minimize the noise. The noise source might be surrounded by an enclosure, or the operators may be located inside a sound-proofed enclosure. In some cases, it may be possible to use a silencer, which would act as a barrier, to reduce the noise. For example, the exhaust from a car is reduced by the muffler, which acts like a silencer. The barrier could also be in the form of earplugs, ear phones, or some other sound-reducing safety device worn by people in the vicinity of the noise. Another technique is to reduce the exposure time to the noise since noise has a cumulative effect on human hearing. In some cases, it may be possible to replace noisy equipment with new equipment that has been specifically designed to produce less noise or to retrofit existing equipment to produce less noise. For example, old combustion air blowers and fans could be replaced by new, quieter blowers and fans (see Chapter 3). Another way to reduce noise is to increase the pipe size and reduce the number of bends in the pipe to reduce the fluid noise caused by the fluids flowing through the pipe. Resonance and instabilities can usually be designed out of a system if they are a problem. Noisy burners can be replaced by quieter burners. Burner noise is a function of the burner design, fuel, firing rate, air–fuel stoichiometry, combustion intensity, and aerodynamics of the combustion chamber. These strategies are discussed next.

The primary source of noise in most industrial combustion systems is from the burner(s). The design of the burner nozzle and the burner tile or quarl are important factors in noise generation in combustion processes. One method to reduce noise from a burner is to use larger exit ports that produce lower gas velocities. However, there are limits to how low gas velocities can be, depending on the fuels used and the burner type. For example, the exit gas velocities in a premix burner must be greater than the flame speed of the oxidizer/fuel mixture or else flashback may result. Fuels containing high concentrations of hydrogen will necessarily require higher exit velocities because of the high flame speed of hydrogen.

Another method for reducing the sound from a combustion process is to add a pipe or tube, often referred to as a quarter wave tube, to the resonance system to cancel out the harmonics causing the noise. In this technique, a specially designed tube is attached to the chamber where sound of a predominant frequency is causing resonance. The quarter wave tube is designed to cancel out this resonance and therefore reduce the noise levels. Figure 1.10 shows a quarter wave tube installed on the side of an exhaust stack for a gas-fired propylene vaporizer that previously exhibited a low-frequency rumble during operation, producing excessive noise levels. The quarter wave tube was built with some length adjustability to determine the best length to maximize noise reduction. The total noise level for the vaporizer without the tube was 95.3 dBA. The total noise level with the tube dropped to 83.8 dBA for a noise reduction of more than 10 dBA. This brought the noise levels within acceptable limits. Note that adding a quarter wave tube is not always a practical option because, depending on the combustion chamber geometry and the frequency of the harmonic, the tube diameter and length may be excessive.

FIGURE 1.10
Quarter wave tube on a propylene vaporizer. (From Baukal, C., Testing safety, in *Industrial Combustion Testing*, C.E. Baukal, ed., Chapter 2, Fig. 2.9, CRC Press, Boca Raton, FL, 2011.)

Other techniques to mitigate noise caused by combustion instability include modifying the[66]

- Furnace stack height
- Internal volume of the furnace
- Acoustical properties of the furnace lining
- Pressure drop through the burner by varying the damper positions
- Location of the pilot
- Flame stabilization techniques

Figure 1.11 shows some of the common silencers used to reduce the source of noise from industrial processes.[67] Figure 1.12 shows a natural draft burner without an air inlet muffler. These burners are often quiet enough that no muffler is needed. Figure 1.13 shows a common type of muffler used on a natural draft burner used in the

Absorptive silencer. This silencer is the most common type and take the form of a duct lined in the interior with a sound-absorptive material.

Reactive expansion chamber. This type reflects sound energy back toward the source to cancel some of the oncoming sound energy.

Reactive resonator. This type functions in approximately the same way as the reactive expansion chamber type.

Plenum chamber. This device allows the sound to enter a small opening in the chamber; that sound which has not been absorbed by the chamber's acoustical lining leaves by a second small opening, generally at the opposite end of the chamber.

Lined bend. Sound energy flowing down a passage is forced to turn a corner, the walls of which are lined with acoustical material. The sound energy is thus forced to impinge directly on a sound-absorbing surface as it reflects its way around the corner. Each successive impingement takes sound energy from the traveling wave.

Diffuser. This device does not actually reduce noise. In effect, it prevents the generation of noise by disrupting high-velocity gas streams.

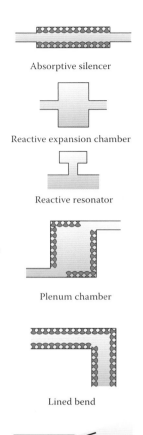

FIGURE 1.11
Silencers. (Adapted from Liu, D.H.F., 6.7 Noise control in the transmission path, in *Environmental Engineers' Handbook*, D.H.F. Liu and B.G. Lipták, eds., Fig. 6.7.8, Lewis Publishers, Boca Raton, FL, 1997.)

FIGURE 1.12
Natural draft burner with no air inlet muffler. (From Baukal, C.E., *Industrial Combustion Pollution and Control*, Fig. 10.16, Marcel Dekker, New York, 2004.)

FIGURE 1.13
Natural draft burner with an air inlet damper. (From Baukal, C.E., *Industrial Combustion Pollution and Control*, Fig. 10.17, Marcel Dekker, New York, 2004.)

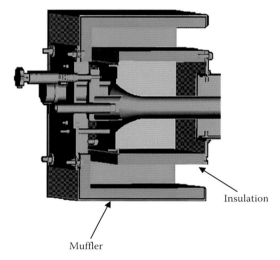

FIGURE 1.14
Typical muffler for a radiant wall-fired natural draft burner. (From Baukal, C.E., *Industrial Combustion Pollution and Control*, Fig. 10.18, Marcel Dekker, New York, 2004.)

floor of refinery heaters where the air inlet has a baffle lined with sound-deadening insulation. Figure 1.14 shows a comparable standard muffler for a radiant wall burner used in the side of ethylene cracking furnaces. Figure 1.15 shows nonstandard mufflers used on floor-fired natural draft burners where the muffler is larger than the burner because of the very-low-noise requirements for the particular application. Figure 1.16 shows a nonstandard muffler for low-noise requirements on wall-fired burners.

Sound transmission mitigation is a technique that involves mitigating the transmission of the sound from

Safety

FIGURE 1.15
Large mufflers on two natural draft burners. (From Baukal, C.E., *Industrial Combustion Pollution and Control*, Fig. 10.19, Marcel Dekker, New York, 2004.)

FIGURE 1.16
Large mufflers on two radiant wall burners. (From Baukal, C.E., *Industrial Combustion Pollution and Control*, Fig. 10.20, Marcel Dekker, New York, 2004.)

the source to the receiver. This can be done in a variety of ways. One rather simple, but not always practical, method is simply to increase the distance between the source and the receiver, which reduces the sound levels at the receiver. Another strategy is to put some type of barrier between the source and the receiver. For example, a concrete wall could be built around the source. People commonly plant trees and shrubs on their property to mitigate some of the sound from their neighbors and from road traffic noise. Figure 1.17 shows enclosed flares with walls around the enclosures to help mitigate the sound produced by the flares. Some type of sound-absorptive material could be placed between the source and the receiver. Different materials have different sound-absorbing characteristics, depending on both the composition and configuration.[68] Another strategy could be to use a medium that is less transmissive for sound. For example, water is less transmissive than air.

Noise generated by the burners in a combustion system may be greatly mitigated by the combustion chamber, which is usually a furnace of some type. The refractory linings in most furnaces generally significantly reduce the noise emitted from the burners. Noise is not commonly considered in many industrial heating applications for a variety of reasons. This is evidenced by the general lack of information available on the subject. It is difficult to predict the noise levels before installing the equipment due to the wide variety of factors that influence noise. Often, there are many other pieces of machinery that are much noisier than the combustion system so that the workers are already required to wear hearing protection. In the future, noise reduction may become more important. In combustion testing, noise may be less of a concern because of the temporary nature of many tests.

FIGURE 1.17
Two enclosed flares. (From Baukal, C.E., *Industrial Combustion Pollution and Control*, Fig. 10.21, Marcel Dekker, New York, 2004.)

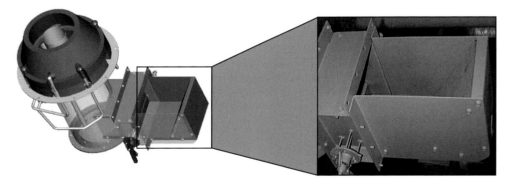

FIGURE 1.18
Typical muffler used on a natural draft burner. (From Baukal, C.E., *Industrial Combustion Pollution and Control*, Fig. 10.23, Marcel Dekker, New York, 2004.)

Mufflers are commonly used on burners to reduce noise levels (see Figure 1.18). Figure 1.19 shows how effective a muffler is at mitigating the noise produced by a burner. These mufflers typically go on the combustion air inlet to the burner and usually have some type of noise-reducing insulation on one or more sides. Reed gives an example of a natural draft burner producing 107 dB of noise before any mitigation.[69] The addition of a primary muffler reduced the noise to 93 dB, and the addition of a secondary muffler further reduced the noise to 84 dB.

Figure 1.20 shows a custom-made cylindrical muffler on the air outlet of a velocity thermocouple, often called a suction pyrometer (see Chapter 8), used to measure higher temperatures. This type of temperature measuring device relies on high air flow rates across a venturi to induce furnace gases to flow through the pyrometer and across a shielded thermocouple. This minimizes the effects of thermal radiation on the thermocouple and produces a more accurate measure of the true temperature. Measurements using bare wire thermocouples can be as much as 200°F (93°C) or more too low. Compressed air is often used as the motive gas to induce furnace gas inspiration. The high exit gas flow rates into the atmosphere can produce relatively high noise levels that are a concern for workers in the vicinity. The cylindrical

Safety

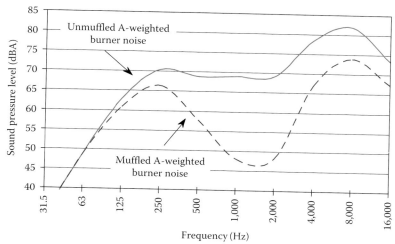

FIGURE 1.19
Sound pressure versus frequency with and without a muffler.

FIGURE 1.20
Cylindrical muffler on a suction pyrometer. (From Baukal, C.E., *Industrial Combustion Pollution and Control*, Fig. 10.25, Marcel Dekker, New York, 2004.)

FIGURE 1.21
Typical ear plugs. (From Baukal, C.E., *Industrial Combustion Pollution and Control*, Fig. 2.6, Marcel Dekker, New York, 2004.)

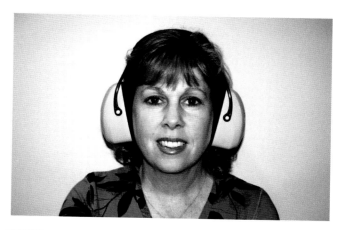

FIGURE 1.22
Typical ear muffs. (From Baukal, C.E., *Industrial Combustion Pollution and Control*, Fig. 10.27, Marcel Dekker, New York, 2004.)

muffler shown in Figure 1.20 reduced noise levels by more than 10 dBA down to an acceptable level.

There are two levels of protection commonly used by industrial workers to reduce noise levels[70]: plugs (see Figure 1.21) and muffs (see Figure 1.22). One or both types may be used, depending on the noise levels. A third type of protection device is a helmet, used commonly by motorcycle drivers, which provides relatively little hearing protection and is rarely used in industry for hearing protection. Therefore, this is not considered further here.

Plugs can reduce noise levels by 5–45 dB, depending on the plug type and sound frequency. They come in a variety of forms including disposable and reusable. Disposable plugs are typically made of some type of moldable material (e.g., foam) that can be inserted into a variety of ear sizes and shapes. These are very inexpensive and are typically bought in large quantities. They are especially convenient for visitors who do not have their own ear plugs. Reusable ear plugs can be cleaned and used multiple times. Custom molded ear plugs are available that are made to fit exactly in a specific person's ears and are designed to be reusable.

Ear muffs are designed to cover the entire ear and typically reduce levels by 5–50 dB, depending on the muff type and sound frequency. These can be used in lieu of or in addition to ear plugs. When both plugs and muffs are worn, noise protection is greater than either individually, but is not additive. Muffs may be more comfortable to some people compared to ear plugs. However, they may also interfere with other personal protective equipment such as hard hats. Special muffs are made to attach to certain types of hard hats where the muffs can be folded down or up as needed (see Figure 1.23).

Convenience and comfort are important factors when choosing appropriate noise protection for a given environment. If too much effort is required to use or maintain the hearing protection devices, or if they are uncomfortable, then they are less likely to be used. Proper training and education are essential to maximize the effectiveness of any hearing conservation program.[71]

While not possible in some cases, a simple way to protect workers is either to increase their distance from the sound source or to put them in a sound-proofed enclosure such as a control room or building. However, it is almost impossible to keep all workers away from high-noise sources all the time, so hearing protection will probably be necessary.

FIGURE 1.23
Ear muffs designed to be used with hard hats. (From Baukal, C.E., *Industrial Combustion Pollution and Control*, Fig. 10.28, Marcel Dekker, New York, 2004.)

1.3.4 High Pressure

There are multiple ways that equipment and personnel might be exposed to high pressures in combustion applications. High-pressure gases are sometimes used, for example, as span gases to calibrate gas analyzers. Specialty fuels may sometimes be supplied from high-pressure cylinders. Proper handling techniques should be used when moving these cylinders around. The cylinders must also be properly stored. For example, oxidizers and flammables are not normally stored together. High-pressure cylinders should be stored away from high-temperature sources. If possible, cylinders should be stored outside, so that if a leak should occur it would not be confined in a building (see Figure 1.24).

Some combustion processes may have the potential to produce high pressures. This can happen if liquids or gases are being heated, such as water in a boiler or hydrocarbon fluids in a process heater. The high pressures may be caused by high temperatures, by fluids being blocked in, by some type of blockage in the system that causes a restriction, or possibly by fluid vaporization from a liquid to a gas which causes a large and rapid increase in volume. In most commercial systems, adequate provisions are designed into the system for pressure relief in case the fluid pressures exceed the design limits. This may not be the case for some systems, so the operator must ensure that adequate precautions are taken so the pressures do not exceed the design limits. This may include the addition of a properly designed relief system and some type of warning system to signal the operators before potentially excessive pressures are reached. Adequate sensors should be provided so the operators can monitor the pressures so corrective action can be taken if necessary. One type of hazardous overpressure condition is a rapid deflagration, which is considered separately next.

1.3.5 Fires

1.3.5.1 Heat Damage

One of the most devastating causes of fires in process plants is process tube rupture (see Figure 1.25). A furnace tube rupture feeds the furnace firebox with an uncontrolled amount of fuel usually resulting in enormous damage and, sometimes, loss of life. As in all safety issues, prevention is preferred to remediation. Although there are numerous safety features of process equipment that can minimize damage when a tube failure occurs, the most desirable, safest, and least costly mode of operation is to prevent the occurrence in the first place.

Furnace operators should be carefully trained to monitor and assess the state of tubes as a regular part of the operational routine. Every occasion of tube overheating, whether it is caused by excessive firing or

Safety

FIGURE 1.24
Pressurized gas cylinders located outside a building where gases are used in a lab inside the building. (From Baukal, C., Testing safety, in *Industrial Combustion Testing*, C.E. Baukal, ed., Chapter 2, Fig. 2.23, CRC Press, Boca Raton, FL, 2011.)

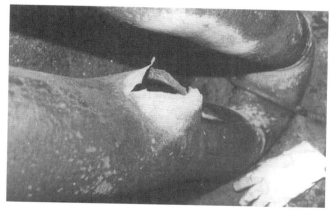

FIGURE 1.25
Tube rupture in a fired heater. (From Sanders, R.E., *Chemical Process Safety: Learning from Case Histories*, Fig. 10.2, Butterworth-Heinemann, Boston, MA, 1999.)

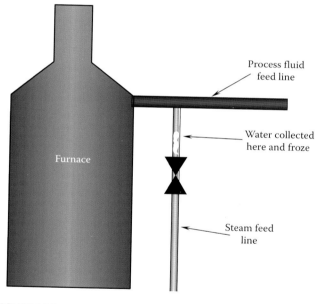

FIGURE 1.26
Trapped steam in a dead-end that can freeze and cause pipe failure. (From Kletz, T., *What Went Wrong: Case Histories of Process Plant Disasters*, 4th edn., Gulf Publishing Company, Houston, TX, 1998.)

inadequate cooling action by the process fluid, should be recorded and assessed to determine the likelihood of tube failure. Tubes are normally designed to last ten or more years under normal operation, but excessive temperatures can shorten the life of the tubes to a few days or less. Coking in the tubes, if not identified and corrected expediently, can cause hot spots on the tubes that will result in premature failure. Another mechanism for tube failure is trapped liquid that freezes and expands (see Figure 1.26).

Prevention of tube failure requires adequate instrumentation and proper furnace design to allow the tube condition to be monitored continuously and accurately. Accurate temperature measurement of tubes can be a difficult problem. Adequate instrumentation is expensive, but inadequate monitoring of tube condition can be dangerous and much more expensive. Temperature measurements of the process fluid should be possible at several points in the tube layout. Furnace wall temperatures should be measurable at several locations in the firebox as well as the convection section. Interlocks should be capable of shutting down the fuel supply if the process fluid temperature is too high or if the process fluid pressure rises above safe levels.

Furnaces should be provided with adequate viewing ports, and operators should be trained to visually assess tube condition during furnace operation. Flame impingement on tubes (see Chapter 13) is the most common cause of tube coking and failure and is usually discovered only by visual observation. When impingement occurs, adjustments in furnace operations should be made immediately. Tubes intended to last 10 years can fail in a matter of hours if they are heated more than 100°F (38°C) above their designed operating temperature. It is important to be able to assess flame impingement visually, since even well-instrumented systems cannot be certain of measuring the tubes at their point of highest temperature. If the furnace is shut down for maintenance, the opportunity should be used to make a careful, close-up inspection of the tubes. Alterations in the furnace structure, layout, or operating procedure should only be undertaken with the advice and consent of an engineer qualified to assess the proposed changes.

If in spite of careful precautions, tube failure occurs, adequate safety procedures should be in place to minimize the hazard and the damage. Process liquid flow should be controllable from a location that is an adequate distance from the furnace. Remote isolation valves should be included in the furnace design and should be inspected and tested regularly. Remote stop buttons should be located at a sufficient distance from the furnace to allow circulation pumps and any other equipment that feeds any form of fuel or oxidizer to the furnace area to be shut down even in a worst case scenario.

Sanders[5] reported two cases of tube failures in process heaters that illustrate the necessity of the earlier guidelines. In one case, a fire occurred during the start-up procedure of a U.S. Gulf Coast plant during a cold winter morning. A natural gas-fired heater was used to provide the heat energy to a heat-transfer fluid. The combustible heat-transfer fluid was to be circulated through piping in a gaseous phase reactor until a start-up temperature of 500°F (260°C) for the process was reached. Once the start-up temperature was reached, the exothermic nature of the reaction was used to sustain the reaction, and the furnace was taken off-line.

After start-up, the process fluid was switched to circulate through the reactor and a cooler to remove excess heat from the reactor. On this occasion, the weather had been unseasonably cold. All of the piping in the system, except for the piping that was inside the furnace, was steam traced and insulated. Apparently, while the system was not in use, the heat-transfer fluid had frozen in the piping inside the furnace. The start-up procedure called for the operator to establish a flow of the heat-transfer fluid through the piping in the furnace, light the furnace to heat the transfer fluid, then switch the flow of the heated fluid through the piping in the reactor.

When the circulation pump was started, a flow could not be established through the furnace. Correctly assuming that the fluid in the furnace had frozen, the operator started a small fire in the furnace in the hopes of thawing the fluid and establishing a flow. This procedure had been used successfully in the past. About 1 h after the operator had ignited the furnace and established a low fuel rate to thaw the pipes, the operations foreman entered the control room and noticed that the natural gas flow to the furnace was much higher than it should be for the thawing operation. The rate was immediately cut to about one-fourth of the existing flow.

Fifteen minutes later, black smoke and fire were reported coming from the heater stack. The fire was quickly extinguished, but inspection of the heater revealed that two tubes had ruptured. Fire investigators could find no one who would admit to increasing the fuel flow to the furnace. The flow recorder for the natural gas supply to the heater was not working properly so neither the exact flow rate nor the times of flow rate change could be ascertained.

An investigation revealed that there were no written instructions for thawing frozen pipes in the heater. The operations crew was not fully aware of the hazards of lighting the furnace before a flow was established in the piping. The heater tubes had carbon buildup which restricted the flow of the heat-transfer fluid. There was evidence of thinning of the tubes in the higher heat flux zones. Since the same technique for thawing the tubes had been used in the past, damage to the tubes may have accumulated over time.

In this case, several common safety precautions were not observed. Adequate instrumentation for measuring the temperature of the process fluid in the heater tubes was not available. The flow recorder for the fuel to the heater was not in working order. An alternative start-up procedure which had apparently been used on more than one occasion was not properly documented. The operators and foreman had failed to adequately oversee the furnace operation during a time in which the furnace was being used for a purpose for which it was not designed.

In the second incident reported by Sanders,[5] a similar tube failure in a furnace caused about $1.5 million in property damage and over $4 million in business interruption. The furnace in this incident performed a function similar to the furnace in the previous incident. A combustible heat-transfer fluid was heated in the furnace, then circulated through gas-phase reactors where solvents were produced. Since the gas-phase reaction is self-sustaining once operating temperature is reached, a single furnace was used to provide a start-up supply of heated fluid to each of five reactors one at a time. The piping and operational procedures for the system were complex. When a reactor was brought online, the operator was to align the valves to allow circulation of the heat-transfer fluid from the heater to the reactor being started, start the circulation pump, and ignite the heater. On this occasion, the operator erred by starting the heater while the heater tubes were isolated from the circulation pump by closed block valves. About 30 min after the heater was fired, a water sprinkler system tripped followed shortly by a heater-flame failure alarm and the rupture of a heat-transfer fluid pipe in the heater. Heat from the resulting fifty-foot flames was spread throughout the unit by 10–12 mile/h winds.

The 15-year-old heater had been relatively well designed to prevent an accident of the type that occurred. Automatic shutdown equipment and alarms were provided to respond to flame failure, high tube-wall temperature, low fuel supply pressure, and high heat-transfer fluid pressure. Although records indicated that some maintenance of the shutdown equipment had been performed, there was no systematic program of inspection and testing that would have assured that the equipment was properly adjusted and would operate dependably in an emergency. The specific system that provided for automatic shutdown due to high tube-wall temperature was equipped with a set point adjustment that allowed the operator to set the temperature at which alarms would sound and automatic shutdown would occur. For some reason, the set point temperature had been set to 1600°F (870°C). The alarm temperature should have been set at 830°F (440°C), and the shutdown temperature should have been set at 850°F (450°C). No reason was ever discovered for the unreasonable and clearly dangerous temperature setting.

Accidents occur due to both human error and equipment failure. Safety instrumentation is designed to prevent human error from creating a dangerous situation. In this case, the safety equipment which would have prevented the human error from producing a destructive incident had been defeated by improper management of the equipment. An appropriate safety program which included routine inspection and testing of the safety equipment would have caught the improper setting and prevented a costly accident.

1.3.5.2 Smoke Generation

Smoke is produced in most uncontrolled fires. Smoke is generated by incomplete oxidation of the fuel caused by insufficient mixing of air and the combustible materials.[72] The smoke contains fine particles made primarily of solid carbon. In many fires, more people die from smoke inhalation than from the heat produced in the fire. There are several potential problems with the smoke. One problem is the elevated temperatures, which can damage the lungs upon inhalation. Another problem is the deposition of smoke particles on the lungs, which can hinder breathing. Smoke can also block or impair vision, which can hinder escape from the fire. Only trained personnel with adequate breathing and eye protection should ever deliberately enter smoky conditions produced by a fire.

Another product of incomplete combustion is carbon monoxide (CO), which is an extremely toxic gas that can quickly kill humans due to respiratory failure. Smoke generation is an indicator of the probability of the presence of CO. CO kills by blocking the ability of hemoglobin in the blood to carry oxygen to the cells in the body. CO has a 300 times greater affinity for hemoglobin than does oxygen. Unfortunately, CO is colorless and odorless so one must be careful to avoid possible situations where it may be present, such as in smoky fires. Fortunately, inexpensive CO detectors are available to warn of its presence (see Figure 1.27).

1.3.6 Explosions

Danger of explosion may come from many sources, but explosions most often occur when the equipment involved is in a state of change such as start-up, shutdown, or maintenance. Since furnaces are made primarily of metals, maintenance often involves welding. The welding process provides an ignition source for any combustible gases or liquids that might remain in the work area. Typically, tests for the presence of flammable gases or liquids are conducted before any maintenance is allowed to commence. However, thorough testing to assure the absence of flammable materials can be complex and difficult.

Two principles should be followed when testing for combustibles in a planned, enclosed maintenance area. First, testing should be done not only in the immediate work area, but also in any connecting plumbing or equipment. Second, testing should be done immediately before the work begins and periodically during the time the ignition source is in use.

Stacks are designed to vent exhaust gases from industrial processes. If the gases are combustible, flare stacks may be used to burn the exhaust gases before they are vented into the atmosphere. In the case of combustible gases, operational procedures should be used which prevent the infiltration of oxygen into the stack where it could mix with the exhaust gases and cause an explosion. Stacks should be built with welded seams rather than bolted joints. Bolted joints, especially joints having surfaces that are not machined, can leak air into the stack in areas of negative pressure. Operational procedures should be used which assure that a continuous flow of gas is moving up the stack. A continuous flow of gas will sweep away small leaks of air and prevent air from moving down into the stack from the top.

Obviously, the best plan is to prevent an explosion from occurring, but appropriate plans should be made in the event one should occur. For example, it is

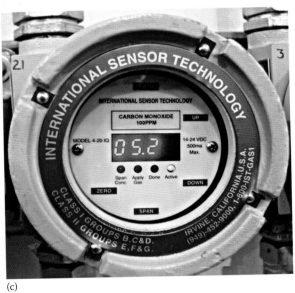

FIGURE 1.27
CO detector: (a) permanent, (b) portable, (c) area monitor.

recommended that all personnel be located away from the furnace during light off in case there is an incident. It may also be advisable to have some type of relief system built into the furnace, for example, pressure relief doors (sometimes referred to as explosion doors) that can open in the case of overpressurization.

1.3.6.1 Explosions in Tanks and Piping

Liquid storage tanks should be thoroughly flushed, cleaned, and tested before any maintenance begins. Yet, even storage tanks that have been thoroughly flushed can have vapor buildup over time from the evaporation of liquid residue in the cracks, seams, and structural members of the tank. It is almost impossible to completely clean tanks that contained heavy oils or polymers. Such tanks can test completely clean with a combustible gas detector yet fill with explosive vapors when maintenance activities heat the residues to the vapor point.

Pipes containing a number of bends, low points, or attached equipment may test clear of flammable materials in a proposed work area, yet contain significant amounts of explosive materials that can be vaporized by welding or migrate to the work area from elsewhere in the piping. Low points or dips in piping are particularly dangerous since they may store liquids that can vaporize and be ignited by maintenance work. Storage tanks, furnaces, pipes, or other metallic containers that have been in contact with an acid can have hydrogen buildup due to the action of the acid on the metal.

1.3.6.2 Explosions in Stacks

Stacks are designed to vent exhaust gases from industrial processes. If the gases are combustible, flare stacks are used to burn the exhaust gases as they are vented into the atmosphere. In the case of combustible gases, operational procedures should be used which prevent the infiltration of oxygen into the stack where it could mix with the exhaust gases and cause an explosion. Figure 1.28 shows the results of an explosion caused by improper purging of a flare stack allowing air to leak in the stack leading to the explosion when fuel gas was introduced back into the stack. Stacks should be built with welded seams rather than bolted joints.

FIGURE 1.28
Flare stack explosion due to improper purging. (From Kletz, T., *What Went Wrong: Case Histories of Process Plant Disasters,* 4th edn., Gulf Publishing Company, Houston, TX, 1998.)

Bolted joints, especially joints having surfaces that are not machined, can leak air into the stack in areas of negative pressure. Operational procedures should be used which assure that a continuous flow of gas is moving up the stack. A continuous flow of gas will sweep away small leaks of air and prevent air from moving down into the stack from the top. If the industrial process does not provide a continuous flow of gas, purge gases should be used to assure a flow velocity of 1–3 in./s (3–8 cm/s). Oxygen sensors should be used to assure that the oxygen content of the exhaust gases does not rise above 5% to avoid creating a flammable mixture. A lower percentage is safer if hydrogen is present in the stack gases.

Flare stacks always include some form of pilot burner or other ignition source. However, other ignition sources exist even in vent stacks where ignition of the stack gases is not intentional. For example, in furnaces that are improperly operated with too little oxygen for combustion, the stack gases can contain unburned fuel. Spectacular incidents have occurred when lightning has ignited stack gases for a distance of 100–200 ft (30–60 m) above the top of a furnace stack. As long as there is no oxygen in the stack itself and the gases are ignited above the top of the stack, such incidents, though disquieting, are not particularly dangerous. Other incidents have occurred in flare stacks in which fuel gases have reached combustion temperature and exploded inside the stack. The only way to assure that stack explosions do not occur is to prevent oxygen from infiltrating and mixing with the stack gases.

1.3.6.3 Explosions in Furnaces

The most common source of furnace explosions is the use of an improper lighting procedure. Even procedures that have been used for years without incident can cause explosions if conditions change. For example, if the fuel source is not completely isolated from the furnace before the ignition source is inserted, a leaky valve can allow enough fuel gas into the furnace to cause an explosion. A safe furnace lighting procedure for a furnace containing piloted burners should include the following generalized steps:

1. Confirm that the furnace fuel lines are completely isolated from the furnace (either by disconnection, blind flange, or a double block, and bleed valve assembly).
2. Confirm that all auxiliary furnace equipment is functioning properly, including all instrumentation and measurement devices; open both the burner air inlet register/damper and the furnace stack damper to the fully open position.

3. Purge the furnace of any combustible or flammable substances; follow the NFPA recommendation of purging the furnace with four furnace volumes of fresh air or inert gas (e.g., N_2 or CO_2).[31]

4. Test the atmosphere inside the furnace to assure that there are no combustibles present.

5. Connect the pilot fuel line to the burner; activate the permanent pilot igniter or insert a portable pilot igniter or premixed ignition torch.

6. Slowly open the pilot fuel control valve to the manufacturer-specified pilot fuel pressure; visually confirm stable ignition of the pilot flame.

7. Reestablish the main burner fuel supply (either by connecting the main fuel line, removing a blind flange, or reversing the double block and bleed valve assembly) to provide a fuel source to the furnace.

8. Slowly open the main burner fuel control valve to supply the burner with the manufacturer-specified ignition fuel pressure; visually confirm stable ignition of the main burner flame.

If the burner ignition attempt is unsuccessful, or if the procedure is aborted prior to successful burner ignition, the burner fuel supply should be immediately disconnected. Subsequent attempts to successfully light the burner must begin at the step 1.

In multiburner furnaces, the operator must be certain to follow the designed lighting procedures. In some furnaces, one burner may be cross-ignited by another burner until all burners are lit. In others, however, the burners may be too far apart for one to be safely lit from another. If the furnace is not designed to have one burner light from another, the full procedure given in the eight steps earlier should be followed for each burner. Although the eight steps given are a safe procedure for a typical furnace, the designed lighting procedure provided by the manufacturer of the furnace should always be followed. The manufacturer of the furnace should preapprove any deviation from the designed procedure.

Another source of explosions in furnaces is improper air management. If the furnace is starved for air, a pulsating huffing sound may result (see Chapter 13). The flame will be unsteady, changing from long to short or wide to narrow. The variations are the result of the available air being completely consumed without burning all of the fuel. The air-starved flame will then be reduced in volume until more air is available. It will then increase in size until the available air is again consumed. This alternating cycle causes the huffing sound.

The correct action, when a furnace is huffing, is to reduce the flow of fuel until there is enough air for full combustion. If the operator incorrectly increases the air without first reducing the fuel flow, the increased air supply may mix with the large volume of unburned fuel already in the furnace causing an explosion.

If low fuel pressure causes the flame to be extinguished in a furnace that burns fuel oil, careful tests for combustible vapors should be made immediately before attempting to relight the furnace. Although furnaces are equipped with automatic valves that shut off the fuel flow when the flame is lost, the piping between the valve and the furnace may still contain fuel that can be vaporized into an explosive source. The vaporization process can take time. Fuel that is still in its liquid state at the time of a test can cause a furnace to test clear of combustibles. Yet, the vaporization process can cause the furnace to be filled with explosive vapors as quickly as 10 min later.

Obtaining an accurate test with a gas detector can be problematic when oil is being used as fuel. Vapors in the furnace can condense back into their liquid state in the tube of the gas detector, preventing the vapors from reaching the detector head and being recorded. If the vaporization temperature of the fuel oil being used is near the ambient temperature, an accurate reading may be hard to obtain. Where inaccurate readings may be suspected, the furnace should be purged until the operator is certain that all unburned fuel oil has evaporated and has exited the furnace.

1.3.7 Flame Instability

There are multiple potential problems that could cause flame instability. The traditional fire triangle shows that fuel, oxygen, and an ignition source are required for a fire. However, these are not enough to guarantee a stable flame. Combustion chemistry is very complicated. The combustion of methane with air includes over 350 chemical reactions and dozens of species,[73] most of which only survive for fractions of a second in the flame before they recombine into more stable species such as CO_2 and H_2O. This chemistry must be allowed to proceed to completion because any interruptions could extinguish the flame or cause the flame to become unstable. For example, Halon fire extinguishers interrupt the chemistry in a flame and cause it to go out. Note that Halon fire extinguishers are being phased out in some locations due to environmental concerns. The velocity of the air and fuel are critical to flame stability. If these velocities are too slow or too fast, the flame can become unstable. If the fuel and air mixing is inadequate, this can produce an unstable flame. The more common causes of flame instability are briefly discussed here.

If the exit velocity from a premixed burner is below the flame speed for the air–fuel mixture, flashback (sometimes referred to as blowback) can occur. This is where the flame travels back inside the burner. This is

Safety

TABLE 1.3
Laminar Flame Speeds

Fuel	Formula	Burning Velocity	
		ft/s	m/s
Ethane	C_2H_6	1.46	0.445
Propane	C_3H_8	1.86	0.566
n-Butane	C_4H_{10}	1.47	0.448
Methane	CH_4	1.42	0.434
n-Pentane	C_5H_{12}	1.40	0.427
Ethylene	C_2H_4	2.23	0.680
Propylene	C_3H_6	2.30	0.702
Hydrogen	H_2	5.58	1.700

Source: Adapted from Glassman, I. and Yetter, R.A., *Combustion*, 4th edn., Academic Press, Burlington, MA, 2008.

particularly an issue for fuels with high flame speeds such as hydrogen. Table 1.3 shows the laminar flame speeds for various fuels.[74] The laminar flame speed for hydrogen is considerably higher than for other fuels, which makes it more likely to flashback. The flame can only travel as far as the air–fuel mixture is flammable. The burner design also dictates how far the flame can travel as shapes inside the burner can act as flameholders or flame arrestors. Figure 1.29 shows a radiant wall burner that has been damaged from a flashback. Some design features that prevent flashback include

- Making the outlet velocity profile as uniform as possible
- Maximizing the air–fuel exit velocity from the burner
- Making the air–fuel mixture fuel lean which reduces the flame speed
- Making the air–fuel mixture inflammable (either above the UFL or below the LFL)
- Promoting laminar flow out of the burner because the laminar flame speed is much slower than the turbulent flame speed

Flashback is normally easy to visually and aurally detect. The general response to flashback is to either shut off the fuel or increase the firing rate so the exit velocity is above the mixture flame speed. Operators should pay particularly close attention to any conditions that might result in flashback. If flashback conditions are likely to occur, appropriate precautions should be taken. These might include using stronger and higher temperature materials in the burner and shielding the equipment in case the burner fails.

If the air–fuel mixture velocity exiting the burner is much higher than the flame speed, liftoff (sometimes referred to as blow off) may occur. Figure 1.30 shows an example of a premix radiant wall burner lifting off. In this type of burner design, the flame exits the burner at a 90° angle and impinges on the refractory. When it is operating correctly, the refractory is uniformly heated and radiates to the process tubes in the furnace. In this photo, the burner is being overfired and the flame is partially lifting off, causing the flower-shaped heating pattern. Liftoff typically occurs when the air–fuel mixture velocity exiting the burner is much higher than the flame speed and there is inadequate flame stabilization in the burner to anchor the flame. There are some flames that are designed to be

FIGURE 1.29
Burner damaged from a flashback. (From Baukal, C., Testing safety, in *Industrial Combustion Testing*, C.E. Baukal, ed., Chapter 2, Fig. 2.24, CRC Press, Boca Raton, FL, 2011.)

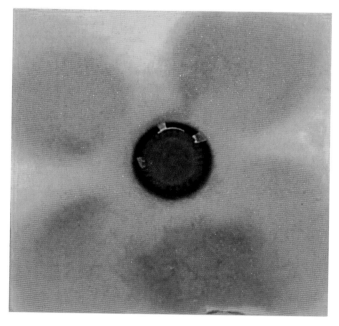

FIGURE 1.30
Premix burner lifting off. (From Baukal, C., Testing safety, in *Industrial Combustion Testing*, C.E. Baukal, ed., Chapter 2, Fig. 2.25, CRC Press, Boca Raton, FL, 2011.)

FIGURE 1.31
Coen iScan® flame detector. (From Baukal, C., Testing safety, in *Industrial Combustion Testing*, C.E. Baukal, ed., Chapter 2, Fig. 2.26, CRC Press, Boca Raton, FL, 2011.)

lifted, but most industrial flames are not lifted because of the possibility of the flame blowing out and reigniting causing an explosion. One indicator of this type of instability is a bouncing or pulsing flame. This phenomenon is commonly referred to as "huffing." This type of low-frequency instability can cause damage to the furnace, supporting structure, and associated equipment due to vibration. The normal procedure to handle huffing is to slowly reduce the fuel flow until the huffing stops and then to make whatever adjustments (e.g., repairing or cleaning the burner) are necessary to prevent this from occurring again. If the cause is due to overfiring the burner, then the burner design may need to be modified to reduce the outlet velocities and prevent liftoff. If the flame were to completely lift off and go out, the fuel should immediately be shut off to prevent an explosion or limit the damage that could occur from an explosion. This is typically done with some type of optical flame detector (see Figure 1.31), such as an ultraviolet flame scanner. Some burner design features that minimize the chance for liftoff include

- Reducing the air–fuel outlet velocities (e.g., reducing the firing rate)
- Designing flameholders into the burner to anchor the flame and reduce the outlet velocities
- Improving the air–fuel mixing (e.g., changing the fuel injection angles)
- Increasing the air–fuel turbulence which increases the flame speed

1.3.8 Environmental

There are many potential environmental problems from combustion processes that can be safety issues (see Vol. 1, Chapters 14 and 15). Incomplete combustion of carbon-containing fuels can produce carbon monoxide (CO) and smoke. CO is usually a regulated pollutant that can asphyxiate people in high enough concentrations. Smoke is also a commonly regulated pollutant that can damage the lungs and block or impair vision. These pollutants are typically more prevalent with liquid and solid fuels. Both can generally be easily controlled with sufficient oxygen to complete combustion, proper mixing of the oxygen with the fuel, and adequate temperatures for the reactions to go to completion. Incomplete combustion of hydrocarbon fuels can produce a wide range of products of incomplete combustion (PICs). In some extreme cases, this could even mean allowing some fuel to go through the combustor unchanged. These PICs are also generally easily controlled with sufficient oxygen, temperature, and mixing.

Another common environmental hazard is the production of nitrogen oxides (NOx) which have many potentially harmful characteristics (see Vol. 1, Chapter 15).[75,76] NO is poisonous to humans and can cause irritation of the eyes and throat, tightness of the chest, nausea, headache, and gradual loss of strength. Prolonged exposure to NO can cause violent coughing, difficulty in breathing, and cyanosis and could be fatal. NO_2 has a suffocating odor and is highly toxic and hazardous because of its ability to cause delayed chemical pneumonitis and pulmonary edema. Intermittent low-level NO_2 exposure may also induce kidney, liver, spleen, red blood cell, and immune system alterations.[77] In addition to the poisoning effect that NOx has on humans, there are also other problems associated with these chemicals. In the lower atmosphere, NO reacts with oxygen to form ozone (O_3), as well as NO_2. Ozone is also a health hazard which can cause respiratory problems in humans. NO_2 decomposes on contact with water to produce nitrous acid (HNO_2) and nitric acid (HNO_3) which are highly corrosive and cause acid rain.

Fuels that contain potentially harmful chemicals may produce effluents that contain hazardous materials. For example, fuels containing benzene could produce benzene in the exhaust under certain conditions. Since benzene is a hydrocarbon, it is generally easy to completely destroy if there is adequate oxygen, temperature, and mixing. Fuels containing chlorine or fluorine can produce dioxins and furans that are toxic. These are not as easy to eliminate from the exhaust, so some type of posttreatment system is often required to minimize these pollutants to acceptable levels.

Appropriate analyzers should be used to detect the emissions from combustion processes. There are two general types of analysis systems: extractive (see Figure 1.32) and in situ (see Figure 1.33). The extractive system pulls a small sample from the exhaust and sends it first to a conditioning system which cools, cleans, and dries the sample, before sending the conditioned sample to the analyzers. These systems have some lag time because they are remotely located from the stack, usually in some type of conditioned building such as the control room. In situ analyzers measure the gas composition at the

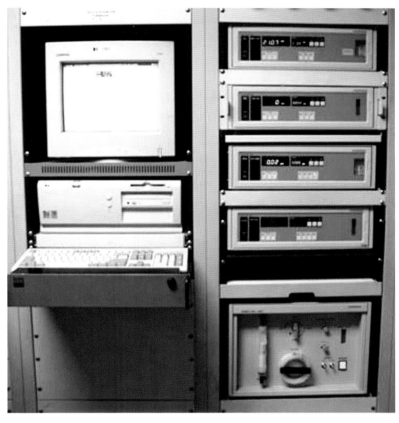

FIGURE 1.32
Typical continuous emission measurement system. (From Baukal, C.E., *Industrial Combustion Pollution and Control*, Fig. 1.20, Marcel Dekker, New York, 2004.)

FIGURE 1.33
Typical in situ analyzer. (From Baukal, C., Testing safety, in *Industrial Combustion Testing*, C.E. Baukal, ed., Chapter 2, Fig. 2.28, CRC Press, Boca Raton, FL, 2011.)

sample location, where the sample is often hot and contains water. These systems provide very rapid response as there is very little delay in getting the measurement. This faster response in generally preferred when an analyzer is being used for safety monitoring purposes.

1.4 Codes and Standards

There is often confusion regarding the differences between codes and standards. The NFPA[78] defines codes and standards in the following manner.

Code—A standard that is an extensive compilation of provisions covering broad subject matter or that is suitable for adoption into law independently of other codes and standards.

Standard—A document, the main text of which contains only mandatory provisions using the word "shall" to indicate requirements and which is in a form generally suitable for mandatory reference by another standard or code or for adoption into law.

1.4.1 NFPA Codes and Standards

The NFPA publishes a variety of codes and standards that address key safety issues related to fire protection. The NFPA website[79] contains a complete listing and description of all available codes and standards. However, the following NFPA codes and standards are particularly important to the safe operation of combustion equipment:

1.4.1.1 NFPA 86: Standard for Ovens and Furnaces, 2011 Edition

NFPA 86[80] is the primary standard that addresses fire and explosion hazards related to the operation and design of fired equipment used for heat utilization. Many of the components of Chapters 5, 6, and 8 can be directly applied to process furnaces in the hydrocarbon and petrochemical industries:

- *Chapter 5: Location and Construction*—The NFPA's recommendations regarding fired equipment and their proximity to personnel, buildings, and external combustible materials. Also addressed are design considerations such as structural integrity, explosion relief, observation port locations, and skin temperature restrictions.
- *Chapter 6: Furnace Heating System*—The furnace heating system refers to both the heating source as well as all associated piping and electrical wiring. The standard provides the NFPA's rules on the design and selection of burners, fuel piping, fittings, valves, and flue ventilation devices.
- *Chapter 8: Safety Equipment and Application*—The NFPA's detailed guidance concerning the design of automated safety systems (burner management systems) and the process conditions (safety interlocks) that should trigger an automated emergency shutdown (ESD). Requirements regarding the design and placement of automated fuel gas safety shutoff valves, high and low fuel pressure switches, flame supervision, excess temperature limit controllers, burner pilots, and flame-proving devices are included. Guidelines for burner pre-ignition procedures and ignition trials are also well documented.

1.4.1.2 NFPA 70: National Electric Code (NEC), Updated Annually

NFPA 70[81] provides "practical safeguarding of persons and property from hazards arising from the use of electricity." The NEC covers the installation of electric conductors and associated equipment in both the public and private sectors, including all electrical wirings associated with fired equipment. The NEC is considered to be law throughout the United States.

1.4.1.3 NFPA 497: Classification of Flammable Liquids, Gases, or Vapors and of Hazardous (Classified) Locations for Electrical Installations in Chemical Process Areas, 2012 Edition

NFPA 497[82] recommends steps to determine the location, type, and scope of hazards presented by electrical installations in operations where flammable or combustible liquids, gases, or vapors are processed or handled. NFPA 497 can be considered a companion standard to NFPA 70: National Electric Code (NEC).

1.4.1.4 NFPA 54: National Fuel Gas Code, 1999 Edition

NFPA 54 sets minimum safety requirements for fuel gas piping systems, fired equipment, flue gas ventilation systems, and related equipment.[31] The NFPA considers fuel gas to include natural gas fuel, manufactured gas, and liquefied petroleum gas (propane/butane). NFPA 54 is an American National Standard, appearing as designation Z223.1.

1.4.1.5 NFPA 58: Liquefied Petroleum Gas Code, 2011 Edition

NFPA 58 provides minimum safety requirements for the design and operation of liquefied petroleum gas (LPG) facilities, including fuel tank locations and piping.[83] NFPA 58 is the basis of LPG law for the United States.

1.4.1.6 NFPA 30: Flammable and Combustible Liquids Code, 1996 Edition

NFPA 30 provides the minimum safety requirements for liquid fuel installations, including requirements for bulk storage tanks, spill control, and emergency relief ventilation.[30] NFPA 30 is accepted as law in most U.S. states.

1.4.1.7 NFPA 921: Guide for Fire and Explosion Investigations, 2011 Edition

NFPA 921 provides guidance for the safe and systematic investigation and analysis of fires and explosions.[84] Topics include fire science, fire patterns, planning and conducting of investigations, and origin and root cause determination.

1.4.2 Additional Standards and Guidelines

In addition to the aforementioned NFPA codes and standards, several voluntary standards and guidelines address the design and operation of combustion devices. These standards and guidelines include

- *European CEN*—The multinational European organization develops standards addressing industrial safety concerns (including fuel handling and combustion) for its national member countries.
- *CSA International*—The independent, not-for-profit organization is the largest standards development organization in Canada. CSA has published many standards addressing combustion and the petroleum refining industry.
- *American Petroleum Institute (API)*—The API publishes a wide variety of standards applicable to combustion processes. *API Standard 521: Pressure-relieving and Depressuring Systems* provides valuable information on pressure relief devices which are an important safety element in most plants. *API Standard 535: Burners for Fired Heaters in General Refinery Service* provides guidelines for the selection and/or evaluation of burners installed in fired process heaters. *API Standard 537: Flare Details for General Refinery and Petrochemical Service* gives recommendations for the design of flare systems which are critical safety elements in refineries and petrochemical plants. *API Standard 560: Fired Heaters for General Refinery Service* also has useful information on safety equipment used in fired heaters.

1.4.3 Industrial Insurance Carriers

Industrial insurance carriers have also developed rigorous standards that must be met by companies wishing to get insurance for their plants using industrial combustion equipment. As might be expected, these standards are rigorous to minimize the chances of an accident that could damage equipment and injure personnel. Two examples are discussed next.

- *Industrial Risk Insurers (IRI)*—IRI provides comprehensive insurance protection for industrial losses due to fire, explosion, hail, lightning, windstorm, smoke, etc. IRI generally requires adherence to NFPA codes and standards. However, the IRI often supplements the NFPA with their own requirements.[85]
- *Factory Mutual (FM)*—The Factory Mutuals consist of three insurance firms, as well as the Factory Mutual Research Corporation. The FM Research Corporation conducts reliability and efficiency testing on a variety of equipment. The "FM Approval" label provides consumers with the confidence that the equipment bearing the label has been rigorously tested and found worthy of use in a fire protection system. The Factory Mutual Approval Guide contains a listing of FM-approved items, as well as the details regarding the application and installation criteria for which the equipment is approved. Similarly to IRI, Factory Mutual will often supplement the NFPA codes and standards with their own requirements.[85]

1.4.4 Testing Laboratories

Underwriters laboratory (UL) is an example of an independent organization that "certifies, validates, tests, inspects, audits, and advises and trains."[86] For example, UL certifies combustion equipment components, such as automatic shutoff valves as safe products. They also offer verification services to verify product performance, such as pollution emissions.

1.5 Accident Prevention

The best safety strategy is to prevent accidents from occurring in the first place, thus preventing equipment damage and personnel injury. Quintana et al. describe the application of a predictive risk analysis model to a large-scale combustion test facility.[87] The model uses continuous data sampling and analysis of an experiment to predict potential hazards that could lead to an incident, a system malfunction or unacceptable risk conditions. Operators are warned of a potential problem so they can take appropriate corrective action. If the proper corrective action is not implemented quickly enough, then the system can be designed to take appropriate action. This type of automated safety analysis system is particularly beneficial in combustion testing because of the fast and dynamic environment that can be present. Some of the more common strategies for preventing accidents are briefly discussed in this section.

1.5.1 Ignition Control

Ignition is the process through which combustion is initiated and occurs when a flammable mixture of fuel and oxidizer comes in contact with a suitable ignition source. Direct contact with a spark or flame is a very common energy source often used for the intentional ignition of industrial combustion equipment. For many types of burners, the ignition source may be in the form of a small premixed pilot burner, a portable electrostatic igniter, or a portable premixed gas torch.

Static electricity is a common potential ignition source in chemical processing plants. An electrostatic charge is formed whenever two dissimilar surfaces move relative to each other. A relevant example is liquid flowing through a pipeline, moving past the walls of the pipe. In this example, one charge is formed on the pipe surface, while another equal but opposite charge is formed on the surface of the moving liquid. When the voltage becomes strong enough, the static electricity will discharge in the form of an electrical spark. The spark can ignite any combustible and flammable materials present. NFPA 77[88] presents detailed explanations of design fundamentals for the prevention of fires and explosions due to electrostatic discharge. Some methods used to prevent unintended ignition include proper grounding of electrical equipment, using nonsparking tools, preventing smoking tobacco products, and keeping electrical devices that are not intrinsically safe (e.g., cell phones) out of areas in the plant where flammables could be present.

1.5.2 General

There are many general precautions that can be taken to prevent an accident during combustion processes. A sample of such precautions is given here. Others may be implemented as appropriate.

Some setups may be temporary, for example, for a one-time experiment, which may require specific precautions to prevent equipment damage and personnel injury. Figure 1.34 shows a photo of a rubber mat over hoses being used for a temporary combustion test. For setups that will be permanent, cables and hoses should be located so they will not be tripping hazards.

For temporary rigs that are commonly used in testing, appropriate precautions should be taken to prevent injuries that could be caused by these types of setups. Figure 1.35 shows safety tape around a portion of a combustion test setup to prevent unauthorized personnel from accidentally entering an area where they could possibly be injured.

Monitoring and control equipment can be used to rapidly stop a process if a problem is detected. The conventional method used in commercial combustion systems is some type of burner management system that includes a flame detector (e.g., Figure 1.31). A furnace camera (Figure 1.36) may be useful for remote monitoring of the burner, flame, and furnace. These cameras are specially designed for use in furnaces and may include air cooling, water cooling, and air purging of the lens. The camera monitor can be located in the control room so the process can be rapidly and remotely stopped. Figure 1.37 shows some emergency stop push buttons in a control room that can be used to rapidly stop a process if necessary. It may also be advisable to have emergency stop pushbuttons at or near the heater. Figure 1.38 shows one of these pushbuttons located next to a sight port on a test furnace so someone watching the testing through the sight port can rapidly stop the test if a problem is detected.

Another example of a precaution that should be taken is to prevent liquid fuel in storage tanks from leaking into the ground and possibly contacting an ignition source that could lead to a fire or explosion, as well as an environmental incident. Figure 1.39 shows an example of a metal containment system at the base of elevated diesel storage tanks. The containment needs to be capable of holding the entire volume of the tanks

FIGURE 1.34
Rubber mat over tripping hazard. (From Baukal, C., Testing safety, in *Industrial Combustion Testing*, C.E. Baukal, ed., Chapter 2, Fig. 2.29, CRC Press, Boca Raton, FL, 2011.)

Safety

FIGURE 1.35
Safety tape around test apparatus. (From Baukal, C., Testing safety, in *Industrial Combustion Testing*, C.E. Baukal, ed., Chapter 2, Fig. 2.30, CRC Press, Boca Raton, FL, 2011.)

FIGURE 1.36
Furnace camera. (From Baukal, C., Testing safety, in *Industrial Combustion Testing*, C.E. Baukal, ed., Chapter 2, Fig. 2.31, CRC Press, Boca Raton, FL, 2011.)

FIGURE 1.37
Emergency stop pushbuttons in a control room. (From Baukal, C., Testing safety, in *Industrial Combustion Testing*, C.E. Baukal, ed., Chapter 2, Fig. 2.32, CRC Press, Boca Raton, FL, 2011.)

FIGURE 1.38
Emergency stop pushbutton next to a sight port on a test furnace. (From Baukal, C., Testing safety, in *Industrial Combustion Testing*, C.E. Baukal, ed., Chapter 2, Fig. 2.33, CRC Press, Boca Raton, FL, 2011.)

FIGURE 1.39
Liquid fuel containment for diesel storage tanks. (From Baukal, C., Testing safety, in *Industrial Combustion Testing*, C.E. Baukal, ed., Chapter 2, Fig. 2.34, CRC Press, Boca Raton, FL, 2011.)

FIGURE 1.40
Example of a cabinet used to store flammables.

plus includes some additional capacity for rain and any other materials that could possibly enter the containment system. Figure 1.40 shows an example of a cabinet specially designed to store flammables to minimize the chance of a fire or explosion caused by leaks from flammables' containers. Some type of monitoring system to detect leaks may be advisable depending on the circumstances, particularly in enclosed locations. As an example, Figure 1.41 shows gas analyzers mounted in a combustion laboratory wall inside a building.

Other potentially hazardous fluids used in the facility should also have suitable protection against leaks getting on the ground (see Figure 1.42). This includes fluids that may not be flammable, but may damage the environment.

Personnel must be properly trained to be able to identify potential hazards, to operate the equipment, and to respond in the event of an accident. Appropriate safety signs (see Figure 1.43) should be used to alert personnel of potential hazards in a particular area.

Safety

FIGURE 1.41
NO, CO, O$_2$, and combustible analyzers in a combustion laboratory. (From Baukal, C., Testing safety, in *Industrial Combustion Testing*, C.E. Baukal, ed., Chapter 2, Fig. 2.35, CRC Press, Boca Raton, FL, 2011.)

FIGURE 1.42
Example of a containment system designed to keep potentially hazardous fluids from leaking onto the ground.

FIGURE 1.43
Example of safety signs on the outside of a building. (From Baukal, C., Testing safety, in *Industrial Combustion Testing*, C.E. Baukal, ed., Chapter 2, Fig. 2.36, CRC Press, Boca Raton, FL, 2011.)

1.6 Accident Mitigation

While every effort should be made to prevent an accident, appropriate plans should also be in place in case there is an accident to minimize the consequences if one should occur. If a fire should occur, a fire alarm system (see Figure 1.44) should be in place to alert personnel and to notify the local fire department. If the fire is not too large and personnel have been trained to extinguish fires, appropriate fire extinguishers (see Figure 1.45) should be located in the vicinity of the combustion equipment.

There may be multiple levels of shutdown depending on the severity of an incident. In a multiburner heater, the lowest level of shutdown might be to shut down a single burner, for example, in the event that burner flames out. Another level might be to shut off all the burners to a single heater. Depending on how the facility is designed, it might be desirable to shut off an entire section of the plant. For example, if a unit in

FIGURE 1.44
Fire alarm system in a combustion test facility. (From Baukal, C., Testing safety, in *Industrial Combustion Testing*, C.E. Baukal, ed., Chapter 2, Fig. 2.37, CRC Press, Boca Raton, FL, 2011.)

FIGURE 1.45
Large portable fire extinguisher. (From Baukal, C., Testing safety, in *Industrial Combustion Testing*, C.E. Baukal, ed., Chapter 2, Fig. 2.38, CRC Press, Boca Raton, FL, 2011.)

the facility contains multiple heaters, it might be desirable to rapidly shut off all the heaters in the unit in the event of a fuel leak in the unit. The highest level of shutdown might shut down the entire plant in the event of a very serious incident. Figure 1.46 shows an emergency pull ring designed to shut down an entire facility in case of a serious incident. These shutdown systems are designed to minimize the impact of an incident that could become much worse if it adversely affects nearby equipment and personnel.

Figure 1.47 shows some examples of safety and medical kits to treat personnel who might be injured in an incident. A well-equipped medical kit should be readily available and personnel properly trained and certified in first aid to treat minor injuries and provide temporary treatment until emergency medical personnel arrive in the event of a serious injury. It may also be advisable to have an emergency shower (see Figure 1.48) available in case personnel are exposed to chemicals or hot fluids.

Appropriate personal protection equipment (PPE) should be used during industrial combustion operations. This may include steel-toed shoes, safety glasses, a hard hat, gloves, ear plugs, and flame-retardant clothes. It may also include a face shield, special glasses for viewing bright objects such as flames in furnaces, and possibly even a special heat-resistant suit (see Figure 1.9). PPE can prevent or minimize personnel injury that could result during operations.

Figure 1.49 shows a photo of a wind sock located near fuel gas piping. This sock can be used to determine which direction leaking chemicals might be taken by the wind in the event of a leak. This is particularly important for toxic substances that could harm personnel and flammable materials which could start or feed a fire. In general, personnel would normally want to go upwind of leaking equipment unless that path would be more hazardous (e.g., they would have to go through a fire to get upwind).

1.6.1 Design Engineering

Several resources are available with general discussions of safety equipment used for industrial combustion applications.[51,88–90] The intent of this section is not

FIGURE 1.46
Emergency pull ring to shut down the entire facility. (From Baukal, C., Testing safety, in *Industrial Combustion Testing*, C.E. Baukal, ed., Chapter 2, Fig. 2.39, CRC Press, Boca Raton, FL, 2011.)

FIGURE 1.48
Emergency shower. (From Baukal, C., Testing safety, in *Industrial Combustion Testing*, C.E. Baukal, ed., Chapter 2, Fig. 2.41, CRC Press, Boca Raton, FL, 2011.)

FIGURE 1.47
Examples of safety and medical kits. (From Baukal, C., Testing safety, in *Industrial Combustion Testing*, C.E. Baukal, ed., Chapter 2, Fig. 2.40, CRC Press, Boca Raton, FL, 2011.)

to detail how equipment should be designed, but to point out the factors that should be considered in the design.

1.6.1.1 Flammability Characteristics

There are three primary parameters used to measure the relative flammability of a substance: the flashpoint (FP), the UFL, and the LFL. The FP is the lowest temperature of a liquid at which it evaporates enough vapor to form an ignitable mixture with the air immediately over the surface of the liquid. The UFLs and LFLs bracket the ignitable concentration range of a gas or vapor mixed with air. Table 1.4 contains experimentally determined FP temperatures, as well as UFLs and LFLs for a wide range of pure component substances in air.[92]

1.6.1.1.1 Liquids

The FP of pure component liquids is usually determined experimentally. However, FP estimates can be obtained

FIGURE 1.49
Wind sock. (From Baukal, C., Testing safety, in *Industrial Combustion Testing*, C.E. Baukal, ed., Chapter 2, Fig. 2.42, CRC Press, Boca Raton, FL, 2011.)

for multicomponent mixtures containing a single combustible species if both the FP and molar concentration of the combustible component are known. Raoult's law is used to determine the vapor pressure of the pure component in the diluted mixture (P^{SAT}), based upon the vapor pressure of the combustible species at its FP temperature (p):

$$p = xP^{SAT} \qquad (1.1)$$

where
P^{SAT} is the vapor pressure of the combustible component present within the mixture
x is the mole fraction of the combustible compoent present within the mixture
p is the vapor pressure of the pure combustible component at its FP

Once the vapor pressure of the combustible component present within the mixture (P^{SAT}) has been calculated, the resulting FP temperature of the mixture can be determined using a vapor pressure versus temperature diagram.[45] Figure 1.50 is a vapor pressure versus temperature diagram for light hydrocarbon fuels.[93]

Experimental methods are recommended for FP determination of multicomponent mixtures involving two or more combustible components.[45]

1.6.1.1.2 Vapors
Similar to the FP for pure component liquids, the UFLs and LFLs are also determined experimentally. For multicomponent gas mixtures, the Le Chatelier equation[94] is used to estimate the UFLs and LFLs of gaseous fuels:

$$LFL_{MIX} = \frac{1}{\sum_{i=1}^{n} y_i / LFL_i} \qquad (1.2)$$

$$UFL_{MIX} = \frac{1}{\sum_{i=1}^{n} y_i / UFL_i} \qquad (1.3)$$

where
LFL is the lower flammability limit for component i (vol%)
UFL is the upper flammability limit for component i (vol%)
y_i is the mole fraction of component i on a combustible basis
n is the number of combustible species present within the fuel mixture

Flammability limit data are often provided at process conditions of 77°F (25°C) and 14.695 psia (1 atm). However, the flammability limit ranges increase dramatically with temperature. The following empirical equations describe the temperature dependency of the flammability limits in air[95]:

$$LFL_T = LFL_{25}\left[\frac{1 - 0.75(T - 25)}{\Delta H_C}\right] \qquad (1.4)$$

TABLE 1.4

Flammability and Ignition Characteristics of Liquids and Gases

Compound	Normal Boiling Point at 14.695 psia (°F)	Flash Point (°F)	LFL (vol% in Air)	UFL (vol% in Air)	Autoignition Temperature (°F)
Acetone	133	−4	2.5	12.8	869
Acetylene	−120	—	2.5	100	581
Acrolein	127	−15	2.8	31	428
Aniline	363	158	1.3	11	1139
Benzene	176	12	1.2	7.8	928
Butane	31	−76	1.9	8.5	549
Carbon monoxide	−313	—	12.5	74	1128
Chlorobenzene	269	82	1.3	9.6	1099
Cyclohexane	177	−4	1.3	8.0	473
Ethane	−127	—	3.0	12.5	882
Ethyl alcohol	173	55	3.3	19	685
Ethylene	−155	—	2.7	36	842
Ethylene oxide	51	−4	3.0	100	804
Ethyl ether	−13	−42	3.4	27	662
Formaldehyde	−2	185	7.0	73	795
Heptane	209	25	1.1	6.7	399
Hexane	156	−8	1.1	7.5	437
Hydrogen	−423	—	4.0	74	—
Isopropyl alcohol	180	54	2.0	12.7	750
Isopropyl ether	45	−35	2.0	10.1	374
Methane	−259	—	5.0	15	999
Methyl acetate	134	14	3.1	16	849
Methyl alcohol	148	52	6.0	36	867
Methyl chloride	−11	—	8.1	17.4	1170
Methyl ethyl ketone	175	16	1.4	11.4	759
Methyl isobutyl ketone	242	64	1.2	8.0	838
Methyl propyl ketone	216	45	1.5	8.2	846
Napthalene	424	174	0.9	5.9	979
Octane	258	55	1.0	6.5	403
Pentane	97	−40	1.4	8.0	500
Phenol	359	174	1.8	8.6	1319
Propane	−44	−155	2.1	9.5	842
Propylene	−54	—	2.0	11.1	851
Propylene dichloride	206	70	3.4	14.5	1035
Styrene	293	88	0.9	6.8	914
Toluene	321	39	1.1	7.1	896
o-Xylene	292	90	0.9	6.7	865
m-Xylene	282	81	1.1	7.0	981
p-Xylene	281	81	1.1	7.0	982

Source: Lide, D.R. (ed.)., *CRC Handbook of Chemistry and Physics*, CRC Press, Boca Raton, FL, 1999.

$$\text{UFL}_T = \text{UFL}_{25}\left[\frac{1 + 0.75(T - 25)}{\Delta H_C}\right] \quad (1.5)$$

where
ΔH_C is the heat of combustion (kcal/mol)
LFL_T is the lower flammability limit at temperature, T (vol%)
LFL_{25} is the lower flammability limit at 25°C (vol%)
UFL_T is the upper flammability limit at temperature, T (vol%)
UFL_{25} is the upper flammability limit at 25°C (vol%)
T is the actual process temperature (°C)

Variation in pressure does not significantly affect the LFL except at very low pressures (below 27 in. w.c. absolute = 50 mm Hg absolute). However, increases in pressure can significantly raise the UFL. The following

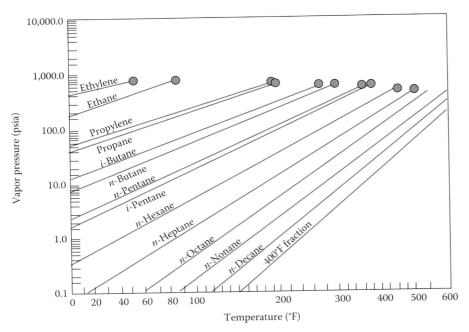

FIGURE 1.50
Vapor pressures for light hydrocarbons. (Adapted from *GPSA Engineering Data Book*, Vol. 2, 10th edn., Gas Processors and Suppliers Association, Tulsa, OK, 1994.)

empirical expression describes the pressure dependence of the UFL in air[96]:

$$UFL_P = UFL_{1atm} + 20.6(\log_{10} P + 1) \qquad (1.6)$$

where
  UFL_P is the upper flammability limit at pressure, P (vol%)
  UFL_{1atm} is the upper flammability limit at pressure, P (vol%)
  P is the actual process pressure (MPa, absolute)

The presence of pure oxygen as the oxidizing agent (as opposed to air) has very little effect on the LFL, as the oxygen concentration of air is in excess of that required for combustion at the LFL. However, the UFL of most hydrocarbon fuels in pure oxygen is increased by approximately 45%–55% when compared to the equivalent UFL in air.[97] Table 1.1 compares flammability limits of several common fuels using both air and pure oxygen as the oxidizing agent.[51]

1.6.1.2 Ignition Control

Ignition is the process through which combustion is initiated and occurs when a flammable mixture of fuel and oxidizer comes in contact with a suitable ignition source. The MIE is the minimum ignition energy required to initiate combustion and can be obtained through a variety of sources: direct contact with a spark or flame, static electricity, autoignition, auto-oxidation, and adiabatic compression. Table 1.5 lists the ignition sources tabulated from over 25,000 fires by the Factory Mutual Engineering Corporation.[98] Table 1.6 contains the MIE required for several common fuels.[99]

In general, the MIE decreases with pressure and increases with inert gas concentration. MIEs for hydrocarbon fuels are relatively low when compared to ignition sources. Walking across a rug can produce an electrostatic discharge of 22 mJ (2.1×10^{-5} Btu). An internal combustion engine's spark plug can generate an electrical discharge energy of 25 mJ (2.4×10^{-5} Btu).[45]

Direct contact with a spark or flame is a very common energy source often used for the intentional ignition of industrial combustion equipment. For process burners in fired heaters, the ignition source may be in the form of a small premixed pilot burner, a portable electrostatic igniter, or a portable premixed gas torch. Flare burners typically use a continuous flare pilot that is ignited by a flame front generator (FFG) (see Volume 3). An FFG is a custom-built device that sends a small burst of flame to the top of a tall flare stack, allowing the operator to ignite the pilot burner from ground level. FFG devices have been proven to safely ignite flare pilot burners a distance of 1 mile (1.6 km) away from the FFG. Regardless of the source, all ignition devices should be designed for the particular equipment and the specific set of process conditions for which they will be used.

TABLE 1.5

Ignition Sources of Major Fires

Electrical (wiring of motors)	23%
Smoking	18%
Friction (bearings or broken parts)	10%
Overheated materials (abnormally high temperatures)	8%
Hot surfaces (heat from boilers, lamps, etc.)	7%
Burner flames (improper use of torches, etc.)	7%
Combustion sparks (sparks and embers)	5%
Spontaneous ignition (rubbish)	4%
Cutting and welding (sparks, arcs, heat, etc.)	4%
Exposure (fires jumping into new areas)	3%
Incendiarism (fires maliciously set)	3%
Mechanical sparks (grinders, crushers, etc.)	2%
Molten substances (hot spills)	2%
Chemical action (processes not in control)	1%
Static sparks (release of accumulated energy)	1%
Lightning (where lightning rods are not used)	1%
Miscellaneous	1%

Source: National Safety Council, *Accident Prevention Manual for Industrial Operations*, National Safety Council, 1974.

TABLE 1.6

Minimum Ignition Energies

Compound	Pressure (atm)	Minimum Ignition Energy (mJ)
Methane	1	0.29
Propane	1	0.26
Heptane	1	0.25
Hydrogen	1	0.03
Propane (mol%) $[O_2/(O_2 + N_2)]*100\%$		
1.0	1	0.004
0.5	1	0.012
0.21	1	0.15
1.0	0.5	0.01

Source: Zabetakis, M.G., Fire and explosion hazards at temperature and pressure extremes, *Proceedings of Chemical Engineering Under Extreme Conditions, American Institute of Chemical Engineers Symposium Series No 2*, pp. 99–104, 1965.

Static electricity is a common ignition source of fires and explosions in chemical processing plants. An electrostatic charge is formed whenever two dissimilar surfaces move relative to each other. A relevant example is liquid flowing through a pipeline, moving past the walls of the pipe. In this example, one charge is formed on the pipe surface, while another equal, but opposite, charge is formed on the surface of the moving liquid. When the voltage becomes strong enough, the static electricity will discharge in the form of an electrical spark. The spark can ignite any combustible and flammable materials present. Crowl and Louvar[45] and the NFPA[100] present detailed explanations of design fundamentals for the prevention of fires and explosions due to electrostatic discharge. Kletz[22] discussed several case histories of fires and explosions ignited by electrostatic discharge.

Table 1.4 contains common autoignition temperatures for a variety of flammable liquids. Autoignition temperatures are dependent upon a number of factors, including fuel vapor concentration, fuel volume, system pressure, presence of catalytic material, and flow conditions. Because the autoignition temperature is a function of so many process variables, it is important that the autoignition temperature is determined experimentally under the conditions that most closely simulate actual process conditions.[45] Figure 1.51 shows the wreckage of an ethylene oxide plant explosion caused by autoignition leading to fire and explosion.[22]

Auto-oxidation is the process of slow oxidation, resulting in the production of heat energy, sometimes leading to autoignition if the heat energy is not removed from the system. The most common example of this potential ignition process is when rags saturated with oils are discarded or stored in a warm area. If allowed to auto-oxidize, the increased temperatures can result in autoignition of the rags, and a damaging fire or explosion can result. Relatively high-FP materials are the most susceptible to the auto-oxidation process, while low-FP materials may often evaporate without ignition. Fuel leaks that saturate thermal insulation or other absorbent materials should be isolated immediately, and the contaminated absorbent should be removed promptly and discarded in a suitable manner.[97]

Adiabatic compression of combustible or flammable materials can result in high temperatures, which in turn may result in autoignition of the compressed fuel. Examples of adiabatic compression include internal combustion engines and gas compressors. The temperature rise associated with the adiabatic compression of an ideal gas can be determined using thermodynamic relationships:

$$\frac{T_2}{T_1} = \left(\frac{P_2}{P_1}\right)^{\gamma-1/\gamma} \qquad (1.7)$$

where
T_2 is the final absolute temperature
T_1 is the initial absolute temperature
P_2 is the final absolute pressure
P_1 is the initial absolute pressure
γ is the specific heat ratio, C_p/C_v

FIGURE 1.51
Ethylene oxide plant explosion caused by autoignition. (From Kletz, T., *What Went Wrong: Case Histories of Process Plant Disasters,* 4th edn., Gulf Publishing Company, Houston, TX, 1998.)

1.6.1.3 Fire Extinguishment

In premixed burners, there is a chance that the fire can flash back into the burner. This occurs when the flame velocity toward the burner is higher than the velocity of the fuel mixture leaving the burner. In order to prevent flashback in premixed burners, flame arrestors (see Figure 1.52) are commonly used.[101] The primary applications for flame arrestors are to protect people and equipment from flashbacks, fires, and catastrophic explosions. A flame arrestor is designed to stop (extinguish) a flame or explosion from further propagation past the arrestor. It is a special type of heat exchanger that cools the flame, thus removing one of the legs of the fire triangle. Time is required to dissipate the heat, so the design and construction of the quenching media are important.[102] In fuel piping systems, the arrestor must be perforated or porous to allow gas flow through it. The technique often used in flame arrestors is to cool the propagating flame or explosion enough to extinguish the fire.

Thermal mass, usually in the form of metal, is used to extract enough energy from the reacting gases that the flame can no longer be supported and is extinguished. Many different arrestor designs are available including gauzes, perforated plates, expanded metal, sintered metal, metal foam, compressed wire wool, loose filling, hydraulic arrestors, stacked plate, and crimped ribbon.[103]

1.7 Safety Documentation and Operator Training

Safety documentation and operator training (see Volume 1, Chapter 17) provide the backbone of a strong safety program and are absolutely essential in order to maintain a safe combustion working environment. Table 1.7 illustrates some of the benefits of a successful process documentation program.[104]

The American Institute of Chemical Engineers (AIChE) Center for Chemical Process Safety publishes several titles that address implementation of process safety documentation.[55,104–106] Safety documentation for combustion-related processes includes design information, PHA reports, standard operating procedures (SOPs), and training documentation. Feedback from each of these documentation elements are linked together as part of a plant's overall process safety program. Figure 1.53 visually describes the documentation feedback linkage suggested by the AIChE.[105]

1.7.1 Design Information

Design information, or process knowledge, refers to all of the documents that pertain to the safe design of the

FIGURE 1.52
Photo of a flame arrestor.

Safety

TABLE 1.7
Benefits of a Successful Process Knowledge and Documentation Program

- Preserves a record of design conditions and materials of construction for existing equipment, which helps assure that operations and maintenance remain faithful to the original intent
- Allows recall of the rationale for key design decisions during inception stage, design, and construction of major capital projects, which is useful for a variety of reasons (i.e., an aid in future projects and modifications)
- Provides a basis for understanding how the process should be operated and why it should be run in a given way
- Offers a "baseline" for use in evaluating a process change
- Records accident/incident causes and corrective action and other operating experience for future guidance
- Protects the company against unjustified claims of irresponsibility and negligence
- Retains basic research and development information on process chemistry and hazards to guide future research effort

Source: Center for Chemical Process Safety, *Guidelines for Process Safety Fundamentals in General Plant Operations*, American Institute of Chemical Engineers, New York, 1995.

combustion system. This set of information can include, but is not limited to,[104,105]

- Process information
- Detailed information regarding the design criteria of the combustion system (i.e., heat load, process flow rates, temperatures, pressures, and fuel composition)
- Process chemistry
- Safe operating ranges for process conditions (flow, temperature, pressure, compositions, etc.)
- The known hazardous effects of deviation from the stated safe operating conditions
- Equipment information
- Process flow diagrams (PFDs)
- Piping and instrumentation diagrams (P&IDs)

- Detailed equipment drawings (e.g., heater, flare, or incinerator drawings illustrating details such as tube locations and burner orientation)
- Electrical wiring schematics and electrical classification data
- Manufacturer's equipment manuals (including design criteria and safe operation recommendations)
- Equipment, valve, and instrumentation specification sheets
- Maintenance manuals
- Automated safety interlock system details

1.7.2 Process Hazard Analysis Reports

Regardless of the evaluation technique chosen, the PHA process (see Section 1.2) puts the combustion system (as well as the entire processing unit) under the microscope, systematically analyzing the entire system piece by piece. In order to ensure that all process safety recommendations are executed, thorough documentation must be conducted in an organized manner. The PHA process is conducted both prior to initial start-up of the system and periodically (for example, every 5 years) during the system's lifetime. This cyclic approach guarantees continuous safety feedback for the operators and engineers entrusted with the safe operation of the combustion system. The AIChE provides detailed guidance on the selection and execution of various PHA evaluation techniques.[55]

1.7.3 Standard Operating Procedures

Day in and day out, SOPs provide operators with clear, detailed, sequenced instructions regarding the safe operation and maintenance of combustion equipment. In addition to providing detailed instructions for the operation of a combustion system, SOPs also assist in the training of both new and existing plant

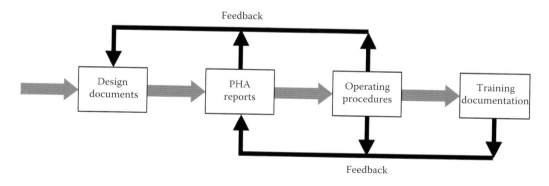

FIGURE 1.53
Safety documentation feedback flowchart. (Adapted from Center for Chemical Process Safety, *Guidelines for Process Safety Fundamentals in General Plant Operations*, Fig. 5.5, American Institute of Chemical Engineers, New York, 1995.)

FIGURE 1.54
Aerial photos of Phillips 66 Incident in Pasadena, TX, on October 23, 1989. (From Sanders, R.E., *Chemical Process Safety: Learning from Case Histories*, Butterworth-Heinemann, Boston, MA, 1999.)

personnel. The AIChE provides guidance regarding the preparation, revision, content, and distribution of SOPs.[105]

As with any set of SOPs, written procedures for combustion systems should include prestart-up, start-up, normal operation, shutdown, and emergency shutdown procedures, as well as additional procedures for preventative maintenance operations. The equipment manufacturer should always be consulted regarding the proper ignition and operation of individual combustion devices. Figure 1.54 shows the charred wreckage of the Phillips 66 plant in Pasadena, Texas where an accident was caused due to improper maintenance procedures.

1.7.4 Operator Training and Documentation

Training is essentially the successful communication and transfer of knowledge and skills (see Volume 1, Chapter 17). Applied to combustion systems, training is the successful communication and transfer of knowledge regarding basic combustion science, safety hazards associated with fires and explosions, and skillful execution of SOPs. In general, training exercises should emphasize the need to complete all tasks in a safe manner. Training is not a substitute for a poor combustion system design. Rather, the design, documentation, and training should complement each other in order to provide a safe working environment.

Employees should be selected for a level of training that is commensurate with the level of exposure they will encounter on the job. New employees should successfully complete their training regimen before they are allowed to perform training-related tasks unsupervised. Training may involve formal (attendance and examination required), informal (safety discussions, demonstrations, seminars), self-study, and on-the-job training methods.[105] Training sessions may include but are not limited to the following topics:

- Start-up operations (including combustion device ignition)
- Normal operations
- Shutdown operations (including combustion device extinguishment)
- Maintenance
- OSHA HAZWOPER (Hazardous Waste Operations and Emergency Response)
- Fuels handling
- Emergency procedures and evacuation
- Confined space entry
- Lockout and Tag-out (hazardous energy sources)
- Hazard communication
- Blood-borne pathogens
- Fire extinguishment

The safety training program must be documented thoroughly to ensure that it is conducted in an acceptable and timely manner. Management and trainers should solicit continuous feedback in order to evaluate and improve program effectiveness. The AIChE provides guidance on the successful management and implementation of a comprehensive safety training program.[104,105]

1.8 Recommendations

Industrial combustion processes have the potential to be very dangerous because of the large quantities of fuels being combusted. New and modified industrial combustion processes can be even more dangerous due to the potential of using unproven equipment designs and configurations. Rigorous safety reviews are strongly recommended to identify potential hazards so appropriate precautions can be taken to prevent equipment damage and personnel injury. There are many potential hazards that could be present during industrial combustion processes, some of which have been discussed in this chapter. Appropriate measures should be taken to prevent accidents, but plans should also be made to minimize the consequences of an accident if one should occur. Safety should be the primary concern for any industrial combustion processes.

1.9 Sources for Further Information

Besides the references cited in this chapter, there are many organizations that have good information on safety. Many of these have general safety information, but some also have specific information related to combustion. Typical organizations are listed below:

American Institute of Chemical Engineers (AIChE)
Center for Chemical Process Safety (CCPS)
3 Park Avenue
New York, NY 10016-5991
(646) 495-1371d
www.aiche.org/ccps/

American Society of Safety Engineers (ASSE)
1800 E Oakton St
Des Plaines, IL 60018
(847) 699-2929
www.asse.org

American National Standards Institute (ANSI)
1899 L Street NW, 11th Floor
Washington, DC 20036
(202) 293-8020
www.ansi.org

Board of Certified Safety Professionals (BCSP)
2301 W. Bradley Avenue
Champaign, Illinois 61821
(217) 359-9263
www.bcsp.com

Occupational Safety and Health Administration (OSHA)
U.S. Department of Labor
Office of Public Affairs, Room N3647
200 Constitution Avenue
Washington, DC 20210
(800) 321-6742
www.osha.gov

National Fire Protection Association (NFPA)
1 Batterymarch Park
PO Box 9101
Quincy, MA 02169-7471
(800) 344-3555
www.nfpa.com

Society of Fire Protection Engineers (SPFE)
7315 Wisconsin Avenue
Suite 1225 W
Bethesda, MD 20814
(301) 718-2910
www.sfpe.org

U.S. Chemical Safety Board
2175 K. Street, NW, Suite 400
Washington, DC 20037-1809
(202) 261–7600
www.csb.gov

References

1. T. Dark and C.E. Baukal, Combustion safety. In *The John Zink Combustion Handbook*. C.E. Baukal, ed., Chapter 10, 1st edn., CRC Press, Boca Raton, FL, pp. 327–349, 2001.
2. C. Baukal, Testing safety. In *Industrial Combustion Testing*. C.E. Baukal, ed., Chapter 2, CRC Press, Boca Raton, FL, pp. 41–61, 2011.
3. I. Faisal and S.A. Abbasi, Major accidents in process industries and an analysis of causes and consequences, *Journal of Loss Prevention in the Process Industries*, 12(5): 361–378, 1999.
4. Center for Chemical Process Safety, *Explosions in the Process Industries*, American Institute of Chemical Engineer, New York, 1994.
5. R.E. Sanders, *Chemical Process Safety: Learning from Case Histories*, Butterworth-Heinemann, Boston, MA, 1999.
6. T. Richardson, Learn from the Phillips explosion, *Hydrocarbon Process*, 70(3), 83–84, 1991.
7. Workplace Health, Safety and Compensation Commission of New Brunswick, Explosion and fire at the Irving Oil Refinery in Saint John, New Brunswick: Interim Report—EDB 98–22, Fredericton, New Brunswick (Canada), Government of New Brunswick, 1998.
8. Anonymous, A thirty year review of large property damage losses in the hydrocarbon chemical process industries, *Loss Prevention Bulletin*, 99, 3–25, 1991.
9. Fire Protection Association, The hydrocarbon processing industry: Fire hazards and precautions, *Fire Protection Association Journal*, 82, 252–264, 1969.
10. F.K. Crawley and G.A. Dalzell, Fire and explosion hazard management in the chemical and hydrocarbon processing industry, *IMechE Conference Transactions: Management of Fire & Explosions*, Vol. 5, London, U.K., pp. 61–72, 1997.
11. T. Kletz, *Learning from Accidents*, 2nd edn., Butterworth-Heinemann, Oxford, U.K., 1994.
12. T. Kletz, *An Engineer's View of Human Error*, 2nd edn., Institution of Chemical Engineers, Rugby, U.K., 1991.
13. Center for Chemical Process Safety, *Guidelines for Preventing Human Error in Process Safety*, American Institute of Chemical Engineer, New York, 1994.
14. R.K. Eckhoff, *Explosion Hazards in the Process Industries*, Gulf Publishing, Houston, TX, 2005.
15. American Petroleum Institute, *Safety Digest of Lessons Learned: Section 2, Safety in Unit Operations*, Publication 758, Washington, DC, 1979.
16. Y. Uehara, Fire safety assessments in petrochemical plants, *Proceedings of the Third International Symposium on International Association for Fire Safety Science*, Edinburgh, U.K., pp. 83–96, 1991.
17. W.P. Crocker and D.H. Napier, Thermal radiation hazards of liquid pool fires and tank fires, *Institution of Chemical Engineers Symposium Series No. 97*, Pergamon Press, Oxford, U.K., pp. 159–84, 1986.
18. Center for Chemical Process Safety, *Thermal Radiation 1: Sources and Transmission*, American Institute of Chemical Engineers, New York, 1989.
19. Center for Chemical Process Safety, *Thermal Radiation 2: The Physiological and Pathological Effects*, American Institute of Chemical Engineers, New York, 1996.
20. M.A. Fry, Benefits of fire and explosion computer modeling in chemical process safety, *Proceedings of the American Chemical Industries Week'94*, October 18–20, Philadelphia, PA, 1994.
21. R.A. Ogle, Explosion hazard analysis for an enclosure partially filled with a flammable gas, *Process Safety Progress*, 18(3), 170–177, 1999.
22. T. Kletz, *What Went Wrong: Case Histories of Process Plant Disasters*, 4th edn., Gulf Publishing Company, Houston, TX, 1998.
23. B. Lewis and G. Von Elbe, *Combustion, Flames and Explosions of Gases*, 3rd edn., Academic Press, New York, 1987.
24. G. Cox, *Fire Safety Science*, Routledge, London, U.K., 1990.
25. J.C. Jones and A Williams, *Topics in Environmental and Safety Aspects of Combustion Technology*, Whittles Publishing, Scotland, U.K., 1997.
26. V.C. Marshall and S. Ruhemann, *Fundamentals of Process Safety*, Institution of Chemical Engineers, London, U.K., 2000.
27. W.O.E. Korver, *Electrical Safety in Flammable Gas/Vapor Laden Atmospheres*, William Andrew Publishing, Norwich, CT, 2001.
28. N.G. Thomson, *Fire Hazards in Industry*, Butterworth-Heinemann, Oxford, U.K., 2002.
29. BP Safety Group, *Safe Furnace and Boiler Firing*, 4th illustrated edn., Institute of Chemical Engineers, Rugby, U.K., 2005.
30. NFPA 30: Flammable and combustible liquids code, National Fire Protection Association, Quincy, MA, 2008.
31. NFPA 54: National fuel gas code, National Fire Protection Association, Quincy, MA, 2009.
32. NFPA 85: Boiler and combustion systems hazard code, National Fire Protection Association, Quincy, MA, 2007.
33. NFPA 86: Standard for ovens and furnaces, National Fire Protection Association, Quincy, MA, 2007.
34. D.R. Stull, *Fundamentals of Fire and Explosion*, American Institute of Chemical Engineers Monograph Series, No. 10, Vol. 73, 1977.
35. W. Bartknecht, *Explosions*, Springer-Verlag, New York, 1980.
36. F.T. Bodurtha, *Industrial Explosion Prevention and Protection*, McGraw-Hill, New York, 1980.
37. F.P. Lees, *Loss Prevention in the Process Industries*, Vol. 1, Butterworths, London, U.K., 1980.
38. G.L. Wells, *Safety in Process Design*, George Godwin Limited, London, U.K., 1980.
39. H.H. Fawcett and W.S. Wood, *Safety and Accident Prevention in Chemical Operations*, 2nd edn., John Wiley & Sons, New York, 1982.
40. Center for Chemical Process Safety, *Guidelines for Engineering Design for Process Safety*, American Institute of Chemical Engineers, New York, 1993.
41. Center for Chemical Process Safety, *Guidelines for Safe Process Operations and Maintenance*, American Institute of Chemical Engineers, New York, 1995.

42. Center for Chemical Process Safety, *Guidelines for Process Safety Fundamentals in General Plant Operations*, American Institute of Chemical Engineers, New York, 1995.
43. American Petroleum Institute (API), *RP 750, Management of Process Hazards*, 1st edn., API, Washington, DC, 1990, reaffirmed 1995.
44. R. Skelton, *Process Safety Analysis*, Gulf Publishing, Houston, MA, 1997.
45. D.A. Crowl and J.F. Louvar, *Chemical Process Safety: Fundamentals with Applications*, Prentice Hall, Englewood Cliffs, NJ, 1990.
46. R. King, *Safety in the Process Industries*, Butterworth-Heinemann, London, U.K., 1990.
47. T.A. Kletz, *Critical Aspects of Safety and Loss Prevention*, Butterworths, London, U.K., 1990.
48. D.P. Nolan, *Handbook of Fire and Explosion Protection Engineering Principles for Oil, Gas, Chemical, and Related Facilities*, Noyes Publications, Westwood, NJ, 1996.
49. M.A. Niemkiewicz and J.S. Becker, Safety overview. In *Oxygen-Enhanced Combustion*. C.E. Baukal, ed., CRC Press, Boca Raton, FL, pp. 261–278, 1998.
50. C.E. Baukal (ed.), *Oxygen-Enhanced Combustion*, CRC Press, Boca Raton, FL, 1998.
51. R.J. Reed, *North American Combustion Handbook, Vol. I: Combustion, Fuels, Stoichiometry, Heat Transfer, Fluid Flow*, 3rd edn., North American Manufacturing Company, Cleveland, OH, 1986.
52. T.H. Pratt, *Electrostatic Ignitions of Fires and Explosions*, Burgoyne, Inc., Marietta, GA, 1997.
53. J.H. Burgoyne, *Reflections on Process Safety, Institute of Chemical Engineers Symposium Series No. 97*, Institute of Chemical Engineers, Pergamon Press, Oxford, U.K., pp. 1–6, 1986.
54. Center for Chemical Process Safety (CCPS), *Guidelines for Chemical Process Quantitative Risk Analysis*, 1st edn., American Institute of Chemical Engineers, New York, 1989.
55. Center for Chemical Process Safety (CCPS), *Guidelines for Hazard Evaluation Procedures*, 2nd edn., American Institute of Chemical Engineers, New York, 1992.
56. D.P. Nolan, *Application of the HAZOP and What-If Safety Reviews to the Petrochemical and Chemical Industries*, Noyes Publications, Park Ridge, NJ, 1994.
57. T. Kletz, *Hazop and Hazan*, 4th edn., Taylor & Francis Group, Philadelphia, PA, 1999.
58. G. Wells, *Hazard Identification and Risk Assessment*, Gulf Publishing, Houston, MA, 1996.
59. G. Wells, *Major Hazards and Their Management*, Gulf Publishing, Houston, MA, 1997.
60. C.E. Baukal, *Industrial Combustion Heat Transfer*, CRC Press, Boca Raton, FL, 2000.
61. A.G. Slavejkov, C.E. Baukal, M.L. Joshi, and J.K. Nabors, Advanced oxygen/natural gas burner for glass melting, *Proceedings of 1992 International Gas Research Conference*, Orlando, FL, November 16–19, 1992, Government Institutes, Inc., Rockville, MD, 1993.
62. P.D. Owens, Health hazards associated with pollution control and waste minimization. In *Process Engineering for Pollution Control and Waste Minimization*. D.L. Wise and D.J. Trantolo, eds., Marcel Dekker, New York, pp. 227–244, 1994.
63. C.E. Baukal and W.R. Bussman, Air infiltration effects on industrial combustion efficiency. In *Fuel Efficiency*. J.K. Bernard, ed., Chapter 4, Nova Science Publications, New York, pp. 101–134, 2011.
64. G. Sams and J. Jordan, How to design low-noise burners, *Hydrocarbon Processing*, 75(12), 101–108, 1996.
65. C.E. Baukal, *Industrial Combustion Pollution and Control*, Marcel Dekker, New York, 2004.
66. W. Bussman and J.D. Jaykaran, Noise. In *The John Zink Combustion Handbook*. C.E. Baukal, ed., CRC Press, Boca Raton, FL, pp. 223–249, 2001.
67. D.H.F. Liu, 6.7 Noise control in the transmission path. In *Environmental Engineers' Handbook*. D.H.F. Liu and B.G. Lipták, eds., 2nd edn., Lewis Publishers, Boca Raton, FL, pp. 496–503, 1997.
68. R. Moulder, Sound-absorptive materials. In *Handbook of Acoustical Measurements and Noise Control*. C.M. Harris, ed., 3rd edn., Acoustical Society of America, Woodbury, New York, 1998.
69. R.D. Reed, *Furnace Operations*, 3rd edn., Gulf Publishing, Houston, MA, 1981.
70. C.W. Nixon and E.H. Berger, Hearing protection devices. In *Handbook of Acoustical Measurements and Noise Control*. C.M. Harris, ed., 3rd edn., Acoustical Society of America, Woodbury, New York, 1998.
71. L.H. Royster and J.D. Royster, Hearing conservation programs. In *Handbook of Acoustical Measurements and Noise Control*. C.M. Harris, ed., 3rd edn., Acoustical Society of America, Woodbury, New York, 1998.
72. J. de Ris, Fire radiation: A review. In *Seventeenth Symposium (International) on Combustion*, Vol. 17, The Combustion Institute, Pittsburgh, PA, pp. 1003–1016, 1979.
73. S.R. Turns, *An Introduction to Combustion: Concepts and Applications*, 3rd edn. McGraw-Hill, New York, 2012.
74. I. Glassman and R.A. Yetter, *Combustion*, 4th edn., Academic Press, Burlington, MA, 2008.
75. C.E. Baukal, Everything you need to know about NOx, *Metal Finishing*, 103(11), 18–24, 2005.
76. C.E. Baukal, NOx 101: A primer on controlling this highly regulated pollutant, *Process Heating*, 15(2), 34–37, 2008.
77. M. Sandell, Putting NOx in a Box, *Pollution Engineering*, 30(3), 56–58, 1998.
78. NFPA, Codes and Standards, http://www.nfpa.org/categorylist.asp2categoryID=124&URL=codes%20/&%20standards, accessed November 7, 2012.
79. NFPA website: http://www.nfpa.org/, accessed November 7, 2012.
80. National Fire Protection Association, *NFPA 86: Standard for Ovens and Furnaces*, 2011 edn., NFPA, Quincy, MA, 2011.
81. National Fire Protection Association, *NFPA 70: National Electric Code (NEC)*, 2011 edn., NFPA, Quincy, MA, 2011.
82. National Fire Protection Association, *NFPA 497: Classification of Flammable Liquids, Gases, or Vapors and of Hazardous (Classified) Locations for Electrical Installations in Chemical Process Areas*, 2012 edn., NFPA, Quincy, MA, 2012.
83. National Fire Protection Association, *NFPA 58: Liquefied Petroleum Gas Code*, 2011 edn., NFPA, Quincy, MA, 2011.
84. National Fire Protection Association, *NFPA 921: Guide for Fire and Explosion Investigations*, 2011 edn., NFPA, Quincy, MA, 2011.

85. J.W. Coons, *Fire Protection Design Criteria, Options, Selection*, R. S. Means Company, Kingston, MA, 1991.
86. UL, What We Do, http://www.ul.com/global/eng/pages/aboutul/whatwedo/, accessed November 6, 2012.
87. R. Quintana, M. Camet, and B. Deliwala, Application of a predictive safety model in a combustion testing environment, *Safety Science*, 38, 183–209, 2001.
88. National Fire Protection Association, *NFPA 77: Recommended Practice on Static Electricity*, 1993 edn., NFPA, Quincy, MA, 1993.
89. J.R. Cornforth (ed.), *Combustion Engineering and Gas Utilization*, E&FN, London, U.K., 1992.
90. IHEA, *Combustion Technology Manual*, 5th edn., Industrial Heating Equipment Association, Arlington, VA, 1994.
91. M.A. Niemkiewicz and J.S. Becker, Equipment design. In *Oxygen-Enhanced Combustion*. C.E. Baukal, ed., CRC Press, Boca Raton, FL, pp. 279–313, 1998.
92. D.R. Lide (ed.), *CRC Handbook of Chemistry and Physics*, 80th edn., CRC Press, Boca Raton, FL, 1999.
93. GPSA, *GPSA Engineering Data Book*, Vol. 2, 10th edn., Gas Processors and Suppliers Association, Tulsa, OK, 1994.
94. H. Le Chatelier, Estimation of firedamp by flammability limits, *Annales des Mines*, 19(8), 388–395, 1891.
95. M.G. Zabetakis, S. Lambiris, and G.S. Scott, Flame temperatures of limit mixtures. In *Seventh Symposium on Combustion*, Butterworths, London, U.K., pp. 484, 1959.
96. M.G. Zabetakis, Fire and explosion hazards at temperature and pressure extremes, *Proceedings of Chemical Engineering Under Extreme Conditions, American Institute of Chemical Engineers Symposium Series,* London, No 2, pp. 99–104, 1965.
97. R.H. Perry, D.W. Green, and J.O. Maloney (eds.), *Perry's Chemical Engineers' Handbook*, 7th edn., McGraw-Hill, New York, 1997.
98. National Safety Council, *Accident Prevention Manual for Industrial Operations*, National Safety Council, 1974.
99. M.G. Zabetakis, *Flammability Characteristics of Combustible Gases and Vapors*, U.S. Bureau of Mines Bulletin 627, USNT AD 701, 576, 1965.
100. National Fire Protection Association, *NFPA 77: Recommended Practice on Static Electricity*, 1993 Edition, NFPA, Quincy, MA, 1993.
101. V.A. Mendoza, V.G. Smolensky, and J.F. Straitz, Don't detonate—Arrest that flame, *Chemical Engineering*, 103(5), 139–142, 1996.
102. V.A. Mendoza, V.G. Smolensky, and J.F. Straitz, Understand flame and explosion quenching speeds, *Chemical Engineering Progress*, 89, 38–41, 1993.
103. H. Phillips and D.K. Pritchard, Performance requirements of flame arresters in practical applications, *Institute of Chemical Engineers Symposium Series*, London, No. 97, Pergamon Press, Oxford, U.K., pp. 47–61, 1986.
104. Center for Chemical Process Safety, *Plant Guidelines for Technical Management of Chemical Process Safety*, Revised Edition, American Institute of Chemical Engineers, New York, 1995.
105. Center for Chemical Process Safety, *Guidelines for Process Safety Documentation*, American Institute of Chemical Engineers, New York, 1995.
106. Center for Chemical Process Safety, *Guidelines for Technical Management of Chemical Process Safety*, American Institute of Chemical Engineers, New York, 1989.

2
Combustion Controls

Rodney Crockett and Jim Heinlein

CONTENTS

2.1 Fundamentals
    2.1.1 Control Platforms 46
        2.1.1.1 Relay System 46
        2.1.1.2 Burner Controller 46
        2.1.1.3 Loop Controller 46
        2.1.1.4 Programmable Logic Controller 47
        2.1.1.5 Distributed Control System 47
        2.1.1.6 Hybrid Systems 47
        2.1.1.7 Future Systems 48
    2.1.2 Discrete Control Systems 48
    2.1.3 Analog Control Systems 48
    2.1.4 Failure Modes 48
    2.1.5 Agency Approvals and Safety 50
        2.1.5.1 Double Block and Bleed for Fuel Supply 50
        2.1.5.2 Unsatisfactory Parameter System Shutdown 51
        2.1.5.3 Local Reset Required after System Shutdown 51
        2.1.5.4 Watchdog Timer to Verify PLC Operation 52
        2.1.5.5 Critical Input Checking to Verify PLC Operation 52
        2.1.5.6 Master Fuel Trip Relay Operation 52
    2.1.6 Pipe Racks and Control Panels 53
2.2 Primary Measurement 54
    2.2.1 Discrete Devices 54
        2.2.1.1 Annunciators 56
        2.2.1.2 Pressure Switches 56
        2.2.1.3 Position Switches 56
        2.2.1.4 Temperature Switches 56
        2.2.1.5 Flow Switches 56
        2.2.1.6 Run Indicators 56
        2.2.1.7 Flame Scanners 57
        2.2.1.8 Solenoid Valves 57
        2.2.1.9 Ignition Transformers 57
    2.2.2 Analog Devices 57
        2.2.2.1 Control Valves 58
        2.2.2.2 Thermocouples 58
        2.2.2.3 Velocity Thermocouples 59
        2.2.2.4 Resistance Temperature Detectors 59
        2.2.2.5 Pressure Transmitters 60
        2.2.2.6 Flow Meters 60
        2.2.2.7 Analytical Instruments 60
2.3 Control Schemes 61
    2.3.1 Parallel Positioning 62
        2.3.1.1 Mechanical Linkage 62
                62

2.3.1.2 Electronic Linkage 63
2.3.1.3 Characterizer Calculations 64
2.3.2 Fully Metered Cross Limiting 65
2.4 Controllers 67
2.5 Tuning 70
References 71

2.1 Fundamentals

This chapter discusses the various control system components, concepts, and philosophies necessary for understanding how control systems work, what the systems are designed to accomplish, and what criteria the controls engineer uses to design and implement a system. The interested reader can find further information on controls in numerous references.[1–11]

The purpose of the control system is to start, operate, and shut down the combustion process and any related auxiliary processes safely, reliably, and efficiently. The control system consists of various physical and logical components chosen and assembled according to a control philosophy and arranged to provide the user with an informative, consistent, and easy-to-use interface.

A combustion system typically includes a fuel supply, a combustion air supply, and an ignition system that all come together at one or more burners. During system start-up and at various times during normal operation, the control system needs to verify or change the status of these systems. During system operation, the control system needs various items of process information to optimize system efficiency. Additionally, the control system monitors all safety parameters at all times and will shut down the combustion system if any of the safety limits are not satisfied.

2.1.1 Control Platforms

The control platform is the set of devices that monitors and optimizes the process conditions, executes the control logic, and controls the status of the combustion system. There are several different types of platforms and several different ways that the tasks mentioned earlier are divided among the types of platforms. Following is a list and a brief description of the most commonly used platforms.

2.1.1.1 Relay System

A relay consists of an electromagnetic coil and several attached switch contacts that open or close when the coil is energized or de-energized. A relay system consists of a number of relays wired together in such a way that they execute a logical sequence. For example, a relay system can define a series of steps to start up the combustion process. Relays can tell only if something is on or off and have no analog capability. They are generally located in a local control panel.

Relays have several advantages. They are simple, easily tested, reliable, and well-understood devices that can be wired together to make surprisingly complex systems. They are modular, easily replaced, and inexpensive. They can be configured in fail-safe mode so that if the relay itself fails, combustion system safety is not compromised.

There are also a few disadvantages. Once a certain complexity level is reached, relay systems can quickly become massive. Although individual relays are very reliable, a large control system with hundreds of relays can be very unreliable. Also, relays take up a lot of expensive control panel space. Because relays must be physically rewired to change the operating sequence, system flexibility is poor.

Today, the durability of the electromagnetic relay has been combined with the flexibility of the programmable logic controller (PLC) to create a programmable relay. Initially, it had a lot of limitations. But they now include such functions as timers, counters, expansion I/O, security protection, and multiple language options. They are programmed with either manufacturer-provided programming software or with the LCD faceplate. The programs can also be simulated and stored using nonvolatile memory. And, they include such options as additional communication and analog inputs. The LCD faceplate can also be used to control the process or provide status information.

2.1.1.2 Burner Controller

A variety of burner controllers is available from several different vendors. They are prepackaged, hardwired devices in different configurations to operate different types of systems. Burner controllers will execute a defined sequence and monitor defined safety parameters. They are generally located in a local control panel. Like relays, they generally have no analog capability.

Advantages of burner controllers include the fact that they are generally inexpensive, are compact, are simple to hook up, require no programming (except dipswitch configuration), are fail-safe, and are very reliable. They are often approved for combustion service by various safety agencies and insurance companies (see Chapter 1).

There are also some disadvantages. Burner controllers cannot control combustion systems of much complexity. System flexibility is nonexistent. If it becomes necessary to change the operating sequence, the controller must be rewired or replaced with a different unit. Controllers also require the use of attached peripherals from the same vendor, so some design flexibility is lost.

2.1.1.3 Loop Controller

A loop controller has the flexibility and capability for implementation of advanced control and batch sequencing. This includes onboard I/O and a vast array of different function blocks used for different types of control strategies including single loop, multiloop, and cascade. They usually include several different networking communication options and optional I/O expansion. They are programmed with either manufacturer-provided programming software or with the faceplate.

Advantages include the fact that they are generally inexpensive, self-contained units, single or multiple loops, security options, and a vast set of function block commands.

There are also some disadvantages. They have limited I/O capability, usually only one loop can be displayed at a time on the faceplate, and communication can be limited.

2.1.1.4 Programmable Logic Controller

A PLC is a small modular computer system that consists of a processing unit and a number of input and output modules that provide the interface to the combustion components. They usually consist of a single processor and I/O, but they can be configured with redundant processors and/or I/O. PLCs are usually rack mounted and modules can be added or changed (see Figure 2.1). There are many types of modules available. Unlike the relays and burner controllers earlier, PLCs have analog control capability. They are generally located in a local control panel.

PLCs have the advantage of being a mature technology. They have been available for over 30 years. Simple PLCs are inexpensive and prices are generally very competitive. They are compact, relatively easy to hook up, and, because they are programmable, supremely flexible. They can operate systems of almost any level of complexity. PLC reliability has improved over the years and is now very good.

Disadvantages of PLCs include having to write software for the controller. Coding can be complex and creates the possibility of making a programming mistake,

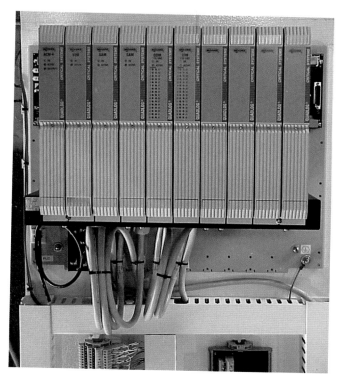

FIGURE 2.1
Programmable logic controller.

which can compromise system safety. The PLC may also have internal or I/O failures that could result in inputs and outputs not responding correctly. Because of this possibility, standard PLCs should never be used as a primary safety device without additional safety precautions. Additional safety functions could include an external watchdog timer, external removing of power to outputs, critical input checking, and redundant input channels per single sensor. Special types of redundant or fault-tolerant PLCs are available that are more robust and generally accepted for this service, but they are very expensive and generally difficult to implement.

2.1.1.5 Distributed Control System

A distributed control system (DCS) is a larger computer system that can consist of a number of processing units and a wide variety of input and output interface devices. Unlike the systems described earlier, when properly sized, a DCS can also control multiple systems and even entire plants. The DCS is generally located in a remote control room, but peripheral elements can be located almost anywhere.

DCSs have been around long enough to be a mature technology and are generally well understood. They are highly flexible and are used for both analog and discrete (on/off) control. They can operate systems of almost any level of complexity and their reliability is excellent.

However, DCSs are often difficult to program and are often restricted when communicating with other hardware platforms. Each DCS vendor has proprietary system architecture, so the hardware is expensive and the software is often different from any other vendor's software. Once a commitment is made to a particular DCS vendor, it is extremely difficult to change to a different one.

2.1.1.6 Hybrid Systems

If one could combine several of the systems listed earlier and build a hybrid control system, the advantages of each system could be exploited. In practice, that is what is usually done. A typical system uses relays to perform the safety monitoring, a PLC to do the sequencing, and either dedicated controllers or an existing DCS for the analog systems control. Sometimes, the DCS does both the sequencing and the analog systems control, and the safety monitoring is done by a fault-tolerant logic system. Most approval agencies and insurers require the safety monitoring function to be separate from either of the other functions.

2.1.1.7 Future Systems

Over the next decade or so, it is expected that embedded industrial microprocessors using touchscreen video interfaces (see Figure 2.2) will continue to gain market share in combustion control. These interfaces will communicate with field devices such as valves and switches via a single communications cable. They will use a digital bus protocol such as Profibus or Fieldbus. These systems are becoming more common on factory floors around the world. Because establishment of a single standard has not yet happened and combustion standards are slow to change, these systems have not yet achieved widespread acceptance in the combustion world.

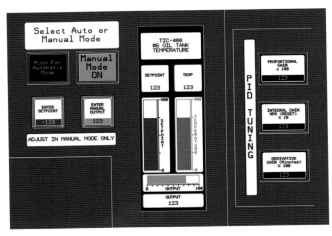

FIGURE 2.2
Touchscreen.

2.1.2 Discrete Control Systems

The world of discrete controls is black and white. Is the valve open or shut? Is the switch on or off? Is the button pressed or not? Is the blower running or not? There are two basic types of discrete devices: (1) input devices (sensors) that have electrical contacts that open or close depending on the status of what is being monitored and (2) output devices, or final elements, that are turned on or off by the control system.

In a typical control system, sensors such as pressure switches, valve position switches, flame scanners, and temperature switches do all the safety and sequence monitoring. These devices tell the control system what is happening out in the real world. They are described in more detail in Section 2.1.3.

The final elements carry out the on/off instructions that come from the control system. These are devices such as solenoid valves, relays, indicating lights, and motor starters. These devices allow the control system to make things happen in the real world and are described in more detail in Section 2.1.3.

The discrete control system does safety monitoring and sequencing. Typically, the system monitors all of the discrete inputs and, if they are all satisfactory, allows combustion system start-up. If a monitored parameter is on when it should be off, or vice versa, the start-up process is aborted and the system must be reset before another start-up is permitted. The system also controls such things as which valves are opened in what order, if and when the pilot is ignited, and if and when main burner operation is allowed. Once the system starts, the discrete system has little to do other than monitor safety parameters. If any of the defined safety parameters are not satisfactory, the system immediately shuts down. Figure 2.3 is a simplified flow diagram showing a standard burner light-off sequence.

2.1.3 Analog Control Systems

The world of analog controls is not black and white—it is all gray. How far open is that valve? What is the system temperature? How much fuel gas is flowing? As transmitters replace switches in basic discrete control systems, the additional diagnostics and flexibility now available for those systems will resolve some of the complex operational issues required today.

There are two categories of analog devices with familiar names: (1) sensors, which measure some process variable (PV) (e.g., flow or temperature) and generate a signal proportional to the measured value, and (2) final elements (e.g., pumps and valves), which change their status (e.g., speed or position) in response to a proportional signal from the control system.

In contrast to the discrete control system, the analog control system usually has few tasks to perform

Combustion Controls

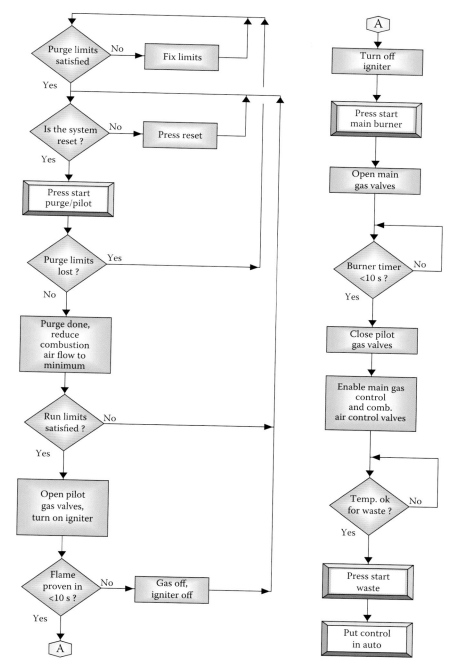

FIGURE 2.3
Simplified flow diagram of a standard burner light-off sequence.

until the system completes the start-up sequence and is ready to maintain normal operation. Most analog devices are part of a control loop. A simple loop consists of a sensor, a final element, and a controller. The controller reads the sensor, compares the measured value to its setpoint (SP) set by the operator, and then positions the final element to make the measured value equal the SP. Figure 2.4 illustrates a simple analog loop.

In this case, the thermocouple transmits the temperature to the controller. If the temperature is higher than the SP, the controller will decrease its signal to the control valve. This will decrease the fuel flow to the burner, thus lowering the temperature. In this way, the loop works to maintain the desired temperature—also known as the SP.

The previous illustration is a good example of a simple feedback system. After the controller adjusts the control valve, the resulting change in temperature is fed back

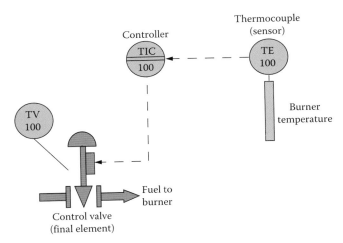

FIGURE 2.4
Simple analog loop. TV, temperature value; TIC, temperature indicator controller; TE, temperature element.

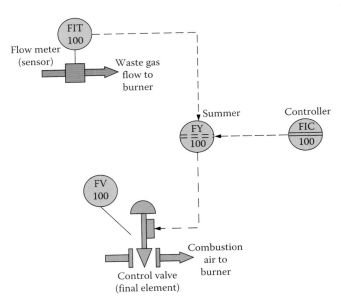

FIGURE 2.5
Feedforward loop. FIT, flow indicating transmitter; FV, flow valve; FY, flow Relay; FIC, flow indicating controller.

to the controller. This way, the controller "knows" the result of the adjustment and can make a further adjustment if it is required. Another good example of feedback takes place whenever one drives a car. If one gets on the expressway and decides to drive at 60 mph (97 km/h), one presses the accelerator and watches the speedometer. Near 60 mph, one begins to ease off the accelerator so as not to overshoot. From then on, one glances at the speedometer every now and then and adjusts one's foot position as necessary.

Feedback alone is not enough, however. What if there is traffic congestion? One begins to slow down in anticipation. This is called feedforward, which occurs when one changes the operating point because some future event is about to happen and one needs to prepare for it. Feedforward is commonly used in combustion control systems. A good example from the world of combustion is waste flow. In a waste destruction system, what happens if the waste flow suddenly doubles? There will no longer be enough combustion air in the system to allow destruction of all the waste. Unburned waste will burn at the tip of the smokestack and clouds of smoke will billow from the stack. Phone calls from irate neighbors will soon begin to accumulate. Using feedforward, as shown in Figure 2.5, the waste flow is measured. When it doubles, the combustion airflow SP immediately increases by a similar amount, avoiding all of the unpleasant consequences listed earlier.

2.1.4 Failure Modes

Almost everything fails eventually. No matter how well the components of a control system are designed and built, some of them will fail from time to time. One of the primary tasks of the controls engineer is to design the control system so that failure of one or even several components will not cause a safety problem with the combustion system.

All components used in a control system have one or more defined failure modes. For example, if a discrete sensor fails, it will most likely cause the built-in switch contacts to fail open. To design a safe system, the controls engineer must choose and install the sensor so that when an alarm condition is present, the switch contacts will open. Thus, the alarm condition coincides with the failure mode. If it did not, the sensor could fail and the control system would still think everything was normal and attempt to keep operating as before—a condition that could be catastrophic.

In addition to sensors, final control elements also have failure modes. The controls engineer can usually select the desired failure mode. If there is an actuated valve that turns the fuel gas supply on and off, the actuator is installed so that the valve will spring closed (fail shut) upon loss of air. In addition, assume there is a solenoid valve that turns air to the actuator on and off. The solenoid valve should be selected and installed to rapidly dump air from the actuator upon loss of electrical power, thus closing the valve. These designs ensure that the two most likely circumstances of component failure enhance system safety.

Construction of a well-designed system ensures that every component that can fail is installed so that component failures do not compromise system safety. At its core, that is what controls engineering is all about.

2.1.5 Agency Approvals and Safety

Worldwide, there are hundreds of private, governmental, and semigovernmental safety organizations. Each ostensibly has the proper implementation of safety (see Chapter 1) at the top of its agenda. Some agencies are concerned with the electrical safety and reliability of the components used in a control system; others are concerned about preventing explosions caused by sparking equipment in a gaseous atmosphere; and still others are concerned with the proper design of control systems to ensure safe operation of various combustion processes. No single organization does all of the things listed.

The design of combustion systems in the United States should include specifications that meet National Fire Protection Association (NFPA) and National Electrical Code (NEC) standards. In accordance with the applicable standards and years of experience in the field, systems should be designed with some or all of the following safety features.

2.1.5.1 Double Block and Bleed for Fuel Supply

This means that there are two fail-closed safety shutoff block valves with a fail-open safety shutoff vent valve located between them, as shown in Figure 2.6. Each of the three safety shutoff valves (SSOVs) in the double-block-and-bleed system has a position switch not shown in the figure. For a system purge to be valid, the block valves must be shut and verified. For burner light-off, the vent valve is shut. After the vent valve position switch confirms that the valve is shut, the two block valves are opened. If there is a system failure, all three of the valves de-energize and return to their failure positions. Note that if the upstream block valve ever leaks, the leakage will preferentially go through the open vent valve and vent to a safe location rather than into the burner. The vent valve is not provided in some applications that include heavier than air gases. An alternate solution could be to provide an automated leak test or valve proving systems.

2.1.5.2 Unsatisfactory Parameter System Shutdown

An unsatisfactory parameter for any critical input immediately shuts down the system. The control system typically receives critical input information as shown in Figure 2.7. The pressure switch PSLL-03073 is wired so that if it fails, the voltage is interrupted to the relay (CR-xx) and the PLC. The PLC will then shut down the system. If either the switch or the relay fails, the system shuts down.

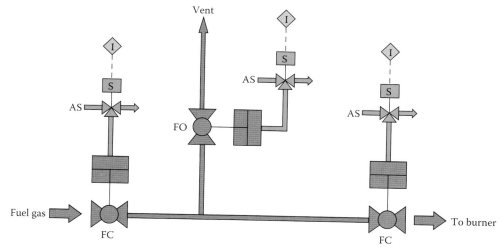

FIGURE 2.6
Double-block-and-bleed system. AS, air supply; FC, fail closed; FO, fail open; pneumatic valve; solenoid valve.

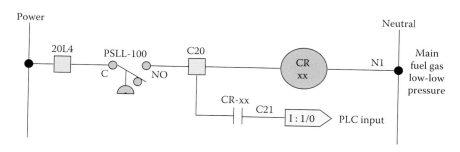

FIGURE 2.7
Fail-safe input to PLC.

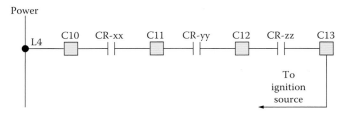

FIGURE 2.8
Shutdown string.

In addition, the relay has another contact in series with all the other critical contacts. If any of these contacts open, the power cuts off to all ignition sources (all fuel valves, igniters, etc.), immediately shutting down the combustion system.

An alternate method could be the use of a critical input checking circuit or single sensor wired to redundant inputs and an external hardwired master fuel trip (MFT) circuit to remove power from all ignition sources.

In Figure 2.8, if there is a failure anywhere in the circuitry, the system shuts down. Even if damage to the PLC occurs, the relays will shut down the system. This is an excellent example of redundancy, fail-safe design principles, and effective design philosophy.

2.1.5.3 Local Reset Required after System Shutdown

After a system shutdown caused by an alarm condition, the system allows a remote restart only after an operator has pressed a reset button located at the combustion system. The operator should perform a visual inspection of the system and verify the correction of the condition that caused the shutdown.

2.1.5.4 Watchdog Timer to Verify PLC Operation

The BMS logic system has many levels of diagnostics to continuously verify the integrity of its internal systems and operation. These diagnostics test for various failures such as (but not limited to) program or data memory failure, program logic corruption, loss of communication with an input or output module, certain failures within the input and output modules, and program execution exceeding a preset time limit. Any failure that is considered critical enough to compromise the system will cause the logic system to stop executing the BMS program logic.

The physical design of the logic system should be such that any time program logic execution stops all outputs revert to the off state. Since the BMS energizes fuel shutoff valves to open them, this assures that all fuel shutoff valves de-energize and close when any of these failures are detected.

One method to meet the requirement of NFPA 85[12] and to include additional redundancy, is to utilize a specialized watchdog timer that is physically separate from the logic system that can directly open the MFT circuit when it detects logic system trouble. The watchdog timer works by continuously monitoring a "strobe" output from the BMS logic system. The BMS continuously alternates the state of the strobe (from off to on and back to off again). The watchdog timer electronics are specifically designed to detect a constant change of state of this strobe input. As long as the strobe changes state, the timer holds its output contact closed. If the BMS logic were to halt for any reason, the strobe from it to the watchdog timer would freeze at some state. The watchdog timer sees this as a failure and opens its contact which is in series with the MFT circuit. Note that although the BMS normally would cause its output to turn off in a failed state, the watchdog timer would detect even a failed-on state as a failure.

2.1.5.5 Critical Input Checking to Verify PLC Operation

In the earlier days of burner management systems (BMS), limit switches and interlocks were usually hardwired in series to the master fuel trip relay (RMFT). This provided a high level of confidence that a limit changing from normal to a trip state would cause the burner to trip by directly causing power to be removed from all of the fuel shutoff valves. Later, as electronic programmable logic devices became popular, these hardwired trip circuits were still used, but the logic system was added primarily to handle sequencing (replacing hardwired timers), alarm detection, and a redundant means of shutting down the system.

Eventually, as logic systems came into more mainstream usage and experience has been gathered to prove their reliability, the hardwired limit circuits have been eliminated. Limit switches and interlocks are now wired directly to logic system inputs. The benefit of this is that much more advanced alarming and diagnostics can be obtained than through the traditional hardwired approach. However, with the elimination of the ability for these limits to directly de-energize the fuel valves (via the RMFT), the logic system must be designed so that it can provide at least the same level of confidence of detecting a limit or interlock changing to a trip state.

Basically with discrete inputs, there are two failure modes—the logic system sees an input as "off" even if it is "on" or vice versa. In the case where the logic system incorrectly sees an input from a limit switch as being in an incorrect position ("off" for an open circuit) when in fact it is satisfied ("on" for a closed circuit), the logic system would simply cause a nuisance shut down of the burner.

However, if such an input were to fail in the "on" state, the logic system could not detect if the limit switch changed to a trip state and the burner could continue to fire under unsafe conditions. So the logic system must continuously test for this particular failure mode.

There are several ways to accomplish this, one example being to use input modules with the logic system that have added functionality specifically for detection of an input that has failed "on."

An alternate approach that has successfully been used for many years is to have the "critical input checking" scheme as described next.

Each input to the logic system, to which a limit or interlock switch is wired, is classified as a critical input (as opposed to noncritical inputs from devices like push buttons or selector switches). In turn, interrogation voltage that powers all of these devices is controlled from a logic system output. Periodically (about every 4 s), the logic system momentarily removes power (for a very short time) from the output, which then should cause all of the critical inputs to also be detected as "off." During the time when the output has been turned off, the logic system verifies that all of these critical inputs are changed states. Any input that did not change state will be marked as having failed.

Power to the critical input circuits is removed about once every 4 s and only for about 300 ms. During the time that power is removed, the BMS logic uses the last known state of each critical input. This 300 ms window is considered short enough that if a limit goes off normal during the power "off" period, the BMS is still able to detect a shutdown condition and shut down the system well within 1 s.

Because of the continuous cycling nature of the logic system output that powers all of the critical input circuits, the output must be a solid-state (triac) type. A relay contact-type output would quickly wear out. Triacs are sensitive to overcurrent damage; therefore, the output circuit includes a "critical input power" fuse. This fuse will blow quickly if a short occurs (most likely to occur when servicing field instrumentation).

2.1.5.6 Master Fuel Trip Relay Operation

The purpose of the MFT circuit (see Figure 2.9) is to provide a way to rapidly shut off all fuels to the burner. NFPA 85 defines the MFT in this way and provides many references to its use according to the type of equipment.

One of the key elements of the circuit is the RMFT itself. The relay is in an energized state during normal operation, but becomes de-energized (to its fail-safe state) in the event that the system must be rapidly shut down. Normally, open contacts on the relay are wired in series with power that feeds all fuel shutoff valves. The relay must be energized for any of the shutoff valves to also be energized and thus open. Another key element is that the RMFT is in a hardwired circuit completely independent of the BMS control system (i.e., PLC- or DCS-based logic system) so that an effective MFT can be initiated regardless of the state of the BMS logic system.

In the past, there have been several variations in the MFT circuit. The BMS logic can be implemented in roughly three ways. The first is a strictly hardwired system of limit switches, relays, and some sort of timing mechanism(s) for sequential control to energize the fuel valves. The second is a hybrid mix of this type of hardwired system, but enhanced with a programmable logic system. And more recently, one in which the logic system directly monitors all limits and controls the fuel shutoff valves. Many of the variations in the MFT circuit were due to or related to these BMS differences.

There are many variations of this concept from the different systems. This is one method that has been

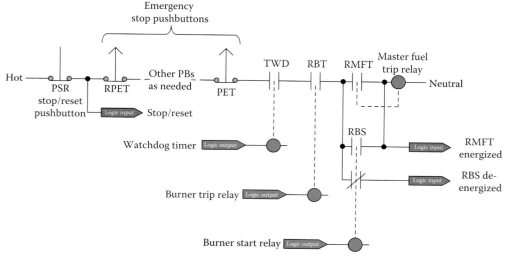

FIGURE 2.9
Master fuel trip circuit.

developed for an MFT circuit in a BMS. This operating narrative incorporates this standardized design.

The MFT circuit utilizes a "self-sealing" relay. That is, it uses a contact from itself to close the circuit around the device that first energizes the circuit. Once energized, the relay remains energized until the circuit is interrupted from an upstream source as described later.

The device that initially closes the circuit is a "burner start relay" (RBS). This relay is controlled by the logic system and is intended to only close momentarily. In fact, failure of the RBS in the "start" position can interfere with the normal operation of the MFT circuit, and so a dedicated signal from the same contact pair as the start contact is wired to a logic system input. If the logic system detects that RBS has indeed failed in the start state, the BMS opens the circuit via the burner trip relay (RBT).

2.1.6 Pipe Racks and Control Panels

For most combustion control systems, two major assemblies comprise the bulk of the system: the pipe rack and the control panel. A pipe rack is shown in Figure 2.10. Sometimes called a skid, the pipe rack is a steel framework that has a number of pipes and associated components attached to it. Usually, most of the combustion process feeds such as air, fuel, and waste have their shutoff and control elements located on the pipe rack. This makes maintenance and troubleshooting more convenient and reduces the amount and complexity of wiring systems required to connect all of the components.

Typical control panels are shown in Figure 2.11a and b. Figure 2.12 shows the inside of the large control panel. The control panel is usually attached to the pipe rack. All of the devices on the pipe rack, as well as the field devices, are electrically connected to the control panel. The control system usually resides inside the control panel. In addition to the wiring, maintenance, and troubleshooting benefits mentioned earlier, another benefit to packaging the control system is far more important—the people who designed and built the system can test and adjust it at the factory. When the control system arrives at the job site, installation consists mostly of hooking up utilities and the interconnecting piping, minimizing the amount of expensive on-site troubleshooting, and tuning.

2.2 Primary Measurement

This section describes a number of different analog and discrete devices used to provide the interface between combustion systems and control systems. All of these devices are available from numerous suppliers around the world. There are vast differences in price, quality, and functionality among the different devices and suppliers.

FIGURE 2.10
Typical pipe rack.

Combustion Controls

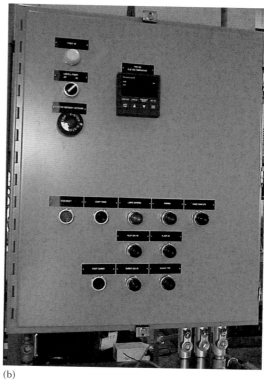

FIGURE 2.11
(a) Large control panel. (b) Small control panel.

FIGURE 2.12
Inside the large control panel.

2.2.1 Discrete Devices

2.2.1.1 Annunciators

An annunciator is a centralized alarm broadcasting and memory device. It can be a physical hardwired device as described later. Or, this function can be included in the existing graphics and/or data logging system. Typically, all of the alarms in a control system are routed to the annunciator. When any alarm is triggered, the light associated with that alarm will flash and the annunciator horn will sound. If the annunciator is a "first-out" type, any subsequent alarms that occur will trigger their associated lights to come on solidly—rather than flashing, as the first alarm does. This is very useful for diagnosing system problems. When a safety shutdown occurs, other alarms are usually triggered, while the system is shutting down. With a first-out capability, the original cause of the shutdown can easily be determined.

2.2.1.2 Pressure Switches

Pressure switches (see Figure 2.13) are sensors that attach directly to a process being measured. They can be used to detect absolute, gage, or differential pressure. The switches generally have a pressure element such as a diaphragm, tube, or bellows that expands or contracts against an adjustable spring as pressure changes. The element attaches to one or more sets of contacts that open or close upon reaching the SP. The devices are used in a number of ways, but in combustion systems, they are usually used to test for high and low fuel gas pressure. They are normally set so the contacts are open when in the alarm condition.

2.2.1.3 Position Switches

Also called limit switches, these sensors attach to or are built into valves, insertable igniters, and other devices. Position switches usually employ a mechanical linkage, but proximity sensors are also quite common. They are adjustable and can tell the control system if a valve is open, closed, or in some defined intermediate position. Position switches are not usually used for alarms. In combustion systems, they are generally used to check valve positions during purges and burner light-off sequences. Position switches are often used with integrated beacons and other visual devices. They are normally installed so that their contacts are closed only when the valve is in the desired safety position.

2.2.1.4 Temperature Switches

Temperature switches are usually attached to auxiliary equipment such as tanks or flame arresters.

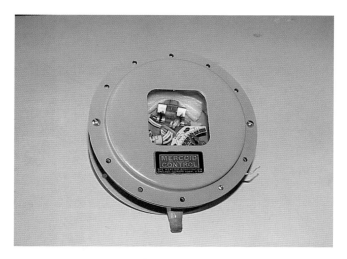

FIGURE 2.13
Two different types of pressure switches.

These sensors generally do not have the range to test for combustion system temperatures, so those applications use other devices. The switches usually use a bimetallic element, where the differential expansion of two different metals generates physical movement. The movement opens or closes one or more sets of contacts. The failure mode of temperature switches is not always predictable. Generally, installation requires open contacts when the switch is in the alarm condition.

2.2.1.5 Flow Switches

Flow switches are sensors that generally insert into the pipe or duct in which flow is measured. Because of the lack of a quantitative readout and the improved reliability of analog transmitters in this service, these devices are becoming less common. Their failure mode is not always predictable. Usual installation requires open contacts when the switch is in the alarm condition.

2.2.1.6 Run Indicators

A run indication sensor shows whether or not a pump or fan is running. It is usually possible to order a motor starter with a built-in set of signal contacts that close when the starter motor contacts are closed. However, that does not always ensure that the pump is running and pumping fluid. A magnetic shaft encoder rotates a magnetic slug past a pickup sensor every revolution and provides positive indication of shaft revolution, but that too does not always ensure that the pump is pumping fluid. It is usually preferable to have a pressure or flow indicator that shows that the system is functioning normally and moving fluid.

2.2.1.7 Flame Scanners

Flame scanners (see Figure 1.31) are crucial to the safe operation of a combustion system. If the flame is out, the fuel flow into the combustion enclosure is stopped and the area is purged before a relight can be attempted. Flame scanners come in two main varieties: infrared and ultraviolet. The name tells which section of the electromagnetic spectrum it is designed to see. Depending on the type of fuel, the correct scanner should be chosen for the application. The detector is either a gas-filled tube or silicon chip that emits as long as the flame continues. When the flame stops, the current stops. There is a 1–4 s delay (usually programmable), to minimize spurious shutdowns, and then the contacts open to designate the alarm condition. Most systems have two flame scanners and both scanners must fail to achieve system shutdown. Use of infrared scanners is desirable if there is a waste stream, such as sulfur, that absorbs ultraviolet light and makes operation of ultraviolet scanners unreliable. Self-checking scanners are usually used. They have output contacts that open on either loss of flame or failure of the self-check. Sometimes, the contacts are part of an amplifier/relay unit located in the control panel, but most newer systems have everything located in the scanner housing, which mounts on the burner end plate. One limitation with flame scanners is the possibility of the power wire to the scanner inducing false flame indications in the signal wire from the scanner. If there are separate wires for scanner power and scanner signal, they must run in separate conduits or shielding to prevent false signals caused by induction.

2.2.1.8 Solenoid Valves

Solenoid valves are turned on or off by the presence or absence of voltage from the control system. A solenoid valve has a relay coil that links mechanically to a valve disk mechanism. Energizing the solenoid causes the linkage to push against a spring to reposition the valve disk. De-energizing the solenoid allows the spring to force the valve to the failure position. The most common types of solenoid valves are two-way and three-way valves. Two-way valves have two positions. They either allow flow or they do not. They are often used to turn pilot gas on and off. Three-way solenoid valves have three ports but still only two positions. If ports are labeled A, B, and C, energizing the valve may allow flow between ports A and B, while de-energizing the valve may allow flow between ports A and C. It is very important to carefully select, install, and test three-way solenoid valves. Three-way solenoid valves typically attach to control valves and SSOVs. In the case of control valves, when the solenoid valve is energized, the control valve is enabled for normal use. When the solenoid valve is de-energized, the instrument air is dumped from the control valve actuator diaphragm, causing the control valve to go to its spring-loaded failure position. For SSOV service, the solenoid valve is hooked up so that when energized, instrument air is allowed to reposition the SSOV actuator away from its spring-loaded failure position. When the solenoid valve is de-energized, the air is dumped from the SSOV actuator, causing the control valve to go to its spring-loaded failure position. The failure modes of the solenoid valve, control valve, and SSOV are coordinated to maximize system safety no matter which component fails.

2.2.1.9 Ignition Transformers

Ignition transformers supply the high voltage necessary to generate the spark used to ignite the pilot flame during system light-off. The type of transformer usually used converts standard AC power to a continuous 6000 V DC. This voltage then continuously jumps the spark gap at the igniter, which is located at the head of the pilot burner. High-energy igniters provide a more intense spark. A high-energy igniter is similar to the transformer mentioned earlier except that a capacitor is included to store energy and release it in spurts, resulting in a more intense spark. Both types of transformers are usually located close to the burner in a separate enclosure and hooked to the igniter using coaxial cable similar to the spark plug wire used in cars.

2.2.2 Analog Devices

2.2.2.1 Control Valves

Control valves are among the most complex and expensive components in any combustion control system. Numerous books document the nearly infinite variety of valves.[13–20] Misapplication or misuse of valves compromises system efficiency and safety. Controls engineers cannot simply pick control valves from a catalog because they are the right size for the line where they will be used. Control valves must be engineered for their specific application. A typical pneumatic control valve is shown in Figure 2.14.

As shown in Figure 2.15, the type of service and control desired determines the selection of different flow characteristics and valve sizes. Controls engineers use a series of calculations to help with this selection process. A typical control valve consists of several components that are mated together before installation in the piping system.

2.2.2.1.1 Control Valve Body

The control valve body can be a globe valve, a butterfly valve, or any other type of adjustable control valve. Usually, special globe valves of the equal percent type are used for fuel gas control service or liquid service. Control of combustion air and waste gas flows generally require the use of butterfly valves—often the quick-opening type. Because the combustion air or waste line usually has a large diameter and because the cost of globe valves quickly becomes astronomical after line size exceeds 3 or 4 in. (8–10 cm), butterfly valves are usually the most economical choice. In Section 2.3, a discussion of parallel positioning describes how controls engineers use a globe valve and a butterfly valve together to work smoothly for system control.

2.2.2.1.2 Actuator

The actuator supplies the mechanical force to position the valve for the desired flow rate. For control applications, a diaphragm actuator is preferred because compared to a piston-type actuator, it has a relatively large pressure-sensitive area and a relatively small frictional area where the stem touches the packing. This ensures smooth operation, precision, and good repeatability. Proper selection of the actuator must take into account valve size, air pressure, desired failure mode, process pressure, and other factors. Actuators are usually spring-loaded and single-acting, with control air used on one side of the diaphragm and the spring on the other. The air pressure forces the actuator to move against the spring. If air pressure is lost, the valve fails to the spring position, so the actuator is chosen carefully to fail to a safe position (i.e., closed for fuel valves, open for combustion air valves).

2.2.2.1.3 Current-to-Pressure Transducer

The current-to-pressure transducer, usually called the I/P converter, takes the 24 V DC (4–20 mA) signal from the controller and converts it into a pneumatic signal. The signal causes the diaphragm of the actuator to move to properly position the control valve.

2.2.2.1.4 Positioner

The positioner is a mechanical feedback device that senses the actual position of the valve as well as the desired position of the valve. It makes small adjustments to the pneumatic output to the actuator to ensure that the desired and the actual positions are the same.

FIGURE 2.14
Pneumatic control valve.

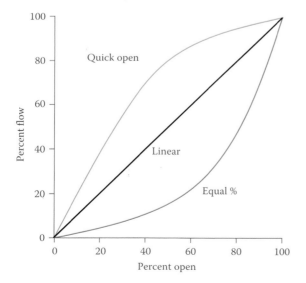

FIGURE 2.15
Control valve characteristics.

Current conventional wisdom states that positioners should be used only on "slow" systems and not on "fast" systems, where they can actually degrade performance. There is no defined border between "fast" and "slow," but virtually all combustion control applications are considered to be "slow," so positioners are almost always used in these systems.

2.2.2.1.5 Three-Way Solenoid Valve

When energized, the three-way (3-way) solenoid valve admits air to the actuator. When de-energized, it dumps the air from the actuator. Because single-acting actuators are generally used, the spring in the actuator forces the valve either fully open or fully closed, depending on the engineer's choice of failure modes when specifying the valve. Obviously, a control valve that supplies fuel gas to a combustion system should fail closed, while the control valve that supplies combustion air to the same system should fail open. In an application in which the failure mode of the valve is irrelevant—and there are some—solenoid valves are not used.

2.2.2.1.6 Mechanical Stops

Mechanical stops are used to limit how far open or shut a control valve can travel. If it is vital that no more than a certain amount of fluid ever enters a downstream system, an "up" stop is set. If it is necessary to ensure a certain minimum flow (e.g., for cooling purposes), a "down" stop is set. In the case of a fuel supply control valve, the "down" stop is set so that during system light-off, an amount of fuel ideal for smooth and reliable burner lighting is supplied. After a defined settling interval, usually 10 s, the 3-way solenoid valve is energized and normal control valve operation is enabled.

2.2.2.2 Thermocouples

Whenever two dissimilar metals come into contact, current flows between the metals, and the magnitude of that current flow, and the voltage driving it, vary with temperature. This phenomenon is called the Seebeck effect. If both metals are carefully chosen and are of certain known alloy compositions, the voltage will vary in a nearly linear manner with temperature over some known temperature range. Because the temperature and voltage ranges vary depending on the materials employed, engineers use different types of thermocouples for different situations. In combustion applications, the K-type thermocouple (0°F–2400°F = –18°C–1300°C) is usually used. When connecting a thermocouple (see Figure 2.16) to a transmitter, the transmitter should be set up for the type of thermocouple employed. Installing thermocouples in a protective sheath known

FIGURE 2.16
Thermocouple.

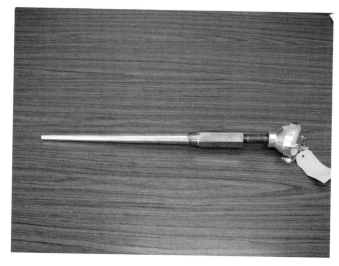

FIGURE 2.17
Thermowell and thermocouple.

as a thermowell (see Figure 2.17) prevents the sensing element from suffering the corrosive or erosive effects of the process being measured. However, a thermowell also slows the response of the instrument to changing temperature and should be used with care.

2.2.2.3 Velocity Thermocouples

Also known as suction pyrometers, the design of velocity thermocouples attempts to minimize the inaccuracies in temperature measurement caused by radiant heat. Inside a combustor, the thermocouple measures the gas temperature. However, the large amount of heat radiated from the hot surroundings significantly affects this measurement. If a thermocouple is shielded from its surroundings by putting it in a hollow pipe as shown in Figure 2.17, the response time is slowed because the thermocouple is now located in a shield-created low-flow zone. Drawing suction on the shield quickly pulls gas in from the combustor and the response time improves.

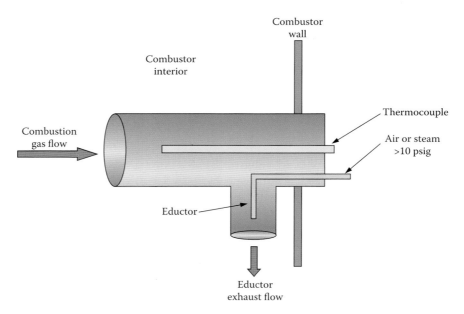

FIGURE 2.18
Velocity thermocouple.

Using velocity thermocouples (see Figure 2.18) provides a high degree of precision in combustion temperature measurement.

2.2.2.4 Resistance Temperature Detectors

The resistance of any conductor increases with temperature. If a specific material and its resistance are known, it is possible to infer the temperature. Similar to the thermocouples described earlier, the linearity of the result depends on the materials chosen for the detector and their alloy composition. Engineers sometimes use resistance temperature detectors (RTDs) in place of thermocouples when higher precision is desired. Platinum is a popular material for RTDs because it has good linearity over a wide temperature range. Like thermocouples, installation of RTDs in thermowells is common.

2.2.2.5 Pressure Transmitters

A pressure transmitter (see Figure 2.19) is usually used to provide an analog pressure signal. These devices use a diaphragm coupled to a variable resistance, which modifies the 24 V DC loop current (4–20 mA) in proportion to the range in which it is calibrated. In recent years, these devices have become more accurate and sophisticated, with onboard intelligence and self-calibration capabilities. They are available in a wide variety of configurations and materials and can be used in almost any service. It is possible to remotely check and reconfigure these "smart" pressure transmitters using a handheld communicator.

FIGURE 2.19
Pressure transmitter (left) and pressure gage (right).

2.2.2.6 Flow Meters

There are many different types of flow meters and many reasons to use one or another for a given application. The following is a list of several of the more common types of flow meters, how they work, and where they are used.

2.2.2.6.1 Vortex Shedder Flow Meter

A vortex shedder places a bar in the path of the fluid. As the fluid goes by, vortices (whirlpools) form and break off constantly. An observation of the water swirling on the downstream side of bridge pilings in a moving stream reveals this effect. Each time a vortex breaks away from the bar, it causes a small vibration in the bar. The frequency of the vibration is proportional to the flow. Vortex shedders have a wide range and are highly accurate, reasonably priced, highly reliable, and useful in liquid, steam, or gas service.

2.2.2.6.2 Magnetic Flow Meter

A magnetic field, a current-carrying conductor, and relative motion between the two create an electrical generator. In the case of a magnetic flow meter, the meter generates the magnetic field and the flowing liquid supplies the motion and the conductor. The voltage produced is proportional to the flow. These meters are highly accurate, are very reliable, and have a wide range, but are somewhat expensive. They are useful with highly corrosive or even gummy fluids, as long as the fluids are conductive. Only liquid flow is measured.

2.2.2.6.3 Orifice Flow Meter

Historically, almost all flows were measured using this method and it is still quite popular. Placing the orifice in the fluid flow causes a pressure drop across the orifice. A pressure transmitter mounted across the orifice calculates the flow from the amount of the pressure drop. Orifice meters are very accurate, but have a narrow range. They are reasonably priced, highly reliable, and useful in liquid, steam, or gas service.

2.2.2.6.4 Coriolis Flow Meter

The Coriolis flow meter is easily the most complex type of meter to understand. The fluid runs through a U-shaped tube that is being vibrated by an attached transducer. The flow of the fluid will cause the tube to try to twist because of the Coriolis force. The magnitude of the twisting force is proportional to flow. These meters are highly accurate and have a wide range. They are generally more expensive and their reliability is not as good as some other types.

2.2.2.6.5 Ultrasonic Flow Meter

When waves travel in a medium (fluid), their frequency shifts if the medium is in motion relative to the wave source. The magnitude of the shift, called the Doppler effect, is proportional to the relative velocity of the source and the medium. The ultrasonic meter generates ultrasonic sound waves, sends them diagonally across the pipe, and computes the amount of frequency shift. These meters are reasonably accurate, have a fairly wide range, are reasonably priced, and are highly reliable. Ultrasonic meters work best when there are bubbles or particulates in the fluid.

2.2.2.6.6 Turbine Flow Meters

A turbine meter is a wheel that is spun by the flow of fluid past the blades. A magnetic pickup senses the speed of the rotation, which is proportional to the flow. These meters can be very accurate, but have a fairly narrow range. They must be very carefully selected and sized for specific applications. They are reasonably priced and fairly reliable. They are used in liquid, steam, or gas service.

2.2.2.6.7 Positive Displacement Flow Meters

Positive displacement flow meters generally consist of a set of meshed gears or lobes that are closely machined and matched to each other. When fluid is forced through the gears, a fixed amount of the fluid is allowed past for each revolution. Counting the revolutions reveals the exact amount of flow. These meters are extremely accurate and have a wide range. Because there are moving parts, the meters must be maintained or they can break down or jam. They also cause a large pressure drop, which can sometimes be important for certain applications.

2.2.2.7 Analytical Instruments

There are many different types of analytical instruments used for very specific applications. Unlike the sensors described previously, these devices are usually systems. They are a combination of several different sensors linked together by a processor of some sort that calculates the quantity in question. Unlike a pressure transmitter, most analytical instruments sample and chemically test the process in question. Because the process takes time, the engineer, when designing the system, must plan for a delayed response from the analytical instrument. A detailed discussion of the design and operation of analytical instruments is beyond the scope of this book; however, a list of several of the more common types and their uses is given later.

2.2.2.7.1 pH Analyzer

Almost any combustion system occasionally requires the scrubbing of effluent or other similar processes. pH monitoring is needed to ensure that the water going into the scrubber is the correct pH to neutralize the acidity or alkalinity of the effluent. The analyzer sends information to a controller that is responsible for opening or closing valves that add alkaline chemicals to the water to raise pH.

2.2.2.7.2 Conductivity Analyzer

Conductivity analyzers are often used in conjunction with pH analyzers. Where the pH analyzer system functions to raise pH, the output from the conductivity analyzer is usually sent to a controller responsible for opening or closing valves that dilute the water to lower pH.

2.2.2.7.3 Oxygen Analyzer

Oxygen (O_2) or combustible analyzers monitor the amount of O_2 or combustibles in the exhaust of a combustion system. The analyzer sends data back to the control system, which uses it to tightly control the amount of combustion air coming into the system. This has the dual result of making the system more efficient and reducing the amount of pollutants that result from the combustion process. Different models have varying methodologies, accuracies, and sample times, but there are two major types: (1) in situ analyzers carry out the analysis at the probe and (2) extractive analyzers remove the sample from the process and cool it before analysis.

2.2.2.7.4 Nitrogen Oxide Analyzer

Nitrogen oxides (NO, NO_2, etc.; see Volume 1, Chapter 15) are one of the main components of smog and are the result of high-temperature combustion. Noxidizers are combustion systems that use an extended low-temperature combustion process designed to minimize the formation of nitrous oxide compounds. Noxidizers use nitrogen oxide (NOx) analyzers. To properly control the process, the NOx analyzer output goes to a controller that controls airflow into the system, minimizing NOx formation.

2.2.2.7.5 Carbon Monoxide Analyzer

Carbon monoxide (CO) is also an undesirable pollutant and is a product of incomplete combustion (see Volume 1, Chapter 14). The output of the CO analyzer (see Figure 1.27) is often used in the analysis of system efficiency or to control airflow to the combustion system.

2.3 Control Schemes

Other chapters of this book present the combustion process and the definition of the terms used to describe it. This section describes methods used to control the process. Generally, controlling the process means controlling the flow of fuels and combustion air.

2.3.1 Parallel Positioning

Designers use analog control schemes to modulate valve position and control fan and pump speeds to achieve the required mix of fuel and O_2 in a combustion system. Simple systems often use parallel positioning of fuel and air valves from a single analog signal.

2.3.1.1 Mechanical Linkage

A common method of parallel positioning is to mechanically link the fuel and air valves to a single actuator. Adjustment of a cam located on the fuel valve supplies the proper amount of fuel throughout the air valve operating range. Figure 2.20 shows the arrangement.

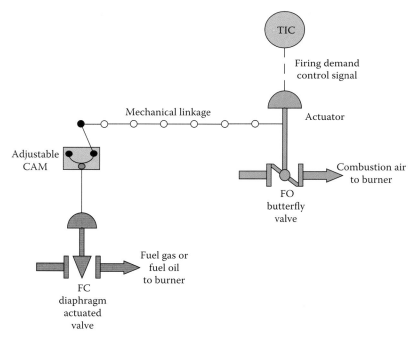

FIGURE 2.20
Mechanically linked parallel positioning. TIC, temperature indicating controller; FC, fail closed; FO, fail open.

In the figure, the temperature indicating controller (TIC) operates an actuator attached to the air valve. A mechanical linkage and an adjustable cam operate the fuel valve in parallel with the air valve. Springs or weights attached to the air valve shaft force a full open position of the air valve if the mechanical connection to the air valve fails. The system uses a fail-closed actuator to ensure that a low fire failure mode results from loss of signal or loss of actuator power.

Mechanical linkage is simple in operation but requires considerable adjustment at start-up to obtain the correct fuel:air ratio over the entire operating range. Predictable flow rates of fuel and air throughout valve position require a fixed supply pressure to the valves and constant load geometry downstream of the valves. Analytical feedback to control fuel gas or combustion air supply pressure can make dynamic corrections for fuel variations, temperature changes, and system errors. Dynamic adjustments should be small, trimming adjustments, rather than primary control parameters.

2.3.1.2 Electronic Linkage

Electronically linked fuel and combustion air valves for parallel positioning have many advantages over mechanically linked valves. Figure 2.21 illustrates the scheme.

In the example, a TIC generates a firing rate demand. The controller applies an output of 4–20 mA to the fuel valve and to a characterizer in the air valve circuit. Electronic shaping of the characterizer output positions the air valve for correct airflow. Predictable and repeatable valve positions require the use of positioners at each valve. Without positioners, valve hysteresis causes large errors in flow rate.

Signal inversion (1 minus the value being measured) is sometimes integral to the characterizer. Signal inversion is necessary because the air valve fails open and the fuel valve fails closed. Safety concerns dictate failure modes. Fuel should always fail to minimum and air should fail to maximum.

Electronically linked parallel positioning works well if properly designed. Good design requires valves with known coefficients throughout valve position and the use of high-performance positioners. Supply pressure of fuel and air to each valve must be constant or repeatable. System load downstream of the valves must be of fixed geometry. Section 2.3.1.3 shows an example of how to calculate and configure a characterizer for the air valve. Figure 2.22 shows a variation of parallel positioning that permits use of the combustion air valve for the multiple purposes of

1. Supplying combustion air during normal operation
2. Supplying quench air when burning exothermic waste
3. Using another heat source requiring quench air

When showing a range of milliampere signals, the first value is the minimum valve position and the second value is the maximum valve position. This convention aids system analysis and is especially useful for complex systems. The TIC output is split-ranged. The top half (12–20 mA) is used for firing fuel gas. When

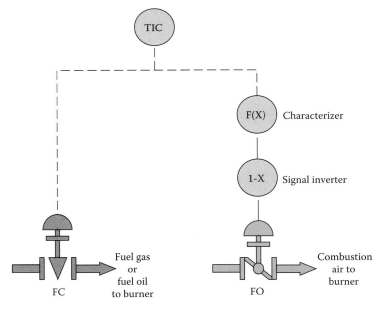

FIGURE 2.21
Electronically linked parallel positioning. TIC, temperature indicating controller; FC, fail closed; FO, fail open.

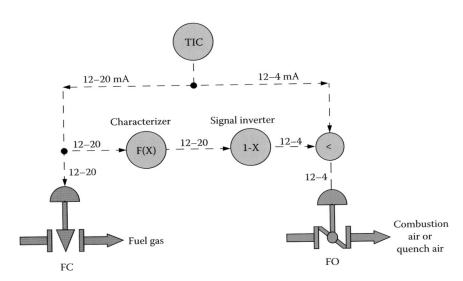

FIGURE 2.22
A variation of parallel positioning. TIC, temperature indicating controller; FC, fail closed; FO, fail open; <, low select.

burning exothermic waste requiring quench air, the temperature controller output decreases, providing low fire fuel at 12 mA, then quench air below 12 mA.

The TIC output is actually 4–20 mA. The description of the action of the receivers uses the term "split-ranged." For example, the TIC applies the entire 4–20 mA range to the fuel gas valve, but the valve is configured to respond only to the partial range of 12–20 mA.

2.3.1.3 Characterizer Calculations

Parallel positioning of a globe-type fuel gas valve and a butterfly-type combustion air valve requires characterizer calculations as described in the following. Figure 2.21 shows the control scheme.

Three general steps are required to define the characterizer:

1. Calculate and graph fuel flow rate versus control signal.
2. Calculate and graph combustion airflow rate versus control signal.
3. Tabulate and graph air valve characterizer.

2.3.1.3.1 Step 1: Fuel Flow Rate versus Control Signal

Predictable and repeatable calculations of fuel gas flow rate versus control valve position require

1. Pressure regulator upstream of control valve to provide constant inlet pressure (varying inlet pressure can be used only if it is repeatable with flow rate)
2. Constant temperature and composition of fuel gas
3. High-quality positioner on the control valve to eliminate hysteresis and to ensure that valve percent open equals percent control signal
4. Knowledge of valve coefficient versus valve position throughout the control valve range, including the pressure recovery factor
5. Fixed and known pressure drop geometry downstream of the control valve
6. Subsonic regime throughout the flow range

Results of Step 1 are shown in Figure 2.23 for a typical fuel gas valve with equal percent trim. Low fire position of the valve is approximately 25% open for many applications. Maximum firing rate occurs between 70% and 80% open, resulting in a near-linear function of flow rate versus valve position throughout the firing range. The linear function is not necessary for configuring a combustion air characterizer, but is useful for the

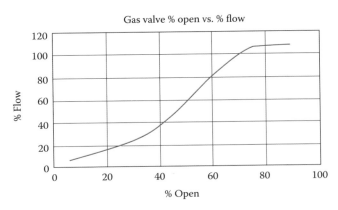

FIGURE 2.23
Fuel flow rate versus control signal.

Combustion Controls

TABLE 2.1
Gas Valve Data

Control Signal TIC (Output %)	Gas Valve (% Open)	Fuel Gas Flow Rate (%)
10	10	10
20	20	16
30	30	25
40	40	39
50	50	58
60	60	83
70	70	100
80	80	107
90	90	109
100	100	110

TABLE 2.2
Data for Characterizer

Control Signal TIC (Output %)	Fuel Gas Flow Rate (%)	Combustion Airflow Rate (%)	Air Valve (% Open)
10	10	10	5
20	16	16	13
30	25	25	22
40	39	39	29
50	58	58	33
60	83	83	41
70	100	100	46
80	107	107	48
90	109	109	49
100	110	110	50

application of a dynamic fuel:air ratio correction to the control circuit.

Use of a positioner on the fuel gas valve establishes equality between percent control signal and percent valve opening. Columns 1 and 2 of Table 2.1 show gas valve data.

2.3.1.3.2 Step 2: Airflow Rate versus Air Valve Position

Calculation of airflow rate versus position that is predictable and repeatable requires

1. Known and repeatable valve inlet pressure versus flow rate
2. Near-constant temperature
3. High-quality positioner on the valve
4. Knowledge of valve coefficient and pressure recovery factor of the air valve at all valve positions
5. Fixed and known flow (pressure drop) geometry downstream of the control valve

Figure 2.24 shows the results of a typical butterfly-type valve calculation for Step 2. The low fire mechanical stop is normally set at approximately 20% open.

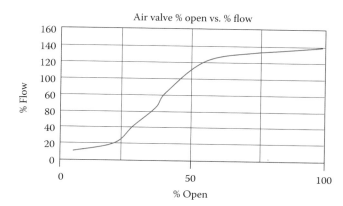

FIGURE 2.24
Typical butterfly-type valve calculation.

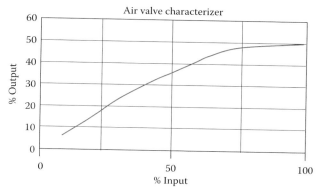

FIGURE 2.25
The required shape of the air valve characterizer.

2.3.1.3.3 Step 3: Air Valve Characterizer

Table 2.2 combines data from Figure 2.24 with data from Table 2.1. Air valve graph data are tabulated in columns 3 and 4. Figure 2.25 is a plot of the data from columns 1 and 4 and represents the required shape of the air valve characterizer. The TIC output signal is the characterizer input and is plotted on the *x*-axis. The characterizer output is the percent open of the air valve and is shown on the y-axis. Many characterizer instruments are available that will model a curved response using straight-line segments. This characterizer is sufficiently defined using three segments.

2.3.2 Fully Metered Cross Limiting

Development of a fully metered control scheme for modulating fuel and air to a burner begins with the electronically linked parallel positioning scheme as previously shown in Figure 2.21. Figure 2.26 adds flow meters and flow controllers.

Flow meters are linear with flow rate. Meter output signal scaling provides the firing rate and air:fuel ratio

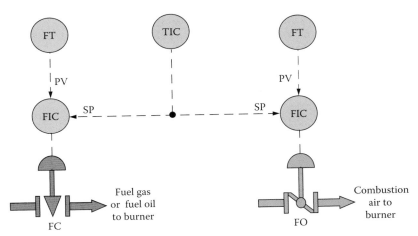

FIGURE 2.26
Fully metered control scheme. FT, flow transmitter; PV, process variable; SP, setpoint; FIC, flow indicating controller; TIC, temperature indicating controller; FC, fail closed; FO, fail open.

required for the application. The combustion air characterizer used for parallel positioning is not required because the transmitters are linear with flow rate.

In the illustration, the TIC output sets the firing rate by serving as SP to each flow controller. Signal inversion, shown as (1 minus parameter value) in the parallel positioning scheme, is not required. Instead, controller output mode is configured to match the valve failure mode.

Controller output mode, reverse or direct acting, defines the change in output signal direction with respect to PV changes. For example, if the controller output increases as the PV increases, the controller mode is direct acting. In combustion control schemes, fail-closed fuel valves require a reverse-acting flow controller, while fail-open combustion air valves require direct-acting flow controllers. From controller mode definitions, it is clear that the TIC should be reverse acting. That is, the TIC output should decrease, reducing the firing rate, in response to an increase in temperature, the PV.

Addition of high and low signal selectors provides cross limiting of the fully metered control scheme, as shown in Figure 2.27. The low signal selector (<) compares demanded firing rate from the TIC to the actual combustion airflow rate and applies the lower of the two signals as the SP to the fuel flow controller. The low signal selector ensures that the fuel SP cannot exceed the amount of air available for combustion.

The high signal selector (>) compares demanded firing rate from the TIC to actual fuel flow rate and applies the higher of the two signals as the SP to the airflow

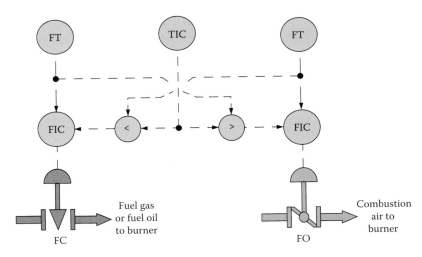

FIGURE 2.27
Fully metered control scheme with cross limiting. FT, flow transmitter; FIC, flow indicating controller; TIC, temperature indicating controller; FC, fail closed; FO, fail open; <, low select; >, high select.

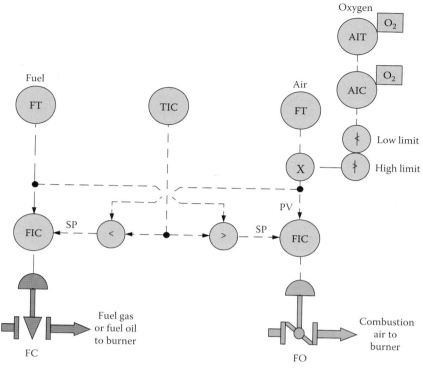

FIGURE 2.28
O_2 trim of airflow rate. AIT, analysis (oxygen) indicating transmitter; AIC, analysis (oxygen) indicating controller; FT, flow transmitter; FIC, flow indicating controller; TIC, temperature indicating controller; SP, setpoint; PV, process variable; FC, fail closed; FO, fail open; <, low select; >, high select.

controller. This ensures that the air SP is never lower than required for combustion of actual fuel flow rate.

Together, the high and low signal selectors ensure that unburned fuel does not occur in the combustion system. Unburned fuel accumulations can cause explosions. Cross limiting by the signal selectors causes airflow to lead fuel flow during load increases and for airflow to lag fuel flow during fuel decreases. This lead/lag action explains why the fully metered cross-limiting control system is often called "lead-lag" control. Whatever the name, the system performs the function of maintaining the desired air:fuel mixture during load changes. The system also provides fuel flow rate reduction in the event airflow is lost or decreased.

It is possible to trim the control scheme using measurement of flue gas O_2 content, as illustrated in Figure 2.28. For most systems, the O_2 signal should be used to "trim," and not be a primary control. Many O_2 analyzers are high maintenance and/or too slow in response to be used as a primary control in the combustion process. As shown, the O_2 controller is utilized for SP injection and provides tuning parameters to help process customization. High and low signal limiters restrict the O_2 controller output to a trimming function, normally 5%–10% of the normal combustion airflow rate.

A multiplication function (X) in the combustion airflow transmitter signal makes the O_2 trim adjustment. The multiplier gives a fixed trim gain. Substituting a summing function for the multiplier would result in high trim gain at low flow rates and could produce a combustion air deficiency.

O_2 trim may be applied to the combustion airflow controller SP rather than the flow transmitter signal. If this technique is used, the airflow signal to the low signal selector must retain trim modification (see Figure 2.29 for the scheme).

Multiple fuels and O_2 sources are accommodated by the cross-limiting scheme, as shown in Figure 2.30. When multiple fuels are used, heating values must be normalized by adjusting flow transmitter spans or by addition of heating value multipliers. Similar methods are used to normalize O_2 content for multiple air sources.

2.4 Controllers

Controllers have historically been called analog controllers because the process and I/O signals are usually analog. Controller internal functions performed within

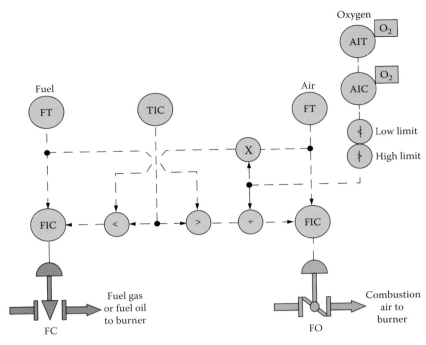

FIGURE 2.29
O₂ trim of air SP. AIT, analysis (oxygen) indicating transmitter; AIC, analysis (oxygen) indicating controller; FT, flow transmitter; FIC, flow indicating controller; TIC, temperature indicating controller; FC, fail closed; FO, fail open; <, low select; >, high select.

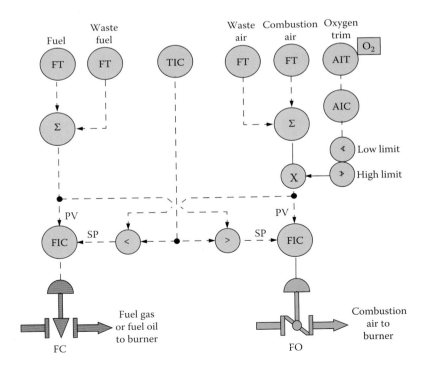

FIGURE 2.30
Multiple fuels and O₂ sources. AIT, analysis (oxygen) indicating transmitter; AIC, analysis (oxygen) indicating controller; FT, flow transmitter; FIC, flow indicating controller; TIC, temperature indicating controller; FC, fail closed; FO, fail open; <, low select; >, high select.

Combustion Controls

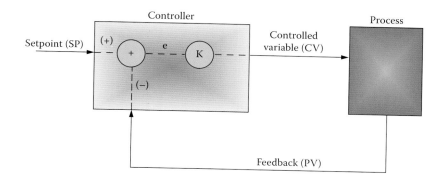

FIGURE 2.31
Controller.

a computer or microprocessor by algorithm are sometimes called digital controllers, although the I/O largely remains analog. Some digital controllers communicate with other devices via digital communication, but for the most part, controllers connect to other devices by analog signals. The analog signal is usually 4–20 mA, DC.

In Figure 2.31, the SP is a signal representing the desired value of a process. If the process is flow rate, the SP is the desired flow rate. SP signals can be generated internally within the controller, called the local SP, or may be an external signal, called the remote SP.

Controller output, called the controlled variable (CV) or the manipulated variable (MV), connects to a final element in the process. In this example of flow control, the final element is probably a control valve. Feedback from the process, called the PV in this example, could be the signal from a flow meter.

The CV is generated within the controller by subtraction of feedback (PV) from the SP, generating an error signal e, which is multiplied by a gain K. The product eK is the controller output (CV):

$$\text{Output (CV)} = (SP - PV)(K) = eK \quad (2.1)$$

This simple controller is an example of the first controller built in the early twentieth century and is called a proportional controller. The output is proportional to the error signal. The proportional factor is the gain K.

Proportional controllers require an error (e) to produce an output. If the error is zero, the controller output is zero. To obtain an output that produces the correct value of the PV, the operator is required to adjust the SP higher than the desired PV in order to create the requisite error signal. Reduction of error by increasing controller gain is limited by controller instability at high gain.

Offset is the term given to the difference between the SP and PV. Correction of offset was the first improvement made to the original controller. Offset correction was accomplished by adding bias to the controller output:

$$\text{Output (CV)} = eK + \text{Bias} \quad (2.2)$$

Bias adjustment required operator manipulation of a knob or lever on the controller, which added bias until SP and PVs were equal. The operator considered the controller "reset" when equality occurred. Each SP change or process gain change required a manual reset of the controller. Figure 2.32 illustrates the proportional controller with manual reset.

Many operators prefer the term "proportional band" (PB) when describing controller gain. PB is defined as

$$\text{Proportional band (PB)} = \frac{100}{\text{Gain}} = \frac{100\%}{K} \quad (2.3)$$

PB represents the percent change in the PV required to change the controller output 100%. For example, a

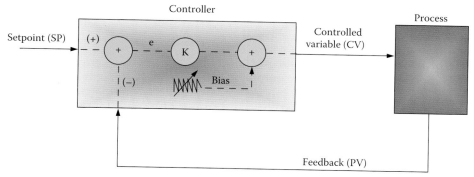

FIGURE 2.32
Analog controller with manual reset.

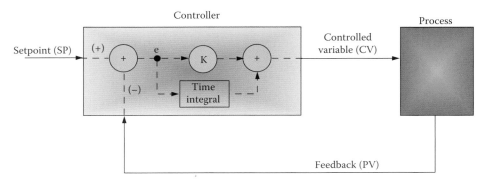

FIGURE 2.33
Analog controller with automatic reset.

controller gain of 1 (K = 1.0) requires a PV change of 100% to obtain a 100% change in controller output. PBs for combustion PVs are generally in the range of 1000%–20% (gain = 0.1–5.0). Flow controller gains are always less than unity. Temperature controller gains vary from 0.1 to 3.0, depending on the process gain. Pressure controllers generally have gains higher than those of flow or temperature controllers. Controller output becomes unstable (oscillatory) when the gain is too high. When instability occurs, the controller operating mode must be changed from automatic to manual to stabilize the process and prevent equipment damage. A reduction of controller gain must occur before a return to automatic mode.

Automatic reset was the next improvement to the process controller. This was a most welcome addition that eliminated the need for manual reset. Automatic reset is a time integral of the error signal, summed with the proportional gain (P) signal to produce the controller output. Integral gain (controller reset) is highest with large errors and continues until the error is reduced to zero. Figure 2.33 illustrates automatic reset.

Automatic reset is the "I" component of a PID (three-mode) controller. P is proportional gain and D is derivative (rate) gain. I is expressed as repeats per minute (RPM) or as minutes per repeat (MPR), depending on the controller manufacturer's choice of terms. Some controllers permit user selection of the term. Controller output is the same regardless of terminology, but the operator must know and apply proper tuning constants. For example, if tuning requires an I of 2 RPM, the operator must enter 0.5 MPR into the controller if MPR is the terminology in use. Integral gain of 0.5 MPR means that automatic reset equal to the P will be applied at the controller output each 30 s. Integral gain is a smooth continuous process that contributes phase lag to the system. Additional phase lag contributes to system instability (oscillation), which prohibits high values of integral gain.

D is a function of how fast the PV is changing. For slow-changing processes, D is of little use. D is not used on flow control loops with head meters or on other loops with noisy PV signals. High noise levels will drive D to instability. D contributes phase lead that can sometimes be beneficial.

Controllers have many modes of operation. P, I, D, automatic, and manual modes have been discussed. Reverse or direct mode is another choice that must be configured to match the process. Reverse or direct describes the change in direction of controller output when the PV changes. Reverse acting means the controller output decreases if the PV increases. An example illustrates how to select reverse or direct.

In this example of flow control, the PV (flow rate) increases when the final element (control valve) opens. In addition, the flow meter output or PV increases with increased flow rate. If the control valve fails closed and opens on increasing signal (increasing controller output), the controller mode must be reverse acting. That is, if PV increases, the controller output must decrease to close the valve and restore flow rate to the correct value. If the control valve fails open (closes on increasing signal), the controller mode must be direct acting. This example illustrates the need to know if each element in a control loop is reverse or direct acting, including transmitters, isolators, transducers (such as I/Ps), positioners, and actuators. Proper selection and configuration of loop elements provide not only proper operation, but also proper failure mode. Reversal of any two elements within a loop will not affect loop response, but failure modes will change.

2.5 Tuning

Many modern controllers have built-in automatic tuning routines. Tuning parameters calculated automatically require a loop upset to enable calculation. Parameters are normally de-tuned considerably from optimum because process gains are often nonlinear. Variable loop gain can also be a problem for manual tuning. A controller tuned at low flow rates or low temperature could become unstable at high flow rates or high temperatures.

Control of most process loops benefits from addition of feedforward components that relieve the feedback controller of primary control. Operation improves if the feedback controller functions as SP injection and error trimming of the feedforward system.

Many processes controlled by a current proportional controller successfully use the following tuning procedure. The process must be upset to produce oscillations of the PV. A graphic recorder should be used to determine when the oscillations are constant and to ascertain the time for one cycle (oscillation):

1. In manual mode, adjust the output to bring the PV near the desired value.
2. Set the rate time to 0 min and set reset time to the maximum value (50.00 min), or set RPM to the minimum value to minimize reset action.
3. Increase gain (decrease PB) significantly. Try a factor of 10.
4. Adjust local SP to equal PV and switch to automatic mode.
5. Increase the SP by 5% or 10% and observe PV response.
6. If the PV oscillates, determine the time for one oscillation. If it does not oscillate, return to the original SP, increase the gain again by a factor of 2, and repeat Step 5.
7. If the oscillation of Step 6 dampens before cycle time is measured, increase the gain slightly and try again. If the oscillation amplitude becomes excessive, decrease gain slightly and try again.
8. Record the current value of gain, and record the value for one completed oscillation of PV.
9. Calculate gain, reset, and rate:
   a. For PI (two-mode controller)
      - Gain = measured gain × 0.5
      - Reset time = measured time/1.2 (MPR)
   b. For PID (three-mode controller)
      - Gain = measured gain × 0.6
      - Reset time = measured time/2.0 (MPR)
      - Rate = measured time/8.0 (min)
10. Enter the values of gain, reset, and rate into the controller.
11. Make additional trimming adjustments, if necessary, to fine-tune the controller.
12. To reduce overshoot: less gain, perhaps a longer rate time.
13. To increase overshoot or increase speed of response: more gain, perhaps shorter rate time.

References

1. J.O. Hougen, *Measurement and Control Applications for Practicing Engineers*, Barnes & Noble Series for Professional Development, CAHNERS Books, Boston, MA, 1972.
2. Combustion control, 9ATM1, Fisher controls, Marshalltown, IA, 1976.
3. Boiler control, Application data sheet 3028, Rosemount, Inc., Minneapolis, MN, 1980.
4. ANSI/ISA-S5.1, *Instrumentation Symbols and Identification*, Instrument Society of America, Research Triangle Park, NC, 1984.
5. ANSI-MC96.1, *Temperature Measurement Thermocouples*, Instrument Society of America, Research Triangle Park, NC, 1984.
6. F.G. Shinskey, *Process Control Systems, Application, Design, and Tuning*, 3rd edn., McGraw-Hill, New York, 1988.
7. M.J.G. Polonyi, PID controller tuning using standard form optimization, *Control Engineering*, March, 102–106, 1989.
8. D.W. St. Clair, Improving control loop performance, *Control Engineering*, October, 141–143, 1991.
9. Controller tuning, Section 11, *UDC 3300 Digital Controller Product Manual*, Honeywell Industrial Automation, Fort Washington, PA, 1992.
10. F.Y. Thomasson, Five steps to better PID control, *Control*, April, 65–67, 1995.
11. API Recommended Practice 556, *Instrumentation and Control Systems for Fired Heaters and Steam Generators*, 1st edn., American Petroleum Institute, Washington, DC, 1997.
12. NFPA 85, *Boiler and Combustion Systems Hazards Code*, National Fire Protection Association, Quincy MA, 2011.
13. P.A. Schweitzer, *Handbook of Valves*, Krieger Publishing Co., Malabar, FL, 1972.
14. R. Greene, *The Chemical Engineering Guide to Valves*, McGraw-Hill, New York, 1984.
15. J.J. Pippenger, *Hydraulic Valves and Controls: Selection and Application*, Marcel Dekker, New York, 1984.
16. J. Hutchison, *ISA Handbook of Control Valves: A Comprehensive Reference Book Containing Application and Design Information*, 2nd edn., Instrument Society of America, Research Triangle Park, NC, 1990.
17. G. Borden and P.G. Friedmann, *Control Valves*, Instrument Society of America, Research Triangle Park, NC, 1998.
18. T.C. Dickenson, *Valves, Piping, and Pipelines Handbook*, Elsevier Advanced Technology, Oxford, U.K., 1999.
19. C. Matthews, *A Quick Guide to Pressure Relief Valves*, Professional Engineering Publishers, London, U.K., 2004.
20. Crane Company Engineering Division, *Flow of Fluids through Valves, Fittings, and Pipe*, Crane Company, Stamford, CT, 2011.

3

Blowers for Combustion Systems

John Bellovich and Jim Warren

CONTENTS

3.1 Introduction 73
3.2 Applications 73
3.3 Types of Blowers for Combustion Systems 73
3.4 Fan Arrangements 74
3.5 Design Considerations 76
    3.5.1 Fan Control 77
    3.5.2 Materials of Construction 83
    3.5.3 Motors and Drives 84
    3.5.4 Couplings and Belts 84
    3.5.5 Bearings and Lubrication 85
    3.5.6 Vibration and Installation 86
    3.5.7 Shaft Seals 87
    3.5.8 Noise Considerations 89
    3.5.9 Filtration 89
3.6 Operational Costs 89
3.7 Inspection and Testing 89
3.8 Maintenance and Troubleshooting 90
References 92
.... 93

3.1 Introduction

Compared to industrial air handling or air conditioning blowers, blowers and fans used in the service of combustion systems have some unique requirements that make design, installation, and operation a little more difficult. An incorrect fan design or material selection can have a disastrous effect on the performance of an entire facility. It is very important to understand the service of the fan as well as its performance in order to properly run a combustion system. The terms "blower" and "fan" are used interchangeably in the context of this book. Blowers and pumps are very similar in the fact that they both push a fluid from one point to another by putting more energy into the system. Pumps push liquids and fans push gases. Fans for combustion systems can be used to introduce air into a combustion zone, prevent smoke formation, move exhaust gases, cool or reoxygenate a stream, purge a stack, and more. The purpose of this section is to introduce the basic concepts of blowers for combustion services and give awareness to the critical issues. The intricacies of blower design are not discussed. To gain more knowledge on the topic, consult the *Fan Engineering Handbook* by the Howden/Buffalo Fan Company.[1] Other valuable resources include any of the Air Movement and Control Association International (AMCA) standards.[2,3] Some good general references related to this topic are also available.[4–8]

3.2 Applications

Typical applications for fans in combustion systems include

- Forced and induced draft fans for fired heaters, burners, boilers, and thermal oxidizers
- Introduction of air for smoke suppression of flare systems
- Cooling or purging of flare systems
- Movement of waste gases into the burner or flare system
- Reoxygenation and quenching of thermal oxidizer flue gases

FIGURE 3.1
A centrifugal fan.

3.3 Types of Blowers for Combustion Systems

There are primarily two types of blower used in the combustion industry, centrifugal and axial. Centrifugal fans (see Figure 3.1), also known as radial fans, are very similar to centrifugal pumps. The air enters the center of the impeller and is ejected radially outward and "pushed" by the impeller blades through the housing and out the exit. Centrifugal fans usually have the same basic configuration with options concerning what type of impeller and drive system is used (see Figure 3.2 for different impeller designs). Other variations of the centrifugal fan will include high pressure fans and multistage turbo blowers. Axial fans are more like an airplane propeller that pushes the air along in the direction of the axis of rotation. Axial fans fall

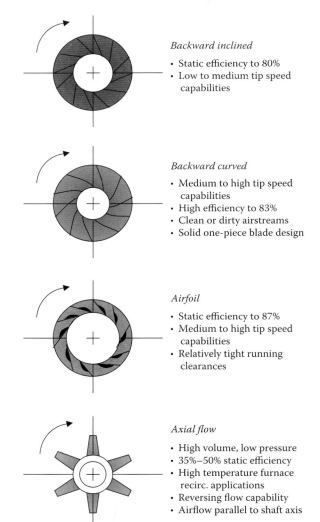

Paddlewheel
- Open design/no shroud
- 60%–65% static efficiency
- Inexpensive design
- Good for high temperature or highly erosive applications
- Medium to high pressure

Radial blade
- Static efficiency to 75%
- High tip speed capabilities
- Reasonable running clearance
- Best for erosive or sticky particulate

Radial tip
- Static efficiency to 75%
- Medium to high tip speed capabilities
- Running clearance tighter than radial blade but not as critical as backward inclined and airfoil
- Good for high particulate airstream

Forward curved (sirrocco)
- Smallest diameter wheel for a given pressure requirement
- High volume capability
- 55%–65% static efficiency
- Often used for high temperatures

Backward inclined
- Static efficiency to 80%
- Low to medium tip speed capabilities

Backward curved
- Medium to high tip speed capabilities
- High efficiency to 83%
- Clean or dirty airstreams
- Solid one-piece blade design

Airfoil
- Static efficiency to 87%
- Medium to high tip speed capabilities
- Relatively tight running clearances

Axial flow
- High volume, low pressure
- 35%–50% static efficiency
- High temperature furnace recirc. applications
- Reversing flow capability
- Airflow parallel to shaft axis

FIGURE 3.2
Fan wheel designs. (Adapted from Robinson Fans, Zelienople, PA.)

Blowers for Combustion Systems

FIGURE 3.3
A vane axial fan.

into two categories, vane axial and tube axial. Vane axial fans (see Figure 3.3) have a set of internal guide vanes, while tube axial fans have none. Which fan is used will depend on the application. Centrifugal fans typically can generate higher pressures than vane axial fans, and the motor can be maintained outside of the fan. Vane axial fans are a high-volume low-pressure fan, are usually lower in cost, and are typically a little more efficient. Unless the vane axial fan is belt driven, the motor is usually inside the fan housing and the unit would have to be disassembled for work on the motor. Some vane axial fans have an extended shaft to move the motor out of the housing, but these are generally for large ventilating applications.

Backward-curved and air foil fans usually offer the highest efficiency. Radial and radial-tipped fans are usually used in applications that have particulate (see Table 3.1). See Figure 3.4 for an example of a blower used for purging. See Figure 3.5 for an example of a centrifugal blower used for a landfill flare application.

FIGURE 3.4
A purge air blower on the side of a combustion chamber.

TABLE 3.1

Relative Characteristics of Centrifugal Blowers

	Radial	Forward-Curved Radial Tip	Backward Curved
Efficiency	Medium	Medium	High
Tip speed	High	Medium	Medium
Size[a]	Small	Medium	Large
Initial cost[b]	Small	Medium	Large
HP curve	Medium rise	Medium rise	Power limiting
Accept corrosion coating	Excellent	Fair to poor	Good (thin coat)
Abrasion resistance	Good	Medium	Medium
Sticky material handling	Good	Poor	Medium
High-temperature capability	Excellent	Good	Good
Running clearance	Liberal	Medium	Minimum req'd
Operation without diffuser	Not as efficient	Must use	Good efficiency
Noise level	High	Medium	Low
Stability/non-surge range[c]	Medium	Poor	Medium
	20%–100%	40%–100%	20%–100%

[a] Size is based on fans at the same speed, volume, and pressure.
[b] Cost is based on fans at the same speed, volume, and pressure.
[c] More a function of operating point along a curve than fan type.

FIGURE 3.5
A multistage high-speed centrifugal blower for a landfill application.

3.4 Fan Arrangements

Arrangements are the AMCA's designations for the basic configurations of fans.[3] There are two considerations that need to be made before finalizing what arrangement is required, installation and maintenance. Arrangements 1 through 3 (as seen in Figure 3.6) require a separate pedestal for the fan housing and the driver. This reduces the up-front equipment cost, but because a separate foundation needs to be poured for the driver, the installation cost is higher. After installation, alignment will be necessary for direct drive units using these arrangements. These arrangements are less often used in the combustion industry and typically used for large motor sizes above 300 hp (224 kW). Arrangements 4, 7, and 8 offer the convenience of mounting the driver on a common frame with the fan for a higher equipment cost. These arrangements are more common. Arrangements 9 and 10 are belt-driven configurations where alignment and installation are simply a matter of preference.

The second factor to consider with the different arrangements is the location of the bearings. The locations of the bearings will determine the extent of maintenance to be done on the machine. From a maintenance

Notes:

SW—single width   DW—double width
SI—single inlet   DI—double inlet

Arrangements 1, 3, 7, and 8 are also available with bearings mounted on pedestals or base set independent of the fan housing.

ARR. 1 SWSI—For belt drive or direct connection. Impeller overhung. Two bearings on base.

ARR. 2 SWSI—For belt drive or direct connection. Impeller overhung. Bearing in bracket supported by fan housing.

ARR. 3 SWSI—For belt drive or direct connection. One bearing on each side and supported by fan housing.

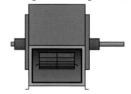

ARR. 3 DWDI—For belt drive or direct connection. One bearing on each side and supported by fan housing.

ARR. 4 SWSI—For direct drive. Impeller overhung on prime mover shaft. No bearing on fan. Prime mover base mounted or integrally directly connected.

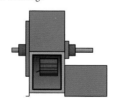

ARR. 7 SWSI—For belt drive or direct connection. Arrangement 3 plus base for prime mover.

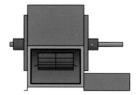

ARR. 7 DWDI—For belt drive or direct connection. Arrangement 3 plus base for prime mover.

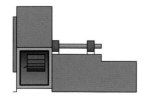

ARR. 8 SWSI—For belt drive or direct connection. Arrangement 1 plus extended base for prime mover.

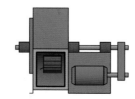

ARR. 9 SWSI—For belt drive. Impeller overhung, two bearings, with prime mover outside base.

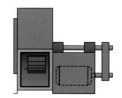

ARR. 10 SWSI—For belt drive. Impeller overhung, two bearings, with prime mover inside base.

FIGURE 3.6
Fan drive arrangements for centrifugal fans AMCA standard 99-2404-03. (Adapted from Robinson Fans, Zelienople, PA.)

FIGURE 3.7
An arrangement 4 fan.

point of view, arrangement 4 (see Figure 3.7) and 5 fans are the simplest. In arrangements 4 and 5, the fan impeller is mounted directly to the motor shaft and there are no bearings other than the motor bearings. Arrangement 5 is similar to arrangement 4 except that the motor is mounted directly to the fan housing. This may be perfectly acceptable for smaller fans. Arrangements 1, 2, 8, 9, and 10 are what is known as overhung, i.e., the wheel is cantilevered on the shaft and the bearings are both on the same side. There is an inboard and outboard bearing. Because both bearings are on the same side of the wheel, the bearings are subjected to uneven wear. Arrangements 3 and 7 have the fan wheel in between the bearings, which results in more even bearing wear and longer life. This arrangement is the best in terms of maintenance of bearings but has a higher associated cost.

3.5 Design Considerations

When selecting a fan for a combustion application, the design of the entire system needs to be taken into account. Location of the fan in the system is very important in determining what kind of fan will be required. If the fan is located at the beginning of the system and "pushing" the air or gas through, then it is considered a forced draft system. If the fan is located at the back end of the system and is "pulling" the flue gas through, then it is considered an induced draft system. A system with both a forced draft fan and an induced draft fan is called a balanced draft system. Once it is determined where the fan is located in the system, then all the process conditions that can affect the fan selection need to be determined. Final selection is based on flow rate, composition of the gas, range of operating temperatures, elevation, and inlet/outlet pressure required. Because the area around a flare system is a harsh environment with a sterile zone that limits access to personnel, vane axial fans are preferred for flares.

TABLE 3.2

Affects of Temperature and Pressure on Volume and Horsepower

	68°F at Sea Level	32°F at Sea Level	100°F at Sea Level	68°F at 3,000 ft ASL	100°F at 3,000 ft ASL
SCFM	10,000	10,000	10,000	10,000	10,000
ACFM	10,000	9,318	10,606	11,170	11,847
lb/h	44,940	48,180	42,360	40,200	37,920
HP	63	58	67	70	75

ASL, above sea level.

Temperature and elevation can greatly affect the performance of a fan. The fan itself should be sized using the maximum flow rate at the highest temperature. Table 3.2 shows the effect that temperature and elevation (pressure) has on a volume of air. The first row shows the flow rate in standard cubic feet per minute (SCFM). This number is always the same because it is corrected back to standard temperature and elevation, in this case 68°F (20°C) at sea level. However, because of the effects of temperature and elevation on volume, blower vendors tend to want to work in actual volumes and not standard or normal volumes. Row two, or actual cubic feet per minute (ACFM), shows that the volume of air, or gas, is proportional to the ratio of the absolute temperatures. As the temperature increases, so does the volume and vice versa. Also note that the volume increases with respect to the elevation, i.e., the atmosphere does not push down on a volume of gas as much at 3000 ft (900 m) above sea level as it does at sea level. So, in the second row, the volume of gas increases to 11,170 ACFM (316 ACMM = actual cubic meters per minute) at the higher elevation at standard temperature and 11,847 ACFM (336 ACMM) at the same elevation and higher temperature. Even though the flow rates are the same on a standard basis, more horsepower (HP) is required to push the increased volume corrected to the higher temperature, and conversely, less HP is required to push the decreased volume at the lower temperature. This can be seen in the last row that shows approximate HP:

$$HP = \frac{ACFM \times (\text{Discharge pressure inches of water column})}{(6354 \times \text{Efficiency})} \quad (3.1)$$

The example in this table uses a we used a discharge pressure of 30 in. (760 mm) water column and a mechanical efficiency of 75%. From this it is plain that as the temperature or elevation increases, so does the required HP.

TABLE 3.3

Effects of Density on Horsepower

	68°F at Sea Level	32°F at Sea Level	100°F at Sea Level	68°F at 3,000 ft ASL	100°F at 3,000 ft ASL
ACFM	10,000	10,000	10,000	10,000	10,000
lb/h	44,940	48,180	42,360	40,200	37,920
lb/ft³	0.075	0.080	0.071	0.067	0.063
HP	63	67	60	56	53

ASL, above sea level.

So, this shows the importance of specifying this information correctly. A fan designed to push 10,000 ACFM (280 ACCM) at sea level will not work adequately at 3,000 ft (900 m) for the same application.

However, Table 3.2 is simply a demonstration of how standard units will not accurately describe the conditions to which a fan may be subjected. Blowers are constant volume machines, meaning that the volume of air or gas pushed through the impeller stays the same. But the density change needs to be considered in the HP equation. As the temperature changes, the density does too (see Table 3.3):

$$HP = \frac{(ACFM \times \text{Discharge pressure inches of water column})}{(6354 \times \text{Efficiency})(\text{Density}_2/\text{Density}_1)} \quad (3.2)$$

The motor should be sized based on the corresponding HP at the maximum flow rate and temperature. Additionally, the blower vendor will need to verify the motor is correctly sized at the low temperature end so that too much electrical current is not pulled by the motor during start-up, causing a breaker trip and shutting the system down.

The final piece of information that is needed to size a blower is the discharge pressure to overcome the system back pressure. Typically, blower manufacturers tend to use static pressure, i.e., the portion of pressure measured with a pressure gauge. The pressure drop must be calculated for each fitting in the system. The components of the pressure drop are as follows:

$$P(\text{total}) = P(\text{velocity}) + P(\text{static}) \quad (3.3)$$

where

$$P(\text{velocity}) = \frac{\rho V^2}{2g_c} \quad (3.4)$$

$$P(\text{total}) = P(\text{velocity}) \times K \quad (3.5)$$

$$K = f \times \left(\frac{L}{D}\right) \quad (3.6)$$

for pipe where L is the length of pipe, D is the inside diameter of the pipe, and f is the friction factor. For additional K values for fittings, see Table 8.4. The friction factor is calculated using

$$f = 64/\text{Re for laminar flow} \quad (3.7)$$

$$f = 0.3164/\text{Re}^{0.25} \text{ for turbulent flow in reasonably smooth piping} \quad (3.8)$$

Re is the Reynolds number and is defined by

$$\text{Re} = \frac{\rho V D}{\mu} \quad (3.9)$$

where
ρ is the density
μ is the dynamic viscosity
V is the average velocity of the fluid in the pipe

For air at standard conditions,

$$\text{Re} = 102.3\, DV \text{ (imperial) (ft, ft/m)} \quad (3.10)$$

$$\text{Re} = 65,970.3\, DV \text{ (SI) (m, m/s)} \quad (3.11)$$

For a full description of calculating pressure drop through ducting (see Volume 1, Chapter 9).

Proper selection of a blower is not accomplished by a single straightforward formula. Experience, usage, and careful evaluation of each application are necessary to ensure proper fan selection. It is always best to work with the blower manufacturers to get the most cost-effective recommendation. When possible, calculate the most accurate pressure drop required so that the fan may be appropriately sized. Be sure to account for future expansion and system change.

With all the information discussed earlier, the blower vendor will generate a performance curve. The performance curve plots the flow rate (volumetric) versus the discharge pressure (static). As a minimum, there are two parts to the performance curve, the fan curve and the system curve. See Figures 3.8 and 3.9 for representative fan curves. The fan curve is a graphical representation of the fan performance for one wheel size. Manufacturers also have what are called multirating charts for each models of fan that show the performance of a particular fan with different sizes of impellers at different speeds.

The fan curve represents the performance of a particular fan. Following the curve all the way to the right would indicate what is called the free discharge, meaning there is no back pressure and the flow rate is at its maximum. Following the curve back to the vertical axis would show the flow rate if the exit of the fan were

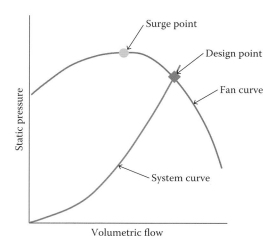

FIGURE 3.8
Basic centrifugal fan curve.

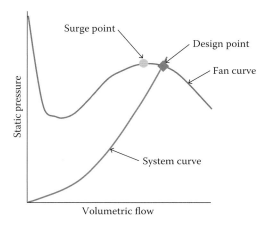

FIGURE 3.9
Basic vane axial fan curve.

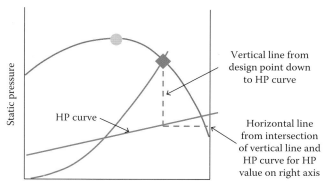

FIGURE 3.10
Basic centrifugal fan curve with HP.

completely blocked off. The more static discharge pressure required to overcome the pressure losses, the less flow rate the fan can push out, and conversely, a reduced static discharge pressure results in a greater flow rate. The system curve shows the possible combinations of discharge pressure and volumetric flow rates for that particular application from no flow up to the design point. The surge point, or peak pressure point, is the location on the curve where operation to the left may result in unstable performance. The pressure will fluctuate in positive and negative pressure pulses that result in loss of flow and greatly increased noise and vibration. Blower surge is similar in effect to pump cavitation, leading to lost efficiency, vibration, and improper performance. It is strongly recommended to choose operating points to the right of the surge point. It is also recommended that the design point should not be within 10% of the flow of the surge point. Some fan curves will also have an HP curve and a value for HP on the right-hand axis. This value is known as the brake horsepower (BHP), and it can be found by following a vertical line down from the design point until it intersects with the BHP curve, then reading along the BHP axis horizontally to obtain the BHP value (see Figure 3.10). The motor, or driver, chosen should supply more HP than BHP value so that the motor is operating within its safety factor. The design fan speed should not be greater than 85% of the maximum safe fan speed. Figures 3.11 and 3.12 are examples of typical fan curves. See Figures 3.13 and 3.14 for more examples of fans in the field.

One final comment about blower curves is that they can be changed by controls (dampers) or by speed. The fan laws are a handy approximation tool that operators can use to change the fan performance if required. The speed change fan laws state the following:

- Volumetric flow varies directly as a function of speed (RPM).
- Static pressure varies as a function of the square of speed (RPM).[2]
- BHP varies as a function for cube of the speed (RPM).[3]

These correlations give users the flexibility to vary the speed to change the system or avoid undesirable operating points. However, all three laws must be applied simultaneously. The speed cannot be changed without affecting all three attributes of flow, pressure, and HP. For example, the flow cannot be increased without the discharge pressure being increased. If the system cannot handle the increased pressure, then increasing the volume should be avoided.

Mathematically, the speed change fan laws can be stated as follows:

Volumetric flow varies directly as a function of speed (RPM):

$$\text{CFM}_{(new)} = \left(\frac{\text{RPM}_{(new)}}{\text{RPM}_{(old)}}\right) \times \text{CFM}_{(old)} \quad (3.12)$$

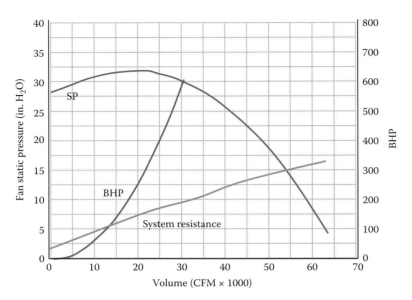

FIGURE 3.11
Forward-tip blade operating curve for 1780 RPM, 70°F, and 0.075 lb/ft³ density. (Adapted from Robinson Fans, Zeleinople, PA.)

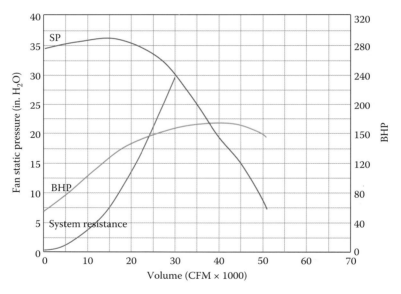

FIGURE 3.12
Backward-curved blade operating curve for 1780 RPM 70°F and 0.075 lb/ft³ density. (Adapted from Robinson Fans, Zeleinople, PA.)

Static pressure varies as a function of the square of speed (RPM)²:

$$SP_{(new)} = \left(\frac{RPM_{(new)}}{RPM_{(old)}}\right)^2 \times SP_{(old)} \quad (3.13)$$

BHP varies as a function for cube of the speed (RPM)³:

$$BHP_{(new)} = \left(\frac{RPM_{(new)}}{RPM_{(old)}}\right)^3 \times BHP_{(old)} \quad (3.14)$$

Changing the speed of the fan actually shifts the fan curve up or down depending upon the change. There is more discussion on this topic in Section 3.5.1. Note: Before increasing the speed of the fan, verify the maximum safe

FIGURE 3.13
One primary and one backup fan in the field with ducting.

FIGURE 3.14
Six-blade vane axial fan in the field.

speed with the manufacturer. Also verify that the existing electrical system can handle the additional load. Be aware that there are critical speeds that must be avoided. The critical speed is a point where the fan operation is close to the natural frequency and increased vibration will occur.

There are many variations of the fan laws, but the speed law referenced earlier is the most useful. A second useful set is for changing the diameter of the impeller. The size change law is as follows:

$$\text{CFM}_{(\text{new})} = \left(\frac{\text{DIA}_{(\text{new})}}{\text{DIA}_{(\text{old})}}\right)^3 \times \text{CFM}_{(\text{old})} \quad (3.15)$$

$$\text{SP}_{(\text{new})} = \left(\frac{\text{DIA}_{(\text{new})}}{\text{DIA}_{(\text{old})}}\right)^2 \times \text{SP}_{(\text{old})} \quad (3.16)$$

$$\text{BHP}_{(\text{new})} = \left(\frac{\text{DIA}_{(\text{new})}}{\text{DIA}_{(\text{old})}}\right)^5 \times \text{BHP}_{(\text{old})} \quad (3.17)$$

Again, all three laws must be applied at the same time. Use this second set with a little caution. There are limits as to how big or small a wheel diameter may be changed. Results will vary with this second set.

Finally, the density change fan laws can be practical and easy to use. Notice that these are directly proportional. They are

$$\text{CFM}_{(\text{new})} = \text{CFM}_{(\text{old})} \text{ (No change)} \quad (3.18)$$

$$\text{SP}_{(\text{new})} = \left(\frac{\text{Density}_{(\text{new})}}{\text{Density}_{(\text{old})}}\right) \times \text{SP}_{(\text{old})} \quad (3.19)$$

$$\text{BHP}_{(\text{new})} = \left(\frac{\text{Density}_{(\text{old})}}{\text{Density}_{(\text{new})}}\right) \times \text{BHP}_{(\text{old})} \quad (3.20)$$

Example 3.1

A chemical manufacturer has a process that now has a reduced amount of product going to it for the rest of the equipment life. The unit was originally designed for 50,000 CFM of air at 10 in. WC static pressure with a speed of 1800 RPM using a 115 hp motor. The new flow rate only needs to be 30,000 CFM. The owner would like to purchase a new motor or VFD to reduce utility cost and flow rate. What will the new motor speed be?

To begin, use the speed change law for flow rate:

$$CFM_{(new)} = CFM_{(old)} \times \left(\frac{RPM_{(new)}}{RPM_{(old)}} \right)$$

$$30{,}000 \text{ CFM} = 50{,}000 \text{ CFM} \times \left(\frac{RPM_{(new)}}{1800 \text{ RPM}} \right)$$

$$RPM_{(new)} = 1080 \text{ RPM}$$

Note that none of the fan laws in each set can be applied individually. The other two laws in the speed change must be applied. The static pressure and HP are calculated as follows:

$$SP_{(new)} = \left(\frac{RPM_{(new)}}{RPM_{(old)}} \right)^2 \times SP_{(old)}$$

$$SP_{(new)} = \left(\frac{1080 \text{ RPM}}{1800 \text{ RPM}} \right)^2 \times 10 \text{ in. WC}$$

$$SP_{(new)} = 3.6 \text{ in. WC}$$

$$BHP_{(new)} = \left(\frac{RPM_{(new)}}{RPM_{(old)}} \right)^3 \times BHP_{(old)}$$

$$BHP_{(new)} = \left(\frac{1080}{1800} \right)^3 \times 115 \text{ hp}$$

$$BHP_{(new)} = 25 \text{ hp}$$

From the previous example, reducing the speed would probably be very beneficial. Because the pressure drop for the blower to overcome was less than 10 in., it can be assumed that the new flow rate will simply follow the system curve and the discharge pressure of 3.6 in. WC is sufficient to overcome the pipe losses. Obviously, 1/4 of the HP would be a significant savings.

Example 3.2

A refinery has a forced draft combustor. But the installed ducting is a different configuration than was initially designed and takes more pressure drop than was expected. In order to generate more static pressure, the owner wants to look at increasing the impeller diameter. The flow rate of the gas is 50,000 CFM at 10 in. WC. The motor HP is 115 and the impeller diameter is 36 in. What size does the new impeller need to be to get 10.5 in. WC?

To find a new diameter, use the size change laws. Again, all three laws have to be applied at the same time.

First, solve for the new diameter using the static pressure formula:

$$SP_{(new)} = \left(\frac{DIA_{(new)}}{DIA_{(old)}} \right)^2 \times SP_{(old)}$$

$$10.5 \text{ in. WC} = \left(\frac{DIA_{(new)}}{36 \text{ in.}} \right)^2 \times 10 \text{ in. WC}$$

$$DIA_{(new)} = 36.9 \text{ in.}$$

The new diameter seems feasible, but the owner should always check with the manufacturer before making such a change. The manufacturer will need to verify if the impeller will fit appropriately in the housing. The new flow rate will be as follows:

$$CFM_{(new)} = \left(\frac{DIA_{(new)}}{DIA_{(old)}} \right)^3 \times CFM_{(old)}$$

$$CFM_{(new)} = \left(\frac{36.9}{36} \right)^3 \times 50{,}000 \text{ CFM}$$

$$CFM_{(new)} = 53{,}845 \text{ CFM}$$

And the HP is

$$BHP_{(new)} = \left(\frac{DIA_{(new)}}{DIA_{(old)}} \right)^5 \times BHP_{(old)}$$

$$BHP_{(new)} = \left(\frac{36.9}{36} \right)^5 \times 115 \text{ hp}$$

$$BHP_{(new)} = 130 \text{ hp}$$

The change in diameter has a great impact on the HP. A 5% change in discharge pressure leads to a 13% change in HP. Assuming the increase in flow rate is acceptable, the owner will also have to decide if he is willing to use the 1.15 service factor for his motor. As a minimum, the owner can run in the service factor while waiting for a new motor to be purchased and installed at a later time.

3.5.1 Fan Control

Most combustion applications require some sort of control of the fan or the air handling system. Control can be accomplished in several different ways. However, before considering what type of control devices are required, the user will want to consider what turndown is required, meaning how far below the design rate will the system need to operate. Additionally, the user needs to consider what the daily, or nominal, operating point will be. With this information at hand, the designer can develop a control scheme that will fit the system. Fans are typically controlled by pressure, temperature, flow rate, and, in some cases, an optical device sensitive to smoke formation.

The simplest and probably the most common form of control is using an inlet or an outlet damper. Although the two devices can certainly function in the same manner, the results are a little different. The outlet damper will change the system curve by putting more or less resistance, or back pressure, on the fan. Figure 3.15 shows how the system curve changes as the outlet damper is opened or closed. Closing the damper pushes the system resistance up and decreases the available flow rate. Note that closing the damper too much may result in undesirable operation around the surge point.

Opening the damper reduces the system back pressure and results in a higher flow rate. The turndown associated with an outlet damper alone is approximately 6 to 1.

But using an inlet damper has a completely different effect on the fan than the outlet damper. Instead of changing the system curve, the inlet damper changes the fan curve as seen in Figure 3.16. Opening or closing the inlet damper pivots and changes the shape of the fan curve along

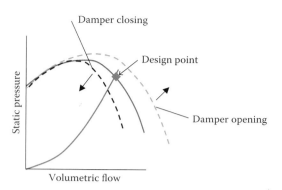

FIGURE 3.16
Inlet damper effects on fan performance.

the system curve, allowing much lower turndown than an outlet damper alone. Note though that the left-hand side of the curve is still anchored at the same spot regardless of how open or closed the damper is. Effectively, the inlet vanes change the blower performance curve so that the HP reduction ratio is actually greater than the flow ratio change. The turndown associated with an inlet damper is around 4–1 but varies depending upon the type of fan.

Dampers can have single-blade, opposing-blade, or parallel-blade configurations. Opposing blades offer the best control configuration because the airflow is evenly distributed across the damper. Parallel- and single-blade dampers are more cost-effective and can work well if the control of the air does not need to be precise, but these dampers can create an uneven airflow distribution in the ducting. Figure 3.17 shows a fan with both inlet and outlet dampers.

Speed control offers even greater flexibility than either an inlet or an outlet damper. Using a variable-frequency drive (VFD) shifts the entire curve up or down proportionally (see Figure 3.18). Using the VFD offers around a 10–1 turndown and the efficiency of the fan stays the

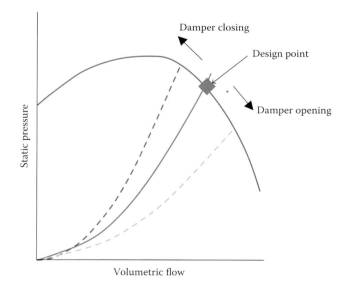

FIGURE 3.15
Outlet damper effects on fan performance.

FIGURE 3.17
Centrifugal fan with inlet and outlet dampers.

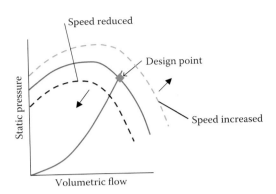

FIGURE 3.18
Speed change effects on fan performance.

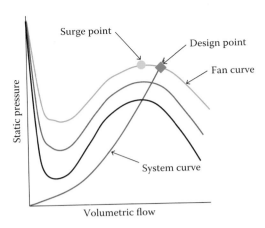

FIGURE 3.19
Variable and controlled pitch change effects on fan performance.

same throughout. Note, however, that it is possible to turn the fan down below the speed required for the internal motor fan to keep the motor bearings cool. Consult your blower vendor before finalizing your design. If using a VFD, ask the vendor to provide performance curves for the minimum rated speed, maximum speed, and three other speeds equally spaced between the minimum and maximum speeds.

Other control options may include two speed motors and multiple blowers depending upon the application. A two-speed motor simply has two sets of windings within that can turn the motor at a high speed and low speed (1/2 the RPM of high speed). Depending upon the application, a two speed may be a good fit. Other variations may be to use multiple blowers. Some flare applications use very large vane axial blower. Bringing blowers on line as required can result in significant cost savings.

One final topic of fan control is variable pitch and variable control impellers in vane axial blowers. The angle of the impeller blades in centrifugal blowers cannot be changed without changing the entire impeller assembly. However, some vane axial blowers come with what is called a variable pitch impeller, meaning if the operator shuts down the unit, the pitch of blades can be changed to alter the fan curve. This has a similar effect on the fan curve as an inlet damper (see Figure 3.19), but the pitch of the blades can only change when there is a problem and a shutdown is required. Figure 3.20 shows a close-up view of these variable pitch blades. Another option is to purchase a controlled pitch impeller, meaning the pitch of the blades can be changed during operation, but with the lower cost of VFDs, this option is becoming less common.

3.5.2 Materials of Construction

The materials of construction of a fan must be suitable for the services of the application. For most air applications, carbon steel is adequate. Low temperatures may require stainless steel or low-temp alloys. Applications where caustic or acidic gases can collect in the fan, such as

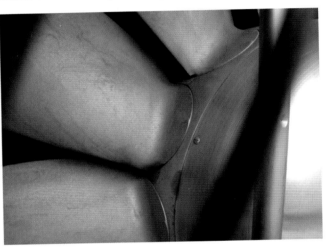

FIGURE 3.20
Close-up of variable pitch blades on a vane axial fan.

an ID fan on a thermal oxidizer, need to have corrosion-resistant alloys on all wetted parts. High-temperature applications, above 250°F (120°C), may require the use of heat flingers, which are heat transfer fins on the shaft, to keep the high temperatures from damaging the bearings. Other applications may use a plastic or FRP material for corrosion resistance. FRP fans should not run at 3600 RPM but should actually run at the lowest possible speed to prevent fatigue cracking and failure. Internal linings should, in general, be avoided as they tend to peel off after years of operation. Using alloys in this case may result in a higher up-front cost but less in maintenance. The cost of an unexpected shutdown is much greater than a little up-front capital cost.

3.5.3 Motors and Drives

Motors for fans and blowers can be very complex, and in fact, an entire book can be written on the subject. In this section the highlights and pitfalls to help in selection, but not the extreme details of motors and drivers are discussed. The most common driver for fans is the electrical induction motor. The most common voltage for combustion applications is around 460 V, but larger voltages can be used to reduce the power supply wiring size. Because of the inefficiencies of our physical world, induction motors do not turn at exactly their rated speed. For instance, an 1800 RPM motor will generally turn at 1780 RPM. This is known as slip. Synchronous motors, meaning a motor that does turn at its rated speed, can be used, but the additional cost may not be justified for small motor sizes. An additional consideration is the frequency of the supply voltage. As stated, an 1800 RPM motor will turn close to the rated speed using a 60 Hz power supply, but the rated speed will be around 1500 RPM using a 50 Hz power supply. This may be a critical factor in developing the proper pressure at the end of the impeller blades.

Motors for centrifugal fans should have a totally enclosed fan-cooled (TEFC) enclosure, meaning the motor has a small fan attached to the shaft that blows cooling air across the motor housing. Because the motor for most vane axial fans is mounted within the air stream, a totally enclosed air over (TEAO) motor is acceptable. However, a belt-driven motor mounted on the outside of a vane axial housing would need to be TEFC.

The motor should be adequately sized for the duty it will see. As discussed earlier, the fan needs to be sized for the flow rate at high temperature, but the motor needs to be adequately sized for the start-up amperage at the required flow rate and low temperature. Start-up current draw can be as much as seven times the full motor current load. Because of all the nuances involved in motor selection, let the blower vendors size the motors, but make sure they include all the options required and that the motor HP and torque are greater than what the fan requires.

Another item to consider is the motor service factor. The motor service factor is important because it tells the user how much extra HP a given motor can output. A motor with a service factor of 1.0 cannot exceed its nameplate HP without damage, but a motor with a service factor of 1.15 can handle loads up to 15% more for short periods of time. The service factor is available in case the user needs to operate the unit above the original design conditions for a short amount of time. The service factor of the motor is not to be used to satisfy the design requirements.

Motors for combustion applications should generally be suitable for a hazardous area. The equipment supplier does not have intimate knowledge of the area the equipment is going into and must sometimes make assumptions. It is the responsibility of the user of the equipment to inform the vendor of area classification requirements. Most applications will meet the NEC Class I, Div 2, Groups C&D, T3 requirements, meaning that equipment is suitable for a hazardous area, but that explosive conditions are not normally present. The comparable IEC equivalent is Class I, Zone 2, IIB, T3. However, it is important to note that depending upon the country in which the equipment is installed, the two may not be interchangeable, meaning where IEC equipment is required, NEC may not be an acceptable substitute and vice versa. Area classification needs to be verified by the local permitting agencies. Also note that an NEC Class I, Div 2 motor may only require a weatherproof terminal box, while IEC motor may require an explosion proof terminal box.

In certain applications where electricity is not available, or even when the possibility of the loss of electricity can create a problem, other drivers are available. Steam turbines offer a fairly efficient way to make use of a common utility in most facilities. Occasionally, end users will have an electric-driven fan and a steam-driven backup in case of an emergency. Steam turbines tend to be more expensive and require a higher level of maintenance compared to electric motors. Another option is an internal combustion motor with a gear box. These options may fit some applications a little better than an electric motor.

3.5.4 Couplings and Belts

Couplings and belts are used to transfer the drive energy from the motor to fan impeller. Non-lubricated, flexible couplings are common today and join the motor drive shaft to the fan shaft (see Figure 3.21). Other than periodic balance checks, they are low maintenance and easy to use. The materials must be suitable for the application. Additionally, couplings protect the motor and the fan shaft by serving as the weak point in the rotating system. A failed coupling is easier to replace than a motor or fan shaft and wheel. Belts are not as common in the combustion industry, but they offer a lower-cost alternative to couplings, and exact alignment is not as important as with a coupling (see Figure 3.22). Belts offer a low-cost alternative to speed changes without using a VFD. However, belt speed settings can only be changed when the fan is shut down. Belts need to be routinely checked for tightness and wear. Additionally, belts tend to slip, i.e., efficiency loss, as

FIGURE 3.21
Close-up of a flexible coupling.

FIGURE 3.22
A belt-driven centrifugal blower.

time goes on. Both belts and couplings need to have some sort of protective guard (OSHA or equivalent) to keep personnel from accidentally contacting moving parts.

3.5.5 Bearings and Lubrication

Bearings and the bearing lubrication system are typically the most critical components of turbo machinery to design and maintain, and blowers are not an exception. When possible, consider eliminating blower bearings and using an arrangement 4 or 5 fan, meaning the impeller is attached directly to the motor shaft, for small applications, less than 50 hp, and almost all vane axial fans. The elimination of the bearings means the system is less complex and requires less maintenance. The motor bearings are all that would need to be lubricated. Some motors come with sealed bearings that do not require adding lubrication. For most services, greased bearings are adequate. Grease bearings do not leak and are easier to maintain. Bearings should be an L-10 grade with approximately 100,000 h of expected run life. Sleeve bearings are adequate for many applications up to 250°F (120°C). Ball and roller bearings can handle temperatures up to 550°F (290°C). Oil lubrication may be required on higher-speed and higher-temperature services. The simplest oil lubrication option is the oil bath, where a pool of oil sits in the lower portion of the bearing housing and lubricates the bearings as they pass through the oil. Oiler systems are available that have an oil reservoir that operators can see the oil level and add lubrication when required (see Figure 3.23). Another option is what is known as an oil mist system that uses compressed air and injects a small amount of oil as needed. The most complicated option is an oil recirculation system, which has a pump and

FIGURE 3.23
Oil-lubricated bearing with reservoir.

reservoir. On critical applications, or high-temperature applications, an oil cooling system may be required. Additionally, heat slingers or shaft coolers may be used to reduce the temperature on the bearings. Oil mist lubrication may be required in severe service applications where the $D \times N$ value is greater than 200,000 where D is the diameter of the wheel in mm and N is the speed in RPM. Heat slingers are simply a set of cooling fins that dissipate heat into the atmosphere by radiation and convection.

If the fan is in critical service, meaning an emergency shutdown will cause much delay and loss of revenue, then it is recommended to monitor the vibration and temperatures of the bearings. Vibration is measured by an accelerometer or switch, while temperature is measured by resistance temperature detectors (RTD). Careful review of the vibration and temperature data can give an indication of potential bearing failure (refer to Table 3.4). If vibration occurs in the unsatisfactory range, then new bearing should be purchased and a shutdown of the blower in question planned. If the vibration is in the unacceptable range, the fan should be shut down to avoid possible catastrophic failure, which can lead to equipment damage and injury to personnel. Bearing temperatures greater than 150°F–170°F (65°C–76°C) indicate that the bearing needs to be regularly monitored to determine if any maintenance is required. Figure 3.24 shows an arrangement 8 fan during maintenance of the bearings.

FIGURE 3.24
Maintenance of arrangement 8 bearings.

3.5.6 Vibration and Installation

Blower foundations should be rigid and approximately five times the weight of the blower itself. See Figure 3.25 for a typical foundation drawing. Units installed on skids, or in critical areas where low noise is required, should be installed with vibration isolators. The skid may need additional stiffening above and beyond what would normally be required to simply support the equipment. A flexible sleeve should be attached to the blower discharge and, depending upon the installation, the inlet, so that vibration will not be carried into the ducting (see Figure 3.26). This is referred to as vibration isolation. The fan vendor should balance the wheel to ISO G 6.3 as a minimum for limit vibration.

Aside from proper blower sizing and overall system design, the single most important time in the life span of the blower will be the initial installation and balancing. An experienced and competent turbo machinery installation specialist can correctly align the motor, coupling, and impeller and can make the difference between replacing the bearings as soon as 2 years or the full lifetime of 8 or 10 years. Vibration readings are typically taken on the bearings, if they exist, and locations on the motor. Any velocity reading less than 0.1 in./s (2.5 mm/s) is considered good. For other ranges, see Table 3.4.

Different kinds of vibration can be indicative of the problem. See Table 3.5 for more information.

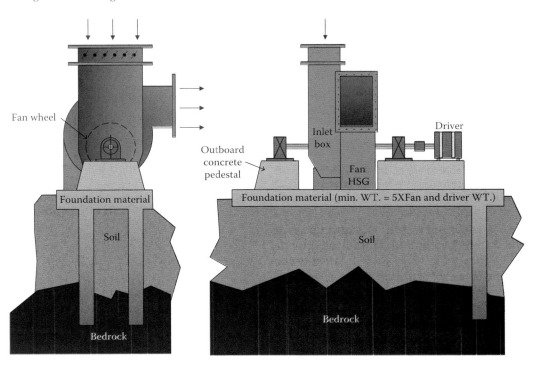

FIGURE 3.25
Fan foundation.

FIGURE 3.26
Inlet and outlet expansion joints for vibration isolation of ducting.

TABLE 3.4

Fan Bearing Vibration Limits

Vibration Limits for Fans (Bearing Measurements)									
Vibration Severity				Displacement Mils Peak to Peak					
Velocity (in./s)		Quality		For Various Fan Speeds (RPM)					
Peak	RMS	Rigid	Flexible	3600	1800	1200	900	720	600
0.025	0.018	Good	Good	0.1	0.3	0.4	0.5	0.7	0.8
0.040	0.028	↓	↓	0.2	0.4	0.6	0.8	1.1	1.3
0.062	0.044	↓	↓	0.3	0.7	1.0	1.3	1.6	2.0
0.100	0.071	↓	↓	0.5	1.1	1.6	2.1	2.7	3.2
0.16	0.11	Satisfactory	↓	0.8	1.6	2.5	3.3	4.1	5.0
0.26	0.18	↓	Satisfactory	1.4	2.7	4.1	5.4	6.8	8.1
0.40	0.28	Unsatisfactory	↓	2.1	4.2	6.3	8.4	10.5	12.6
0.62	0.44	↓	Unsatisfactory	3.3	6.6	9.9	10.3	16.5	19.8
1.00	0.71	Unacceptable	↓	5.3	10.6	15.9	21.2	26.5	31.8
1.56	1.10	↓	Unacceptable	8.0	16.0	24.8	33.1	41.4	49.7

Source: Jogensen, R., *Fan Engineering*, Buffalo Forge Company, Buffalo, NY, 1983.

TABLE 3.5

Fan Vibration Diagnostic Clues

Amplitude	Frequency	Phase	Possible Cause
Steady—radial largest	1 × RPM	Steady—single reference mark	Unbalanced (see text for clues to correcting)
Axial largest	1 or 2 × RPM	1, 2, or 3 reference marks	Misalignment or bent shaft (check with dial indicators)
Unsteady	1 × RPM	Unsteady	Resonance (check for impeller or other flexibility using modal-analysis techniques)
Unsteady	2 × RPM	2 reference marks	Looseness (check bolts, keys, etc.)
Unsteady	RPM/2	Unsteady	Oil whip (uncommon for fans)
Low	60 or 120 Hz	1 or 2 rotating marks	Electrical
Erratic	Many × RPM	Erratic	Faulty antifriction bearings
Erratic	1 or 2 × belt RPM	Unsteady	Defective belts (Check by freezing with strobe.)
Low	Blade-passing	Steady	Aerodynamic (check cutoff clearance)

Source: Jogensen, R., *Fan Engineering*, Buffalo Forge Company, Buffalo, NY, 1983.

3.5.7 Shaft Seals

Because the shaft on centrifugal fans has to pass through the housing on at least one side of the housing, shaft seals are required to keep air or gases from escaping through these openings. On combustion air blowers, some amount of leakage is acceptable and expected. Felt and graphite seals do the job well and only need to be replaced when overhauling the fan. On other critical applications where escaping gases can cause injury or death due to high temperatures, acids and caustics, or toxins, a substantial shaft seal is required. Shaft seals can be a carbon ring seal, or packing glands. Note that to obtain a seal, the seal material needs to be in contact with the shaft. This contact causes friction, and as a result of this friction, heat builds up in the shaft and bearings. Other options may include a purged seal, where nitrogen or another inert gas is continuously injected into the seal to maintain a positive pressure, preventing gases from escaping the blower. Bellows seals and spring seals are also available depending upon the application. Bellows seals offer a much greater amount of protection but have a higher up-front cost and greatly increased amount of maintenance.

3.5.8 Noise Considerations

Blower noise considerations can be a big issue, especially if the equipment is in an area where personnel are continuously exposed to the rotating machinery in question. Typically, blower noise should be less than 85 dBA at a distance of less than 3 ft (1 m), in an area where workers perform their duties in an 8 h period. OSHA allows for higher levels depending upon the circumstances, but 85 dBA is considered the industry norm. Inlet silencers are available for fresh air applications. For applications that require further noise suppression, a noise-insulating enclosure, sometimes called an attenuator, may be installed, but these can be fairly expensive. The higher the speed or pressure, the more noise a fan will generate.

When considering noise, blower vendors typically discuss it in terms of ducted and unducted. Ducted means that the blower is connected to a duct and that the duct will have some noise-suppression properties. Applications where fresh air is drawn into the system will typically have an unducted inlet. So, the unducted noise numbers should be used for noise estimates. Occasionally, lagging and noise insulation will have to be installed on the blower and ducting to prevent noise transmission through the steel to meet certain noise requirements. Noise in general is discussed in some detail in Volume 1, Chapter 16.

3.5.9 Filtration

Filtration is rarely used for combustion applications. However, if there is a large amount of particulate in the area, such as sand, then a filter may be considered. Installations in dusty or sandy climates may require filtration. Filtration of particulate can prevent abrasion wear on the fan impeller. Filtration can be as simple as a replaceable corrugated filter, much like the air conditioning filter in a house, to the very complex pulse jet bag filters. Other options are electrostatic dust collectors and cyclonic settling drums. The options on filtration can be very complex, and in some cases, the filtration system may cost more than the blower itself. Carefully consider if filtration is required.

3.6 Operational Costs

When considering operational costs, the user must understand where the unit will operate the majority of the time. Never assume that the latest technology is what is best for the application without reviewing other possibilities. For instance, take the example of Table 3.6 showing possible recommended control schemes for a flare. The users must understand where the system is going to operate most of the time, then it might be beneficial to evaluate if an inlet damper or VFD would save money. If the unit was a batch process that was full on when used and completely off when not, then a single-speed blower would be sufficient. Further, if the normal operating point was only half of the smokeless capacity, then a two-speed motor might be lower cost than a VFD and still function approximately the same. However, if normal rates are below 50%, then a VFD would be justified and substantial operating cost savings can be achieved.

Table 3.7 shows the potential relative savings using different control strategies. The base case assumes no

TABLE 3.6

Control Options Relative to Design Rate

Normal Operation	Control
100% of design rate	Single-speed blower
75%–100% of design rate	Single-speed blower with damper
50%–75% of design rate	2-speed blower with damper
25%–50% of design rate	VFD
0%–25% of design rate	VFD with damper

TABLE 3.7

Potential Controls Cost Savings

Control Type	Flow Rate	% Operation	HP	Cost ($)	Savings ($)
Base case	150,000	100	1275	1,232,167	—
Outlet damper	110,000	73	1095	1,073,363	158,799
Inlet damper	110,000	73	880	862,614	369,553
VFD	110,000	73	521	510,707	721,460

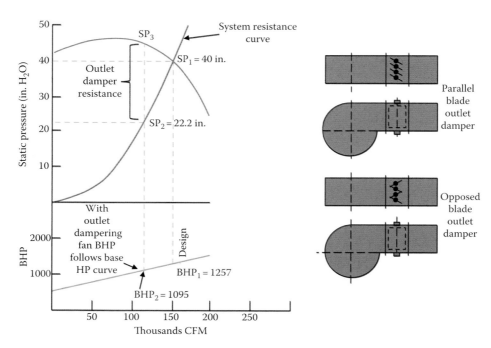

FIGURE 3.27
Outlet damper fan curve with HP.

damper control, while the others assume the process flow required is 110,000 CFM (3,100 CMM). All other aspects are the same and the cost of electricity is $0.15/kWh. The calculation is

$$\text{Annual Power Cost} = \text{HP} \times 0.746 \text{ kW/hp} \times 365 \text{ days/year}$$
$$\times 24 \text{ h/day} \times 0.15 \text{ \$/kWh}$$

HP is obtained by the HP curves in Figures 3.27 through 3.29.

Another consideration is the blower efficiency. Use the fan wheel that offers the highest efficiency. Centrifugal blowers should have a mechanical efficiency of 60%–70%, while vane axial blowers should be in the 60%–80% range. Fans with 50% efficiency or less should be reevaluated.

3.7 Inspection and Testing

The level of inspection and testing required will depend upon three things, the criticality of the blower in service, the size of the blower, and the level of comfort required by the owner. The most basic inspection check would be a visual and dimensional check. This visual and dimensional check should verify all of the connection flange dimensions and the anchor bolt hole dimensions. Note that for off-the-shelf blowers, a dimensional check may be all that is available. Material certifications may also be required. Discuss the quality control and documentation requirements with the blower vendor before procuring services. Additionally, the basic inspection may require an impeller balancing certificate with reference to the applicable balancing standard.

Fans that are designed for low or high temperatures or hazardous and toxic gases may require additional inspection of the welds. The type of testing required will depend upon the material of the components to be welded and the weld material. Radiographic examination uses x-ray exposure to take a "picture" of the weld metal to see if there are any flaws within the weld and the surrounding area. However, the shape of the piece to be welded must be somewhat flat to use this process. Other methods available for testing of welds include ultrasonic testing to look for cracks or imperfections below the surface, liquid penetrant examination to look for surface cracks, and magnetic particle examination to find defects in alloy materials and welds. Additionally to testing, weld procedures and other documentation may be required to verify proper techniques. It is important to discuss inspection requirements with the fan vendor before purchasing any equipment.

Some applications such as fans handling toxic or hazardous gases may require a pressure test of the housing. This is done in order to verify that there are no leaks in the housing. Pressure testing may be accomplished using water, known as a hydro test, or air, known as a

Blowers for Combustion Systems

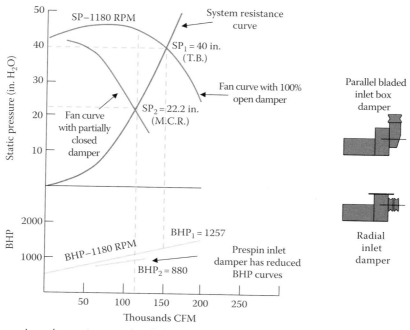

FIGURE 3.28
Inlet damper fan curve with HP.

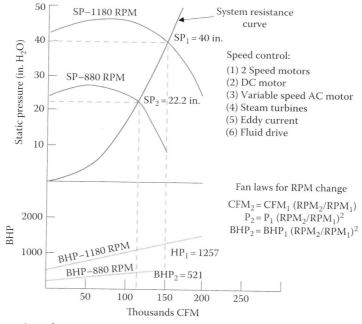

FIGURE 3.29
Speed control fan curve with HP.

pneumatic test. The test pressure required is usually 1.5 times the working pressure of the fan for a minimum of 30 min.

The mechanical run test is performed to verify if the fan and drive system are correctly balanced and is usually performed for a minimum of 2 h. During the 2 h, the bearing vibration and temperature are measured at regular intervals. Mechanical run tests are generally not performed on fans less than 30 hp. Sound tests may be performed as well to determine the overall noise level. The sound meter should be calibrated for the A scale and noise measurement taken at 3 ft (1 m) from the blower and at multiple points around the blower to find the maximum level.

For critical fans and large fan applications, a performance test may be required. A performance test is what the fan manufacturer would do in order to develop a fan curve for a particular fan. The fan is connected to a test chamber that has a flow straightener, a pitot tube and manometer setup to measure the total and velocity pressure, and a set of dampers to adjust the back pressure on the chamber and fan.

A final note concerning quality control and fan vendors is that it is recommended to discuss in detail any requirements concerning API 5607 or 673.8 These standards contain a varying degree of design and inspection requirements that some vendors are simply not prepared to meet. API 560 and 673 should not be applied to small fans, less than 30 hp (22 kW), unless all parties are in agreement. The documentation in these standards is stringent. Additionally, they can require a torsional and lateral analysis of the shaft. Applying these standards can greatly increase the cost of a blower. Depending upon the degree of the requirements, the price may be four times as much as a standard unit, or even more.

3.8 Maintenance and Troubleshooting

Fan maintenance is extremely important. The most common items in need of attention are the fan bearings. The speed of the fan and the shaft size determine the size of the bearings and consequently the maintenance requirements. Grease bearings should be lubricated routinely according to the manufacturer's instructions. Lubrication of the bearings is extremely important and should be carried out regularly. Even automatic lubrication systems should be routinely checked in order to catch problems before a failure occurs. During shutdowns, bearings, if not too worn, need to be cleaned and repacked with the proper grease, or oil. Bearings that need to be replaced should be overhauled or the entire assembly replaced. The wheel, coupling, and motor should be checked for balance, stress cracking, and excessive vibration. The fan housing should be checked and debris removed.

Often fans may develop problems in the field despite the best maintenance efforts. Table 3.8 provides a general list of common problems and possible solutions.

TABLE 3.8

Common Fan Problems and Possible Solutions

Problem	Probable Cause
Hot bearing	Recheck temperature measurement (do not trust "hand feel")
	Over-lubrication
	Wrong lubricant (check MFG recommendations)
	Improper internal clearances
	Misalignment
	High ambient temperature, no ventilation around bearings
	Sunlight
	Excessive thrust loading
	Improper shimming
	High vibration
	Bearing race turning inside housing
	Moisture in bearing
Excessive vibration	Accumulated material on impeller
	Worn or corroded impeller
	Bent shaft
	Impeller or sheaves loose on shaft
	Motor out of balance
	Impeller out of balance
	Sheaves eccentric or out of balance
	Bearing or drive misalignment
	Loose or worn bearings
	Weak or resonant foundation
	Foundation unlevel
	Structure not cross braced
	Fan operating in unstable system condition
Excessive HP draw	Fan speed higher than design
	Gas density higher than design
	Impeller rotating in the wrong direction
	Fan size or type not appropriate for the application
Insufficient airflow	Duct elbows near fan inlet or outlet
	Restricted fan inlet or outlet
	Impeller rotating in the wrong direction
	Fan speed lower than design
	System resistance higher than design
	Dampers shut
	Faulty ductwork
	Dirty or clogged filters and/or coils
	Inlet or outlet screens clogged
Excessive airflow	System resistance less than design
	Fan speed too high
	Filters not in place
	Air registers or grilles not installed
	Improper damper adjustment
Inoperative fan	Blown fuse
	Broken belts
	Loose sheave
	Motor too small
	Wrong voltage
Excessive noise	Imperfection in bearing element
	Improper clearance
	Internal wear of bearing parts

References

1. R. Jorgensen, *Fan Engineering Handbook*, Buffalo Forge Company, New York, 1983.
2. AMCA Publication 200-95 (R2007), *Air Systems*, Air Movement and Control Association International, Inc., Arlington Heights, IL, 2007.
3. AMCA Publication 201–02 (R2007), *Fans and Systems*, Air Movement and Control Association International, Inc., Arlington Heights, IL, 2007.
4. R.W. Fox and A. McDonald, *Introduction to Fluid Mechanics*, John Wiley & Sons, Inc., New York, 1985.
5. E.A. Avallone and T. Baumeister III, *Marks Standard Handbook for Mechanical Engineers*, 9th edn., McGraw-Hill, New York, 1987.
6. H.L. Gutzwiller, Fan performance and design, Robinson Industries, Inc., Zelienople, PA, January 2000.
7. API Standard 560, Section 11, *Fired Heaters for General Refinery Service*, 3rd edn., API Publishing Services, Washington, DC, May 2001.
8. API Standard 673, *Centrifugal Fans for Petroleum, Chemical, and Gas Industry Services*, 2nd edn., API Publishing Services, Washington, DC, January 2002.

4

Metallurgy

Wes Bussman, Garry Mayfield, Jason D. McAdams, Mike Pappe, and Jon Hembree

CONTENTS

- 4.1 Introduction
- 4.2 Background 96
  - 4.2.1 Carbon Steel 96
    - 4.2.1.1 Composition of Carbon Steel 96
    - 4.2.1.2 Iron Oxide Scale 96
  - 4.2.2 Stainless Steel 97
    - 4.2.2.1 Brief History 97
    - 4.2.2.2 Chromium Oxide Scale 97
    - 4.2.2.3 Types of Stainless Steel 97
    - 4.2.2.4 Metals Commonly Used in the Process Burner and Flare Industries 98
- 4.3 Common Forms of Metal Failure in the Combustion Industry 98
  - 4.3.1 Intergranular Corrosion 99
  - 4.3.2 Stress Corrosion Cracking 99
  - 4.3.3 Breakaway Corrosion 100
  - 4.3.4 Sulfidation Attack 100
  - 4.3.5 Hot Corrosion 101
  - 4.3.6 Chlorination Attack 101
  - 4.3.7 Carburization Corrosion 102
    - 4.3.7.1 Exposure to Carbonaceous Environments 102
    - 4.3.7.2 Metal Dusting 102
  - 4.3.8 Cryogenic Service 103
- 4.4 Material Selection 103
  - 4.4.1 Process Burners 105
    - 4.4.1.1 Air Plenum 105
    - 4.4.1.2 Fuel Delivery System 105
    - 4.4.1.3 Flame Holders/Stabilizers 106
  - 4.4.2 Process Flares 106
    - 4.4.2.1 Flare Burner (Typical) 106
    - 4.4.2.2 Pilots 108
    - 4.4.2.3 Moleseals and Flare Riser 108
- 4.5 Examples and Case Histories 108
  - 4.5.1 Process Burners 109
    - 4.5.1.1 Formation of Oxide Scale on Heater Process Tubes 109
    - 4.5.1.2 Ruptured Process Tubes 109
    - 4.5.1.3 Process Tube Distortion 109
    - 4.5.1.4 Burner Damage 112
    - 4.5.1.5 Control Valve Damage 112
    - 4.5.1.6 Oil Gun Damage 113
    - 4.5.1.7 Corroded Orifice Spud 113
    - 4.5.1.8 Damaged Premixed Burner Tip 114
  - 4.5.2 Process Flares 117
    - 4.5.2.1 Flame Retention Segments 117
    - 4.5.2.2 Pilot in an Enclosed Flare 117

      4.5.2.3 Pilot on an Elevated Flare 118
      4.5.2.4 Air-Assisted Flare 118
      4.5.2.5 Steam-Assisted Flare 119
4.6 Welding 119
    4.6.1 Types of Welding Processes 119
    4.6.2 Welding Carbon Steel...120
    4.6.3 Welding Stainless Steel 124
4.7 Nondestructive Testing..125
    4.7.1 Introduction..125
    4.7.2 Liquid Penetrant Testing..125
    4.7.3 Magnetic Particle Testing 126
    4.7.4 Radiographic Testing..128
    4.7.5 Ultrasonic Testing...129
    4.7.6 Positive Material Identification/Alloy Verification 130
    4.7.7 Metallographic Replication 130
References..131

4.1 Introduction

Equipment failure can create safety hazards and loss of production in industry. The premature failure of combustion equipment can occur in a variety of ways; however, a common source of failure is attributed to the destruction of the equipment's material. Common sources that cause the destruction of the metallurgy in combustion equipment stem from one or a combination of the following factors: (1) continuous or frequent flame impingement, (2) gasses containing sulfur, (3) temperature of process gasses, (4) atmospheric corrosive environments, and/or (5) design and fabrication. The purpose of this chapter is to discuss these major sources.

4.2 Background

4.2.1 Carbon Steel

4.2.1.1 Composition of Carbon Steel

Iron is one of the most abundant materials found on earth. Pure iron has little practical use because it is soft; however, with the addition of carbon and other alloying elements and the proper processing, a new material called steel is formed. Steel can be harder, stronger, and more useful than pure iron. Commercial carbon steel consists of about 0.03%–1.5% carbon by weight (wt%). A small change in the amount of carbon added to iron can have a big impact on its properties. Table 4.1[1,2] shows the classification of iron–carbon alloys. Low-carbon and mild-carbon steels are two of the most widely used grades of steel and are characterized by low strength/high ductility and are easily machined, formed, and welded. These steels cannot be hardened by heat-treating. Common uses of low-carbon steel include pipes, chains, wires, building structures, bridges, cars, ships, nails, rivets, screws, refrigerator bodies, flare stacks, burner plenums, air registers, and heater structures. Medium-carbon steel is harder than low- and mild-carbon steel and is hardened by heat treatment. Typical uses of medium-carbon steel include crank shafts, axles, machine parts, connecting rods, and rollers on conveyer belts. High-carbon and very high-carbon steels are the hardest of all carbon steel materials. Typical uses include springs, masonry nails, files, knives, ax heads, punches, and tool steel for cutting.

Low-, mild-, and medium-carbon steels (0.05–0.45 wt% carbon) are commonly used in process burner and flare applications. A couple of these standard steel alloys include SA-36, SA-106, and SA-516. Table 4.2 shows the composition ranges of each of the metals.[3]

TABLE 4.1

Classification of Iron–Carbon Alloys

Material	% Weight of Carbon in Iron
Wrought iron	0–0.05
Steel	
Low-carbon steel	0.05–0.15
Mild-carbon steel	0.15–0.3
Medium-carbon steel	0.3–0.45
High-carbon steel	0.45–0.75
Very high-carbon steel	0.75–1.5
White cast iron	2.0–4.5
Malleable iron	2.5–3.0
Gray iron	3.0–4.0
Ductile iron	4.0–4.5

Metallurgy

TABLE 4.2

Composition Ranges of a Few Standard Carbon Steel Alloys

Element	SA-36	SA-106	SA-516 (70 Grade)
Carbon (C)	0.29 maximum	0.25–0.35	0.31 maximum
Silicon (Si)	0.4 maximum	0.1	0.15–0.30
Manganese (Mn)	0.60–1.2	0.27–1.06	0.85–1.2
Copper (Cu)	0.2 minimum	—	0.3
Phosphorus (P)	0.04 maximum	0.48	0.035
Sulfur (S)	0.05 maximum	0.058	0.04
Iron (Fe)	Balance	Balance	Balance

4.2.1.2 Iron Oxide Scale

There is an old saying that "rust never sleeps."[4] We commonly see reddish-brown rust on many things such as steel bridges, buildings, automobiles, and metal fences. Rust is created when iron, water, and oxygen from the ambient air react with each other to form *iron oxide* (Fe_2O_3).

When iron oxide forms on the surface of a metal, it is sometimes beneficial because it can act as a protective barrier to help slow down the rate of corrosion. The problem, however, is that when iron oxide is formed on the surface, it is sparsely packed and does not adhere to the surface very well. The reason is because the iron atom is much smaller than the oxygen atom, which does not allow the molecules to neatly pack together. Figure 4.1[5] is an electron microscope image of corroded carbon steel. Notice the small islands; this is the iron oxide that has formed and is attached to the surface. The height of these islands is about 1/1000 of an inch (0.025 mm). This photograph shows two interesting things: (1) iron oxide does not cover the entire surface of the steel leaving some areas exposed to the corrosive environment and (2) iron oxide appears to be poorly attached to the steel.

4.2.2 Stainless Steel

4.2.2.1 Brief History

In the early nineteenth century, scientists began to look for ways to prevent steel from rusting. In 1821 a French scientist, Pierre Berthier, found that if chromium (a very hard and brittle material) was mixed with carbon steel, it became corrosion resistant to some acids.[6] He suggested using it to replace silverware eating utensils. Unfortunately, at that time they were not able to eliminate all carbon from the chromium that made the steel very brittle and of no real practical use. In the 1890s, Hans Goldschmidt of Germany discovered a way to produce carbon-free chromium. However, an English metallurgist named Harry Brearley is sometimes considered the "inventor" of stainless steel (SS) because he was the first to industrialize the SS alloy.

In 1914, Brearley experimented with different amounts of chromium in iron and found that about 13 wt% was required to give good corrosion resistance. In 1914, he

FIGURE 4.1
Electron microscope image of corroded carbon steel. (From Sutter, L., Research and professional page, Michigan Tech Transportation Institute, Houghton, MI, www.tech.mtu.edu/~llsutter/rust.html)

produced and commercialized eating utensils that contained about 13 wt% chromium. Today, eating utensils have about 18 wt% chromium and 10 wt% nickel.

About the time Brearley was commercializing his SS, a company in Germany called Krupp Iron Works began similar developments. They developed a unique SS with a mixture of chromium and nickel (21 wt% chromium and 7 wt% nickel). Nickel alters the atomic structure of the steel, making it more ductile, stronger, creep resistant, elastic, and easier to weld compared to iron–chromium alloys. Over the years, industry has been adding various amounts of chromium, nickel, and other elements such as molybdenum, aluminum, and silicone to help improve the protective barrier.

4.2.2.2 Chromium Oxide Scale

During the development period of SS, people did not realize why chromium prevented steel from corroding. Today, how chromium helps prevent corrosion is better understood.

When chromium is added to steel, the oxygen in the air first reacts with the chromium instead of the iron to produce *chromium oxide*. The chromium and oxygen atoms are

FIGURE 4.2
Electron microscope image of chromium oxide layer on SS surface. (Courtesy of P. Hou, Berkeley Lab, Berkeley, CA.)

similar in size so they pack tightly together on the surface of the steel. This chromium oxide scale is invisible to the eye because it is only a few atoms thick (about 1 millionth of a millimeter). This scale is called the *passive film*.

The passive film adheres to the surface of the steel and protects the steel from further corrosion. If the metal is scratched, the passive film will quickly form, protecting it once again from corrosion. In order for the passive film to form effectively, the steel needs at least 10.5 wt% chromium.

Figure 4.2 is an electron microscope image of SS that has been scratched. Notice the chromium oxide scale has been removed, but next to the scratch, the scale is firmly attached. The steel will repair itself with a new layer of chromium oxide. In order for it to do this, oxygen must be present. If oxygen is not present, the passive film will not develop and leave the steel exposed to further corrosion.

4.2.2.3 Types of Stainless Steel

Stainless steels are defined as iron-based alloys containing a minimum of 10.5% chromium.[7] There are hundreds of separate and distinct alloy steels commercially available, each one with its own unique properties. The *American Iron and Steel Institute* (AISI) has developed a *unified numbering system* (UNS) to identify many of these steels. In addition to the AISI types, there are many more proprietary SSs worldwide, for example, Hastelloy.

AISI SSs are identified by a numbering system that falls into three series: 200, 300, and 400 series. The 200 and 300 series are referred to as *austenitic* SSs. Austenitic steels are nonmagnetic in the annealed condition and cannot be hardened by heat-treating; heat-treating will tend to soften them. The 300-series SSs consist of a chromium–nickel–iron composition. Other elements are added to these compositions to help improve corrosion resistance and strength. For example, types 321 and 347 contain small amounts of titanium in order to help prevent high-temperature corrosion from *carbide precipitation* (see Section 4.3.1). Types 316 and 317 contain small amounts of molybdenum to help improve corrosion and oxidation resistance. The 200-series SSs are similar to the 300 series except manganese replaces some of the nickel to help give them more strength.

The 400 series fall into the ferritic and martensitic group SSs. The ferritic SSs are chromium–iron alloys. They are magnetic and cannot be hardened by heat treatment. Types 430 (14%–18% chromium) and 446 (23%–27% chromium) are a couple of common variations in the ferritic group. The martensitic group has a higher carbon:chromium ratio than the ferritic group. Martensitic SSs are magnetic and can be hardened by heat treatment. Type 410 is the most common in this group and is often used for heat exchange service.

4.2.2.4 Metals Commonly Used in the Process Burner and Flare Industries

The 300-series SSs are typically used in process burner and flare applications. Typical of this group is 304 SS (see Table 4.3), also known as 18-8 SS because it contains about 18 wt% chromium and 8 wt% nickel. Relative to the other SS in the 300 series, 304 SS is typically the least cost with relatively little chromium and nickel. On the other end of the 300-series SS spectrum, in terms of chromium and nickel content, is 310 SS. Series 310 SS contains about 5 wt% more chromium and about two times more nickel than 304 SS and is about twice as expensive. As a general rule, SS prices are dictated by how much nickel they contain; that is, the more nickel, the higher the cost. Both 304 and 310 SSs are commonly used in combustion equipment and are excellent steels where occasional flame impingement (see Chapter 13) occurs.

In the middle of the 300-series SS spectrum, in terms of chromium and nickel content and cost, is 309 SS. Series 309 SS has about the same amount of chromium, but about 4–10 wt% less nickel than 310 SS. The lower nickel content helps reduce sulfidation corrosion (discussed later) and is sometimes used when the process gas contains sulfur. Other common SSs used in the process burner and flare applications include 316 and 321.

Metallurgy

TABLE 4.3

Metals Commonly Used in the Process Burner and Flare Industry

	Weight Percent of Element					
Material	Nickel	Chromium	Iron	Carbon	Other	Relative Cost
Alloy 800	32.5	21	Balance	0.05	Aluminum, titanium	20.1
310 SS	19–22	24–26	Balance	0.25	—	9.4
309 SS	12–15	22–24	Balance	0.2	—	7.8
321 SS	9–12	17–19	Balance	0.08	Titanium	6.7
316 SS	10–14	16–18	Balance	0.08	Molybdenum	6.1
304 SS	8–10	18–20	Balance	0.08	—	4.9
A-36 Carbon	0	0	99	0.26	Copper, sulfur, phosphorus, manganese	1.0

Occasionally, end users request process burner and flare equipment to be fabricated of a special grade, patent-protected, high-nickel alloy. Although some of these alloys have lost their patent protection, they are still known by their original trade name. For example, alloy 600 is commonly associated with Inconel and alloy 800 with Incoloy. Relative to carbon steel and 300-series SS, high-nickel alloys are relatively expensive (see Table 4.3).

4.3 Common Forms of Metal Failure in the Combustion Industry

4.3.1 Intergranular Corrosion

When SS is exposed to thermal cycling (repetitive heating and cooling), it can degrade through a mechanism called intergranular corrosion. Intergranular corrosion initiates within the steel at locations called the grain boundaries. These grain boundaries form during the manufacturing process of SS.

As molten steel begins to cool and transform to a solid state during the manufacturing process, individual grains in the steel form as referred to as dendritic structure as shown in Figure 4.3.[8] These grains resemble the look of pine trees. The space that exists between grains is usually just a few atoms wide and is referred to as a grain boundary. At the grain boundary, the SS is rich in chromium.

When SS is heated to a temperature range of 1110°F–1650°F (600°C–900°C) and slowly cooled, the carbon in the steel migrates to the grain boundaries: this is referred to as *carbide precipitation*. Once inside the grain boundary, the carbon combines with the chromium and forms a brittle material called chromium carbide. As thermal cycling continues, more chromium is depleted from the alloy forming more chromium carbide at the grain boundaries. The alloy can become brittle at the grain boundaries and start to form cracks as shown in Figure 4.4[9] (referred to as embrittlement).

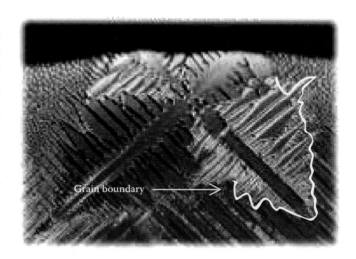

FIGURE 4.3
Magnified view of SS showing grain boundary where the steel is rich in chromium content. (From Bondhus, Crystals, grains, and cooling, Picture 1—Dendrite crystal, Monticello, MN, http://www.bondhus.com/metallurgy/body-3.htm)

FIGURE 4.4
Magnified view showing intergranular corrosion.

FIGURE 4.5
Magnified view showing crack propagating across the grain boundary due to stress corrosion cracking. (From Metallurgical Technologies Inc., Mettalography/microstructure evaluation, Mooresville, NC, http://www.met-tech.com/metallography.htm)

4.3.2 Stress Corrosion Cracking

Stress corrosion cracking occurs when steel is under stress in a corrosive environment; both of these have to be present in order to qualify as *stress corrosion cracking*. Stresses can come from residual stress during manufacturing, such as welding or bending or from thermal expansion due to flame impingement or from the pressure induced by the process gas. When stress corrosion cracking occurs, a rapid or sometimes catastrophic failure occurs.[10] Figure 4.5[11] shows a magnified view of a crack propagating across the grain boundary due to stress corrosion cracking.

4.3.3 Breakaway Corrosion

As mentioned earlier, the presence of chromium in SS helps protect the alloy from corrosion. Chromium protects the steel by reacting with the oxygen to create a thin, chromium oxide barrier on the surface. When SS is heated and then cooled, both the steel and oxide scale will grow and shrink due to thermal expansion and contraction; however, the rate of thermal growth between the steel and the oxide scale is different. At 68°F (20°C), the coefficient of thermal expansion for SS and chromium oxide (Cr_2O_3) is 17.3×10^{-6} and $6.5 \times 10^{-6}/K$[12,13]; this difference leads to spalling of the oxide scale causing it to detach from the surface of the steel as shown in Figure 4.6.

When the oxide scale detaches, fresh metal is exposed to the environment; if oxygen is present, a new oxide scale forms. As thermal cycling persists, the oxide scale is continually removed and reformed; this cyclic event

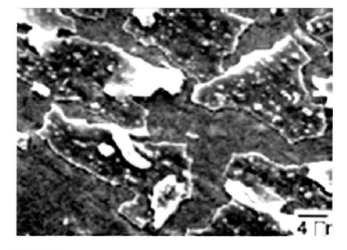

FIGURE 4.6
Magnified view showing removal of the oxide scale from an alloy surface due to thermal cycling. (Courtesy of P. Hou, Berkeley Lab, Berkeley, CA.)

progressively depletes the chromium content in the alloy and is referred to as *cyclic oxidation*. Given sufficient scale removal, enough chromium may be lost to cause the alloy to lose its heat-resistant properties.[14] Eventually, after a substantial amount of chromium is depleted, oxygen in the air begins to react with the iron and nickel causing rapid deterioration of the steel known as *breakaway corrosion*.

Table 4.4 shows the generally accepted service temperature of several 300-series SS grades for intermittent (thermal cycling) and continuous service temperature in air.[15] Notice that the continuous service temperature is

TABLE 4.4
Generally Accepted Service Temperatures for Several 300-Series SSs

Material	Intermittent Service Temperature	Continuous Service Temperature
304	1600°F (870°C)	1700°F (925°C)
316	1600°F (870°C)	1700°F (925°C)
309	1800°F (980°C)	2000°F (1095°C)
310	1900°F (1035°C)	2100°F (1150°C)

higher than the intermittent temperature. The reason is thermal cycling can result in breakaway and intergranular corrosion, which lowers the service temperature.

4.3.4 Sulfidation Attack

Sulfurous gases, such as hydrogen sulfide (H_2S) and sulfur oxides (SO_2, SO_3), are found in many applications in industry, including combustion atmospheres and petrochemical processing.[16,17] From a process gas stream, sulfurous gases are one of the most common contributors to corrosion damage in the process burner and flare industry.

Verma[18] has shown that the presence of chromium is the most important alloying element in resisting sulfidation attack because it forms a protective layer of chromium oxide over the metal surface. Howes[19] showed that high-nickel alloys generally increase susceptibility of sulfidation attack. When the nickel diffuses through the chromium oxide scale, it reacts with the sulfur environment to form nickel sulfides on top of the oxide scale. One particular nickel sulfide that forms on top of the protective chromium oxide scale is $Ni-Ni_3S_2$. This particular nickel sulfide melts at 1175°F (635°C). The molten nickel sulfide can easily destroy the chromium oxide scale and lead to catastrophic *sulfidation attack*.

Figure 4.7[20] shows a cross-sectional view of a component made of alloy 800H that failed due to sulfidation attack. The chromium oxide scale broke down creating iron-rich sulfides on the outer surface. The photograph also shows chromium sulfide and depleted chromium regions lying below the iron-rich oxide scale.

Field experience has shown that high-nickel alloys are not the best choice for sulfur applications; in general, the higher the nickel content, the more sensitive the alloy is to sulfidation corrosion. SSs and iron-based alloys are preferred over high-nickel alloys because nickel is prone to forming the low-melting nickel sulfide that leads to sulfidation attack.[16]

4.3.5 Hot Corrosion

Equipment located near coastal areas appears to suffer more extreme high-temperature corrosion damage than those located inland. The reason is because near coastal areas, the air is laden with sea salt that contains

8000 H, 1200°F–1400°F, sulfidation attack

1. Iron and chromium oxide scale (57Fe-18Ni-16Cr-5S-4K)
2. Iron and chromium oxide scale (80Fe-8Ni-12Cr)
3. Nickel sulfide (71Ni-2Cr-27S)
4. Depleted chromium (72Ni-28Fe)
5. Chromium sulfide (40Cr-10Fe-50S)

FIGURE 4.7
Magnified view showing the cross section of an alloy suffering from sulfidation attack. (Reprinted from Lai, G., *High Temperature Corrosion of Engineering Alloys*, ASM International, Materials Park, OH, 1990, Figure 7.2b. With permission.)

very corrosive elements. Studies have found that mist and fog can consist of 5% salt water solution by volume and can be carried as far as 5 miles (8 km) inland from the coast. When the salt-laden air comes in contact with steel and is heated, a very corrosive environment is produced known as *hot corrosion*.[21,22]

The most corrosive elements of salt water include (1) chlorine, (2) sodium, and (3) sulfur. The mechanism of hot corrosion is complex and not fully understood. What appears to happen is that sodium and sulfur react to form sodium sulfate. At a temperature of 1470°F (800°C), the sodium sulfate melts and adheres to the surface of the metal.[20] The sodium sulfate diffuses into the chromium oxide protective scale and destroys it, leaving the metal vulnerable for attack; this is referred to as the incubation period. Chromium is the best alloy to improve resistance to hot corrosion.

The next stage is referred to as accelerated corrosion attack. In this stage the sodium sulfate reacts with chromium in the steel and forms chromium sulfide; this depletes the steel of chromium, leaving steel vulnerable for attack by other corrosion mechanisms, one of these being *chlorination attack*.

4.3.6 Chlorination Attack

In hot environments, chlorination attack occurs when the chlorine from the salt water in the air reacts with the iron in the steel to form iron chloride. Iron chloride melts at a temperature of 577°F (303°C) leaving internal voids in the metal. Both chromium and nickel are good for resistance to chlorination attack.

Chlorine in the air can also corrode equipment at ambient temperatures. During winter months, many cities put salt on the roads to help melt ice and snow. Usually, not long after the roads are salted, pot holes start to form. This deterioration is sometimes caused by the chloride in the salt. The chloride corrodes the carbon steel rebar embedded within the concrete. When rust scale forms, it takes up about three to seven times the volume of the original steel causing the concrete to expand, crack, and break away resulting in potholes.[23] A similar phenomenon occurs to the paint on combustion equipment located in coastal areas. That is, salty air environments more readily create rust on equipment causing expansion that flakes the paint off and exposes the steel. Figure 4.8 shows paint flaking off process equipment located along the Texas coastline. Keeping fresh coats of paint on carbon steel helps protect it from further corrosion.

4.3.7 Carburization Corrosion

For thousands of years, man has hardened steel by the process called *pack hardening*. This process involved packing charcoal around low-carbon steel and heating it for a period of time. After the heating process, the steel was dropped into water to quench. This process allows carbon to migrate into the steel, making the surface extremely hard while maintaining a core that retained the toughness of low-carbon steel.[24] Today, the surface of low-carbon steel is heated in furnaces that contain a carbon atmosphere. As the steel is heated, the carbonized gas ingresses into the surface of the steel. This process is referred to as *gas carburization* and is commonly used to manufacture shafts, files, saw blades, etc.

Although gas carburization is a useful manufacturing process, it is one of the major methods of high-temperature corrosion in the petrochemical industry.[20] In a survey, Moller and Warren[25] found that carburization was a major mode of process tube failure in ethylene and olefin pyrolysis furnaces.

There are two different corrosion processes caused by carburization: (1) internal carbide formation in alloy steels exposed to carbon-containing atmospheres (carbonaceous environments) and (2) metal dusting.

4.3.7.1 Exposure to Carbonaceous Environments

Carburization can occur when alloys are exposed to temperatures of about 1550°F–1700°F (840°C–930°C)[20] in atmospheres containing carbon monoxide (CO) and hydrocarbon gases such as methane (CH_4) and propane (C_3H_8). At high temperatures, the carbon molecules diffuse through the pores and cracks of the chromium oxide protective barrier.[26] Once past the chromium oxide barrier, the carbon combines with the alloy, forming metal oxides making the steel brittle. As a general rule, when the carbon content of the steel reaches about 1% by weight, the steel becomes brittle at room temperature[24]; this, however, depends on the microstructure and carbon content of the steel. When the steel becomes brittle, it loses its tensile strength and becomes more susceptible to cracking. The carbon also ties up the chromium which reduces the oxidation resistance of the metal.

Metal exposed to carbonaceous environments can also suffer damage in another way; this is not referred to as carburization corrosion. When soot (carbon) from a flame collects on metal, it can penetrate deep into small cracks in weld joints or surface defects. As the soot deposits continue to collect and grow, they tend to pry open the cracks resulting in larger cracks and cavities.[24]

FIGURE 4.8
Corrosion of carbon steel process equipment located along the Texas coastline.

4.3.7.2 Metal Dusting

Metal dusting (also referred to as carbon attack, carbon rot, and catastrophic carburization[27]) is a carburization process that disintegrates alloys to a powdery black residue of metal particles. The attack results in metal wastage through pitting and/or thinning of the material as shown in Figure 4.9.[28] Metal dusting is particularly conducive to a stagnant atmosphere containing carbon monoxide (CO), carbon dioxide (CO_2), hydrogen (H_2), water vapor (H_2O), or hydrocarbon gases at temperatures between 800°F and 1650°F (430°C–900°C).[20]

The mechanism of metal dusting for iron-based alloys is different than for nickel-based alloys.[29] For both cases, metal dusting begins with saturation of carbon into the steel. For iron-based alloys, the carbon decomposes the metal into iron particles and powdery carbon. For nickel-based alloys, the carbon decomposes the metal into metal particles and graphite. Carbon steels are usually uniformly thinned by metal dusting. More highly alloyed materials usually display damage through small pits that increase in size over time.[16] For more details on metal dusting, the reader is referred to reference [30].

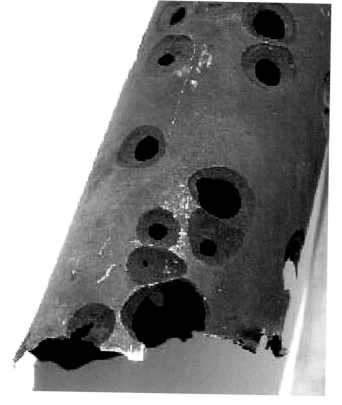

FIGURE 4.9
Metal dusting on an inlet tube of a heat exchanger unit. The material is alloy 800. (From Hrivnak, I. et al., *Kovove Materialy*, 43, 290, 2005.)

4.3.8 Cryogenic Service

It is not uncommon for equipment to be exposed to extremely low temperatures. For example, in many parts of the world, such as the northern parts of Canada, the ambient temperature can get as cold as −40°F to −50°F (−40°C to −46°C). Also, at times, equipment must be designed to handle process fluids at very cold temperatures. For example, in ethylene and liquid methane flare applications, the waste gas temperature can be as low as −240°F (−151°C). Under these extremely cold conditions, it is important that equipment manufacturers select the proper metal. If the metal is not properly selected, catastrophic failure could result.

One may be familiar with the experiment of freezing a rose to a temperature of −320°F (−196°C) by dipping it in liquid nitrogen. When the rose is dropped on a solid surface, it shatters like glass; steel behaves in the same manner. At very low temperatures, many steels become brittle and loses its ability to absorb energy. One of the first American disasters caused by embrittlement of steel occurred on January 15, 1915, in Boston.[31] The Purity Distilling Company used a distilling tank 58 ft (18 m) high and 90 ft (27 m) diameter to contain molasses. On the day of the disaster, the ambient temperature reached a chilly 2°F (−17°C). One thought is that the cold steel became brittle and fractured. Newspapers reported 15 ft (5 m) waves of molasses moving at a speed of 35 mph (56 km/h). The disaster killed 21 people, injured 150, and destroyed buildings and trains. Today, we have a much better understanding of how alloys behave under cold conditions. A standard method for measuring the brittleness of an alloy is the Charpy impact test.

The Charpy impact test requires a rectangular piece of material, with a machined notch; this notch acts as a point of crack initiation. A pendulum, designed with a weight on one end, is released and allowed to impact the specimen causing a sudden load[32] (see Figure 4.10). This test measures the amount of energy the material will absorb during the impact. If the specimen is ductile, it will absorb a lot of energy; if it is brittle, it will break similar to glass and not absorb much energy. Figure 4.11 shows results of two carbon steel specimens that went through the Charpy impact test at a temperature of −50°F (−46°C). Clearly, both specimens failed at this temperature.

Figure 4.12 shows results of Charpy impact tests for high-carbon steel, low-carbon steel, and austenitic SS (300 series).[33] Notice at warm temperatures, all three alloys are ductile corresponding to a large amount of energy absorbed. However, as the temperature is reduced below about 32°F (0°C), the carbon steel begins to lose its impact strength and becomes more brittle; this region is referred to as the transition phase. As the temperature is further reduced, carbon steel absorbs almost no energy from the impact indicating that the steel is brittle.

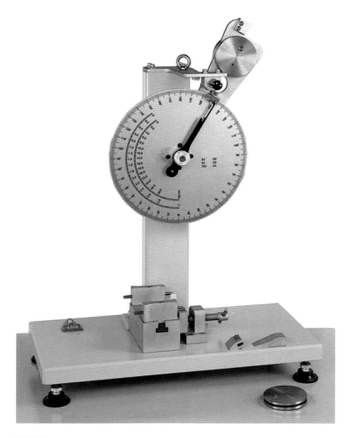

FIGURE 4.10
Charpy impact test apparatus.

FIGURE 4.11
Charpy impact test specimens of carbon steel at −50°F (−10°C).

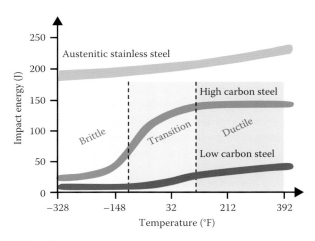

FIGURE 4.12
Results of Charpy impact tests for high-carbon steel, low-carbon steel, and austenitic SS.

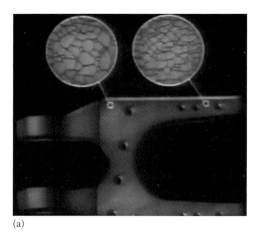

(a)

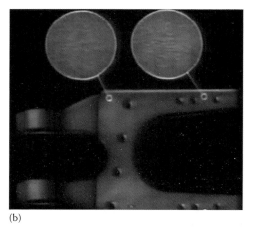

(b)

FIGURE 4.13
Normalizing of a forged carbon steel specimen. (a) Magnified view of grain structure before normalizing and (b) after normalizing. The normalizing process involved heating the specimen to 1600°F and then allowing it to cool in still air. (From United States Office of Education Training Film, Elements of tempering, normalizing and annealing, Produced by the Division of Visual Aids, U.S. Office of Education, 1945.)

Metallurgy

The transition from ductile to brittle for a low-carbon steel occurs in about the same temperature region as for high-carbon steel; however, within this temperature range, the transition from ductile to brittle does not occur for most austenitic SSs. Austenitic SSs have a different crystal structure than carbon steel, allowing the crystals to more easily slide over one another; this keeps the steel more ductile at low temperatures.

As a general rule, austenitic SS is good to temperatures as low as −320°F (−196°C), while some carbon steel is typically good to about −20°F (−29°C). Carbon steel can be used at lower temperatures if it is normalized. Normalizing is a process that requires heating the steel to a temperature of about 1625°F (885°C) (depends on how much carbon is in the steel) and letting it cool in air at room temperature. The normalizing process alters the structure of the steel by aligning and refining the grains as shown in Figure 4.13.[34] As a general rule, the temperature at which normalized carbon steel can no longer be used is about −50°F (−46°C). Below these temperatures, SS should be used.

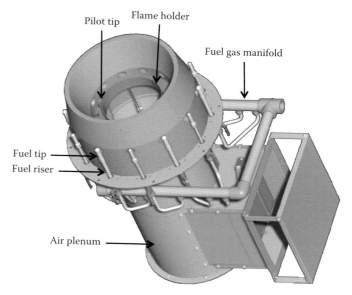

Burner component	Typical fabrication material
Air plenum	Carbon steel
Gas manifold	Carbon steel
Fuel gas riser	Carbon steel
Fuel gas tips	Stainless steel
Pilot tips	Stainless steel
Oil gun	Carbon steel
Flame holders	Stainless steel

FIGURE 4.14
Process burner schematic.

4.4 Material Selection

4.4.1 Process Burners

Process burners (see Figure 4.14) are designed to provide constant, reliable flames for heating various processes. As such, certain components are subject to constant heat from the flames they produce. These components must be fabricated out of materials that can withstand the high temperatures and potential oxidizing environments present for long periods of time. Other components of the process burner are designed to deliver fuel gas or fuel oil to the appropriate place in the burner. These components must be designed to resist corrosion or erosion from the fuels present within them.

The materials selected for construction of the various burner components depend upon whether that component needs to resist corrosion, heat (or, more specifically, oxidation), cold (embrittlement), and erosion; hold a structural shape; or any combination of these requirements.[35] The following sections detail metal selection considerations for various major portions of process burners.

4.4.1.1 Air Plenum

The burner air plenum, or register, is the housing that accepts air from the atmosphere or preheated air system and delivers it to the combustion reaction through the throat of the burner. Typically, most process burner air plenums are fabricated out of painted or galvanized carbon steel. These materials hold the required structural shape well at an optimized cost. Painting, galvanizing, or other surface coatings are needed to protect the surface of the plenum from corrosion, from rain, or from airborne salts in a marine environment. Some burner plenums may be subjected to high internal temperatures from preheated air. The common approach to burner design for this situation is to internally insulate so that the elevated temperatures do not require a plenum material upgrade. Insulation inside plenums is typically a soft material, such as mineral wool, and can be held in place with pins and keepers, or galvanized or SS liners. Galvanized liner materials may not be allowed in high-temperature processes using high-alloy tubes. This is due to the concern of accidental contact of the galvanizer with the tubes. Even small amounts of the galvanizer being deposited on the high-alloy process tubes in these services can degrade their strength and/or creep resistance, resulting in reduction of expected service life, or even causing catastrophic failure.

For cold environments where the burner will be subjected to extreme low ambient air temperatures, API 535 and 560 do not provide guidance.[35,36] Some manufacturers have adopted a minimum ambient operation limit of −40°F (−40°C). Below this temperature, special normalized carbon steel or alternatively 300 series (typically 304) is needed for the plenum to have sufficient embrittlement resistance. Using 300-series SS provides the structural temperature rating needed and extends that rating below the −50°F (−46°C) minimum possible with normalized carbon steel. An additional benefit of using SS is that it does not need to be painted.

4.4.1.2 Fuel Delivery System

4.4.1.2.1 Gas Manifold

The gas manifold is the connection point for the gas fuel to the burner. It handles the distribution of this fuel to the various risers and tips of the burner. Most process burner fuel gas manifolds are fabricated from painted carbon steel. However, when the fuel gas contains sulfur components in concentrations greater than 100 ppm (by weight) and the fuel is at a temperature higher than 300°F (150°C), API 560—"Fired Heaters for General Refinery Service"—recommends that the material of this part of the burner fuel delivery system be upgraded to 316L SS to resist internal sulfidic corrosion.[36] Some end users have required an upgrade to SS on fuel gas manifolds because they have installed fuel gas filtering and/or coalescing systems and have chosen to run SS piping all the way to the burners to reduce the possibility of scale forming in the lines downstream of the filter.

4.4.1.2.2 Fuel Gas Risers

Fuel gas risers connect the gas tips of the burner to the fuel gas manifold system. Fuel gas risers are also generally fabricated out of carbon steel. Risers that are located inside the burner air plenum and are subjected to combustion air temperatures greater than 700°F (370°C) should be upgraded to a 300-series SS to prevent oxidation and scaling due to heat. Like the fuel gas manifold, combinations of sulfur concentrations greater than 100 ppm (by weight) and heat from the fuel itself, or from the combustion air, are reasons to recommend a metal upgrade to 316L SS.[36] Some end users have required an upgrade to SS on burner risers, especially when they have specified SS on the burner manifold. In services with higher-temperature fireboxes, such as ethylene cracking (pyrolysis) or steam methane reforming, burner risers may need to be upgraded to a 300-series SS at a minimum and may need to be upgraded to the more heat-resistant 310 SS in order to provide sufficient scaling resistance in these applications.

4.4.1.2.3 Fuel Gas Tips/Pilot Tips

Fuel gas tips are used to direct the flow of gas to the appropriate place within the burner. These tips are usually in the direct radiation zone of the burner flame and are subject to high levels of heat. Counteracting this heating from flame radiation is the cooling from the flow of fuel gas flowing through the tip and also, in some cases, the flow of combustion air around the tip. Typical metallurgy for fuel gas tips that are not highly cooled by incoming air is 310 SS (or CK-20, its cast equivalent). The 310 SS alloy has been shown to be the most economical choice for long-term operation because it has the highest oxidation resistance for its price. Table 4.5[37] provides a listing of alloys and their oxidation resistance.

As can be seen in Table 4.5, even higher nickel-based alloys like alloy 800 will eventually oxidize away at 2000°F (1100°C) and higher. Burner tips made of alloys with more nickel content than 310 SS have been investigated in some of the highest-temperature operating furnaces. While the temperature resistance is higher, it has been found that there is no significant increase in the life of the burner tips and the cost-to-benefit ratio is not high.

4.4.1.2.4 Oil Guns

Oil guns are present in burners designed to fire oil. They are the connection point for the incoming oil and atomizing media. Oil guns in process burners are typically constructed using cast or ductile steel with carbon steel components. Some atomizers and other components are machined from brass. Oil gun tips are often machined out of 416 SS.[36] When abrasives such as catalyst fines are present in the incoming fuel oil, the oil gun tip should be upgraded to a harder tool steel and the atomizer body should be upgraded to nitride-hardened alloy to resist erosion from the hard particles in these oils and provide a reasonable operational life. When the mass percentage of sulfur in the fuel oil is greater than 3%, it is recommended that the atomizer be upgraded to 303 SS to resist corrosion.

4.4.1.3 Flame Holders/Stabilizers

Flame holders (or stabilizers) are devices in the airstream of the burner that are designed to provide a stable continuous anchor point for the burner flame. A majority of these devices create a bluff body in the airflow that forms a recirculation zone within which the flame attaches, keeping the burner flame in close proximity to the flame holder. Flame holders are generally cooled by the incoming airstream, which provides for the dissipation of a portion of the heat loading from the close proximity flame. This allows flame holders to be fabricated out of 310 SS and have an expected operational life of many years.

In some instances, burners are stabilized by a swirler in the throat instead of a bluff body. The swirler creates a recirculation vortex zone, stabilizing the burner flame. This zone is located away from the face of the swirler and the burner throat, which in many cases allows the swirler to be fabricated out of 310 SS without sacrificing expected life.

4.4.2 Process Flares

If a flare tip fails, it may not safely or effectively dispose of the waste gas, which could result in a costly plant

TABLE 4.5

Estimated Corrosion Rate for Oxidation

Material of Construction	Corrosion Rate—Mills (0.001 in.) per Year (mpy) Maximum Metal Temperature °F (°C)												
	900–950 (482–570)	951–1000 (511–538)	1001–1050 (538–566)	1051–1100 (566–593)	1101–1150 (594–621)	1151–1200 (622–649)	1201–1250 (649–677)	1251–1300 (677–704)	1301–1350 (705–732)	1351–1400 (733–760)	1401–1450 (761–788)	1451–1500 (788–816)
Carbon steel	2	4	6	9	14	22	33	48	—	—	—	—
1¼ Cr	2	3	4	7	12	18	30	46	—	—	—	—
2¼ Cr	1	1	2	4	9	14	24	41	—	—	—	—
5 Cr	1	1	1	2	4	6	15	35	65	—	—	—
7 Cr	1	1	1	1	1	2	3	6	17	37	60	—
9 Cr	1	1	1	1	1	1	1	2	5	11	23	40
12 Cr	1	1	1	1	1	1	1	1	3	8	15	30
304SS	1	1	1	1	1	1	1	1	1	2	3	4
309SS	1	1	1	1	1	1	1	1	1	1	2	3
310SS/HK[a]	1	1	1	1	1	1	1	1	1	1	1	2
800H/HP	1	1	1	1	1	1	1	1	1	1	1	2

Material of Construction	Corrosion Rate—Mills (0.001 in.) per Year (mpy) Maximum Metal Temperature °F (°C)												
	1501–1550 (816–843)	1551–1600 (844–871)	1601–1650 (872–899)	1651–1700 (899–927)	1701–1750 (927–954)	1751–1800 (955–982)	1801–1850 (983–1010)	1851–1900 (1011–1038)	1901–1950 (1038–1066)	1951–2000 (1087–1093)	2001–2050 (1094–1121)	2051–2100 (1122–1149)	2101–2150 (1149–1177)
Carbon steel	—	—	—	—	—	—	—	—	—	—	—	—	—
1¼ Cr	—	—	—	—	—	—	—	—	—	—	—	—	—
2¼ Cr	—	—	—	—	—	—	—	—	—	—	—	—	—
5 Cr	—	—	—	—	—	—	—	—	—	—	—	—	—
7 Cr	—	—	—	—	—	—	—	—	—	—	—	—	—
9 Cr	60	—	—	—	—	—	—	—	—	—	—	—	—
12 Cr	50	—	—	—	—	—	—	—	—	—	—	—	—
304SS	6	9	13	18	25	35	48	—	—	—	—	—	—
309SS	4	6	8	10	13	16	20	30	40	50	—	—	—
310SS/HK[a]	3	4	5	7	8	10	13	15	19	23	27	—	37
800H/HP	3	4	6	8	10	13	17	21	27	33	41	50	60

Source: American Petroleum Institute, *Damage Mechanisms Affecting Fixed Equipment in the Refining Industry*, API Recommended Practice 571, Washington, DC, December 2003.

[a] Same composition as CK-20.

shutdown. The premature failure of flare tips can often be directly attributed to corrosion damage (see Chapter 14). The major factors that influence the corrosion damage of a flare tip are service and operational conditions, metal selection, design, and/or fabrication.[38] The metallurgy that makes up a typical flare system usually consists of several types of steel. Usually, the flare tip is made of relatively expensive SSs. As a general rule of thumb, the further the steel is away from the heat of the flame, the less expensive the steel. This section will provide a general overview of the material used for various components of a flare system.

4.4.2.1 Flare Burner (Typical)

When making a material selection for a flare tip, one must consider cost, ease of fabrication, and corrosion resistance. The key parameter for flare tip material selection, however, is corrosion from flame impingement. It is important to understand the application of the flare and the possible thermal exposure. A flare tip that is exposed to continuous or frequent flame impingement from internal or external burning will ultimately suffer high-temperature corrosion damage regardless of the material however, proper material selection can enhance the service life.

The material for the heat-affected zone of most large flare tip applications is 310 SS. Series 310 SS gives good overall protection against low-intensity or occasional flame impingement; however, like any material, it will ultimately fail with repeated, intense impingement. Some flares are designed so that they rarely experience direct flame impingement. For these applications, certain components of the flare can be made of 304, 316, or 321 SS or similar cast alloys. From time to time, other materials including those with high nickel content are used. While the use of high-nickel alloy materials can be intriguing, the added cost of such material is often not justified by the gain, if any, in flare tip service life. The lower section of the flare tip is made of either carbon steel or a lower-grade SS such as 304. If this section is made of carbon steel, it is typically painted.

In addition to exposures to high temperatures and reducing atmospheres, the flare tip can also be subjected to chemical attack from the waste gas stream. The most common contributors to corrosion from the waste gas stream are hydrogen sulfide and sulfur oxides. The presence of sulfur in the waste gas stream can cause severe corrosion damage to a flare tip. Some material suppliers have recommended using high-nickel alloys for flares burning a waste gas with sulfur present. Field experience, however, has shown that high-nickel alloys are not always the best choice for sulfur applications. The combination of high temperature, reducing atmosphere, and sulfur can quickly corrode high-nickel alloys.

Atmospheric corrosion is typically not a predominant mechanism that leads to the failure of a flare tip; however, it can contribute to a reduced service life. Flare tip installations on offshore platforms or near the shore experience salty atmospheres. This atmosphere can lead to a chemical reaction between the oxide scale and the salt resulting in a breakdown in the protective scale. Material suppliers may recommend using high-nickel alloys in salty environments. Experience with flare tips exposed to salt water atmospheres, however, indicates that high-nickel alloys are not required. Exposure to heat and sulfur are usually more detrimental than salt water exposure.

Flare tips installed in various plants can experience corrosion due to the gas fumes in the atmosphere. However, the concentration of these corrosive gases in the atmosphere is typically extremely low and not a major contributor to corrosion.

4.4.2.2 Pilots

According to the API 537, Section 5.3.3.3, the fuel gas line feeding the pilot should be made of SS; however, carbon steel is commonly used. Carbon steel is used because it is less expensive than SS. API 537 recommends using SS because rust scale on the inside of carbon steel pipe can dislodge and plug the pilot orifice spud (which usually has one or more very small holes) or the strainer; SS significantly helps reduce this problem. API 537 states that if stainless pipe is used, it must be properly supported. This is mentioned because the tensile strength of SS is less than that of carbon steel, therefore requiring more supports along the height of the flare stack. API 537 also states that if carbon steel is used for the pilot fuel line, it is a good idea to run separate lines to each pilot; this helps reduce risk in a couple of ways: (1) reduces the velocity of the gas in the pipe that helps prevent scale from flowing up the pipe and plugging the orifice and (2) allows the user to turn off one pilot at any time for maintenance.

4.4.2.3 Moleseals and Flare Riser

Moleseals and flare risers are commonly made of carbon steel and painted on the outside with an inorganic zinc-rich coating. It is important that the moleseal has proper drainage. If not, rainwater can collect on the inside of the seal and cause corrosion damage, especially along coastal areas. Moleseals need to be inspected for corrosion damage anytime the flare is taken out of service. It is also important to check for corrosion at the moleseal cap. There have been cases during sudden flare gas releases that the cap was blown off due to corrosion damage. If a cap is blown off, the moleseal is no longer effective and can result in a flashback into the flare system.

4.5 Examples and Case Histories

4.5.1 Process Burners

4.5.1.1 Formation of Oxide Scale on Heater Process Tubes

When iron and/or chrome oxide (oxidation) form on the outer surface of process tubes, the appearance resembles that of tree bark as shown in Figure 4.15. Typically, the oxide scale is about three to seven times the thickness of the base metal from which it formed. When the oxide scale is heated, it can appear to look like the tube is "on fire." Generally, the scale on the outer surface of process tubes should be removed because it can reduce the heat transfer to the process fluid.

Ceramic tube coatings (see Figure 4.16) have been used successfully to help reduce or eliminate the deterioration of process tubes. Ceramic coatings tend to slightly insulate the tubing; this reduces the heat transfer in the radiant section that forces more of the heat into the convection section. When considering ceramic coatings, one must carefully evaluate any possible shifts in heat load.[39]

4.5.1.2 Ruptured Process Tubes

Figure 4.17 shows a process tube failure after long-term overheating. Notice the swelling or bulge on the tube at rupture location. When process tubes overheat, they lose their tensile strength and will sometimes create a bulge in the hot zone prior to rupturing. When tubes finally rupture, they typically flare outward and create a very sharp edge along the opening seam; this is typically referred to as a "fish mouth"- or "fish lip"-type failure.

Figures 4.18 and 4.19 show the end result of continuing to fire burners inside a furnace after the flow of the process fluid has stopped; a process tube ruptured and fire occurred within minutes. The fire overheated the tubes in the radiant section of the heater causing them to bend and fall out of their hangers and land on top of the burners.

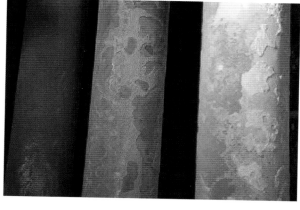

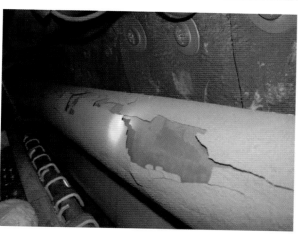

FIGURE 4.15
Oxide scale formed on the outer surface of process tubes.

FIGURE 4.16
Heater process tubes coated with ceramic.

Figure 4.20 shows a failed process tube after long-term overheating. Notice the cracks in the area of the ruptured seam. These cracks may indicate that the tube suffered embrittlement from carbide precipitation (formation of internal carbides) due to carburization.

Figure 4.21 shows a ruptured steam methane reformer (SMR) tube. The tube failure was due to a leak that went undetected. After the tube failed, it caused a fire within the furnace that destroyed a dozen or more other process tubes. These tubes are a made of a high-alloy steel that reside inside the furnace, which operates at a temperature of about 2100°F (1150°C) at normal conditions. It is interesting to note that this particular failure caused the tube to crack along the length of the tube creating jagged edges; in many cases, when tubes rupture, they will create a bulge and peel back. The tape, located on the bottom of the tube in photograph (b), was used to hold the catalyst inside while the tube was cut out and removed from the heater.

Figure 4.22 shows a photograph of a heater process tube that has ruptured due to severe external corrosion damage. Notice the tube wall is thin and metal scale on the heater floor just below the location of the rupture. The pitting on the external surface of the tube may indicate that the tube suffered from carburization corrosion, in particular, metal dusting.

Process fluid flow inside heater tubes typically consists of a single-phase liquid or a two-phase gas–liquid mixture. When two-phase flow occurs, the flow pattern is complex. Depending on the velocity of the gas and liquid, several different flow regimes can occur as illustrated in Figure 4.23. Heaters are usually designed so that the process fluid is allowed to wet the entire inner surface of the tube wall; this improves heat transfer to the process fluid and helps prevent tubes from overheating. If stratified flow occurs, the tubes are only wetted near the bottom. Since the heat transfer coefficient of a gas is about ten times lower than for a liquid, the

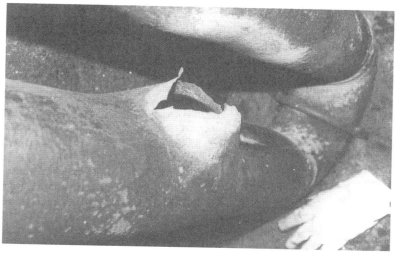

FIGURE 4.17
Rupture process tube after long-term overheating. (From Sanders, R.E., *Chemical Process Safety: Learning from Case Histories*, Fig. 10.2, Butterworth-Heinemann, Boston, MA, 1999.)

FIGURE 4.18
Ruptured process heater tube.

(a)

(b)

FIGURE 4.19
(a) Process tubes lying on top of a burner. (b) Tubes pulled from the radiant section of the heater.

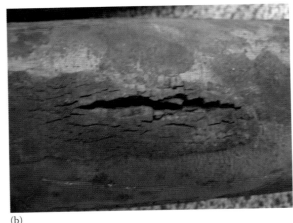

FIGURE 4.20
(a) Failure of a process heater tube. (b) Close-up view of ruptured tube.

FIGURE 4.21
Large crack in a process tube used in a SMR. (a) SMR furnace and (b) cracked tube.

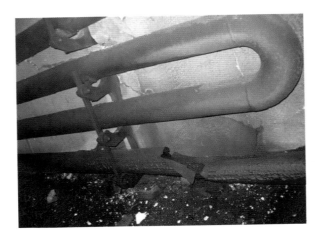

FIGURE 4.22
External corrosion of a heater process tube.

tube wall temperature can become much higher near the top. Figure 4.24 shows a 10 in. (25 cm) diameter vacuum heater tube damaged due to stratified flow. Notice the pipe is not damaged near the bottom where the liquid was flowing; however, the pipe suffered oxidation corrosion near the top where the gas was flowing.

4.5.1.3 Process Tube Distortion

Figure 4.25 shows a process tube that was overheated causing it to distort. As the process tube overheated, it lost much of its tensile strength and began to sag due to high-temperature creep. Figure 4.26 shows how the rupture strength of low-carbon steel and 300-series SS varies as a function of temperature.

4.5.1.4 Burner Damage

The damage to the burner shown in Figure 4.27 was caused by a failure of a support system on the hot coke angle riser in

Metallurgy

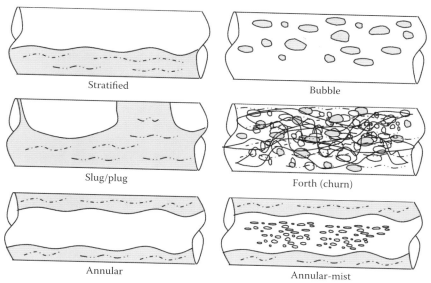

FIGURE 4.23
Various two-phase flow regimes that can occur in process heater tubes.

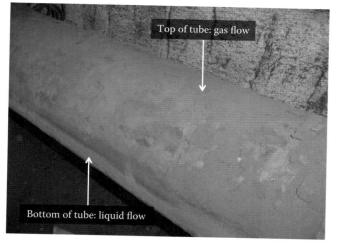

FIGURE 4.24
A vacuum heater tube showing signs of oxidation due to stratified two-phase flow.

FIGURE 4.25
Sagging process tube due to high-temperature creep.

the coker reactor upstream; it was unrelated to the burner. The riser failure caused the coker feed to be operated in a modified way. A fire broke out above a grid that melted, causing coke to fall onto the burner; this caused the heater to behave similar to a blast furnace causing the burner to be destroyed. The burner and pilot materials were upgraded; however, if the same scenario occurred again, the burner would be destroyed regardless of the material used.

4.5.1.5 Control Valve Damage

Sulfide vapors at ambient temperatures can be extremely corrosive to equipment. Figure 4.28 shows corrosion damage to a control valve in high hydrogen sulfide service located on an offshore platform.

4.5.1.6 Oil Gun Damage

Figure 4.29 shows an oil gun tip that failed due to stress corrosion cracking. The buildup of the iron oxide scale on the outer surface of the tip indicates that the oil gun suffered cyclic oxidation from thermal cycling. The heating and cooling of the tip caused carbide formation resulting in embrittlement and failure.

4.5.1.7 Corroded Orifice Spud

Furnace flue gas can contain very acidic gaseous effluents such as sulfur compounds. If the temperature of the flue gas drops below the dew point, the water vapor can condense resulting in a precipitating, highly corrosive sulfuric acid that can severely damage equipment.

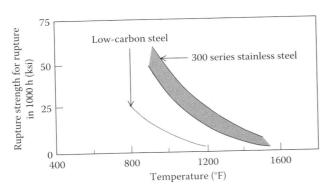

FIGURE 4.26
Rupture strength of low-carbon steel and 300-series SS at various temperatures. (Adapted from The Stainless Steel Information Center, Specialty steel industry of North America, http://www.ssina.com/index2.html)

FIGURE 4.28
Damage to a control valve due to sulfidation corrosion.

FIGURE 4.29
Damaged oil burner fuel nozzle.

For example, Figure 4.30 shows a fuel orifice spud that was used on a venturi-style eductor to entrain and recirculate furnace flue gas into a burner. This particular example shows the corrosive wear on the top of the orifice spud after approximately 10 years in service. As the fuel jet exits the orifice, a recirculation zone is created just above the face of the spud. When the acidic flue gas comes in contact with the cooler metallic surface, it condenses creating a highly corrosive environment resulting in damage. Also, notice that the threads are also gone due to corrosion damage.

4.5.1.8 Damaged Premixed Burner Tip

Figure 4.31 is a photograph of a center fuel gas burner tip showing signs of damage due to high temperature

FIGURE 4.27
Damage to a burner caused by a failure of a support system upstream in the coker reactor.

Metallurgy

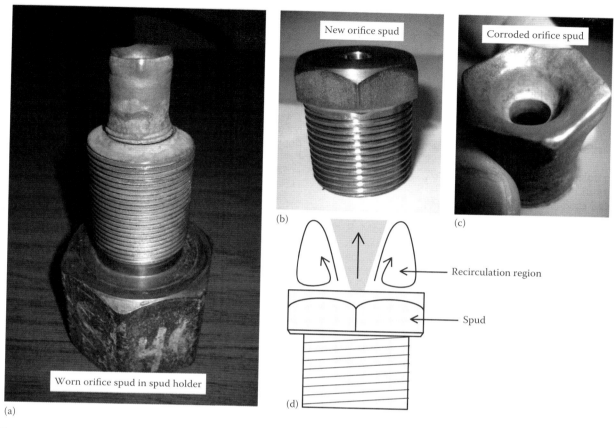

FIGURE 4.30
Orifice spud in service for 10 years (a) spud, (b) new orifice spud, (c) corroded orifice spud, and (d) recirculation zone at the outlet of a spud.

FIGURE 4.31
Center fuel gas burner tip showing signs of corrosion damage due to high-temperature oxidation: (a) in operation and (b) out of service.

oxidation. Notice the scale buildup along the outer edges of the burner tip. During the plant turnaround, the damaged tip was removed and replaced with a new gas tip as shown in Figure 4.32. The same metallurgy was used because the tip lasted for many years in operation.

Figure 4.33 shows erosion–corrosion damage of a pressure swing absorption (PSA) burner gas tip. This tip was in service for about 5 years in a hydrogen reformer furnace that operated above 1800°F (980°C). The tip was fabricated with HF alloy. HF alloy is an iron–chromium–nickel alloy with high strength and

FIGURE 4.32
(a) Corroded burner tip and (b) new burner tip.

FIGURE 4.33
Photographs of erosion–corrosion damage of a tip from a PSA burner gas.

Metallurgy

corrosion resistance in the temperature range from 1000°F to 1600°F (540°C to 870°C).[40] The fuel composition consisted of the following (volume percent): 13% methane, 43% carbon dioxide, 0.78% water vapor, 5.06% nitrogen, 10.99% carbon monoxide, and 26.55% hydrogen. The cutaway view of the tip clearly shows the erosion–corrosion damage. Notice that the steel is much thinner near the outlet of the fuel ports. The thinning metal is attributed to a combination of high-velocity fuel gas and carburization/metal dusting. This damage substantially increased the fuel port area. Failure to replace this burner tip could have eventually altered the performance of the burner in a detrimental way.

4.5.2 Process Flares

4.5.2.1 Flame Retention Segments

The purpose of a flame retention ring is to prevent the flame from lifting off and blowing out at high waste gas flow rates. A flame retention ring typically consists of a number of segments located around the inside perimeter of the flare tip at the exit as shown in Figure 4.34. As their purpose would indicate, it is expected that the flame retention segments will be exposed to direct flame contact.

A flare tip designed with 52 flame retention segments was placed in service on a waste gas stream that contained small amounts of hydrogen sulfide. The flare manufacturer recommended that the segments be made of 310 SS. However, the user requested that the flame retention segments be made of Incoloy 800. After a relatively short period of operation, an inspection revealed that 48 of the segments were completely deteriorated, while the other four segments suffered little deterioration. An analysis showed that the segments made of Incoloy 800 were completely destroyed and that the four good segments were made of 310 SS. The investigation determined that the tip was built on a rush basis and the 310 SS segments had been used, with the customers knowledge, to complete the flare tip on schedule. X-ray analysis showed the presence Fe_3O_4 in both good and bad segments and the presence of sulfides by qualitative chemical analysis indicated sulfidation had occurred. It is believed that internal burning, due to the capping effect from the perimeter steam nozzles (see Chapter 14), may have created a corrosive sulfur atmosphere. This particular example demonstrates flare tip corrosion damage due to both improper operation and metallurgical selection. High-nickel alloys are not the best application for flares burning a waste gas stream with sulfur present. The literature shows that nickel sulfides can form and easily destroy the protective oxide scale and lead to catastrophic sulfidation attack.

4.5.2.2 Pilot in an Enclosed Flare

A pilot operating on refinery fuel gas had difficulty maintaining a stable flame after only 2 months in service. An inspection of the pilot tip revealed a metal sulfide scale and loss of tip material as shown in Figure 4.35. Upon investigation, it was determined that the refinery fuel gas contained 2% hydrogen sulfide and the pilot

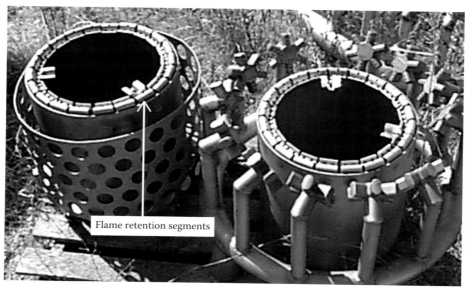

FIGURE 4.34
Flame retention segments located around the inside perimeter of the flare tip.

FIGURE 4.35
Scale and loss of a pilot tip due to sulfidation attack.

was operated in such a manner that the flame, at the tip, produced a reducing atmosphere. The problem was corrected by changing the operation of the pilot to produce an oxidizing flame. There was no change in the metallurgy of the pilot tip. The pilot has now been in service for several years without any further reports of failure or damage.

4.5.2.3 Pilot on an Elevated Flare

Flare pilots are typically positioned around the perimeter of a flare tip and are used to ignite the waste gas stream. Most pilots operate with an air–fuel mixture near stoichiometric conditions and can last for many years without suffering substantial corrosion damage. However, the service life of a pilot can be substantially reduced if the flame from the flare engulfs the pilot for long periods of time.

A flare tip with three pilots was installed at a natural gas treatment plant. The flare was burning gas continuously at a low flow rate (approximately 1–1.5 ft/s = 0.3–0.5 m/s exit velocity). Due to a predominant wind direction, the flare experienced external burning on one side approximately 75% of the time over an 18 month period. When the flare tip was inspected, it was found that the pilot that was positioned on the downwind side of the flare tip suffered extreme high-temperature corrosion damage to the windshield as shown in Figure 4.36. The other two pilots on the flare were not exposed to a continuous external burning and did not suffer any damage. The extent of damage seriously reduced the ability of the pilot to function properly. This problem was eliminated by staging to a smaller tip at low waste gas flow rates. Figure 4.37 shows another example of a flare pilot that suffered severe high-temperature corrosion damage due to external burning for long periods of time.

4.5.2.4 Air-Assisted Flare

Air-assisted flares use air from a fan as a supplemental energy source to help improve the smokeless performance. At low waste gas flow rates, wind and air from the blower can induce internal burning by creating recirculation zones inside the arms (waste gas side) of the flare tip. When internal burning occurs in this type of an air flare, heat causes a buildup of stresses and crack formations at or near the weld seams. In turn, these cracks allow waste gas to enter into the air side of the flare causing burning to occur on both sides

Metallurgy

FIGURE 4.36
(a) External burning on flare tip and (b) pilot that was positioned on the downwind side of the flare tip suffered extreme corrosion damage to the windshield.

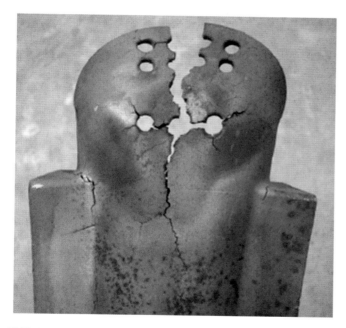

FIGURE 4.37
Pilot that was positioned on the downwind side of the flare tip suffered extreme corrosion damage to the windshield.

of the gas tip. When burning occurs on both sides of the tip, it can quickly deteriorate, regardless of the metallurgy.

An air-assisted flare was put into service burning a typical refinery waste gas. After 12 months in service, the flare was inspected. Inspection revealed cracks in or near many of the weld seams and severe high-temperature corrosion damage to the 310 SS arms as shown in Figure 4.38. Studies show that a tip of the same or similar design would be susceptible to damage from internal burning. The manufacturer solved the air-flare problem through the development of new tip designs. Tips of the new design, which normally use 310 SS, have shown significantly improved service life.

4.5.2.5 Steam-Assisted Flare

Steam-assisted flares use steam as a supplemental energy source to help improve the smokeless performance. At low waste gas flow rates, ambient wind can cause the flame to pull down and burn on the external surface of the flare tip. Figure 4.39 shows the deterioration of a lower steam ring on a steam-assisted flare that was exposed to a flame for a long period of time. Figure 4.40 shows deterioration to the inlet of the lower steam eductor tube. Figure 4.41 shows the upper steam deterioration of the upper steam spider.

4.6 Welding

4.6.1 Types of Welding Processes

The most utilized welding processes for manufacturing process burner and flare equipment include the following: gas metal arc welding (GMAW), flux core arc welding (FCAW), gas tungsten arc welding (GTAW), shielded metal arc welding (SMAW), and submerged

FIGURE 4.38
(a) Internal burning on an air-assisted flare and (b) failure of an air-assisted flare due to stress corrosion cracking.

arc welding (SAW). Common uses for these welding processes are listed in the following:

1. GMAW-S (short circuit)
   a. Nonstructural attachment welds
   b. Root passes in piping
2. FCAW-GS (gas shielded) and SMAW
   a. Structural welding
   b. Piping fill and cap passes
   c. Attachment welds
   d. Plate welding
3. GTAW
   a. Pipe welding
   b. Attachment welds
   c. Plate welding
4. SAW
   a. Large-diameter (OD > 8 in., 20 cm) pipe welding
   b. Plate welding (thickness > 3/16 in., 0.48 cm)

4.6.2 Welding Carbon Steel

Each of the welding processes listed earlier fuses two or more pieces of steel together by heating the metal to a liquid state. The temperature required to achieve a liquid state is highly dependent on the amount of carbon in the steel as demonstrated in Figure 4.42.[2] This plot is referred to as binary iron–carbon phase diagram. It shows, for example, that a carbon steel containing 2% by weight carbon must reach a temperature greater than about 2600°F (1400°C) to become liquid.

When metals are heated above transformation temperatures, they undergo microstructural changes.

Metallurgy

FIGURE 4.39
(a) Inspection by helicopter of steam leaking from underneath a steam-assisted flare tip and (b) close-up view showing a ruptured lower steam ring.

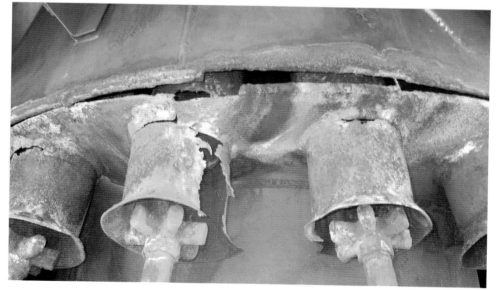

FIGURE 4.40
Deterioration to the inlet of the lower steam eductor tube due to flame impingement for an extended period of time.

FIGURE 4.41
Deterioration of a steam spider due to flame impingement for an extended period of time.

Microstructural changes during welding are of primary concern because it can alter the properties of the material. Depending on a variety of factors such as carbon content, the metal may have to be post weld heat-treated in order to return it to its prewelded state. Fortunately, low-carbon and some mild-carbon steels are not readily heat-treatable and do not typically require post weld heat treatment. Figure 4.43[41] shows the time–temperature-transformation (TTT) curve for a medium-carbon steel (AISI 1040).

Martensite is a hard, brittle material that is undesirable in most load-bearing, pressure-containing structures. It can be formed in low- and mild-carbon steels and in some medium-carbon steels if quenched too quickly (on the order of 1–5 s). Low-carbon and mild-carbon steels typically do not form martensite when they are slowly cooled (i.e., air cooled). Carbon-alloyed steels can have a cooling time on the order of 60 s and still form martensite. Figure 4.44[25] shows a TTT diagram for a medium-carbon steel alloy (AISI 4340). The composition of AISI 4340 is 0.38%–0.43% carbon, 0.6%–0.8% manganese, 0.035% max phosphorus, 0.04% max sulfur, 0.7%–0.9% chromium, 0.2%–0.3% molybdenum, 1.65%–2% nickel, 0.15%–0.3% silicon, and balance iron.

Rapid air cooling of welds from 1200°F to 500°F (650°C to 260°C) is possible and martensite formation can occur. The majority of carbon steel used in manufacturing of process burner and flare equipment is of the low and mild carbon type. When welded, these steels do not appreciably change upon cooling.

While time and temperature determine the as-welded microstructure, carbon content determines the maximum hardness and alloy content determines the hardenability—how easily the steel can be hardened. Figure 4.45[42] correlates maximum hardness to carbon content.

It is important that welders be qualified in accordance to code and that they pass a test for each type of weld that they perform. These welds should be tested to ensure that the as-welded metal meets certain

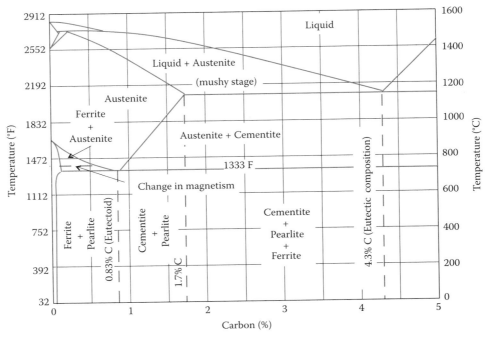

FIGURE 4.42
Binary iron–carbon phase diagram. (Adapted from Jefferson, T.B. and Woods, G., *Metals and How to Weld Them*, 2nd edn., James F. Lincoln Arc Welding Foundation, Cleveland, OH, 1962.)

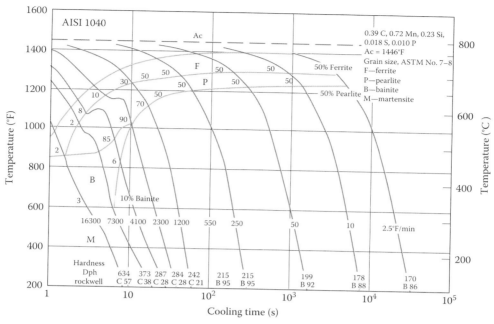

FIGURE 4.43
Typical TTT diagram for medium-carbon steel. (Adapted from Vander Voort, G.F. (Ed.), Additional steels I-T and CCT diagrams, *Atlas of Time-Temperature Diagrams for Irons and Steels*, ASM International, Materials Park, OH, 1991.)

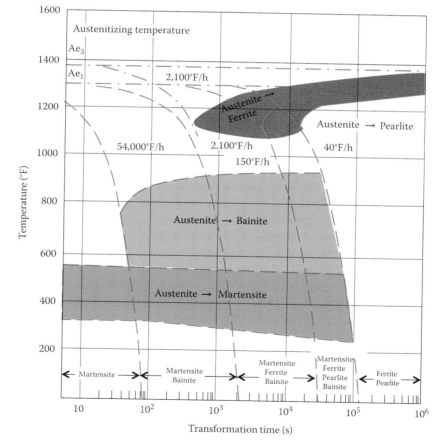

FIGURE 4.44
Typical TTT diagram for medium-carbon alloy steel AISI 4340. (Adapted from Moller, G.E. and Warren, C.W., Presented at *Corrosion/81*, National Association of Corrosion Engineers, Houston, TX, Paper no. 237.)

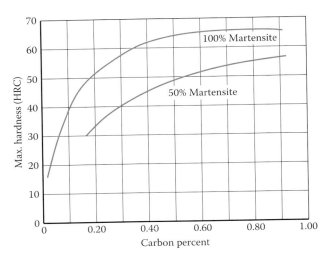

FIGURE 4.45
Maximum hardness versus % carbon by weight. (Adapted from Kearns, W.H. (Ed.), *Welding Handbook Metals and Their Weldability*, 7th edn, Vol. 4, American Welding Society, FL, p. 36+, 1982.)

requirements. Specimens can be tested for ductility, tensile stress, hardness, and impact. Figure 4.46 shows several welded specimens that were tested for bending and tensile stress. Section 9 of the ASME Boiler and Pressure Vessel Code provides a step-by-step process on how a weld should be made and how it should be properly tested. There are several volumes of books that explain the recipe for a variety of welds, such as speed of the weld, number of passes, and type of filler.[42]

4.6.3 Welding Stainless Steel

SSs follow a similar set of principles to those discussed for carbon steel. Additional considerations for welding include section thickness, weld restraint, ferrite content, and heat input among others. For austenitic SSs, ferrite content can be detrimental and beneficial. Ferrite increases the risk of $M_{23}C_6$ carbide formation, which can cause embrittlement as discussed earlier.[26] Fully austenitic SS welds have the highest potential for weld solidification/cracking. Welds that form as ferrite–austenite and ferrite are the least susceptible to solidification cracking. Figure 4.47[26] shows the susceptibility to cracking as a measure of chrome–nickel equivalency ratios and solidification mode. Figure 4.47 shows that welds containing ferrite are less likely to crack. The austenite–ferrite mixtures require that cracks follow a nonplanar path; thus, crack growth becomes more difficult. Table 4.6[43] provides a general estimate of solidification mode based on ferrite concentration; ferrite number (FN) is equivalent to percent ferrite.

The majority of materials used to fabricate process burner and flare equipment can be readily welded using GMAW, GTAW, FCAW, SMAW, and SAW. For carbon steels, one should consider thickness, cooling rate, hydrogen entrainment, and weld joint rigidity. Austenitic SSs typically utilized in process burner and flare equipment are readily welded using the same processes as carbon steels. Ferrite content of austenitic SSs, as a general rule, should be held between 3% and 10%

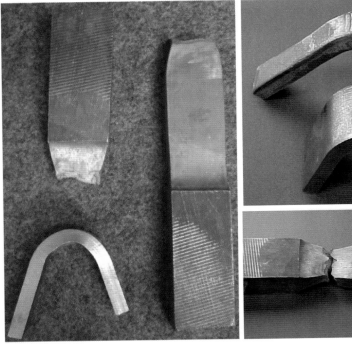

FIGURE 4.46
Tensile and bend specimens.

Metallurgy

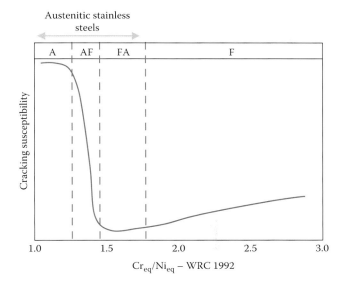

FIGURE 4.47
Solidification cracking susceptibility (A, austenite; AF, austenite–ferrite; FA, ferrite–austenite; F, ferrite).

TABLE 4.6
Solidification Mode Based on FN

Ferrite Number (FN)	Solidification Mode
0	Austenitic (A)
0–3	Austenite–ferrite (AF)
3–20	Ferrite–austenite (FA)
20+	Ferrite

Source: Data taken from Lippold, J.C. and Kotecki, D.J., Austenitic stainless steels, in *Welding Metallurgy and Weldability of Stainless Steels*, John Wiley and Sons, Inc., Hoboken, NJ, 2005.

by weight, except for those alloys such as 310 that form a fully austenitic microstructure. Construction codes (e.g., ASME B31.3, ASME section VIII) provide a good source for determining what factors should be considered for a given set of circumstances; these codes in combination with welding codes (e.g., ASME section IX, AWS D1.1) provide excellent guidelines for developing and testing welding procedures.

4.7 Nondestructive Testing

4.7.1 Introduction

Nondestructive testing (NDT) is a collection of processes for examining and evaluating an item without affecting its future usefulness. Other names are sometimes used interchangeably for NDT; this may be associated with a specific industry or region. Several names associated with this process include the following:

(1) NDT, the term used by most U.S. national codes; (2) nondestructive examination or evaluation (NDE); or (3) nondestructive inspection (NDI). Regardless of the name, the processes and methods are the same.

Procedural requirements for NDT are guided by ASME section V NDT, a mandatory supporting code to the ASME B&PV Codes. Personnel qualifications are guided by SNT-TC-1A, a recommended practice.

The purpose of this section is to provide an overview of the most common NDT methods applied to equipment produced and used in the process burner and flare industries. These methods include the following:

1. Surface and near-surface methods
    a. Liquid penetrant testing (PT)
    b. Magnetic particle testing (MT)
2. Full volumetric methods
    a. Radiographic testing (RT)
    b. Ultrasonic testing (UT)
3. Metallurgical methods
    a. Positive material identification/alloy verification (PMI) or (PMI/AV)
    b. Metallographic replication

4.7.2 Liquid Penetrant Testing

Figure 4.48 shows a flare undergoing PT. The liquid penetrant process relies on capillary action. A brilliant-colored liquid dye is applied on the metal surface at the area of interest. Through capillary action, the dye is drawn into micro flaws. After the penetrant is allowed to reside on the metal for a certain amount of time, the surface is cleaned of all visual penetrant. The metal surface is then dusted with a powder of contrasting color that wicks the liquid remaining in the flaw. A flaw, typically referred to as an *indication* (see Figure 4.49), appears as a brilliant color (usually red) against a contrasting background (usually white). Individual flaws can be categorized by the physical size and amount of dye absorbed. This information allows the technician to qualify and quantify the indication for acceptance or rejection. It should be noted that prior to testing, it is important that the metal surface is free of oil, grease, dirt, and paint.

Two types of water- or solvent-removable penetrant materials are commercially available: (1) brilliant contrasting colors that can be visibly examined under white light or (2) fluorescent colors that are observable under a black light. Service providers of liquid penetrant NDE are guided by ASME B&PV Code section V latest revision and SNT-TC-1A latest revision.

The liquid penetrant process is a simple, portable, low investment cost requiring minimum operator

FIGURE 4.48
Penetrant testing the connection welds of a flare.

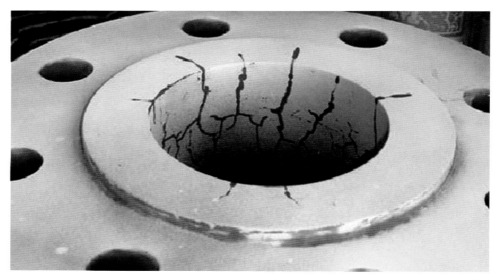

FIGURE 4.49
Dye penetrant indication example.

skill, qualifications, and technical procedures. It produces reliable results on most surface types including ferrous or nonferrous metals and ceramics and is not affected by the size or the orientation of a flaw provided it is open to the surface. This method also reveals most manufacturing flaws including cracks, laps, porosity, casting shrinkage, and laminations provided they are at, or open to, the surface. There are, however, several limitations and drawbacks to using this method: (1) flaws must be open to the surface, (2) exceedingly rough or coarse surfaces make cleaning of excess penetrant and subsequent evaluation difficult, (3) permanent records produced by transferring to transparent tape are inconclusive without accompanying photographs, and (4) safety concerns are present with flammable materials in the presence of an ignition source and/or confined space.

4.7.3 Magnetic Particle Testing

The magnetic particle process is based on the principle of flux leakage at breaks in a magnetic field. The magnetic field is produced by passing electric current through (direct magnetism) or by applying an external magnetic field (indirect magnetism) to the object under examination. The most common method in the process burner and flare industry is to pass an electric current through the object using an AC yoke (see Figure 4.50). Next, iron particles are introduced on the surface of the object. The magnetic field causes the iron particles to become active and agglomerate around the flux leakage region thereby identifying the flaw. The orientation of a potential flaw to the magnetic lines of force is critical for detection. A properly prepared and executed procedure that accounts for the positioning of the yoke or prods relative to potential flaws is essential for reliable detection. Service providers of magnetic particle NDE are guided by ASME B&PV Code section V latest revision and SNT-TC-1A latest revision.

Two distinctly different types of equipment are typically used for inducing the magnetic current: (1) AC yoke that is free from arc strikes and overheating local areas of the subject under examination and (2) DC prod that produces a stronger magnetic field with deeper penetration at the risk of localized arc burns.

Two types of metallic particles are typically used for MT: (1) dry iron particles color dyed for visual examination under white light by contrasting with the subject under examination and (2) solution-suspended iron particles fluorescent dyed that are observable under a black light.

The magnetic particle process is relatively simple, requiring portable equipment with low investment cost and moderate operator skill. It is able to detect surface and slightly subsurface flaws. The depth of flaw detection depends on several factors including intensity of the magnetic field and orientation/size of the flaw. It reveals most manufacturing flaws including cracks, laps, porosity, casting shrinkage, and laminations provided they are at or near the surface.

The magnetic particle process, however, has some limitations and drawbacks: (1) it is applicable only to ferromagnetic material, (2) flaws must be open or near the surface, (3) exceedingly rough or coarse surfaces inhibit probe contact and iron particle surface activity reducing the sensitivity of the test, (4) the surface must be clean, free of oil, grease, dirt, and paint to provide confidence of flaw detection, (5) permanent records produced by transferring to transparent tape are inconclusive without accompanying photographs, (6) damage by arc burning is a significant concern and potential for destroying the usefulness of an item when examining finely prepared surfaces and heat-treated components via the DC prod method, and (7) it cannot be safely used in a combustible atmosphere.

FIGURE 4.50
Magnetic particle testing of gas piping using the AC yoke dry method.

4.7.4 Radiographic Testing

The radiographic process is based on differing materials having a differential absorption of penetrating radiation. These variations in absorbed energy are transferred to a recording media as varying shades of gray. Recording media are traditionally specialty films of transparent plastic base with silver emulsion coating. These films are processed with developer and fixer solutions, followed by drying. Digital, filmless imaging is being used more in the industry; however, it has not overtaken traditional film as the recording media of choice. Figure 4.51 shows RT of flare component welds and a radiographic film image of a weld.

To create an image on the recording media, the item under evaluation is prepared for access from both sides. Often the radiation passes through an undesirable layer before the actual subject item; this is referred to as double-wall technique. The preferred method is referred to as the single-wall technique. In this technique, the radiation passes only through the thickness of the item under consideration.

The single-wall technique requires radiation to enter from one side, pass through the object, and have the energy broadcast onto a recording medium placed on the opposite side that is in direct contact with the object. The radiation penetrating the material produces a uniform image, provided the material has the same energy absorption throughout. However, it will produce different shades of gray as the amount of absorbed energy varies throughout the material; the various shades of gray identify the differences in the material density. For example, a flawless steel plate radiograph will show a uniform and featureless gray field. If an anomaly is introduced to the steel plate, representing a material or weld flaw such as porosity or slag inclusion, the image in that area will darken and take the shape of the area having differential absorption. If the material density is increased by thickness, such as a protrusion in a weld bead or actual difference in material types (tungsten inclusion left during the welding process), the image in that area will be lighter than that of the base plate.

Two types of radiation sources are commonly used for industrial radiography: (1) gamma radiation resulting from the decay of an artificially produced unstable isotope—the most common industrial radiography isotopes are iridium 192 and cobalt 60, and (2) x-ray radiation resulting from the conversion of electrical energy into x-radiation in a vacuum tube. Service providers of radiographic NDE are guided by ASME B&PV Code section V latest revision and SNT-TC-1A latest revision.

RT provides full volumetric examination and a permanent record of the results; this makes it the preferred nondestructive test method. It is also able to detect surface and subsurface flaws with equal reliability provided they are in a parallel path to the radiation. Welded butt joints are ideally suited for radiographic evaluation. Reliable and effective flaw sizing, orientation, and depth for evaluation activities are possible with the proper technique and experience. Most manufacturing flaws including cracks, laps, porosity, welding penetration, casting shrinkage or core shift, and thickness variations will be revealed by a properly executed radiographic technique. This technique is applicable to virtually any material.

RT has several limitations and drawbacks: (1) cost of equipment, radiation safety, and personnel training and qualifications are among the highest of the nondestructive processes; (2) certain flaws, particularly planar, are undetectable unless oriented in a manner that presents a portion of the flaw in a parallel path to the radiation, (3) welded joints exceeding approximately 30° from a normal plane are difficult and impractical for radiography; (4) exceedingly rough or coarse surfaces may mask or confuse the radiograph interpretation; and (5) it requires a radiographic vault or large area cleared of personnel to meet safety regulations and prevent unintentional radiation exposure.

(a)

(b)

FIGURE 4.51
(a) Radiographic testing of flare component welds using an iridium 192 radioactive source in a containment vault and (b) radiographic film image of a weld as viewed in a prescribed darkened room.

4.7.5 Ultrasonic Testing

The ultrasonic process is based on the principle that sound travels at different speeds in different materials and reflects back at interfaces of unlike materials. For example, sound travels one particular speed through steel; however, if it encounters a different material such as weld slag, the sound wave will travel at a different speed and will partially reflect some of the sound energy back. If the sound wave encounters a void in the metal, the sound energy will be fully reflected.

When a beam of high-frequency acoustic energy (sound wave) is directed into an item, it will travel at a known speed until an interface is encountered. This interface may be an element of the item such as a bolt-hole, a machined groove, the outer edge of the item, or a flaw somewhere in between. In either case, the sound is reflected back and relevant information is processed and displayed on a screen, typically the sound amplitude versus time. The sound amplitude provides the information necessary to help determine the type and orientation of flaws. The time information is used to help locate the depth of the flaws.

The two most common methods of UT are straight beam and angle beam. For the straight-beam method, the acoustic signal enters the item and travels directly to the point where an interface is encountered and reflects back. For the angle-beam method, the acoustic signal enters the item at a prescribed angle and travels through the item in a zigzag pattern. Figure 4.52 shows an ultrasonic, angle-beam apparatus (in the foreground) with the resulting signal. Service providers of ultrasonic NDE are guided by ASME B&PV Code section V latest revision and SNT-TC-1A latest revision.

UT provides full volumetric examination with a permanent record when the device is linked to a data downloader; this makes UT the preferred nondestructive method for configurations impractical with radiography. Welded joints exceeding approximately 30° from a normal plane are readily accessible using UT; this makes UT the preferred choice for corner joints such as nozzle projections and right-angle structural configurations. UT is also able to detect surface and subsurface flaws with equal reliability and is not affected by flaw orientation. Reliable and effective flaw sizing, orientation, and depth for evaluation activities are routine with the proper technique and experience. Most manufacturing flaws including cracks, laps, porosity, welding penetration, casting shrinkage or core shift, and thickness variations will be revealed by a properly executed ultrasonic technique. Straight-beam processes can be relatively simple and applied with a minimum of expense by an instrument referred to as T-gauge. This instrument is used primarily to measure material thickness and excels at demonstrating thickness of corroded or repaired areas and providing material thickness of otherwise inaccessible plates and shapes. UT is also applicable to virtually any material that will support the transducer.

UT has several limitations and drawbacks, however, such as (1) cost of equipment, accessories, and personnel training and qualifications; (2) reliable flaw detection is dependent on the operator skill and technique; and (3) exceedingly rough or coarse surfaces must be prepared to provide a relatively smooth surface.

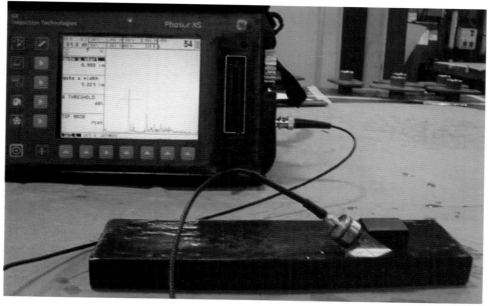

FIGURE 4.52
An ultrasonic, angle-beam apparatus (foreground) with the resulting signal.

4.7.6 Positive Material Identification/Alloy Verification

PMI is based on the principle that every element, when bombarded with radiation, predictably releases an exact amount of energy intensity. This principle allows an analyzer to be used to determine the element and the percentage of the element in an alloy. The analyzer recognizes the first few thousandths of material as representing the entire material volume. Therefore, it is important that the alloy surface is clean. Sometimes the alloy surface is abraded to present uncontaminated material representative of the entire alloy for analysis. Figure 4.53 shows a handheld analyzer being used on a process burner riser pipe to confirm the material type.

Elements that can be identified by a portable analyzer include titanium, chromium, manganese, nickel, copper, molybdenum, cobalt, niobium, tungsten, vanadium, iron, zinc, selenium, zirconium, silver, tin, hafnium, tantalum, gold, lead, and bismuth. Service providers of portable PMI NDE must rely on self-documented technical procedures.

PMI analyzers are extremely portable and can give immediate results. They can be used on most alloy materials with an accuracy of approximately ±10%. These analyzers, however, have several limitations and drawbacks: (1) cost of equipment; (2) radiation safety; (3) personnel training and qualifications are required and are among the highest of the nondestructive processes; (4) it cannot be safely used in a combustible atmosphere; and (5) it cannot read carbon content; if carbon content is desired, refer to equipment designation optical emission spectroscopy (OES).

4.7.7 Metallographic Replication

Metallographic replication is a sampling technique that records the topography and microstructure of a prepared metal surface obtained from an item without destructive consequences. Production of a replica begins with polishing a selected location in the area of interest. Next, an acid-etched thin plastic film is used to record the area. This recorded film is magnified, photographed, and printed. Evaluation of the records then proceeds as though the samples were produced as an actual metallic sample.

Metallographic replication is a valuable tool in creating a baseline for equipment conditions. The replication can be repeated throughout the life of the equipment, and conclusions, with resultant actions, can be drawn from changes occurring in the microstructure.

Service providers of metallographic replication are guided by ASTM E 1351 latest revision. Actions that

FIGURE 4.53
Positive material identification using a handheld analyzer on a process burner riser pipe to confirm the material type.

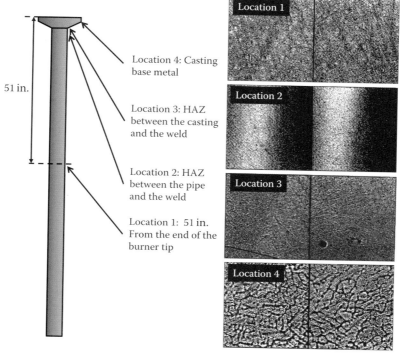

FIGURE 4.54
Metallographic replicas taken from a flare tip during fabrication.

result from a replication program may include the following: (1) revised welding procedures that are designed to further control the welding temperatures, (2) addition or deletion of certain modifying elements to the welding consumables, and (3) rapid cooling or solution annealing where size and configuration allows.

Metallographic replication has several limitations: (1) replication is not equal to an actual mounted sample (although it does provide an understanding of the characteristics of the subject under evaluation), (2) accessibility by the provider of the replica service is mandatory to ensure a properly prepared surface and repeatability in the future, and (3) accurate maps of the replica locations are essential.

Austenitic SS alloys commonly used for fabricating process burner and flare equipment are susceptible to a condition known as sensitization. Sensitization can occur during the welding process. Sensitization of austenitic SS is generally described as the depletion of chromium at the grain boundaries resulting from heating in the range of 930°F to 1560°F (500°C to 850°C). When an alloy becomes sensitized, it raises concerns of the susceptibility to intergranular corrosion. During the welding process, the potential for sensitization is increased as the cooling rates are slowed and the level of carbon in the alloy is increased. The carbon is a necessary constituent of the SS to ensure that the material maintains strength at elevated temperatures; therefore, total elimination of sensitization is probably not practical or even achievable. If the alloy becomes sensitized, it cannot be concluded that intergranular corrosion is inevitable. If sensitized, it is a concern and can be reduced or controlled by monitoring of heat and rapid cooling where possible.

Metallographic replication is commonly used to evaluate the presence of sensitization occurring from the welding process by examining the microstructure. Figure 4.54 shows metallographic replicas taken from a flare tip during fabrication.

References

1. Granet, I., *Modern Materials Science*, Reston Publishing Co. Inc., A Prentice-Hall Company, Reston, VA, 1980.
2. Jefferson, T.B. and Woods, G., *Metals and How to Weld Them*, 2nd edn., James F. Lincoln Arc Welding Foundation, Cleveland, OH, 1962.
3. Bardes, B.P. (Ed.), *Metals Handbook, Vol. 1, Properties and Selection: Irons and Steels*, 9th edn., American Society for Metals, Metals Park, OH, 1978.
4. Fisher, S.M., Sulfidation: Turbine blade corrosion, http://www.aviationpros.com/article/10378159/sulfidation-turbine-blade-corrosion, accessed December 2010.
5. www.tech.mtu.edu/~llsutter/rust.html
6. Cobb, H.M., *The History of Stainless Steel*, Copyright 2012, ASM International, Materials Park, Ohio, www.asminternational.org

7. American Iron and Steel Institute, *The Role of Stainless Steel in Industrial Heat Exchangers*, Washington, DC.
8. http://www.bondhus.com/metallurgy/body-3.htm
9. http://www.corrosion-club.com/intergr.htm
10. Chandler, H., *Metallurgy for the Non-Metallurgist*, ASM International, Cleveland, OH, 2006.
11. www.corrosion-doctors.org/Forms/images
12. Pang, X., Gao, K., Yang, H., Qiao, L., Wang, Y., and Volinsky, A.A., Interfacial microstructure of chromium oxide coatings, *Advanced Engineering Material*, 9(7), 594, 2007.
13. Pang, X., Gao, K., Yang, H., Qiao, L., Wang, Y., and Volinsky, A., Interfacial microstructure of chromium oxide coatings, *Advanced Engineering Materials*, 9(7), 594–599, 2007.
14. Allegheny Ludlum, Technical data blue sheet—Stainless steel, http://www.alleghenyludlum.com/ludlum/Documents/309_310.pdf, accessed December 2010.
15. The Stainless Steel Information Center, Specialty steel industry of North America, http://www.ssina.com/index2.html, accessed December 2010.
16. Elliott, P., Choose materials for high-temperature environments, *CEP*, February 2001.
17. Rolled Alloys, Inc., RA 353 MA Alloy Data Sheet, http://www.rolledalloys.com/products/nickel-alloys/ra330
18. Verma, S.K., Presented at *Corrosion/85*, National Association of Corrosion Engineers, Houston, TX, Paper no. 336.
19. Howes, M.A.H., High temperature corrosion in coal gasification systems, Final Report, GRI-8710152, Gas Research Institute, Chicago, IL, August 1987.
20. Lai, G.G., *High Temperature Corrosion of Engineering Alloys*, Materials Park, OH, ASM International, 1990.
21. Rapp, R.A., Hot corrosion of material, *Pure and Applied Chemistry*, 62(1), 113–122, 1990.
22. ASM International, Hot corrosion in gas turbines. In *High Temperature Corrosion and Materials Application (#05208G)*, Chapter 9, pp. 249–258, www.asminternational.org
23. Whiteway, P., Building better bridges, *Nickel Magazine*, September 1998, http://www.nidi.org/nickel/0998/1-0998n.shtml
24. Carburization—The temper of iron for files, http://www.rolledalloys.com/trcdocs/heatresist/CARBURIZATION.pdf, accessed December 2010.
25. Moller, G.E. and Warren, C.W., Presented at *Corrosion/81*, National Association of Corrosion Engineers, Houston, TX, Paper no. 237.
26. Naogljičenje, K.T. and Koksanje, U.K., Carburization, carbide formation, metal dusting and coking, *MTAEC 9*, 36(6), 297, 2002, ISSN 1580-2949.
27. Kelly, J., Metal dusting causes HRA headaches, http://industrialheating.com, accessed December 2010.
28. Hrivnak, I., Caplovic, L., Bakay, G., and Bitter, A., Metal dusting of inlet tube made of alloy 800, *Kovove Materialy*, 43, 290–299, 2005.
29. Baker, B.A. and Smith, G.D., Alloy selection for environments which promote metal dusting, *Corrosion 2000*, NACE International, Houston, TX, Paper no. 00257.
30. Al-Meshari, A.I., Dusting of heat-resistant alloys, PhD dissertation, Department of Materials Science and Metallurgy Metal, University of Cambridge, Hughes Hall, Cambridge, U.K., October 2008.
31. Meier, M., The ductile to brittle transition, Department of Chemical Engineering and Materials Science, University of California, Davis, CA, September 2004.
32. University of Cambridge, The Ductile-Brittle transition, DoITPoMS teaching and learning packages, http://www.doitpoms.ac.uk/tlplib/ductile-brittle-transition/printall.php, accessed December 2010.
33. Ricart, J.B., Dennehy, M., Fredriksson, L., Holland, G., Otte, W, Petit, P., Puype, H., Cryogenic vaporization systems—Prevention of Brittle fracture of equipment and piping, European Industrial Gases Association AISBL, Document 133/06/E.
34. http://www.kuleuven.ac.be/bwk/materials/Teaching/master/wg02/l0100.htm
35. API Recommended Practice 535, *Burners for Fired Heaters in General Refinery Service*, 2nd edn., American Petroleum Institute, Washington, DC, 2006.
36. API 560, *Fired Heaters for General Refinery Service*, 4th edn., American Petroleum Institute, Washington, DC, August 2007.
37. API Publication 581, *Risk Based Inspection—Base Resource Document*, 2nd edn., American Petroleum Institute, Washington, DC, 2000.
38. Bussman, W., Franklin, J., Schwartz, R., Corrosion of flare tips, Presented at *NACE 97*, Chicago, IL, 1997.
39. Romero, S., Delayed coker fired heater design and operations, *Rio Oil and Gas 2010*, Brazilian Petroleum, Gas and Biofuels Institute (IBP), Rio de Janeiro, Brazil, IBP2710-10.
40. Kubota Metal Corporation, http://www.kubotametal.com/alloys/heat_resistant/HF.pdf, accessed December 2010.
41. Vander Voort, G.F. (Ed.), Additional steels I-T and CCT diagrams, *Atlas of Time-Temperature Diagrams for Irons and Steels*, ASM International, Materials Park, OH, 1991.
42. Kearns, W.H. (Ed.), *Welding Handbook Metals and Their Weldability*, 7th edn, Vol. 4, American Welding Society, FL, 1982, p. 36+.
43. Lippold, J.C. and Kotecki, D.J., Austenitic stainless steels. In *Welding Metallurgy and Weldability of Stainless Steels*, John Wiley and Sons, Inc., Hoboken, NJ, 2005.
44. Sutter, L., Research and professional page, Michigan Tech Transportation Institute, Houghton, MI, www.tech.mtu.edu/~llsutter/rust.html
45. Bondhus, Crystals, grains, and cooling, Picture 1—Dendrite crystal, Monticello, MN, http://www.bondhus.com/metallurgy/body-3.htm
46. Metallurgical Technologies Inc., Metallography/microstructure evaluation, Mooresville, NC, http://www.met-tech.com/metallography.htm
47. Lai, G., *High Temperature Corrosion of Engineering Alloys*, ASM International, Materials Park, OH, 1990.
48. United States Office of Education Training Film, Elements of tempering, normalizing and annealing, Produced by the Division of Visual Aids, U.S. Office of Education, 1945.
49. American Petroleum Institute, *Damage Mechanisms Affecting Fixed Equipment in the Refining Industry*, API Recommended Practice 571, Washington, DC, December 2003.

5
Refractory for Combustion Systems

Jim Warren

CONTENTS

5.1 Introduction
5.2 What Are Refractories? 133
5.3 Monolithic Refractory Products 134
    5.3.1 Hydraulically Bonded Castables 134
    5.3.2 Chemically (Phos) Bonded Castables 135
    5.3.3 Chemically Bonded Plastics 135
5.4 Brick Refractory Products 136
    5.4.1 Characteristics of Refractory Brick 136
    5.4.2 Refractory Brick Installation 137
5.5 Soft Refractory Products 137
5.6 Refractory Materials: Chemical and Physical Properties 139
    5.6.1 Typical Refractory Systems 141
    5.6.2 Refractory Anchoring Systems 141
        5.6.2.1 Primary Function 142
        5.6.2.2 V-Anchors 142
        5.6.2.3 Footed Wavy V-Anchor 142
        5.6.2.4 Double-Hooked V-Anchor 144
        5.6.2.5 Tined Anchors 145
        5.6.2.6 Anchor Distance below Refractory Surface 145
        5.6.2.7 Anchor Spacing 145
        5.6.2.8 Steel Fiber Reinforcing 145
        5.6.2.9 Other Refractory Anchoring Systems 145
5.7 API-936 Considerations 146
    5.7.1 Surface Preparation 147
    5.7.2 Installer Certification 148
    5.7.3 Curing, Drying, and Firing 148
    5.7.4 Repairs to Existing Refractory Lining Systems 149
    5.7.5 Inspection of Existing Refractory Lining Systems 149
    5.7.6 Shipping Refractory Equipment to Tropical Environments 149
    5.7.7 Laboratory Testing 150
Reference 150
                                                     150

5.1 Introduction

The history of manufacturing involving high heat and refractory technology began with the discovery of fire. Nature provided the first refractory materials as crucibles of rock where metals were softened and shaped into primitive tools. Refractory can now be seen in everyday life such as in a fireplace as shown in Figure 5.1. Modern industrial refractory materials are customized, high-temperature ceramics designed to withstand the destructive and extreme service conditions needed to manufacture chemicals, cement, glass, metals, petroleum, and other essentials of contemporary life.[1]

In the early 1980s, several things happened that were a matter of concern to the operators of hazardous waste incinerators. The Resource Conservation and Recovery Act (RCRA) was enacted, and the proper handling and disposal of industrial waste became the responsibility of

FIGURE 5.1
Example of everyday refractory.

the waste producers. This established a "cradle to grave" responsibility for the disposal of hazardous wastes.

Temperatures and residence times in incinerators were increased to insure complete combustion of volatiles and combustion products. The refractories required to contain the incineration process had to be upgraded to handle the increased volume and demand of the incineration systems. They also changed how these incineration systems were operated to extend the refractory life of the equipment. This increase in incineration systems and the harsher conditions that refractories face created a need for longer service life and higher quality refractory materials.

Depending on the type of waste to be disposed, and emission restriction, refractory requirements for combustion equipment vary considerably. The criteria for selecting suitable refractory materials should usually be confined to the following:

1. *Temperature*. The refractory must be able to withstand the maximum heat inside the combustion chamber. The limiting temperature shall be the runaway temperature excursions under the most severe operating conditions.
2. *Slag*. The selected refractory shall be dense and chemically compatible with any slag likely to form on it. Usually, the more chemically compatible a refractory is to slag with which it contacts, the more resistant it is to slag attack.
3. *HCl, Cl_2, HF, Br, and SO_x in flue gas*. The presence of these chemicals in the flue gas can cause an attack of the lime (CaO) in castables. This can cause refractory failures even at very low temperatures.
4. *Thermal shock*. All possible operating conditions shall be evaluated, and the most suitable refractory material for preventing damage due to rapid temperature changes shall be selected.
5. *Corrosion*. Refractory materials shall be selected and designed to resist corrosion attack and to also maintain the steel temperature above the acid dew point at all operating conditions.
6. *Differential thermal expansion*. Refractory design shall take into consideration the differential thermal expansion between different temperature zones and different refractory materials.

5.2 What Are Refractories?

Refractories are used for the containment of substances at high temperatures in incinerators, furnaces, reactors, and other process units. Refractories provide the necessary protective linings for these vessels. Some common materials used to make refractories include alumina, andalusite, fireclays, bauxite, chromite, dolomite, magnesite, silica, silicon, zirconia, and numerous others.

Raw material type is one means by which refractories are classified: fireclay, high alumina, silica, basic, and others. These raw materials are available from a variety of sources throughout the world. Figure 5.2 shows several examples of raw materials. As a rule, the properties of any refractory are dependent upon the chemistry and purity levels of the raw materials.

The bonding phase holds together the refractory grains and is also used to differentiate between refractory types. A refractory's bond is a function of additives to the raw materials and subsequent heat treatment. Most bonds can be described as a glue phase—a material totally surrounds the refractory grains and holds them together (see Figures 5.3 and 5.4). Most of the refractory properties are determined by the bonding system. There are several common bonding phases that give refractory its strength. The two basic phases are chemical bonds and ceramic bonds. The chemical bond begins to form around 500°F (260°C). Ceramic bonds usually form above 1800°F (982°C).

Refractories may be further divided into "hard" and "soft" categories, which apply to their state when ready for service. Hard refractories can be further categorized as castables, plastics, or bricks. Castables and plastics are referred to as monolithic linings.

5.3 Monolithic Refractory Products

Monolithic refractory products (hard refractories) are unshaped materials as compared to firebrick, tile, and other prefired shapes. They are represented in three forms: hydraulically bonded castables, chemically bonded castables, and chemically bonded plastics.

FIGURE 5.2
Several raw materials used in making refractory.

FIGURE 5.3
Drawing representing glue phase.

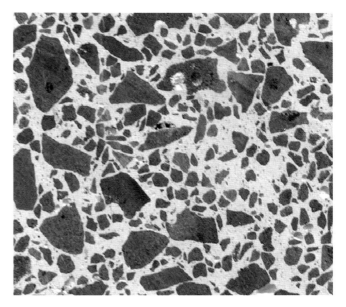

FIGURE 5.4
Photograph demonstrating glue phase.

5.3.1 Hydraulically Bonded Castables

- Dry mixes containing calcium aluminate cement, fired and/or calcined aggregates, and proprietary mixtures
- Tempered with water at the time of placement
- Installed by casting, troweling, hand packing, and pneumatic gunning in a wide range of thicknesses
- They are supplied in a wide range of densities and other physical properties

5.3.2 Chemically (Phos) Bonded Castables

- Shipped dry requiring water only
- High density, strength, and erosion resistance
- Installed by ramming, hammering, or hand packing in thin applications

5.3.3 Chemically Bonded Plastics (See Figure 5.5)

- Water or other wetting agents included at time of manufacture
- Will not achieve a set until fired
- Installed by ramming or hammering in thin and thick applications as shown in Figure 5.6

FIGURE 5.5
Chemically bonded plastic pieces.

FIGURE 5.6
Installation of plastic.

FIGURE 5.7
Plastic refractory anchoring system.

The primary function of the refractory anchoring (reinforcing) system, pictured in Figure 5.7, is to hold the refractory system tightly against the steel substrate it is intended to protect. This will insure the following:

- Protect the steel substrate from high temperatures.
- Protect the steel substrate from abrasion.
- Protect the surface from corrosion.

This is discussed in more detail under Refractory Systems (Section 5.6.1).

5.4 Brick Refractory Products

Brick refractory is available in a wide variety of compositions ranging from high alumina content aluminosilicates to magnesites. The binding material in brick can be calcium cement based or phosphoric acid based. A brick lining is held in place by gravity and/or compressive forces resulting from proper placement (as in the construction of an arch). The lining must be installed in the vessel by skilled craftsmen and requires more time to install, especially if special shapes have to be assembled by cutting individual bricks.

Because of its high density and low porosity (good penetration resistance to molten or refractory attacking materials), brick typically offers the best abrasion and corrosion resistance of any refractory. However, the high density results in the brick usually being heavy

(120 lb/ft³ = 1900 kg/m³ or more), and the lower insulating value results in greater lining thicknesses required to achieve the same thermal resistance. An additional consequence of a thicker lining is that a larger, more expensive combustion chamber shell is needed to maintain the same inside vessel diameter.

5.4.1 Characteristics of Refractory Brick

Premanufactured refractory shapes are commonly referred to as "firebrick." Figures 5.8 and 5.9 show some common refractory brick shapes. Some of their attributes are shown in the following:

1. Shipped ready to install.
   a. Integrity of the brick is not compromised at the jobsite
   b. Quantities are controlled with minimal waste
2. Prefired
   a. The free water is removed, and the risk of explosive spalling is eliminated
   b. Permanent shrinkage is complete
   c. The true refractory bond is completed
3. Premanufactured without significant irregularities
   a. Installed without significant waste
   b. Long shelf life for surplus
4. Quality controlled at point of manufacture
   a. No mixing required in the field
   b. Hand installed

Some of the challenges of refractory brick construction are the long lead time required for shipment. Inventories are kept low due to the lack of demand for some products and shapes. The installation schedule is also longer because the firebrick is hand installed, piece by piece, and is quite labor intensive (see Figure 5.10). This is affected more by the particular application and is a function of custom cutting at termination points, corners, and expansion joints (see Figure 5.11).

There is also the problem of availability of skilled craftsmen to install the firebrick. This is dependent on the location of the installation as most installations are performed at site to minimize any damage to the lining during transit. Many of these skilled craftsmen are retiring, and there is very little apprenticeship training available to replace this aging workforce.

There is a broad array of refractory firebrick classifications available. These are differentiated by insulating firebrick, general duty firebrick, and high-alumina firebrick.

Insulating firebrick is lightweight with excellent insulating properties, normally used as a backup lining in dual-layer applications. It is not the best choice for corrosion or abrasion applications if it is being considered as a hot face lining.

General duty firebrick has good insulating properties and is used as a hot face lining for lower duty applications.

High-alumina firebrick is a dense firebrick, usually in excess of 120 lb/ft³ (1900 kg/m³), that resists extremely high temperatures. These are usually in excess of 90% alumina and are usually the hot face lining in thermal oxidizers in applications where abrasion and corrosion is a concern.

5.4.2 Refractory Brick Installation

There are several problems that can be encountered when installing firebrick. When using experienced installers, most of these problems can be avoided. This is discussed further in the API-936 installation (Section 5.7). Some of the installation problems that can occur are identified immediately below, but their prevention measures are discussed in later sections.

Hacking, which is exposed edges of a brick offsetting due to incorrect brick combinations in a ring, must be managed, or the lining integrity may be suspect. There is also concern about incorrect mortar joints, usually being applied too thickly. If the lining is not installed properly, a sagging of the lining may occur, which can indicate a separation of the backup lining. This can be avoided by properly keying the brick with hydraulic rams if necessary and may require the use of what are called pogo sticks, as seen in Figure 5.12, to insure the brick is properly seated. Sometimes, bricks are cut to use less than ½ the length. The last two bricks should have been cut at ¾ length to avoid a weak brick.

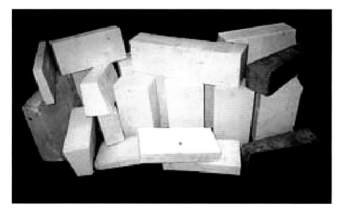

FIGURE 5.8
Common firebricks.

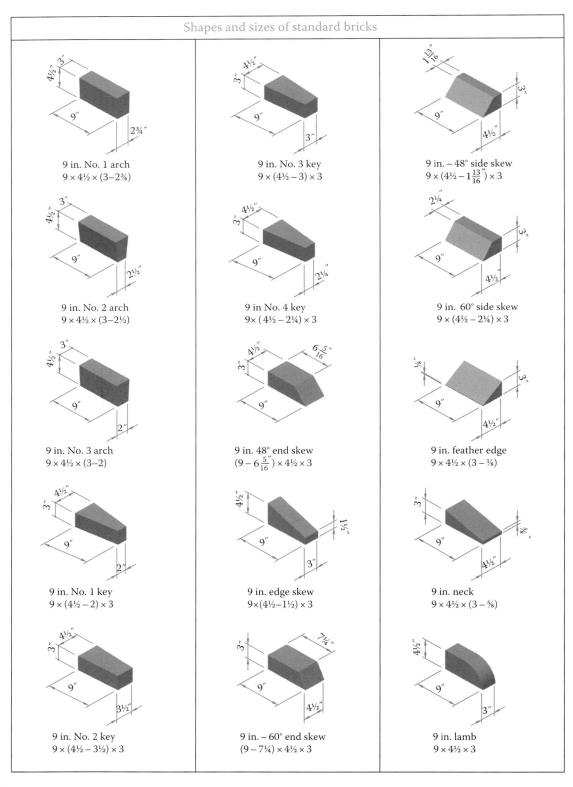

FIGURE 5.9
Shapes and sizes of standard bricks. (Adapted from Harbison-Walker.)

FIGURE 5.10
Example of labor-intensive brick installation.

FIGURE 5.11
Firebricks cut and installed in a circular pattern.

FIGURE 5.12
Use of pogo sticks to ensure brick is properly seated.

The expansion joints also need to be inspected to insure they are installed properly, and there are a sufficient number.

5.5 Soft Refractory Products

Ceramic refractories, which are called "soft refractories," are composed of ceramic fibers formed into a blanket, a soft block module, bulk, paper, vacuum-formed shapes, or a hard board. Figure 5.13 shows examples of these and other forms of ceramic refractories. They remain soft when in service. The blanket, as seen in Figure 5.14, and board are usually held in place with stainless steel or other high-temperature alloy anchors or pins. The blanket is easily installed by impaling it over the anchor against the steel shell with the pin protruding through the blanket. Self-locking washers

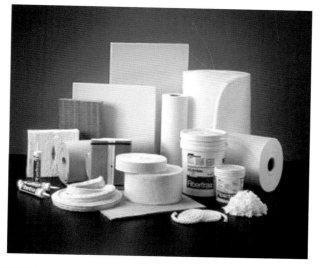

FIGURE 5.13
Example of many different types of ceramic refractory.

FIGURE 5.14
Refractory blanket.

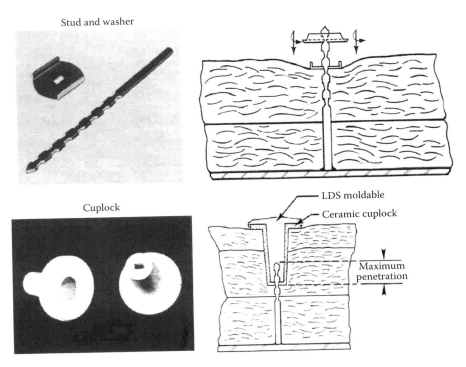

FIGURE 5.15
Ceramic refractory anchoring components.

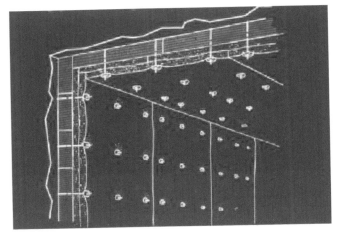

FIGURE 5.16
Illustration of anchors penetrating through refractory.

are then placed on the pins to keep the material from coming loose. This anchoring technique can be seen in Figures 5.15 through 5.17.

Soft refractories are much lighter (usually less than 12 lb/ft^3 = 190 kg/m^3), are much better insulators, and can be heated rapidly without fear of damage due to thermal shock. Soft refractories have limited temperature limits to 2600°F (1427°C). They are susceptible to erosion and do have poor resistance to alkali liquids and vapors. Ceramic refractory is very cost effective in certain applications.

FIGURE 5.17
Photograph of installed ceramic refractory.

5.6 Refractory Materials: Chemical and Physical Properties

An understanding of the important properties of refractories is fundamental to the development, improvement, and selection of refractory linings. Figure 5.18 shows how these are related. Important properties of refractories which can be determined most readily are the following:

1. Density or bulk density
   a. Weight per unit volume of the refractory
   b. Indicated as "dried" density (after drying to 225°F = 110°C) or fired density (after firing to 1500°F (815°C)
2. Compressive (cold crushing) strength
   a. The ultimate strength of the material in compression
3. Modulus of rupture
   a. The flexural strength of the material in compression
4. Permanent linear change
   a. The percent change in length from the dried to the fired state
5. Thermal conductivity
   a. The rate of heat flow through a refractory material at a given mean temperature
   b. Used to predict skin temperatures, heat loss, and lining interface temperatures
6. Chemical analysis
   a. The average chemical composition of the refractory material per the material manufacturer.
   b. In some cases, a refractory material will not perform properly under given conditions. Always consult the material manufacturer when in doubt.

5.6.1 Typical Refractory Systems

A single-component lining reinforced with V-Anchors is the most common refractory lining system in petrochemical applications. It may serve primarily as a heat barrier for the steel shell in which case the refractory lining would be a lightweight insulating castable. If moderate mechanical strength and insulating ability are needed, the refractory lining would be a medium weight insulating castable. If the intent is to protect the steel shell from process erosion, and still act as insulation, the refractory lining would be a semi-insulating dense castable refractory with good physical strength and abrasion resistance.

A multiple-component lining, as shown in Figure 5.19, is used when the service environment demands a refractory lining with good insulating value, a high temperature use limit, and good strength and durability. While a single-component system is not able to provide all of these requirements at a given time,

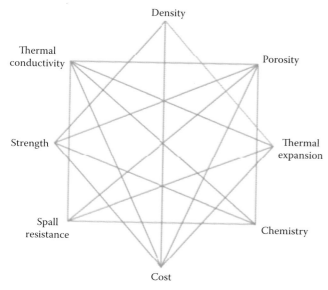

FIGURE 5.18
Relationship of different refractory physical properties.

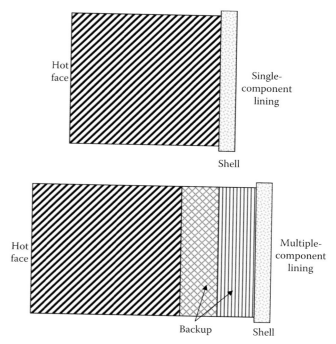

FIGURE 5.19
Difference between single- and multiple-component lining.

the two (or more) component system is able to do so. The insulating castable backup lining provides good insulating value, reducing heat flow to the shell, while the hot face lining is able to withstand higher temperatures and mechanical abuse.

5.6.2 Refractory Anchoring Systems

The importance of a properly designed and installed refractory lining reinforcing system cannot be over emphasized. If the reinforcing system fails, the entire refractory lining system will fail regardless of the quality of the installation of refractory materials. When considering anchors for a refractory lining system, anchor alloy, style, spacing, orientation, and weld attachment are all important considerations. The alloy of the anchor must be compatible with the environment in which it will function and compatible with the steel surface to which it will be attached. Table 5.1 shows the maximum use temperature for several different types of steel. Anchor style must be appropriate for the refractory lining system being supported and must be spaced and oriented in an appropriate manner. Figures 5.20 and 5.21 show several different styles of anchors and anchor positioning. Figure 5.22 shows anchor spacing.

5.6.2.1 Primary Function

To hold the refractory system tightly against the steel substrate it is intended to protect, thereby

- Protecting the substrate from high temperatures
- Protecting the substrate from abrasion
- Protecting the substrate from corrosion

5.6.2.2 V-Anchors

- Usually used in monolithic linings.
- Occasionally used in dual-layer linings.

TABLE 5.1

Maximum Use Temperature of Various Grades of Stainless Steel

Maximum Temperature of Metallic Anchors (°F)	Type of Steel Required
500	Carbon steel
1500	304 SS
1500	316 SS
1600	309 SS
1700	310 SS
2000	Inconel 601

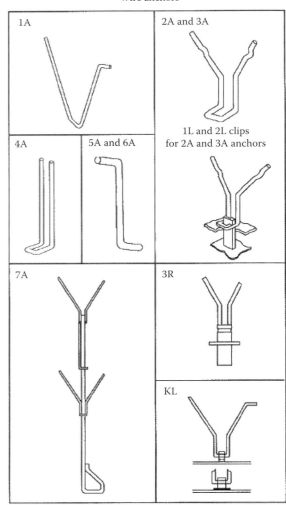

Grades of steel required for high temperature service

Type of Steel	Color Code	Maximum Temp of Metallic Components (°F)
Carbon steel	Blue	1000
304 SS	No color	1600
309 SS	Red	1650
310 SS	Yellow	1700

FIGURE 5.20
Several different wire anchors and components.

- Some typical types are footed wavy V-anchors (preferred), V-anchors, footed V-anchors, double-hooked vee, steerhorns, split-pin vee, and wavy vees (see Figure 5.23). See Figure 5.24 for an example of footed wavy V-anchors.

V-anchors are fabricated from stainless steel round bar, and typical bar diameters vary from a minimum of 3/16 in. to a maximum of 5/16 in.

Three of the more popular styles of V-anchors are discussed:

Refractory for Combustion Systems

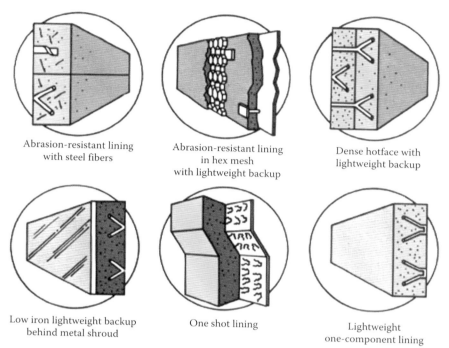

FIGURE 5.21
Examples of different anchoring.

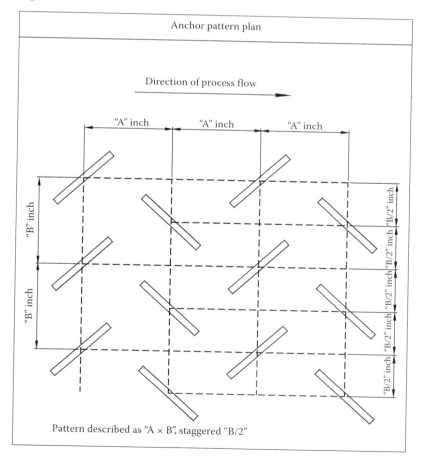

FIGURE 5.22
Typical anchor spacing.

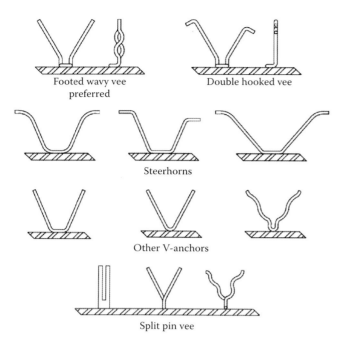

FIGURE 5.23
Several different vee-type anchors.

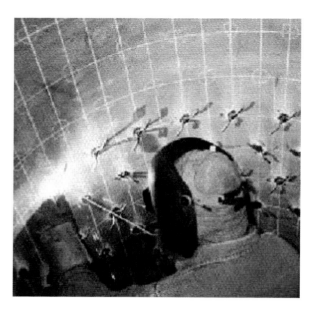

FIGURE 5.25
Anchor welding.

requiring one leg of the anchor to be hammered down after welding, then bent back in place. The weld, seen in Figure 5.25, is then checked for fusion. Sometimes, fillet welding a V-anchor does not pass this test due to the quality of the weld. The footed anchor is more expensive and takes longer to install as it is a manual operation, but has a much higher success rate of passing the hammer test, insuring the integrity of the anchoring system. Certainly, the decision to use this type of anchor is dependent on the application and the quality of the installation, but if a weld fails this test, one may be required to remove and reinstall thousands of anchors. Figure 5.26 shows an application where these are used extensively. The footed anchor provides much more

FIGURE 5.24
Example of footed wavy V-anchor installation.

5.6.2.3 Footed Wavy V-Anchor

A footed wavy V-anchor is the preferred anchor for castable linings from 2 to 8 in. (5 to 20 cm) in thickness. The foot at the bottom of the anchor allows for the installation of more weld metal insuring that the anchor is secured to the steel. Most API-936 installations require a hammer test to be conducted on the anchor,

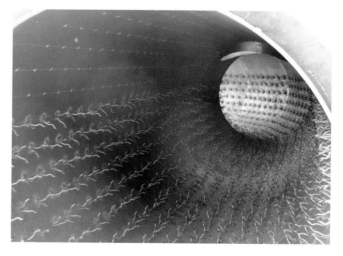

FIGURE 5.26
Extensive use of wavy anchors.

weld penetration. The wavy vee, or crimped zigzag legs, provide additional locking ability, which helps to keep the refractory lining tightly against the steel surface.

5.6.2.4 Double-Hooked V-Anchor

Double-hooked V-anchors provide more positive retention of the refractory monolith than the footed wavy V-anchor because of the hooked anchor ends. However, there is a tendency for voids to occur behind the hooked anchor ends especially when refractory installation is by pneumatic gunning. In addition, as the anchor arms expand longitudinally toward the refractory lining surface on heat up, the hooks are pushed sideways through the refractory lining which tends to create shear planes near the refractory surface. Plastic tips are sometimes placed on the end of the anchor to minimize this occurrence. Double-hooked anchors can be purchased with or without wavy arms and with or without a footed leg and must be installed by manual welding.

5.6.2.5 Tined Anchors

Tined anchors are available in 2 tine and 3 tine configurations and can be installed by either manual or automatic stud welding. Tined anchors are installed straight after which the legs are bent to the desired included angle which is normally between 60° and 90°. The center tine of the 3 tine anchor acts as a lining thickness depth gauge.

5.6.2.6 Anchor Distance below Refractory Surface

In most instances, anchor tips should be located between ½ and 1 in. (1.3 and 2.5 cm) beneath the refractory surface, as shown in Figure 5.27. In years past, anchor height was based on extending a minimum of 2/3–3/4 the distance through the hot face lining, not on the combined lining thickness. This can change based on the application, but most of the time, the anchor height is just below the hot face of the lining.

5.6.2.7 Anchor Spacing

The distance between anchors needs careful consideration.

Edges, roofs, bullnoses, and areas where vibration, mechanical movement, or gravity impose loads on the lining need more anchoring than a straight wall or floor. Standard spacing for various areas is suggested in Table 5.2. Anchors are usually welded in a square pattern (as near as possible in some cases), but alternative patterns, such as a diamond pattern, are also suitable in many installations.

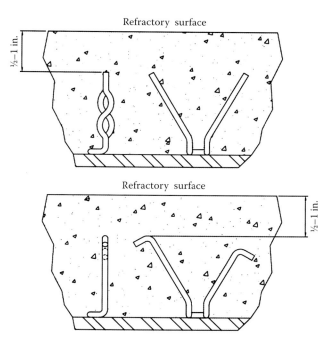

FIGURE 5.27
Anchor distance below refractory surface.

TABLE 5.2
Standard Spacing for Various Anchors

			Suggested Anchor Centers			
	Lining Thickness		Wire and Rod		Ceramic	
Location	(in.)	(cm)	(in.)	(cm)	(in.)	(cm)
Walls and slopes	2–4	5.1–10.2	6	15.2	—	—
	4–8	10.2–20.3	9	22.9	—	—
	8–12	20.3–30.5	12	30.5	18	45.7
	12–16	30.5–40.6	—	—	18	45.7
	>16	>40.6	—	—	24	61
Roofs and bullnose	4–8	10.2–20.3	7	17.8	12	30.5
	>8	>20.3	10	25.4	12	30.5
Floors	2–4	5.1–10.2	12	30.5	—	—
	4–9	10.2–22.9	18	45.7	—	—
	>9	>22.9	20	50.8	—	—

The tines are rotated 90° from neighboring anchors. The average anchor spacing is between 6 and 9 in. (15–23 cm), but this varies depending on the lining thickness and location.

5.6.2.8 Steel Fiber Reinforcing

The use of stainless steel wire fiber reinforcing in gunned and vibration cast linings is very common within the petrochemical industry (see Figure 5.28). Wire fibers come in three basic forms—melt-extracted fiber, cold drawn, and

FIGURE 5.28
Example of steel fibers.

FIGURE 5.30
Example of chopped fibers.

chopped. Most specifications require installation to be 3% by weight of the refractory material.

5.6.2.8.1 Melt-Extracted Fiber

Melt-extracted fiber is somewhat inconsistent in nominal diameter and length. This is the most common fiber in use today.

5.6.2.8.2 Cold-Drawn Fiber

Cold-drawn fiber is simply chopped stainless steel wire. It is therefore consistent in diameter and length and is significantly stiffer than melt-extracted fiber (see Figure 5.29). Because of its stiffness, it does not easily pass through conventional refractory gunning equipment and is therefore primarily used in cast applications.

5.6.2.8.3 Chopped Fiber

Chopped fiber is manufactured from thin stainless steel strip which is passed through specialized shears. It is rectangular in cross section and, like toe cold-drawn fiber, is relatively consistent in cross section and length (see Figure 5.30). It can be used for gunning and casting.

5.6.2.9 Other Refractory Anchoring Systems

5.6.2.9.1 Ceramic (Refractory) Anchor

When process-operating temperatures approach 2000°F (1100°C), refractory brick anchors are often used. These anchors are fired refractory shapes usually about 4 in.^{2} (26 cm^2) in cross section and formed with corrugated sides. Anchor length is adjusted to place the end face of the anchor at the surface of the refractory lining. They are attached to the steel shell by heavy alloy clips, tees, or other retainers which have been welded or bolted to the steel casing. Because the anchor is refractory, it operates well at elevated temperatures where alloy anchors could fail.

5.6.2.9.2 Y-Anchors

Y-anchors are, basically, no more than V-anchors that have been attached to straight studs. They are usually used in dual-layer lining applications where the "V" portion of the anchor rests in the hot face layer. They can also be used in single-layer applications.

5.6.2.9.3 Hexmesh

Hexmesh remains the most commonly used form of reinforcement for thin erosion resistant linings, usually used in linings, less than 2 in. (5 cm) in thickness, for an abrasive service. The completed panels appear to be a series of contiguous hexagons which are about 2 in. (5 cm) wide. Sometimes, a tab is punched and bent toward the center of the hex. These tabs are referred to as lances, and hexmesh having lances is referred to a lance grid. The lances provide additional anchoring.

Several unitized anchors have been developed over the years as alternative anchors to hexmesh. The advantage of alternative (unitized) anchors over hexmesh is

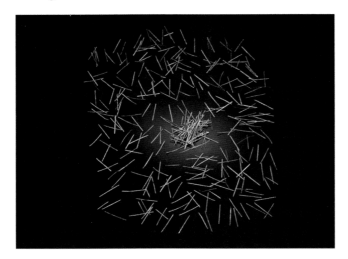

FIGURE 5.29
Example of cold drawn fibers.

Refractory for Combustion Systems

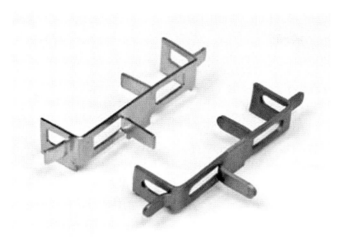

FIGURE 5.31
Example of curl anchors.

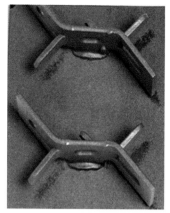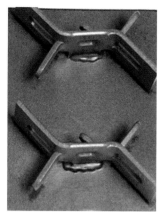

FIGURE 5.32
Example of K-bar anchors.

that alternative anchors act independently, whereas hexmesh panels act continuously. They are also used in irregular or small-diameter applications where rolling and shaping of hexmesh can be time consuming and expensive. The absence of any alternative anchor from this list does not suggest its unacceptability in any way.

5.6.2.9.4 Curl Anchor

The curl anchor is a rectangular "C" in plan and has four tabs which are punched through and bent outward (see Figure 5.31).

5.6.2.9.5 Flexmesh

Flexmesh is essentially hexmesh where the clinches have been replaced by continuous rods. The Flexmesh design allows free form rolling to very small diameters, as small as 6 in. (15 cm). Hexmesh cannot be used in this application.

5.6.2.9.6 K-Bar

As its name implies, the K-bar looks like a "K" in plan. It has good angularity for ramming refractory against and can be attached through automatic stud welding (see Figure 5.32).

5.6.2.9.7 S-Bar

S-bars, possibly the oldest of the alternative anchors, have had widespread use as an alternative for hexmesh in thin erosion linings. The "S" bar is roughly "T" shaped in profile and "S" shaped in plan and has tabs punched in the "T" for additional anchoring (see Figure 5.33).

5.6.2.9.8 Tacko Anchor

The Tacko anchor is round in plan with three semicircular cutouts and three resulting welding feet.

FIGURE 5.33
Example of S-bar anchors.

5.7 API-936 Considerations

It does not matter how well a lining is installed. It does not matter how good the design is. It does not matter how good the materials are. The only thing that matters is what is inside. What is on the walls is what counts. If it is not good inside, then nothing else matters.

The American Petroleum Institute (API) recommended practice 936, second edition in February of 2004. The API-936 Refractory Personnel Certification Exam has been accepted by the process industry as the industry standard to insure that the refractory lining is installed correctly.

Everything discussed previously is critical to the integrity of the refractory lining, but if it is not installed properly and supervised closely, a failure could easily occur such as the one shown in Figure 5.34.

The API-936 starts with the purchase of materials requiring testing at the factory to make sure the

FIGURE 5.34
Example of refractory failure.

5.7.2 Installer Certification

The installer chosen may not be competent to perform the work as specified. This is why API-936 installer certification is often required. The installer is then required to demonstrate their ability to do the work in advance of the actual refractory installation by prequalifying their work on separate panels (see Figure 5.35). If the installer is not qualified, they should demonstrate their abilities on the ground and not in a vessel such as the one shown in Figure 5.36. The installer uses owner specifications

FIGURE 5.35
Prequalifying work.

proper materials are received and checking the QC at the factory before it is shipped to site. Upon receipt of the materials in the field, the materials are tested again to make sure what is about to be installed in the equipment matches what was purchased. Strict storage requirements are followed, and the 12-month shelf life of the materials is not exceeded. API-936 installations require contracting with an API-certified refractory installation inspector to sign off on the materials and supervise the installation of the refractory. This supervision requires testing a sample from each pallet, qualifying the installer, site conditions, and anchor installation to insure what is inside the vessel is correct. After the API-936 installation is complete, documentation will be supplied that insures at a much higher level that no shortcuts were taken.

5.7.1 Surface Preparation

Surface preparation often by grit blasting (sand blasting) prepares the steel surface for refractory installation. In most cases, an SSPC-SP-6 or SP-7 surface condition is appropriate for refractory installation. Surfaces should be free of chemical contamination and free of any loose scale or dirt that could affect the ability of the refractory to remain tightly against the shell during the setting process. Surface preparation is a judgment call to insure the surface for anchor welding and the ability for the refractory to stick to the shell.

FIGURE 5.36
Example of vessel having refractory spray installed.

FIGURE 5.37
Example of adequate personnel and equipment.

and agreed-upon written installation procedures. The adequacy of the installer personnel is confirmed, and the adequacy of the installer equipment is confirmed (see Figure 5.37).

5.7.3 Curing, Drying, and Firing

Curing for hydraulically bonded refractories is that period of time between the introduction of mixing water to the dry refractory material and the completion of all hydraulic and chemical reactions that occur as the material becomes refractory concrete (see Figure 5.38). Water must remain within the concrete for the curing period by water spraying or by sealing with an approved concrete curing agent. It is best to remain above 60°F (16°C) for the curing period. The material must not be allowed to freeze during the curing period. It is also recommended to maintain temperatures less than 100°F (38°C) during the curing period.

Drying is the removal of all free water from the refractory concrete. This is the water that was not chemically combined during curing. This is usually considered complete when the refractory mass has soaked for a period of time at 250°F (110°C). This is sometimes monitored by placing thermocouples on the external steel shell holding the steel temperature to at least 212°F (100°C), insuring the removal of any free water. This is required when the equipment will possibly be stored outside in a cold environment and known as *freeze protection*.

Firing is final heating to the desired elevated temperature. It involves specific temperature rates and holding times. This is usually performed during unit start-up. New equipment is sometimes fired prior to shipment to the field when required, but after final firing, the lining becomes somewhat brittle and may experience surface cracking that causes big concerns to the untrained eye. Fiber and steel needle reinforcement can sometimes be added to minimize this cracking. It is a function of the geometry of the equipment whether this reinforcement is required or not. Larger vessels are more prone to this problem.

5.7.4 Repairs to Existing Refractory Lining Systems

For any area of refractory lining found to be defective, the full thickness of the lining in the defective area, down to the steel, shall be removed. For refractory linings 3 in. (7.5 cm) thick or more, the minimum repair shall be approximately 1 ft² (0.03 m³) and shall expose at least three refractory anchors. For 1 in. (2.5 cm) thick refractory linings in hexmesh, the minimum repair will be 1 hexmesh opening. Adjacent refractory lining damaged during refractory removal must also be repaired. Any hexmesh damaged during repairs must be repaired or replaced.

If the total amount of repairs in any piece of refractory lined equipment exceeds 25% of the lining, the entire lining shall be replaced. The method of repair and curing shall be the original installation. Additional refractory anchors may be required if deemed necessary by the refractory inspector. Pointing of cracks or inserting mortar or refractory material in the crack using a tuck-pointing trowel is not an acceptable repair. Unless directed otherwise, repairs must be thermally dried when individual repairs exceed 1 ft² (0.03 m³) or when the sum of all repairs exceeds 1% of the refractory lined area.

5.7.5 Inspection of Existing Refractory Lining Systems

Inspections of existing refractory lining systems is required to determine the condition of existing refractory lining systems with respect to serviceability and required repairs. Visual inspection of the entire refractory surface is required to check for cracking (see Figure 5.39), spalling, lamination, and bad joint work. Inspection should be made for erosion of refractory anchorage and the refractory surface. Hammer testing is required of the entire refractory surface based on

FIGURE 5.38
Example of poured refractory in a mold.

FIGURE 5.39
Cracked refractory.

sound and feel. Inspection for lamination, separation from the anchoring system or substrate, soft materials, and inadequate joint work could be performed.

5.7.6 Shipping Refractory Equipment to Tropical Environments

When refractory equipment is shipped to tropical environments exposed to high humidity, and curing is not complete, alkali hydration is possible, especially in lightweight, low-density products. This appears as spalling.

Alkali hydrolysis, also known as carbonation, is the formation of calcium carbonate caused by the reaction of lime in cement and carbon dioxide in the atmosphere. The hydrolysis reaction breaks down the cement bond, which creates a volume expansion that weakens the refractory lining surface. This weakened surface is friable and can peel off in ¼–1 in. (6–25 mm) layers depending on the severity.

High porosity and alkali content make lightweight castables susceptible to alkali hydrolysis, which can occur in unprotected linings exposed to weather conditions, especially rain. This reaction does not occur in protected linings, such as insulating linings that are protected by a solid structure on the back and a dense refractory on the front.

Three steps must be followed to minimize/prevent the alkali hydrolysis reaction:

1. Cast and cure material at warmer temperatures to develop stable cement hydrates, which are more resistant to alkali hydrolysis (higher than 70°F = 21°C is preferred).

2. Dry the material out as soon as possible after the 24 h cure time. Drying will remove excess water and convert cement hydrates to more stable phases. The dryout temperature should be in the 500°F–750°F (260°C–400°C) range on the hot face to allow heat to penetrate the material and the temperature to reach at least 230°F (110°C) partway into the lining. The lining does not have to be completely dried.

3. Keep the material dry by covering it with plastic. Do not use surface sealants because they will break down over time and trap water inside the lining. The trapped water can act as a catalyst to promote the hydrolysis reaction.

5.7.7 Laboratory Testing

Many people ask, "Why laboratory testing?" It increases cost and extends the installation time. Just because the material is new does not mean the materials are of high quality. Certification of material prior to refractory shipment from the manufacturer insures quality. The parties are in agreement on the acceptable physical properties prior to order placement. Making sure the physical properties of the materials shipped are as specified is fundamental to insure a quality lining.

The following four tests are recommended for laboratory testing:

1. Density—Checking weight per unit volume. This reflects on the insulating ability of a refractory system.
2. Compressive strength—This is a measure of ultimate strength and ruggedness.
3. Permanent linear change—Projects permanent shrinkage and expansion.
4. Abrasion resistance—A guide to the expected resistance to abrasion in service.

Installer certification works hand in hand with the material certification making sure that the refractory installer can achieve quality on the wall where it counts if they have good material.

Reference

1. C.A. Schacht, *Refractories Handbook*, CRC Press, Boca Raton, FL, 2004.

6
Burner Design

Richard T. Waibel, Michael G. Claxton, and Bernd Reese

CONTENTS

- 6.1 Introduction 152
- 6.2 Combustion 152
- 6.3 Burner Design 153
  - 6.3.1 Metering: Fuel 153
    - 6.3.1.1 Gas Fuel 154
    - 6.3.1.2 Liquid Fuel 154
  - 6.3.2 Metering: Air (Combustion O_2) 156
    - 6.3.2.1 Natural Draft 156
    - 6.3.2.2 Forced Draft 157
  - 6.3.3 Air Control 157
  - 6.3.4 Mixing Fuel/Air 158
    - 6.3.4.1 Entrainment 159
    - 6.3.4.2 Co-Flow 160
    - 6.3.4.3 Cross Flow 160
    - 6.3.4.4 Flow Stream Disruption 160
  - 6.3.5 Maintain (Ignition) 160
  - 6.3.6 Mold (Patterned and Controlled Flame Shape) 161
  - 6.3.7 Minimize (Pollutants) 162
- 6.4 Burner Types 162
  - 6.4.1 Premix and Partial Premix Gas 162
  - 6.4.2 Raw Gas or Nozzle Mix 163
  - 6.4.3 Oil or Liquid Firing 164
    - 6.4.3.1 High-Viscosity Liquid Fuels 164
    - 6.4.3.2 Low-Viscosity Liquids 164
  - 6.4.4 High Intensity (KEU Combustor) 164
  - 6.4.5 Conventional Process Heater Application High Intensity 167
- 6.5 Configuration (Mounting and Direction of Firing) 168
  - 6.5.1 Conventional Burner, Round Flame 168
  - 6.5.2 Flat Flame Burner 168
    - 6.5.2.1 Wall Fired 169
    - 6.5.2.2 Freestanding 169
  - 6.5.3 Radiant Wall 169
  - 6.5.4 Downfired 170
- 6.6 Materials Selection 170
- References 171

6.1 Introduction

What is a burner? In its simplest form, a burner is a device used to provide heat to a process system. More definitively, it is a device which provides a controlled exothermic oxidation reaction.

Using this definition, one could argue that a wooden torch is a burner—even if there is some question as to the defined condition of a controlled exothermic reaction condition being met. In an industrial setting, it is necessary to assume that the device is not itself consumed by the reaction. See Ref. [1] for a wide range of discussions on industrial burners. As a reasonable approach, a burner is a device that provides three basic design functions:

1. A burner must provide for controlled mixing of the reactant (fuel) and the oxidizing agent (in most cases air).
2. A burner must provide a stable and self-renewing ignition source (normally the heat generated by the combustion process).
3. A burner should provide for a controlled region of reaction, a controllable flame shape.

The modern concept of controlling the flame pattern is a direct outgrowth of continuous hydrocarbon and petrochemical processes. Processing of naturally occurring crude oils and organic by-products was initially accomplished through "batching." In this method, the necessity of controlling the actual flame to precise dimensions was not critical. If, at the end of a single batch process, there were carbon residues or undesirable deposits, the fire was extinguished and the vat cleaned. With the advent of tubed, continuous throughput process furnaces, it was no longer economically desirable, and in most cases not physically possible, to perform mechanical cleaning. This made it necessary to reduce or eliminate the carbon residues and deposits generated by localized overheating of the fluid being processed. Direct conductive heat transfer from flame impingement (see Chapter 13) is a major source of this localized overheating.

6.2 Combustion

Combustion as a controlled process is often considered a "black art." While there are substantial amounts of experience-based information and design "rules of thumb" involved, there are very real chemical and physical laws involved. The layman's view of flames and combustion stability is actually soundly based in the principles of chemistry, chemical reactions, and fluid flow (see Volume 1 for detailed discussions of these topics).

Combustion has been defined as a relatively fast exothermic gas-phase chemical reaction. It can occur in either flame or non-flame mode. For the purposes of this discussion on burner design, only flame mode will be considered and is defined as a dual-reactant flame. This flame is an exothermic reaction propagating at subsonic velocity through the mixture of the two reactants. The two reactants are, of course, the fuel and the oxidant. The oxidant is usually atmospheric air. However, the combustion O_2 can be from a number of alternative sources, including the following:

- *Ambient atmospheric air* is the predominant source.
- *Preheated atmospheric air* is commonly used to improve the thermal efficiency and typically requires a forced-draft application.
- *Turbine exhaust gas* (TEG) is the high-temperature, reduced O_2 content exhaust gas from a gas turbine. In general, this is a good source for combustion O_2; however, low oxygen content with low-temperature exhaust gas can result in an oxidant that cannot support combustion.
- *Diesel engine exhaust* is used periodically as an alternative to TEG. The major problem with this as a source of combustion O_2 is in the low temperatures associated with lowered oxygen levels and the pulsing flow associated with internal combustion exhaust.
- *Kiln and dryer off-gas* is seldom seen but can be a usable source for combustion O_2. Again, attention must be paid to the O_2/temperature relationship.
- *Oxygen-enriched streams* are not currently common in typical industrial combustion applications; however, they do exist in specialized processes[2] and are being actively investigated for more common refining and chemical processing applications.

Each of these sources for combustion O_2 have been designed for by burner designers and utilized by industry. For the purposes of this chapter, the majority of the discussion is limited to the use of ambient atmospheric air. "Special" considerations within a burner design that are particular to one or more of the alternative sources are periodically noted.

The temperature versus %O_2 (gross) relationship with respect to sustainable combustion is particularly

Burner Design

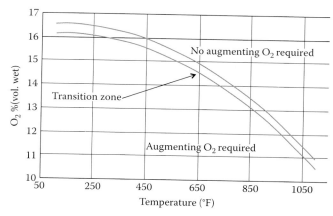

FIGURE 6.1
Graph of sustainable combustion for methane.

unforgiving. Figure 6.1 shows the relationship of natural gas combustion in a combustion O_2 stream varying in oxygen and temperature, as determined through testing at the John Zink research facility (see Chapter 8). Practical application does not allow for operation near the theoretical limit.

Practical writings on combustion often include discussions of the 3 Ts of combustion: time, temperature, and turbulence. The 3 Ts are simply a condensation of fluid flow and chemical reaction principles:

1. *Turbulence* is the interaction between two fluid streams required to achieve intermixing of the two.
2. *Temperature* is the required energy for the initiation of a chemical reaction—oxidation.
3. *Time* is the period for the reaction to reach completion.

In other words, mix the fuel with the air, and initiate the oxidation reaction in the mixture by heating to its ignition temperature. The result is a flame with a specific volume dictated by the time for the reaction to complete.

The major differences in burner designs come from the specific requirements addressed by that design. Mixing and reaction rates, or flame lengths and diameters, are directly related to the 3 Ts. They are also dependent on the fuels themselves. Variations on fuel composition, direction and distribution of the fuel in the air stream, and port velocity of the fuel can and will produce different results. The flame generated by a combustion system on one fuel will not necessarily duplicate the flame generated by that same system with another fuel. Their general flame shape, length, and diameter or width may be within acceptable tolerances; however, they will not be completely identical. The wider the variance in the properties of the fuels, the greater will be the deviation in flame properties.

6.3 Burner Design

Specialization of processes, and furnace designs to meet the demands of those processes, has resulted in the necessity of specialized burners. A burner is designed to provide stable operation and an acceptable flame pattern over a specific set of operating conditions. In addition, there may be a specified maximum level of pollutant emissions that can be generated through the combustion process. The American Petroleum Institute (API) gives some guidelines for burners used in fired heaters.[3] Specifications of operating conditions include

- Specific types of fuels
- Specific range of fuel compositions
- Maximum, normal, and minimum heat release rates for each fuel
- Maximum fuel pressures available for each fuel
- Maximum atomizing medium pressures available for liquid firing
- Fuel temperature for each fuel
- Oxidant source, either ambient air, heated ambient air, exhaust gases, or oxygen rich
- Available combustion air pressure loss, whether from a forced (positive) or induced (negative) system
- Combustion oxidant source temperature
- Furnace firebox temperature
- Furnace dimensions for flame size restrictions
- Type of flame (configuration or shape)

To provide acceptable operation, the burner must be designed to perform the 5 Ms:

1. *Meter* the fuel and air into the flame zone.
2. *Mix* the fuel and air to efficiently utilize the fuel.
3. *Maintain* a continuous ignition zone for stable operation over the range.
4. *Mold* the flame to provide the proper flame shape.
5. *Minimize* pollutant emissions.

6.3.1 Metering: Fuel

Typically, the furnace operating system is able to monitor only the total flow of fuel to a furnace. A typical process heater has multiple burners installed to provide the proper heat distribution. The fuel system must then be designed to ensure that the fuel is properly distributed to all burners. Uniform fuel pressures to each burner are critical to the proper operation of the burners. The burner designer then ensures that each burner takes the correct amount of fuel from the system. Controlling the proper amount of fuel flow is accomplished through a system of metering orifices designed for each burner. These ports are specifically designed to act as metering and limiting orifices, passing a specified and known amount of fuel at a given fuel pressure.

6.3.1.1 Gas Fuel

The system of burner tips and ports provided by the burner designer allows him to provide the operator with a capacity curve that specifies heat release versus pressure for a given fuel composition and temperature. For gaseous fuels, in compressible or incompressible flow, the calculations for the mass flow through a given orifice are dependent on

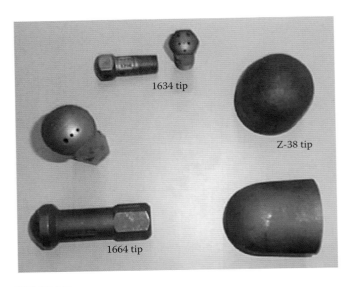

FIGURE 6.2
Typical raw gas burner tips.

- P_o, the fuel pressure immediately upstream of the orifice
- P_a, the downstream pressure (generally atmospheric pressure)
- T_o, the fuel temperature upstream of the orifice
- K, the fuel's ratio of specific heats, which is dependent on the composition of the fuel (this is a factor used in calculating the compressibility of the fuel gas)
- A, the area of the port(s)
- C_d, the discharge coefficient, which depends on the design of the orifice port(s)

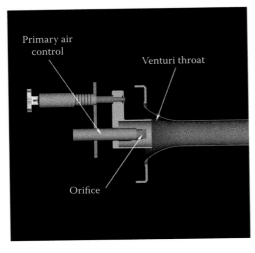

FIGURE 6.3
Typical premix metering orifice spud and air mixer assembly.

The fuel metering orifices for gaseous fuels in raw gas burners are typically installed at the point where the flame is formed. This is generally located in a region of the burner tile often termed the "burner throat." The fuel injector tips with the fuel ports can be centered in the burner throat or located on the periphery. Figure 6.2 shows typical raw gas burner tips.

For premix or partial premix burners, the metering orifice serves dual functions. It is generally a single port located at the entrance of a venturi eductor. The fuel gas discharging from the orifice is used to entrain air for combustion. A typical premix metering orifice spud and air mixer assembly is shown in Figure 6.3.

Capacity curves are generally presented in terms of heat release versus fuel pressure. The heat release is calculated from the mass (or volume) flow of the fuel, multiplied by the heating value per unit mass (or volume) of the fuel. Figure 6.4 presents a typical gas fuel capacity curve.

6.3.1.2 Liquid Fuel

Liquid fuels must be vaporized in order to burn. Burners designed for firing liquid fuels include an atomizer designed to produce a spray of small droplets, which enhances the vaporization of the fuel (see Volume 1, Chapter 10 for a discussion of oil atomization). The design of the atomization system will have a significant impact on the liquid fuel flow metering of the burner. With liquid fuels, the metering design is more complicated because of the need to "mix" the oil with an atomizing medium. This results in fuel metering orifices and atomizing media (typically steam or air) metering orifices in combination with orifices designed to flow the mixture. Figures 6.5 through 6.7 depict typical liquid fuel

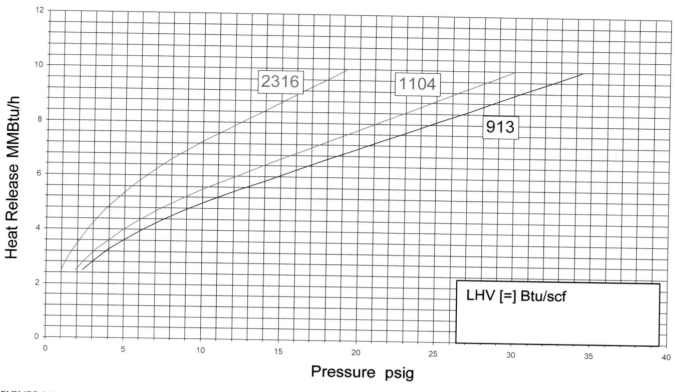

FIGURE 6.4
Typical gas fuel capacity curve.

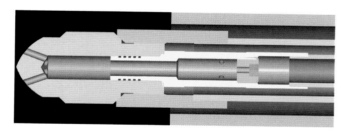

FIGURE 6.5
Internal mix twin fluid atomizer (EA style).

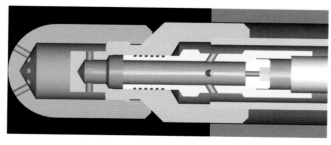

FIGURE 6.6
Internal mix twin fluid atomizer (MEA style).

atomizer/spray tip configurations for various liquid fuel systems. Based on the burner designer's knowledge of the oil metering and atomization system, the designer is capable of providing the operator with a capacity curve that specifies heat release versus fuel pressure for a given liquid fuel.

Many factors are important in the operation of a liquid fuel atomization system. Variations in any of a number of factors must be considered when designing the system of ports for the metering of flow for a liquid fuel, including the following:

- Temperature/viscosity relationship of the fuel
- Atomization medium temperature
- Fuel/atomizing medium pressure ratio
- Temperature/vaporization relationship of the fuel

Because capacity curves are generally presented in terms of heat release versus fuel pressure, the heat release is calculated from the mass (or volume) capacity of the system, multiplied by the heating value per unit mass (or volume) of the fuel. Figure 6.8 presents a typical liquid fuel capacity curve.

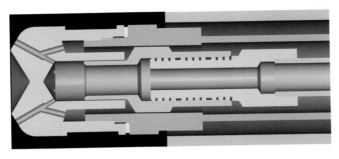

FIGURE 6.7
Port mix twin fluid atomizer (PM style).

6.3.2 Metering: Air (Combustion O_2)

The burner designer must ensure that the required air flow for the design operating conditions can be achieved with the air pressure loss that is available. The design of the burner throat is critical for two independent reasons. First, there is the requirement of achieving the proper flow of combustion O_2 to meet the demands of the fuel. Second, the burner throat is important in controlling the pattern of the flame.

In conventional burners, the primary air flow metering point and the point at which the bulk of the air pressure loss is utilized are designated as the "throat."

In most burners, especially raw gas or nozzle mix and liquid fuel burners, it is the point at which the fuel is first injected into the air stream. In a premix or partial premix burner, it is the point at which the fuel/air mixture is injected and, if necessary, secondary or completion air is introduced. In those cases, it is the location at which the initiation of combustion—or ignition—occurs. Figure 6.9 illustrates the throat of a typical raw gas burner. In some of the more recent low-emissions burners, air distribution is more critical than air turbulence. In those cases, the major air pressure loss may not occur at the same point as where the fuel is introduced or the flame is initiated.

6.3.2.1 Natural Draft

Many burners installed in refinery or petrochemical process heater service utilize the natural draft available in the furnace as the motive force for drawing combustion air through the burner (see Chapter 12). This negative pressure is typically less than 1 in. water column (2.5 mbar) at the burner location. This furnace internal negative pressure is dependent on the height of the radiant firebox, the temperature of the radiant firebox, and the level of draft (negative pressure)

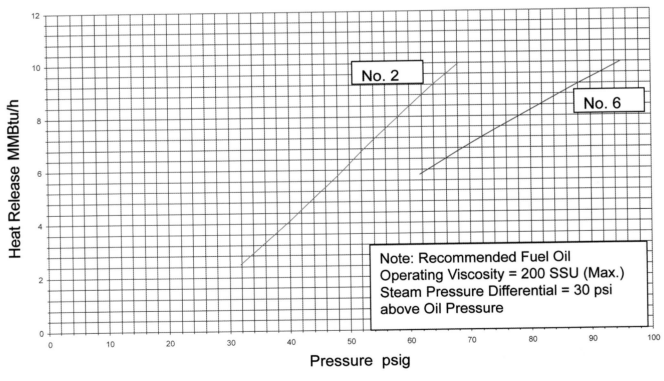

FIGURE 6.8
Typical liquid fuel capacity curve.

FIGURE 6.9
Typical throat of a raw gas burner.

maintained at the top of the radiant firebox. For burners required to operate under a natural-draft application, the burner throat is designed to pass the correct amount of combustion air required at the design heat release, utilizing only the available negative pressure provided by the furnace. In some cases, the natural draft can be enhanced by an induced draft fan; however, the available pressure still rarely exceeds 1 in. w.c. (2.5 mbar) of draft.

6.3.2.2 Forced Draft

Some burner applications provide the combustion O_2 stream at a positive pressure. These are generally applications either with preheated air, turbine exhaust gas, or another alternative oxidant source. Applications requiring smaller flame volumes can also utilize increased air pressure drop for enhanced mixing. If the application operates on forced draft at all times, the available pressure loss is often in the range of 4–10 in. w.c. (10–25 mbar). With this level of available pressure loss, the flame dimensions are considerably smaller due to the increased turbulent mixing in the flame. In addition, it is often possible to control the combustion O_2 stream flow over a wider heat release range. This enhanced mixing and improved control allows operation with low excess O_2 over a wider range of fuel input, resulting in improved combustion efficiencies across the heat release range.

In some forced-draft applications, it is specified that the burner must also operate in an ambient air, natural-draft operating mode. For such a dual-operational mode specification, it is most common that the natural-draft case is the limiting design. Burners designed for natural-draft and forced-draft operation seldom require forced-draft pressures greater than 2 in. w.c. (5 mbar). The exception to this is when the forced-draft operation uses the lowered oxygen concentrations of some of the alternative oxidant sources. For this design condition, the increased mass flow of the combustion O_2 source, with oxygen content lower than air, along with the increased temperature of that stream, generally increase the required burner pressure to levels greater than 2 in. w.c. (5 mbar).

6.3.3 Air Control

In conventional and most of the earlier low-emissions burners, a majority of the pressure loss is expected to be utilized within the burner throat. Industry groups such as the American Petroleum Institute (API),[3,4] process design companies, and some furnace manufacturers provide burner design guidelines that define the use of the available airside pressure drop at that point.

Depending on the level of accuracy desired, there are limitations to the practical range of control, especially when applied to burners designed for natural-draft applications. The percentage of available draft utilized for throat metering can leave little room for the design of an adequate control mechanism.

Example 6.1

Natural-draft burner
Refinery fuel gas: 1,200 Btu/scf, 0.70 sp.gr. (47 MJ/N m³, 11,000 kcal/N m³)
Maximum design duty: 10.0 MMBtu/h (2.9 MW, 2.5 MMkcal/h)
Normal duty: 8.0 MMBtu/h (2.3 MW, 2.0 MMkcal/h)
Minimum design duty (turndown): 2.0 MMBtu/h (0.59 MW, 0.50 MMkcal/h)
Draft: 0.5 in. w.c. (125 Pa, 1.25 mbar, 12.7 mm H_2O)
20% Excess air design, 10% excess air operation
Design to meet API 560 and 535

The burner shall be selected to use no less than 90% of the maximum draft available for the maximum specified heat release (API 560, Section 10.1.12).[4]

At design condition, minimum of 90% of the available draft with the air register fully open shall be utilized across the burner. In addition, a minimum of 75% of the airside pressure drop with the air registers full open shall be utilized across burner throat (API 560, Appendix A).[4]

Dampers and burner registers shall be sized such that the air rate can be controlled over a range of at least 40%–100% of burner capacity (API 535, Section 8.2).[3]

Assume that, through the efforts of the burner designer, the burner actually requires between the specified 90% and 100% of the available draft and that the specified 75% is measurable as static loss across the throat. Table 6.1 indicates the

- Operating heat release (column 1, Heat Release)
- Desired operational excess air (column 3, Percent X-Air)
- Required pressure loss across the throat (column 4, Throat Drop)
- Required pressure loss across the register or air control device (column 5, Control Drop)
- Ratio percentage of register or damper opening required (column 7, Percent Control Open)

Table 6.1 shows the theoretical percentage of air control opening based on no change in control flow coefficient and 3% leakage at the full closed position. Table 6.2 is based on 6% leakage. These tables assume that the variable orifice, the air control mechanism, has a constant discharge coefficient (see Volume 1, Chapter 9) throughout its full range of operation. This is a simplification and is not totally correct in practical application.

These charts (Tables 6.1 and 6.2) provide insight into another common fallacy in burner specification—the error in excess design capacity. Many who specify combustion equipment seek to provide excess capacity as a cushion against reduced operation with burners out of service, upset operation, process design contingencies, and future desired capacity. The fallacy is in the ability to control excess air and to utilize the available pressure drop for the normal operation. The air and fuel pressures are both reduced in the mixing zone and the resulting flame quality will suffer.

6.3.4 Mixing Fuel/Air

Mixing is a general term used to describe the function of bringing the fuel (reactant) into close molecular proximity with the oxygen source (oxidant). The higher the level of turbulence and shear between the streams, the more uniform the fuel and air mixture and the more rapid and complete the combustion reaction. It is generally accepted that the higher the level of mixing, the more intense and more complete and better the combustion. However, there are a number of conditions and factors that should be considered, including flame shaping and control of pollutant emissions (see Volume 1, Chapters 14 and 15) that lead to some exceptions to this rule. Intimate mixing does not always produce the most desirable results.

Designing a combustion system for special applications (e.g., high inert composition fuels) can also set

TABLE 6.1

Theoretical Air Control Opening, Based on No Change in Control Flow Coefficient and 3% Leakage at the Full Closed Position

Heat Release (MMBtu/h)	Percent Design (%)	Percent X-Air (%)	Throat Drop		Control Drop		Total dP (in. w.c.)	Percent Control Open	
			90% Total	100% Total	90% Total	100% Total		90% Total	100% Total
			(in. w.c.)		(in. w.c.)			% (w/3% Leakage)	
10.0	100	15	0.338	0.375	0.162	0.125	0.50	86.3	Full open
10.0	100	10	0.309	0.343	0.191	0.157	0.50	75.9	83.8
9.0	90	10	0.250	0.278	0.250	0.222	0.50	59.4	63.1
8.0	80	10	0.198	0.220	0.302	0.280	0.50	47.7	49.6
7.0	70	10	0.151	0.168	0.349	0.332	0.50	38.6	39.6
6.0	60	10	0.111	0.124	0.389	0.376	0.50	31.0	31.6
5.0	50	10	0.077	0.086	0.423	0.414	0.50	24.5	24.8
4.0	40	20	0.059	0.065	0.441	0.435	0.50	20.7	20.9
3.0	30	40	0.045	0.050	0.455	0.450	0.50	17.6	17.7
2.0	20	80	0.033	0.037	0.467	0.463	0.50	14.7	14.8
0	0	3	0.00031	0.00034	0.49969	0.49966	0.50	Full closed	Full closed

TABLE 6.2

Theoretical Air Control Opening, Based on No Change in Control Flow Coefficient and 6% Leakage at the Full Closed Position

Heat Release (MMBtu/h)	Percent Design (%)	Percent X-Air (%)	Throat Drop		Control Drop		Total dP (in. w.c.)	Percent Control Open	
			90% Total	100% Total	90% Total	100% Total		90% Total	100% Total
			(in. w.c.)		(in. w.c.)			% (w/6% Leakage)	
10.0	100	15	0.338	0.375	0.162	0.125	0.50	84.8	Full open
10.0	100	10	0.309	0.343	0.191	0.157	0.50	74.4	82.3
9.0	90	10	0.250	0.278	0.250	0.222	0.50	57.9	61.6
8.0	80	10	0.198	0.220	0.302	0.280	0.50	46.2	48.1
7.0	70	10	0.151	0.168	0.349	0.332	0.50	37.1	38.1
6.0	60	10	0.111	0.124	0.389	0.376	0.50	29.5	30.1
5.0	50	10	0.077	0.086	0.423	0.414	0.50	23.0	30.1
4.0	40	20	0.059	0.065	0.441	0.435	0.50	19.2	23.3
3.0	30	40	0.045	0.050	0.455	0.450	0.50	16.1	19.4
2.0	20	80	0.033	0.037	0.467	0.463	0.50	13.2	16.2
0	0	6	0.0012	0.0014	0.4988	0.4986	0.50	Full closed	Full closed

limitations on the desired level of mixing. All combustion reactions have a rate at which they will proceed and a minimum temperature that is required to initiate or sustain that reaction. The introduction of inert components to the fuel can generate two conditions that will affect burner stability. First, the inert components slow the reaction. This slowing of the flame speed can result in the stabilization point being translated out of the desired position. Second, the inerts introduce a heat absorption component that narrows the flammability limits and reduces the flame temperature. Quenching of the flame, to extinction, is an important consideration when inert components are present in the fuel or the combustion O_2 stream.

Most combustion O_2 streams with greater inert compositions than ambient air are the result of a combustion process. As such, they are typically available at an elevated temperature. This elevated thermal energy will often partially offset the potential quenching effect of the elevated inert levels. However, there are limits, as discussed in 6.1.

An effective form of emissions control is through the delaying of the combustion process. Highly mixed fuel and combustion O_2 in the proper proportions will generate the maximum flame temperature. This will result in the formation of a large quantity of nitrogen oxides (NOx), which are highly regulated pollutants (see Volume 1, Chapter 15). Certain burner designs will utilize a reduced level of fuel and air mixing, or a delayed mixing, extending the reaction zone to achieve reductions in these emissions.

In the case of a dual-reactant flame, there are only two sources from which to obtain the energy required for mixing. First is the relatively high-mass, low-velocity combustion O_2 stream and second is the low-mass, relatively high-velocity fuel stream. Each stream will contribute energy to the work required to mix in proportion to its mass and its velocity.

The intimate mixing of two or more dissimilar fluid streams under flowing conditions occurs in a turbulent shear zone defined by the intersection of the two streams. This region can be described as the dynamic interaction of each stream's mass and velocity, or momentum. This surface of intersection will vary in its turbulence in proportion to the magnitude of the shear forces developed.

Four basic mechanisms are available for the development of fuel and air mixing:

1. Entrainment
2. Co-flow mixing
3. Cross flow mixing
4. Flow streamline disruption or eddy formation

6.3.4.1 Entrainment

Entrainment is an effective demonstration of the law of conservation of momentum. One stream, usually the fuel, is utilized to draw the other lower velocity stream into its flow stream. This motive stream's energy and momentum must be conserved. As the velocity of the primary jet is dissipated, the mass of that stream must increase by the entrainment of the secondary stream to satisfy the conservation of momentum.

A well-designed premix burner primarily utilizes entrainment as its fuel/air mixing mechanism. By intimately mixing fuel and oxidant prior to the distribution tip, the highest possible uniform mixing of the two streams is achieved.

6.3.4.2 Co-Flow

When the streams are effectively parallel in directional flow, the amount of shear force developed is proportional to only the differential mass velocity of the streams. At high-velocity differentials, the interface of these streams becomes a turbulent region in which the components of the streams become intermixed. However, in low-velocity differentials, as in all transitional or laminar boundary layers, the relative thickness of this interface is small in comparison with the total cross section. Co-flow mixing with low-velocity differential conditions is—by its dynamic configuration—slow mixing.

6.3.4.3 Cross Flow

The shear energy generated between two flowing streams is greater anytime the streams are intersecting. The work required to redirect the combined flow of the streams results in turbulent intermixing of those streams. The included angle of intersection of the streams, the relative differential velocity of the streams, and the mass densities of the streams are all factors in the resulting direction and the turbulence generated for the mixing of the streams.

Included angles of intersection closer to perpendicular result in higher levels of mixing. However, quick mixing is not always the most important function being sought. Flame shape, stability, and emissions are all primary functions that can be affected by the rate and direction of combustion.

6.3.4.4 Flow Stream Disruption

Additional turbulence can be developed through the disruption of the "path" of a flowing fluid. Strategically located obstructions (e.g., bluff bodies) and sudden expansions (e.g., tile ledges) provide forced changes in flow streams, generating turbulence. If these disruptions are located in a region where both fuel and air streams are present, mixing will occur.

6.3.5 Maintain (Ignition)

The most important function that a burner performs is to provide for the continuous and reliable ignition of the fuel and air passing through the burner over a specified range of operating conditions. Each burner is designed to provide a specific location in which a portion of the fuel and air is continuously introduced in near-stoichiometric proportions at velocities at or below the mixture flame speed, thereby allowing for a continuous "ignition zone." At this location, a continuous flame is maintained and ignites the fresh fuel/air mixture as it is introduced. The ignition zone is designed to operate over the specified range of operation of the burner. The flame from this ignition zone is then used to ignite the remainder of the fuel and air mixture.

In natural-draft burners, the ignition zone is often situated in the eddy formed downstream of a step or ledge in the burner tile (see Figure 6.10) or in the wake of a bluff-body "flame stabilizer or flame holder" (see Figure 6.11) located in the center of the combustion air stream. Forced-draft burners can utilize similar techniques or other aerodynamic methods, such as a swirler (see Figure 6.12), to generate reverse flow zones that recirculate hot combustion products that assist in providing continuous ignition zones.

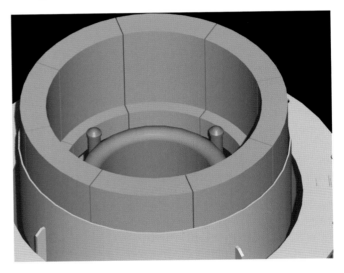

FIGURE 6.10
Ledge in the burner tile.

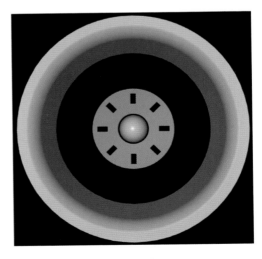

FIGURE 6.11
Flame stabilizer or flame holder.

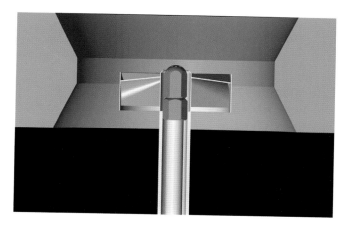

FIGURE 6.12
Swirler.

Safe ignition of the fuel/air mixture depends on the fuel and combustion O_2 source compositions. These factors are primary in affecting the flammability limits, ignition temperatures, and required stoichiometry of the mixture. The rate of reaction (flame speed) is controlled by the stoichiometry of the mixture and combustion characteristics of the fuel. The ability to sustain a single point of ignition for a fuel/air mixture is controlled by both the stoichiometry and the velocity of the mixed stream. Wide ranges in capacity, excess air requirements, and fuel compositions make strict dependence on the mixed stream for stability unreliable. The development of low-velocity zones and recirculating eddies ensures a single point for the kindling of the flame over a broad range of operating conditions. Each burner is designed for a limited range of fuel (and "air") compositions and rates, and one should not attempt to operate the burner outside this range.

6.3.6 Mold (Patterned and Controlled Flame Shape)

The design of combustion equipment, in the form of burners, for the hydrocarbon processing industry (HPI), the chemical processing industry (CPI), and to a major extent the power generation industry (PGI) has critical restrictions on the shape, size, and consistency of the flame generated. In fact, all burners providing energy to a process will be restricted in flame qualities by the design of the chamber, or furnace, into which it is fired.

The furnace chamber and burner flame in every process furnace are each designed to provide for the efficient transfer of heat to the process load. Flame size and shape for the HPI is especially critical due to the sensitivity to overheating of the hydrocarbons being processed. The rate of heat transfer to the process tubes must be limited to prevent overheating of the process tubes leading to the formation of carbon or coke inside the process tubes. As a result, there are generally strict guidelines for the flame dimensions. Typical specifications for flames include maximum flame lengths and widths. The number, heat release, and layout of the burners in the furnace are designed to provide the proper heat transfer pattern.

Patterning of the air flow through approach distribution, tile throat sizing and shape, and the tile exit configuration provides the most reliable method for flame pattern control. Introduction of the fuel into the established air flow streams provides the primary function in a raw gas burner. The proper flame pattern is generated by the combination of fuel injection pattern provided by the fuel injectors and the burner tile and flame holder which controls the air flow. The fuel injectors are also often called *spuds* or tips. The injectors have fuel injection ports that introduce the main portion of the fuel into the air stream in a manner that generates the desired flame pattern or shape. In conjunction, the air stream must be shaped in an appropriate manner by the air flow passages provided by the shape of the tile and flame holder. In many cases, the flame shape is round or brush shaped and acceptable in length and diameter (see Figure 6.13).

FIGURE 6.13
Round-shaped flame.

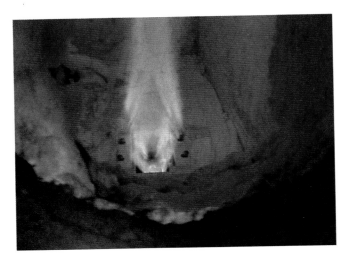

FIGURE 6.14
Flat-shaped flame.

In this case, the burner tile is typically round, and the fuel is injected symmetrically. Some furnaces require a fan-shaped flame, often termed a flat flame. In that case, the burner tile is generally rectangular, and the fuel is injected in a manner producing a flame that is essentially rectangular or "flat" rather than round (see Figure 6.14).

6.3.7 Minimize (Pollutants)

Most societies have come to the point at which the environment is of foremost concern. Therefore, the governments of most nations and localities are very critical of any source that puts certain undesirable materials into the air, soil, or water. Environmental regulations limiting air pollution have direct impact on the design of combustion equipment.

The challenge presented to burner design by these restrictions comes from the thermochemical reactions that form the regulated emissions. Emission control issues are discussed in Volume 1, Chapters 14 and 15.

6.4 Burner Types

Burners are typically classified based on the type of fuel being burned. A subdivision in burner type often includes the method of combustion O_2 supply. Therefore, there can be as many as eight basic design criteria:

1. Gas—premix and partial premix, natural-draft and/or low-pressure drop air
2. Gas—raw gas or nozzle mix, natural-draft and/or low-pressure drop air
3. Gas—raw gas or forced nozzle mix, forced-draft high-pressure drop air
4. Liquid—natural-draft and/or low-pressure drop air
5. Liquid—forced-draft high-pressure drop air
6. Solid fuel—forced-draft high-pressure drop air (typically)
7. Combination gas and liquid—raw gas and oil (typically), natural-draft and/or low-pressure drop air
8. Combination gas and liquid—raw gas and oil (typically), forced-draft high-pressure drop air

The final two designs listed are simply extensions of other designs. The combination of any two or more burners is simply a matter of basing the design on the most difficult of the fuel types and adapting the other fuel distribution systems to the base design.

6.4.1 Premix and Partial Premix Gas

Premix is a term applied to burners that educts part or all of the total air required for combustion. This type of burner provides for intimate mixing of the fuel and combustion O_2 prior to the ignition zone.

In a premix burner, the motive energy is supplied by the low-mass, high-potential-energy fuel. The fuel gas is metered through one or more orifices at the entrance to a venturi or mixer. The entrained air stream is made available at zero, or virtually zero, velocity at the same location. The conversion of the potential energy, pressure of the fuel stream, to kinetic energy, jet velocity, is achieved within the zone of air supply. The "free jet" of the fuel immediately begins to expand and decelerate. Conservation of momentum requires that the reduction in velocity be balanced by an increase in mass of the moving stream. This additional mass is entrained air.

Burners designed with a venturi typically require fewer adjustments to the air control. The utilization of the mass and velocity of the fuel results in that portion of the air educted being proportional to the gas flow. Therefore, with the reduction in fuel mass flow, there is a resulting reduction in the entrained air. The efficiency of the venturi and the restriction imposed by the fuel/air distribution nozzle are the limits to the capacity (volume) of air that can be educted.

Another benefit to this design is in the volume of the flame generated. If the majority of the air is initially intimately mixed with the fuel, the resulting flame volume of this premixed burner will be much smaller than that of any other low-air pressure drop burner. Conversely, if only a small portion of the combustion air is premixed, the flame will actually be larger than a raw gas burner. This is a result of the reduced secondary air mixing energies. By utilizing the majority of the available energy of the fuel stream to achieve the primary premix, the remaining energies for any required secondary air mixing are reduced to only that available from the air due to the draft loss.

Further benefits provided by this style of burner lie in the fuel metering configuration. Because all of the fuel is metered through a single or minimal number of orifices, the size of those orifices will be maximized for the conditions. Larger-diameter orifices, as long as they do not jeopardize the function of the burner, are a benefit because they minimize the chances of plugging from dirty fuels (see Chapter 13).

One of the basic limitations to this type of burner is founded in the burning characteristics of fuels. Each fuel chemical compound has its "rate of reaction" with oxygen. This rate is dependent on concentration levels of the fuel, the oxygen, and any inert components. This rate is also dependent on the temperature of the mixture. Another way to describe this rate is in terms of "flame speed" or the velocity at which the flame will propagate. The design of the distribution tip/system into the ignition and combustion zones is dependent on this same burning characteristic of the design fuel. Changes in the firing rate or the fuel will result in a change in the volume, velocity, and burning characteristics at the distribution tip. If the fuel/air mixture is within combustible limits, and the velocity is not maintained above the flame propagation speed, the result is a translation of the flame front back along the fuel/air flow streams. In the worst case, this flashback condition may result in flames being translated back through the premix tip distribution ports to inside the premix burner tip, causing thermal damage to metallic mechanical and structural parts.

6.4.2 Raw Gas or Nozzle Mix

Raw gas (nozzle mix) burners are designed to introduce the fuel and air separately into the combustion zone. They provide all of the mixing of the fuel and combustion O_2 at or after the ignition point of the burner. Typically, these burners provide for the major, or metering, pressure drop for both the fuel and air immediately prior to the ignition zone.

With all of the fuel maintained as segregated from the combustion O_2 until mixed in the ignition zone, there is no possibility of flashback. Therefore, the raw gas style of design can effectively handle a wide range of fuels without concern for equipment damage or personnel safety. Turndown on a raw gas burner is limited only by the method of stabilization, the flammability limits of the fuel, and the controllable and safety lower pressure limits of the fuel system.

With all of the air being supplied through and metered by the throat of the burner, it becomes necessary to adjust the air flow for efficient operation. Independent of whether the burner is designed for natural draft, low forced draft, high induced draft, or high forced draft, the ability to efficiently combust the fuel and transfer the heat is directly related to the amount of excess air. Therefore, it is important to control the air with every change in fuel flow or fuel composition, or there will be a related change in efficiency.

Another problem intrinsic to the raw gas design condition comes from the fuel distribution system, especially in burners designed for natural draft and/or low-combustion O_2 pressure drop. Patterning of the flame, distribution of the fuel into the combustion air stream, and mixing of the fuel with the oxygen will typically require multiple points of fuel injection. This breaking up of the fuel flow into multiple metering orifices while maintaining high potential energy, pressure, can require small orifices. These small orifices are highly subject to fouling problems due to fuel quality and foreign material in the piping.

Natural draft on the low pressure drop, combustion O_2 side often requires a burner design utilizing quiescent zones for stabilization. These low-flow, low-pressure zones can be generated through the use of flow stream disruption. Cones or bluff bodies located in the throat of the burner are a common form of flow stream disruption. Flow disrupters or shields around fuel tips and ledges or sharp changes in tile profile are also common. In all cases, these mechanisms are basically designed to prevent a portion of the combustion O_2 from leaving the designed ignition zone prior to the introduction of fuel. They provide "pockets" of a continuously renewing combustible mixture at a location in which the velocity is lower than the flame propagation speed.

High combustion O_2 pressure drop designs have another available form of flame stabilization—vortex recirculation (see Volume 1, Chapter 9). Through the utilization of swirl mechanisms and tile profiles, the combustion O_2 stream can be forced into a vortex. The physical properties of a vortex provide a low-pressure zone on the interior of the rotation that actually generates a backflow to the point of origin. The shape and strength of the vortex is determined by the flow and pressure loss through the swirler, the flow around the perimeter of the swirler, and the profile of the surrounding air throat. By injecting fuel into the swirling O_2 stream, a portion of the fuel and oxygen is continuously recirculated to the centralized ignition zone.

6.4.3 Oil or Liquid Firing

Oil firing is more complicated than fuel gas firing because the oil must be "atomized" and vaporized before it can be properly burned. Atomization involves producing a relatively fine spray of droplets that will vaporize quickly (see Volume 1, Chapter 10). Industrial oil burners typically utilize twin fluid atomizers and employ steam as the atomizing medium. Compressed air, or even high-pressure fuel gas, can be utilized in some applications rather than steam.

Some *internal mix twin fluid atomizers* form a gas/liquid emulsion which then issues from the fuel tip. This type of atomizer is shown in Figures 6.5 and 6.6. The oil issues from a single port at the entry to the atomizer section. Steam is injected into the oil stream through multiple ports, forming the emulsion. This oil and steam emulsion then travels to the chamber between the atomizer section and the exit ports. Multiple exit ports are placed on the tip, which allows for enhanced mixing of the spray and the combustion air, flame shaping, and stabilization. Internal mix atomizers often utilize a steam pressure maintained at 10–30 psig (70–210 kPa) above the oil pressure over the operating range of the burner. The oil pressure may be 100–120 psig (700–800 kPa) at maximum firing rate.

Port mix twin fluid atomizers (see Figure 6.7) use the atomizing media to form a thin liquid annulus on the inner surface of the port. The high velocity of the atomizing media pushes the liquid along the length of the port shearing and stretches the liquid into a thin film. These atomizers generally operate with a constant steam pressure, typically 120–150 psig (800–1000 kPa) over the entire operating range of the burner. The oil pressure may be 100–150 psig (700–1000 kPa) at maximum firing rate.

In each case, internal mix or port mix, droplets are formed as the liquid forms a sheet as it exits the port, and this sheet then breaks up as it expands. The size of the droplets formed depends on the relative velocity of the liquid sheet and the surrounding media. It also depends on the viscosity, surface tension and density of the liquid, size of the port, momentum of the fuel, and ratio of fuel to atomizing media.

6.4.3.1 High-Viscosity Liquid Fuels

The bulk of the industrial liquid fuel (see Volume 1, Chapter 3) fired is high-viscosity or heavy fuel oil. Heavy oils include No. 6 fuel oil, bunker C oil, residual fuel oil, pitch, tar, and vacuum tower bottoms. Most heavy fuel oils are the residue from the oil refining process and are considered a low-value by-product. This viscous oil must be heated to be efficiently atomized. Typically, the best atomization requires the viscosity of the oil to be in the range of 100–250 SSU. This often requires heavy oil to be heated to the range of 200°F–250°F (90°C–120°C). However, a heavy pitch may require a temperature of 600°F (320°C) to achieve the proper viscosity for atomization.

Heavy fuel oils have high boiling points and are therefore difficult to vaporize. Many heavy oil burners have special refractory tiles that redirect the intense flame radiation back to the root of the spray to enhance vaporization rates and help stabilize the oil flame. Some other heavy oil burners utilize swirlers to promote internal recirculation of the hot combustion products back into the flame root to enhance heating of the spray (see Figure 6.15).

6.4.3.2 Low-Viscosity Liquids

Low-viscosity liquid fuels include light oils such as No. 2 fuel oil, naphtha, and by-product waste liquids containing a variety of hydrocarbon by-products such as alcohols. These liquids are also typically fired using twin fluid atomizers. However, care must be taken to avoid overheating low-boiling-point fuels. Vaporization of the liquid within the oil gun can cause disruption in flow within the atomizer and unsteady operation. Many light liquids require compressed air rather than steam atomization. Because light oils generally have lower boiling points, they are easier to vaporize, and their flames can often be stabilized using simple bluff-body stabilizers rather than swirlers or special tiles.

6.4.4 High Intensity (KEU Combustor)

The John Zink KEU combustor is a functional unit of burner and burner chamber for thermal oxidizers (TO) (see Volume 3, Chapter 9). The combustor is designed with three concentric stainless steel cylinders. At the combustor inlet, oxygen-rich waste air or ambient air is supplied through a tangential nozzle, generating an air rotation (vortex) (see Figure 6.16). This rotating air is forced through the outer and inner passages of the concentric cylinders and ensures intensive cooling of the burner chamber wall. The pressure drop across the burner is a function of the intensity of the vortex which will be adjusted during commissioning via the manual flap.

Waste gases, waste liquids, and the auxiliary fuel are injected via the so-called "organ set" in the burner center (see Figures 6.17 and 6.18). In the case of burning explosive waste gases, a flame flashback can be prevented through the addition of a specially proven safety device (dynamic flame arrester).[5–12]

The centrifugal forces of the rotating air create a low-pressure zone in the burner center, which results in an outer stream rotating forward with an inner stream

Burner Design

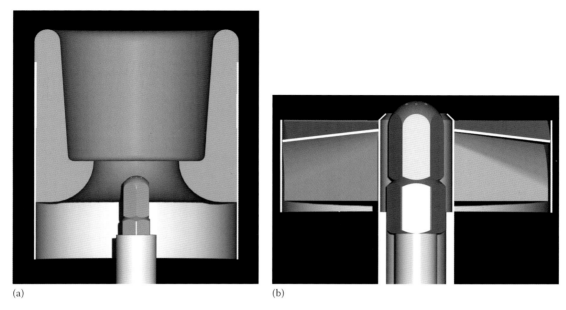

FIGURE 6.15
(a) Regen tile and (b) swirler.

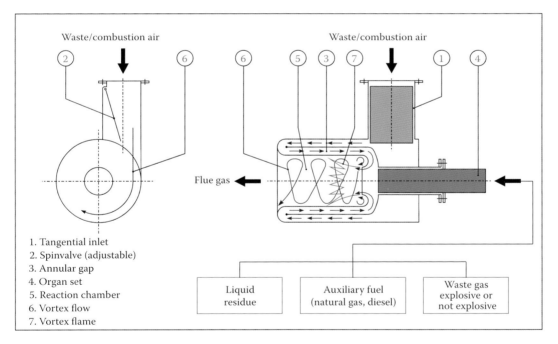

1. Tangential inlet
2. Spinvalve (adjustable)
3. Annular gap
4. Organ set
5. Reaction chamber
6. Vortex flow
7. Vortex flame

FIGURE 6.16
Combustor schematic.

flowing backward axially (see Figure 6.19). This pattern is called supercritical vortex flow.[13]

The combustor has no mechanical flame holders such as a cone or tile. The flame is stabilized by the backflow of active, energy-rich flue gas compounds (radicals) from the flame front back into the flame root. This creates a short rotating flame which fills the burner chamber (see Figure 6.20).

All reactants must pass the flame front. Bypassing is not possible. The intensive mixing of the reactants leads to a homogeneous temperature profile across the reaction chamber and ensures an excellent burnout result,

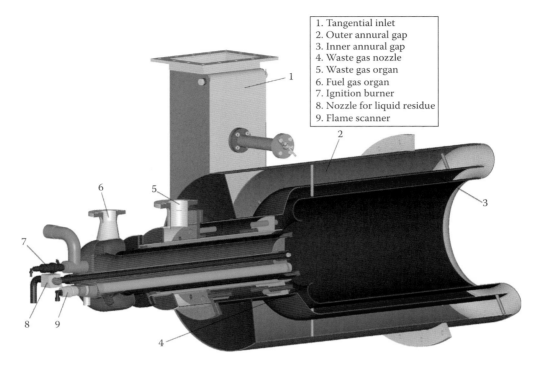

1. Tangential inlet
2. Outer annural gap
3. Inner annural gap
4. Waste gas nozzle
5. Waste gas organ
6. Fuel gas organ
7. Ignition burner
8. Nozzle for liquid residue
9. Flame scanner

FIGURE 6.17
View combustor with organ set.

FIGURE 6.18
Front view to the organ set of a combustor.

even at a low-reaction temperature. Prevention of hot flame zones diminishes the formation of NOx.

The standard combustor type has a maximum pressure drop of 12 in. w.c. (30 mbar) on the combustion air side with an air side turndown of three to one (3:1). If a higher air side turndown rate, up to five to one (5:1), is required, a forced-draft combustor in combination with a pressure control loop can be used without requiring a change in pressure drop. The turndown rate for the fuel is nine to one (9:1).

The combustor typically runs with excess air (above stoichiometric ratio) at temperatures between 1436°F and 2192°F (780°C and 1200°C). Depending on the waste composition and operation mode, the emission limits can be reached at a residence time between 0.6 and 1.0 s. The combustor series allows sizing with total flow rates between 370 and 33,600 scfm (600 and 54,000 N m³/h).

If waste liquid firing is required, double or multiple atomizing lance systems are used. A fine atomizing spray pattern is important for a good and efficient burnout. Since the waste liquids often contain particles, lances with outer mixing nozzles are the preferred solution. Depending on the waste liquids and their characteristics, compressed air or superheated steam is used as the atomization medium.

The John Zink KEU combustor (see Figure 6.21) is a proven burner system for multipurpose thermal oxidation in the chemical and pharmaceutical industries. For these applications, the combustor provides a high level of reliability and stable combustion during quick changes in flows, oxygen content, and input of waste. In the case of burning halogenated hydrocarbons, the highly efficient combustion of the combustor avoids the formation of dioxins.[14]

These benefits are the reason why the European Commission listed the combustor system as BAT (best available technique) for the manufacture of organic fine chemicals.[14]

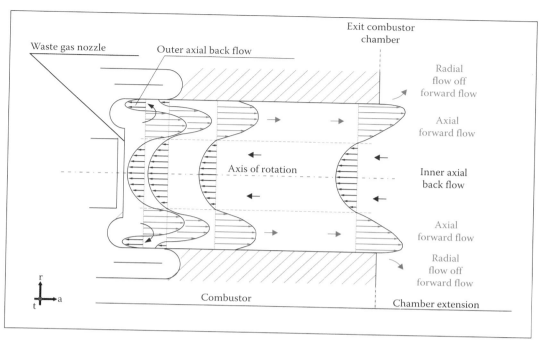

FIGURE 6.19
Axial velocity within combustion chamber (cold air flow model).

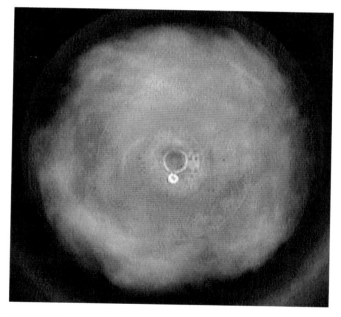

FIGURE 6.20
Combustor flame.

FIGURE 6.21
Standard combustor normal—FD combustor short.

6.4.5 Conventional Process Heater Application High Intensity

High-intensity, high-air pressure drop combustor-type burners are available for process heaters. In the conventional design of these burners, the combustion chamber is not constructed of concentric high alloy metal shells. The most common construction utilizes high-temperature refractory. The typical combustor is designed for the fuel and excess air specified for operation. The majority of the air pressure drop is utilized to develop the swirling air pattern and the stabilizing air vortex, similar to the KEU design. However, the swirl is typically imparted directly prior to the burner throat. The fuel(s) are introduced into the focus of the vortex, allowing the central recirculation zone to provide the burner stability.

6.5 Configuration (Mounting and Direction of Firing)

Burners can be mounted in the furnace or heater floor to fire vertically upward, in the heater wall to fire horizontally, or in the roof to fire vertically downward. The major consideration required for burner design is to ensure proper support for the burner tile. Burner blocks for floor-mounted service can typically be simply placed on the furnace steel or burner mounting plate. Burner blocks for horizontal mounting must be supported by tile case assemblies and must be held in place so that they do not move if subjected to vibration. Roof-mounted tiles must have support surfaces cast into them so that they can hang from steel supports in the furnace roof.

6.5.1 Conventional Burner, Round Flame

The following figures illustrate the different types of burners that are typically used in refinery and petrochemical furnaces. The round flame burner is the most universal design and used in many applications. Figure 6.22 illustrates a typical raw gas conventional burner. This burner is used where NOx emissions are not a primary concern and a short flame is desired.

Figure 6.23 shows a typical premix gas burner. A premix round flame burner is useful when a short heater does not have enough draft to supply the required combustion air. The premix burner uses the fuel jet as a motive force to allow the burner to pull in part or all of its combustion air.

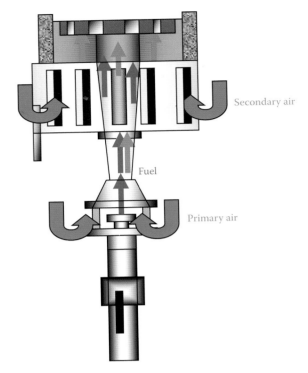

FIGURE 6.23
Typical premix gas burner.

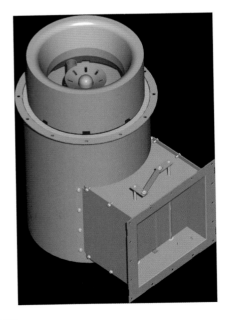

FIGURE 6.22
Typical conventional raw gas burner.

Figure 6.24 shows a typical round/conical flame combination oil and gas burner. This burner can be used to burn gas or liquid fuels. This versatility is desirable to applications with liquid fuels or where gas fuels may be in short supply at various times throughout the year and an alternative fuel must be fired to maintain a process.

Figure 6.25 shows a typical round/conical flame high-intensity combination oil and gas burner. This burner is used in applications where a high-heat release per burner is required but a short flame length is required.

6.5.2 Flat Flame Burner

Some applications require a flat- or fan-shaped flame due to the close proximity of process tubes or due to the fact that the burner is fired along a wall or across the floor. Figure 6.26 shows a typical staged-fuel flat flame gas burner. This burner produces a freestanding flame and is used in applications where the process tubes are close to the centerline of the burner.

The advantage in using this particular burner is that, because of fuel staging, the NOx emissions are significantly reduced over that of a non-staged burner. Based on fired capacity, one can utilize single or multiple primary and staged injectors to effect the proper fuel distribution to produce the desired flame pattern. A flat flame burner can be fired in two basic configurations—freestanding or wall adjacent (or across the floor) fired.

Burner Design

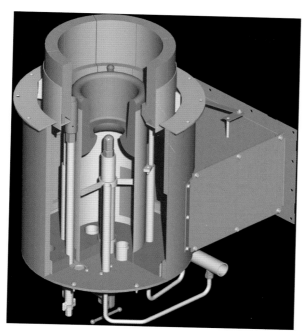

FIGURE 6.24
Typical round flame combination burner.

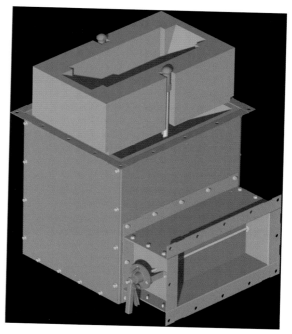

FIGURE 6.26
Typical staged-fuel flat flame burner.

6.5.2.1 Wall Fired

A wall or floor-fired flat flame burner is installed and fired against a refractory wall or floor. By directing the fuel jets toward the refractory surface, the flame heats the refractory, which in turn radiates heat to the process tubes. A typical example of wall-fired flat flame burners is the "double fired coil design" used in ethylene furnaces and some coker furnaces. Other furnace designs call for flat flame burners firing on either side of a center wall or firing across the floor to a center dividing wall.

6.5.2.2 Freestanding

A freestanding flat flame burner is used in applications where it is necessary to fire a burner between two sets of process tubes. Close spacing between the two sets of process tubes often requires the use of a flat flame burner. A staged-fuel, flat flame is shaped by firing staged fuel jets opposite one another into an elongated air flow pattern. This makes the flame shape into a fan. The flame thickness is typically less than or equal to the burner tile width.

6.5.3 Radiant Wall

Radiant wall burners are used in some cracking furnaces and hydrogen reformers (see Volume 1, Chapter 2 and Volume 3, Chapter 6). These burners are mounted through the furnace wall and produce a flat circular disk of flame adjacent to the wall with minimal projection into the furnace. There are typically several rows

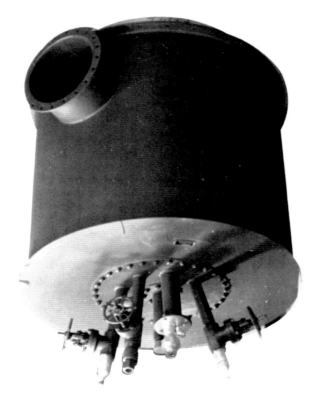

FIGURE 6.25
Typical round flame, high-intensity combination burner.

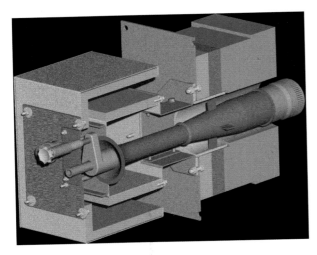

FIGURE 6.27
Typical radiant wall burner.

of burners, with burners spaced in a grid pattern on the walls. The burners uniformly heat the walls, which radiate heat to the process tubes on the centerline of the furnace. Figure 6.27 shows a typical radiant wall burner.

Many applications, such as reformers and cracking furnaces, require high furnace temperatures in the radiant section of the heater to produce the required high-process temperatures. Multiple, small capacity burners distributed across a radiation wall provide the very uniform heat transfer to the process tubes that is needed. Additionally, in such applications, a floor-mounted flat flame burner that fires adjacent to the refractory wall can be utilized in conjunction with radiant wall burners.

6.5.4 Downfired

A downfired burner is a burner that is fired vertically into a furnace from the ceiling (see Figure 6.28). Both round pencil-shaped flames and freestanding flat flame burners have been applied to this style of furnace. The most common applications produce a flame that is round/conical in shape. These burners are utilized in applications such as hydrogen reformers or ammonia production. Downfired burners are often dual-fuel burners firing a process off-gas and a makeup fuel. In the case of hydrogen reforming, the makeup fuel is typically the same as the process feed. This fuel can be natural gas, heavy or light naphtha, gas oil, diesel oil, or a propane/butane gas fuel.

6.6 Materials Selection

Burner materials are selected for strength, temperature resistance, and in many cases corrosion resistance. Carbon steel is used for most metal parts unless

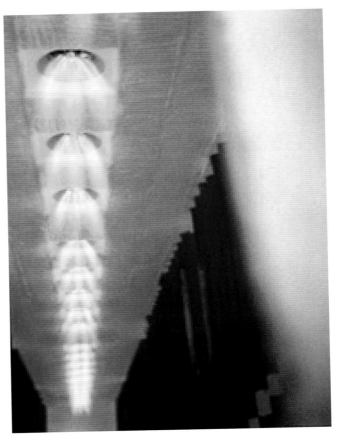

FIGURE 6.28
Downfired burners in a hydrogen reformer furnace.

temperature or corrosion considerations require more resistant alloys. See Chapter 4 for a general discussion on metallurgy including some specific discussions on the use of various metals in burners.

Cast iron or carbon steel can be utilized for fuel gas manifolds. The fuel gas riser material between the manifold and the fuel injector tip is generally carbon steel for ambient air service and 304 SS for parts in contact with air temperatures higher than 700°F (370°C). If H_2S is present in the fuel gas and preheated air over 300°F (150°C) is used, 316L stainless steel may be required for the gas piping passing through the windbox.

Fuel injector tips for premix burners may be cast iron. Cast steel or cast stainless steel may be required for fuel gases containing appreciable levels of hydrogen (typically >50 vol%).

Raw gas burners typically use 300 series stainless steel gas tips. Tips for oil burners are normally 416 SS. For oils containing erosive particles, hardened tungsten containing tool steel is generally used. Atomizers for oil service are normally brass or 303 SS for oils containing sulfur. Other oil injector parts are carbon steel, although nitride-hardened parts can be used for erosive liquids.

Flame stabilizer cones and swirlers are normally 300 series stainless steel.

Mineral wool is commonly used for noise reduction in plenums and as insulation in preheated air service up to 1000°F (540°C). Ceramic fiber insulation is used for higher temperature applications.

High-strength, low-alloy structural steel (ASTM A 242/A 242M) or 304 SS is used for other metal parts subject to >700°F (370°C) air preheat.

Burner block material is typically 55%–60% alumina refractory with a 3000°F (1650°C) service temperature. Primary oil tiles may require 90% alumina refractory if the combined vanadium and sodium in the oil is greater than 50 ppm by weight.

References

1. C.E. Baukal (Ed.), *Industrial Burners Handbook*, CRC Press, Boca Raton, FL, 2004.
2. C.E. Baukal (Ed.), *Oxygen-Enhanced Combustion*, CRC Press, Boca Raton, FL, 1998.
3. American Petroleum Institute, Burners for Fired Heaters in General Refinery Services, API 535, 1st edn., Washington, DC, July 1995.
4. American Petroleum Institute, Fired Heaters for General Refinery Services, API 560, Washington, DC, November 1996.
5. IBExU Institut für Sicherheitstechnik GmbH, *Prüfbericht IB-03-3-776/1. 2004. Über die Prüfung der strömungsüberwachten Einrichtung KEU EX-Organ IIB3 DN als Flammendurchschlagsicherung für brennbare Flüssigkeiten und Gase der Explosionsgruppe IIB3 mit einer Normspaltweite von ≥0,65 mm*, IBExU, Freiberg, Germany, 2004.
6. IBExU Institut für Sicherheitstechnik GmbH, *Prüfbericht IB-03-3-776/2. 2004. Über die Prüfung der strömungsüberwachten Einrichtung KEU EX-Organ IIB3 DN als Flammendurchschlagsicherung für brennbare Flüssigkeiten und Gase der Explosionsgruppe IIC mit einer Normspaltweite von < 0,50 mm*, IBExU, Freiberg, Germany, 2004.
7. IBExU Institut für Sicherheitstechnik GmbH, *Prüfbericht IB-03-3-776/3. 2004. Über die Prüfung der strömungsüberwachten Einrichtung KEU GPR-Organ IIB3 DN als Flammendurchschlagsicherung für brennbare Flüssigkeiten und Gase der Explosionsgruppe IIB3 mit einer Normspaltweite von ≥0,65 mm*, IBExU, Freiberg, Germany, 2004.
8. IBExU Institut für Sicherheitstechnik GmbH, *Prüfbericht IB-03-3-776/4. 2004. Über die Prüfung der strömungsüberwachten Einrichtung KEU GPR-Organ IIB3 DN als Flammendurchschlagsicherung für brennbare Flüssigkeiten und Gase der Explosionsgruppe IIB3 mit einer Normspaltweite von <0,50 mm*, IBExU, Freiberg, Germany, 2004.
9. IBExU Institut für Sicherheitstechnik GmbH, *EG-Baumusterprüfbescheinigung gemäß Richtlinie 94/9/EG, Anhang III.2004, Nummer: IBxU04ATEX2227 X*, IBExU, Freiberg, Germany, 2004.
10. IBExU Institut für Sicherheitstechnik GmbH, *EG-Baumusterprüfbescheinigung gemäß Richtlinie 94/9/EG, Anhang III.2004, Nummer: IBxU04ATEX2228 X*, IBExU, Freiberg, Germany, 2004.
11. IBExU Institut für Sicherheitstechnik GmbH, *EG-Baumusterprüfbescheinigung gemäß Richtlinie 94/9/EG, Anhang III.2004, Nummer: IBxU04ATEX2229 X*, IBExU, Freiberg, Germany, 2004.
12. IBExU Institut für Sicherheitstechnik GmbH, *EG-Baumusterprüfbescheinigung gemäß Richtlinie 94/9/EG, Anhang III.2004, Nummer: IBxU04ATEX2230 X*, IBExU, Freiberg, Germany, 2004.
13. O. Carlowitz and R. Jeschar, *Entwicklung eines variablen Drallbrennkammersystems zur Erzeugung hoher Energieumsetzungsdichten*, Brennstoff Wärme Kraft, 32, Nr. 11, 1980.
14. European Commission, Integrated pollution prevention and control. reference document on best available techniques for the manufacture of organic fine chemicals, Chapter 4, pp. 240, 2006.

7
Combustion Diagnostics

Wes Bussman, I.-Ping Chung, and Jaime A. Erazo, Jr.

CONTENTS
7.1 Pressure Management
    7.1.1 Manometer 173
    7.1.2 Bourdon Tube Gauge 173
        7.1.2.1 Design of the Bourdon Tube Gauge 174
        7.1.2.2 Common Failure Mechanisms 174
        7.1.2.3 Calibration of Pressure Gauges 175
        7.1.2.4 Selection 176
        7.1.2.5 Installation 176
7.2 Flow Measurement 176
    7.2.1 Orifice Meter 177
        7.2.1.1 Description 177
        7.2.1.2 Upstream Flow Conditioners 177
        7.2.1.3 Calculating the Mass Flow Rate 178
        7.2.1.4 Accuracy of Flow Measurements 178
    7.2.2 Venturi Meter 179
    7.2.3 Turbine Flow Meter 179
    7.2.4 Vortex Flow Meter 180
    7.2.5 Magnetic Flow Meter 180
    7.2.6 Ultrasonic Flow Meter 181
    7.2.7 Thermal Mass Meter 182
    7.2.8 Positive Displacement Meter 182
    7.2.9 Pitot Tube 183
    7.2.10 Averaging Pitot Tube 184
7.3 Advanced Diagnostics 185
    7.3.1 Fourier Transform Infrared Spectroscopy 185
    7.3.2 Phase Doppler Particulate Anemometer 186
    7.3.3 Liquid Planar Laser-Induced Fluorescence 186
References 188
.... 189

7.1 Pressure Management

7.1.1 Manometer

One of the simplest and most useful pressure measuring devices is the manometer. There are several variations of the manometer; however, in this section, two types will be discussed: U-tube and inclined manometers. In its most basic structure, the manometer takes the form of a U-shaped tube which is partly filled with some liquid, such as water, oil, or mercury, as shown in Figure 7.1.

The differential pressure in a U-tube manometer can be expressed as

$$P_d = \gamma h \quad (7.1)$$

where
    P_d is the differential pressure (lb_f/ft^2)
    γ is the specific weight of the fluid (lb_f/ft^3)
    h is the height of the liquid (ft)

To demonstrate how Equation 7.1 is used, consider the following example.

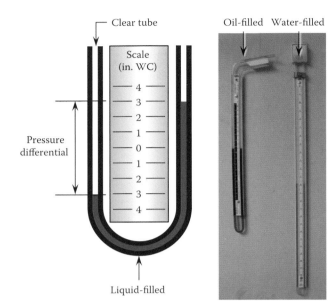

FIGURE 7.1
U-tube manometer.

Example 7.1

A U-tube manometer, filled with mercury, is connected to a pressurized vessel. The difference in the height of mercury reads 60 in. (a) What is the pressure in units of pounds per square inch (psi)? (b) What would the difference in height be if the manometer was filled with water?

The specific weight of water and mercury is 62.4 lb_f/ft^3 and 847 lb_f/ft^3, respectively.

(a) $P_d = \gamma h$

$$= 847 \frac{lb_f}{ft^3} \times 60 \text{ in.} \frac{1 \text{ ft}}{12 \text{ in.}} \times \frac{1 \text{ ft}^2}{144 \text{ in.}^2} = 29.4 \text{ psi}$$

(b) $\dfrac{h_{mercury}}{h_{water}} = \dfrac{P_d/\gamma_{mercury}}{P_d/\gamma_{water}}$

$$h_{water} = h_{mercury}\left(\frac{\gamma_{mercury}}{\gamma_{water}}\right)$$

$$= 60 \text{ in.} \times \frac{847}{64.2} = 814.4 \text{ in.}$$

This example illustrates that mercury-filled U-tube manometers are reasonable to use in pressure ranges as high as 30 psig; at this pressure, a 5 ft (60 in.) manometer is required. A water-filled manometer, however, is not reasonable at this pressure because it would require a U-tube manometer over 68 ft (21 m) long.

7.1.2 Bourdon Tube Gauge

In 1849, Eugene Bourdon, a French watchmaker and engineer, patented an instrument used to measure pressure. This instrument is called the Bourdon tube gauge and is one of the most common pressure measurement instruments used in industrial plants. The purpose of this section is to discuss the design, failure, and calibration of these gauges.

7.1.2.1 Design of the Bourdon Tube Gauge

Bourdon tube gauges are available in a variety of sizes (see Figure 7.2). The diameter of the dial faces can vary from less than 1 in. (2.5 cm) to as large as 3 ft (1 m), with the largest typically used for extreme accuracy.[1] Bourdon tube gauges are also available in a variety of pressure ranges: from 5 psig (140 kPa) full scale to as much as 10,000 psig (69,000 kPa) full scale.[2]

The Bourdon gauge consists of a flattened, thin-walled metal tube bent into the form of an arc or "C" shape as shown in Figures 7.3 and 7.4.[3] One end of the tube is fixed, securely sealed, and bonded to a threaded connection called a "socket." The other end is tightly sealed and free to move. At atmospheric pressure (zero gauge pressure), the tube is undeflected, and for this condition, the gauge pointer is calibrated to read zero pressure.[4] However, when pressure is applied into the fixed end of the tube, the effect of the forces tends to uncoil or straighten the tube causing the end to move in a slightly curved path. The end of the Bourdon tube does not move a great distance within its pressure range: typically 0.125–0.25 in. (0.31–0.63 cm).[1] Using a gear and linkage system (called the "movement"), the motion of the tube is magnified and converted into rotary movement of the indicating needle. The indicating needle rotates around a circular dial calibrated for pressure.[5]

The indicating needle is commonly fixed to a spiral spring which tightens up as more pressure is applied. The purpose of the spiral spring is to assist the tube and

FIGURE 7.2
The Bourdon pressure gauge.

Combustion Diagnostics

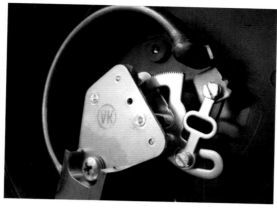

FIGURE 7.3
Internal view of the Bourdon pressure gauge.

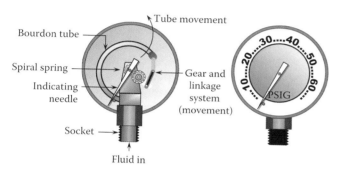

FIGURE 7.4
Basic components of a Bourdon pressure gauge. (Adapted from Hydraulic Energy Transmission and Control, Hill Learning Systems, 2001–2007.)

pointer to quickly respond to a drop in pressure after reading some higher value.

The most important part of the gauge is the Bourdon tube. Bourdon tubes are made of many materials: beryllium copper, phosphor bronze, and various alloys of steel and stainless steel.[6] Beryllium copper is typically used for high-pressure applications. Most gauges in air, light oil, or water applications use phosphor bronze. Stainless steel alloys usually add cost to the gauge if specific corrosion resistance is not required.

The socket is usually made of brass, steel, or stainless steel. The most common socket size is usually ¼ in. and ½ in. NPT tapered pipe threads. For dial sizes smaller than 4.5 in. (11 cm), a ¼ in. (6.4 mm) NPT is common; however, in some cases, a 1/8 in. (3.2 mm) NPT will be used. For pressure-gauge dial sizes smaller than 2 in. (5 cm), a 1/8 in. (3.2 mm) NPT is common.

The movement mechanism can be made from a variety of material such as nickel silver, brass, or stainless steel; brass is a very common material. It is important that the material used allows for a friction-free assembly. The movement mechanism is protected within an enclosure commonly made of brass, aluminum, steel, or plastic. Most lenses, protecting the indicating needle and dial, are made of plastic instead of glass for safety reasons due to possible breakage.

7.1.2.2 Common Failure Mechanisms

Since a typical plant will use many pressure gauges, they may not always receive the proper attention to maintenance. Three common reasons for gauge failure are harsh environments, vibrations, and water condensation.

Harsh environments such as corrosive and dirty fluids can cause a pressure gauge to fail. Bourdon tubes can be made from a variety of materials and are most often manufactured from 316 stainless steel, phosphor bronze, Monel, and Inconel.[7] The selection of material depends on the fluid for the application. If the application consists of a highly corrosive fluid, then a diaphragm seal can be used with the gauge; the seal prevents the fluid from coming in contact with the Bourdon tube. Diaphragm seals not only protect a pressure gauge from corrosion attack, but also prevent dirt and scale from clogging the tube. The diaphragm seal will typically reduce the accuracy of the gauge.

If a gauge is used in a vibrating environment, it can cause excessive motion of the movement mechanism causing it to wear and fail prematurely. Vibrating environments also make it difficult to read the pressure gauge accurately because the indicating needle oscillates. To help prevent this problem, pressure gauges are often filled with a dampening fluid such as glycerin or a silicon fluid. The fluid helps cushion the tube and movement mechanism against damage from sudden impact and vibration. Many industrial plants have adopted liquid-filled gauges as a standard.

Excessive motion of the movement mechanism can also occur if rapid pressure changes occur within a

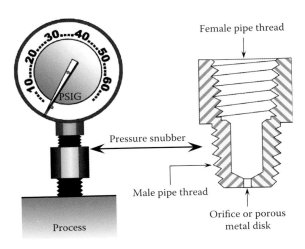

FIGURE 7.5
Pressure snubber. (Adapted from O'Keefe Controls Co., Trumbull, CT, 2003, http://www.okcc.com/PDF/Pressure%20Snubber%20pg.42.pdf)

process. Pressure snubbers are commonly used to protect pressure gauges from the system pressure shock and surges and can prolong the life of a gauge. The pressure snubber is placed in line between the gauge and the process as shown in Figure 7.5[8] and is typically designed with an orifice or a porous metal disk to help dampen pressure fluctuations.

7.1.2.3 Calibration of Pressure Gauges

To insure accuracy, it is customary to calibrate pressure gauges periodically. Depending on their use and importance to a given process, gauges might be calibrated monthly, quarterly, yearly, or every several years. Deadweight testers are the primary standard for accurate pressure calibration. A typical oil-type deadweight tester is shown in Figure 7.6.

FIGURE 7.6
An oil-type deadweight tester.

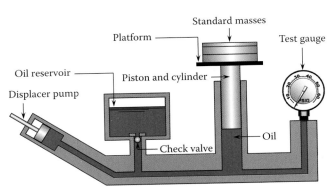

FIGURE 7.7
Basic components of a deadweight tester. (Adapted from Miller, R.W., *Flow Measurement Engineering Handbook*, McGraw-Hill, Inc., New York, 1989.)

The basic components of an oil-type deadweight tester are shown in Figure 7.7. These testers consist of an accurately honed piston of known cross-sectional area inserted into a cylinder. Known weights (standard masses) are placed on a freely rotating platform attached to the top of the piston; the platform is designed to freely rotate to reduce friction between the piston and cylinder. When the oil pump supplies enough pressure to raise the weights, the force exerted by the oil pressure over the piston area is balanced by the force of the weights.

When using a deadweight tester, it is important to apply corrections for local gravity, elevation, buoyancy, or thermal expansion of the piston, since these errors may be significant.[9] Other sources of error include weights that are dirty, corroded, or chipped or using a tester that is not properly leveled.[10]

7.1.2.4 Selection

When selecting a pressure gauge, it is recommended to use one that has a range of twice the normal working pressure. The maximum operating pressure, in all cases, should not exceed 75% of the maximum range of the gauge. If gauges are used in applications where pulsations are encountered, the pressure should be limited to 67% of the maximum range of the gauge.

The ambient temperature range for most standard dry or silicone-filled gauges is about −40°F to 140°F (−40°C to 60°C). The ambient temperature range for glycerin-filled gauges is 4°F to 140°F (−20°C to 60°C). As a general rule, the error caused by temperature change is ±0.3% per 18°F (10°C) rise or fall, respectively, based on the temperature of the gauge.

7.1.2.5 Installation

Bourdon gauges are commonly used to measure the static pressure of a fluid flowing in a pipe. To reduce the potential for error, it is important that static pressure taps are designed correctly. The following are several important design considerations for static pressure taps:

- The hole should be drilled perpendicular to the pipe.
- The hole should be free of burrs.
- The edge of the hole should be square with the inner pipe surface with no rounded corners or no concave or convex surface surrounding the hole.
- The drilling of the hole should be done after all fittings for attaching the pressure gauge have been welded to the pipe.

Ferron[11] states that countless static pressure measurements are in error because of poor pressure tap construction. Brunkalla[12] suggests a ¼ in. (6.4 mm) diameter hole for pipe sizes greater than 2 in. (5 cm). Departure from the recommended hole size, inclination, or edge condition results in errors of −0.5% to +1.1% of the velocity or dynamic pressure.[9]

FIGURE 7.8
Orifice plate.

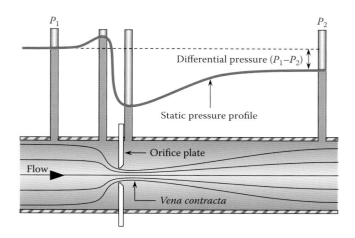

FIGURE 7.9
Illustration showing static pressure drop through an orifice metering run.

7.2 Flow Measurement

This section discusses three common types of differential-producing flow meters: the orifice meter, the venturi meter, and the nozzle meter.

7.2.1 Orifice Meter

7.2.1.1 Description

A differential-producing flow meter is a device used to create varying static pressure within a flow stream that can be used to determine the flow rate of the fluid. These devices have been used throughout history for over 1000 years.[9] Today, differential flow meters are the most common and reliable flow meters used in industry. In this section, two types of differential-producing flow meters are discussed: the orifice meter and the venturi meter.

Orifice meters have been in commercial use since the early 1900s and are the most common differential-producing flow meter used today.[13,14] The orifice meter consists of a flat, thin plate with a circular hole machined into the center of it as shown in Figure 7.8 and is held in place with a holding device such as a pipe flange.

When a fluid flows through an orifice plate, the static pressure within the pipe varies as illustrated in Figure 7.9. Notice that just upstream of the orifice plate, the static pressure reaches a maximum value. When the fluid passes through the bore of the plate, it accelerates. As the fluid jet exits the other side of the plate, it continues to accelerate and decrease in cross-sectional area; hence, the minimum flow area is actually smaller than the area of the orifice. The location where the jet reaches its minimum cross-sectional area is referred to as the *vena contracta* and is where the velocity is the highest and the static pressure the lowest. As the fluid continues to flow downstream, it decelerates, where some, but not all, of the static pressure is recovered.

Pressure taps are required on each side of the orifice plate to allow the measurement of the pressure drop across the plate when the fluid is flowing. Knowing the differential pressure, the ratio of the orifice plate bore to pipe inside diameter (beta ratio), and fluid density, the flow rate can be calculated.

The differential pressure depends on the location of the pressure taps. Pressure taps are located in four general arrangements as illustrated in Figure 7.10: flange taps, D and $D/2$ taps, corner taps, and pipe taps. Flange taps are typically placed on the pipe flange and are

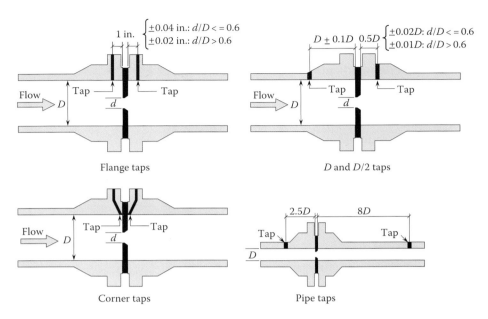

FIGURE 7.10
Common pressure-tap arrangements for orifice metering runs.

located 1 in. (2.5 cm) upstream and 1 in. (2.5 cm) downstream from the face of the plate to centerline of the tap. D and D/2 taps are located one pipe inner diameter upstream and one-half downstream from the face to the flange. Corner taps are located at the face of the orifice plate. Pipe taps are located 2.5 pipe inner diameters upstream and 8 diameters downstream from the face of the orifice plate; this arrangement measures the pressure drop at the point of full pressure recovery.

7.2.1.2 Upstream Flow Conditioners

When fluid flows through elbows, tees, and valves, a swirling flow with an uneven velocity profile develops. For accurate flow measurement, the flow upstream of a differential pressure meter should be free of these disturbances. In order to eliminate these disturbances, a long length of straight pipe is required. For example, swirl flow inside a pipe can require 50–100 diameters of straight pipe to eliminate the spin.[15]

Swirl-free conditions are defined as where the angle of the swirl is less than 2°.[15] An acceptable velocity profile condition exists when the velocity at each point within the pipe cross section agrees to within ±5% of the velocity of a very long straight length (over 100 diameters) of similar pipe.

Flow conditioners are commonly placed upstream of the orifice metering device to help straighten the flow profile and reduce swirl, thus allowing shorter lengths of upstream pipe. In general, there are three standard types of flow conditioners: tube bundle conditioner, Zanker straightener, and Sprenkle straightener.[15] The tube bundle conditioner consists of a number of parallel, thin-walled tubes connected together. The tube bundles do a good job of eliminating swirl but do a poor job of straightening the velocity profile. The Zanker straightener consists of a plate with holes of a certain size located just upstream of a number of parallel, thin-walled square channels connected together; one channel for each hole. The Sprenkle straightener consists of the three perforated plates, in series, spaced one pipe diameter apart. The plates are held together by studs or bolts around the perimeter of the plates. Research and development efforts in this area are ongoing in order to gain better insight into the best design and location of flow conditioners upstream of a differential metering device.

7.2.1.3 Calculating the Mass Flow Rate

The following equation represents the theoretical or ideal mass flow rate ($\dot{m}$) through an orifice metering run, that is, an incompressible, inviscid flow:

$$\dot{m} = A_o \sqrt{\frac{2\,dP\,\rho}{1-\beta^4}} \qquad (7.2)$$

where
dP is the pressure drop measured across the orifice plate
ρ is the density of the fluid
A_o is the area of the orifice bore
β is the ratio of pipe diameter to orifice bore diameter (D/d)

To determine the actual mass flow rate, this equation is multiplied by two correction terms: gas expansion factor (Y_1) and discharge coefficient (C_d). When a gas flows through an orifice plate, its pressure drops causing the gas to expand and reduce its density. To correct for the density change, the equations are multiplied by a gas expansion factor. The gas expansion factor is based on laboratory tests and is defined as follows:

$$Y_1 = 1 - (0.41 + 0.35\beta^4)\frac{dP}{kP_1} \qquad (7.3)$$

where
  k is the ratio of specific heat of the gas (isentropic coefficient)
  P_1 is the static pressure upstream of the orifice plate

The discharge coefficient (C_d) corrects the ideal theoretical equation for the effects of the approaching velocity profile and the assumption that no energy is lost between the pressure taps. The discharge coefficient is determined from laboratory experiments and varies with tap arrangement, diameter of pipe, and diameter of orifice bore. For a complete listing of discharge coefficient equations, see Miller.[9] The final form of the actual mass flow rate through an orifice is the theoretical mass flow rate multiplied by the gas expansion factor and orifice discharge coefficient and is written as

$$\dot{m} = A_o\, Y_1\, C_d \sqrt{\frac{2\, dP\, \rho_1}{(1-\beta^4)}} \qquad (7.4)$$

The discharge coefficient is a function of the upstream velocity in the pipe; therefore, since the upstream velocity is not known, the solution to this equation requires an iterative procedure.

Other correction factors can be applied to Equation 7.4. For example, for applications where the fluid temperature is very high or low, it is important to include the thermal-expansion factor. Pipe and orifice plate bore diameters are typically provided by the manufacturer and are based on room temperature. As the temperature of the orifice plate and pipe vary from room temperature, they expand and contract. To correct for the expansion and contraction of the material, a thermal-expansion factor is applied to the dimensions; these new dimensions are then used in the calculation.

7.2.1.4 Accuracy of Flow Measurements

It is not practical to list all factors that affect the flow accuracy through an orifice metering system; however, the following list provides a few factors that commonly contribute to errors:

- Incorrect information as to the bore of the orifice plate
- Incorrect information as to the inner diameter of the pipe
- Orifice plate inserted in line backward
- Orifice plate bore damaged
- Excessive pipe roughness
- Leakage around orifice plate flanges
- Orifice plate bore not properly centered inside pipe
- Flow disturbances upstream of orifice plate due to elbows, etc.
- Differential tap in the wrong location in relation to orifice plate
- One of the differential taps plugged with debris
- Liquids accumulating in pipe
- Instrumentation not zeroed
- Inaccurate temperature reading

7.2.2 Venturi Meter

The orifice meter is a simple and accurate device for measuring flow rates; however, the pressure drop for an orifice meter can be quite large.[4] A meter that operates on the same principle as the orifice meter, but with a much smaller pressure drop, is the venturi meter. A photograph of a venturi meter is shown in Figure 7.11.

The venturi meter is designed with gradual upstream converging cone with an angle of 15°–20° as illustrated in Figure 7.12. Downstream of the converging cone, the fluid enters into a throat section. The flow rate is determined based on the pressure difference between the upstream side of the converging cone and the throat section. After leaving the throat, the fluid enters into a diverging section designed with a total (included) angle of 7°–15°. Due to the gradual upstream contraction, there is no vena contracta (like for the flow through an orifice plate) resulting in a low pressure drop; again, this makes the venturi meter suitable where only small pressure heads are available.

The equation for the mass flow rate through a venturi meter is the same as for the orifice meter, Equation 7.2. The discharge coefficient for a venturi meter is typically larger and typically varies from 0.95 to 0.975. For the orifice meter, the discharge coefficient typically varies from 0.6 to 0.7. For more information on the design and discharge coefficients of a venturi meter, see ASME MFC-3M-1989.[15]

FIGURE 7.11
Cutaway view of a venturi meter. (Photo courtesy of Aplitex, http://www.directindustry.com/prod/aplitex-sl/venturi-differential-pressure-flow-meters-50657-361558.html)

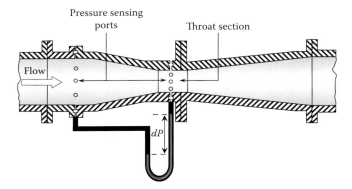

FIGURE 7.12
Schematic of a venturi meter.

7.2.3 Turbine Flow Meter

Turbine flow meters, when installed and calibrated properly, are capable of providing the highest accuracies attainable by any currently available flow sensor for both liquid and gas volumetric flow measurement.[16] The turbine meter consists of a cylindrical housing containing a balanced, free-spinning rotor designed with a series of curved blades that mounts on the centerline of the pipe as illustrated in Figure 7.13. The number of blades typically varies from about 6 to 20 or more depending on the design with a pitch angle of 30°–45°.[16] As fluid passes over the blades, it forces the rotor to spin.

The rotational speed of the rotor, or turbine, increases linearly with flow velocity: within ±0.5% over a wide flow range 10:1–20:1.[9] The turbine speed is typically measured

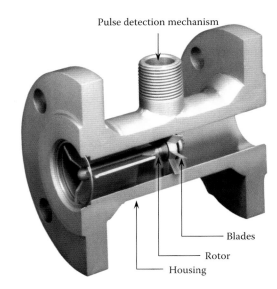

FIGURE 7.13
Turbine meter. (Photo courtesy of FMC Technologies, Houston TX.)

by detecting the pulse of each blade electrically, mechanically, or optically; each pulse represents a certain volume of fluid. The flow rate of the fluid is determined by integrating the total number of pulses over a period of time. The accuracy of these meters is better than 1% over a wide range of flow rates.[4] Accuracy is compromised, however, with blade wear, bearing friction due to wear, and when a vapor enters the line for liquid flows.[9]

7.2.4 Vortex Flow Meter

When a fluid flows past a bluff body, the wake downstream will form rows of vortices that shed continuously from each side of the body. Figure 7.14 shows

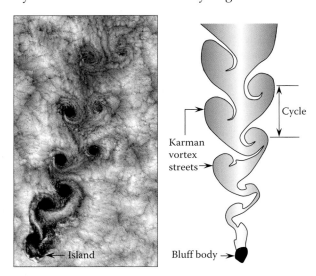

FIGURE 7.14
NASA satellite image of clouds off the Chilean coast near the Juan Fernandez Islands showing von Karman vortex streets and figure drawn for greater clarity.

a satellite image of clouds blowing past an island. Notice the unique swirling vortices downstream of the island. These repeating patterns of swirling vortices are referred to as Karman vortex streets: named after the fluid dynamicist Theodore von Karman. Vortex shedding is a common flow phenomenon that causes car antennas to vibrate at certain wind speeds and also led to the collapse of the famous Tacoma Narrows Bridge in 1940. Each time a vortex is shed from the bluff body, it creates a sideways force causing the body to vibrate. The frequency of vibration is linearly proportional to the velocity of the approaching fluid stream and is independent of the fluid density.

The frequency of vortex shedding or vibration of the bluff body is determined as follows:

$$f = \frac{St\, V_{average}}{d_{body}} \quad (7.5)$$

where
St is the Strouhal number (dimensionless)
$V_{average}$ is the average upstream velocity
d_{body} is diameter of the bluff body

For flow over a cylindrical body, the Strouhal number is constant for Reynolds numbers from 10,000–1,000,000: based on the diameter of the bluff body. As the Reynolds number increases, the wake becomes more complex and turbulent, but the alternate shedding can still be detected at a Reynolds number of 10,000,000.[17] Knowing the Strouhal number, frequency of vibration, and the diameter of the bluff body, the average velocity can be determined.

Vortex flow meters are designed to partially obstruct the fluid stream using a bluff body in order to create vortex shedding (see Figure 7.15). The frequency of the vortex shedding is measured by various methods and converted to an average velocity. Today, most Vortex meters operate accurately over a wide range of flows and are available for line sizes from ½ to 16 in. (1.3 to 41 cm). The main advantage of this meter is that it has no moving parts, but it does create a pressure drop comparable to other obstruction-type meters[9]: turbine meter, orifice plate, etc.

7.2.5 Magnetic Flow Meter

All the flow meters mentioned so far require that some sort of obstruction, such as a turbine or orifice, be placed within the flow stream. Magnetic flow meters, however, do not obstruct the flow of the fluid. The magnetic flow meter operates on the basic principle of magnetic induction; that is, when a conducting material moves through a magnetic field, it produces an electromagnetic force as illustrated in Figure 7.16. It is interesting to note that the basic principle of the magnetic flow meter was investigated by Faraday in 1832; however, it was not practically used until a century later.[9]

When a current is applied to the coil, a magnetic field is produced that is at right angles to the pipe. As the conductive liquid flows through the pipe, an electrical voltage is induced and measured by two electrodes mounted flush to the pipe wall. This induced voltage is proportional to average velocity of flow and therefore to the volume flow rate. In order for a magnetic flow meter to work, the fluid must be conductive; therefore, they are not suitable for gas flow.[18]

It is important that the strength of the magnetic field remain constant; this is accomplished with a

FIGURE 7.15
Vortex meter. (Photo courtesy of Aalborg Instruments & Controls, Inc., Orangeburg, NY.)

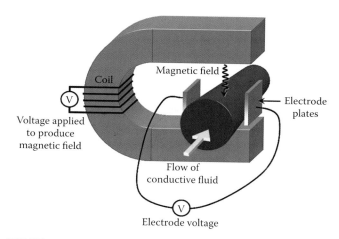

FIGURE 7.16
Principles of magnetic induction. (Adapted from Precision measuring PD 340 flow transmitter, http://www.instruments-gauges.co.uk/pd340.html)

microprocessor built into the transmitter controls. The microprocessor also detects the voltage of the electrodes and converts the signal to a fluid flow rate. The main advantages of the magnetic flow meter are that the output signal varies linearly with the flow rate and that the meter does not obstruct the flow. The major disadvantages are its relative high cost and that it is not suited for gas flow.[4]

7.2.6 Ultrasonic Flow Meter

There are two types of ultrasonic flow meters: time-of-flight and Doppler. The time-of-flight meter sends pulses of high-frequency sound waves diagonally across the pipe as illustrated in Figure 7.17.[19] The time required for the sound wave to reach the opposite wall depends on whether it is moving with the flow or against the flow; if the wave is moving with the flow, it will travel faster than the wave moving against the flow. The time difference is a measure of the fluid flow rate. The ultrasonic flow meter illustrated in Figure 7.17 is referred to as a single-path meter because only one beam path crosses the pipe. Some ultrasonic flow meters send sound waves along several paths to improve accuracy as illustrated in Figure 7.18.

The Doppler flow meter sends out sound waves similar to the time-of-flight meter, but the waves are reflected back to a detector by small particles or bubbles moving with the fluid. The sound wave reflected back from the fluid will have a slightly lower frequency than the transmitted sound wave due to the Doppler effect. The difference between transmitted frequency and the reflected frequency is used to determine the flow rate. Commercially available Doppler flow meters require that the liquid contains at least 100 parts per million of 100 μm or larger suspended particles or bubbles.[20]

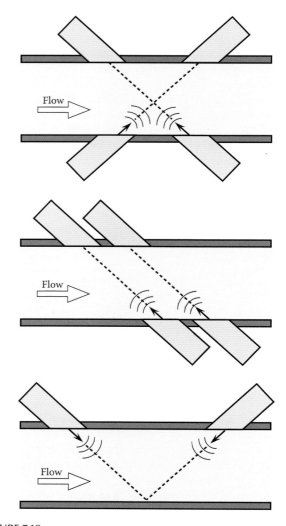

FIGURE 7.18
Time-of-flight ultrasonic flow meter: multipath type. (Adapted from Miller, R.W., *Flow Measurement Engineering Handbook*, McGraw-Hill, Inc., New York, 1989.)

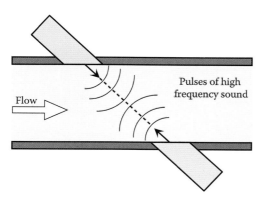

FIGURE 7.17
Time-of-flight ultrasonic flow meter: single-path type.

7.2.7 Thermal Mass Meter

A typical thermal mass meter is shown in Figure 7.19. The operation of these meters is based on the principle that the rate of heat absorbed by a flowing fluid is directly proportional to the mass flow rate.[21] In general, there are three types of thermal mass meters: constant temperature, constant power, and energy balance.

Energy balance thermal mass meters require one heating element located between two temperature sensors as illustrated in Figure 7.20.[22] Although several design variations exist, their operation is basically similar. As the fluid flows past the heating element, the fluid absorbs heat. This heat is carried downstream where it is transferred to the downstream temperature sensor. The temperature difference between the upstream and downstream sensor is detected. This output signal is

FIGURE 7.19
Thermal mass meter.

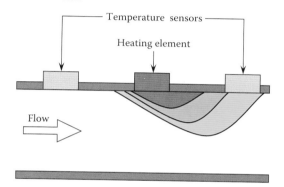

FIGURE 7.20
Energy balance type thermal mass meter.

FIGURE 7.21
Positive displacement meter.

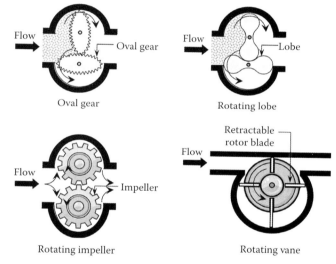

FIGURE 7.22
Types of positive displacement meters. (Adapted from Universal, Flow Monitors (UFM), Positive Displacement Flowmeters, Hazel Park, MI, http://www.flowmeters.com/ufm/index.cfm?task=positive_displacement)

then converted into a mass flow rate. These meters typically have a turndown ratio of 10:1, while the constant temperature and constant power meters have a turndown ratio of 1000:1 and 100:1, respectively.

7.2.8 Positive Displacement Meter

Positive displacement flow meters, or PD meters, are used more than all other flow measurement devices[9]; a typical PD flow meter is shown in Figure 7.21. Millions of PD meters are used daily to meter gasoline, natural gas, and water into our homes. Although there are many different PD meter designs commercially available, they all share the same principle of operation: they measure the volume of fluid by separating the flow into known volumes and then counting, repeatedly, the filling and discharging of volumes passing through the meter over time.

The pumping of individual parcels of fluid is accomplished using rotating parts, pistons, or diaphragms that form moving seals between each other and/or the flow meter body.[23] A few of these mechanical units designed with rotating parts is illustrated in Figure 7.22. As the flow rate of the fluid increases, these parts turn or reciprocate proportionally faster and are sensed by mechanical registers or electronic transmitters. The rate of revolution or reciprocation determines the volume of fluid passing through the meter. It is important that these mechanical units have tight tolerances to prevent fluid from passing through the flow meter without being measured; this is referred to as slippage. Due to these tight tolerances, filters are typically installed upstream to prevent dirt from entering the meter; this helps reduce mechanical wear and prevent plugging. Typically, filters are used to remove particles larger than 100 μm as well as gas bubbles from the liquid flow.[24] Mechanical wear can result in substantial flow error; therefore, PD meters should be periodically calibrated.

Since PD meters only measure the volume of fluid that passes through, they are rarely used as flow rate meters; however, the average flow rate can be determined by measuring the volume of fluid that passes through in a given amount of time. It should also be mentioned that PD meters create a considerable pressure drop which should be considered for any application.

7.2.9 Pitot Tube

A Pitot tube is an instrument used to measure the velocity of a flowing fluid. The Pitot tube was invented by French engineer Henri Pitot in the early 1700s. Today, it is widely used in industrial applications; refer to Volume 1, Section 9.4.4 for a more complete discussion.

The static reading on a Pitot tube is accurate to 0.5% for Mach numbers up to 0.5. For Mach numbers between the values of 0.5 and 0.7, the error increases to 1.5%. Above a Mach number of 0.7, the error can increase as much as 10% due to the formation of shock waves on and around the tip of the probe. Above a Mach number of 1.0, both the total and static readings vary significantly for the actual values.[25]

Although Pitot-static tubes are not extremely sensitive to their angle with respect to the flow streamline (angle of attack), it is good practice to position them as parallel as possible to the flow stream. Errors, as high as 2%, can result if the angle of attack varies from 5° to 30°.

If the probe is positioned too close to a wall, the fluid can accelerate between the probe and wall decreasing the static pressure on one side. It is recommended that the probe be positioned at least 10 probe diameters away from the wall.

If the volume or mass flow rates are required in round or rectangular ducts, then one must take measurements at various locations across the duct. It is important that the Pitot traverse be conducted at least 8 duct diameters downstream or 2 diameters upstream of an obstruction such as fans, elbows, dampers, expansions, etc. Figure 7.23 shows the locations for a Pitot tube traverse in a round or rectangular duct, based on centroids of equal area, for determining the volume or mass flow rate.[26]

After measuring the velocity at each point in the duct using the locations provided in Figure 7.23, the total volume and mass flow rate of the fluid can be determined as follows:

$$Q = A_{duct} \frac{1}{n} \sum_{i=1}^{n} V_i \qquad (7.6)$$

$$\dot{m} = \rho Q \qquad (7.7)$$

where
Q is the volume flow rate in the duct
A_{duct} is the duct cross-sectional area
n is the total number of points traversed
V_i is the velocity at each measurement point
$\dot{m}$ is the mass flow rate in the duct

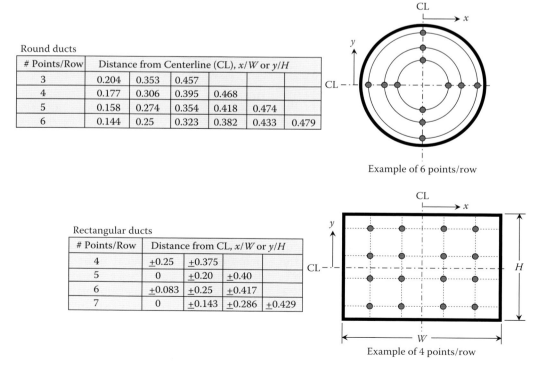

Round ducts

# Points/Row	Distance from Centerline (CL), x/W or y/H				
3	0.204	0.353	0.457		
4	0.177	0.306	0.395	0.468	
5	0.158	0.274	0.354	0.418	0.474
6	0.144	0.25	0.323	0.382	0.433	0.479

Rectangular ducts

# Points/Row	Distance from CL, x/W or y/H			
4	±0.25	±0.375		
5	0	±0.20	±0.40	
6	±0.083	±0.25	±0.417	
7	0	±0.143	±0.286	±0.429

FIGURE 7.23
Locations for a Pitot tube traverse in a round or rectangular duct, based on centroids of equal area.

Combustion Diagnostics

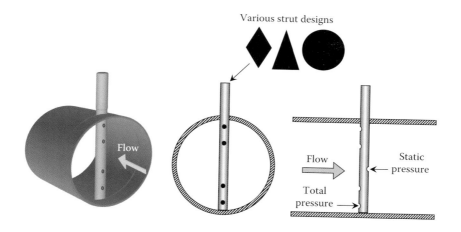

FIGURE 7.24
Averaging Pitot tube.

7.2.10 Averaging Pitot Tube

The averaging Pitot tube is widely used as a flow measurement device in industry. In general, an averaging Pitot tube consists of a hollow tube or strut, with various cross-sectional designs, that spans a flow stream as illustrated in Figure 7.24. The strut is designed with several holes strategically placed along its length so that the average of the total and static pressure is measured. The difference in the average, total, and static pressure provides the mean velocity across the pipe; the volume flow rate of the fluid is determined by multiplying the mean velocity by the cross-sectional area of the pipe. A single Pitot tube measurement will give a similar reading if it is located at a point in the pipe cross section where the flowing velocity is near the average velocity.[27]

The strut can consist of various designs such as a cylinder, diamond, or triangular cross section. The position of the pressure ports is critical to the accuracy of the device. If the ports are not located in the proper location, then the average velocity profile will be in error. It is also important that the velocity profile is symmetric; if skewed to one side, significant errors can result. Typically, as the pipe or duct size increases, the number of total pressure ports and static ports increases.

Averaging Pitot tubes are commonly used to measure the flow rate of fluids in ducts and large pipe as shown in Figure 7.25. The advantages of the averaging Pitot tube are that they are easy to install, relatively inexpensive, and do not take a large pressure drop.[9] The disadvantage is their poor performance with dirty fluids.

7.3 Advanced Diagnostics

Sampling combustion products is not only necessary for emission control, but also useful for the evaluation of efficiency or effectiveness of combustion systems. For example, NOx analysis in exhaust gas uses a sampling probe to pull the exhaust gases through a chemiluminescence detector to determine the pollutant emission concentration. This is a continuous emission monitoring system, which is discussed in Volume 1, Chapter 14. Other measurements, such as CO probing in the exhaust gas and heat flux mapping inside the furnace, are useful for the determination of combustion efficiency and furnace heat transfer efficiency.

In addition, to fully understand the combustion process and conduct combustion research, combustion diagnostic systems are required. Traditional approaches for combustion diagnostics typically use a mechanical probe inserted into a region of interest. For example, a thermocouple probe is used for temperature measurements in the flame. A Pitot tube or hot wire anemometry is used for the flow velocity measurements. Those are invasive techniques. Invasive techniques have the

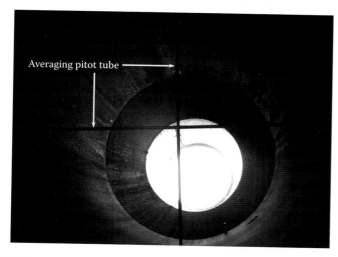

FIGURE 7.25
Photograph of an averaging Pitot tube located inside a large duct.

disadvantage of disturbing the system being probed. Disturbance of the system may cause measurement errors. Recently, many noninvasive techniques have been developed to overcome this disadvantage of invasive techniques. These new systems generally use either optics or laser instruments.

Many advanced optics or laser instruments have been developed and used for academic combustion research. Examples include coherent anti-Stokes–Raman scattering (CARS) for temperature and species concentration measurements,[28,29] laser Doppler anemometry (LDA) for flow field velocity measurements,[30] particle image velocimetry (PIV) for velocity vector mapping,[31] and laser-induced incandescence (LII) for soot mass concentration and particle size measurements,[32] among others. Some of them are commercially available and can be applied to industrial applications. The scope of this chapter is to discuss only the commercially available techniques or easy-to-apply industrial tests with which the authors have had experience.

7.3.1 Fourier Transform Infrared Spectroscopy

Fourier transform infrared (FTIR) spectroscopy is a technique to identify chemical species and their concentrations. This instrument introduces an infrared beam through a sample and measures the spectral frequencies and the intensity absorbed. The absorption spectrum is unique for each infrared-active gas. Each molecule has certain natural vibration frequencies, which are in infrared regions. The infrared regions have wavelengths in the ranges of 0.8–2.5 μm (near infrared), 2.5–50 μm (mid infrared), and 50–800 μm (far infrared).

There are many methods to analyze species spectrum and intensities in the infrared region. The FTIR, which uses an interferometry technique, is one of them. Interferometry is a technique to combine two or more waves and diagnose the properties of their interference. Most interferometers employ a beam splitter which takes the incoming infrared beam and divides it into two optical beams. One beam reflects off a mirror which is fixed. The other beam reflects off another mirror which is on a mechanism moving a very small distance (typically a few millimeters) away from the beam splitter. In the FTIR, a coherent IR light is guided through an interferometer and the sample to perform a Fourier transform as illustrated in Figure 7.26. Since the information at all frequencies is collected simultaneously, the process of FTIR analysis is very fast. The FTIR also has the advantage of averaging multiple samples resulting in a sensitivity improvement.

Figure 7.27 shows an FTIR instrument which can simultaneously measure 250 different organic compounds at concentrations in ppm range. The instrument

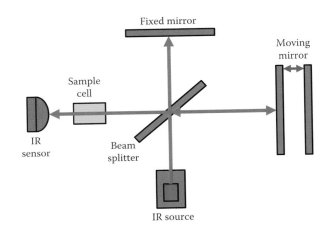

FIGURE 7.26
A simple FTIR spectrometer layout.

FIGURE 7.27
Photograph of an FTIR system.

is useful for measuring a chemical's destructive removal efficiency (DRE), identifying combustion intermediate by-products, and analyzing gas compositions.

7.3.2 Phase Doppler Particulate Anemometer

The phase Doppler particulate anemometer (PDPA)[33,34] is an instrument for spray combustion applications. Spray combustion is discussed in detail in Volume 1, Chapter 10 and Volume 3, Chapter 2. As discussed in these references, flame shape and pollutant emissions are dependent on droplet evaporation rates, droplet size, and droplet velocity. Droplet size and velocity measurements in the past were conducted using high-speed cameras with short-duration lighting and double-image photography.[35] These methods only provided limited amounts of data. New optical techniques are available to obtain more accurate

Combustion Diagnostics

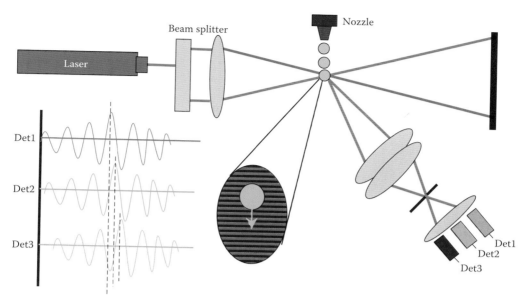

FIGURE 7.28
Schematic of phase Doppler particle anemometer (PDPA).

measurements of droplet size and velocity in a liquid fuel spray. The PDPA is one of them.

PDPA uses a Doppler signal and the interference phase shift to characterize the particle velocity and size simultaneously.[36] Usually, an optical instrument has the drawback that the beam intensity is easily attenuated by other objects, such as droplets, aerosols, or solid particles. PDPA is only dependent upon the wavelength of scattered light and the measurement scale, which can overcome this disadvantage.

The schematic of a PDPA system is shown in Figure 7.28. The figure shows that laser light splits into two equal intensity beams. When the two beams refocus, they interfere and generate a fringe pattern, where the fringe pattern is magnified at the center of the schematic. As droplets flow through the fringe pattern, the Doppler signal is scattered. Three detectors collect the different phase Doppler signals and present them as one sample point. The temporal frequency of the fringe pattern is the Doppler difference frequency, which is a function of the optical geometry, laser wavelength, and particle velocity. Therefore, the Doppler signal is used for velocity measurement, while the spatial frequency of the fringe pattern (related to the drop size) is used for size measurement.

All the droplets passing through the sample volume contribute to the measurements. The sample volume is the cross-sectional area of the intersection of the two laser beams. A typical PDPA measurement contains a large number of sample points, and the information is presented statistically to yield the mean droplet size, size distribution, standard deviation, and velocity distribution. The measurements made for the oil atomizer tests presented here contain 10,000 sample points.

The PDPA, as with any instrument, has its limitations and measurement uncertainties. The operator should understand these limitations to ensure data collected are in an accurate region. For more detailed discussions on the PDPA uncertainties, refer to other technical papers listed in the references.[37,38]

The PDPA has been used to investigate numerous topics of interest and applications both in industry and academia.[39] The authors have used the PDPA to develop oil guns and conduct atomization research by using water as the fluid and air as atomizing medium. Oil guns generally have multiple exit ports. The PDPA model used has a small capacity. In order not to saturate the data, only one of the exit ports was used in the measurement. This was accomplished by dividing the spray chamber into two internal volumes as illustrated in Figure 7.29. One volume allows the single port being measured to discharge its spray to the PDPA probe head, and the other volume collects the spray from the other ports then discharges it to the drain.

A typical PDPA measurement is shown in Figure 7.30. In the figure, the oil gun is operated at different mass ratios. The mass ratio is defined as a ratio of the atomizing medium flow to the fluid flow (i.e., an air-to-water mass ratio). As discussed in Volume 1, Chapter 10, the atomization quality is dependent on the mass ratio. (A larger mass ratio yields a better atomization.) The PDPA measurements indicate that the high mass ratio sample generates smaller droplets than the low mass ratio sample. If one transfers the number distribution

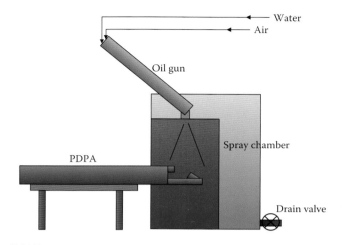

FIGURE 7.29
Schematic of oil gun and spray chamber by using PDPA for droplet size measurements.

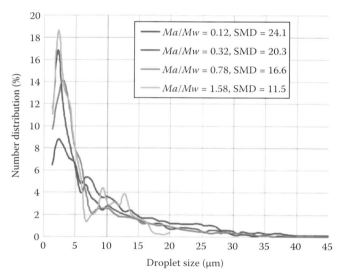

FIGURE 7.30
Typical PDPA droplet size measurements. The oil gun is operated at different mass ratios resulting in different Sauter mean diameters (SMD).

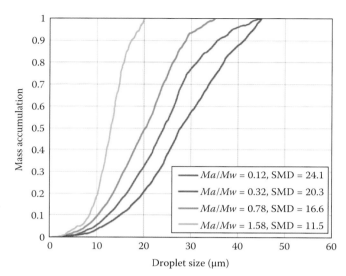

FIGURE 7.31
PDPA mass accumulation measurements at different mass ratios.

data to mass accumulation data, as indicated in Figure 7.31, the atomization quality at different mass ratio conditions is clearly distinguished. In the high mass ratio case (i.e., $Ma/Mw = 1.58$), the droplet size distribution is within 20 µm; while in the low mass ratio case (i.e., $Ma/Mw = 0.12$), the droplet sizes are much larger than 20 µm.

The PDPA measurements are very useful for atomization research and new oil gun development.

7.3.3 Liquid Planar Laser-Induced Fluorescence

A liquid planar laser-induced fluorescence (LPLIF) is an optical method to observe liquid mass distribution. LPLIF is usually applied to spray applications. According to the referenced works,[40,41] the laser-induced fluorescence signal, based on a simple two-level model with weak excitation, is a function of the laser energy, sample volume, number density of the species, and the thermodynamic state of the species (i.e., temperature and pressure). For cold spray studies (i.e., non-combustion spray), the detected fluorescent signal is simply proportional to the density of dyne molecules in the flow, because no chemical reaction occurs, and temperature and pressure are constants.[42]

The LPLIF uses a laser and a dye which can be excited by laser light to emit fluorescent light. A sharp cutoff filter is installed to filter out the laser-scattered light but allow the fluorescent light to pass. For example, fluorescein dye dissolved in distilled water as the fluorescent medium can be excited by an Ar-ion laser. A typical multiline Ar-ion laser emits 488, 514, and 528 nm laser lights. The fluorescent light frequencies emitted from fluorescein dye are continuous, ranging from 488 to 600 nm. A sharp cutoff filter at around 550 nm can filter out the laser-scattered light and allow fluorescent light to pass. The detailed experimental setup and procedures can be found in the reference by Chung et al. (1997).[43] Different images for spray fluorescent light and typical scattered light were demonstrated as shown in Figure 7.32. The intensity of scattered light (right image) depends strongly on the viewing angle and the droplet size. The LPLIF image (left image) overcomes those disadvantages and is more representative of the liquid mass distribution.

A patternator as mentioned in Volume 1, Chapter 10, can also be used to measure the liquid distribution in the spray. However, the liquid flux measured by the

FIGURE 7.32
Spray images for liquid planar laser-induced fluorescence (LPLIF) (left) and scattered light (right).

patternator is only one cross section with a fixed distance from the origin of the spray. The LPLIF provides the whole spray image, and collecting data is much faster and more convenient. This method saves a great deal of time if several measurements are required.

References

1. Betts, D., *How Products Are Made*, 1994, Gale Group, www.encyclopedia.com/doc/1G2-2896500079.html
2. Zigrang, D.J., *Elements of Engineering Measurements*, Department of Mechanical Engineering, University of Tulsa, Tulsa, OK, 1985.
3. HETaC, Screen Shots from Program, 2012, www.hetac-fluidpower.com/screen_shot_gallery.html
4. Roberson, J.A. and Crowe, T., *Engineering Fluid Mechanics*, Houghton Mifflin Co., Dallas, TX, p. 45, 1980.
5. Cornforth, J.R., *Combustion Engineering and Gas Utilization*, British Gas plc, London, U.K., 1992
6. Encyclopedia.com, Pressure Gage, 2012, www.encyclopedia.com/doc/1G2-2896500079.html
7. Jankura, R., Seven steps to pressure gage selection, Dresser Instrument, Internet website
8. O'Keefe Controls Co., www.okcc.com/PDF/Pressure%20Snubber%20pg.42.pdf
9. Miller, R.W., *Flow Measurement Engineering Handbook*, McGraw-Hill, Inc., New York, 1989.
10. Improving deadweight tester accuracy, www.transcat.com/PDF/DeadweightTesters.pdf
11. Ferron, A.G., Construction of static pressure taps, Technical Memorandum, Alden Research Lab, Alden, MA, September 1986.
12. Brunkalla, R.L., Effects of fabrication technique on the discharge coefficient for throat tap nozzles, ASME Paper 84-JPGC-PTC-10, October 1985.
13. Buckley, B., Fundamentals of orifice metering, help.intellisitesuite.com/ASGMT%20White%20Papers/papers/002.pdf
14. Smith Metering, Inc., Fundamentals of orifice metering, www.afms.org/Docs/gas/Fundamenatls_of_Orifice.pdf
15. ASME MFC-3M-1989, *Measurement of Fluid Flow in Pipes Using Orifice, Nozzle, and Venturi*, The American Society of Mechanical Engineers, New York.
16. Wadlow, D., Turbine flowmeters, www.sensors-research.com/articles/turbines.html
17. White, F.M., *Viscous Fluid Flow*, McGraw-Hill, St. Louis, MO, p. 10, 1991.
18. Precision measuring PD 340 flow transmitter, www.instruments-gauges.co.uk/pd340.html
19. Roxar, Why and how to measure flare gas, www.flowmeterdirectory.com/flowmeter_artc/flowmeter_artc_02021401.html
20. Introduction to ultrasonic Doppler flow meters, www.omega.com/prodinfo/ultrasonicflowmeters.html
21. Cross Instrumentation, A flow measurement primer—Thermal mass flow meters, www.crossinstrumentation.com/ga/mfg/Flow/tmf_meters.html
22. Flow measurement, www.en.wikipedia.org/wiki/Flow_Measurement
23. Universal Flow Monitors, Positive Displacement Flowmeters, 2012, www.flowmeters.com/ufm/index.cfm?task=positive_displacement
24. efunda, Positive Displacement Flowmeters, 2012, www.efunda.com/designstandards/sensors/flowmeters/flowmeter_pd.cfm
25. United Sensor, Pitot-static pressure probes: for measuring total and static pressures of a moving fluid, *Bulletin* 1, 6–83, 2011.
26. Flow Kinetics LLC, Using a Pitot static tube for velocity and flow rate measurement, www.flowmeterdirectory.com/flowmeter_artc/flowmeter_artc_02111201.html
27. www.en.wikipedia.org/wiki/Annubar
28. Eckbreth, A.C., *Laser Diagnostics for Combustion Temperature and Species*, Gordon and Breach Publishers, Amsterdam, the Netherlands, 1996.
29. Lackner, M., *Lasers in Chemistry*, Wiley-VCH, Weinheim, Germany, 2008.
30. Durst, F., Melling, A., and Whitelaw, J.H., The application of optical anemometry to measurements in combustion systems, *Proceedings of the Combustion Institute*, 14, 699–706, 1973.
31. Reuss, D.L., Adrian, R.J., Landreth, C.C., French, D.T., and Fansler, T.D., Instantaneous planar measurements of velocity and large-scale vorticity and strain rate in an engine using particle-image velocimetry, *SAE Technical Paper Series*, Paper No. 890616, 1989.
32. Will, S., Schraml, S., and Leipertz, A., Two-dimensional soot-particle sizing by time-resolved laser-induced incandescence, *Optics Letters*, 20, 2342–2344, 1995.
33. Bachalo, W.D., Method for measuring the size and velocity of spheres by dual-beam light-scatter interferometry, *Applied Optics* 19(3), 363, 1980.

34. Bachalo, W.D. and Houser, M.J., Phase/Doppler spray analyzer for simultaneous measurements of drop size and velocity distribution, *Optical Engineering*, 23(5), 583, 1984.
35. Beér, J.M. and Chigier, N.A, *Combustion Aerodynamics*, Halstead Press Division, John Wiley & Sons, Inc., New York, 1973.
36. Sankar, S.V. and Bachalo, W.D., Response characteristics of the phase Doppler particle analyzer for sizing spherical particles larger than the light wavelength, *Applied Optics*, 30, 1487–1496, 1991.
37. Widmann, J.F. and Presser, C., A benchmark experimental database for multiphase combustion model input and validation, *Combustion and Flame*, 129, 47–86, 2002.
38. Schneider, M. and Hirleman, E.D., Influence of internal refractive index gradients on size measurements of spherically symmetric particles by phase Doppler anemometry, *Applied Optics*, 33, 2379–2387, 1994.
39. Erazo, J.A., Parthasarathy, R.N., and Gollahalli, S.R., Atomization and combustion of canola methyl ester biofuel spray, *Journal of Fuel*, 89, 3735–3741, 2010.
40. Bechtel, J.H., Dasch, C.J., and Teets, R.E., *Laser Applications*, Academic Press, New York, 1984.
41. Hiller, B. and Hanson, R.K., Simultaneous planar measurements of velocity and pressure field in gas flows using laser-induced fluorescence, *Applied Optics*, 27(1), 33–48, 1993.
42. Igushi, T., McDonnell, V.G., and Samuelsen, G.S., An imaging system for characterization of liquid volume distributions in sprays, *Proceedings of the International Conference on Liquid Atomization and Spray Systems*, Worcester, MA, 1993.
43. Chung, I.P., Dunn-Rankin, D., and Ganji, A., Characterization of a spray from an ultrasonically modulated nozzle, *Atomization and Sprays*, 7(3), 295–315, 1997.

8
Burner Testing

Jaime A. Erazo, Jr. and Thomas M. Korb

CONTENTS

8.1 Introduction 191
8.2 Burner Testing 191
    8.2.1 Benefits 192
    8.2.2 Drawbacks 192
    8.2.3 Burner Testing versus CFD 192
    8.2.4 Testing Parameters and Measurements 192
8.3 Burner Testing Equipment and Methodology 193
    8.3.1 Test Furnaces 193
    8.3.2 Air Delivery Systems 193
    8.3.3 Instrumentation and Control 194
    8.3.4 Fuel Flow and Composition 195
    8.3.5 Flue Gas Analysis 195
    8.3.6 Flue Gas Temperature and Pressure 196
8.4 Special Equipment 197
    8.4.1 Heat Flux 197
    8.4.2 CO Probe 197
    8.4.3 Noise 198
    8.4.4 Unburned Hydrocarbons, Particulate Matter, and Oxides of Sulfur 198
8.5 Test Fuel Selection 198
8.6 Test Procedure 199
8.7 Conclusions 199
References 201

8.1 Introduction

Many different types of burners are used in industrial and commercial applications.[1] Some examples of burners are boiler burners (see Volume 3, Chapter 3) used to produce steam and process burners (see Volume 3, Chapter 1) used to refine petroleum. Each industry has its own unique requirements and challenges to be addressed in the burner design. For more information about the various industries in which burners are used and their design, the reader is also referred to the other two volumes of this handbook series.

The purpose of this chapter is to discuss burner testing. Due to the variety of industries in which burners are utilized and the numerous, and often unique, performance requirements of a burner, this chapter cannot cover burner testing methodology in detail for every industry application (see Ref. [2] for a discussion of testing a variety of different burner types). However, there are many universal concepts which are applicable in the industries where burners are used. For this reason, this chapter focuses on general concepts and refers the reader to other references within this handbook which are more industry specific. To illustrate these concepts, examples from the refining/petrochemical industry which uses process burners are provided. For more information about process burners and their use in refining applications, the reader is referred to Refs. [3,4].

8.2 Burner Testing

Burner tests are typically conducted for one of the following reasons:

- Research and development purposes
- Commercial testing to validate burner performance against customer specifications
- Resolution of existing applications issues

Each type of test has its own objectives. For example, during a research and development test, a detailed test matrix will be executed to completely characterize the capabilities of a burner. This information may then be used to validate burner CFD models (see Volume 1, Chapter 13) and to develop burner-specific performance prediction models used in the production design engineering process. On the other hand, a commercial test focuses on the operating conditions specific to a customer's needs. A customer may be replacing burners to meet more stringent emissions regulations. As new processes are introduced, the fuel gas composition may change dramatically which may require testing the burner to confirm flame dimensions and emissions. In some cases, a customer may require a custom design to meet unique requirements; in these cases, a burner test is usually requested to confirm performance. In all cases, testing provides verification of the design and performance of the burner.

While these performance requirements may vary widely from application to application, there are a number of burner attributes that are considered to be universal. These attributes can be summarized as safety (see Chapter 1), performance, and maintenance. A burner must be designed to safely burn all of the supplied fuel with a stable flame under all design operating conditions. The burner should also be able to meet flame dimension and emission requirements for the entire design operating range. The burner design should allow limited maintenance work to be performed on the burner while the other burners in the same heater are still in operation without putting personnel or equipment at risk. Regardless of what type of burner test is being conducted, these attributes are an important indicator of the success of a burner test.

8.2.1 Benefits

A dedicated test furnace allows for burner experiments that are difficult if not impossible to do with the actual production furnaces. The controlled and isolated environment of a test furnace facilitates the ease and safety of a demonstration. Fuel sources can be quickly isolated and other precautions can be taken to improve the safety of the operation. Common instabilities such as flashback, liftoff, and even acoustic coupling can be demonstrated. The reproduction of these conditions is invaluable to the customer as the flame appearance, fuel pressures, and air register settings can be documented to avoid this behavior in the field. This information can then be relayed to the operators and technicians who work with the burners on a daily basis.

Burner research and development is greatly facilitated, since a test furnace will typically accommodate just a single burner. The burner can, therefore, be studied in greater depth as there is more operating control and better access for instrumentation. Bench-scale testing can provide some insight into the behavior of a full-scale burner, but there is no substitute for reproducing the flame that will be present in the furnace. Complicated scaling and extrapolations are avoided, and real-time data are produced.

8.2.2 Drawbacks

In spite of the benefits of a full-scale test, there are some real limitations. Actual furnaces typically utilize many burners to provide the heat needed for the process. A single burner test cannot predict if there will be burner-to-burner interaction in a furnace (see Chapter 13). Burner-to-burner interaction occurs when burners are tightly spaced together and their flames interact. What typically occurs is that if the flames are close enough to one another, they will coalesce into a single large flame. This flame may be much larger than each individual flame, which presents the possibility of flame impingement and increased pollutant emissions. A quick demonstration of this can be accomplished by holding the flames of two candles close to one another. When brought close enough, the two flames coalesce into one large flame.

The burner test facility will have a limited number of types of furnaces available. It will never have an identical furnace into which the burners will be installed. The flue gas recirculation patterns and heat sinks will be significantly different. It is often the case that the burner will still behave slightly differently in the field than in the test furnace due to these reasons and others. Therefore, it can be a challenge to extrapolate the test results to actual furnace conditions and predict how the burner performance may change as a function of these differences.

8.2.3 Burner Testing versus CFD

The CFD modeling of combustion processes has greatly advanced in recent years (see Volume 1, Chapter 13).

The development of a range of different combustion models for diffusion flames combined with accurate gray or non-gray radiation models has led to good successes predicting burner performance in actual furnaces. Premixed flame modeling remains difficult, however. Pollutant emissions are also very difficult to predict using CFD, due to the large differences in reaction rates between the main combustion reactions and the formation of NOx (see Volume 1, Chapter 15). Another disadvantage of CFD is that the near-burner region needs to be finely resolved, typically leading to meshes with large amounts of cells. For that reason, the convergence times needed to resolve full-scale burner flames are still substantial compared to the speed of a burner test. This becomes much more evident when transient modeling is attempted to simulate flame instabilities or ignition.

So while it has some shortcomings, the great advantage of CFD is that it greatly facilitates the extrapolation of the test results to the actual furnace. Where the burner test provides insight in ignition, stability, and emissions, a CFD model of the actual furnace can predict whether flame interactions lead to rollover or impingement on process tubes. So it is the tandem of burner testing and CFD modeling that will yield the most complete burner performance prediction.

8.2.4 Testing Parameters and Measurements

During a burner test, a wide variety of operating parameters are being controlled, measured, and recorded to quantify the burner performance characteristics. Operating parameters such as pressures and temperatures are recorded to verify that the burner is operating within design specifications. Flue gas analysis is conducted, and flame dimensions are recorded to monitor the combustion process in the furnace. Each industry has testing parameters specific to its needs, but in general, the following parameters are common in most applications and may be measured and recorded during a burner test:

- Fuel flow rate (heat release) and composition
- Fuel temperature and pressure
- Furnace draft and burner airside pressure drop
- Combustion air temperature and humidity
- Flue gas analysis (CO, O_2, NOx, SOx, UHC, and particulate matter)
- Furnace temperatures
- Radiant heat flux profile
- Visible flame length and width/diameter
- Noise measurements

8.3 Burner Testing Equipment and Methodology

A significant investment in specialized equipment is required to properly perform full-scale burner tests. The following is a list of some of the equipment that are discussed in detail in subsequent sections. An aerial view of an industrial combustion testing facility is shown in Figure 8.1:

- Test furnaces
- Fuel measurement and delivery
- Air delivery systems
- Instrumentation and control
- Special equipment

8.3.1 Test Furnaces

Selecting the correct test furnace for a burner test is essential to producing valid data that are representative of the actual application for the burner. Every industry will have many different furnace designs available to achieve a certain process requirement. A well-equipped burner vendor for a particular industry will have a wide variety of test furnaces available to simulate real-world applications. Some of the most critical features to be considered when selecting a test furnace are as follows:

- Test furnace size and geometry
- Burner mounting location and direction of firing
- Single versus multiburner testing capabilities
- Furnace cooling and temperature control

Test furnaces are tailored to different burner designs. The direction a burner will fire is an important consideration in determining the furnace to be utilized for a test. A poor selection of test furnace can result in data that do not accurately represent the intended application of the burner. All furnaces need sufficient space to accommodate the flame produced from the burner. Upfired and horizontally fired, freestanding burners need to be tall/wide enough to prevent flame impingement and/or propagation of the flame into the stack. Terrace wall firing requires a test furnace with a sloped wall to approximate terrace-type furnaces that are typically used in the steam-methane reforming industry.

Multiburner tests involve two or more burners simultaneously firing in a furnace. These types of tests are common in ethylene applications where one

FIGURE 8.1
Aerial view of an industrial combustion testing facility.

or more wall-fired burners on the furnace floor and several radiant wall-fired burners are firing simultaneously. Thus, the furnace needs to be tall enough to approximate the dimensions of the cracking furnace. It is also common to test the radiant wall burner individually prior to the multiburner test. For this reason, small box type furnaces are available to accommodate the small firing rate of a horizontally installed radiant wall burner.

Furnace cooling and temperature control are important for a number of reasons. A burner test can be compromised by operating at temperatures significantly higher or lower than the design temperature specified by the client. Temperatures that are too low can negatively impact burner stability or overpredict CO production. Depending on the temperature range in which a furnace is designed to operate, testing at a significantly higher furnace temperature could lead to excessive NOx production.

The furnace temperature can be regulated through the use of insulation, cooling tubes, and/or a water jacket. Insulation such as refractory and high temperature mineral fiber is used to protect the furnace steel from high temperature furnace gases. Refractory (see Chapter 5) comprised of oxides of aluminum, silicon, and magnesium is used to line the interior of a furnace and to withstand direct flame impingement. Cooling tubes can be used to cool the flue gases, and the extent of cooling can be controlled by the flow rate of cooling fluid, such as water, through the tubes or by insulating the tubes themselves. Similarly, water jackets are also used to provide cooling. Photographs of several different test furnaces representative of the process burner industry are shown in Figures 8.2 through 8.4.

8.3.2 Air Delivery Systems

Air delivery systems for burners are typically classified as either natural draft or forced draft. Owing to the buoyant forces of the hot gases inside the furnace, the pressure inside the furnace at the burner level is lower than the pressure outside the furnace. This pressure differential, known as draft, is the driving force used to supply air to the burner in natural draft systems (see Chapter 12). Natural draft burners have an air inlet that is open to ambient air.

Forced draft systems use a fan or blower (see Volume 3) to provide air to the burner by way of a plenum or air duct. The air being supplied to the burner can be either at ambient temperature or preheated, with the amount of preheating depending on the application. Pitot tubes

FIGURE 8.2
Test furnace used primarily for ethylene applications.

FIGURE 8.3
Test furnace capable of simulating terrace wall-fired heaters.

are used to measure the static pressure just before the inlet to the burner. The difference in static pressure between the duct and the floor of the furnace provides the pressure drop across the burner.

Forced draft burners utilize the air pressure to provide a superior degree of mixing between fuel and air. Also, with forced draft systems, air control can be better maintained allowing the operator to realize economic savings by operating the furnace at lower excess air rates over a wider firing range. Figure 8.5 shows an example of a mobile air preheater used during forced draft testing with or without preheated combustion air.

8.3.3 Instrumentation and Control

The instrumentation and control system of the furnace controls and measures important parameters such as fuel flow rate, flue gas composition, and furnace temperature. Following is a list of instrumentation and control equipment functions that are discussed:

- Fuel flow and composition
- Analysis of flue gases
- Flue gas temperature and pressure

8.3.4 Fuel Flow and Composition

Accurately metering the flow rate of each individual component used to make up a fuel blend is necessary to measure the heat release of the burner. There are many ways to measure flow: differential pressure, magnetic, mass, oscillatory, turbine, and insertion flow meters, just to name a few (see Chapter 7 and Volume 1, Chapter 9). For purposes of burner testing, the differential pressure flow meter is discussed. Even limiting this discussion to differential flow meters, there are still several different methods of measurement available. Measuring

FIGURE 8.4
Vertical cylindrical furnace for freestanding, upfired burner tests.

FIGURE 8.5
Self-contained, portable combustion air heater and blower used for testing forced draft, preheated air burner designs.

the differential pressure across a known orifice plate is the most commonly applied method.[5] The advantage of using orifice plates is that they are versatile and can be changed to match a flow rate and fuel to be metered. Also, there is a significant amount of data concerning fuel flow via an orifice plate. Finally, there are no moving parts to wear out. A drawback to orifice plates is that they are precision instruments with an accuracy that is generally determined by the condition of the plates: the flatness of the plate, the smoothness of the plate surface, the cleanliness of the plate surface, the sharpness of the upstream orifice edge, the diameter of the orifice bore, and the thickness of the orifice edge.[6] Another drawback is loss of accuracy when measuring flow rates of dirty fuels. While dirty fuels are a way of life for many industrial facilities, test fuels are clean (no liquid or solid particles in the gaseous fuels), and this concern is minimized.

The Coriolis meter is commonly used to measure liquid flow rates. The Coriolis meter utilizes the Coriolis effect to directly measure the mass flow rates of liquids. The meter is equipped with a specially shaped vibrating tube through which the liquid flows. When a fluid flows through the tube, it alters the manner in which the tube vibrates. The mass flow rate of the fluid is then calculated based on this response. The Coriolis meter is a more expensive means of measurement, but this is often offset by its degree of accuracy and its low maintenance requirements.

8.3.5 Flue Gas Analysis

The measurement of NOx, CO, and O_2 concentration in the test furnace flue gas (see Volume 1, Chapters 14 and 15) is necessary to evaluate the performance of the burner. A sample of flue gas is continuously extracted from the furnace using a water-cooled sample probe installed in the furnace stack. The sample flows from the sample probe through heat traced tubing to a sample conditioner inside a data building. The conditioner dries the sample and removes any soot or particulate matter. The sample then flows to a series of emission analyzers where the actual concentration of each species of interest is measured. The methods commonly used to measure each component are briefly reviewed in the following sections.

There are a few common methods of measuring oxygen concentration in the gas phase. Electrochemical sensors and paramagnetic cells are typically used to measure oxygen concentration on a dry and wet basis, respectively. Carbon monoxide (CO) is most commonly measured using a nondispersive infrared (NDIR) technique. A gas sample flows between an infrared radiation source and an infrared detector. Carbon monoxide absorbs infrared radiation, and the difference in intensity is directly proportional to the concentration of CO in the gas sample.

Oxide of nitrogen (NOx) is most commonly measured using chemiluminescence. This method is capable of measuring oxides of nitrogen from subparts per million to 5000 ppm. The principle of operation of these analyzers is based on the reaction of nitric oxide

(NO) with ozone. The sample is drawn into a reaction chamber where it is reacted with ozone generated by an internal ozonator. This reaction produces a characteristic luminescence where the intensity is proportional to the concentration of NO. The luminescence is detected by a photomultiplier tube which in turn generates a proportional electronic signal. The electronic signal is processed by the microcomputer into an NO concentration reading. To measure the NOx (NO + NO_2) concentration, the NO_2 is transformed to NO before reaching the reaction chamber. This transformation takes place at a temperature of approximately 1160°F (625°C). Upon reaching the reaction chamber, the converted molecules along with the original NO molecules react with ozone. The resulting signal represents the total NOx.

8.3.6 Flue Gas Temperature and Pressure

A suction pyrometer (also known as a suction thermocouple or velocity thermocouple) is widely considered to be the preferred method for obtaining gas temperature measurements in the harsh environment of an operating furnace. If a bare thermocouple is introduced into a hot furnace environment for the measurement of gas temperature, measurement errors may arise due to radiative exchange between the thermocouple and its surroundings. A suction pyrometer is typically comprised of a thermocouple recessed inside a radiation shield (see Figure 2.18). An eductor rapidly aspirates the hot gas across the thermocouple. This configuration maximizes the convective heat transfer to the thermocouple while minimizing radiation exchange between the thermocouple and its surroundings, assuring that the temperature is nearly that of the true gas temperature.

Gas pressure is measured using gauges and pressure transducers. Differential pressures are measured using calibrated manometers in conjunction with differential pressure transducers.

8.4 Special Equipment

8.4.1 Heat Flux

Several techniques have been developed to measure heat flux levels at different locations within a furnace. The instruments designed to successfully obtain heat flux data in the hostile environment of a full-scale furnace are typically water-cooled probes, which are inserted through a furnace port at the location of interest. The probes may utilize radiometers that measure radiant heat flux levels. The sensing element is typically composed of a thermopile-type sensor that produces a voltage proportional to the temperature difference between the area of the element that is exposed to heat transfer from the furnace and the area that is cooled and kept at a relatively constant temperature per the element design. Common designs utilize a plug-shaped thermopile element with the exposed face at one end and the opposite end cooled by contact with a heat sink. Figure 8.6 is a

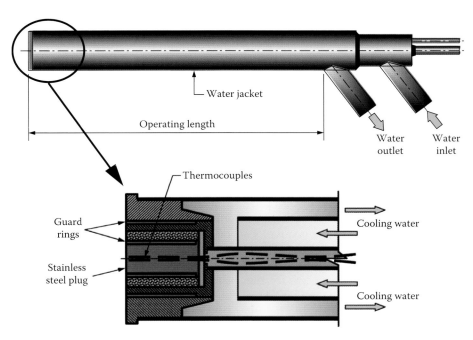

FIGURE 8.6
Schematic of heat flux probe.

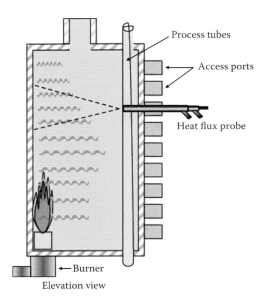

FIGURE 8.7
Schematic of heat flux probe mounted in a test furnace.

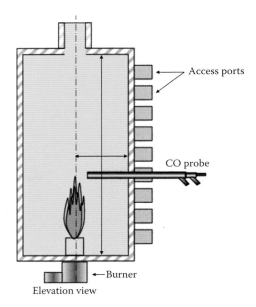

FIGURE 8.8
Schematic of a CO probe mounted in a test furnace.

schematic of a typical heat flux probe. The probe utilizes a crystal window, gas screen, or a mirrored ellipsoidal cavity to negate convective heat transfer to the sensor. A radiometer is also often equipped with a gas purge in an effort to keep the crystal clean and free from fouling. Critical parameters to consider when using a heat flux meter include ruggedness, sensitivity, calibration method, and view angle of the instrument.

Since heat transferred to the furnace tubes is the desired information, the heat flux is placed in the same plane as the hot face of the tubes. The heat flux is measured opposite the fired wall via access ports along the vertical axis of the furnace (Figure 8.7). Once the data are collected, they are plotted on a curve and compared to the desired heat flux profile.

8.4.2 CO Probe

Spatially resolved measurement of local CO concentration (CO probing) can be used to measure flame envelope much more accurately than visual observation. The CO probing typically defines the flame edge as 2000 ppm of CO. In most applications, the industrial burners produce turbulent flames; therefore, CO probe data must be averaged over a period of time. A CO probe is a water-jacketed probe that is inserted into an operating furnace and connected to a sample line that feeds into emissions monitors. Using ports similar to the access ports for the heat flux, the CO probe is traversed across the furnace at different elevations to measure flame width and height (Figure 8.8). Near-flame data can be collected, and concentrations of oxygen, carbon monoxide, and oxides of nitrogen can be measured.

8.4.3 Noise

Noise emissions are becoming increasingly important (see Volume 1, Chapter 16). With some refineries located near populated areas, it is important to keep noise to a minimum. Burner testing is usually conducted on a single burner, and noise emissions are usually measured at a distance of approximately 3 ft (1 m) from the burner air inlet. Data collected during the test include an overall A-weighted sound pressure level and the sound level at each octave band, ranging from 31.5 to 8000 Hz. When collecting noise data, it is important to measure it with and without the burner operating, in order to correct the measurements for background noise.

8.4.4 Unburned Hydrocarbons, Particulate Matter, and Oxides of Sulfur

Combustion processes can create pollutant emissions other than carbon monoxide and oxides of nitrogen. Unburned hydrocarbons (UHCs) is a term describing any fuel or partially oxidized hydrocarbon species that exit the stack of a furnace. The cause for these emissions is typically due to incomplete combustion of the fuel from poor mixing or low furnace temperature. A low-temperature environment can be created by operating the furnace at a reduced firing rate or turndown. Particulate matter (commonly referred to as soot) is often produced from fuel-rich regions in diffusion flames. Soot becomes smoke if the rate of formation of soot exceeds the rate of oxidation of soot in the flame. Oxides of sulfur are formed when sulfur is present in the fuel.

8.5 Test Fuel Selection

The customer fuel cannot always be reproduced exactly at a test facility due to the sheer number of components that could be present and the cost associated with supplying and maintaining such an inventory. To circumvent this problem, a test fuel is blended to simulate the customer fuel and reproduce its fluid transport and chemical properties.

Three properties of the simulated test fuels that need to be matched are the isentropic coefficient (ratio of specific heats), molecular weight, and lower heating value (LHV). The properties need to be closely approximated to reproduce the fuel pressure versus heat release relationships. The Wobbe number may also be calculated to determine similarity in fuels. The chemical properties that need to be matched are the adiabatic flame temperature, inert content, olefins and hydrogen content, and LHV. These properties need to be closely approximated so that the combustion process produces similar flame heights, pollutant emissions, and flue gas temperatures.

As an example, Table 8.1 illustrates a refinery gas. The test fuel chart shown in Figure 8.9 is a general procedure for composing test fuel blends. By following this flow chart, a test fuel can be composed that will accurately simulate both the fluid transport and chemical properties of the actual fuel.

Based on the fuels available for blending, the hydrogen content is matched, propylene is used to substitute the ethylene content, and a natural gas/propane mixture is used to simulate the methane and ethane balance as described in Figure 8.9. By holding the hydrogen content fixed at 38%, natural gas and propylene are balanced to obtain a match of the LHV and molecular weight.

TABLE 8.1

Example of a Refinery Fuel Gas Composition

Fuel Component	Formula	Volume%
Methane	CH_4	8.13[a]
Ethane	C_2H_6	19.9[a]
Propane	C_3H_8	0.30
Butane	C_4H_{10}	0.06
Ethylene	C_2H_4	32.0[b]
Propylene	C_3H_6	0.78
Butylene	C_4H_8	0.66
1-Pentene	C_5H_{10}	0.07
Benzene	C_6H_6	0.12
Carbon monoxide	CO	0.22
Hydrogen	H_2	37.8[c]

[a] Balance of fuel is primarily methane and ethane.
[b] Level of olefins in the fuel.
[c] Hydrogen content.

TABLE 8.2

Comparison of RFG and Simulated Test Fuel

Property	Refinery Fuel	Test Fuel
LHV (Btu/scf)	1031	1026
HHV (Btu/scf)	1124	1121
Molecular weight	18.09	18.38
Specific heat ratio at 60°F	1.27	1.26
Adiabatic flame temperature (°F)	3481	3452
Wobbe index	1422	1407

This balance is further refined by including the adiabatic flame temperature. A test fuel blend of 34% natural gas, 28% propylene, and 38% hydrogen would be acceptable to simulate the refinery fuel gas (RFG) illustrated in Table 8.1. Table 8.2 gives a side-by-side comparison of the fuel properties.

8.6 Test Procedure

Before a burner test starts, all physical dimensions of the burner have to be verified and recorded, in particular, the number and arrangement of fuel ports. The test configuration has to be verified as well, such as the burner spacing and test furnace insulation pattern. Each day of testing must begin with a calibration of the flue gas analyzers.

A test procedure will vary from burner test to burner test due to difference in application. However, there are several test conditions that are routinely tested to determine the general performance of a burner.

The testing of a burner begins with a cold furnace light-off. This is an important test point as it demonstrates burner stability during start-up conditions. Also important during light-off/warm-up period is the production of carbon monoxide. Increasingly stringent emissions regulations are limiting the amount of carbon monoxide that can be emitted during start-up. The light-off point is usually tested with natural gas or with some other purchased start-up fuel, such as butane. If a pilot is present, its performance may also be verified. Pilot stability can be confirmed by increasing the furnace draft, closing and opening the air registers, increasing the airside pressure drop across the burner or increasing/decreasing the pilot fuel pressure. This phase of the burner test is also used for verification of the flame scanner ability to check the flame of either the pilot or the main burner. Once the flame scanner operation has been confirmed, it can be used to verify the burner ignition times and extinction times, whenever these

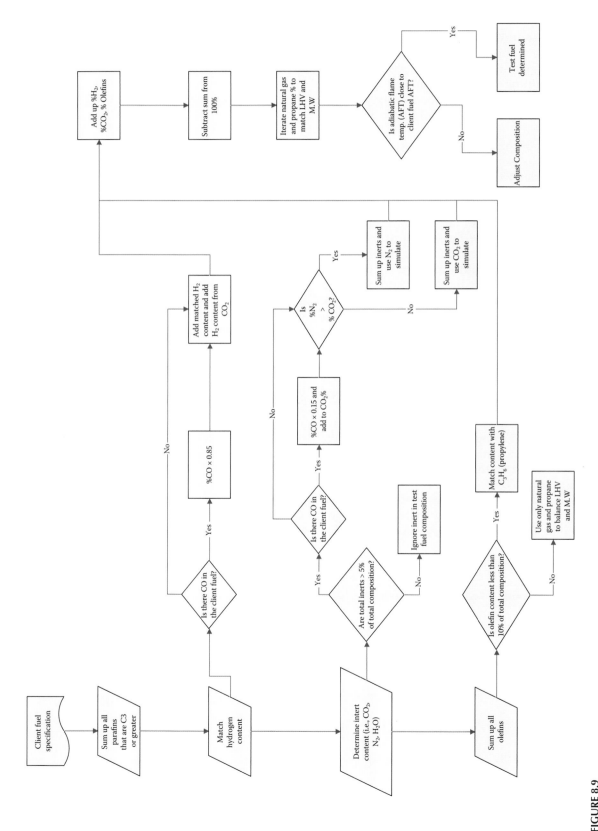

FIGURE 8.9
Test fuel selection process flow diagram.

are a part of the burner specification (e.g., if EN746-2 specifications, apply[6]).

Design heat release or maximum capacity is demonstrated to confirm the burner will provide the heat release required. This test point is often the condition at which the burner will operate for most of its lifetime. Therefore, documentation of flame dimensions and pollutant emissions at these conditions is important. Carbon monoxide break is a condition that is achieved by increasing the fuel flow rate until the furnace is almost depleted of excess air. This is conducted to simulate how a burner will respond in a situation where excess air suddenly decreases. Turndown ratio or minimum heat release is demonstrated by reducing the heat release of the burner. These points are repeated if more than one fuel composition is to be tested. Certain fuels such as start-up fuels are only used during light-off and warm-up periods and therefore do not require testing at every condition.

8.7 Conclusions

Burner testing is an invaluable tool in validating client requirements, research, and development of new products and characterizing burner models to develop predictive tools. Burner testing is an intensive experimental undertaking that requires careful selection of test furnace and setup instrumentation. Testing must be completed in a detailed and disciplined approach in order to collect reliable data. In all cases, burner testing tailors individual burners to meet the stringent demands of furnaces for any application. By testing burners, combustion engineers, clients, and end users can be assured that the product to be utilized will be thoroughly developed and meet requirements regarding safety, reliability, operability, and maintenance.

References

1. C.E. Baukal (ed.), *Industrial Burners Handbook*, CRC Press, Boca Raton, FL, 2004.
2. C.E. Baukal (ed.), *Industrial Combustion Testing*, CRC Press, Boca Raton, FL, 2011.
3. API 560, *Fired Heaters for General Refinery Service*, ANSI/API Standard 560, 4th edn., API Publishing Services, August 2007.
4. API 535, *Burners for Fired Heaters in General Refinery Services, Recommended Practice*, 2nd edn., API Publishing Services, January 2006.
5. Spitzer, D.W., *Flow Measurement: Practical Guides for Measurement and Control*, Instrument Society of America, Research Triangle Park, NC, 1991.
6. European Standard EN 746-2, *Industrial Thermoprocessing Equipment: Part 2: Safety Requirements for Combustion and Fuel Handling Systems*, API Publishing Services, March 1997.

9

Flare Testing*

Charles E. Baukal, Jr. and Roger Poe

CONTENTS

9.1 Introduction 203
9.2 Literature Review 205
    9.2.1 Large-Scale Flare Test Facilities 205
    9.2.2 Field Testing of Flares 205
9.3 Large-Scale Flare Test Facility 205
    9.3.1 System Description 206
    9.3.2 Flow Control System 207
    9.3.3 Data Acquisition System 211
        9.3.3.1 Thermal Radiation 212
        9.3.3.2 Noise 214
        9.3.3.3 Flare Conditions 215
9.4 Flare Pilot Test Facility 215
9.5 Sample Experimental Results 216
    9.5.1 Hydrostatic Testing 216
    9.5.2 Cold Flow Visualization 217
    9.5.3 Ground Flare Burner Interactions 217
    9.5.4 Unassisted Flare 217
    9.5.5 Air-Assisted Flares 217
    9.5.6 Steam-Assisted Flare 219
    9.5.7 Water-Assisted Flare 219
9.6 Summary 221
References 224

9.1 Introduction

ANSI/API Standard 537 (API 537) defines a flare as a "device or system used to safely dispose of relief fluids in an environmentally compliant manner through the use of combustion."[1] Flares (see Volume 3, Chapter 11) are used to destroy unwanted gases in a wide variety of applications from petroleum production, to refineries and petrochemical plants, to downstream product handling.[2] These unwanted gases may include off-spec product or vented streams that may be flammable, toxic, or corrosive.[3] In each case, the flare system must separate liquids from gases (if necessary), ignite the gases, and provide the stable combustion necessary for destruction while minimizing smoke, thermal radiation (see Volume 1, Chapter 7), and noise (see Volume 1, Chapter 16).[4] A flare is an important interface between a plant and its surroundings. Industry, the public, and government regulators are increasingly challenged by the need for factual knowledge of flare performance. Reed notes that flares are the most critical element in the safe operation of a process plant.[5] Chaudhuri and Diefenderfer state that flare system design is very important to the economics of plant operation.[6]

Flares can be classified by geometry (elevated or ground) or by mixing enhancement (steam-assisted, air-assisted, pressure-assisted, or non-assisted).[7] Seebold[8] lists flare height, flame stability, purge requirements,[9] ignition reliability, materials of construction, and liquid

---

* This chapter is adapted from C.E. Baukal, J. Hong, R. Poe, and R. Schwartz, Large-scale flare testing, Chapter 28 in *Industrial Combustion Testing*, C.E. Baukal (ed.), CRC Press, Boca Raton, FL, 2011.

removal as important elements in flare design. Schwartz and Kang list reliable burning, hydraulics, liquid removal, air infiltration, flame radiation, smoke suppression, and noise/visible flame as important design considerations.[4] API 537 lists the following as high-level safety and operating goals for a flare:

- To provide safe,[10] reliable, and efficient discharge and combustion of hydrocarbons with a high reaction efficiency
- To ensure the discharged hydrocarbons burn with stable combustion over the entire defined operating range
- To ensure a continuity of the flare flame under severe weather conditions
- To ensure that ground level concentrations of specified compounds do not exceed environmental limits
- To ensure that the back pressure does not exceed the maximum allowable
- To ensure that velocity throughout the flare burner does not exceed the maximum specified
- To ensure the opacity limit at the smokeless flow rate range does not exceed that allowed
- To ensure the flare radiation intensity and noise levels do not exceed the maximum allowable

Other important aspects of flare design include determining the plume rise to ensure the plume does not affect people inside or outside the plant,[11] ensuring noise levels are acceptable to the plant personnel and surrounding community, and ensuring the flare can handle the range and composition of waste gas flows that will be sent to it. These goals are very difficult to verify in operating flare systems and can generally only be proven with an appropriate experimental facility. Failure to meet these goals can lead to significant problems.[12]

Durham et al. note that a flare test facility can be used to study variables that are critical to flare performance including the following[13]:

- Flame types (momentum vs. buoyancy dominated)
- Flame length, shape, and stability
- Radiation
- Smokeless burning rate
- Nozzle size and number for a given flow, composition, and pressure
- Nozzle location and spacing
- Flame stabilization

- Integration of high- and low-pressure streams for dual-pressure designs
- Gas discharge velocity
- Effects of wind and low-pressure zones on burner life

Other critical issues include the flame stabilization location and acoustic issues that could be high or low frequency. A flare test facility can also be used to validate computational fluid dynamics (CFD) (see Volume 1, Chapter 13) predictions.[14] Noise[15] and acoustic instability[16,17] can also be parameters of interest in flare design and operation. Choi notes that flare performance testing may be part of the process for buying flares.[18] Because of the critical safety nature of flares, reliability[19] is an important element that often can only be thoroughly examined through testing.

Straitz notes that poor design is one of the major factors leading to poor flare performance.[20] API 537 provides many details and guidelines for designing flare systems. One of the major challenges for developing these guidelines is collecting operating data from full-scale flare systems. In a review article, Brzustowski notes the challenges of scaling many of the important factors related to flare design and operation such as flame length, thermal radiation, and noise.[21] He writes, "full-scale testing is almost out of the question in many aspects of flaring… Full-scale tests will be exceedingly rare." He is referring to field or in-plant testing which can be very expensive and can interrupt plant production. Cassidy and Massey also note the difficulty of carrying out accurate and dependable tests on in-plant flares.[22] This chapter focuses on testing large-scale flares. The interested reader can consult Ref. [23] for a discussion of testing small-scale flares.

There are some unique challenges for collecting test data from large-scale flares compared to other combustion systems. The first is the extremely large scale. Some of the largest flares are capable of burning over 1 million lb/h (0.5 million kg/h) of a fuel such as propane, which translates to a firing rate of 20 billion Btu/h (6000 MW). In many cases, the waste gas composition and flow rate may be unknown because of the nature of various waste streams being vented to the flare and the difficulty in measuring such flows.[24] The lower heating value, flammability limits, and stoichiometric ratio can all vary widely with the waste gas composition (see Volume 1, Chapter 3). Another problem is the intermittent operation of flares. The highest flow rates normally only occur during some type of emergency upset condition. These are very rare events that generally cannot be predicted in advance. The location of flares is often a problem as the tips may be elevated hundreds of feet above the ground. Add to that is the fact that flare flames are usually not enclosed, that is, there is no combustion chamber. This makes it very difficult, for example, to make emissions measurements.[25–27] The weather plays an important role in flare operation and is also very unpredictable. Flares must be capable of operation in both high-wind and high-rain conditions. High winds can significantly impact flare performance, and high rains have the potential to extinguish them. The combination of elevated location, no combustion chamber, and unpredictable weather conditions makes it extremely difficult to instrument large-scale flares.

9.2 Literature Review

Relatively few studies have been reported in the literature on large-scale flare testing. The distinction between small- and large-scale testing is somewhat arbitrary. Here, laboratory-size tests, generally conducted indoors, are considered small. Full-scale or near full-scale size flares being tested in an industrial test facility or in the field are considered large scale. For example, field testing of a relatively small, but industrial size flare would be considered large scale.

Large-scale flare testing is categorized in this section into facilities testing and field testing. Test facilities can be used to precisely measure a wide range of parameters under somewhat controlled conditions. Fuel flow rates and compositions can be controlled, but ambient conditions such as wind speed and direction and ambient air temperature, pressure, and humidity generally cannot be controlled in large-scale tests. Field testing is where a flare installed in a production site is tested. There is usually much less control over field tests, with fewer measurements being made. Because of the importance of a flare for plant safety, field tests are generally only practical in very limited circumstances. This literature search section is not intended to be exhaustive but is designed to give the reader examples of various types of larger scale tests that have been conducted.

9.2.1 Large-Scale Flare Test Facilities

Schwartz and coworkers discuss the use of a large-scale flare pilot test facility to develop new flare pilots and monitoring system.[28,29] Hong et al. showed that large-scale testing was an integral part of designing a new type of steam-assisted flare which has many important improvements including significantly reduced steam consumption.[30] D'Amico and Nazzaro describe full-scale flare testing of air-assisted flares.[31] They studied smokeless performance and measured or calculated combustion efficiency, radiation, temperature, and noise. They also measured a wide range of other parameters such as fuel and air flows, flame length, and ambient conditions such as wind speed and direction and air temperature and humidity. Some of these experiments and others that have been conducted in large-scale test facilities are discussed later in more detail in Section 9.5.

9.2.2 Field Testing of Flares

Oenbring and Sifferman describe full-scale field tests of a 16 in. (41 cm) diameter flare in an operating gas plant and in a refinery.[32] Gas flow rates, flame lengths, and flame angles measured from various positions around the flares were reported. However, gas composition, wind speed and direction, noise, and radiation were not reported and do not appear to have been measured. Radiation was calculated using several different calculation procedures. McMurray presented flame radiation data collected from field tests conducted over a 10-year period.[33] Flame length, flame shape, position, and radiation were used to correlate flare radiation as a function of a wide range of parameters such as type, size, orientation, waste gas characteristics, and ambient conditions. Blackwood tested the emissions from two very large flares using open-path Fourier transform infrared technology.[34] Different tracers were introduced into each stack to distinguish between the two plumes which were located fairly close together. A range of emissions were measured. Strosher conducted a 5-year study to investigate the emissions of diffusion flares used to burn solution gases at oil-field battery sites in Canada.[35] The study also included laboratory testing as well. The focus of the study was determining emissions where 199 volatile and semi-volatile hydrocarbons were identified. Schwartz et al. discuss successful field testing of a new acoustic flare pilot monitoring system in an olefins plant in the Gulf Coast.[36]

9.3 Large-Scale Flare Test Facility

A large-scale flare test facility (see Figure 9.1) will be used as an example to illustrate the concepts discussed in this chapter.[37,38] A large maximum flaring capacity allows many flares to be tested at full scale (see Figure 9.2).

Today's industries demand increasingly larger smokeless capacities from flares. Although methods exist for estimating the performance of large flares, full-scale testing is still the most reliable method due to the complexity of the process. Flare design is particularly challenging because of the wide range of operating conditions including fuel composition and flow rates (see Volume 3, Chapter 11). It is not typically feasible to test new flare designs in the field. This is because operating conditions

FIGURE 9.1
World-class flare test facility at John Zink Company, LLC, in Tulsa, OK.

FIGURE 9.2
An air-assisted flare undergoing testing. (From Hong, J. et al., Flare test facility ready for the challenge, Presented at *the John Zink International Flare Symposium*, Tulsa, OK, June 11–12, 2003.)

cannot be modified as desired in the field. One would have to wait for an actual emergency condition to occur with whatever gases happened to be in the system at the time. Discovery of a design flaw in the field during operation would be a big problem as the next time there is a planned shutdown of the flare could be in 5 or more years when the next turnaround is scheduled. Correcting a problem on an operating flare is often impossible without shutting down the unit being protected by the flare which is very expensive and undesirable.

Many flares have some type of assisting media such as steam or air to increase the smokeless capacity.[39] Figure 9.2 shows an example of an air-assisted flare firing propane. The large ducts on the ground, attached to the base of the flare, are connected to two large blowers, one on each side, and provide the assist air. A flare test facility must have the capability of providing large quantities of steam or air for assisted flares.

Some types of flares do not require any assisting media because they are designed to use the available energy from the waste gas pressures. Figure 9.3 shows a comparison of a Coanda-type flare operating at a normal condition and at a condition where the gas flow is higher than the maximum design flow rate. The Coanda effect is no longer present when the maximum design flow rate is exceeded. The flame becomes bushy, and the sound level increases dramatically. This is an example showing the importance of characterizing flare performance to avoid improper application. A flare test facility must be capable of supplying the higher gas pressures typically encountered in these applications.

9.3.1 System Description

A block diagram of the facility is shown in Figure 9.4 where the overall flow is from left to right. Starting at the upper left of the figure, a wide range of fuels are available for testing. Standard fuels include propane, propylene, butane, natural gas, or blends of these. The first three fuels are stored as liquids and vaporized to gases for use in testing. Other fuels such as ethylene can be supplied in temporary storage vessels as needed. Inert gases such as nitrogen and carbon dioxide may also be part of the waste gases that need to be simulated. For lower flow rates, the fuels can be sent directly to a given flare. To achieve higher flow rates, a small and/or large storage vessel is/are filled with gases at an elevated pressure to increase the available

FIGURE 9.3
Pressure-assisted flare utilizing the Coanda principle: (a) normal operation, (b) "breakaway" condition. (From Hong, J. et al., Flare test facility ready for the challenge, Presented at *the John Zink International Flare Symposium*, Tulsa, OK, June 11–12, 2003.)

hydraulic capacity. A compressor can be used to pressurize the tanks and also to circulate the gases to ensure that blends are well mixed. Sample taps are available on the blend tanks, so samples can be taken to measure the blend composition. The fuel then flows through a computer-controlled flow regulation and metering station before going to the flare. Multiple different size metering runs significantly increase the available flow range. A variety of flare testing venues are available, depending upon the flare size and application. Nitrogen is used to purge the lines between tests for safety and to ensure there is no fuel contamination for succeeding tests. Nitrogen is also used to create fuel blends with reduced heat content.

The bottom left side of Figure 9.4 depicts the arrangement of the steam supply system used for steam-assisted flare testing. A high-pressure boiler generates steam that can be stored in a holding vessel to increase the available capacity for tests requiring high steam flow rates. A complete metering and control system regulates and measures the steam flow rates.

Figure 9.5 shows the fuel processing system. The large gas storage vessel (V-114) is located in the center of the picture with the gas flow controls on the right. The LPG vaporizer (with the twin exhaust stacks) is located behind the large storage vessel in the upper part of the photo with the small gas storage vessel to its right. Gas blends are produced by sequentially supplying the appropriate gases in the desired quantities to the fuel storage vessels. The mixture is then circulated with an LPG compressor (to the left of the large storage vessel) until the blend is well mixed. Blends are sampled and tested to verify the proper composition.

Appropriate notification systems have been designed to alert appropriate personnel. The test facility is located only a few miles from Tulsa International Airport. When very large flares are tested, the airport is contacted, and permission must be received from them to run the test, because of the possible impact of a large flame on aircraft traffic. The fire department is also contacted in case they receive calls from concerned citizens about a large fire. Before this procedure was put into place, the fire department sometimes came to the test facility during testing of a large flare prepared to put out a fire. Personnel at the test facility are notified of an impending flare test by a series of horn blasts.

9.3.2 Flow Control System

The flow controls for the new flare test facility were particularly challenging for several reasons. One was the wide range of fuels and flow rates that it would have to handle. The system is capable of flowing less than 100 lb/h (45 kg/h) up to more than 150,000 lb/h (70,000 kg/h). Since the fuel supply system is not capable of continuously

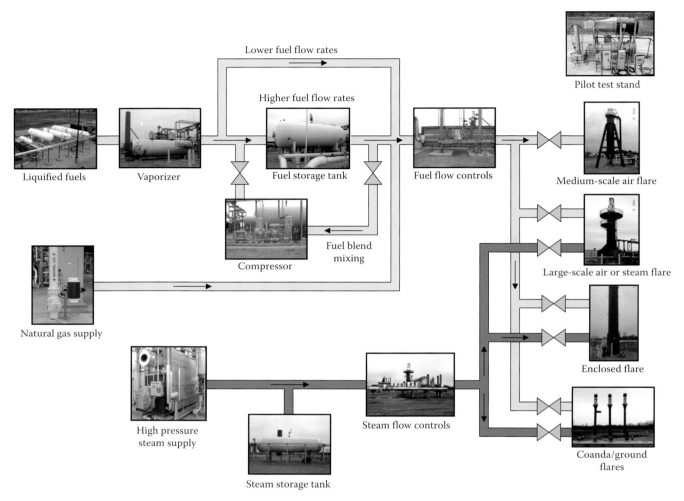

FIGURE 9.4
Block diagram of flare test facility. (From Hong, J. et al., *Chem. Eng. Prog.*, 102(5), 35, 2006.)

FIGURE 9.5
Fuel processing system. (From Hong, J. et al., Flare test facility ready for the challenge, Presented at *The John Zink International Flare Symposium*, Tulsa, OK, June 11–12, 2003.)

flowing the higher flow rates, a pressurized storage vessel is charged with the desired fuel composition at an elevated pressure (above that required to flow the desired amount of fuel). During the actual testing, the fuel is withdrawn from the tank at the desired flow rate. Because of the high flow rates and fuel costs, a high flow rate flare test may only last a matter of minutes. Therefore, a challenge for the flow control system was the extremely fast response time required. The system must reach the desired flow rate quickly and then maintain that rate as the pressure rapidly declines in the gas storage vessel. Depending on the desired flow rate, a test may be completed in less than a minute, so the flow needs to be constant during that time, despite the declining fuel tank pressure. Another challenge is synchronously recording data and video images on a single test record.

Figure 9.6 shows the inside of a specially constructed control building dedicated to flare testing. The building is reinforced concrete with special impact-resistant glass to protect the operators in the unlikely event of an incident during the testing and development of flare designs. The operator seated to the right is controlling the test of an air-assisted flare firing outside the far left window. The control station and operator are shown on the right in the photo. The operator can manually or automatically control the fuel flow rate using the custom-designed hardware and software systems. Flow rates can be varied during a test as required including ramping the flows up and down.

The system was constructed with a wide array of safety devices and instrumentation. There are redundant safety systems to minimize the possibility of an undesired incident. The system can be run with a single operator if desired. An extensive set of procedures was developed based on a comprehensive process hazards analysis (PHA) of the system (see Chapter 1). An environment, health, and safety (EH&S) review is conducted prior to the start of any testing program.

Figure 9.7 shows a typical computer screen from the customized software control system. The left side of the screen shows the current wind speed and direction and the ambient temperature, pressure, and humidity. The right side of the screen shows pressures and flow rates (in this case, zero, as no test was in progress during the screen capture). Some of the other available functions are shown at the bottom of the screen. The software system is designed to systematically and safely guide an operator through a test.

Testing a large flare tip at a very high flow rate can be challenging. For instance, to flow hundreds of thousands of lb/h (kg/h) of a blended fuel to a flare tip typically involves compressing fuels in certain proportions into a storage tank and then depressurizing the tank in a relatively short period of time. The reason

FIGURE 9.6
A test in progress as viewed from the control room. (From Hong, J. et al., *Chem. Eng. Prog.*, 102(5), 35, 2006.)

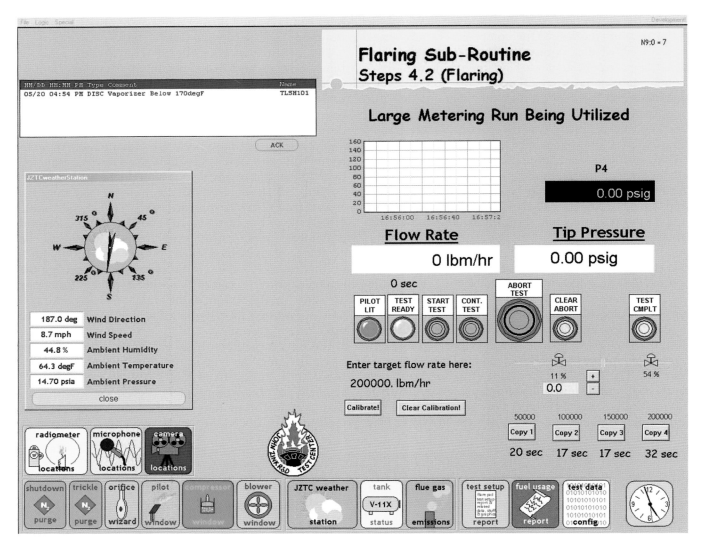

FIGURE 9.7
Typical flare test control screen. (From Hong, J. et al., Flare test facility ready for the challenge, Presented at *The John Zink International Flare Symposium*, Tulsa, OK, June 11–12, 2003.)

behind this batch operation is the extreme high flow rate and the physical limitations on equipment such as vaporizer capacity for LPG (propane, propylene, butane, ethylene, etc.). During the batch operation, the pressure of the tank drops rapidly, and the temperature of the fuel blend also drops noticeably due to the Joule-Thomson cooling effect. When properly planned, short tests can yield valuable data and can be cost effective at the same time.

To maintain a constant mass flow rate to the flare tip, flow control valves must continually be opened to compensate for the decreasing tank pressure. Historically, this was done manually by skilled operators who watched a pressure gauge connected to the flare tip fuel plenum and continuously opened a valve in the fuel supply line in an attempt to keep the tip pressure constant. This process had severe shortcomings. It was a manual feedback control dependent on the operator's skill. During a short duration test, the operator tended to perform an overshoot-correction-undershoot-correction cycle, resulting in oscillation around the target flow rate (see Figure 9.8).

For the upgraded system, a sophisticated computer control algorithm was designed and implemented to quickly reach the target flow rate and then maintain that flow rate within a very tight range. This system includes multiple orifice runs, each consisting of an orifice meter and two actuated valves, one upstream of the orifice meter and one downstream. The algorithm that drives these two valves includes a cascade of feedback controls and a feed-forward control to dramatically improve the response time. This control system is capable of controlling flow rates ranging from 10^1 to 10^5 lb/h (kg/h) of fuel.

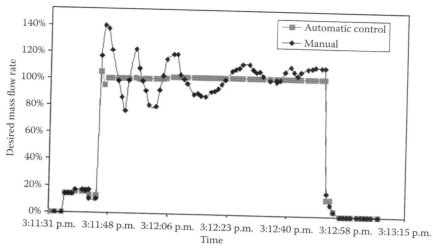

FIGURE 9.8
Graph of a typical flow control history for manual and automatic operation. (From Hong, J. et al., *Chem. Eng. Prog.*, 102(5), 35, 2006.)

9.3.3 Data Acquisition System

At the John Zink Company, various flare designs are tested comprehensively to determine performance parameters such as flame stability, flame length, smokeless capacity, purge rate required, blower horsepower or steam requirements for assisted flares, tip longevity, radiation, and noise. All relevant data are recorded for each test in a single record. The data acquisition system consists of three computers (see Figure 9.9):

1. A computer that provides the human–machine interface (HMI) for the operator. This computer collects general data such as ambient conditions, fuel temperature and flow rates, tip pressure, radiation fluxes and radiant fraction, and

FIGURE 9.9
Flare test control center. (From Hong, J. et al., Flare test facility ready for the challenge, Presented at *The John Zink International Flare Symposium*, Tulsa, OK, June 11–12, 2003.)

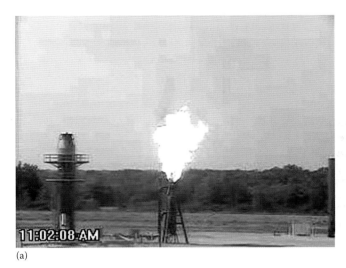

FIGURE 9.10
(a) Upwind and (b) crosswind views of a flame during a flare test. (From Hong, J. et al., Flare test facility ready for the challenge, Presented at *the John Zink International Flare Symposium*, Tulsa, OK, June 11–12, 2003.)

locations of radiometers and microphones. This computer is used to control the actual operation during a flare test.

2. A computer that records the digital videos from multiple cameras positioned to view the flare tests from various directions (see Figure 9.10). These cameras (with pan-tilt-zoom capabilities) are operated from the control room. Videos record multiple frames per second and can be played back simultaneously using a multiplexer.

3. A computer that records the acoustic data is connected to the high-speed noise signal processor (the so-called front end from Bruel and Kjaer Company). Three different computers are used to ensure that the various software packages (HMI, video recording, and noise recording) have enough computational resources. It is important to have these three computers time synchronized. For example, in order to determine the exact pressures at which the flame becomes detached from the Coanda nozzle (see Figure 9.3), the video was visually inspected to locate the transition from an attached flame to a detached flame, and the time stamp on the video was used to identify the flare tip pressure corresponding to the "breakaway" phenomenon from the general data file.

9.3.3.1 Thermal Radiation

Thermal radiation is one of the most important considerations in flare designs.[40,41] Figure 9.11 shows an example of a temperature profile called a thermogram of a flare flame. The stack height of a flare is often designed

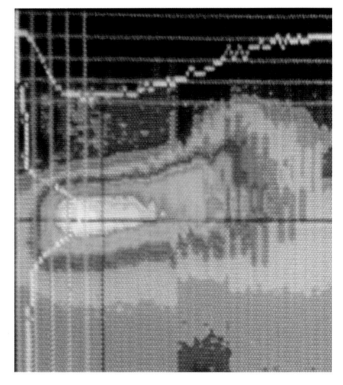

FIGURE 9.11
Thermogram of a flare flame.

to be just tall enough to meet certain radiation criteria at specified locations on the ground. Excessive radiation loads can damage equipment and injure personnel.[42,43] The flare tip design can have a large impact on the radiation characteristics of the flare. Effective tip designs reduce the radiation fluxes from the flame, thus making it possible to use a shorter flare stack and reduce the

Flare Testing

overall cost of the flare system. McMurray argues that the best source of flame radiation data comes from testing.[33]

Before a test is started, the coordinates of the flare and the radiometers used to measure radiation are determined by utilizing a laser range finder to measure distances to three fixed objects with known coordinates and a technique called "triangulation." Multiple radiometers are used to measure various radiant fluxes simultaneously. A photo of the radiation measurement system is shown in Figure 9.12. The measured radiant fluxes, through sophisticated mathematical analysis, are used to determine the coordinates of the effective *epicenter* of the flame, and the radiant fraction, which is defined as the fraction of heat release from combustion that is emitted as thermal radiation.[44] Solar radiation is subtracted from the radiation measurements as appropriate.

If the flare test is expected to produce significant levels of radiation, a flare radiation model is run to estimate the radiation at ground level that could damage equipment and injure personnel. Areas may be blocked off to prevent personnel from entering areas that may have potentially high levels of thermal radiation. In some cases, the grass around the flare may need to be sprayed with water to prevent it from catching on fire, particularly during the summer when the ambient temperatures are high and there has not been any rain for some time.

Numerous calculation methods have been proposed for estimating the radiation from a flare. Schwartz and White have shown that a wide range of predictions is possible depending on which model is used and what assumptions are made.[45,46] Overestimating radiation results in the design of a flare stack that is taller and more costly than necessary. Underestimating radiation means the radiant flux at the ground will be higher than desired and potentially dangerous to personnel in the area during a flaring event. Many designers have relied on the historic models due to their lack of salient data from their flares operating under a variety of conditions. DeFaveri et al. used small-scale experiments to develop a flare radiation model.[47] The problem with such tests is scaling them up several orders of magnitude to actual production flare sizes. A large-scale test facility can be used to compare prediction methods with actual data. The effects of wind on flame radiation can also be studied as flare flames in low-wind conditions are more vertical, while flames under high-wind conditions may be nearly horizontal (see Figure 4.36a).

Figure 9.13 shows a plot of constant radiation lines (isoflux lines) at ground level for a high-pressure flare such as that shown in Figure 9.3. This plot was generated using measurements from an array of radiometers located at various distances and angles from the flare.

FIGURE 9.12
Radiation measurement system (foreground) in use during a flare test. (From Hong, J. et al., Flare test facility ready for the challenge, Presented at *The John Zink International Flare Symposium*, Tulsa, OK, June 11–12, 2003.)

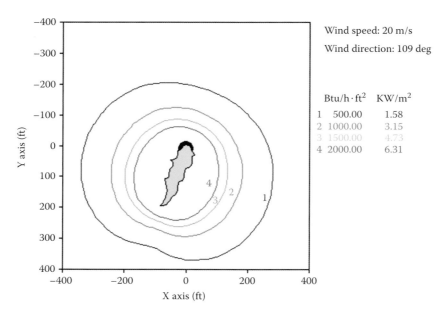

FIGURE 9.13
Isoflux profiles for typical radiation measurement. (From Hong, J. et al., *Chem. Eng. Prog.*, 102(5), 35, 2006.)

9.3.3.2 Noise

Noise (see Volume 1, Chapter 16) is an important emission from a flare that must be adequately controlled to protect personnel in the vicinity of a flare event.[48] To study the effects of noise from flares, a sophisticated sound measurement system was designed for the flare test facility. It consists of state-of-the-art Brüel & Kjær technology and is powered by the latest Pulse Platform. This measurement system (see the photo in Figure 9.14) consists of five stages.

FIGURE 9.14
Sound measurement system in use during a flare test. (From Hong, J. et al., *Chem. Eng. Prog.*, 102(5), 35, 2006.)

The first stage is the transducer where ultrahigh precision microphones (foreground in the picture) are used. The microphones are equipped with a transducer electronic data sheet (TEDS) that enables the sound system to recognize each microphone for ease of identification and communication. In this stage, the sound pressure from the noise source (flare) is picked up by different microphones that are positioned around the source at predetermined distances, heights, and directions. The use of multiple microphones facilitates resolving the directivity of the noise source (flare) and enables verifying analytical models for flare noise.[49,50]

The second stage is the preconditioning circuit in which all signal conditioning such as pre-amplification and anti-aliasing filtration are carried out for all microphones. The third stage is the data logging system where the analog-to-digital (A/D) conversion is completed and the data for signals from all microphones are simultaneously recorded. The fourth stage is the digital signal processing (DSP) in which all signal processing for spectrum analysis such as fast Fourier transform (FFT) and constant percentage band (CPB) are accomplished simultaneously for all microphones. The fifth stage is the post-processing and control platform in which all the sound measurement conditions are set through a powerful graphical user interface.

The duration of the measurements, the microphones to be used, the type of data recorded, and the type of spectrum analyzers are among the numerous measurement conditions that can be varied before and after taking the sound measurements. An example of the sound pressure record from two microphones at different locations, for a typical flare test, is shown in Figure 9.15. The spike at 0 s and again at about 10 s are from a safety horn alerting everyone in the area of an impending flare test. In this particular example, there was a rapid rise in the sound level at the start of the test followed by a steady decline as the fuel flow rate was reduced as dictated by the given test plan.

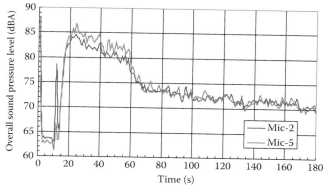

FIGURE 9.15
Typical overall sound level as a function of time. (From Hong, J. et al., *Chem. Eng. Prog.*, 102(5), 35, 2006.)

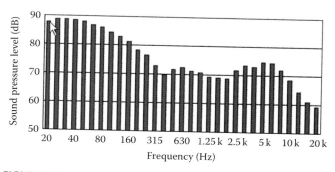

FIGURE 9.16
Average sound profile for a given time window. (From Hong, J. et al., Flare test facility ready for the challenge, Presented at *the John Zink International Flare Symposium*, Tulsa, OK, June 11–12, 2003.)

A one-third octave band analysis for the sound signal from microphone No. 5 during the time period from 20 to 40 s is shown in Figure 9.16. Notice that there are three peaks in the spectrum analysis. These peaks correspond to combustion roar and to the noise generated from the assist air and fuel jets.

9.3.3.3 Flare Conditions

A number of other data points may be collected during a given test, depending on the type of flare and the intended application. These include, for example, metal temperatures (see Chapter 4 for a discussion of metals used in flares) in and around the exit of the flare tip to observe potential overheating problems.

Flares are often tested with noncombustible gases such as air to study flow patterns. One example is the use of smoke generators in an air-assisted flare. Another uses handheld Pitot probes to measure velocities. For the enclosed ground flare, a video camera is used to visually observe the combustion process. A gas analysis system can be used to measure the exhaust emissions (see Volume 1, Chapter 14).

Flare performance can be significantly affected by ambient wind speeds. Wind can cause a flame to bend and change the radiation flux on the ground. In addition, high wind speeds can significantly impact flare performance. Data that are continuously recorded include wind speed and direction and ambient air temperature. Data, such as relative humidity and ambient pressure, that are essentially constant during a test are recorded only once for that test.

9.4 Flare Pilot Test Facility

A test stand (shown in Figure 9.17) was constructed specifically for testing flare pilots (see Volume 3, Chapter 12).[29] These are critical safety devices that

FIGURE 9.17
Flare pilot test stand. (From Hong, J. et al., Flare test facility ready for the challenge, Presented at *The John Zink International Flare Symposium*, Tulsa, OK, June 11–12, 2003.)

must remain lit under severe weather conditions such as heavy wind and rain. The test stand is capable of simulating wind speeds over 160 mph (260 km/h) blowing against both the pilot and the pilot mixer and simulating rain at over 30 in./h (76 cm/h). This test rig was used to develop the WindProof[51] and the InstaFire pilots (both patented) that can remain lit when exposed to those extreme weather conditions.[29]

9.5 Sample Experimental Results

This section gives a sample of some large-scale flare tests. The first few examples are where the flares were not firing. The last few are examples of flares firing. They demonstrate the wide range of types of tests that may be done to determine the performance of flares.

9.5.1 Hydrostatic Testing

In some cases, hydrostatic testing is done to evaluate the maximum pressures that can be introduced to a flare tip. Figure 9.18 shows a senior John Zink engineer carefully taking measurements of a single test element.

FIGURE 9.18
Hydrostatic flare tip test. (From Baukal C. et al., Large-scale flare testing, Chapter 28 in *Industrial Combustion Testing*, C.E. Baukal (ed.), CRC Press, Boca Raton, FL, 2011.)

Pressure was hydrostatically increased until the element began to fail. This scenario was repeated several times to ensure the continuity of the data taken. This element was tested to pressures of more than 20 times the typical working pressure of the unit.

9.5.2 Cold Flow Visualization

In some cases, cold flow visualization can be used to inexpensively display patterns in a flare. The researcher may be looking for flow uniformity, for any recirculation zones, for mixing lengths, entrainment, or other related fluid dynamics. For example, fluid flow uniformity may be important to ensure the resulting flare flame is uniform and not leaning. CFD[52,53] (see Volume 1, Chapter 13) can be used as well as cold flow visualization (see Volume 1, Chapter 11), where both can be important tools to ensure there are no flow inversions in a given tip design. Smoke can be injected into the air stream of an air-assisted flare to help validate flow patterns predicted by CFD. Cold flow visualization can be a powerful tool to validate CFD predictions and to see flow patterns relatively quickly and inexpensively.

9.5.3 Ground Flare Burner Interactions

Ground flares (see Figure 9.19), sometimes referred to as multipoint flares, may be used for a variety of reasons. If a very large turndown is required, rows of burners can be staged on as required. If elevated flare flames are not desirable, this type of flare design is an option. Flare radiation can also be shielded with some type of fence. Design tools have been developed to predict the performance of these ground flares, for example, to ensure proper confinement and accurately predict radiation. Single burner testing will suffice to determine turndown capabilities, baseline flame shape, and a range of smokeless performance. Cross lighting capabilities, total system flame height, and combined radiation confirmation can only be accomplished by more expensive multi-burner testing. Figure 9.20 shows a row of ground flare burners being tested to determine flame height. This is compared to previous measurements of single and dual burners. Single and dual burner tests may underestimate the flame height compared to a row of burners. This is important when designing the height of the flare fence.

Figure 9.21 shows another example of a test for the type of ground flare burners commonly used in the array shown in Figure 9.19. These are full-scale burners being tested to determine flame heights and centerline-to-centerline distances for cross lighting. A range of different molecular weight fuels are tested to determine the performance over a range of conditions that might be encountered in the field.

9.5.4 Unassisted Flare

Another type of flare which uses the Coanda principle (see Figure 9.3) is referred to as an Indair flare. The Coanda effect occurs within a specific pressure range.[54] Full-scale testing can be used to determine the minimum waste gas flow to initiate the Coanda effect. These flares are spring-loaded, so the minimum pressure is needed to determine the proper spring opening pressure. This keeps the tip velocity sufficiently high to negate the smoke typically associated with excessively low turndown pressures. This type of burner is routinely performance tested at application pressures to ensure smoke or sooting is not an issue during normal operation. Figure 9.22 shows multiple Indair burners being tested to determine centerline spacing for proper cross lighting. Cross light timing may also be important as the longer it takes to light flare burners, the more unburned waste gas that could potentially be emitted into the atmosphere.

9.5.5 Air-Assisted Flares

Air-assisted flares (see Figure 9.2) use one or more blowers to supply combustion air to the flare flame, in addition to the ambient air around the flare that is entrained into the flame. The blowers increase the smokeless capacity of the flare. Figure 9.23 shows a series of photos of an air-assisted flare test where the blower goes from off to full on. The flame goes from heavily smoking to smokeless with the addition of the blower.

One safety test that is sometimes done is to simulate a power failure for an air-assisted flare. The blower on an air-assisted flare is designed to increase the smokeless capacity of the flare. A blower failure test is designed to demonstrate that the air flare will still safely handle the waste gases going to the flare with the blower off, although smoke will likely be produced at lower flow rates with the blower off compared to when the blower is in operation. In particular, the blower failure test is

FIGURE 9.19
Typical ground flare array.

FIGURE 9.20
Photo of a row of ground flares, with flame heights compared to those previously measured for single and dual burner tests.

FIGURE 9.21
Multiple ground flare burners being tested to determine flame heights and cross lighting distances.

FIGURE 9.22
Multiple Indair® flare test to determine minimum operating pressure range and tip spacing for cross lighting.

to make sure fuel gas exiting the flare tip does not flow backward in the air duct toward the inlet of the air blower, which could lead to dangerous consequences. This hazardous reverse flow pattern sometimes arises due to the mechanical design of the flare tip. Figure 9.24 shows a high flow rate of propane going to an air-assisted flare with the air blower off. While smoke was generated, the test demonstrated the flare safely operated when the blower was off.

9.5.6 Steam-Assisted Flare

Steam-assisted flares are similar to air-assisted flares in that additional air is entrained into the flare to increase the smokeless capacity.[55] In steam-assisted flares, steam is used to entrain air into the flare, in addition to the ambient air surrounding the flare which is entrained into the flame. Figure 9.25 shows a series of photos of a steam-assisted flare as the steam flow goes from zero to the required flow rate for smokeless operation.

Over-steaming of steam-assisted flares can impact both the stability and the effectiveness of a flare. Figure 9.26 shows the testing of a steam-assisted flare to determine the range of stability when significantly over-steamed. Over-steaming sometimes occurs in a plant when the steam control is manual, so the flare should be stable in case this condition does occur.

9.5.7 Water-Assisted Flare

Figure 9.27 shows a flare on an offshore oil production platform. A particular challenge for these flares is the close proximity to personnel, where it is generally not feasible to elevate the flare high enough to minimize thermal radiation loads on the platform. This is due to the structural issues associated with elevating a flare high above a platform that itself is elevated above the floor of the ocean. One technique that has proven to significantly reduce both radiation and noise levels from flares is water injection.[56] Large volumes of water are not always readily available in a land-based plant, but salt water is very plentiful around offshore platforms. Figure 9.28 shows full-scale testing of water injection in a flare designed for use on an offshore platform.[49] Figure 9.29 shows the results of these tests and how dramatically water injection can reduce radiation from a flare. Figure 9.30 shows how this water injection reduced noise by 13 dBA compared to the same flare with no water injection. Full-scale testing was used to determine the optimum water injector design and injection flow rates.

FIGURE 9.23
Air-assisted flare test: (a) blower off, (b) blower starting up, (c) blower speed increasing, and (d) blower at full speed and flare smokeless.

FIGURE 9.24
Air flare blower failure test.

9.6 Summary

A flare test facility can be used for many purposes. One is to develop new designs with enhanced performance. Another is to characterize the performance of existing technologies, particularly for new conditions that may not have been previously encountered. This includes extending them to new applications such as increasing the operating range of an established design. A well-equipped flare test facility can also be used to validate both proprietary equipment-specific empirical prediction models as well as computational fluid dynamic mathematical models.

In the past, flares have been designed using narrowly defined empirical and semi-analytical models that sometimes produced disparate results. This was due primarily to the limited amount of credible and comprehensive experimental data from industrial-scale flares tested under precisely measured conditions. The large-scale flare test facility discussed here was designed and built to correct that problem and provide data for greatly improving flare analysis techniques leading to new and improved designs. While field testing may be done under certain limited circumstances, this is not generally practical and typically only provides limited data under a narrow range of operating conditions.

FIGURE 9.25
Effectiveness of steam in smoke suppression: (a) no steam, (b) starting steam, and (c) smokeless.

FIGURE 9.26
Steam-assisted flare test experiencing over-steaming conditions.

Flare Testing

FIGURE 9.27
Radiation from an offshore flare. (From Baukal, C.E., *Industrial Combustion Pollution and Control*, CRC Press, Boca Raton, FL, 2004.)

FIGURE 9.28
Testing a water-assisted flare.

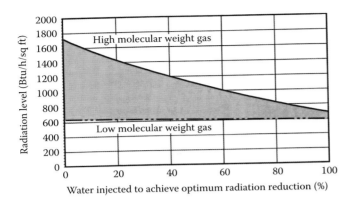

FIGURE 9.29
Radiation reduction by water injection. (From Leary, K. et al., *Oil Gas J.*, 100(18), 76, 2002.)

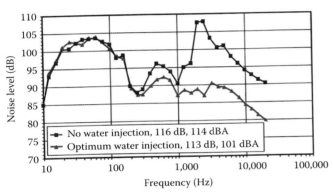

FIGURE 9.30
Noise reduction by water injection. (From Leary, K. et al., *Oil Gas J.*, 100(18), 76, 2002.)

References

1. API Standard 537, *Flare Details for General Refinery and Petrochemical Service*, 2nd edn., American Petroleum Institute, Washington, DC, December 2008.
2. R. Schwartz, J. White, and W. Bussman, Flares, Chapter 20 in *The John Zink Combustion Handbook*, C.E. Baukal, (ed.), CRC Press, Boca Raton, FL, 2001.
3. K. Banerjee, N.P. Cheremisinoff, and P.N. Cheremisinoff, *Flare Gas Systems Pocket Handbook*, Gulf Publishing Co., Houston, TX, 1985.
4. R.E. Schwartz and S.G. Kang, Effective design of emergency flaring systems, *Hydrocarbon Engineering*, 3(2), 57–62, 1998.
5. R.D. Reed, *Furnace Operations*, 3rd edn., Gulf Publishing, Houston, TX, 1981.
6. M. Chaudhuri and J.J. Diefenderfer, Achieving smokeless flaring—Air or steam assist? *Chemical Engineering Progress*, 91(6), 40–43, 1995.
7. F.A. Akeredolu and J.A. Sonibare, A review of the usefulness of gas flares in air pollution control, *Management of Environmental Quality*, 15(6), 574–583, 2004.
8. J.G. Seebold, Practical flare design, *Chemical Engineering*, 91(25), 69–72, 1984.
9. H. Husa, How to compute safe purge rates, *Hydrocarbon Processing*, 43(5), 179–182, 1964.
10. D. Shore, Making the flare safe, *Journal of Loss Prevention in the Process Industries*, 9(6), 363–381, 1996.
11. G. Fumarola, D.M. DeFaveri, E. Palazzi, and G. Ferraiolo, Determine plume rise for elevated flares, *Hydrocarbon Processing*, 61(1), 165–166, 1982.
12. J.F. Straitz, Flare case histories demonstrating problems and solutions, *Process Safety Progress*, 25(4), 311–316, 2006.
13. E.B. Durham, D.D. McClain, and R.E. Schwartz, Profitable flare selection, *Hydrocarbon Processing*, 4(5), 88–90, 1999.
14. D. Castiñeira and T.F. Edgar, CFD for simulation of steam-assisted and air-assisted flare combustion systems, *Energy and Fuels*, 20, 1044–1056, 2006.
15. D. Banks, Combustion pulsation and noise, *Process Heating*, 14(8), 30–34, 2007.
16. G.J. Nathan, P.J. Mullinger, D. Bridger, and B. Martin, Investigation of a combustion driven oscillation in a refinery flare. Part A: Full scale assessment, *Experimental Thermal and Fluid Science*, 30(4), 285–295, 2006.
17. M. Riese, R.M. Kelso, G.J. Nathan, and P.J. Mullinger, Investigation of a combustion driven oscillation in a refinery flare. Part B: Visualisation of a periodic flow instability in a bifurcating duct following a contraction, *Experimental Thermal and Fluid Science*, 31(8), 1091–1101, 2007.
18. D.W. Choi, How to buy flares, *Hydrocarbon Processing*, 79(12), 76–78, 2000.
19. J.F. Straitz, Improve flare design, *Hydrocarbon Processing*, 73(10), 61–66, 1994.
20. J.F. Straitz, Improve flare safety to meet ISO-9000 standards, *Hydrocarbon Processing*, 75(6), 109–114, 1996.
21. T.A. Brzustowski, Flaring in the energy industry, *Progress in Energy and Combustion Science*, 2(3), 129–141, 1976.
22. J. Cassidy and L. Massey, Flare system evolution, *Hydrocarbon Processing*, 8(9), 76–78, 2003.
23. C. Periasamy and S.R. Gollahalli, Flare experimental modeling, Chapter 29 in *Industrial Combustion Testing*, C.E. Baukal (ed.), CRC Press, Boca Raton, FL, 2011.
24. L. Gilmer, C.A. Caico, J.J. Sherrick, G.R. Mueller, and K.R. Loos, Flare waste gas flow and composition measurement methodologies evaluation document, Report prepared for the Texas Commission on Environmental Quality, 2006.
25. B.C. Davis, Flare efficiency studies, *Plant/Operations Progress*, 2(3), 191–198, 1983.
26. J.H. Pohl, J. Lee, R. Payne, and B.A. Tichenor, Combustion efficiency of flares, *Combustion Science and Technology*, 50, 217–231, 1986.
27. C.E. Baukal, *Industrial Combustion Pollution and Control*, CRC Press, Boca Raton, FL, 2004.
28. R.E. Schwartz, New products enhance flare stability, *World Refining*, 12(3), 20–22, 2001.
29. R.E. Schwartz, J. Hong, and J.D. Smith, The flare pilot, *Hydrocarbon Engineering*, 7(2), 65–68, 2002.
30. J. Hong, C. Baukal, M. Bastianen, J. Bellovich, and K. Leary, New steam assisted flare technology, *Hydrocarbon Engineering*, 12(7), 63–68, 2007.

31. R. D'Amico and M.I. Nazzaro, No smoking, *Hydrocarbon Engineering*, 13(7), 47–51, 2008.
32. P.R. Oenbring and T.R. Sifferman, Flare design: Are current methods too conservative? *Hydrocarbon Processing*, 59(5), 124–129, 1980.
33. R. McMurray, Flare radiation estimated, *Hydrocarbon Processing*, 64(11), 175–181, 1982.
34. T.R. Blackwood, An evaluation of flare combustion efficiency using open-path Fourier transform infrared technology, *Journal of the Air and Waste Management Association*, 50(10), 1714–1722, 2000.
35. M.T. Strosher, Characterization of emissions from diffusion flare systems, *Journal of the Air and Waste Management Association*, 50(10), 1723–1733, 2000.
36. R.E. Schwartz, Z. Kodesh, M. Balcar, and B. Bergeron, Improve flaring operations, *Hydrocarbon Processing*, 81(1), 59–62, 2002.
37. J. Hong, R. Schwartz, C. Baukal, and M. Fleifil, Flare test facility ready for the challenge. Presented at *the John Zink International Flare Symposium*, Tulsa, OK, June 11–12, 2003.
38. J. Hong, C. Baukal, R. Schwartz, and M. Fleifil, Industrial-scale flare testing, *Chemical Engineering Progress*, 102(5), 35–39, 2006.
39. O.C. Leite, Smokeless, efficient, nontoxic flaring, *Hydrocarbon Processing*, 70(3), 77–80, 1991.
40. API RP-521, *Guide for Pressure-Relieving and Depressuring Systems*, 4th edn., American Petroleum Institute, Washington, DC, March 1997.
41. W. Bussman and J. Hong, Flare radiation, Chapter 30 in *Industrial Combustion Testing*, C.E. Baukal (ed.), CRC Press, Boca Raton, FL, 2011.
42. K. Buettner, Heat transfer and safe exposure time for man in extreme thermal environment, ASME paper 57-SA-20, *Proceedings of the Heat Transfer Division ASME Semi-Annual Meeting*, San Francisco, CA, June 9–13, 1957.
43. G. Fumarola, D.M. De Faveri, R. Pastorino, and G. Ferriaolo, Determining safety zones for exposure to flare radiation, *Institution of Chemical Engineers Symposium*, Harrogate, England, No. 83, pp. G23–G30, 1983.
44. J. Hong, J. White, and C. Baukal, Accurately predicting radiation from flare stacks, *Hydrocarbon Processing*, 85(6), 79–81, 2006.
45. R.E. Schwartz and J.W. White, Flare radiation prediction: A critical review, Presented at *30th Annual Loss Prevention Symposium of AIChE*, February 29, 1996, Paper 12a.
46. R.E. Schwartz and J.W. White, Predict radiation from flares, *Chemical Engineering Progress*, 93(7), 42–49, 1997.
47. D.M. DeFaveri, G. Fumarola, C. Zonato, and G. Ferraiolo, Estimate flare radiation intensity, *Hydrocarbon Processing*, 64(5), 89–91, 1985.
48. J.G. Seebold, Flare noise: Causes and cures, *Hydrocarbon Processing*, 51(10), 143–147, 1972.
49. W.R. Bussman and D. Knott, Unique concept for noise and radiation reduction in high-pressure flaring, *Proceedings of the Annual Offshore Technology Conference*, Houston, TX, May 1–4, Volume 1, pp. 721–732, 2000.
50. J.G. Seebold and A.S. Hersh, Control flare steam noise, *Hydrocarbon Processing*, 51, 140, 1971.
51. J. Hong, J.D. Smith, R. Poe, and R.E. Schwartz, Ultrastable flare pilot and methods, U.S. Patent 6,702,572, 2004.
52. C.E. Baukal, V.Y. Gershtein, and X.Y. Li, *Computational Fluid Dynamics in Industrial Combustion*, CRC Press, Boca Raton, FL, 2001.
53. D. Castiñeira and T.F. Edgar, Computational fluid dynamics for simulation of wind tunnel experiments on flare combustion systems, *Energy and Fuels*, 22(3), 1698–1706, 2008.
54. D.G. Gregory-Smith and A.R. Gilchrist, The compressible Coanda wall jet—An experimental study of jet structure and breakaway, *International Journal of Heat and Fluid Flow*, 8(2), 156–164, 1987.
55. G.K. Selle, Steam-assisted flare eliminates environmental concerns of smoke and noise, *Hydrocarbon Processing*, 75(11), 117–118, 1996.
56. K. Leary, D. Knott, and R. Thompson, Water-injected flare tips reduce radiated heat, noise, *Oil and Gas Journal*, 100(18), 76–83, 2002.

10

Thermal Oxidizer Testing

Bruce C. Johnson and Nathan S. Petersen

CONTENTS
10.1 Introduction 227
10.2 Equipment and Facility Design Objectives 229
10.3 Test Data Accuracy 230
10.4 Thermal Oxidizer Equipment Testing 230
    10.4.1 Burners 231
    10.4.2 Thermal Oxidizer Chambers 231
    10.4.3 Waste Gas Injection Methods and Configurations 232
    10.4.4 Test Equipment Sizing 232
10.5 Simulating Thermal Oxidizer Input Streams 232
    10.5.1 Combustion Air 234
    10.5.2 Quench Medium 234
    10.5.3 Burner Fuel 234
    10.5.4 Simulating Waste Streams 234
        10.5.4.1 Endothermic Waste Gas Streams 235
        10.5.4.2 Exothermic Waste Streams 236
        10.5.4.3 Aqueous Wastes 237
        10.5.4.4 Special Components 238
10.6 Instrumentation 238
    10.6.1 Chemical Species Analysis 238
        10.6.1.1 Analytical Methods and Instrumentation 238
        10.6.1.2 Chemical Sample Collection 240
    10.6.2 Flow Measurements 241
    10.6.3 Temperature Measurement 242
    10.6.4 Pressure Measurement 242
10.7 Conclusions 242
References 244
    244

10.1 Introduction

With ever-increasing demands on TO equipment, i.e., to improve destruction removal efficiencies (DREs), lower NOx emissions (see Volume 1, Chapter 15), reduce fuel usage, and increase uptime performance, having a facility to test equipment and processes has become increasingly important (see Volume 3, Chapter 8 for a discussion on thermal oxidizers). Seeker has noted "The challenge to waste combustion technology of the future is to burn the ever changing waste streams in a cost-effective, fuel efficient manner without creating other emissions or operating problems."[1] Test facilities are used to determine emissions from previously untested wastes or at untried operating conditions. These data can then be used by companies and regulatory agencies to help establish operating conditions and/or permits for new systems. Detailed discussions of a variety of aspects of industrial combustion testing are given in Ref. [2].

Thermal oxidizer test facilities (see Figure 10.1) also provide a place to simulate and work on problems that can be sometimes encountered in the start-up of new one-of-a-kind commercial systems or when the waste stream compositions or flow rates in operating systems have changed. Solving problems in commercially

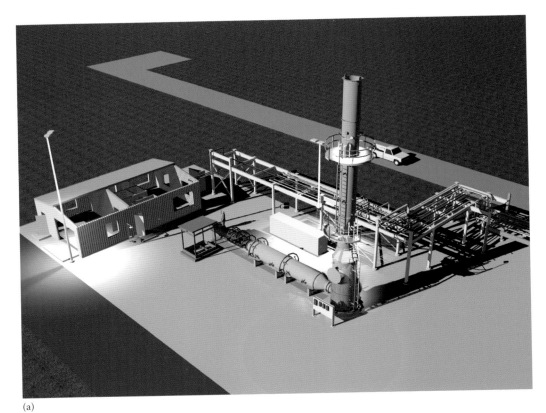

FIGURE 10.1
Thermal oxidizer test facility: (a) drawing and (b) photograph. (From Johnson, B. et al., *Hydrocarbon Engineering*, 15, 71, 2010.)

operating systems without having the benefit of test facility data often results in a lengthy and costly method of problem solving. Some of the difficulties that can be encountered in trying to solve commercial system problems result from the inability to make equipment changes, since the system is required to stay online, and not being able to significantly alter operating conditions since the desired operating conditions may fall outside the range of the equipment's capability or operating permit.

A test facility can also be used to study conditions that may occur infrequently (i.e., emergency conditions in a plant where performance of the system needs to be proven). This includes testing potentially dangerous conditions, where numerous redundant safety systems need to operate correctly to ensure safe operation of the TO system. The test facility can also be used for testing thermal oxidizer control systems.[3]

A test facility is also important for optimizing existing technologies and for developing new technologies.[4,5] It is generally very difficult to do product development in operating units. A test facility can be used to very carefully control specific parameters to study their effects, and test facility equipment is fitted with numerous sample points, pressure taps, viewports, and temperature-measuring locations to better understand the effects of equipment and/or operating parameter changes. Testing results can also be used to identify additional areas for research and development.

A large-scale test facility can be effectively used to train plant operators in a controlled environment, without the time constraints and distractions associated with starting up equipment to meet production demands. Various conditions can be simulated for training purposes in a TO test facility that are not easily obtained in a TO unit operating in a plant. For example, multiple start-ups and shutdowns can be easily performed in a test facility where this would not be practical in an operating commercial system.

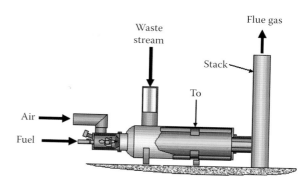

FIGURE 10.2
Schematic of a common configuration of a horizontal thermal oxidizer.

turbulence. In addition to waste destruction, the effects of the three Ts on other parameters such as CO and NOx emissions can also be carefully studied in a test facility. A schematic of a horizontal thermal oxidizer is shown in Figure 10.2.

A properly designed test facility TO system should be equipped with numerous sample ports, pressure taps, thermocouples, gas analyzers, or any other instrument that may provide information that can be used to characterize the performance of the equipment. The facility should be designed to be modular to allow components to be used for multiple purposes and also allow for the insertion of additional components that may be needed from test to test. The utilities, control room (see Figure 10.3), and instrumentation also needs to be designed to allow for simple additions and change outs that may be required.

Additionally, the equipment should be designed to have capabilities to operate over a very wide range of conditions such as throughput and temperature. The materials (e.g., refractory) need to be capable of operating over a wide temperature range and have durability

10.2 Equipment and Facility Design Objectives

The primary purpose of the test facility is to have equipment capable of creating surrogate waste streams similar to those commonly encountered in industry and destroy them under controlled conditions with adequate instrumentation to quantify parameters that affect the process. Destruction of waste streams is achieved by proper mixing, sufficient temperature, and adequate residence time in the TO. These parameters are often referred to as the three Ts of combustion: time, temperature, and

FIGURE 10.3
Thermal oxidizer test facility control room.

to handle significant thermal cycling. Provisions should be in place to allow for a wide variety of fuels and diluents if they are needed. The supply of these components and their metering devices needs to be adequate to allow for careful controlling of test conditions so that high-quality test data can be obtained.

The John Zink Test Facility (see Figure 10.1) has successfully used a horizontal TO system on top of a long metal plate to allow for easy modification of equipment configurations. The horizontal vessel connects directly to the stack, and additional vessels and thermal expansion are accommodated by sliding the equipment on top of the metal plate. Flexible piping is used on all of the connections to the system to also allow for thermal expansion and modularity.

Stack design should be carefully considered when designing a TO test facility since the stack can be the single most costly equipment component and usually requires the most elaborate and costly foundation. Some of the major considerations that need to be addressed in determining the stack size and design include the following: Will the stack ever be used to test natural draft burners? What is the maximum firing rate that could ever be expected? And what is the required/desired stack height? Other stack features to consider include providing additional stack nozzles so a single stack could potentially be used with other pieces of equipment and leaving free space around the stack so other equipment could be either temporarily or permanently attached to it.

Other system design considerations include providing a concrete test pad that is large enough and strong enough to allow for the future addition of equipment and to permit the maneuvering of a crane and fork truck for installing and changing out system components. The addition of future equipment could possibly include a second TO or adding components to an existing TO such as a quench system, venturi scrubber, packed column absorber, or electrostatic precipitator.

With a TO test facility, important operating parameters can be evaluated using a multitude of equipment configurations. For example, in order to study the effects of turbulence on DRE, different types of burners may be used, the pressure drop of the burner can be changed, the TO vessel length-to-diameter ratio (L/D) can be changed, or other geometric features of the TO could be investigated such as the addition of a choke.

10.3 Test Data Accuracy

One of the most important aspects of using a test facility to obtain experimental data is having the ability to check and verify the accuracy of the data being collected. In a test facility with properly calibrated instrumentation, mass and energy balance closures within a few percent are frequently achieved. When field data are obtained from commercial thermal oxidizers, the quality of the data is often questionable. This is because there is seldom time to obtain steady-state conditions, means to verify the accuracy of the instrumentation, or even complete data to close a mass and energy balance around the process to see if the measured data appear to be correct.

In commercial systems, the accuracy of instruments such as fuel and air flowmeters is frequently poor due to the type of meter, placement of the meter, initial setup (mol.wt., temperature range, pressure, etc.), or even as the result of erroneous scaling factors. Attempts made to determine what instrument or if an instrument is in error by making mass and energy balance calculations are also usually not productive since all of the necessary parameters are usually not available to close a mass and energy balance around the system. For example, waste gas flow volumes are often not measured since they may contain aerosols, solids, or tar-like constituents that can plug or coat flowmeters, causing them to fail or to have poor accuracy.

Even if a waste gas flow rate is accurately measured, the composition of the waste is often constantly varying, so closing a mass and energy balance is still not possible as the real-time composition of the gas is still needed and is almost never measured.

When collecting data in a TO test facility, many things can be done to ensure that the instruments used and the data collected are highly accurate. In the case of fuel gas flow measurements, multiple methods of measurements in series can be used. For example, at the John Zink TO Test Facility, three different fuel gas flowmeters were installed in series in order to determine that the original, brand new, properly installed, and calibrated gas flowmeter was in error by approximately 10%.

Once it is determined that a highly accurate fuel flow rate is available, the fuel rate can be used to check the accuracy of the air flowmeter if an oxygen analyzer is available. This is done by measuring the flue gas oxygen content while operating the TO at a fixed fuel flow and airflow rate. A mass and energy balance is then calculated using the measured fuel flow and adjusting the airflow rate (if need be) to get the calculated oxygen value to match the measured oxygen value. The difference between the measured and calculated airflow, if there is one, is the error in the air flowmeter. This method has been found to be very effective in determining the accuracy of air flowmeters because the accuracy of a good extractive-type oxygen meter is very high (within a few hundredths of a percentage) if it is frequently calibrated with certified calibration gases.

When performing any testing in a thermal oxidizer test facility, it is highly recommended that a mass and energy balance is calculated for each test condition before the test is started. Not only does it help to ensure that the test data are accurate, but it allows measurement errors to be quickly recognized so the problem can be immediately fixed rather than finding out later and having to repeat an entire series of tests. Repeating tests can often entail significant expense when it is considered that a TO test system may nominally fire 5×10^6 Btu/h (1.5 MW) of fuel and that it often has to operate up to 24 h just to ensure that the system is at thermal equilibrium before testing can be started. Even more costly is the labor required to repeat the test and the disruption to other scheduled tests.

FIGURE 10.4
Cutaway view of a thermal oxidizer test burner capable of both fuel and air staging.

10.4 Thermal Oxidizer Equipment Testing

10.4.1 Burners

Testing and developing burners specifically for use in thermal oxidizers is a very important function of a TO test facility since burners used in other applications, such as process burners and boiler burners, are not necessarily well suited in thermal oxidizers. One of the main differences in TO burners is that they fire into a combustion chamber with very hot refractory walls, whereas with many other types of burner applications, the combustion chamber walls are relatively cool since heat is being extracted through the chamber walls.

Since TO combustion chamber walls are hot, to maximize the destruction of wastes, some NOx reduction technologies that have been developed for burners in processes with cooled walls are not effective with TO burners. The technology employed in cooled wall chambers primarily involves internally recycling flue gases taken from the cooler side walls and inspirating the gases into the burner throat to reduce NOx (see Volume 1, Chapter 15). However, other techniques can be used to reduce thermal NOx in TO burners. Some of these techniques include fuel gas staging, air staging, partial premix of air and fuel, using numerous small fuel gas tips or ports, and external flue gas recirculation.

Another difference between thermal oxidizer burners and other types of burners is that they are frequently required to fire with much higher quantities of excess air. The high amounts of excess air are required because wastes downstream of the burner may require significant quantities of air for combustion. Also, high excess air may be required to control the combustion chamber temperature since heat is not withdrawn through the chamber walls. In addition, thermal oxidizers sometimes employ burners that are designed to operate under substoichiometric conditions (insufficient combustion air for complete combustion).

Figure 10.4 is a cutaway view of an R&D thermal oxidizer test burner that is equipped with a number of NOx reduction methods. The test burner NOx reduction methods include primary, secondary, and tertiary fuel gas injection in addition to an internal fuel gas ring in a secondary air annulus around the main burner to allow testing of premixed fuel and air. All of the fuel gas delivery methods contained in the test burner would not typically be used simultaneously. External valving is used to direct fuel gas to various combinations of delivery points and at different percentages.

In addition to low NOx emissions, other desirable features of a TO burner include highly turbulent mixing of waste gases or waste liquids, stable operation, high turndown, quiet operation, reasonable cost, and long life.

Highly turbulent mixing is especially desirable in TO waste gas and waste liquid burners since turbulence ensures that the wastes are rapidly mixed with hot combustion products and that wastes are contacted with the available oxygen to ensure waste burnout. Much of the turbulence in the TO system is generated with the combustion airflow pattern. This is one reason forced draft burners are frequently used on thermal oxidizers. Testing different geometries and pressure drops of burners can be used to develop improved equipment, determine optimum operating parameters for existing equipment, or characterize existing equipment. The relative turbulence created by combustion air entry for a given burner geometry can be correlated to the airside pressure drop of the burner. Therefore, test burners should be equipped with appropriate pressure taps to enable airside pressure drop measurements.

When developing new burners, documenting the performance of existing burners, or improving commercial burner designs, special attention should be paid to the construction of the test burner. The test burner should be constructed to facilitate modifications, have sufficient spare nozzles and connections, and incorporate sight glasses.

Test burners are preferably equipped with separate fuel lines with valves to each set of fuel tips. This allows control of the fuel splits to each set of tips without changing tip drillings. Instrumentation should also be in place to determine the fuel flow to each set of tips. This can be done by placing flowmeters in each tip's fuel line or installing pressure gauges that measure the fuel tip pressure. Even with independent flow control to each set of fuel tips, it is desirable for the fuel tips to be easily removable to enable experimentation with different drillings and port angles.

Testing different fuel tip and fuel flow configurations makes it possible to quantify minor species pollutants (NOx, CO) for different configurations that may be used in thermal oxidizer designs. Testing these different configurations at a range of flow rates and fuel splits provides information regarding the operating and stability window for the equipment.

10.4.2 Thermal Oxidizer Chambers

Oxidation reactions of the waste occur primarily in the thermal oxidizer chamber, and the three Ts of combustion can be evaluated by measurements made in this chamber. Numerous nozzles should be incorporated into the chamber design to enable measurement of gas concentrations and temperature throughout the chamber. The effect of residence time on waste destruction is quantified by sampling at various locations along the axial length of the combustion chamber. The effect of turbulence and effectiveness of mixing can be evaluated by measuring gas concentrations and temperature at various radial positions at each axial location. Gas concentration and/or temperature stratification can occur in both the vertical and horizontal directions. Therefore, sampling in both planes is preferable. The gas sampling probe and thermocouple should be designed to have adequate length to have the capability to probe from the vessel centerline to the vessel wall. Care should be taken to seal the sample port after the probe has been inserted to prevent fresh air leaking into the vessel and corrupting the measurements.

Care should also be taken when inserting and retracting the probes to prevent injury. The probes are long and get very hot inside the oxidation chamber. Furthermore, conditions may exist where hot gases blow out of the sample port. Adequate personal protection equipment and proper procedures need to be followed when taking samples.

Permanent gas sample lines and thermocouples should also be installed at the exit of the thermal oxidizer chamber to allow for reliable operation of the thermal oxidizer.

10.4.3 Waste Gas Injection Methods and Configurations

There are numerous methods and equipment configurations for injecting waste gases into a thermal oxidizer. This is one of the reasons for designing a modular test unit. Waste gas properties can vary widely for different applications. The waste gas injection system should be designed to best accommodate the desired process conditions. For example, a waste stream with a heating value near 900 Btu/scf (35 MJ/m^3) available at 5 psig (35 kPa) could be injected essentially the same as a burner fuel, whereas a waste stream with a heating value of 1 Btu/scf (39 kJ/m^3) available at 5 in. w.c. (1.2 kPa) might be injected around a manifold downstream of the burner combustion chamber. The composition of the waste stream can also affect how the process is designed.

After it has been decided if the stream should be introduced through a burner tip or injector into the thermal oxidizer vessel, the type of injector has to be decided. The numbers of injectors, injection angle, and injection locations are critical decisions that have to be made in the design. For example, lean waste gases may be introduced through an annulus around the burner or through multiple tangential nozzles downstream of the burner. Many other types of configurations beyond these, including staging configurations, can also be considered.

The waste injection system can affect both the burner and waste gas combustion performance. It is very difficult to predict the quantitative effects of how a waste stream interacts within the burner's combustion zone without test data. In some cases, waste gases interacting with the burner flame may be beneficial in achieving reduced pollutant emissions. In other cases, the waste gas may act to quench or reduce the rate of oxidation reactions in the flame and inhibit destruction of the waste gases. Testing can be a very valuable tool when evaluating how best to introduce waste gas into a thermal oxidizer.

10.4.4 Test Equipment Sizing

Most of the time, tests are conducted on scaled-down versions of actual equipment since fuel and equipment costs would be very impractical testing full scale. Testing as close to full scale as possible should be attempted where feasible since combustion processes consist of mixing, heat transfer, and chemical reaction processes that do not scale with geometry. For example, scaled-down thermal oxidizers will have a greater surface area per internal volume than a larger vessel with the same L/D ratio. The increased surface area per unit volume leads to an unproportionally high heat loss in the smaller vessel, which affects the temperature profile in the vessel. The higher temperatures in the upstream end of a scaled-down test unit can produce higher NOx emissions and higher destruction efficiencies than would be observed in a larger system.

Velocity matching should always be considered in test equipment sizing, especially at the injection points. The mixing characteristics in the combustion process are strongly tied to velocity. Matching the velocities makes it possible to simulate fluid shearing of two mixing streams, mixing due to fluid impingement, and inertial mixing characteristics.

Matching pressure drops should also be done where it is possible since pressure drop energy is frequently utilized to generate turbulence. For example, the pressure drop taken in burner spin vanes or bluff bodies generates flow patterns that are very effective in mixing burner combustion products into waste streams.

Thermal oxidizer design does not rely on turbulence alone to maximize mixing characteristics. The arrangement of the injection geometries of air, gas, and wastes and their relative proximities to each other needs to be carefully evaluated in designing the test system. Some basic qualitative understanding (the more quantitative the understanding, the better) of the mixing and reaction phenomenon is essential to understanding how to best scale down the test system. Even though selecting the injection geometries in the test system is grounded in scientific principles, it is still very much an art. The unknown factors associated with these decisions may, in fact, be driving the purpose for the test in the first place.

Many times in the scale-down, the number of injectors and their relative locations and flow rates cannot be maintained. Selecting one over the other can be quite challenging. Understanding the processes that are most representative of the performance will dictate the decisions that are made.

For example, scaling down a burner with multiple-staged fuel tips will require selecting the number of tips, distance between tips, and flow to each tip. The burner tip performance, especially minor pollutant emissions (e.g., NOx), is strongly related to the specific flames formed by the tips. Maintaining tip spacing and individual tip flow is more likely to simulate the performance of the full-scale burner since the flamelets from each of the orifices will be comparable to the full-scale burner. One of the uncertainties associated with the scale-down is the effect that burner throat diameter has on the data. The mixing distance from the fuel tips to the center of the burner will be smaller in the test burner than in the full-scale burner even though each tip flow and tip spacing may be identical to the full-scale burner design. This is an unavoidable consequence of scale-down, which is why testing at near full scale may be desirable.

Many times, thermal oxidizer burners are designed with a single burner tip in the center of a bluff body cone. Simulating the minor species pollutant emissions and flames out of a single, central fuel tip will require a different approach since the flow through the tip is necessarily smaller in the scaled-down burner. A test of this type is more practically scaled by matching velocities and pressure drops in the burner, which also scales the flow rates accordingly.

Deciding whether to maintain waste injector number or distance between injectors has to be considered in designing waste injection nozzles. Further complicating this decision is deciding the injector positions relative to the burner. The test equipment should generally try to mimic the location of waste injection relative to the burner flame since this can greatly affect the performance. Equipment that is designed for wastes to pass through the burner throat or around the perimeter of the burner may be introduced at these positions as would be done in the full-scale unit. Equipment that is designed for wastes to be injected downstream of the flame will have injectors positioned at or beyond the expected flame length, which will require more thought about how the flame shape scales.

Sometimes waste gases are introduced through an annulus around the perimeter of the burner tile. Fluid shearing between the waste gas and combustion gases along with turbulence generated in the throat of the burner dictates the mixing performance. One approach to simulate this effect on a scaled-down unit is to match velocities through the waste gas annulus and burner throat to recreate a similar fluid shearing interface condition. Pressure drop through the test burner throat should also be matched to re-create similar mixing energy from the full-scale burner throat. The relative dimensions of the thickness of the waste gas layer injected out of the annulus, the mixing distance across the burner throat, and the flame length will not scale proportionally, which will introduce some uncertainties to the test. These effects could result in achieving more complete mixing near the end of the test burner flame, where this same mixing condition may be achieved closer to the middle of the flame in the full-scale system, which can affect NOx emissions, destruction efficiency, etc.

Systems that are designed with waste injection systems intended to allow a certain level of burnout between injection points may be scaled on distance and/or vessel volume between the injection points. Scaling on volume alone may be inadequate because a certain mixing distance out of the injector is required to get the waste gases into the hot combustion gases before the residence time for burnout "starts." The level of waste gas penetration into the center of the vessel can also be difficult to scale since smaller systems will have a shorter distance to travel to the vessel centerline, which can affect the results. Matching the effective residence time between injection points and mixing efficiencies becomes more and more difficult with each level of scale-down, which is another reason for testing near full scale on certain tests.

CFD modeling (see Volume 1, Chapter 13) can also be used in some instances to assist in making equipment scaling decisions since equipment scaling is largely dictated by mixing processes. There is no one size fits all relation or correlations that can be used to make these decisions. A case by case analysis has to be done. The following is a list of relations that can be used as a starting point in making the scaling decisions:

1. Velocity matching
2. Pressure drop matching
3. Fixed distance spacing
4. Scaled distance spacing
   a. Diameter ratio relations (e.g., number of injector diameters between injectors)
   b. L/D relations
5. Relative volume matching

Other factors that must be considered in sizing a test system relate to the information that needs to be collected. For example, the test thermal oxidizer vessel should be long enough to collect data for a range of residence times. Sufficient sight ports, pressure taps, sample ports, and thermocouples must be provided to allow information to be collected regarding temperature, mixing, and pressure profiles, in order to provide additional insight into the performance of the system. Fabrication limitations can also play a role in designing the test system.

10.5 Simulating Thermal Oxidizer Input Streams

Waste gas properties have a profound impact on operating conditions. Testing needs to be conducted using waste properties as similar as possible to the real waste. Waste streams are unique to the location or plant in which they are created and cannot be easily transported to a test facility. Process streams can be simulated by blending each of the pure components together to closely match the process stream, but this is often impractical due to the number of compounds needed, equipment requirements, and economic considerations. Therefore, a priority of physical and chemical properties for the process stream needs to be developed so a simpler, cost-effective surrogate can be used in place of the process stream for the test. The following discussion gives guidance on what properties are important and how they may be prioritized for different process streams. Figure 10.5 shows a metering skid where multiple components can be independently metered and mixed for simulating an input stream.

10.5.1 Combustion Air

Thermal oxidizers can be designed to operate with fresh ambient air, preheated air, or even oxygen-containing waste gas streams as the combustion air source. The air source composition and temperature can have a profound effect on the operation of the system. Therefore, it is important to conduct testing with a representative combustion airstream.

The physical properties of the test combustion air source should be matched as closely as is practical to the actual air source that will be used in the field. The most important properties of the combustion airstream are the temperature and oxygen content. The concentration of inerts, especially those with relatively high heat capacities (e.g., CO_2, H_2O), should also be factored into how a given test should be conducted.

10.5.2 Quench Medium

Thermal oxidizers are usually designed as adiabatic chambers. Heat recovery equipment is common, but it is almost always located downstream of the thermal oxidizer chamber. Therefore, quench fluids are used to control the temperature of the flue gases in the thermal oxidizer chamber by direct cooling. The most common quench fluids are liquid water, steam, or air.

The quench fluid affects the composition and volume of the resulting flue gases. Therefore, the tested quench fluid should be the same as the actual quench fluid of interest.

10.5.3 Burner Fuel

Thermal oxidizers are generally operated with natural gas as the primary fuel source. However, different types of fuels such as liquified petroleum gas (LPG), fuel oil, or refinery gas can be used. Most natural gas blends that are used across the world are similar enough that the performance of the thermal oxidizer will not be significantly affected by conducting a test with any local natural gas supply to simulate another.

FIGURE 10.5
Test facility metering skid. Multiple components are independently metered in each line and mixed in a header to be delivered into the test equipment.

Surrogates of LPG can be made by blending mixtures of propane, butane, and natural gas. The surrogate LPG fuel should closely match the heating value and molecular weight of the actual fuel.

Liquid fuels can be simulated using mixtures of common petroleum oils (such as No. 2 or No. 6) and water. Liquid fuel requires atomization to allow for efficient combustion. The atomization method has a very large effect on the combustion process. Therefore, it is critical to test the liquid under atomization conditions that will be used in the commercial unit. Matching the injection geometry, spray angle, pressure drop, and atomization fluid to liquid fuel ratio is very important in simulating liquid fuels since the atomization affects combustion processes and pollutant formation. The most important fuel characteristics to match for liquid fuels are heating value, density, and viscosity.

Refinery gas blends may contain significant quantities of hydrogen, inert gases, methane, and/or heavier hydrocarbons. Refinery gas blends can usually be simulated by matching the H_2 concentration and inert concentration (using N_2 or CO_2) into blends of natural gas to match the heating value of the actual fuel. Refer to testing recommendations by Baukal.[6,7] for additional details on simulating refinery fuels. See Volume 1, Chapters 3 and 4 for more detailed discussions of fuels.

10.5.4 Simulating Waste Streams

There are multiple combustion properties that need to be considered when attempting to simulate waste streams. Matching all of the combustion properties is usually not possible or practical. Therefore, identifying the most critical concern in the ability to oxidize the waste should be done so that a priority can be given to matching the most important combustion properties. In some situations the TO may be limited by mixing; in other situations it may be limited by temperature, or secondary pollutant formation may be the primary concern.

There are many waste gas properties that may be considered in selecting an appropriate test surrogate. Some bulk properties frequently considered in simulating waste gases are

1. Heating value
2. Oxygen content
3. Inert content
4. Stoichiometric oxygen requirement

Matching the heating value makes it possible to recreate similar gas volumes and temperatures associated with the combustion of the actual waste gases. The concentration of oxygen in the waste gases is expected to have a significant impact on combustible burnout and also minor pollutant formation (e.g., NOx). The inert constituents in the waste gas dilute the concentration of the combustion reactants and cool the flue gases, which will have an impact on the combustion reaction rates in the thermal oxidizer. The oxygen demand of the waste gas determines the minimum quantity of combustion air that must be supplied to complete oxidation of the waste gas. The consumption of oxygen by the waste gas also affects the flue gas oxygen concentration, which has an effect on waste burnout and minor pollutant emissions. Most of the time, it is possible to recreate bulk properties using mixtures of simple hydrocarbons (e.g., methane, propane), nitrogen, carbon dioxide, and/or steam.

Simulating the waste gas combustible constituents requires a little more thought. The ability to destroy or oxidize waste is strongly tied to chemical kinetics. Matching the bulk properties alone usually does not adequately address this issue. Commonly tabulated combustion properties can be used to help make qualitative assessments of chemical kinetics. Some of these commonly available properties are autoignition temperature, flame speed, and flammability limits. These parameters are useful because they can be compared for different compounds to gain insight into which compounds may be most difficult to combust. However, they cannot be used to generate quantitative correlations because they are inherently oversimplified. Furthermore, these parameters vary from source to source.[5]

The autoignition temperature (sometimes referred to as minimum ignition temperature) is "the lowest temperature at which combustion of a given fuel can start."[8] The autoignition temperature is analogous to the activation energy of the combustion reaction since it gives insight into the energy barrier required to begin the combustion reaction. In general, compounds with higher ignition temperatures are more difficult to combust than those with lower ignition temperatures. However, looking at autoignition temperature alone is an oversimplification. For example, hydrogen with a relatively high autoignition temperature near 930°F (500°C) is considered easier to burn than hydrocarbons with lower autoignition temperatures.

Flame speeds are reported for multiple fuels.[8] The flame speeds in these tables are measured by metering stoichiometric fuel and air into a premixed Bunsen burner at different flow rates (under laminar flow conditions) until the burner flashes back. These flame speeds are analogous to the reaction rate constants of the combustion reaction. It should be noted that these reported flame speeds may not be observed in different burner geometries and configurations (especially those that have turbulent flows).

Flammability limits, another commonly tabulated combustion property, are not so easily associated with a single chemical kinetic property. In general, fuels with a wider flammability limit burn more easily than those with narrow limits because combustion with fuels having narrow flammability may be easily starved for air (when flammability limit is exceeded) or quenched (when below the flammability limits).[9] Note that evaluating flammability limits for mixtures and temperatures other than ambient temperature is not a simple process. Bureau of Mines publication "Limits of Flammability of Gases and Vapors" can give some insight into the complexity of and methods used to determine flammability limits over a wide range of conditions.[10]

Blending surrogates to match autoignition temperature, flame speed, and flammability limits is not feasible since these properties are not well defined for mixtures. Selecting a "limiting" surrogate can be an effective approach. The "limiting" aspect can be arguable depending on the properties of the waste gas. Many times, getting a quantitative relationship between a surrogate and the actual waste gas cannot be accomplished. Selecting a surrogate that is expected or known to give conservative estimates of the thermal oxidizer performance is a reasonable approach to overcome this challenge. Some general waste gas categories are presented in the following text to give some indication on approaches that can be taken.

10.5.4.1 Endothermic Waste Gas Streams

Hazardous vent streams typically contain trace quantities of toxic compounds or pollutants in a diluent such as air, nitrogen, or carbon dioxide. These streams are commonly described as "endothermic" (even though they have a heating value) because a significant amount of energy is required to initiate and sustain oxidation of the combustible species in the waste. The primary concerns when treating these waste streams are to maximize DRE and minimize energy consumption.

One of the most important physical properties to consider in a hazardous vent stream is the heat capacity of the stream since it will determine how much heat must be supplied to get the mixture to the ignition temperature. Combinations of air, nitrogen, carbon dioxide, and steam should be metered to match the concentrations in the vent stream of interest or at least mimic the heat capacity. Matching the actual diluent components should be done where it is practical since they can participate in the chemistry of oxidation reactions (e.g., O_2 content, OH radical formation).

Many times the pollutants of interest are toxic to breathe such as H_2S or benzene. Testing with hazardous components is strongly discouraged due to the safety of personnel. There are many less toxic surrogates that may be used to evaluate whether the target components can be oxidized in the system. Research sponsored by

Thermal Oxidizer Testing

TABLE 10.1

Incineratability of Several Common Hazardous Air Pollutants

Rank of Difficulty to Incinerate	Compound Name	Compound Formula
1	Acetonitrile	C_2H_3N
2	Tetrachloroethylene	C_2Cl_4
3	Acrylonitrile	C_3H_3N
4	Methane	CH_4
5	Hexachlorobenzene	C_6Cl_6
6	1,2,3,4-Tetrachlorobenzene	$C_6H_2Cl_4$
7	Pyridine	C_5H_5N
8	Dichloromethane	CH_2Cl_2
9	Carbon tetrachloride	CCl_4
10	Hexachlorobutadiene	C_4Cl_6
11	1,2,4-Trichlorobenzene	$C_6H_3Cl_3$
12	1,2-Dichlorobenzene	$C_4H_4Cl_2$
13	Ethane	C_2H_6
14	Benzene	C_6H_6
15	Aniline	C_6H_7N
16	Monochlorobenzene	C_6H_5Cl
17	Nitrobenzene	$C_6H_5NO_2$
18	Hexachloroethane	C_2Cl_6
19	Chloroform	$CHCl_3$
20	1,1,1-Trichloroethane	$C_2H_3Cl_3$

the EPA[11] was done to evaluate the thermal stability of 20 common hazardous waste compounds. Table 10.1 ranks the compounds in the study according to the difficulty of incinerating the compound. Notice that methane (CH_4) is more difficult to oxidize compared to most of the compounds studied and, for the most part, is considerably less toxic. Therefore, simulating the hazardous species with an equivalent concentration of methane creates a more difficult stream to oxidize. Others have reported that using CO or THC (e.g., natural gas) does not provide a direct correlation to organic destruction, but does act as a conservative indicator of destruction efficiency.[1]

Testing experience at the John Zink Test Facility has indicated that the oxidation of carbon monoxide in endothermic waste streams is also more difficult compared to most other hydrocarbons. Using carbon monoxide as a test surrogate is also possible, but it is more toxic than methane.

Methane (contained in natural gas) and carbon monoxide also make good surrogates because they are relatively easy to obtain, handle, and meter. Furthermore, common CEM analyzers detecting CO and unburned hydrocarbons (UBH) can be used to measure their presence for determining DRE.

Table 10.2 illustrates an example of a test surrogate that may be used to simulate a hazardous vent stream.

TABLE 10.2

Simulation of Hazardous Vent Stream

Hazardous Vent Stream Composition	Surrogate Vent Stream Composition
79 vol% CO_2	81 vol% CO_2
4 vol% O_2	19 vol% air (4 vol% O_2, 15 vol% N_2)
12 vol% N_2	200 ppmv CH_4 (natural gas)
5 vol% H_2O	
150 ppmv chlorobenzene	
50 ppmv propane	

10.5.4.2 Exothermic Waste Streams

An exothermic waste stream contains enough chemical energy to generate temperatures high enough to sustain continuous oxidation without adding additional energy. However, they often require an auxiliary energy input to initiate the ignition of the waste gas. In some cases, these types of waste streams can present difficulty in achieving oxidation due to equipment mixing limitations, or poor droplet vaporization (with liquid fuels). Testing conducted with exothermic waste streams may be needed to identify optimum burner firing rates (e.g., minimum required ignition energy), operating conditions, and/or equipment configurations for systems destroying these wastes.

Waste stream properties having the largest effect on operating conditions are the heating value, stoichiometric oxygen requirement, and properties that may affect mixing and dispersion (especially for liquid wastes). Matching these properties as closely as practical will help ensure that the results from the test equipment are representative of commercial equipment results. To a lesser extent, matching the inert components should be attempted when practical with emphasis on components with the highest heat capacities such as H_2O and CO_2.

Table 10.3 illustrates how an exothermic waste gas stream may be simulated with a simpler surrogate waste stream.

TABLE 10.3

Simulator of Exothermic Waste Stream

Exothermic Waste Gas Properties	Surrogate Waste Gas Properties
88 vol% CO_2	94.6 vol% CO_2
6.5 vol% H_2O	5.4 vol% C_3H_8
2.3 vol% CH_4	
1.6 vol% C_3H_8	
1.2 vol% C_5H_{12}	
0.4 vol% C_8H_{18}	
LHV = 126 Btu/scf	LHV = 125 Btu/scf
Air demand = 1.3 mol air/mol waste gas	Air demand = 1.29 mol air/mol waste gas

10.5.4.3 Aqueous Wastes

Aqueous wastes contain hydrocarbons within a liquid water stream. These streams may be endothermic or exothermic depending on the concentration of oxidizable compounds. Proper atomization of the liquid stream is probably the most critical aspect of treating aqueous wastes. This is because atomized droplets must be small enough to be completely evaporated within the thermal oxidizer chamber and still allow for sufficient residence time to oxidize the combustible compounds. The fluid nozzle, pressure drop, and atomization fluid (e.g., air or steam) demand (if dual fluid atomization is used) should be matched to that used in the field to obtain a meaningful simulation.

The composition of the test fluid should be selected to match the heating value, viscosity, and density of the waste stream of interest.

10.5.4.4 Special Components

In some cases, a species cannot be simulated using a simpler surrogate. A particular combustion property of the species may be the limiting case in achieving good combustion. For example, conjugated hydrocarbons (e.g., benzene, butadiene, propylene) have a tendency to form smoke. The double bonds in these species make them prime precursors for forming soot. Simulating the smoking tendency cannot be done with simple hydrocarbon fuels such as methane or propane. The actual compound responsible for the smoke formation may be needed for such a test. At least, a species of similar chemical structure should be used to collect meaningful data.

Components with fuel-bound nitrogen are another common species that cannot be simulated with simple hydrocarbon fuels. This is not to say that the actual chemical-bound species must always be used for the test. Hydrogen cyanide (HCN) is obviously not desirable for testing due to its high toxicity. Ammonia or pyridine could be used to provide the necessary fuel-bound nitrogen for the test. The surrogate fuel containing fuel-bound nitrogen should be blended in the appropriate amount to match the nitrogen weight composition of the actual fuel.

10.6 Instrumentation

Any testing campaign requires measurements to evaluate the thermal oxidizer's performance. The set of instrumentation used for the test should be carefully chosen. Each instrument will have certain capabilities and limitations that will affect the quality of the test data. The instrument that is used needs to be capable of making measurements over the entire testing range. Furthermore, the instrument needs to have suitable accuracy for the expected measurement range. The durability and maintenance requirements of instruments will also affect the selection. See Chapter 7 for a discussion of common diagnostic equipment used in combustion testing.

10.6.1 Chemical Species Analysis

10.6.1.1 Analytical Methods and Instrumentation

There may be many types of chemical species of interest for a given test. The most common types of analyzers associated with combustion equipment are contained in a rack or cabinet as seen in Figure 10.6. The typical analyzers found in one of these cabinets are a paramagnetic oxygen analyzer, infrared absorption carbon monoxide analyzer, UBH analyzer, and chemiluminescent NOx analyzer. Commercial sulfur dioxide analyzers based on infrared or ultraviolet absorption are also readily available to fit into one of these racks. The analyzers found in these cabinets contain sample conditioners to remove water vapor from the stream since the analyzers cannot handle condensing water. The water vapor is removed by condensation in chillers and further removed with desiccants. Since water vapor is removed prior to any measurements, the chemical species concentrations are measured on a dry basis rather than the true concentration at the sample point. Mass and energy balances are required to determine how the analyzer dry measurements and true wet measurements relate to each other. Small portable analyzers equipped with multiple electrochemical sensors are also available for combustion flue gas analysis. Figure 10.7 depicts a common portable combustion gas analyzer.

Other species that may be of interest require more specialized equipment that is more difficult to use and/or interpret results. A Fourier transform infrared (FTIR) spectrometer, as can be seen in Figure 10.8, can be used to detect a wide array of components. The FTIR requires that the species of interest be "IR-active." Most polar and polyatomic molecules are "IR-active." An FTIR absorption spectrum is useful for characterization of gases since each molecule has its own unique absorption spectra sometimes referred to as a "fingerprint." The absorption spectrum is also useful for quantification since the molecule concentration is proportional to absorption. Many species may be measured simultaneously with an FTIR, but the accuracy and detection limits will vary depending on the absorption spectra. Nonlinear spectral fitting can be used to quantify species with overlapping absorption spectra.

Gas chromatography (GC) can also be used to measure component concentrations in a mixture. A GC separates components in a mixture inside a column by differences in polarity. Each species in the mixture has a slightly different molecular polarity, which causes

Thermal Oxidizer Testing

FIGURE 10.6
Commercial test rack with the following components from top to bottom: CO meter, oxygen meter, NOx meter, total hydrocarbons (THC) meter, and gas conditioning system.

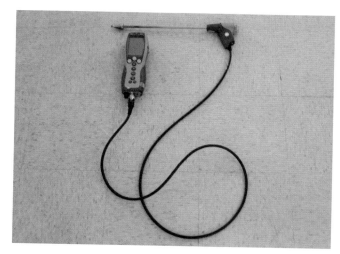

FIGURE 10.7
Portable combustion gas analyzer. A small handheld probe is inserted into the combustion gases and the extracted gases are analyzed using electrochemical cells installed in the instrument.

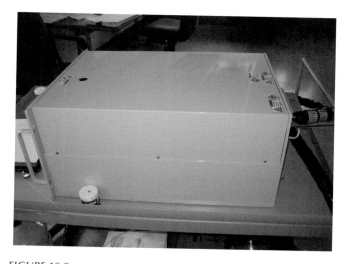

FIGURE 10.8
FTIR gas analyzer. Sample gases are continuously flowed through the instrument while it repeatedly scans the absorption spectrum over infrared wavelengths. This unit can be programmed to identify target species and quantify them.

the molecules to adsorb and desorb from the column surfaces at different rates, which affects their retention times in the column. As the separated gases exit the column, they must be detected. Different types of detectors can be used to detect the presence of target species. Flame ionization (destructive) and thermal conductivity (nondestructive) detectors are the most common detectors used with GC columns. The column type, GC oven temperature, sample quantity, carrier gas flow rate, and other analytical parameters must be optimized to separate and analyze a given mixture. In general, GC is most useful for the characterization and analysis of hydrocarbon mixtures that are easily vaporized.

Several other types of analytical instruments are available for chemical characterization and analysis such as electrochemical sensors (such as the in situ O_2 analyzer shown in Figure 10.9) or wet chemistry and laboratory methods. Some of the most important factors affecting the selection of chemical analytical equipment are

- Measurement accuracy and sensitivity
- Background (matrix) interference
- Chemical interference
- Response time
- Ease of instrument operation
- Ease of data analysis
- Durability of instrumentation
- Cost of instrumentation and/or analysis

10.6.1.2 Chemical Sample Collection

Chemical species measurements can be made from extracted samples or nonintrusively. A special concern of nonintrusive measurements is the interference from nontarget species. Electrochemical sensors exhibit cross sensitivities to other gases and can artificially increase the reported value of interest. Optically based measurements may similarly experience interference from other gas molecules and/or particulates. Extractive samples have the advantage of being able to be conditioned to make the sample more amenable to the desired measurement method. Even though measurement interferences can be controlled by extracting and conditioning a sample, the desired measurement may be corrupted by post-extractive degradation or reactions. Figure 10.10 shows an extractive sample probe connected to a flexible sample line that can be used to extract samples from a vessel at multiple locations and depths.

Some of the most important factors affecting the design or selection of a sample collection system are

- Material compatibility at sampling point (temperature rating, corrosion and fouling resistance, possible reactivity with target compounds)
- Instrumentation compatibility with sample (condensation, or chemical attack to sensitive components, temperature limitations)
- Timescale required for data acquisition
- Sample interactions or reactions with the collection system

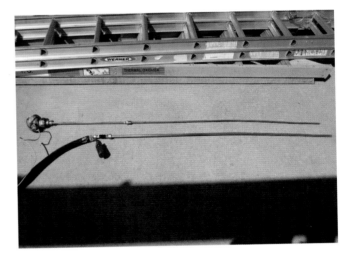

FIGURE 10.9
In situ oxygen analyzer. The oxygen analyzer in this unit is not actually located in the stack. However, the sensing element is very close to the sample location to minimize lag time. Flue gases are rapidly educted across an electrochemical sensor that measures oxygen concentration in real time.

FIGURE 10.10
Extractive sample probe. The tube shown in the bottom of the figure is used to draw flue gases out of the test unit. The extracted gases can be conditioned and analyzed. The gas samples usually take 10–30 s to be conditioned and analyzed using the equipment shown in the figure.

- Possible interference or reactions with other species in the sample
- Durability of equipment
- Extraction (lag) time
- Ease of operation

10.6.2 Flow Measurements

Flow measurement is critical during testing since the flow rates need to be metered accurately to carefully control the experimental conditions. The flow measurements are also essential in the analysis and characterization of the test data. There are several methods for measuring flow. Regardless of the method, care needs to be taken in setting up the flow measurement. The flow should be extremely uniform at the point of measurement. Flow straighteners may be needed to eliminate secondary flow patterns in the pipe or duct. The flow is preferably uniform and uninterrupted for 20 pipe diameters upstream of the measurement location and also 10 diameters downstream. Additional instrumentation for density corrections (temperature, pressure) to the flow measurement may also be required.

Orifice plate flowmeters, as seen in Figure 10.11, are good devices for measuring flows in gases or liquids. Orifice plates are relatively cheap and are very reliable and accurate. The measurement range for a given orifice is limited, but different diameter orifice plates can be easily changed out. Orifice plates take higher pressure drops than other flowmeters, which may make them unsuitable for fluids only available at low pressure drops.

Combustion and quench air sources are generally not available at pressures above 1 psig (7 kPa) due to increased power requirements and high equipment costs of higher pressure fans and blowers. Therefore, flowmeters that do not require much pressure drop, such as thermal mass flowmeters or Pitot tube arrays, are often employed. These flowmeters are generally less accurate than orifice plate flowmeters, but they tend to give reproducible results of reasonable accuracy. These flowmeters are usually capable of measuring flow over wider range compared to an orifice plate flowmeter. Thermal mass flowmeters, as seen in Figure 10.12, have probes with sensing elements on the end of the probe. The sensing element measures heat transferred between two points on the end of the probe and correlates this to mass flow. Since a single probe is used, the measurement is determined locally at the probe position and the overall flow in the duct must be inferred. The duct geometry and gas composition must be known to determine the flow rate using these sensors. Since the overall flow is inferred, irregularities in the flow pattern greatly reduce the accuracy of the measured value since it is assumed to be uniform.

There are many other types of flowmeters that can be used such as venturi meters, vortex shedder flowmeters, and Coriolis flowmeters. Any of these may be preferred depending on the application. Fluid flow can also be measured indirectly using chemical analysis and mass and energy balance calculations. Airflow measurements are commonly made this way since an oxygen analyzer is generally more accurate than a thermal mass flowmeter.

FIGURE 10.11
Orifice plate meter. The orifice plate is located between the flanged connections behind the pressure transmitter. The flanges are designed with pressure taps that connect to the pressure transmitter, which monitors the pressure drop across the orifice. The calibrated data acquisition systems continually record the flow.

FIGURE 10.12
Thermal mass flowmeter. The meter's probe is inserted into a duct of known dimensions. Flowing gases pass through the sensing element at the bottom of the probe. The sensing element consists of two rods. The heat transfer between these rods is correlated to flow. An electrical transmitter and digital display is located in the head of this instrument.

Some considerations to keep in mind in selecting flowmeters are

- Range of flow measurement
- Possible particle deposition, fouling, or corrosion due to fluid
- Accuracy
- Reliability
- Ease of calibration or modification

10.6.3 Temperature Measurement

Thermal oxidizer temperatures are most commonly measured using ceramic-sheathed thermocouples as seen in Figure 10.13. These thermocouples give reliable, reproducible results and can be used in most instances when they are placed properly. To minimize measurement errors, the thermocouple should not be positioned near luminous flames or cool surfaces.

Suction pyrometers can be used to obtain more accurate measurements of the flue gas temperature. A suction pyrometer recesses the thermocouple in an extractive probe to shield it from radiation exchange with the rest of the vessel. Flue gas is rapidly educted into the extractive probe to maximize the convective heat transfer of the flue gases to the thermocouple (while minimizing radiative exchange with the rest of vessel).

There are multiple types of thermocouples that can be used. The most common thermocouples used in combustion applications are type K, type S, and type B. The temperature rating for type S and type B are higher than type K, but are significantly more expensive and not quite as accurate. The thermocouple sheathing material and length also need to be selected to be compatible with a given test.

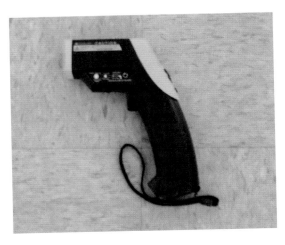

FIGURE 10.14
Handheld radiometer. This radiometer is equipped with the HeNe laser used for sighting. The radiometer quantifies the temperature of the sight point. Note that the emissivity of the material at the sight point must be known.

In cases where the environment is too harsh for a thermocouple, the temperature can be measured optically using a pyrometer or radiometer. Additionally, an infrared thermograph can be used to map temperature. Handheld radiometers, as seen in Figure 10.14, are useful for measuring shell temperatures. Optical measurements usually require some knowledge of the optical properties (emissivity, reflectivity) of the medium being measured.

10.6.4 Pressure Measurement

Pressure and flow measurements are very useful for the characterization and operation of test equipment. Pressures in fuel lines are generally several psi, whereas pressures in the combustion chamber and air duct may be on the order of inches of water column. The expected pressure range will determine which measuring device is best suited for the application.

One of the simplest methods to measure pressure is with a water-filled manometer when the pressure drop is less than 40 in. w.c. (10 kPa). Higher pressure drops can be measured, but may be impractical because of the length of the manometer. When the pressure drop is more than 1 in. w.c. (0.25 kPa), a u-tube manometer, as seen in Figure 10.15, is well suited. When the pressure drop across the manometer is less than 1 in. (0.25 kPa) w.c., an inclined manometer, as seen in Figure 10.16, is well suited. The only difference between a U-tube manometer and inclined manometer is the angle of the water column. The inclined manometer forces the liquid to travel a longer distance for a given pressure,

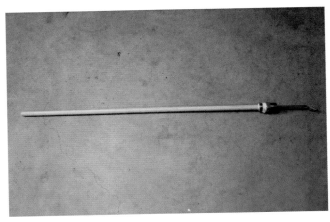

FIGURE 10.13
Ceramic-sheathed thermocouple. A stainless steel type K thermocouple is sheathed with a ceramic thermowell. This is the most common instrument for measuring temperature in a thermal oxidizer.

FIGURE 10.15
U-tube manometer. The tubes on each side of the manometer are connected to pressure taps (or one is open to atmosphere). The difference in liquid columns is measured with a ruler positioned between them.

FIGURE 10.16
Inclined manometer. The tubes on each side of the manometer are connected to pressure taps (or one is open to atmosphere). The difference in liquid columns is measured with a ruler calibrated for the inclination angle.

FIGURE 10.17
Bourdon tube pressure gauge. The pressure force exerted on the spring moves the needle on a calibrated dial displaying pressure.

which makes short water columns easier to measure. Handheld battery-operated digital manometers can be similarly utilized. These manometers are very durable and very simple to operate, but do not permit automatic data acquisition.

Higher pressures are commonly measured with Bourdon tube gauges as seen in Figure 10.17. Bourdon tube gauges can be designed to measure vacuum pressures or positive pressures. Bourdon tube gauges display the pressure by displacement of a needle. Typically, they do not provide electronic outputs. Bourdon tube gauges should be periodically removed and calibrated on a test stand to ensure their accuracy.

Diaphragm pressure transmitters are most commonly used when an electrical signal is desired for data acquisition. These devices are also durable and reliable, but require additional calibration and maintenance compared to a fluid manometer or Bourdon tube gauge.

Placement and orientation of the pressure taps need to be considered in the equipment design to allow for pressure measurements. The pressure taps should be positioned perpendicular to the flow of fluid, otherwise the velocity head of the fluid will alter the measurement. Placing ball valves between the pressure taps and measurement device is commonly done (especially when high pressures are involved) so that the pressure measurement device can be safely taken out of service for calibration or replacement. The pressure measurement device is preferably located near the pressure tap, but is not essential. The diameter of the tubing connecting the pressure tap to the measurement device should be of adequate diameter to minimize the response time. In most cases, ¼ in. (6.4 mm) tubing is adequate.

10.7 Conclusions

There are unlimited configurations of thermal oxidizers that can be tested. Each configuration and test objective will have its own requirements for equipment and measurements. Designing the equipment for maximum flexibility and modularity makes it possible to test the vast majority of configurations. Simulating input streams with simpler streams makes the test easier and usually safer to conduct. Equipping the test equipment with the proper instrumentation allows the test to be carefully controlled and provides accurate data for analysis.

The thermal oxidizer test facility can be used for many objectives such as research and development, equipment characterization, customer demonstrations, and training purposes. The information in this chapter is an introduction to the most common approaches to thermal oxidizer testing and, hopefully, gives insight and considerations for carrying out less common test methods.

References

1. Seeker, W. R., Waste combustion, *Twenty-Third Symposium (International) on Combustion*, Orléans, France, Pittsburgh, PA: The Combustion Institute, 1990, pp. 867–885.
2. Baukal, C. E. (ed.), *Industrial Combustion Testing*, Boca Raton, FL: CRC Press, 2011.
3. Johnson, B., Baukal, C., Petersen, N., Karan, J., and Sidek, S., Putting technology to the test, *Hydrocarbon Engineering*, 15(10), 71–78, 2010.
4. Johnson, B. and McQuigg, K., The effects of operating conditions on emissions from a fume incinerator, *Proceedings of the 1994 International Incineration Conference*, Houston, TX, 1994.
5. Johnson, B. C., Petersen, N. S., and Colannino, J., State-of-the art low NOx thermal oxidation, *International Conference on Thermal Treatment Technologies and Hazardous Waste Combustors*, San Francisco, CA, 2010, Paper #66.
6. Baukal, C. E. (ed.), *The John Zink Combustion Handbook*, Boca Raton, FL: CRC Press, 2001.
7. Baukal, C. E. (ed.), *Industrial Burners Handbook*, Boca Raton, FL: CRC Press, 2004.
8. Reed, R. J., *North American Combustion Handbook Second Edition*, Cleveland, OH: North American Manufacturing Co., 1978.
9. Schnelle, K. B. and Brown, C. A., *Air Pollution Control Technology Handbook*, Boca Raton, FL: CRC Press, 2002.
10. Coward, H. F. and Jones, G. W., *Limits of Flammability of Gases and Vapors*, Bureau of Mines Bulletin 503, Washington, DC: United States Government Printing Office.
11. Dellinger, B., Torres, J. L., Rubey, W. A., Hall, D. L., and Graham, J. L., Determination of the thermal decomposition properties of 20 selected hazardous organic compounds, *EPA-600/S2-84-138*, Washington, DC: United States Environmental Protection Agency, October 1984.

11

Burner Installation and Maintenance

William Johnson, Mike Pappe, Erwin Platvoet, and Michael G. Claxton

CONTENTS

- 11.1 Introduction ... 246
  - 11.1.1 Burner Tile and Why It Is Important ... 246
    - 11.1.1.1 Metering the Combustion Air ... 246
    - 11.1.1.2 Mixing the Air and Fuel ... 246
    - 11.1.1.3 Maintaining Stability ... 246
    - 11.1.1.4 Molding the Flame ... 246
    - 11.1.1.5 Minimize Emissions ... 247
  - 11.1.2 Burner Gas Tips ... 247
- 11.2 Preinstallation Work ... 247
  - 11.2.1 Receiving, Handling, Storage ... 247
  - 11.2.2 New Heater ... 247
  - 11.2.3 Existing Heater ... 247
  - 11.2.4 Safety ... 248
- 11.3 Burner Installation ... 248
  - 11.3.1 Tile Installation ... 248
  - 11.3.2 Mounting the Burner ... 248
  - 11.3.3 Inspection of Key Components ... 250
  - 11.3.4 Air Registers and Dampers ... 254
  - 11.3.5 Fuel Piping ... 254
  - 11.3.6 Electrical Connections ... 255
- 11.4 Burner Maintenance ... 259
  - 11.4.1 Gas Tip Cleaning ... 262
    - 11.4.1.1 Gas Tip Maintenance Recommended Tools ... 264
    - 11.4.1.2 Cleaning Procedure ... 266
    - 11.4.1.3 Corrective/Preventive Actions ... 267
  - 11.4.2 Premix Gas Burners ... 270
  - 11.4.3 Burner Tile ... 271
  - 11.4.4 Flame Stabilizer ... 273
  - 11.4.5 Air Registers and Dampers ... 278
  - 11.4.6 Oil Burner Maintenance ... 280
    - 11.4.6.1 Oil Gun ... 281
    - 11.4.6.2 Oil Gun Insert Removal ... 286
    - 11.4.6.3 Z-56 "Quick Change" Oil Gun ... 289
    - 11.4.6.4 Disassembly ... 289
    - 11.4.6.5 Inspection ... 289
    - 11.4.6.6 Assembly ... 289
  - 11.4.7 Pilots ... 290
    - 11.4.7.1 Electrical Connections ... 290
- References ... 294

11.1 Introduction

The proper installation and maintenance of process burners is absolutely critical for the safe and efficient operation of a process heater. The burners are the "heart" of the entire process system. If the burners do not function correctly, the unit will suffer and process requirements will be unachievable. Before starting a burner installation, the most important item is to review the burner drawing and the "Operating and Installation Manual" that came with the burner. The manual will include detailed instructions that will make the installation easier and help produce the desired end results. The burner drawing will show tile dimensions and gas tip orientation details with required tolerances.

The modern process burner has evolved from a fairly simple fuel and air mixing device to a sophisticated piece of combustion technology over the last thirty years.[1] Burners no longer simply provide heat to a process inside the heater. Today's burners must perform and achieve results that were unheard of 30 years ago.

The key components of a burner are the tile and the gas and oil tips. The installation and maintenance of these critical components are the technology behind the overall burner design.

11.1.1 Burner Tile and Why It Is Important

The burner tile provides the basis for the overall burner operation including metering the combustion air, mixing the fuel and air, maintaining stability, molding the flame, and minimizing emissions (see Figures 11.1 and 11.2). Each of these important functions to burner performance will be examined next.

11.1.1.1 Metering the Combustion Air

The tile throat is sized to allow a certain amount of air into the combustion reaction at a given pressure drop. If the throat is too small or too large, the burner will not operate with the required amount of excess air.

11.1.1.2 Mixing the Air and Fuel

Normally the tile will have channels or zones to aid in the mixing of the air and fuel. Proper mixing is essential to good combustion.

11.1.1.3 Maintaining Stability

Once the combustion reaction has been initiated, flame stability becomes the priority. The burner tile

(a)

(b)

FIGURE 11.1
Burner tiles: (a) flat shaped and (b) round shaped.

FIGURE 11.2
Burner tile with intricate design features.

uses ledges or shoulders (bluff bodies) to provide low-pressure areas to aid in the stability. The tile also provides a hot "chamber" around the flame to aid in the stabilization of the reaction.

11.1.1.4 Molding the Flame

The burner tile helps to shape the flame for the correct pattern in the heater. Burners can be designed for a flat flame or a round flame depending on the heater configuration.

11.1.1.5 Minimize Emissions

The burner tile plays a vital role in reducing the NOx (see Volume 1, Chapter 15) and CO (see Volume 1, Chapter 14) emissions from the combustion reaction. The tile design provides contours or paths that allow the fuel gas and flue gas to combine to reduce the formation of NOx. The strategic location of "ledges" or low-pressure areas helps to anchor the flame and lower CO emissions.

11.1.2 Burner Gas Tips

The gas tips, along with the tile, are the most critical components of a burner.[2] The technology used to develop the gas tip design is in fact the "black magic" behind a successful burner. There are several burner and heater problems related to improper orientation, location, and fouling of gas tips.

The main purpose of the gas tip(s) is to inject the fuel in such a manner that it mixes with the combustion air and provides the proper flame envelop for the heater configuration. The drilling pattern and number of fuel ports on the tip are designed to provide the desired flame pattern and assist in maintaining a stable and safe combustion reaction (see Figure 11.3).

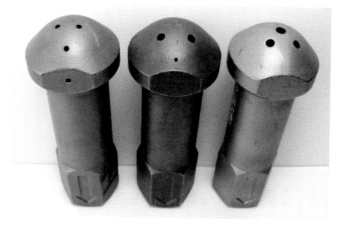

FIGURE 11.3
Gas tips with different drill patterns.

11.2 Preinstallation Work

11.2.1 Receiving, Handling, Storage

The burner assembly and burner tile are typically shipped separately unless there are special circumstances, such as the burner being installed horizontally, in which case the tile could be installed on the burner. The burner should be unpacked and inspected to ensure that all parts are in accordance with the bill of materials included with the burner. Burner tiles and crates are designed to be picked up and transported by a forklift. Refer to the shipping documents for weights and dimensions of the crates and refer to the burner drawings for weights and dimensions of the individual pieces of equipment.

Missing parts or parts that appear to be incorrect or damaged should be immediately reported to the burner manufacturer for correction. The bill of materials will list the main burner parts, such as the burner assembly and burner tile, and other miscellaneous parts. Confirm that any preassembled parts are as shown on the burner drawing.

11.2.2 New Heater

Some preparations to the heater are required so that burners can be installed properly. If the furnace is new, then the refractory and the heater steel will be in good condition. The surfaces will be plumb and level, which will make the installation typically easier than in a revamp situation. The installer must compare the heater manufacturer's and burner manufacturer's drawings for the required cutout in the heater steel and the burner mounting bolt pattern, including bolt circle and bolt size. These are compared to the field measurements of the same dimensions. If differences are discovered, they must be resolved before proceeding with the installation work. The heater steel is checked for interferences with the burner steel and for flatness. Any necessary work should be completed before attempting burner installation.

11.2.3 Existing Heater

When retrofitting burners to existing heaters, the burner installation is often more complicated. Older furnaces often have problems with warped or corroded casing steel or structural supports. Examination of the existing heater steel where the burners will be mounted is required when retrofitting with new burners. Corrosion or any deformation of the heater casing should be corrected before installing the new burners. Cutting out

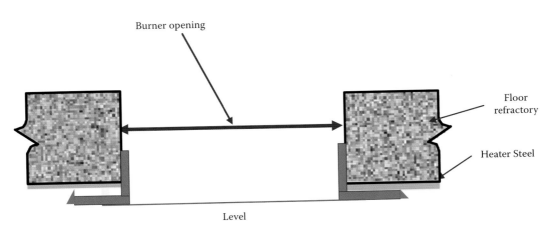

FIGURE 11.4
Burner mounting sleeve.

sections of warped steel and replacing it with new sheet metal may be required depending on the severity of the damage. If replacement of faulty heater sections is not possible, then shim materials or special mounting sleeves may be required in order to install a burner so that it is plumb and level (Figure 11.4). If a burner is not installed on an even plane, it may cause the burner flame to lean to one side.

11.2.4 Safety

All areas around the heater should be free of clutter and debris. The furnace should be checked and permitted for confined space entry.

When working inside the heater during installation and maintenance, the following minimum safety procedures should be put in place before entering the heater:

- Follow all required lockout and tag-out procedures at the jobsite.
- Disconnect and blind all fuel supply lines.
- Place "Man in Heater" signs at all entrances, fuel control panels, supply valves, and fuel line blind locations.
- Provide a person at the heater entry point at all times to notify others that personnel are in the heater.
- Open all burner air inlets and furnace stack dampers to full open position when personnel are inside heater.
- Purge heater for not less than five (5) furnace air volume changes or 15 min (whichever is longer) before entry. Confirm safe atmosphere with hydrocarbon meter.

- If heater has been in operation, make sure the refractory has cooled and that the furnace internal temperature is below 100°F (40°C).
- Sufficient lighting inside the furnace is required to check installation tolerances and proper burner orientation.

11.3 Burner Installation

11.3.1 Tile Installation

The burner tile should be stored in a dry location out of the weather until the tile is ready for installation. Tile that is left outside will absorb moisture from the air. This can be disastrous during start-up since the tile can actually explode from absorbed moisture turning into steam.

A burner tile can withstand high temperatures that are common in furnaces. The tile is normally designed for temperatures between 2600°F and 3000°F (1400°C and 1650°C). The burner tile is usually either round or rectangular depending upon the flame shape needed by the process and furnace configuration. The round type (see Figure 11.5) provides a conical flame pattern and the rectangular tile (see Figure 11.6) provides a flat or fan-shaped flame.

The burner tile may come in several pieces that have to be put together in the field (see Figure 11.7). When putting the tile together, measurements must be made to maintain the burner throat dimensions. A preliminary fit-up installation using a "dry" run without mortar to check the basic fit and overall dimensions should be considered (see Figure 11.8). If this initial check proves satisfactory, then the tile can be installed with

Burner Installation and Maintenance

FIGURE 11.5
Round burner tile.

FIGURE 11.7
Burner tile sections.

FIGURE 11.6
Rectangular or flat flame burner tile.

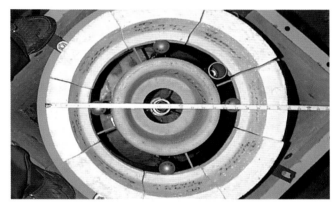

FIGURE 11.8
Dry "fit" to check dimensions.

mortar as required. Normally the tile will sit on the heater steel. Do not put any ceramic blanket or gasket material under the tile unless the drawing specifically calls for this.

Air-setting high-temperature mortar is used to seal the joints on multiple-piece tiles during installation (see Figure 11.9). All joints should be swept clean of mud or debris before being mortared into place. Just a thin layer of mortar is sufficient as too much mortar between the joints can cause the dimensions to be outside the manufacturer's tolerances (see Figure 11.10). Tolerances for the burner tile installation are usually listed on the burner drawing, or in the operating manual.

When doing refractory work inside the heater or on the burner tile, it is necessary to cover the gas tips (see Figure 11.11). This keeps refractory and mortar from falling on the tips and plugging the ports. Do not forget to remove the tape before closing the heater!

The burner tile may have notches or grooves to allow the tile to clear bolt heads and allow the tile to sit flat on the mounting surface.

Some tiles may also have orientation holes or notches so that the tile can only be installed in a certain position. It is important to check that keyways or notches actually fit over the bolts properly (see Figure 11.12). When placing the burner tiles into place, it is important

FIGURE 11.9
Burner tiles with mortared joints.

FIGURE 11.10
Applying mortar to burner tile.

FIGURE 11.11
Gas tips protected with masking tape during installation.

to remember that the burner tile is fragile and should be handled with care (see Figure 11.13).

11.3.2 Mounting the Burner

Each burner installation can present unique mounting requirements. Careful attention to details is required for a successful installation. Mounting the burner usually consists of moving the burner into place and raising or lifting it into place on the heater steel and then bolting it to the mounting bolts. In the case of heavy burners, lifting devices may be necessary (see Figure 11.14). Forklifts, scissor lifts, and chain hoists are commonly used (see Figure 11.15). Depending upon the accessibility at the heater, different kinds of lifting devices are used.

If access to the burner mounting area is not possible using a forklift, then other rigging options may

FIGURE 11.12
Tile with groove-clearing bolt heads.

FIGURE 11.13
Handle burner tile with caution.

FIGURE 11.14
Crane-lifting burner.

FIGURE 11.15
Burner stand and forklift.

FIGURE 11.16
Lifting device for burner installation.

be required. These options include chain hoists, winches, or special lifts (see Figure 11.16). Lifting lugs could be welded to the heater floor externally, and using pulleys or block and tackle cable hoists, the burner can be hoisted into position for bolting (see Figure 11.17).

It may be necessary to lift the burners into place from inside the heater. This requires the use of an "A"-frame-type hoist mounted inside the heater or some other rigging device to pull the burner into place (see Figure 11.18). Some burners may require a gasket between the burner and the heater steel. In this case the gasket must be installed before the burner is actually bolted into place. Burner mounting can come in different orientations. Most burners are mounted in the floor of the heater; they can be mounted on the sides or even in the roof in the case of a down-fired reformer.

Burners are mounted on the heater steel in a variety of ways. The design of the heater, the heater manufacturer's practices, and the process to which the heater is applied all impact how the burners are mounted. Some mounting options include

- Floor mounted firing vertically upward
- Side or end mounted firing horizontally
- Down fired in the roof of the heater
- Multiple burners in a common air plenum

The burner is commonly attached to the heater steel in one of three ways:

1. The air register mounting flange can be bolted directly to the furnace steel casing (see Figure 11.19).
2. The burner front plate is bolted to a common air plenum, which is attached to the furnace steel casing. The plenum may be provided for noise abatement or for distribution of preheated air from a forced draft system (see Figure 11.20).
3. The burner comes with an integral plenum or "P-box" for noise reduction, or forced draft operation is bolted to the heater steel (see Figure 11.21).

Complete a dimensional check of the burner mounting plate and the heater steel mounting location to identify any problems. If the burner is plenum mounted, the air register and plenum depths should be checked prior to attempting assembly to identify any possible problems.

When installing new burners, an expansion joint is required around the burner tile (see Figure 11.22). This expansion joint is usually ceramic blanket that is stuffed or wrapped around the burner tile. This allows the burner tile to expand as it heats up without coming in contact with the heater floor refractory.

FIGURE 11.17
Various lifting techniques.

The burner manufacturer will show where the expansion gap between the burner tile and the heater refractory is needed on the burner drawings (see Figure 11.23).

When retrofitting burners, the burner manufacturer must know the correct thickness of the heater floor. This is so the burner tile can be installed to the correct height above the floor (see Figures 11.24 and 11.25).

Burners can be mounted in the horizontal position. Burners that are horizontally mounted are on the heater sidewalls or end walls. These horizontally fired burners have expansion joints as well; however, the burner tiles are usually mortared into place on the bottom half of the tile with leveling grout, and the top half only has an expansion joint (see Figure 11.26). This allows horizontally fired burner tiles to expand in an upward direction.

Burners mounted on the heater floor have their weight supported by the heater steel. Burners mounted horizontally have their weight supported by the heater steel and the wall refractory (see Figure 11.27). A burner tile case can be used to take the load off of the wall refractory if required. The tile in a roof-mounted burner is designed to be hung from the roof steel of the heater as shown in Figures 11.28 and 11.29. Correct mounting of any type of burner is very important (see Figures 11.30 and 11.31).

11.3.3 Inspection of Key Components

After the burner is installed, it should be inspected for proper alignment and orientation of the gas tips, tile, diffuser cones, and pilots (see Figures 11.32 and 11.33). The burner tile relationship with the gas tips is critical on many burners (see Figure 11.34), and the burner drawing should be referred to during final inspection (see Figure 11.35).

11.3.4 Air Registers and Dampers

The air register or damper controls the amount of combustion air the burner receives and must operate freely (see Figures 11.36 and 11.37). Free movement of the air register or damper should be tested after installation by opening and closing to verify if they open and close properly. In forced draft systems, the burner air inlets are bolted to large air ducts to deliver the combustion air to the burners. It is important that the main air duct mates up with the individual burner air inlets so the burner air damper does not hang up in the ductwork. Installing burners should include operator access to the air control handles.

Air control can be automated so that the excess air in the furnace can be kept at optimal levels without

FIGURE 11.18
Cables and hoist used to mount burners.

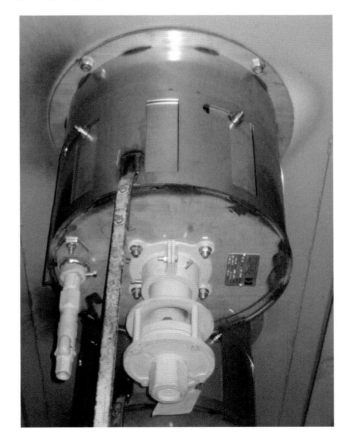

FIGURE 11.19
Burner mounted to heater floor.

operators having to continually adjust individual air registers. One way to accomplish this is by using a jackshaft system (see Figure 11.38). The jackshaft system consists of an actuator that is mounted to a large shaft with individual burner dampers connected to it.

A jackshaft system is particularly useful when there are a large number of burners mounted in a straight row. All the burners in that row can be adjusted simultaneously to keep the excess oxygen at the set point. An alternative to the jackshaft, used in smaller furnaces and vertical cylindrical furnaces, is for each burner damper to be connected to an actuator so the air control can be automated.

11.3.5 Fuel Piping

Care must be taken when connecting the fuel piping at the fuel inlet to the burners. The burner is not designed to provide structural support for the piping, and therefore the piping must not create a load on the burner. Most burners are designed to allow removal of gas risers and inserts while the furnace is in service. This design feature generally results in the riser assemblies being easy to remove, making them susceptible to external loading. Maintaining a zero load ensures that stress from the piping, at the burner manifold or pilot connections, does not create gas tip location problems or mechanical problems at the connection point.

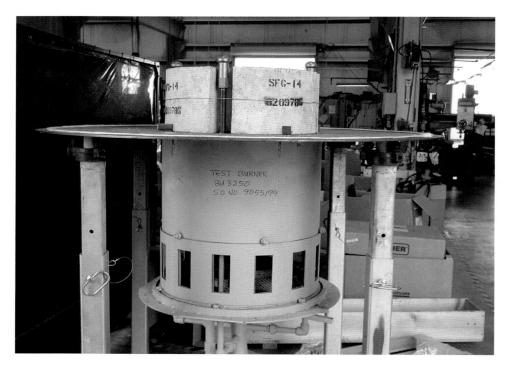

FIGURE 11.20
Burner designed to be mounted in a common plenum.

FIGURE 11.21
P-box-type burner with an integral plenum.

Burner Installation and Maintenance

FIGURE 11.22
Expansion joint between tile and floor.

Also ensure that the piping does not interfere in any way with the operator's access to the burner viewports or operating functions such as air register adjustment.

In cases where flexible metal hoses are used for the burner fuel connections, they should be made of stainless steel to meet the required temperature and pressure ratings and provide reasonable durability. The hoses should have a braided armor covering to resist impact

FIGURE 11.23
Open area around tip for future removal.

FIGURE 11.24
Floor is (incorrectly) higher than gas tips.

FIGURE 11.25
Gas tips above floor as designed.

FIGURE 11.26
Horizontally mounted burner.

FIGURE 11.27
Horizontal burners in service.

FIGURE 11.28
Down-fired burners located on top of the heater.

FIGURE 11.29
Down-fired burners in operation.

FIGURE 11.30
Horizontally mounted burner.

and rough handling. Care should be taken in handling and installation to avoid sharp bends or kinks in order to prevent hose failure.

The piping system should be free from debris. A procedure for "blowdown" of the piping before connecting to the burners is highly recommended. Fuel strainers or filters are also recommended to keep the burner gas tips clean. Fuel systems with liquid carryover should use coalescing-type filters. Systems with dry gas can use strainers with a 32–42 mesh element. This provides sieve openings from 0.0197 to 0.0139 in. (0.50 to 0.35 mm) in diameter. Since the smallest fuel orifice is normally 0.0625 in. (1.6 mm) in diameter, these filters will catch most of the particles in the fuel. A check of the burner manufacturer's data will confirm the smallest fuel metering orifice diameter.

If the fuel contains significant amounts of butane, pentane, or heavier hydrocarbons, it may be necessary to insulate and heat trace the lines as shown in Figure 11.39. This is especially true in colder climates where liquid condensation can be a severe problem. If the piping is to be insulated, it must be installed with the necessary clearances to allow the covering to be installed. Another measure that can be taken is installing the burner fuel lines on top of the header to reduce gas tip fouling (see Figure 11.40).

Oil firing units present many problems compared to the normal gas system. In a heavy oil system, both the oil and the atomizing steam lines must be fully insulated and heat traced. The temperature of the oil, at the burners, is the most important item to achieve proper combustion. The atomizing steam should be dry with approximately 40°F–50°F (4°C–10°C) degrees of superheat. If the steam is saturated then there should be steam traps at all of the low points in the system.

11.3.6 Electrical Connections

Electrical connections on burners include the igniters, flame detection devices, and air register actuators (see

FIGURE 11.31
Ultralow-NOx radiant wall burners.

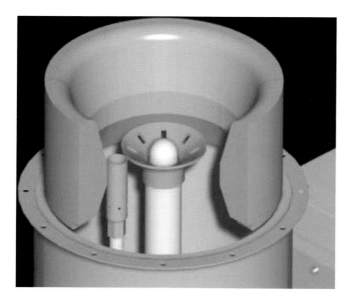

FIGURE 11.32
Gas tip/diffuser cone position.

Figures 11.41 and 11.42). The igniters can have a sparking device that lights the burner pilot and then the pilot lights the main burner. These sparking devices have high energy potential, and if personnel come in contact with the ignition rod, they can receive a very painful jolt. Flame rods will always have voltage present with the system energized. If an electrical device is used, power should be shut off with lockout and tag-out complete before any maintenance is attempted.

The Operating and Installation Manual should be consulted for the type of wiring to be used for each connection. The wiring system should include separate conduits for the ignition wiring and the flame sensing wiring.

Flame scanners are normally mounted on a "swivel"-type connection that allows the sighting point to be adjusted slightly. This line of sight can be adjusted before the scanner is installed on the burner. Moving the connection and looking through the sight pipe will confirm the correct line of sight to

Burner Installation and Maintenance

FIGURE 11.33
Radiant wall tip location.

FIGURE 11.34
Gas tip location (before mounting in a heater).

FIGURE
Final ... pection.

... ain burner. The literature that comes ... ner will have detailed information to ... installation and troubleshooting of the ...

Burner Maintenance

Because of the environment that they are in, burners require routine maintenance. The burners reside in a very harsh place of high temperatures and dirty fuel and air. Burner maintenance is essential to keep the burners performing safely, efficiently, and in compliance with air quality regulations. Routine inspections of the burners are necessary to discover potential burner problems before they become major problems that could create an unsafe operation (see Figures 11.43 and 11.44). Burner maintenance programs help to keep the burners in a safe working condition that will lengthen the lifetime of the burner as well as the heater. The majority of burner maintenance will involve cleaning the fuel tips and checking that they are installed properly. The remaining time will be on items such as repairing air registers to ensure they work, replacing cracked and eroded burner tile or flame retention devices, such as cones or swirlers. The pilots will also require maintenance to keep them working properly.

Keeping the fuel piping and gas tips clean will solve many burner problems. However, the air side can also cause problems. If a burner becomes starved for air due to blocked airways, or closed air doors, the burner performance suffers greatly. Operating a burner without enough combustion air is a serious condition that can quickly lead to a dangerous situation.

It is vital to keep the burner airways free of debris. The air registers or air dampers must be operable and be free of loose refractory or other debris. The burner tile, which serves as the air orifice for the burner, must be in good condition to allow combustion air to flow as designed through the burner throat.

FIGURE 11.36
Dual-blade air dampers.

FIGURE 11.37
Rotary-type air register.

FIGURE 11.38
Jackshaft system to control the combustion air.

FIGURE 11.39
Fuel piping insulated and steam traced.

FIGURE 11.40
Burner fuel lines taken from the top of the header to reduce gas tip fouling.

11.4.1 Gas Tip Cleaning

Keeping the burner fuel tips clean will maximize the burner performance.[2] Each gas tip performs a function that works in tandem with the other tips on the burner to achieve desired flame shapes, heat release, and emission levels (see Figures 11.45 through 11.47). If one or more gas tips become plugged, the burner loses efficiency, the fuel header pressure goes up, and the burner will have poor flame quality. The mechanic must know how to remove the tips for cleaning, clean them, and reinstall them properly. Burner drawings, such as the one seen in Figure 11.48, and data sheets, as seen in Figure 11.49,

Burner Installation and Maintenance

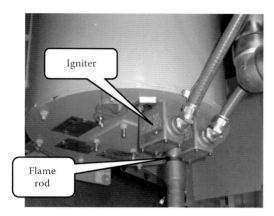

FIGURE 11.41
Example pilot conduit boxes.

FIGURE 11.42
Burners with flame scanners.

must be available for proper burner maintenance. These documents show the proper gas port sizes and the correct way to install the gas tips.

Burners are usually designed so that the gas tips can be removed without having to remove the entire burner from the heater when something goes wrong with one of the tips, such as a tip overheating (see Figure 11.50). The fuel to the burner to be cleaned needs to be shut and blinded. Then each individual gas tip can be disconnected and pulled out for inspection. The gas tip should be checked for coking, corrosion, or plugging (see Figures 11.51 and 11.52). If the gas tip is damaged, it must be replaced. If the ports in the gas tip are not the correct size, then the gas tips need to be replaced with a new one. If the gas tip is in good condition, but it is plugged, then cleaning and reinstalling it is all that is necessary.

FIGURE 11.43
Visual inspection of the burner.

FIGURE 11.44
Recording operating information.

FIGURE 11.45
Premix and raw gas tips.

11.4.1.1 Gas Tip Maintenance Recommended Tools

- Correct burner general arrangement drawing
- The burner manufacturer's operating manual
- Proper PPE
- Proper MTD drill sizes, as shown on the burner drawing and/or data sheet for these sizes

NOTE: Have several of each size along with *properly* sized replacement tips.

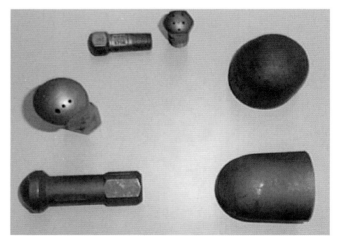

FIGURE 11.46
Raw gas burner tips.

- Tripod pipe vise and/or a sturdy worktable w/pipe vise (see Figure 11.53)
- Good pipe wrenches and other required hand tools
- High-temperature (1800°F = 980°C) anti-seize pipe thread lubricant

Burner Installation and Maintenance

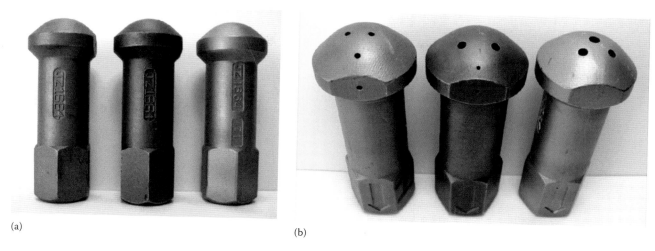

FIGURE 11.47
Gas tips may look similar, but they are not the same. (a) Front view and (b) view of holes drilled.

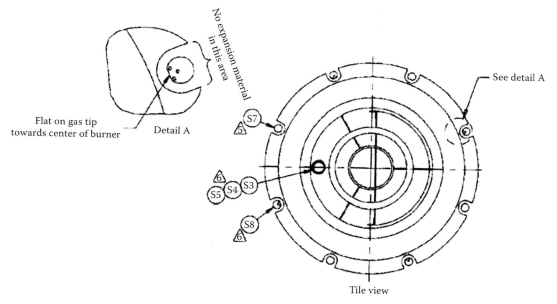

FIGURE 11.48
Burner drawings will show number of gas tips and proper orientation.

- Steam, air, and water source for cleaning and cooling risers and tips
- Tap drill T-handle

11.4.1.2 Cleaning Procedure

1. Remove fuel gas manifold, fuel gas risers, and tips (primary and secondary) from the burner assembly (see Figures 11.54 and 11.55). *The tips will be hot; cool them in a water bath before removing them from the riser.*

2. Remove the tips from the risers. If tips or risers are damaged during their removal, then the tip, riser, or both should be replaced before putting the burner back in service.

> **NOTE**
> It is critical that tip height and tip orientation be correct when installing new risers and/or tips. See burner drawings for tip height and orientation.

Combination gas tip drilling:			
Staged ignition ports:	(2)	~	#38
Primary firing port:	(1)	~	#31
Staged firing port:	(1)	~	1/8
Staged gas tip drilling:			
Staged ignition ports:	(2)	~	#38
Staged firing port:	(1)	~	1/8
Pilot orifice drilled:	(1)	~	1/16
Pilot pressure required	7–10 psig		

FIGURE 11.49
Example of gas tip drilling information on burner documentation.

FIGURE 11.52
Coke buildup in tip.

FIGURE 11.50
Overheated gas tip.

FIGURE 11.51
Plugged gas tip.

> **CAUTION**
>
> The risers should be one continuous length of pipe. No couplings should be used—they eventually leak causing damage to the equipment!

3. After removing the tips, apply steam, air, or water to flush clean the fuel gas manifold and risers (see Figure 11.56). While this is being done, tap on the manifold and the risers with a hammer to dislodge any buildup in these parts. Once the scale/buildup has been removed, you may stop the flushing. Ensure the items are dry and free from plugging or debris before use. If the debris cannot be removed, replace the item with "in kind" parts.

4. Inspect the tips for cracking, deformation of the ports, buildup inside the tip, and proper drill sizes. See burner drawing/data sheet for these drill sizes.

5. Use steam, air, or water to remove buildup inside the tip. If buildup remains, then replace the tip with "in kind" parts.

6. Once tip is free of buildup on the inside, then take the proper size bit and check ports for coke buildup and sizing. Twist the bit into the ports by hand only. You may use a T-handle to do this. The drill bits should be tight in the ports. If the ports are one (1) drill size larger than noted, then we recommend you replace the tip. This twisting of the bits should remove the coking if any is present.

FIGURE 11.53
Pipe vise for gas tip maintenance.

FIGURE 11.54
Floor-mounted burners in a vertical cylindrical furnace.

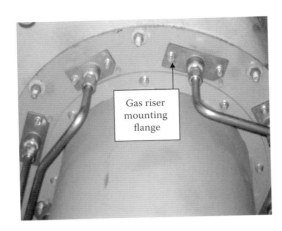

FIGURE 11.55
Gas riser mounting flange.

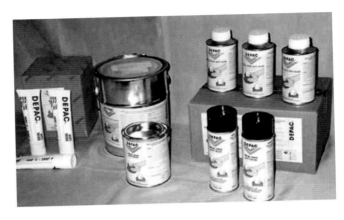

FIGURE 11.57
Thread lubricant.

- If tip has ports that are eroded, replace it.
- If ports are one (1) drill size larger than noted on drawings, replace the tip (see Figure 11.58).
- Remove fouling in the ports with the correct size drill bit turned by hand (see Figure 11.59).
- Do not use a power drill to remove material in the ports (see Figure 11.60).
- When cleaning burner tips, always clean *ALL* tips associated with that burner.
- Always consult burner drawings/data sheets for proper tip heights, drillings, and orientation.

FIGURE 11.56
Using steam to blow out the gas riser/tip assembly.

CAUTION

Do not use a power drill for this. This can deform the drillings and ruin the tip. If the coking cannot be removed without damage to the tip or changing the port size, replace the tip.

7. Once the tips have been properly cleaned and sizing confirmed, reinstall on the risers, being careful to use high-temperature anti-seize lubricant on all the threads (see Figure 11.57).

11.4.1.3 Corrective/Preventive Actions

- If buildup cannot be removed from manifold, risers, or tips, replace them.
- If tip is cracked, replace it.

FIGURE 11.58
Checking ports with correct sized drill bit.

Burner Installation and Maintenance

FIGURE 11.59
Using a drill bit to clean a gas port.

FIGURE 11.60
T-handle used to manually clean gas tips.

After manually cleaning out the fuel orifice, a good practice is to steam out the gas tip and riser assembly again. Depending on how dirty the riser is, it may be best to clean the riser separately, removing pipe scale and residue before replacing the gas tip. The steam exiting each individual port will tell you if every port is clean.

11.4.2 Premix Gas Burners

Maintenance of premix burners, such as the one seen in Figure 11.61, includes periodically cleaning the fuel orifice, checking the gas tip for integrity, lubricating the primary air door, checking the venturi mixer body for damage or plugging, and making sure the secondary air register is not full of debris or inoperable (see Figure 11.62).

There are two common types of fuel orifice spuds that may be in the premix burner. One is the single-port fuel

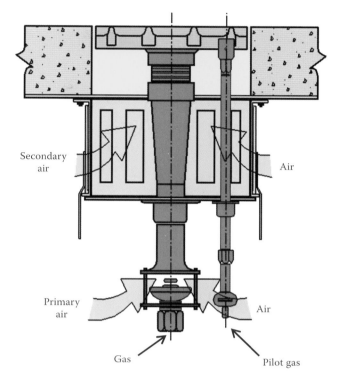

FIGURE 11.61
Diagram of a HEVD premix burner.

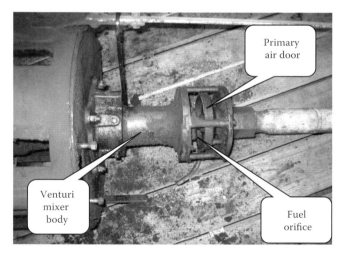

FIGURE 11.62
View of primary air door assembly including fuel orifice.

orifice (see Figure 11.63) and the other is the QD (quiet design) type with multiple ports (see Figure 11.64). In both cases, the ports need to be clean and sized properly (see Figure 11.65). The drillings in the fuel port determines the amount of fuel the burner will receive, while the burner tip will shape the flame pattern.

The gas tip on premix burners shapes the flame. The flame height and width must fit into the firebox such that the flame does not impinge on the heater tubes

FIGURE 11.63
Single-port orifice spud.

FIGURE 11.64
QD orifice spud.

FIGURE 11.65
Plugged QD orifice spud.

FIGURE 11.66
JZV premix gas tip.

FIGURE 11.67
LPM radiant wall gas tip.

in the radiant section or the convection section of the heater. Drilling angles and port sizes direct the fuel and air mixture to achieve the desired flame shape. The gas tips must be removed and checked internally for coke or oily buildup that may block or partially block the gas tip.

The spider gas tip is very common with premix burners, but there are other premix tip designs that provide a flame shape that is suited for that particular application (see Figures 11.66 and 11.67).

Burner Installation and Maintenance

FIGURE 11.68
Burner damaged from flashback.

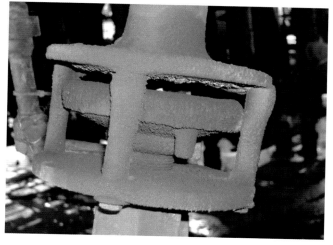

FIGURE 11.69
Dirty mixer/primary air door.

FIGURE 11.70
Spider with severe oxidation from overheating.

The use of penetrating oil on the threaded nipples may be needed in order to unthread the gas tip without damaging the threads. The burners are usually in service for long periods of time and the gas tips can become frozen and difficult to remove.

The burner gas tip, orifice spud, air door, and mixer can be cleaned with steam, solvents, and hand tools (see Figures 11.68 through 11.71).

Large premix burners can be noisy so mufflers may be installed around the primary air inlet to reduce the noise (see Figures 11.72 through 11.74). The muffler gaskets and sound proofing material within the muffler can become worn and will need to be replaced over time (see Figure 11.75). All the primary air for the burner must pass through the muffler.

11.4.3 Burner Tile

Burner tile maintenance is required when the burner tile becomes broken, cracked, or falls apart. The

FIGURE 11.71
HEVD spider with internal fouling.

FIGURE 11.73
HEVD burner with muffler installed.

FIGURE 11.72
HEVD burner without a primary muffler.

FIGURE 11.74
Primary air muffler removed for inspection.

Burner Installation and Maintenance

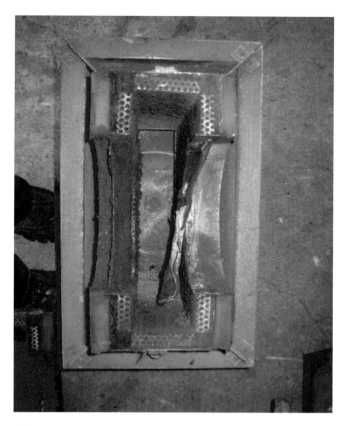

FIGURE 11.75
Worn gasket and dirty insulation.

FIGURE 11.76
Burner tile in good condition.

burner tile is a refractory material and can be damaged if something falls on them. Burner tile can be eroded by steam or fuel streams over time. The heat from the combustion can eventually damage the burner tile. Periodic inspection of the burner tiles is necessary to keep the burners working as designed, safely, and in compliance.

Burner tiles will normally show some small cracks after firing. As long as the cracks do not affect the stability, flame shape, or burner performance, they can be ignored. These small cracks can be repaired by applying mortar patches, but the patches usually do not last long. In most cases replacing the burner tile will require a heater shutdown for entry. Figures 11.76 through 11.85 show burner tiles.

Replacing burner tile requires that the mechanics have the manufacturer's drawings available to maintain critical dimensions. Using excessive amounts of mortar can alter the dimensions of the burner tile. It only requires a thin layer of mortar in most cases to seal the burner tile pieces adequately (see Figure 11.86). The mortar used should be rated for a temperature of 3000°F (1650°C). The mortar should be mixed with water to the same consistency as honey or syrup. The mortar should be easy to spread with a trowel, but it should not run like water.

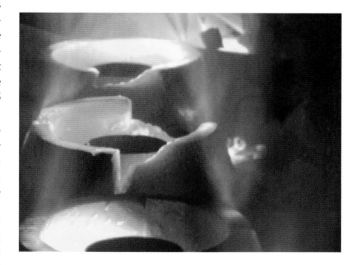

FIGURE 11.77
Burner tile with large broken pieces.

FIGURE 11.78
Tile with large crack.

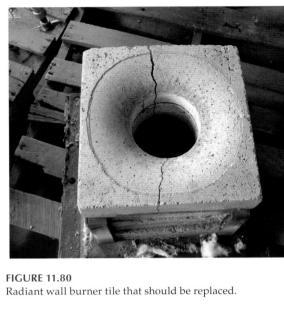

FIGURE 11.80
Radiant wall burner tile that should be replaced.

FIGURE 11.79
Tile crumbling and coming apart.

FIGURE 11.81
Vanadium attack on burner tile.

Burner Installation and Maintenance

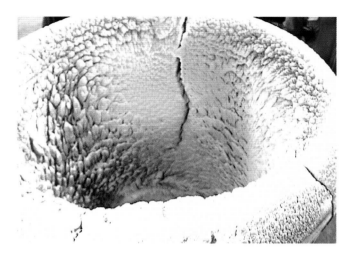

FIGURE 11.82
Catalyst buildup on regen oil tile.

FIGURE 11.83
Inspection of burner tiles inside a furnace.

FIGURE 11.84
Small cracks can be repaired.

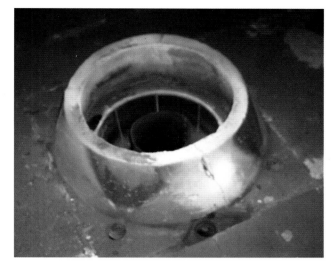

FIGURE 11.85
Tile with large cracks should be replaced.

FIGURE 11.86
Applying a thin layer of mortar.

The height of the tile installed in the burner opening is shown on the manufacturer's drawing. The tile height is related to the thickness of the adjacent heater refractory. Some burner models have the tile height the same as the refractory thickness, while others may not. The installer and the inspector must confirm that the tile height and heater refractory are in accordance with the drawings. If the tile height is not correct, this could result in poor burner performance (see Figures 11.87 and 11.88). It is always recommended to have a ½ in. (13 mm) layer of ceramic blanket as an expansion joint surrounding the burner tile. This will allow the burner tile (see Figure 11.89) to expand slightly as the burner heats up without disturbing the integrity of the tile or the heater refractory. Horizontally fired burners may require an expansion joint on the top 180° and leveling grout on the lower 180°.

11.4.4 Flame Stabilizer

Flame stabilizers are used in burners to create low-pressure zones that the flame attaches to (see Figure 11.90). Flame holders or diffuser cones are other names for flame stabilizers. The flame holder is basically a bluff body that creates a pressure drop that is used to help the flame anchor solidly and maintain ignition. Without the flame holder, the flame can be wild and easy to extinguish. The flame stabilizer is positioned to maximize turbulence needed for mixing the fuel and air. The burner drawing will show where the flame holder should be and what tolerance is recommended. Some flame holders are adjustable, and others are not. Flame stabilizers should be periodically inspected for proper position and condition.

The flame holders can be a variety of shapes and sizes (see Figures 11.90 through 11.94). They can be circular

FIGURE 11.87
Checking gas tip orientation.

FIGURE 11.88
Final tile installation.

FIGURE 11.89
Multiple section burner tile.

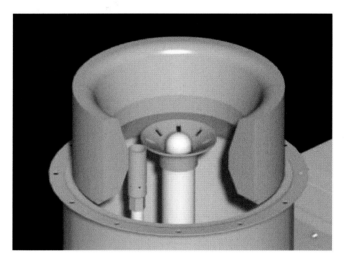

FIGURE 11.90
Diffuser cone used for stabilizing the flame.

FIGURE 11.91
Burner tile ledge used to stabilize the flame.

Burner Installation and Maintenance

or rectangular depending on the burner type. There are burners that rely on the burner tile to serve as a flame holder. In all cases, the flame holder or flame stabilizer is an important part of the burner that must be in good condition for the burner to perform as designed. If the flame holder becomes worn or damaged, it should be replaced to maintain acceptable burner performance (see Figures 11.95 and 11.96).

When replacing the flame holder, it may be important to maintain the orientation of the flame holder. It is positioned to maximize burner performance and the drawings should be checked to see where the flame holder is located (see Figure 11.97).

FIGURE 11.92
Diffuser cones used for flame stability.

FIGURE 11.93
Swirler used to stabilize oil flames.

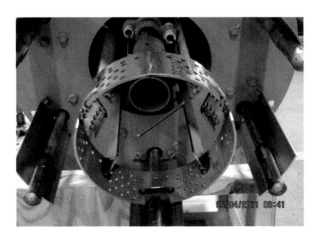

FIGURE 11.94
Flame deflector ring with stabilizing tabs.

FIGURE 11.95
Diffuser cone in good condition.

FIGURE 11.96
Damaged diffuser cone.

FIGURE 11.97
Diffuser cone location too low.

FIGURE 11.98
Rotating-type air registers.

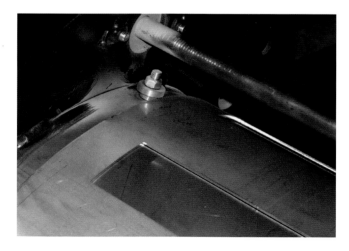

FIGURE 11.99
Rotary air register with E-Z Roll bearings.

11.4.5 Air Registers and Dampers

Air control for the burners is achieved by adjusting the air dampers or air registers on the burners (see Figure 11.98). The air registers must have free movement so that operations can adjust the air to the burners for efficiency and flame quality. Over time the air registers can become stuck in one position. These must be cleaned and repaired so they are operable. Debris can build up on rotating air registers or the registers can become warped from heat. Remove them by unbolting them from the heater plate to clean them or replace them.

The E-Z Roll-type air register offers machined rollers that help keep the register operable. These are especially good when using preheated combustion air or oil firing (see Figure 11.99). Figures 11.100 through 11.103 show some common damper types and air control handles.

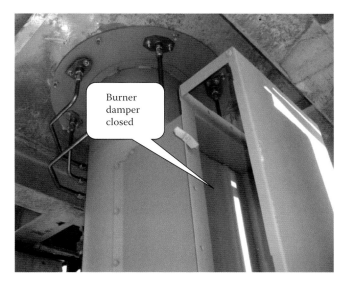

FIGURE 11.100
Burner damper shown in closed and open position.

FIGURE 11.101
Locking air control handle with 18 positions.

FIGURE 11.102
Burners with air handles all set at the same position.

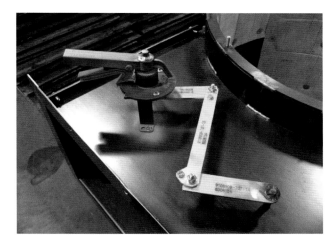

FIGURE 11.103
Damper linkage for dual-bladed opposed motion design.

Burner dampers can be provided with bearings for smooth and positive operation under adverse conditions (see Figure 11.104).

Damper shafts can be lubricated with a high-temperature bearing lubricant. Air registers and vane-type air doors should be lubricated with a dry Teflon-type spray.

11.4.6 Oil Burner Maintenance

The typical oil burner is more complicated than the standard gas-only burner. Because the oil burner is normally supplied as a dual fuel burner, the configuration contains both oil and gas firing capabilities (see Figure 11.105). This combination design utilizes a center-fired oil gun and multiple gas tips located

FIGURE 11.104
Bearings for smooth damper operation.

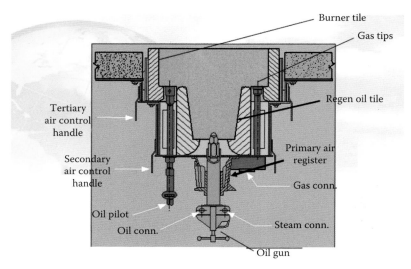

FIGURE 11.105
Combination oil and gas LoNOx burner.

around the oil tile. Some designs will have a concentric dual gun arrangement with a center oil gun and an outer gas gun.

The main components, of an oil burner, that require basic maintenance are

- Secondary tile
- Regen or primary oil tile
- Diffuser cone or swirler
- Air register or damper
- Oil gun assembly

The secondary tile and primary tile are cast from high-temperature refractory (see Figures 11.106 and 11.107). Basic maintenance includes checking the dimensions for proper installation and repairing small cracks in the

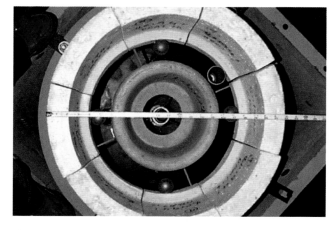

FIGURE 11.106
Secondary and primary (regen) tiles.

Burner Installation and Maintenance

FIGURE 11.107
Regen tile and one section of secondary tile.

surface. The secondary tile normally comes in several pieces that have to be installed as a single tile. The joints between the individual pieces should be coated with a thin layer of high-temperature air-setting mortar.

The stability of the oil fire is accomplished by using a regen tile (see Figure 11.108), a swirler, or a diffuser cone. The type of burner and the physical characteristics of the oil determine which type of device is used (see Figures 11.109 and 11.110). Diffuser cones and swirlers should be replaced if they become damaged or severely oxidized.

The combustion air control is achieved by using slotted registers, vanes, or dampers. Because there can be oil spillage and fouling when firing oil, the standard rotary air register can be given an upgrade to maintain its operability (see Figure 11.111). One way to achieve more reliable operation is to use a design that has bearings or rollers to help the movement of the outer register cylinder.

Another option is to use a multiple vane-type air register. This design provides very smooth and reliable operation. Since the vanes are normally curved, this design will add a degree of "spin" to the combustion air (see Figure 11.112).

A very good option, for oil firing, is to select a burner with an integral plenum box that uses a multiple vane-type damper to control the combustion air (see Figure 11.113).

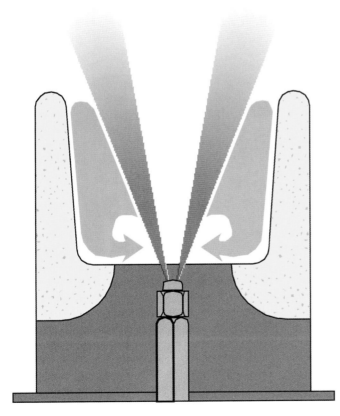

FIGURE 11.108
Regen tile used to stabilize the oil flame.

FIGURE 11.109
Oil and steam spray at the oil tip.

FIGURE 11.110
Regen oil tile with an oil gun in the center.

FIGURE 11.111
Typical rotary-type air registers.

FIGURE 11.112
Vane-type air register.

FIGURE 11.113
Integral plenum box with inlet air damper and muffler.

Air registers and dampers can be lubricated with a high-temperature graphite- or silicone-type lubricant.

11.4.6.1 Oil Gun

The majority of oil guns currently in operation in the refining and petrochemical industry use steam or air as the atomization medium. The use of mechanical or high-pressure atomization is very rare and limited to specific types of equipment.

This section will be describing the design and maintenance for steam-atomized oil guns. The procedures and parts are basically the same for air-atomized oil guns. Figures 11.114 through 11.127 show some typical oil gun components.

FIGURE 11.116
Oil gun bodies.

FIGURE 11.114
Oil gun insert and oil body receiver (with red caps).

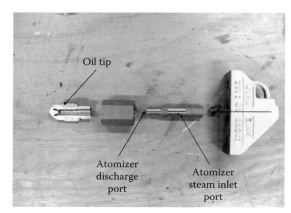

FIGURE 11.117
Oil gun parts.

FIGURE 11.115
Oil body receiver with copper gaskets for sealing.

FIGURE 11.118
EA-/SA-type oil tip.

Burner Installation and Maintenance

FIGURE 11.119
Oil tip stamped with "864."

FIGURE 11.121
MEA-type oil tip.

FIGURE 11.122
MEA oil gun parts.

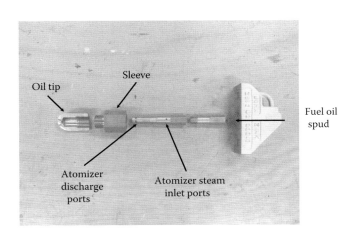

FIGURE 11.120
MEA oil gun parts.

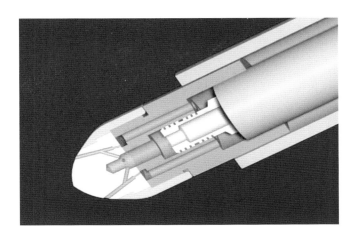

FIGURE 11.123
HERO oil tip with dual atomizing design.

FIGURE 11.124
HERO oil tip, sleeve, and collar.

FIGURE 11.125
HERO oil tips and atomizers.

Oil gun maintenance requires basic hand tools and the oil gun drawing:

1. Cooling water bath
2. Stationary vise
3. Wire brush
4. 12 or 18 in. (305 or 460 mm) pipe wrench
5. 1 1/2 in. open (38 mm)-end wrench
6. 1 5/8 in. open (41 mm)-end wrench
7. 3/4 in. open (19 mm)-end wrench
8. 1/4 in. (6 mm) Allen wrench (male hex drive)
9. Standard flat-bladed screwdriver
10. Copper, bronze, or nickel high-temperature pipe thread lubricant with applicator brush
11. Drill gauges—contact burner manufacturer for specific sizes

FIGURE 11.126
HERO oil gun inserts.

FIGURE 11.127
Applying high-temperature anti-seize to oil gun threads.

11.4.6.2 Oil Gun Insert Removal

The normal-type oil gun is a Quick Change design that allows the oil gun insert to be removed without disconnecting the oil and steam piping. The oil gun guide tube, which provides the tip adjustment, does not have to be removed with the insert assembly for cleaning.

11.4.6.3 Z-56 "Quick Change" Oil Gun

When loosening the oil body receiver clevis, be careful. There could be some oil spray due to the residual pressure between the oil body and the oil body receiver. It is highly recommended that a full purge of the oil side be completed prior to insert removal. This will minimize the possibility of oil spills when the gun is removed. Swing the clevis away and retract the insert.

11.4.6.4 Disassembly

1. Using the cooling water bath, quench the oil tip and sleeve for 2–3 min until they are cool to the touch.
2. Position the oil body in the vise and tighten.
3. Using the 1 5/8 in. (41 mm) end wrench as a backup on the oil gun sleeve, remove the oil tip using the 1 1/2 in. (38 mm) wrench.
4. Using the pipe wrench, remove the steam tube/sleeve assembly from the oil body.
5. Using the pipe wrench as a backup on the oil tube, remove the atomizer assembly using the 3/4 in. (19 mm) end wrench.
6. Holding the atomizer body with the 3/4 in. (19 mm) end wrench, remove the oil spud using either the Allen wrench or the slotted screwdriver.

> **NOTE**
>
> Dependent upon the materials of construction, the type of drive supplied on the oil spud may be either style. A quick inspection from the threaded end of the atomizer assembly will determine the proper tool.

11.4.6.5 Inspection

11.4.6.5.1 Oil Tip

External inspection of the port region will reveal much about the condition of the oil gun. Visually, carbon or oil buildup around the exit ports is an indicator of three (3) possible problems: tip inserted too far, low atomizing medium pressure or flow, and erosive or corrosive action on tip. The oil tip should be stamped with numbers that show the size, the number of ports,

FIGURE 11.128
Atomizer with labyrinth seals and steam ports.

and the angle of the ports. For instance, an 864 tip is a size 8 oil gun with 6 exit ports at 40° total angle.

After cleaning all foreign materials from the face of the oil tip, inspect the exit ports for wear. Visual erosion, egg-shaping, of the exit ports will cause disruption of the flame pattern and should be accepted as reason for replacement.

11.4.6.5.2 Atomizer

External inspection of steam ports should show them free of any foreign material, and the exit port should be concentric and should not show signs of erosion or corrosion, while the labyrinth seal should be clean and with no scoring in the longitudinal direction (see Figure 11.128).

11.4.6.5.3 Oil Spud

Oil spud should be free of foreign materials and not eroded or corroded.

11.4.6.5.4 Tip/Atomizer

The atomizer labyrinth seal should insert into oil tip with a minimum amount of tolerance; new atomizers often require a twisting motion to achieve insertion. If the atomizer/tip fit is not "firm," steam bypassing of the atomizing chamber is possible. Check the atomizer in a new tip to determine if it is the tip or the atomizer that is worn.

11.4.6.6 Assembly

1. Lightly lubricate the threads of the oil spud and install into atomizer.
2. Lightly lubricate the pipe threads on the 3/8 in. (9.5 mm) oil tube, and using the pipe wrench for backup, screw atomizer assembly firmly onto oil tube.

FIGURE 11.129
Checking atomizer location in sleeve.

3. Lightly lubricate labyrinth seal of atomizer taking care to not load up labyrinths or fill exit port or steam ports.
4. Lightly lubricate threads on 1 in. SCH 40 steam tube and reinstall in the sleeve. This procedure should result in the atomizer end being 1/4 ± 1/8 in. (6 ± 3 mm) inside of the sleeve (see Figure 11.129).
5. Lightly lubricate the threads on the oil tip and install onto oil gun making sure that the beveled portion of the tip firmly seats on the beveled portion of the sleeve.

The oil gun is now ready to be installed into the burner.

11.4.7 Pilots

Burner pilots are used to light the main burner in process heaters. They should be maintained to ensure they are reliable. There are a wide variety of burner pilots available that have varied levels of sophistication. A basic burner pilot is fairly simple (see Figures 11.130 and 11.131). There are no electrical components or auxiliary equipment needed to operate the basic pilot. The maintenance items on burner pilots are usually cleaning the fuel orifice and checking that the pilot tip is in good condition. The pilot orifice is subject to plugging; the adjustable air door, if there is one, may get jammed; or the pilot tip could become distorted or cracked over time. Fuel pressure to the pilot is listed on the burner drawing or data sheets, and the gas regulator should be set accordingly.

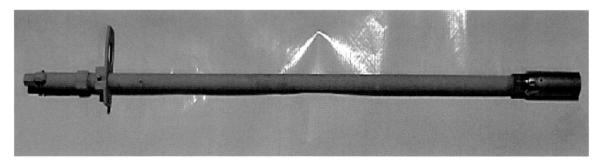

FIGURE 11.130
ST-1S high-stability burner pilot.

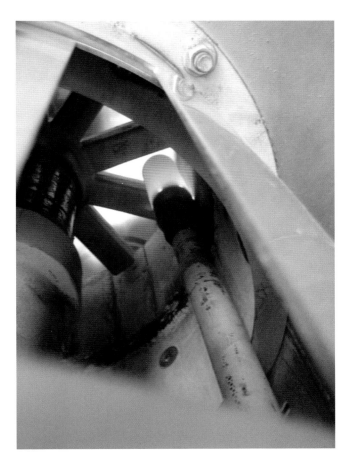

FIGURE 11.131
Pilot shield glowing in normal operation.

A common fuel pressure for John Zink pilots is in the 5–15 psig (35–105 kPa) range when operating on natural gas.

A burner pilot usually can be shut off and removed for maintenance with the burner on line. The pilot is held in place by a pilot boss that is bolted to the front plate of the burner. Set screws on the pilot boss hold the pilot in place. The fuel orifice can be removed from the pilot mixer body for cleaning and inspection (see Figures 11.132 and 11.133).

Electric pilots are designed to provide push button lighting of pilots. A high-voltage transformer is used to create a spark where the fuel and air exit the pilot tip. The ignition rod must be clean and dry and the ceramic insulators must be in good condition.

Normal supply power is 120 VAC to operate the pilot shown in Figure 11.134. High-tension ignition cable is connected from the ignition transformer to the pilot conduit junction box at the pilot.

The ignition rod is assembled by using threaded sections of rod connected with nuts and couplings. There must be an air gap from the ignition rod to the pilot tip so a spark will occur and light the pilot. Pilots may also be fitted with flame rods to detect presence of flame at the pilot. Flame rods must also be clean and dry to function. Flame detection is achieved by the rectification of the electrical current within a flame. Special relay panels that use the direct current in the flame are required to display "flame on" or "flame off" status (see Figures 11.135 through 11.138).

Pilots may have a combination of electric ignition and flame detection. Care must be taken to ensure the power is off before any maintenance is performed. Insulators are required to keep the electrical circuits for these options from shorting out. They must be kept dry, clean, and in good condition. Any cracked insulators should be replaced. Figures 11.139 and 11.140 are examples of electric ignition and flame rods added to the pilot.

Recent versions of process burner pilots are designed to eliminate the ceramic insulators and to be more reliable (see Figures 11.141 through 11.143). Older-style pilots can be prone to problems with condensation that causes the electric igniters and flame rods to ground out.

Another recent development is the use of a high-energy ignition system instead of the traditional transformer (see Figures 11.144 and 11.145). These units work

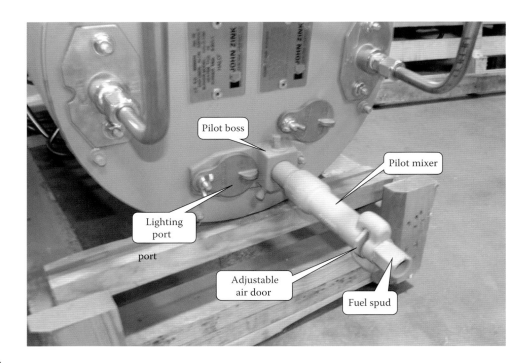

FIGURE 11.132
Pilot parts.

FIGURE 11.133
Checking pilot orifice.

Burner Installation and Maintenance

293

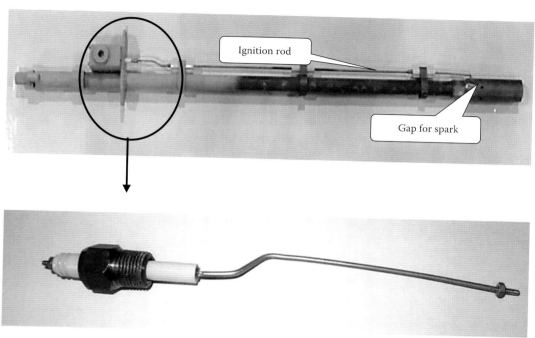

FIGURE 11.134
Pilot ignition rod assembly.

FIGURE 11.135
Typical relay panel.

FIGURE 11.136
Pilot "on" green light illuminated.

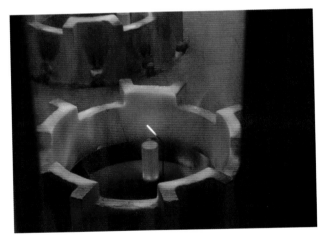

FIGURE 11.137
Operating pilot with flame rod.

on the principle of high capacitance discharge and provide a spark that is immune to condensation and water vapor. Some of these units will actually ignite under water. The John Zink AG-2-HE is a good example. Another example is the ST-1SE FR pilot shown in Figures 11.146 and 11.147.

11.4.7.1 Electrical Connections

Electrical connections on burners include the igniters, flame detection devices, and air register actuators when present. The igniters will have a sparking device that lights the burner pilot and then the pilot lights the main burner. These sparking devices have high energy potential, and if personnel come in contact with the ignition rod, they can receive a very painful jolt. Flame rods will always have voltage with the system energized (see Figure 11.148).

If an electrical device is used, power should be shut off with lockout and tag-out complete before any maintenance is attempted.

The Operating and Installation Manual should be consulted for the type of wiring to be used for each connection. The wiring system should include separate conduits for the ignition wiring and the flame sensing wiring.

Flame scanners are normally mounted on a "swivel"-type connection that allows the sighting point to be adjusted slightly (see Figures 11.149 and 11.150). This line of sight can be adjusted before the scanner is installed on the burner. Moving the connection and looking through the sight pipe will confirm the correct line of sight to the pilot or main burner. The literature that comes with the scanner will have detailed information to help with the installation and troubleshooting of the equipment.

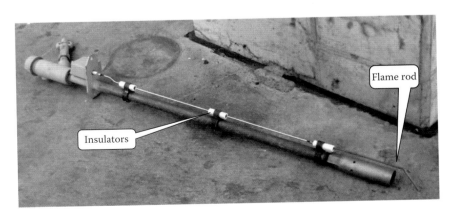

FIGURE 11.138
Pilot with exposed flame rod.

Burner Installation and Maintenance

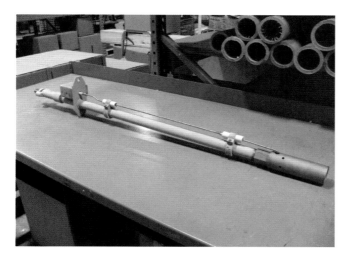

FIGURE 11.139
ST-1SE pilot with external igniter and ceramic insulators.

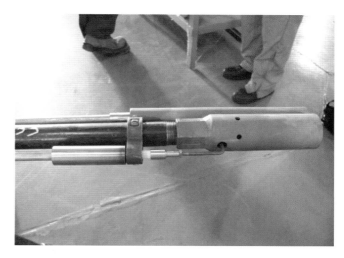

FIGURE 11.142
New ST-1SE-FR pilot tip.

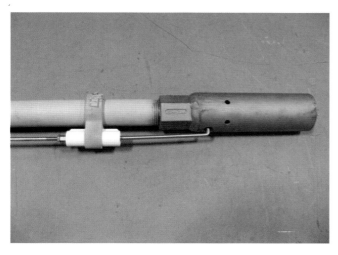

FIGURE 11.140
Close-up view of ceramic insulator and ignition rod.

FIGURE 11.143
New ST-1SE-FR pilot with enclosed flame rod.

FIGURE 11.141
New-style pilot with no exposed insulators.

FIGURE 11.144
New pilot with an internal high-energy igniter system.

FIGURE 11.145
End view of the internal high-energy exciter.

FIGURE 11.146
ST-1SE-FR pilot.

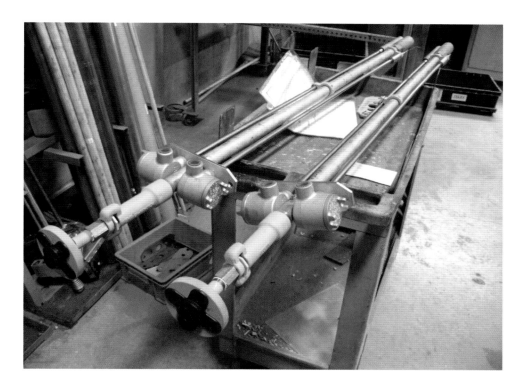

FIGURE 11.147
ST-1SE-FR pilots.

FIGURE 11.148
Meter showing voltage with power on.

FIGURE 11.149
Smart flame scanner mounted on a swivel connector.

FIGURE 11.150
Power supply for a smart scanner.

References

1. C.E. Baukal (ed.), *Industrial Burners Handbook*, CRC Press, Boca Raton, FL, 2004.
2. C.E. Baukal, B. Johnson, K. Vaughn, and M. Pappe, Burner fuel tips, *Process Heating*, 17(2), 11–13, 2010.

12
Burner/Heater Operations

William Johnson, Erwin Platvoet, Mike Pappe, Michael G. Claxton,
Richard T. Waibel, and Jason D. McAdams

CONTENTS

12.1 Introduction 300
12.2 Heater Monitoring for Improved Performance 300
    12.2.1 Heater Draft 301
        12.2.1.1 Definition 301
        12.2.1.2 Draft Sampling 301
        12.2.1.3 Trip and Alarm Settings 302
    12.2.2 Excess Air and Excess Oxygen 303
        12.2.2.1 Definition 303
        12.2.2.2 Excess Oxygen Sampling 303
        12.2.2.3 Oxygen Analyzers 304
        12.2.2.4 Dry versus Wet Oxygen Measurement 306
        12.2.2.5 Tunable Diode Lasers 308
        12.2.2.6 Excess Oxygen versus CO Measurement 308
        12.2.2.7 Alarm and Trip Settings 308
    12.2.3 Fuel Measurements 309
        12.2.3.1 Fuel Flow 309
        12.2.3.2 Fuel Pressure 309
        12.2.3.3 Selecting Trip and Alarm Points 309
        12.2.3.4 Increasing the Fired Duty 310
        12.2.3.5 Fuel Temperature 311
    12.2.4 Combustion Air Measurements 312
        12.2.4.1 Combustion Air Temperature 312
        12.2.4.2 Combustion Air Flow 312
        12.2.4.3 Combustion Air Pressure 314
    12.2.5 Flue Gas Temperatures 314
        12.2.5.1 Standard Thermocouples 314
        12.2.5.2 Suction Pyrometer (Velocity Thermocouple) 314
    12.2.6 Process Tube Temperature 314
    12.2.7 Process Fluid Parameters 315
    12.2.8 Heater Operation Monitoring 316
12.3 Heater Operations 317
    12.3.1 Operating Strategy and Goals 317
    12.3.2 Heater Safety 317
        12.3.2.1 Start-Up Procedures 317
        12.3.2.2 Heater Shutdown Procedure 317
        12.3.2.3 Emergency Procedures 320
    12.3.3 Heater Combustion Control 320
        12.3.3.1 Target Draft Level 321
        12.3.3.2 Target Excess Air Level 321
    12.3.4 Heater Turndown Operation 322
        324

12.4 Visual Inspection inside the Heater 324
    12.4.1 Flame Pattern and Stability 325
    12.4.2 Process Tubes... 326
    12.4.3 Refractory and Tube Support Color 326
    12.4.4 Burner Tile and Diffuser Condition... 326
    12.4.5 Air Leaks... 327
12.5 External Inspection... 328
    12.5.1 Stack Damper 328
    12.5.2 Burner Block Valves.. 328
    12.5.3 Pressure Gauges... 329
    12.5.4 Heater Shell or Casing Condition 329
    12.5.5 Burner Damper Position 329
    12.5.6 Burner Condition 330
References..

12.1 Introduction

Process heaters are fundamental pieces of equipment needed in every refinery, petrochemical, and process plant in the world. The generation of heat is the basic step required to start the reaction and separation processes for literally thousands of units. This generation of heat is normally accomplished by the oxidation of fossil fuels and the resulting release of hot products of combustion ("flue gas") into a fired heater. The burning of fuel can produce a flame that is almost mesmerizing in its vivid color and dynamic movement in the flame zone. On the other hand, a flame that is out of control, unstable, or running out of air can rapidly become a monster that can be devastating to equipment and operating personnel. The safe operation of the process heater has to be the number one priority.

There are four basic combustion rules that govern the operation of process heaters used in the refining and petrochemical industries:

Rule Number 1: Keep the flames inside the firebox.

Rule Number 2: Keep the flames off the process tubes.

Rule Number 3: Keep the process inside the tubes.

Rule Number 4: Purge the heater before light-off.

FIGURE 12.1
Flames outside the heater.

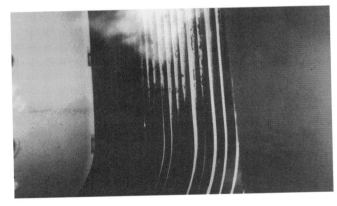
FIGURE 12.2
Flame impingement on process tubes.

Figures 12.1 through 12.4 show what it looks like when these rules are not followed.

This chapter describes how a process heater should be monitored and controlled in order to satisfy these four rules. The two most important items to be monitored are the excess oxygen in the flue gas and the draft inside the heater. The excess oxygen level tells the operator if the combustion is within safe operating limits and how efficiently it is operating. A very common problem in process heater operation is a false indication of excess oxygen in the flue gas. This error is normally the result of air leakage into the heater upstream of the oxygen analyzer and/or the location of the oxygen analyzer itself. The other critical item is the measurement of draft

FIGURE 12.3
Leak in a process tube.

FIGURE 12.4
Failure to purge the heater before light-off lead to an explosion.

or the pressure inside the firebox. Since the vast majority of all process heaters operate under a negative pressure, controlling the draft is essential to safe operation.

12.2 Heater Monitoring for Improved Performance

12.2.1 Heater Draft

12.2.1.1 Definition

Draft is defined as the negative pressure of the flue gas measured at any point within the heater (see Figure 12.5).

The hot flue gas within the heater is less dense than the atmospheric air outside the heater. This density difference results in a force ("buoyancy") that tends to create an upward acceleration of the flue gas. The flue gas will rise until it reaches a point where it has the same density as the air or where its flow is impeded by a restriction. This displacement of the flue gas causes air to move into the heater through the burners or other openings. The hotter the flue gas and/or the colder the surrounding air, the greater the difference in densities and the greater the draft or negative pressure within the heater. This explains why some natural draft heaters can handle higher duty and process rates in the winter or night time when ambient temperatures are lower (creating more draft) compared to summer or daytime operation when ambient temperatures are higher (creating less draft).

Draft loss is the pressure drop of air or flue gas, as it flows through the burners, ducting, air heaters, firebox, tube banks, and the stack. In burner terminology, the draft loss across the burner is the pressure drop of the combustion air as it flows through the burner air control device, the burner housing, and the throat of the burner tile. In a natural-draft heater, the burner draft loss is the difference between the pressure in the firebox at the burner elevation and the atmospheric pressure at that elevation. In a forced-draft (FD) heater, the burner draft loss is the difference between the positive pressure in the combustion air duct and the negative pressure in the firebox at the burner location. Balanced-draft units utilize both a FD fan and an induced-draft (ID) fan. The FD fan provides just enough pressure to move the combustion air through the burners into the firebox. The ID fan removes the flue gas from the system. A point that is often confusing or misunderstood is the heater draft during balanced-draft operation. Many people believe that a balanced-draft unit should provide a positive pressure in the firebox. This is not the case! Whether the unit is a natural-draft heater or a balanced-draft heater, the firebox and flue gas system remain under a negative pressure. The burner tile

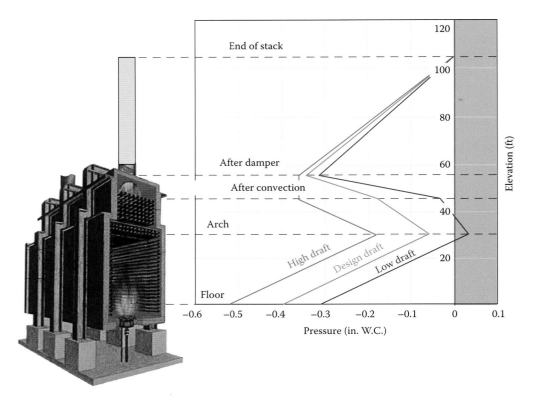

FIGURE 12.5
Heater draft profile.

throat is the "balance" point. On the air side of the tile throat, the combustion air should be under a positive pressure. On the combustion side of the tile throat, the flue gas is under a negative pressure.

12.2.1.2 Draft Sampling

As stated in Section 12.1, the majority of fired heaters should operate with a negative pressure, or draft, throughout the flue gas path. This negative pressure should be measured at specific points.

The critical draft measurement is at the location of the highest absolute pressure within the heater. This typically occurs at the arch or top of the radiant firebox since this is normally where the flue gas leaves the radiant section and enters the convection section (see Figure 12.6). The draft is the lowest at this point due to downstream flow resistances from the convection section and stack damper. Maintaining a slight negative pressure at this point ensures a negative pressure throughout the system. The target draft at this location is normally −0.10 inH_2O (−2.5 mmH_2O).

Another point for draft measurement is at the location of the burners. This is to ensure that at the burner level, an adequate pressure differential exists to supply the necessary combustion airflow.

The third important location for draft measurement is at the flue gas exit from the convection section.

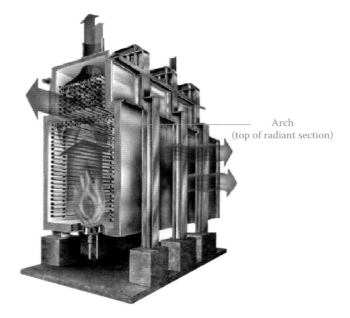

FIGURE 12.6
Draft and O_2 location.

By combining this measurement with the draft value at the top of the firebox, you can determine the draft loss across the convective tube bank. This can help identify the occurrence of damage or excessive fouling in the convection section.

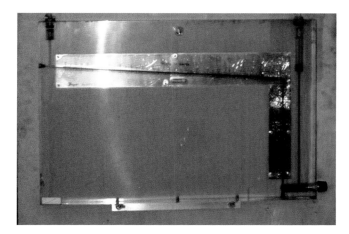

FIGURE 12.7
Inclined manometer.

FIGURE 12.8
Magnehelic draft gauge.

FIGURE 12.9
Draft transmitter.

Draft can be measured with an inclined manometer (see Figure 12.7) or with a Magnehelic dial gauge (Figure 12.8). If an inclined manometer is used, then the proper fluid must be used in the measuring tube. The normal manometer oil has a specific gravity of 0.826. Using water will result in a lower draft reading (less negative) than actually exists. Draft transmitters (see Figure 12.9) should be used to provide remote draft indication to the control room. Since draft levels are typically less than 1.0 inH_2O (25.4 mmH_2O), it is important to design the sensing lines for a draft transmitter so that water condensation from the flue gas cannot collect in these lines and cause erroneous draft readings.

The draft at the top of the radiant firebox is the only location that requires constant monitoring. In a properly tuned heater, the draft within the firebox and convection section should always be a higher negative number than this value. The draft at the firebox roof is controlled by adjusting the damper in the stack or, if an ID fan is provided, by adjusting the fan inlet damper or speed.

12.2.1.3 Trip and Alarm Settings

The heater should have an alarm on the draft at the arch. This is essential to keep the heater from going to positive pressure. The majority of heaters use −0.05 inH_2O (−1.3 mmH_2O) as the alarm setting. The selected points should be confirmed based on the actual heater operating history. Figure 12.10 shows arch refractory that failed and fell to the heater floor due to positive pressure.

12.2.2 Excess Air and Excess Oxygen

12.2.2.1 Definition

Excess air is defined as the amount of air above the stoichiometric requirement for complete combustion of the fuel, expressed as a percentage (see Volume 1, Chapter 4).

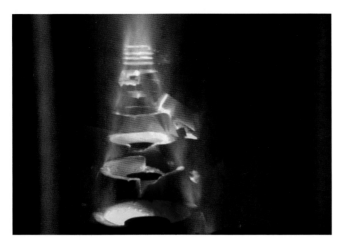

FIGURE 12.10
Arch refractory failed due to positive pressure at the arch.

For example, if the stoichiometric air is expressed as 1.0, then 1.15 would be 15% excess air. The excess oxygen is the amount of oxygen in the incoming air not used during combustion and is related to a percentage of excess air as shown in Figure 12.11. It should be clear that in discussions on the completion of the combustion reaction, excess air and excess oxygen are not interchangeable terms although they are related to each other. As an example, a flue gas containing 3% excess oxygen, on a dry basis, is equivalent to 15% excess air on most fuels. Excess oxygen is relatively easy to measure and is used as a substitute for the actual measurement of the combustion airflow. If there is an adequate amount of excess oxygen in the flue gas, good fuel/air mixing at the burner, and a stable flame observed in the firebox, the operator can be reasonably assured that combustion is complete.

Some people may ask why operate a heater with excess air? Since the extra air actually lowers the heater thermal efficiency, why not run a heater at stoichiometric conditions. There are several reasons why the heater should be operated with excess air:

- Swings in process feed rates can result in higher firing rates.
- Changes in fuel composition can change the combustion air requirements.
- Changes in the ambient air temperature or humidity will affect the amount of oxygen in each cubic foot of air.
- Changes in wind speed and direction can affect the amount of air a natural-draft burner receives.
- Formation of carbon monoxide (CO) or unburned hydrocarbons is difficult to control at stoichiometric conditions.
- Mixing of the fuel and air at the burner is seldom perfect. Having excess air ensures a surplus of air to help overcome this.

For all of these reasons, operating with excess air is the correct mode of operation for the majority of process heaters. The key to achieving high efficiency (see Volume 1, Chapter 12) is to run the excess air at the lowest practical level for the type of burners and the normal process operating conditions.

12.2.2.2 Excess Oxygen Sampling

Since excess oxygen is an indication of the state of the combustion reaction, it is best to sample the flue gas in the radiant firebox where combustion is taking place. The location that best satisfies this requirement for controlling the combustion reaction is at the flue gas outlet from the radiant section. That means near the top of the radiant section for an up-fired heater or at the bottom of the radiant section for a down-fired unit. Cabin type furnaces may require more than one sample point to be sure the combustion is uniform among the burners (see Figures 12.12 and 12.13).

With the heater radiant and convection sections operating under negative pressure, any openings in those zones will allow tramp air into the heater (see Figure 12.14).

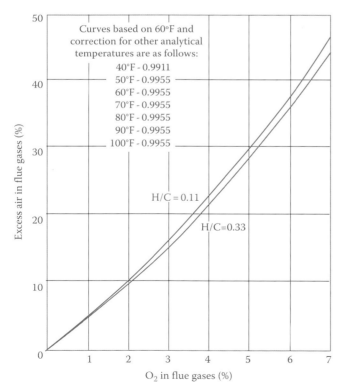

FIGURE 12.11
Excess air indication by oxygen content.

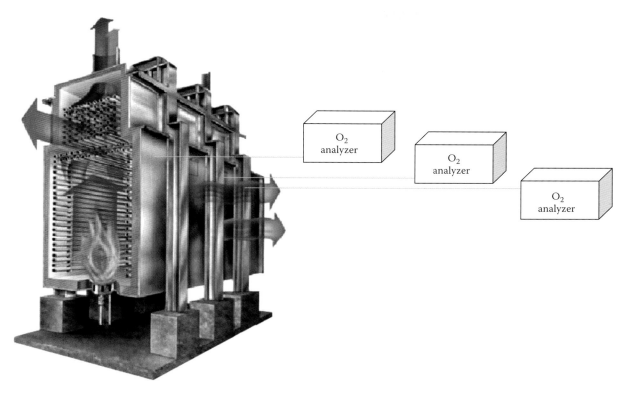

FIGURE 12.12
Multiple oxygen analyzer locations (e.g., for a larger heater).

FIGURE 12.13
Large balanced draft cabin heater.

FIGURE 12.14
Air leakage around process tube.

This tramp air, which does not come through the burners, does not participate directly in the combustion process. The oxygen analysis used to determine the excess air cannot differentiate between air that enters via the burners and air leaking into the heater.

The amount of tramp air is typically low in a well-maintained and sealed radiant section. Convection sections usually have more tube penetrations that provide possible sources of air infiltration. Figure 12.15 shows another possible source of tramp air. If the operator depends on an excess oxygen reading taken in the stack, the reading may indicate an excess oxygen level that is higher than what exists at the burner level. Thus, the adjustments to reduce excess oxygen that the operator would typically

FIGURE 12.15
Open inspection port allowing tramp air into the heater.

FIGURE 12.16
Oxygen probe extending into heater.

make, such as closing burner air registers, could lead to insufficient combustion oxygen at the burners. This could result in unburned hydrocarbons and CO (see Volume 1, Chapter 14) in the flue gases exiting the radiant section.

The unburned hydrocarbons and CO may lead to a condition known as "afterburning." Afterburning is the term given to combustion that occurs downstream of the zone where the combustion was intended to take place. When unburned combustibles leave the burner area or radiant section due to inadequate airflow through the burners, they can burn wherever they come in contact with oxygen within the heater. Afterburning will continue until the flue gas has cooled below approximately 1200°F (650°C). Because the amount of air leaking into the heater is typically greater at the convection section, this is where afterburning is most likely to occur.

Additionally, flow studies and computer simulations have shown that air entering a heater through casing openings tends to stay near the firebox walls as it flows to the exit. If the flue gas sample is drawn from a probe located further into the center of the heater, the sample is more representative of the combustion in the firebox. Therefore, it is recommended to sample through a probe that extends 18 in. (46 cm) or more from the wall into the flue gas (see Figure 12.16). For most heaters, the sample probe should be 310 SS tubing. If the heater has a bridgewall temperature (BWT) of 1800°F (980°C) or greater, then a water-cooled probe must be used to get an accurate measurement of CO.

There will still be requirements to sample the flue gas in the stack before it exits the heater to the atmosphere. To calculate the heater efficiency or to confirm emissions data, the flue gas must be sampled after the last convective tube bank. To calculate the heater efficiency, refer to API RP 560 "Fired Heaters for General Refinery Service" for the recommended procedure.[1]

The justification for controlling the excess oxygen in the heater is shown in Figure 12.17 for natural gas. The information required to determine the annual savings achievable includes the heat release per burner or the total heat release for the heater, the operating excess oxygen, the stack temperature, and the cost of the fuel.

12.2.2.3 Oxygen Analyzers

Conventional oxygen measurement is achieved using extractive sample analyzers where the analyzer can be located away from the sample point. Prior to the analysis, the flue gas must be conditioned to remove all water so that it does not damage the analyzer components. The disadvantage of this method is a slower response to changes in furnace oxygen content and a correspondingly slow change in excess air control, resulting in inefficient combustion control.[2]

Close-coupled extractive analyzers, as shown in Figure 12.18, or in situ oxygen probes containing zirconium oxide sensors provide a much faster measurement of the oxygen concentration in the flue gas. The zirconium oxide sensors work on the basis of the Nernst equation, requiring that the sensor cell be maintained at a high operating temperature of 1300°F (700°C). This temperature is most often accomplished using a heater, but non-heated versions are available that use the heat of the flue gas itself if the temperature is sufficiently high (at least 1025°F = 550°C). The in situ version places the zirconium oxide sensor directly in the flue gas stream. True in situ type analyzers can only measure the oxygen in the flue gas. They are not capable of measuring combustibles. Some types could be limited to applications where the flue gas temperature is less than 1250°F (677°C) due to the type of probe material. As safety precautions, installation of flame arrestors in the flue gas

Burner/Heater Operations

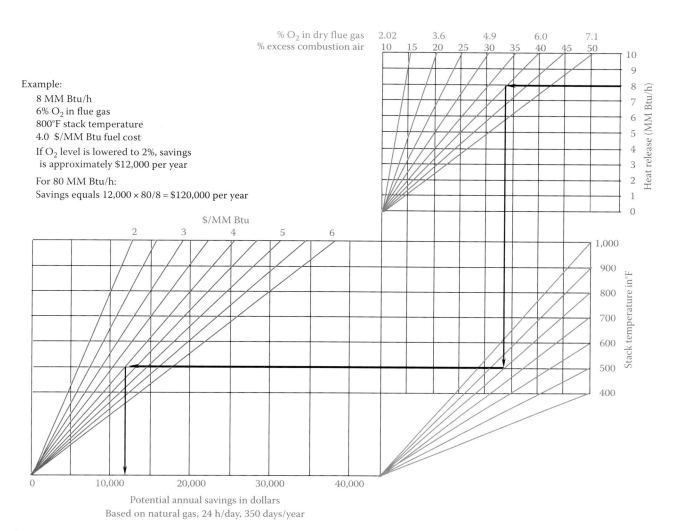

FIGURE 12.17
Cost of operating with higher excess oxygen levels (natural gas).

FIGURE 12.18
Close-coupled extractive flue gas analyzer.

sample line and software controlled probe deactivation on loss of power should be considered. The flame arrestor is designed to stop the propagation of a flame back into the fired equipment if a flammable mixture is present in the flue gas. Since the heated sensor is above the auto-ignition temperature of hydrocarbons, it could act as an ignition source.

The close-coupled extractive method extracts the gas to the sensor located directly outside of the fired equipment or stack (see Figure 12.19). This is the predominant type of analyzer being used in refining and petrochemical plants around the world. These analyzers have the capability of measuring oxygen and combustibles. Whichever type of analyzer is being used, it should be calibrated on a regular basis. The calibration gas should have the correct ratio of oxygen and combustibles as specified by the manufacturer with the balance being nitrogen (N_2). The calibration gas should

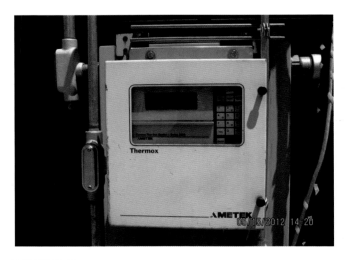

FIGURE 12.19
Flue gas analyzer data panel.

be checked and removed from service when it is six to nine months old. Span gas that has been allowed to sit around for too long will not give accurate calibration results. This is especially important on calibration gases containing nitric oxide (NO) for measuring NOx in the flue gas.

12.2.2.4 Dry versus Wet Oxygen Measurement

Oxygen analyzers can report the excess oxygen concentration based on the wet flue gas or the dry flue gas. A normal flue gas will contain about 15%–30% water vapor on a volume basis, from the combustion of the hydrogen in the fuel. A dry gas analyzer will remove this moisture from the sample. The result is a higher percentage of oxygen, due to the reduction in sample volume. For most fuels, the difference between wet and dry oxygen readings is 0.5%–1.0% with the dry reading always being higher than the wet reading.

Many environmental regulations require sampling and reporting on a dry basis, which is a reason why the dry basis extractive sampling analyzers are often installed. Most portable analyzers, as seen in Figure 12.20, are essentially a dry reading because they use a desiccant or a knockout filter to remove the water from the sample.

12.2.2.5 Tunable Diode Lasers

The disadvantage of probe analyzers is that they provide information about a single point in the heater only. This is usually sufficient in many cases such as vertical cylindrical heaters, but single point information is often insufficient for large cabin heaters. Flue gas recirculation and stratification patterns can cause significant variation of oxygen concentration along the

FIGURE 12.20
Portable flue gas analyzer.

length of the firebox. A single measurement point is not necessarily representative of the true average oxygen concentration.

Tunable diode laser analyzers have originally been developed to measure components in the Earth's upper atmosphere. Recently, a number of commercial applications have been developed, including the use of tunable diode laser spectrometers (TDLS) to monitor oxygen, CO, and other chemical compounds. The system consists of a laser mounted outside the firebox, aiming a beam at a receiver across the firebox at distances up to 100 ft (30 m). The reported value is an average oxygen concentration along the path length. The measurement is fast, typically around 5 s. The system does not provide a potential ignition source like the Zirconium sensors.

12.2.2.6 Excess Oxygen versus CO Measurement

As noted earlier, there are very sophisticated laser analyzers on the market that can be used to monitor the flue gas. For fired heater application, the predominant use is to measure the CO in the flue gas. The goal is to increase the heater efficiency by reducing the excess air until trace amounts of CO are detected in the flue gas. The key benefit in using a CO analyzer is that the normal tramp air, due to leakage, will not affect the reading as it does with excess O_2 measurement. The normal set point for excess O_2 is around 1.0% wet basis. The CO set point is usually 100 ppm. Operating at 1.0% excess O_2 will increase efficiency by about 0.8%–1% compared to operation at 3% with a 400°F (200°C) stack temperature. Another good point is that the formation of NOx drops significantly from 3% to 1% excess oxygen.

Most of these units are located in the stack where the effect of air leakage in the convection section could reduce the concentration of CO in the flue gas. It is possible with this type of arrangement, that the burners could be low on air and generating more CO than 100 ppm. If the CO burns out in the convection section, the analyzer reports that all is good. Another caution is that heaters that have a steam coil, boiler feed water coil, or combustion air preheat coil cannot operate at excessively low excess oxygen levels due to convective heat transfer requirements. On natural-draft heaters, the system uses the stack damper to control both the excess O_2 and the draft. There has to be an alarm to be sure the stack damper does not create positive pressure at the arch due to closing too far.

12.2.2.7 Alarm and Trip Settings

The heater should have adequate alarms for excess oxygen at the arch. These are essential to keep the heater from running out of air. The majority of heaters use 0.5%–1.0% excess oxygen as the alarm setting. A trip should be initiated anytime the excess oxygen goes to zero or the combustibles level spikes to 5000 ppm or greater. The selected points should be confirmed based on the actual heater operating history, flame patterns, and burner stability.

12.2.3 Fuel Measurements

12.2.3.1 Fuel Flow

The fuel flow (or rate of heat release) is one of the most important controlled variables in a process heater. The board operator should be aware of the maximum design heat release or the maximum heat release that has been proven by successful operation. As operation approaches this maximum fired duty, it is important to monitor the excess oxygen, draft, combustibles, and tube metal temperatures.

The amount of fuel, to the burners, is normally controlled by the process outlet temperature. If the outlet temperature starts to drop, the fuel flow control valve will open allowing more fuel to the burners. Common control valve loops use cascade control techniques to minimize the effect of pressure and composition fluctuations in the fuel supply system. Often, fuel control valves are incorporated into a lead-lag control system with the air controls of a heater. This system prevents the input of additional fuel before the air to burn it is present in the heater. The fuel control valve should be carefully placed and sized so that it reacts quickly without imposing an excessive pressure drop on the fuel supply to the burners. It should also be able to control over the full turndown range of the burners. A poorly sized valve may encourage operation with the valve bypass open, thereby reducing the effectiveness of any burner management system.

In order to know the total heat input, knowledge of the fuel heating value is required. The heat content of the fuel gas should be measured by analyzing the fuel gas components. If large changes in heating value can be expected, it is recommended to use a Wobbe index meter.

12.2.3.2 Fuel Pressure

The pressure of the fuel, whether gas or liquid, is a major energy source used within a process burner to achieve the required mixing of the fuel and the air. The design fuel gas pressure for most gas-fired process burners will typically be 15–30 psig (1–2 barg) at the burners when firing at the maximum design heat release with the design fuel composition. Figure 12.21 shows an example of a typical fuel pressure gage that would display this.

For liquid fuels, the design fuel pressure may be 100–150 psig (7–10 barg) at the burner at maximum design firing conditions. Higher fuel pressures allow a greater range of heat release, known as turndown. Turndown is the ratio of maximum heat release to the minimum heat release. For liquid-fired burners, the atomization medium pressure

FIGURE 12.21
Fuel pressure gauge.

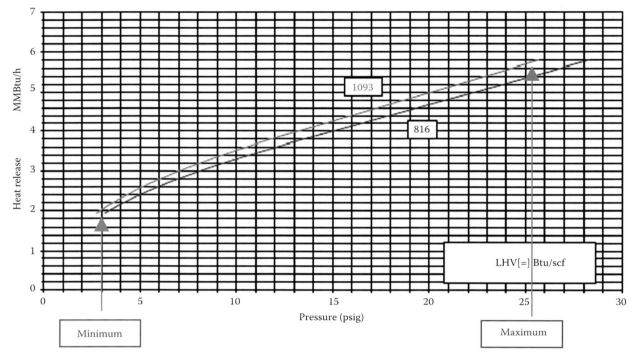

FIGURE 12.22
Maximum and minimum operating pressure.

will depend on the type of oil gun used and possibly on the available pressure of the oil. Either a constant atomization medium pressure of about 70–250 psig (5–17 barg) or a differential pressure controlled to 15–30 psig (1–2 barg) above the oil pressure is typical. The pressures of both the oil and any atomization medium should be monitored at a point downstream of any control valve and close to the burner block valves. The pressure of the oil and atomization medium must be checked with the individual burner block valves fully opened.

The fuel gas pressure at pilot burners will typically be regulated at a constant pressure, usually 5–15 psig (0.35–1.0 barg), depending on the pilot model employed.

The burner manufacturer normally provides a curve of the fuel pressure versus heat release (or fuel flow) for each burner and for each specified fuel composition. This curve (see Figure 12.22) will indicate the maximum and minimum fuel pressures. Operation outside these pressures should not be attempted unless adequate burner performance has been proven by field observations or testing.

12.2.3.3 Selecting Trip and Alarm Points

One common problem for heater operators is to determine fuel pressure alarm and trip points in order to avoid a burner operating problem. The objective is to set limits that allow safe operation under normal operating conditions without causing nuisance shutdowns. See Figure 12.23 for some basic guidelines. This task is made more difficult because the majority of fired heaters use multiple fuels. Different fuels require different pressures at the burners to satisfy the process duty.

The burner fuel capacity curves show the relationship between design operating fuel pressure and heat release for each of the specified fuels. These curves should be used to help determine the set points for actuating alarm points for the heater operation. They are based on burners that are in as designed condition with the

☐ Alarms (determined by end user)
  ☐ Maximum design pressure
  ☐ Minimum design pressure
☐ Shutdowns (determined by end user)
  ☐ Maximum design pressure + (3 to 5 psig)
  ☐ Minimum design pressure − (0.5 to 1 psig)

Notes: 1. Based on clean, properly designed & installed tips
2. Air is usually limiting factor on maximum firing rate
3. Confirm alarms & shutdowns at actual operating conditions (e.g., fuel composition, draft, etc.).

FIGURE 12.23
Fuel pressure guidelines.

proper gas tip port sizes, tip position, and tip orientation. All ports must be clean and free from debris and foreign material that can cause plugging. Gas tips that are not positioned correctly or fouled can have a dramatic impact on the flame pattern and the stability of the burner.

The first step in determining the fuel pressure alarm points is to review the burner design information. Be sure that the burner heat releases and the fuel compositions shown on the capacity curve and burner data sheet are still applicable to your current operation. Figure 12.24 is an example of typical burner capacity curves.

Select the normal fuel, and use the maximum and minimum pressures as the alarm points. After selecting the fuel pressure alarm points, the flames should be verified visually to confirm that the pattern and stability are acceptable. The excess oxygen should be closely monitored while making any changes to the burner firing rates. Operation that includes several fuels makes this process more difficult due to the possibility of multiple alarm points.

Another option, in addition to using the burner capacity curves, is to consider burner testing at the manufacturer's facility. This testing, under controlled conditions, will help to establish maximum and minimum stability points for each of the design fuels.

The high-fuel pressure heater trip point should normally be set at the fuel pressure corresponding to a 10% increase above the maximum duty. Since most heaters operate with 15% excess air, this would still allow for operation with excess oxygen in the flue gas.

The low-pressure heater trip point should be set below the minimum pressure shown on the capacity curve. Real-world operating practices have shown that burners should not be operated below 1.0 psig (0.07 barg) for most applications.

The difficult part of setting trip points is allowing normal operation under safe conditions without having nuisance shutdowns.

The operating and safe practice guidelines established by individual owners should always be consulted when determining heater trip points.

12.2.3.4 Increasing the Fired Duty

Anytime that the fired duty of a burner is increased, both the fuel pressure and the excess air must be evaluated as to their compatibility with the proposed changes in firing rate. A burner is designed to allow a specific amount of air to enter the tile throat at a certain pressure loss. As the firing rate is increased above the design duty, the chance of operating in a fuel-rich condition increases. During such a state of fuel-rich operation, it is possible for a combustible mixture to ignite or explode in areas of the heater where oxygen may be present, due to air infiltration through cracks and openings in the heater casing. If this happens inside a process heater, there could be severe damage to the equipment and

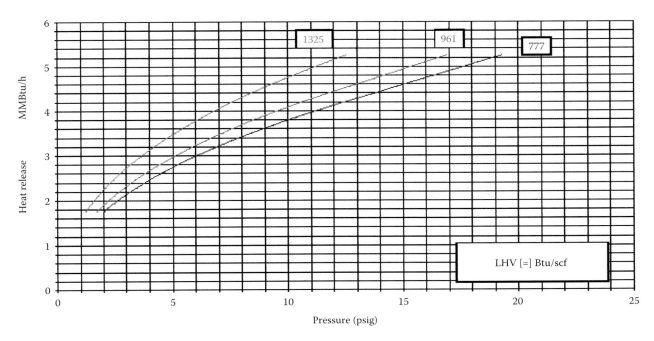

FIGURE 12.24
Typical burner capacity curves.

also injury or loss of life to operating personnel. Before the duty is increased, the operators must confirm the presence of excess oxygen in the firebox in the range of 4.0%–5.0% (wet basis). After the increase in firing rate is achieved, the stability of the burner flames must be visually confirmed.

Each case should be analyzed with regard to burner type and condition, fuel composition, heater configuration, and overall operating conditions. A thorough knowledge of the burner and heater operating manuals is essential before operating the unit. **It is the responsibility of the owner to operate the heater and burners in a safe and reliable manner.**

12.2.3.5 Fuel Temperature

Fuel gas temperature requirements can vary significantly based on the fuel gas composition. Most noncondensable hydrocarbon components and hydrogen can be at any value below the auto-ignition temperature without problem. Most condensable, light hydrocarbon components will require some level of heating of the fuel to prevent liquid condensation and carryover which can result in carbon formation, burner tip fouling, and coke buildup on the tile. Some unsaturated components must be handled with care. These components can crack or polymerize at elevated temperatures, resulting in tip fouling. Typically, the fuel gas temperature can range from ambient to 200°F (95°C). The gaseous fuel temperature must be specified for burner orifice and tip sizing, and any change that varies the density of the gas will affect the pressure and volume of fuel flowing at a given pressure. Normally, any changes will cause corrective action by the fuel control valve and a burner pressure increase or decrease (see Section 12.2.3.1).

If low ambient and fuel temperatures occur together with the presence of higher molecular weight components in the fuel gas, expect liquid hydrocarbon condensation in the fuel line. If proper liquid knockout facilities are not included in the fuel gas system, the liquid can extinguish burners.

Liquid hydrocarbon fuels must be atomized (broken up into small droplets) to be efficiently burnt. The most important factor in achieving good atomization is the viscosity of the liquid. Adjusting the temperature of heavier liquid fuels is the most common method used to control the oil viscosity. Raising the temperature will result in a less viscous liquid. Most burner manufacturers require a fuel oil viscosity ranging from 20 to 43 cSt (approximately 100–200 SSU) at the atomizer for proper burner operation. In general, fuel oils with a specific gravity of greater than 0.93 (~7.75 lb/gal/928 kg/m^3/20.5°API) usually have viscosity levels above 20 cSt at 38°C (>100 SSU at 100°F). Fuel oils with a specific gravity of greater than 0.95 (7.91 lb/gal = 948 kg/m^3/17.6°API) usually have viscosity levels above 50 cSt at 38°C (>240 SSU at 100°F). These oils, and more viscous oils, are required to be heated. That heat must be retained all the way from the fuel heater to the burner which requires that the fuel lines be well insulated to maintain the oil temperature between the fuel heater and the burners. Damaged or missing insulation should be repaired or replaced. Figure 12.25 is a typical plot of viscosity versus temperature for many types of fuels. The viscosity measured at two temperatures can be plotted and the line extended to the desired viscosity to determine the required fuel oil temperature.

12.2.4 Combustion Air Measurements

12.2.4.1 Combustion Air Temperature

The burner designer sets the tile throat area in the burner on the basis of an expected pressure drop under design conditions. The design conditions that set the air density include atmospheric pressure, air temperature, and relative humidity. A higher air temperature requires a larger volume of combustion air to supply the same amount of oxygen required to combust a set amount of fuel. If the burner draft loss is limited, the amount of oxygen flowing through the tile decreases, effectively reducing the capacity of the burner.

Relative humidity can also have a surprisingly large effect on burner and heater capacity. A relative humidity of 50% at 50°F (10°C) ambient temperature corresponds to 0.6 vol% H_2O in the air. Air with 50% relative humidity at 104°F (40°C) contains 3.6 vol% H_2O. This means that in addition to 3% more air being required, the water in the humid air also requires more heat to raise it to the flue gas temperature.

A significantly lower than design air temperature can lead to an increase in airflow and excess oxygen, leading to a reduction in efficiency. In either case, an adjustment of the stack or fan dampers or the burner air registers must be made.

Typically, the natural-draft air temperature is the ambient air temperature that can vary, in extreme, from −30°F to 130°F (−35°C to 54°C). This full temperature range does not normally occur at a single geographical location, but some can get close. A more typical maximum ambient temperature range of 120°F (49°C) causes little problem in the operation of the burners. Today, many heaters have air preheat systems installed to reduce the amount of fuel required. Air preheat systems are generally designed to recover unused thermal energy contained in the stack flue gas and transfer it back into the combustion air. An air preheat system will normally require both a FD blower and an ID fan

Burner/Heater Operations

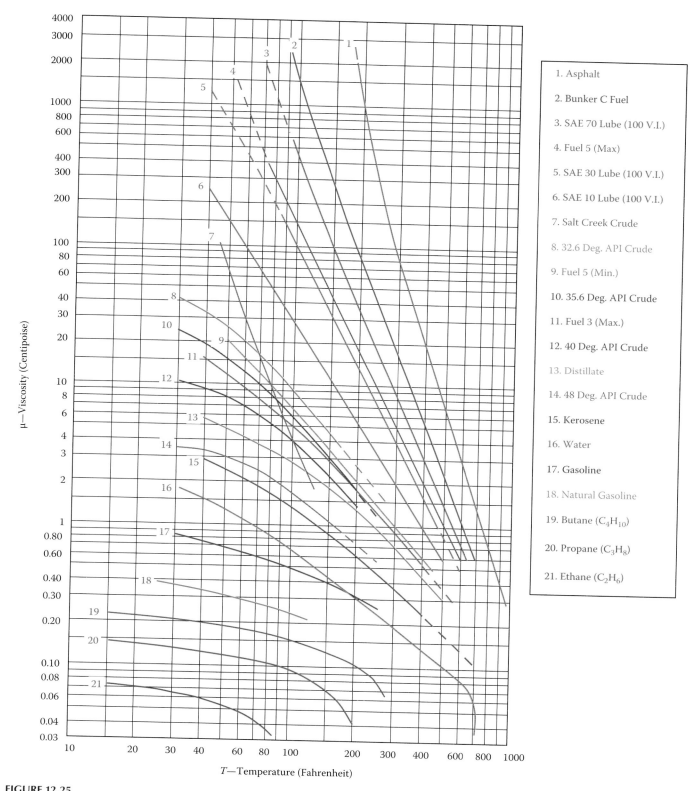

FIGURE 12.25
Viscosity versus temperature for a range of hydrocarbons.

to control the combustion air and flue gas flows while maintaining the required negative pressures within the heater. Typical preheated air temperatures range from 300°F to 850°F (150°C to 450°C). Some processes use turbine exhaust gas (TEG) as the oxygen source for combustion. In this case, the TEG stream could have a temperature as high as 1200°F (650°C).

12.2.4.2 Combustion Air Flow

Combustion airflow is typically difficult to measure. In natural-draft heaters, it is not measured at all. In forced-draft systems, there are several methods, but many problems exist due to uneven velocity profiles inside the ducts and leakages. Since the completeness of combustion is measured by excess oxygen in the firebox, it is normally not necessary to directly measure the airflow rate.

12.2.4.3 Combustion Air Pressure

A low static pressure in the combustion air can be used as a FD fan trip indicator. This measurement is also required to determine the pressure drop across the burners in a FD or balanced-draft application. This measurement should be taken with a pitot tube instead of an open-ended probe. A pitot tube will eliminate any velocity impact on the reading. Be sure that the sample tubing is connected to the static side of the pitot tube. A pressure transmitter should be placed downstream of the control dampers and the air preheater.

12.2.5 Flue Gas Temperatures

The flue gas temperatures of primary interest are the temperature of the flue gas leaving the radiant section and entering the convection section (i.e., the BWT) and the temperature of flue gases leaving the convection section and entering the stack.

The BWT is indicative of the radiant section heat transfer performance and the degree of fouling of the radiant section tubes. The BWT will rise as fouling occurs in or on the radiant section tubes, lowering the heat transfer from flue gas to the process fluid. It will also rise as excess oxygen in the firebox decreases or as fired duty increases. The excess oxygen and duty can be adjusted, but removing fouling deposits in the radiant section tubes usually requires a heater shutdown.

The stack gas temperature is a rough measure of the overall heater efficiency. As the exit temperature rises, the heater efficiency decreases. A rising stack temperature is most commonly caused by any of a number of factors, either singly or more commonly as the result of a combination of those factors. High excess air through the burners, radiant tube fouling resulting in loss of radiant efficiency, afterburning due to high levels of combustibles entering the convection section, and convection tube fouling resulting in lowered convective heat transfer will all result in an overall reduction in furnace efficiency and raised stack temperature. Refer to Section 12.2.2.2 for the relationship between furnace stack temperature, excess oxygen, and overall heater efficiency.

12.2.5.1 Standard Thermocouples

Great accuracy in these temperature readings is not highly important, but observation of trends will give significant information. This lower importance on quantitative data allows these temperatures to be commonly measured using thermocouples inserted in fixed thermowells. The radiant, convective, and conductive heat transfer to and from a thermowell in these services, combined with the typical arrangements in a heater, can result in readings that may be as much as 150°F (66°C) lower than the actual flue gas temperatures.

12.2.5.2 Suction Pyrometer (Velocity Thermocouple)

If a high level of accuracy for these temperatures is required for a test run, for debottlenecking studies or heater efficiency studies, these temperatures should be measured with a suction pyrometer, also known as a velocity thermocouple, as shown in Figure 12.26.[3] The velocity thermocouple uses either air or steam to educt through the flue gas into the thermocouple tube, which acts as a radiation shield. This gives an accurate measurement of the flue gas and eliminates the radiation loss.

FIGURE 12.26
Velocity thermocouple.

12.2.6 Process Tube Temperature

The process heater tube materials lose strength and become more subject to the effects of high-temperature creep from the stresses caused by their internal pressure as the temperature of the tubes increases.

To avoid process tube ruptures, providing for satisfactory tube life and safe operation, the tube-wall or tube-skin temperatures must be held within design limits. The industry design standard reference for defining the allowable tube-skin temperatures in a heater is API Standard 530.[4] The effects of high temperatures can be seen in Figures 12.27 through 12.29.

Tube metal temperatures can be measured in several ways. Most general-service process heaters will have tube-skin thermocouples attached to the radiant tubes at selected locations. These thermocouples are typically radiation shielded. They should be located on the fire side of the tubes and approximately one-half to two-thirds of the estimated flame length. This shielded style of tube-skin thermocouple has been found to give the greatest accuracy with acceptable life. The thermocouples are typically connected to a data collection system to provide a record of the tube external surface temperatures at the selected points that can be trended over time.

Infrared thermography (or "IR scanning") is another method to check furnace tube-skin temperatures. Infrared thermography provides a digital image of the heater firebox interior where the temperature of the tubes and refractory are represented by colors or contrasts.

This technique requires special equipment, appropriate training, firing with gas fuel (so that the flame wavelengths can be filtered out of the detected information), and fairly generous sight port sizing. This is not

FIGURE 12.28
Process tubes and tube hangers.

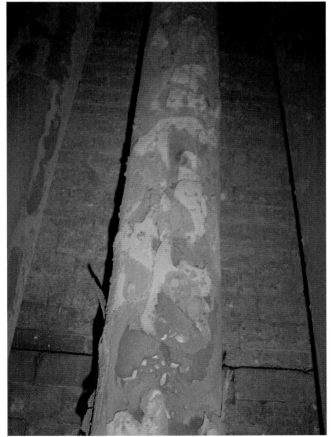
FIGURE 12.29
Tube scale due to flame impingement.

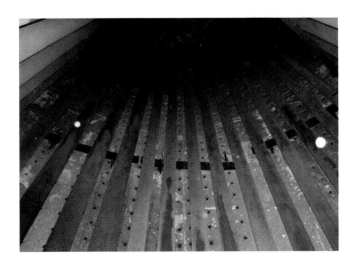
FIGURE 12.27
Process tube fouling due to flame impingement.

(yet) practical as a continuous monitoring technique. It is generally limited to routine inspections to identify hot tube areas caused by flame impingement, internal process tube fouling, or erratic hot flue gas flow patterns. It can be used for monitoring the progress of internal fouling, confirming the accuracy of tube-skin thermocouples, and identifying problem burners. However, since external scaling of heater tubes due to oxidation can simultaneously raise the outermost surface temperature and change surface characteristics of the tube, it may take a trained eye to determine if high temperatures detected during IR scanning are due to internal or external fouling or due to a change in surface emissivity.

Tube-skin temperature can also be measured at specific spectral bands using infrared pyrometers. These are handheld optical instruments which are aimed at the selected tube and measure the infrared energy entering the instrument. The flue gas contains H_2O and CO_2 which both have the capability to radiate heat energy in the infrared spectrum. They radiate in very specific discrete bands because of their limited number of excitation states. This means that the flue gas is transparent in so-called "windows" in the spectrum. If the pyrometer wavelength is chosen in such a window, for example, at 3.9 μm, the measurements become insensitive to the flue gas influences. In addition, the smaller the wavelength, the less dependent the emitted radiance becomes on variations in emissivity.

Using an assumed or measured tube metal emissivity, the instrument electronically converts the incident energy to an indicated metal temperature using Planck's law (see Volume 1, Chapter 7). However, it must be remembered that the infrared energy entering the instrument comes not only from the tube but also from the refractory. If the emissivity is set to 0.85 (typical for an alloy cracking coil tube), the pyrometer assumes all incident radiation comes from the tube and converts it to an equivalent temperature. However, 15% of the incident radiation is actually reflected radiation from the refractory and should not be considered as being emitted by the tube itself. Without any correction, the calculated tube skin temperature is too high. Using a refractory emissivity (typically 0.5–0.6) and a measured refractory temperature, this reflected fraction can be calculated and subtracted from the incident radiation in order to get the most accurate result.

The heaters commonly monitored with pyrometers are those in high-temperature pyrolysis and reforming services where tube-skin temperature is a critical process parameter. Frequency of the measurements is a function of coil design, operation severity, and total run length between decokes and can vary between once a shift to once a week.

High tube temperatures due to internal fouling can often be visually observed. A reddish or silvery spot on the tube is an indication of localized overheating. When hot spots are identified, operation of the heater should be modified to keep the tube from becoming hotter. Also, tube cleaning should be considered.

There are three generally applicable techniques for avoiding further increases in the tube temperature. One option is to increase the process flow in the pass containing the hot tube. The increased flow rate increases the convection heat transfer inside the tube and removes heat from the tube wall more rapidly. This should be undertaken with knowledge of the impact on the other passes, where flow is reduced. The temperatures of tubes in these passes can be expected to rise and must therefore be carefully observed.

The second option is to reduce the radiating effectiveness and temperature of the hot gases in the radiant zone by increasing the excess air. The radiating capability of the hot gases is proportional to its temperature to the fourth power and inversely proportional to the concentration of symmetrical molecules such as oxygen and nitrogen. As the concentration of nitrogen and oxygen increases while the flue gas temperature drops, the energy radiated from the gases to the radiant tubes decreases. Because hot spots usually occur in the radiant zone or firebox, this can be an effective action, particularly if firing gas fuel. The impact of this action on the remainder of the heater is an increase in the amount of the flue gases entering the convection section. The convection section process temperatures and heat absorption will rise. Duty will be shifted from the radiant section to the convection section.

The second option will be of limited effectiveness if the heat transfer to the hot spot is largely radiation from a flame. In such a case, the most effective action is to reduce or eliminate the heat input from that flame by throttling or closing the burner valve. This action should be reported to management because it has the effect of possibly shifting the hot spot to another location. Section 12.5.2 discusses the problems and limitations involved in throttling burner valves.

12.2.7 Process Fluid Parameters

The process fluid flow rate, pressure drop, and outlet temperature are measured with conventional instruments. In the majority of process heaters, the flow rate through each individual pass is monitored. In some cases, the overall flow is measured, and the flow to individual passes is governed by restriction orifices. In other designs, where the addition of measuring elements and control valves would impose an uneconomical pressure drop, careful distribution manifold design equalizes pass flows as much as possible.

The flow of the process fluid cools the tubes and maintains an acceptable metal temperature. A reduction in flow will lead to a rise in the tube-skin temperature and, often, a phase change or chemical degradation in the fluid. An unexpected formation of vapor in a normally liquid-filled tube pass will increase the pressure drop and cause further reduction in flow and cooling capability with an associated increase in tube temperature.

Chemical degradation of hydrocarbons can cause a phase change, which may devalue a product. More frequently, decomposition of hydrocarbons will cause formation of a solid layer of carbonaceous material, referred to as coke, on the tube internal surface. Due to its low thermal conductivity, this layer impedes the conduction of heat from the tube to the process fluid and causes a local increase in temperature. The pressure drop in the pass may increase measurably due to the reduction in cross-sectional area. Ultimately, the tube metal temperature rises to a level where decoking is needed before the tube fails.

In order to minimize differences between tube temperatures, the pass flows should be nearly equal and not vary by more than 10% when the heater is firing at greater than 75% of the rated heat release. At lower heat releases, the heat flux is unlikely to cause the problems mentioned earlier.

Thermocouples are typically installed in the outlets of each pass. These can be used to check the pass flow controls and indicators. If the flow in a pass is reduced, without a change in the pass heat absorption, the outlet temperature will rise. Conversely, if the flow to a pass is increased, its outlet temperature will drop. If a change in pass outlet temperature occurs without a corresponding and appropriate change in the indicated pass flow, the operator should immediately investigate. An instrument problem is likely, and a flow interruption or reduction, with potential tube overheating, is possible.

12.2.8 Heater Operation Monitoring

Every heater should have established control points on certain operating variables that are selected to ensure safe operation and maximum heater utilization. Operators must ensure that these control limits are not exceeded. Items that may be considered limits will include some of those listed in Table 12.1 and should be monitored frequently.

12.3 Heater Operations

12.3.1 Operating Strategy and Goals

The goal of the heater operating strategy should be to achieve a heater operation which is first and foremost safe. Second, it must be in compliance with safety and environmental regulations. Third, the operating strategy should ensure that the heater is operated within the design operational and mechanical envelope. Finally, the objective should be to achieve an operation that meets the process demands in a stable and efficient manner.

12.3.2 Heater Safety

Various codes and standards[5-7] exist that define minimum requirements and recommendations for the safe operation of process heaters. Safety instrumented systems (SIS) perform specific control functions to failsafe or maintain safe operation of a process when unacceptable or dangerous conditions occur. The SIS is independent from all other heater control systems. The SIS can be of varying levels of complexity, dependent on the results of the risk and hazard assessments such as process hazard analysis (PHA), layers of protection analysis (LOPA), or event trees. The implementation of these standards and safety systems has significantly improved the safety performance of petrochemical plants and refineries in general and fired heaters specifically. However, incidents still occur. Start-up of fired equipment is especially critical since many elements of the SIS are not fully functional or need to be bypassed during the start-up phase due to the absence of process flow and flue gas flow. Accumulation of unburned hydrocarbons in the firebox during light-off sequences must be prevented at all times and requires operator training, adherence to procedures, good communication between board and field operators, and properly maintained instrumentation and equipment.

12.3.2.1 Start-Up Procedures

Each plant will need to develop start-up and operating procedures for its particular heater and operations. These procedures should be followed for start-up and shutdown to ensure a safe and efficient operation. The following procedures are only intended as a guide to help develop start-up and operating procedures for a plant.

The start-up and operating procedures are intended to ensure avoidance of an explosive mixture occurring

TABLE 12.1

Some Typical Control Limits

Process flow	The process flow rate should be monitored and trended. The process flow is an indication of the amount of work the heater is to perform. If the process flow is at design conditions, then the burners can be expected to be operating at design conditions. If the process flow is above design conditions, then the burners can be expected to be operating above design conditions.
Pass flow	Pass minimum flow limits are established to avoid maldistribution of process flow through the tubes in each pass. Flows that are below the limit can cause coking of interior tube surfaces or overheating and tube failure.
Tube metal temperature	Tube maximum temperature limits are established to protect against tube overheating and failure.
Stack temperature	Stack gas temperature low limits are intended to protect against condensation and corrosion. High-temperature limits are set to help protect air preheaters and stack materials from failure. This temperature can also be used to trend the heater to determine efficiency changes and identify convection section heat transfer problems.
Bridgewall temperature	Bridgewall or radiant section exit flue gas temperature limits may protect against overfiring and overheating of convection section tubes and supports. The burner manufacturer bases his emissions guarantee on the BWT.
Process inlet and outlet temperature and pressure	Check and monitor the process inlet and outlet temperature limits to help prevent overfiring, protect process quality, and avoid unwanted reactions. Inlet temperatures can be used to identify upstream problems, such as heat exchanger fouling. Check the process inlet and outlet pressure to maintain the equipment integrity and to help identify internal tube deposits.
Draft targets	Draft targets are set to maintain negative pressures within the heater and avoid safety and structural problems while ensuring conditions that allow adequate airflow through the burners.
Excess oxygen targets	Excess oxygen, measured in the flue gas leaving the firebox, will provide guidance on adjusting burner registers (or dampers) to reach the excess oxygen target level. These adjustments must also consider the draft target. Trends in excess oxygen can indicate changes in fuel composition, tube leaks, and deterioration of heater operation.
Fuel pressure, temperature, and flow rate	Check the fuel pressure at the burners. Compare the fuel pressure to the burner capacity curve and to the fuel flow measured by the heater instruments. Fuel flow should be recorded and trended. An increase in fuel flow that brings no increase or even a decrease in process outlet temperature with a constant process flow indicates a "fuel-rich" condition where there is not enough air in the furnace to burn all the fuel. The fuel temperature should be monitored if the fuel contains large amounts of butane or heavier hydrocarbons. Reducing the fuel temperature on a heavy fuel gas can cause liquid hydrocarbons to form in the fuel stream. If the fuel is heavy oil, the temperature at the burner is critical to good atomization and combustion of the liquid fuel.
Stack emissions	Stack emissions, such as NOx, SOx, CO, and combustibles, are monitored to help ensure compliance with regulations and operating permits. A change in the levels may indicate an improperly adjusted burner, inadequate airflow to the burners, unstable flames, or a change in fuel composition.

in the firebox and to establish stable flames and process flows. Start-up procedures will vary with the degree of automation and instrumentation on the heater. The procedures should address the issues listed in the following:

1. Fuel system and burner preparation
   a. Fuel line blinds must be removed.
   b. The fuel lines should be blown out to remove scale and debris. This has to be done before the fuel line is connected to the burners to avoid blowing scale and trash into the burner tips.
   c. The fuel lines and valves should be pressure tested to identify leaks.
   d. The fuel gas knockout vessel should be drained of all liquids.
   e. The circulation in the fuel oil system must be established and confirmation obtained that the heat tracing and atomizing medium systems are functional.
   f. Confirm adequate pressure in the fuel and atomizing medium supply systems.
   g. Confirm that the burners are properly installed, with the tips properly positioned and oriented and no blockage in the fuel ports or the airflow passages (see Figure 12.30).
   h. Isolation valves in the fuel systems must be closed, including those at each burner.
   i. Open and close the burner air registers to confirm operability over the full range. Leave them set at the 100% open position(s) for purging the heater.

2. Check the heater for readiness
   a. Visually check that there is no debris from maintenance or construction left inside the heater. All access doors, header boxes, sight doors, and explosion doors should be closed at this time.
   b. Stroke the stack damper over the full range to confirm operability, and leave it fully open.

FIGURE 12.30
Checking burner gas tip location.

3. Establish flow through the process tubes
   a. Multi-pass heaters with vertical radiant tubes and liquid feed require a special procedure that involves filling one pass at a time. This is to ensure that full and stable flow is reached in each pass during operation. Heater manufacturer guidelines and instructions should be followed to ensure that each pass is flowing properly.
   b. During the filling process, check the operation of the pass flow indicators, valves, and controllers. If any pass does not have the correct flow indicated or if flow fluctuates, analyze and correct the problem before beginning the purge and burner lighting steps. It is often helpful to place the pass flow control valves on manual operation until the flow reaches about 75% of the normal flow value. This will overcome the reset windup problems that occur with some control systems and valves with minimum flow stops.
4. Purge any accumulated combustibles from the firebox
   a. Use steam or fan-supplied air to purge the firebox for at least five volume changes or fifteen minutes, whichever takes longer. Avoid excessive steaming; a long exposure to steam and condensate can damage the furnace refractory.
   b. Check the operation of the draft gauge during the purge period to ensure that the draft gauge is operable and reading the draft within the heater.

5. Light the pilot burners
   a. First, check that all pilot and main burner individual block valves are closed.
   b. Remove any blinds that are in the pilot fuel system, and confirm that the pilot gas pressure regulator is set accurately and in accordance with the required pressure for the pilot to work.
   c. Reduce the purge flow 75%, and sample the firebox in several locations with a combustibles analyzer or a LEL meter, as seen in Figure 12.31.
   d. Close the burner air register or damper to a position of 15%–25% open. Close the stack damper to 25% open. Light all of the pilots before attempting to light a main burner. If the pilots fail to light, refer to Chapter 13.
6. Light the main burners
   a. Visually recheck that all main burner individual fuel valves are closed. Check the fuel control valve for proper operation. Place the fuel control valve in the "manual" mode on initial start-up because operations will be below the automatic controller range. Set the fuel control valve to 15% open, open the main fuel supply block valves, and open one individual burner block valve on a burner with a stable lighted pilot. The burner should light immediately. If it does not light within five seconds, close the fuel valve and refer to Chapter 13. Continue lighting main burners that have stable pilot flames, manually

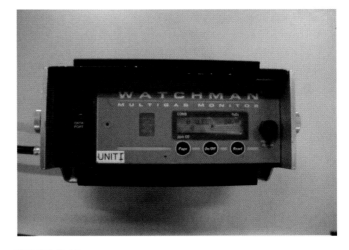

FIGURE 12.31
Typical lower explosion limit (LEL) meter.

opening the fuel control valve after each to maintain fuel pressure above minimum firing pressure shown on the burner capacity curve. Failure to do this may cause flame instability and potential loss of flame, requiring a re-purge of the heater if all the pilots and burners are extinguished. Light burners in a pattern that distributes active burners evenly throughout the firebox.

  b. If operating in the natural-draft mode, increase the fuel rate slowly to warm the heater, and establish the draft that allows more air to enter the heater. Failure to establish adequate draft can cause a fuel-rich, potentially explosive mixture to develop in the firebox. Check the draft at the top of the firebox radiant section to ensure that it is slightly negative at all times.

  c. Visually do a frequent check for stable flames on both main and pilot burners.

  d. When the process outlet temperature warms to within the range of the temperature controller, place the fuel control valve on automatic operation.

  e. Typical allowable warm-up rates for heaters vary from 100°F to 200°F (55°C to 110°C) per hour for heaters with plug headers and up to 350°F (177°C) per hour for heaters with fully welded coils. The lower rates for plug headers will avoid excessive thermal stresses that can cause leaks at plugs or header attachments. The earlier temperatures are flue gas temperatures measured at the bridgewall.

  f. Warm-up rates may be limited by the cure-out procedure for new refractory. The heater manufacturer or installer of the material will recommend a firing procedure to carefully dry and cure the material.

7. Periodic checks

  a. Periodically during the start-up, check the fuel and pass flows, the individual pass outlet temperatures, fuel pressure, draft at the arch, flue gas oxygen content at the radiant section outlet, and firebox and tube-skin temperatures.

  b. Draft should be monitored frequently by the operator when heat input is increasing. As the heat input is increased, the burner air registers (or dampers) and the stack damper will need to be opened. The combustion air should always be increased or decreased on the "lead/lag" principle. This means that the combustion air should be increased before the fuel when firing rates are being raised and the fuel should be decreased before the combustion air when rates are being lowered.

  c. Place the pass flow controllers on automatic operation at about 75% of normal flow (see step 3). Watch the pass outlet temperatures closely. A high or uneven pass outlet temperature can be due to a low pass flow, flame impingement, or uneven firing of the burners. This must be analyzed and corrected quickly to avoid damage or curtailment of the run length due to internal tube fouling. The pass flow indicators may not be reliable; the outlet temperature may be the best information. If flow stoppage or reduction is suspected or a major pass outlet temperature discrepancy cannot be corrected, quickly extinguish the main burners until the problem is resolved. The pilots can usually be kept lit under these circumstances, thereby avoiding having to purge the heater again.

  d. Be aware that without process flow, the tube-wall temperature will rise to the firebox temperature after the main burners are extinguished. Opening the stack damper(s) fully will increase the flow of cooling air. If the process flow is interrupted, do not reintroduce fluid into the tubes until they have cooled to below 800°F (425°C).

12.3.2.2 Heater Shutdown Procedure

Shutdown of the heater is far simpler. Gradually reduce the heat input, taking burners out of service in order to hold fuel pressure on the active burners above the minimum shown on the burner capacity curves. Do not close the registers on burners taken out of service. Purge the contents from the tubes when they have cooled and flames are extinguished. Close all fuel and atomizing medium valves. Open the stack damper and the burner air registers to increase the flow of air and the rate of cooling of the heater. Install line blinds in the fuel and process lines as required by safe practice.

12.3.2.3 Emergency Procedures

An emergency can be defined as an off-design condition that, if not properly handled, will result in major damage to or destruction of equipment and personnel injury or death. Emergency procedures most commonly address the problem of rupture of a tube containing a flammable material while the heater is operating.

A more common and more easily handled event is an incident in which unburned combustibles collect in an active firebox due to errors in fuel and air handling. In the latter case, the appropriate procedure is to gradually reduce the fuel flow rate, without increasing the air supplied to the firebox, until complete combustion (as represented by the presence of some target amount of excess oxygen and elimination of CO in the flue gas) is reestablished. *Do not open the burner air registers or the stack damper when the heater is out of air as this could cause an explosion and possible personnel injury or worse.* When the heater has stabilized, the firing rate can be increased to satisfy the process requirements.

Some events that can result from a tube rupture and for which the proper mitigating actions must be developed include the following:

- Melting or vaporization of the tubes and supports
- Detonation in the firebox
- Convection section collapse into the radiant section
- Heater collapse due to support structure failure
- Flaming oil pool spreading to other areas, putting additional equipment at risk
- Explosive vapor cloud forming around the heater and possibly igniting
- Damage to a stack used by several heaters
- Rapid shutdown of heater and unit causing leaking flanges due to thermal shock
- Rapid depressuring from high pressures causing upset of catalyst beds and distillation column trays

Some actions are almost always appropriate and should be considered when developing emergency procedures to deal with a tube rupture. These include the following:

1. Shut off process flow to the heater or to the ruptured pass to eliminate the source of uncontrolled hydrocarbon entering the heater.
2. Always leave the stack damper in the current position.
3. If conditions are safe to do so, close the burner air registers to minimize the air entering the heater.
4. Turn on firebox smothering (snuffing) steam to cool the fire.
5. Activate firewater monitors and hoses to quench any spilled oil, to protect adjacent equipment with fogging sprays, and to cool the structure and stack to avoid possible collapse.

Aim at containing the fire inside the heater. Slowly, using steam or nitrogen, purge the contents of tubes in the failed pass into the firebox without losing containment. The tubes in the other passes should be purged of contents in the normal manner to avoid a rupture in another pass adding fuel to the conflagration. If the failed pass contained hydrogen, allow the contents to dissipate into the firebox without purging. A hydrogen fire may become so hot that tubes melt or vaporize, and it is important to give the smothering steam a chance to cool the firebox.

It is likely that there will be unburned hydrocarbon gases or vapors in the heater firebox. Beware of the instinctive reaction to cut off the fuel to the burners. The combustion air freed by such action could result in an explosive mixture in the heater, resulting in violent destruction. Avoid creating explosive mixtures in the affected heater and in any ducting or stack common with other heaters. These other heaters, if any, should be kept firing at low rates and with the absolute minimum excess air until they can be shut off and their air registers or plenum dampers closed. Then, introduce smothering steam to them as well.

If valves are available, isolate the heater to minimize the amount of flammables that can flow from other equipment. Begin depressurizing the plant as soon as possible to minimize the amount of fuel that is available.

Keep the smothering steam flow on and the combustion airflow blocked until the heater cools to below 600°F (300°C). The burner plenum damper or air registers can now be opened to increase the rate of cooling.

12.3.3 Heater Combustion Control

12.3.3.1 Target Draft Level

A target draft is established at the point of highest flue gas pressure within the heater. The target value is selected to minimize air leaking into the heater and to provide adequate differential pressure or draft loss at the burner level for necessary airflow across all burners. Fired heater data sheets, most commonly, define this lowest (closest to zero or positive pressure) draft as the draft at the arch, or the top of the radiant firebox. The most common design level for draft at this location is 0.1 inH_2O (2.5 mmH_2O). This is not a design limit; it is the reference value from which draft losses are calculated and flue gas passages sized by the heater designer. This is also the location at which the operator normally monitors and controls the heater draft performance. Each heater should have facilities to monitor the draft available at the arch or at the location where the draft (or negative pressure) within the heater is at a minimum.

Efficient furnace operation generally minimizes excess air. This requires minimizing air leaking into the heater through paths other than the burners. Minimizing the arch level draft minimizes the differential pressure between the outside air and the negative pressure within the heater. This, in turn, minimizes the driving force for air passing through any non-burner openings into the heater. Air leaking into the heater is an important consideration when setting a draft target. The more negative the draft, the higher the volume of air that can leak into the heater. The less negative the draft, the less air leaks into the heater.

Burners require a differential pressure between their air supply and the negative pressure within the furnace. This pressure drop drives the combustion air through the burner air registers and tile throat. The negative pressure within the radiant firebox, or available draft, decreases with height—the draft available at the floor of the firebox will be greater than that at the roof. This is shown in the typical static draft profile (Figure 12.5). Vertically fired burners located at the floor, or horizontally fired burners located near the floor, should be designed for the draft available at their location. In furnaces where the burners are end wall or side wall mounted at different elevations, burners located higher on the walls will have less draft available. In such a case, the draft target in the heater should be adjusted to satisfy the highest elevation burners at the design heat release, plus a small margin for ambient temperature, fuel composition, and operational changes.

The consequence of too much draft in the firebox is excessive air leaking into the heater and lower heater efficiency. The additional air adds to the flue gases and increases the draft loss across the convection section and stack. Too little draft can restrict the airflow through burners, sometimes enough to cause flame instability, flames to exceed the designed flame envelope (flame impingement), and even the formation of CO. If the firebox pressure becomes positive at the arch, hot gases will flow out through openings, potentially damaging the heater casing, weakening refractory anchors, creating unsafe conditions, and restricting heat release.

12.3.3.2 Target Excess Air Level

It is not possible to recommend a single target excess air level for all fired heaters. The condition of the heater, the type and composition of the fuel fired, the type of heater draft (natural, forced, induced, or balanced), the level and variability of process operation, and the ambient conditions all affect the achievable target.

The target excess air level can be established by following a structured procedure. The recommended procedure for establishing the optimum excess air level targets for a natural-draft heater is given in the following. The procedures for forced- or balanced-draft heaters are similar. This same procedure can be used to adjust a heater to the maximum combustion efficiency or to optimize the heater performance. The fuel flowmeter, the process outlet temperature indicator or recorder, the draft gauge (and controller if used), the flue gas oxygen and combustibles analyzer, and the CO analyzer should all be calibrated and in good working order.

Begin the procedure with all possible burners in service and firing equally on one fuel, satisfactory flame appearance, correct fuel temperature and pressure, steady process operation, and all potential limits (such as tube metal temperature) monitored and recorded. Close the air registers (or dampers) on all out-of-service burners. Equalize the airflow to all active burners by adjusting all registers to the same opening. Check the oxygen and CO levels of the flue gas, and ensure that the draft available at the firebox roof is on target.

Next, slightly close the air registers equally on the active burners or the common plenum air supply damper. The draft will increase because the incoming air and the flue gas amounts are lowered, reducing the losses from flue gas flow through the heater. Measure the excess oxygen and CO in the flue gas. Observe the flame condition, and monitor the other instruments for satisfactory operation with no approaching limits. Readjust the draft to the target value with the stack damper.

Continue to close the burner air registers or plenum dampers in slight increments while holding the targeted draft level with stack damper adjustments. Once the target excess oxygen level is achieved and the target draft has been maintained, the heater is "tuned." More sophisticated heater instrumentation and control systems can safely allow lower excess oxygen levels than would be suitable for simple automatic or manual systems. The target excess oxygen level should provide safe and steady heater operations with good flame patterns (no flame impingement), flame stability, tube-skin, and firebox temperatures within the limits set by the design and operating engineers of the heater. Tables 12.2 and 12.3 indicate typical excess air values that should be achieved with this procedure.

With the target draft and the target excess oxygen established, the operator is now ready to make any adjustments needed to keep the heater operating as efficiently as possible. The flowcharts in Figure 12.32

TABLE 12.2

Typical Excess Air Values for Gas Burners

Type of Furnace	Burner System (%)
Natural draft	15–20
Forced draft	10–15

TABLE 12.3

Typical Excess Air Values for Liquid Fuel Firing

Operation	Fuel	Excess Air (%)
Natural draft	Naphtha	15–20
	Heavy fuel oil	20–25
Forced draft	Naphtha	10–15
	Heavy fuel oil	15–20

(natural draft) and Figure 12.33 (balanced draft) will guide the operator through the necessary adjustments on the heater to achieve set targets on a continual operating basis. For example, the operator can control the preset targets as follows:

- The target draft has been determined to be −0.10 inH$_2$O (2.5 mmH$_2$O) from the previous discussion.
- The target excess oxygen has been determined to be 3.0%.
- In a natural-draft heater, the stack damper and the burner air register are adjusted to control the draft and the excess oxygen.

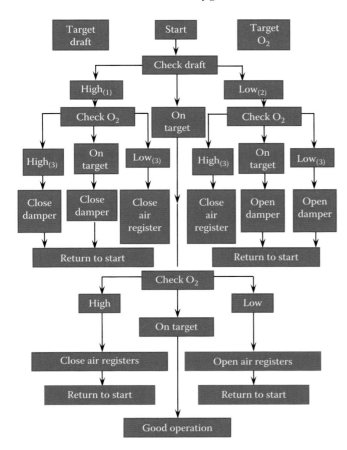

FIGURE 12.32
Natural-draft furnace adjustments.

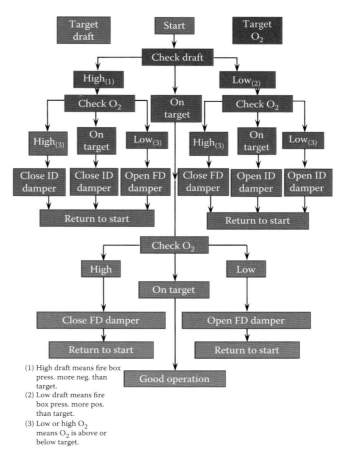

(1) High draft means fire box press. more neg. than target.
(2) Low draft means fire box press. more pos. than target.
(3) Low or high O$_2$ means O$_2$ is above or below target.

FIGURE 12.33
Balanced draft furnace adjustments.

- In an ID system, the inlet damper on the ID fan and the burner air register are adjusted to control the draft and the excess oxygen.
- In a FD system, the inlet damper on the FD fan and the stack damper are used to control the excess oxygen and the heater draft.
- In a balanced-draft (forced/induced) system, the inlet damper on the FD fan and the damper on the ID fan are used to adjust the excess oxygen and the draft.

For the natural-draft example, the operator begins in the "START" box of Figure 12.32.

The arch draft is measured on the heater at −0.15 inH$_2$O (−3.8 mmH$_2$O). The draft is above the target of −0.10 inH$_2$O (−2.5 mmH$_2$O), and the logic box indicates "HIGH." The excess oxygen is measured on the heater at 5%. The chart indicates that if the excess oxygen is also above the target of 3%, then the excess oxygen is in the "HIGH" box. The flowchart indicates that the corrective action required is to close the stack damper on the heater. When the stack damper is closed, the draft within the heater goes from −0.15 to −0.10 inH$_2$O (−3.8 to −2.5 mmH$_2$O). The logic chart indicates a return to "START." The draft (pressure) is now on target, so go to box "ON TARGET." The excess oxygen measured in the field is 4.0%. The excess oxygen is still above the target of 3%. The operator goes to the "HIGH" box. The corrective action indicated is to close the air register or damper on the burner and return to "START."

The draft is measured again and determined to be −0.13 inH$_2$O (−3.3 mmH$_2$O) above the target. The logic chart indicates to check the excess oxygen. The new excess oxygen reading is 3.5%. The logic chart indicates to close the stack damper and return to "START." Return to "START" and measure the draft. The new draft reading is 0.10 inH$_2$O (2.5 mmH$_2$O). The draft is on target; thus, the logic chart indicates to measure the excess oxygen. The excess oxygen reads 3.0% and is on target. The logic chart indicates "Good Operations." The tuning of the natural-draft heater has been completed. Tuning a balanced-draft heater is basically the same process. The difference is that the FD fan and the ID fan are used to control the excess oxygen and the draft at the arch of the heater (see Figure 12.33).

While the operator should be encouraged to operate to the excess oxygen target as determined earlier, there may be conditions where it is desirable to operate with excess oxygen well above the target. For example, in the case of high-radiant tube metal temperatures, increasing the excess oxygen reduces the radiating effectiveness of the firebox gases, lowers the radiation to the radiant tubes, and lowers the heat flux and temperature of the tubes. The effect is to increase the amount of flue gas, increasing the amount and temperature of gas flowing to the convection section. Duty is shifted from the radiant to the convection section. Similarly, for heaters where waste heat steam generation occurs in the convection section, trouble with the plant steam boilers may make it desirable to generate more steam in the waste heat generation coils. Increasing the excess oxygen (excess air) above the target value will increase convection section available heat and mass flow and generate increased heat transfer in the convection section.

12.3.4 Heater Turndown Operation

During normal- or design-level heater operation, the operator strives for even heat distribution in the firebox by equalizing the air and fuel to all burners while maintaining stable flames and acceptable flame quality. The same goals apply to turndown operation. When operating at reduced heat-release levels, the operator may remove burners from service in a selected pattern, closing air registers and fuel valves, so as to maintain adequate fuel pressure well above any low-pressure trip limits. Burner flames should be stable with a well-defined shape due to a higher excess air levels.

Burners are turned off in a pattern that maintains an even heat release throughout the firebox. The heater designer arranges burners so that each radiant tube pass "faces" the same number of burners. It is wise to maintain this practice when reducing the number of active burners. This helps to ensure that each pass receives the same amount of heat and avoids having to overly bias pass flow rates to equalize outlet temperatures. In natural-draft heaters with diffusion flame burners, the design excess oxygen can be held nearly constant with air register and stack damper adjustments from 100% capacity to approximately 75% capacity. The excess oxygen level will begin to increase as the heater (burner) heat release is reduced beyond that point. The majority of natural-draft heaters will normally operate with 30%–50% excess air at turndown conditions. Reducing the air further usually results in poor flame quality or stability problems associated with cold fireboxes.

Some FD burners may be capable of maintaining low excess air levels at reduced firing rates. This is due to a higher air side pressure drop which provides more turbulence for better mixing at turndown conditions.

In natural-draft heaters with premix burners, the air entrained by the fuel corresponds with the fuel gas pressure as the heater duty is reduced. If the premix burner is a small radiant wall type that uses the fuel pressure to induce 100% of the combustion air, the excess oxygen is reduced as the fuel pressure is lowered. Larger premix burners that have secondary air registers will see an increase in excess oxygen at lower rates if the secondary register is not partially closed.

At low firebox temperatures, burners may emit more CO. The colder recirculated flue gas quenches the combustion reactions as it is entrained back into the flame zone. The amount of CO emitted depends predominantly on the fuel composition, burner design, and the firebox floor temperature (which is indicative of the recirculated flue gas temperature).

12.4 Visual Inspection inside the Heater

Much of successful heater operation depends on frequent and knowledgeable visual checks of the equipment. In most applications, visual inspection of the heater and flames as often as necessary to ensure safe and optimum operations is considered good practice. Visual observations should also be made after significant load changes, atmospheric disturbances, fuel gas composition changes, and utility system upsets. A checklist of what to observe should be developed for each heater. The visual inspection of a heater should include the inside of the heater, the outside of the heater, and observation of the performance data of the heater.

Sight doors, mounted on the casing, provide visual access to the firebox interior. The operator can monitor the burner flames, the temperature and condition of tubes, and the temperature and condition of the refractory. The sight doors should ideally be sized and located so that all burners for the full length, or diameter, and of all radiant section tubes can be observed. If the sight doors are not adequate to fully observe the inside of the firebox, additional sight doors should be added to ensure good visual observations inside the heater.

Prior to opening a sight door, the draft at the door elevation should be checked to ensure a negative pressure. While opening, a rag should be held at the door to confirm the flow of air into the heater. Under positive pressure, hot gas will exit through an opened sight door and will blow the rag outward. *Be careful! If the flue gas is hot enough, the rag may burst into flames!* Without this safeguarding procedure, these hot gases could cause injury to the unprotected observer. Proper protective gear should be worn when opening sight doors for inspecting heaters.

12.4.1 Flame Pattern and Stability

The flame pattern from a burner is developed jointly by the heater designer and the burner manufacturer. Based on the specified and/or jointly agreed upon flame pattern, the burner manufacturer selects the gas or oil tip drilling pattern, the type of diffuser (if applied), and the tile shape to achieve that desired flame. The fuel tip drilling patterns can be varied based on testing and experience to obtain short, bushy flames or long, narrow flames depending on the requirements of the process heater. There are many different types and shapes of fuel tips. Each fuel tip is drilled with a given pattern to meter the fuel and to inject the fuel into the combustion air. The tile shape, diffuser (if applied), and the fuel tip drilling pattern are all variables that the burner manufacturer can manipulate to obtain a predictable flame pattern.

The burner tile shape and condition are critical to obtaining the desired flame shape. Round tile shapes provide round flame patterns; rectangular tile shapes provide flat flame patterns. Missing tiles, or poorly maintained tiles with holes and cracks, can cause a poor flame pattern. Substitution of tiles with ones of a different design can restrict airflow, result in poor air/fuel mixing, or cause flame instability because a tile ledge or other feature is missing or incorrectly located. Variation of the fuel composition, slight variations in combustion airflow from burner to burner, flame interactions, furnace gas circulation current strength and pattern, and air leakage all act to vary the flame pattern between burners.

The flame patterns within the heater should be visually inspected often. Any change in shape or dimensions of an individual flame should be noted. The flames from all active burners of the same size should be uniform because they all have the same heat release, the same airflow across the burner, and the same fuel flow at each burner. Any flames that are visually different or unusual, either in dimension or stability, should be investigated and any problem corrected. Chapter 13 discusses several visual indications of flame problems, along with appropriate corrective actions.

Flame patterns are designed to stay within an envelope that is a safe distance from the process tubes and the refractory. API RP 560 cites some standards and recommendations as to burner-to-tube and burner-to-refractory. Some users apply greater clearances between burners and between burners and tubes than API RP 560 recommendations. The burners installed on existing heaters many years ago may not comply with current API RP 560 recommendations. If the installation of burners into an old heater is to comply with API RP 560, the number of burners and burner spacing may need to be evaluated. More burners at a lower heat release will provide smaller flames. Fewer burners at a larger capacity may be able to be moved further from the tubes and/or refractory.

Good flame patterns alone do not protect against localized tube overheating. Flue gas circulation flows within the firebox and uneven firing of burners will affect the distribution of heat to tubes.

A check into the firebox should show stable flames and even flame patterns all active burners, as seen in Figure 12.34. The size, color, and shape of the flames should be the same because all active burners have the same fuel flow (heat release) and draft (combustion airflow). Any flames that are unstable, any uneven flame

FIGURE 12.34
Uniform flames and refractory color.

FIGURE 12.35
Uneven flame patterns.

patterns, or any flames impinging on tubes indicate a problem that needs to be corrected. Figure 12.35 is an example of this. See Chapter 13 for a full discussion on these problems.

12.4.2 Process Tubes

The process tubes should be periodically checked visually for evidence of localized hot spots, tube displacement, and process leaks. Tube hot spots may appear as red or silver spots and indicate temperatures approaching or exceeding the mechanical limit of the tube material. The immediate cause of the hot spot is usually a fouling deposit on the internal tube wall. This deposit may be the result of flame impingement, overfiring, uneven distribution of active burners, or concentrated heat input from flue gas circulation currents. Hot spots need to be continually monitored; if allowed to continue, they will ultimately cause tube failure. The process tube maximum temperature limits should be known to the operator and monitored to protect against tube overheating and failure.

Tubes that are out of position may bow due to overheating or because of loss of a tube support or guide. Overheating may be caused by overfiring, by concentration of active burners in the firebox, or by flame impingement. The bowing may be accompanied by internal tube fouling. Look for metal parts from a broken support or guide on the heater floor.

A process tube leak may initially show as a small flame or wisp of smoke at the tube surface. Flames or smoke may be traveling through the tube sheet into the firebox from a header box. The process tube leak acts as fuel and consumes oxygen. This may result in long flames from the burners, due to the reduction in excess oxygen, and afterburning in the convection section. The oxygen analyzer will often show a steep change to lower oxygen levels, and the measured CO levels will increase.

Dark smoke from the stack, when burning gas, could indicate a process tube failure or incomplete combustion of the fuel. Black rolling smoke, when burning oil, indicates that the burners are short of air. The operator must check the excess oxygen in the firebox and the flame appearance to determine whether the problem is a tube failure or poor combustion at one or more of the burners.

12.4.3 Refractory and Tube Support Color

The refractory is held in place by refractory anchors of various designs. The anchor design, spacing, and attachment to the furnace casing are critical to obtaining a satisfactory refractory installation and service life. Most refractory failures can ultimately be traced to anchor failure. The basic causes for refractory anchor failures are positive pressure in the convection section causing anchors to overheat, corrosion due to positive pressure and CO_2 attack on the anchors, or inadequate anchor spacing (too far apart) to adequately hold the refractory.

Refractory failure may first be evident by debris on the heater floor. With severe refractory failure, hot areas and discolored paint on the outside of the casing may be observed. In furnace casing overheating, accompanied by discolored paint and excessive heat on the exterior furnace casing, the casing can be cooled with a low-pressure steam spray or water flood until repairs can be completed.

When viewing the firebox refractory, color can be an indication of even heat distribution within the firebox. Refractory color can also be a very rough measure of the refractory temperature. Dark streaks on the refractory provide evidence of air leaking into the firebox and cooling the refractory.

The color of the refractory and tube supports can be an indication of high temperatures in the firebox. Tube supports should appear uniform in color if the heat distribution is uniform within the heater. If some tube supports are visibly hotter than other tube supports, there is uneven heat distribution within the heater. If the refractory is not uniform in color, there may also be uneven heat distribution within the heater. Uneven firing on active burners, poor burner air distribution, flame impingement, or flue gas circulation patterns are all suspects and should be checked.

12.4.4 Burner Tile and Diffuser Condition

Broken burner tiles, burner tiles that have deteriorated from chemical attack, improper installation of the tile, and burner diffuser cone conditions should

FIGURE 12.36
Severe tile damage.

be noted. Tile or diffuser cone in poor condition can lead to poor flames and possible impingement. The diffuser cone can normally be replaced while the heater is in service, but tile replacement may require a heater shutdown. Figure 12.36 is an example of tile damage.

12.4.5 Air Leaks

The casing is normally designed for airtight construction. There are typically seams, sight doors, openings for tubes and manifolds, and doors for maintenance access in addition to the burner openings. Because the heater operates under a slight negative pressure, there is the possibility of air leaking into the heater through all of these non-burner-related openings. It is desirable, for best operating efficiency and proper burner operation, to have all air enter the heater through the burners.

Therefore, all doors should be kept tightly closed during operation, and other openings should be minimized or sealed as tightly as possible. Figure 12.37 shows an open sight door that would allow air to enter the heater.

Openings capable of allowing air into the heater can be located by smoke testing using smoke generators (or smoke bombs), usually placed inside the idle firebox. In many cases, cracks and seams identified in this manner, or alternatively through IR thermography, as seen in Figure 12.38, can be successfully sealed using commonly available high-temperature silicone caulk.

In more severe cases where the leak is due to damaged or warped furnace steel casing, the damaged plates should be repaired or replaced.

FIGURE 12.37
Open sight door.

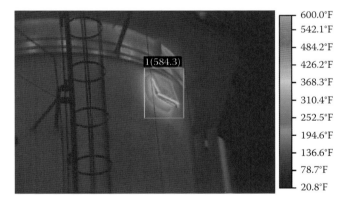

FIGURE 12.38
Thermal image of an open explosion door.

When observing inside the heater, look for air leaks around sight doors and other areas where air leakage into the heater is possible. In hot fireboxes, air leaks may show themselves as a dark outline around sight doors and access doors or dark lines on the refractory surface (see Figure 12.39).

As air leakage into the heater increases, the operating cost increases. Also, as air leakage increases, the likelihood of poor flame patterns increases due to the inability to control air through the burners. Idled burners should have air registers completely closed. Air leaks should be sealed when identified.

FIGURE 12.39
Air leaks at sight door and access door.

12.5 External Inspection

12.5.1 Stack Damper

The stack damper position should be noted to determine if there is sufficient control of the draft within the firebox. If the stack damper is a single-blade damper and is over 75% open, there is probably very little capacity left to increase the draft (negative pressure) within the firebox. However, if this is a multiblade damper, there is probably sufficient control remaining to control the targeted draft and get more capacity out of the heater.

Most furnaces are equipped with an external indicator in the stack damper shaft. Check the arrow indicating the stack damper position. Does it match the position shown in the control room? Does the damper move when the actuator is adjusted, or is it stuck? Most dampers control adequately over a range of positions between 20% and 80% open. Compare the present position to this range to get an idea of the amount of draft control available.

12.5.2 Burner Block Valves

All burners should be installed with manual block valves on the individual fuel line (and atomization medium line, if provided) to each burner. Good operating practice is to keep these valves fully open when the burner is in operation. The purpose of these burners is to "block" fuel flow, fully closed, when the burner must be taken out of service. These valves should not be used to throttle the amount of fuel to the burner during normal operation. The need to throttle an individual burner is an indicator of other problems. Short-term operation with the block valve partially closed can be used as an aid to operation until that time the cause can be determined and corrected. The fuel to the burners should be regulated with the main fuel control valve. If the operator uses individual burner block valves to control flame patterns at selected burners, an uneven distribution of heat within the firebox may result. One possible consequence is overheating of tubes at some locations, causing coking of the tube contents and potential tube failure. It is not uncommon for an operator to react to a high tube temperature alarm by throttling the block valve on the burner closest to the alarming tube-skin thermocouple. While this stops the alarm and reduces the temperature at the alarming thermocouple location, the total fuel input does not change. With the lowering of fuel flow on the burner with the throttled valve, the fuel pressure and input on the other active burners will increase. The result of this is, although the tube temperature which was high will decrease, all other tube temperatures will rise in response to the increased firing of the other burners. The new locations of high tube-wall temperature likely are not at one of the few thermocouples, and tube overheating may go undetected until failure or severe fouling occurs.

If block valves on burners are partially closed, the fuel pressure on the burners varies, depending on valve positions. If the heater controls reduce the fuel flow, one or more burners may experience a fuel pressure below the stable limit and may extinguish. The low-pressure alarm and trip instruments on the burner fuel manifold will not sense the same low pressure as at the burner(s) and will not protect the heater as expected with the burner management control system.

Check the positions of the individual burner block valves. Observe those not fully open or closed as in Figure 12.40. Most heaters should be operated with these valves either fully open if the burner is in operation or fully closed if the burner is out of service. Some special heater designs allow throttling of individual valves to control the tube temperatures. However, in most heaters, the throttling of these valves can result in poor flames as well as compromised safety systems and tube temperature monitoring.

12.5.3 Pressure Gauges

The pressure gauges should be located on the fuel line to provide a pressure reading to indicate the amount of fuel flowing to each burner. The fuel pressure should be recorded and monitored in the control room, and, if possible, it should be trended. The local pressure gauges should be in good working condition to give accurate pressure readings.

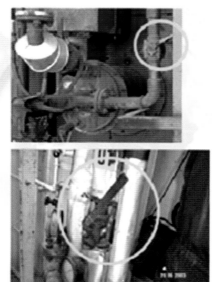

FIGURE 12.40
Examples of pinched block valves.

12.5.4 Heater Shell or Casing Condition

The heater shell or casing should be inspected periodically to determine if there are any hot spots developing on the shell or casing (see Figure 12.41). This is often easiest to detect when it is nighttime and dark. Hot spots on the heater shell are an indication of refractory failure within the firebox. If left uncorrected, the heater efficiency will decrease because of the higher shell heat loss, and casing failure can lead to the heater being shut down.

Hot spots on the casing are the result of internal refractory failure, possibly due to positive pressure in the firebox. Check the draft in the heater, and, if necessary, make adjustments to regain a negative draft at the arch. The hot spot can be cooled with steam or water spray.

FIGURE 12.41
Heater casing oxidation.

12.5.5 Burner Damper Position

The burner damper position or air register position should be at the same opening for all burners in operation. They should be equally open on the active burners and closed on all burners out of service.

12.5.6 Burner Condition

The general condition of the burner should be observed. A burner in good condition can be operated at very low excess oxygen levels if the heater is in good condition and sealed properly. Burners in poor condition will require higher excess oxygen levels to operate safely and should be removed and repaired. Check if the air inlet is unobstructed. Sight and lighting doors need to be closed to prevent too much air going into the burner. The pilot air door should be open, and the pilot mixer and fuel orifice should be unobstructed.

References

1. API Standard 560, *Fired Heaters for General Refinery Service*, American Petroleum Institute, Washington, DC.
2. B. Corripio et al., Recognizing in situ oxygen analyzers as ignition sources in ethylene furnaces, Presented at the *2007 Ethylene Producers' Conference*, Houston, TX, 2007.
3. R.D. Reed, *Furnace Operations*, 3rd edn., Gulf Publishing, Houston, TX, 1981.
4. API Standard 530, *Calculation of Heater-Tube Thickness in Petroleum Refineries*, American Petroleum Institute, Washington, DC.
5. API Recommended Practice 556, *Instrumentation, Control, and Protective Systems for Gas Fired Heaters*, Petroleum Institute, Washington, DC.
6. EN 746-2, Industrial thermoprocessing equipment—Part 2: Safety requirements for combustion and fuel handling systems.
7. NFPA 8502, Standard for the prevention of furnace explosions/implosions in multiple burner boilers.

13
Burner Troubleshooting

William Johnson, Erwin Platvoet, Mike Pappe, Michael G. Claxton, and Richard T. Waibel

CONTENTS

13.1 Introduction 333
13.2 Failure to Light Burners 334
    13.2.1 Pilot Fails to Ignite 334
        13.2.1.1 Indications of the Problem 334
        13.2.1.2 Effect on Operation 334
        13.2.1.3 Corrective Action 334
        13.2.1.4 Flame Rod Testing 334
        13.2.1.5 Recommendations and Conclusions 336
        13.2.1.6 Field Test Procedure 338
    13.2.2 Main Burner Fails to Light Off 338
        13.2.2.1 Indications of the Problem 338
        13.2.2.2 Effect on Operation 338
        13.2.2.3 Corrective Action 338
13.3 Flame/Flue Gas Patterns 339
    13.3.1 Long Flames 339
        13.3.1.1 Indications of the Problem 339
        13.3.1.2 Effect on Operation 339
        13.3.1.3 Corrective Action 339
    13.3.2 Leaning Flames 339
        13.3.2.1 Indications of the Problem 340
        13.3.2.2 Effect on Operation 340
        13.3.2.3 Causes and Corrective Action 341
    13.3.3 Irregular/Nonuniform Flames 341
        13.3.3.1 Indications of the Problem 342
        13.3.3.2 Effect on Operation 342
        13.3.3.3 Cause and Corrective Action 342
    13.3.4 Pulsating Flames/Burners Out of Air 343
        13.3.4.1 Indications of the Problem 344
        13.3.4.2 Cause and Effect on Operation 344
        13.3.4.3 Corrective Action 344
    13.3.5 Flame Lift-Off 345
        13.3.5.1 Indications of the Problem 345
        13.3.5.2 Effect on Operation 345
        13.3.5.3 Corrective Action 346
    13.3.6 Flashback 346
        13.3.6.1 Indications of the Problem 346
        13.3.6.2 Effect on Operation 346
        13.3.6.3 Corrective Action 347
    13.3.7 Flame Impingement on Tubes 347
        13.3.7.1 Indications of the Problem 348
        13.3.7.2 Cause and Effect on Operation 348
        13.3.7.3 Corrective Action 349
.... 351

13.3.8　Burner Spacing/Flame Interaction 352
　　13.3.8.1　Indications of the Problem 352
　　13.3.8.2　Cause and Effect on Operations 352
　　13.3.8.3　Corrective Action 353
13.3.9　Low-Temperature Operation (<1300°F [700°C])/Flame Stability/CO Formation 353
　　13.3.9.1　Indications of the Problem 354
　　13.3.9.2　Causes 354
　　13.3.9.3　Corrective Action 355
13.3.10　High Stack Temperature 355
　　13.3.10.1　Indication of the Problem 355
　　13.3.10.2　Effect on Operations 355
　　13.3.10.3　Causes and Corrective Actions 357
13.3.11　Overheating of the Convection Section 357
　　13.3.11.1　Indications of the Problem 357
　　13.3.11.2　Effect on Operation 357
　　13.3.11.3　Corrective Action 357
13.4　Fuel Gas Problems 357
　13.4.1　Burner Fuel Pressure/Impact on Operation 357
　　　13.4.1.1　Fuel Gas Tip Problems 357
　　　13.4.1.2　Fuel Composition 358
　　　13.4.1.3　Wrong Gas Tips 358
　　　13.4.1.4　Fired Duty Has Changed 359
　　　13.4.1.5　Fuel Flow Measurement Is Incorrect 359
　　　13.4.1.6　Fuel Pressure Is Incorrect 359
　　　13.4.1.7　Operation outside Design Fuel Pressure Range 360
　　　13.4.1.8　Running the Heater Out of Air/Flame Impingement 361
13.5　Oil Firing Problems 361
　13.5.1　Effect on Operations 362
　13.5.2　Oil Combustion 362
　　　13.5.2.1　How Does Oil Burn? 363
　　　13.5.2.2　Viscosity and Temperature 363
　　　13.5.2.3　Steam Atomization 363
　　　13.5.2.4　Contaminants in the Oil 364
　13.5.3　Oil System 364
　　　13.5.3.1　Heating and Storage 364
　　　13.5.3.2　Recirculation System 365
　　　13.5.3.3　Heat Tracing and Insulation 365
　　　13.5.3.4　Pressure Gauges 365
　13.5.4　Steam System 365
　　　13.5.4.1　Insulation 366
　　　13.5.4.2　Checking Steam Traps 366
　　　13.5.4.3　Superheated Steam 366
　　　13.5.4.4　Pressure Indication and Control 366
　13.5.5　Smoke Emission from the Stack 366
　　　13.5.5.1　Indication of the Problem 366
　　　13.5.5.2　Effect on Operation and Equipment 367
　　　13.5.5.3　Corrective Action 367
13.6　Emissions 367
　13.6.1　Nitrogen Oxides 368
　　　13.6.1.1　Burner Type 369
　　　13.6.1.2　Firebox Temperature 370
　　　13.6.1.3　Fuel Composition 370
　　　13.6.1.4　Excess Air 371
　　　13.6.1.5　Combustion Air Temperature 371
　　　13.6.1.6　Relative Humidity

    13.6.2 Sulfur Oxides..371
    13.6.3 Carbon Monoxide 371
    13.6.4 Combustibles/Volatile Organic Compounds/Unburned Hydrocarbons 371
    13.6.5 Particulate Matter 371
13.7 Combustion Air Issues...372
    13.7.1 How Ambient Weather Conditions Can Affect Burner Performance..................................372
        13.7.1.1 Effect on Operations 372
        13.7.1.2 Corrective/Preventive Actions...373
    13.7.2 Burner Sizing..373
    13.7.3 Forced-Draft System 374
        13.7.3.1 Selecting the Correct Size Burner..374
        13.7.3.2 Designing for Uniform Air Distribution 375
13.8 Summary...375
References...375

13.1 Introduction

Diagnosing and solving problems with burners on heaters in the hydrocarbon and petrochemical industries often seems to be as much an art as a science. It is a basic scientific assumption that the principles of physics, chemistry, fluidics, hydraulics, and combustion (see Volume 1) do not change. Yet the myriad variables in a typical refinery or petrochemical operation sometimes make it appear that the equipment has a personality. The complexity of the sciences, multiplied by the many-staged processes in a typical plant, causes problems to occur that were not, and could not have been, anticipated by the design engineers. Moreover, although scientific principles remain the same, equipment changes with use. Parameters that may have been designed correctly may change with time.

With all of the complexity of conditions that may occur, it is still the operator's job to "keep it running." When production suffers because of the inability of equipment to operate at required capacity, costs go up, and profit or product margins decrease. Frequently, the operator of the heater must be trained and use knowledge of the equipment and process unit to make adjustments that bring operations back to the required capacity desired by plant management.

It is essential that troubleshooting be done in a systematic, well-organized fashion. Effective and safe troubleshooting involves four basic steps[1]:

1. Recognizing and discussing the problem with the unit operators
2. Collecting field data and observing the operation of the unit
3. Developing a theory as to the cause of the problem
4. Identifying solutions to resolve the problem and taking the appropriate corrective action

The initial indication that a problem exists may come from controls and instruments on the furnace or from direct observation of conditions within or outside the furnace. Changes in process temperature, process pressure drop, excess oxygen, draft, fuel pressure, stack temperature, and emission levels are typically first observed with instruments. Changes in the noise being emitted from the heater or flame patterns, flame impingement, oil spillage, flame instability, and flashback are usually observed directly during inspection of the heater.

When a problem is noted, it is necessary to evaluate its likely effect on the process or product being produced. Some solutions may require the heater to be shut down for the problem to be resolved. Then, plant management must determine the economic value of meeting contracts for products versus operating the equipment until it fails and has to be shut down for repairs at a higher cost. On the other hand, if the problem might result in large fines from controlling governmental agencies, significant damage to equipment, or danger to personnel, then a solution must be employed immediately—whatever the cost.

If a solution is going to be employed, a careful study of symptoms being exhibited by the furnace should be conducted, and a probable cause for the symptoms should be identified. It is important that the possible causes be carefully identified lest an attempted solution escalate the seriousness of the problem rather than solve it.

Once a cause has been determined, standard procedures should be followed to solve the problem. All personnel involved should be aware of the problem, the planned corrective actions, the ways that safety is addressed, the expected results, and the proper action to take should the problem worsen or not be solved. During normal operations, care should be taken to make incremental changes and adjustments to parameters controlling the combustion process. However, under some conditions, a change must be made quickly and with confidence to prevent additional operating problems.

The following chapter sections describe typical problems, their indicators, their likely effects on operations, and their causes and standard solutions. Additional references are available on troubleshooting.[1–6]

13.2 Failure to Light Burners

Lighting the pilots and main burners during start-up can be easy or can be very stressful. The heater start-up is normally at the end of a turnaround. People are tired and under pressure to get the unit on line. If the burners fail to light, then the entire unit is at a stop until this critical issue is resolved.

Heater refractory work during a shutdown can result in plugged burner tips or blocked tile throats. A careful inspection of the burners should be made after all work has been completed inside the heater. Some of the key items to check before attempting a light-off are the following:

- Heater floor and burner tile are level and installed as shown on the burner drawing.
- The gas tips are properly located with the correct *orientation* for proper burner operation. This is critical for safety and flame stability.
- The pilot tips are located correctly for reliable light-off of the main burner. Also check the ignition rod and flame rod locations for proper operation.
- Look for any damage or debris in the burner tile that could affect operation.
- Confirm available fuel pressure for the burner and pilots.
- Visual check of any noticeable air leakage areas inside the heater.
- Inspect the burner air registers/dampers for proper operation.

A very important item is to be sure that operations have the burner *Installation and Operating Manual* before trying to light the pilots. The manual has information on installation, operation, maintenance, and troubleshooting for the pilots and burners.

13.2.1 Pilot Fails to Ignite

13.2.1.1 Indications of the Problem

Pilots are usually small premix gas burners designed to ignite the main burner. According to API 535, the minimum heat release for pilots is 65,000 Btu/h (19 kW). Because pilot burners have small ports, they are susceptible to plugging. Many users use a clean, dedicated, and reliable pilot gas fuel to eliminate this problem. The pilot flame should remain stable over the full design range of the main burner (API 535). Some conditions, such as a heater trip, may cause the pilots to go out. In a problem situation, although the operator follows the appropriate purge and ignition procedures, the pilot may fail to light or may go out once lit.

13.2.1.2 Effect on Operation

If the pilot does not light upon heater start-up, the corresponding main burner is not placed in operation, or the main burner must be lit using a handheld torch. The latter is a less satisfactory method because a large gas supply valve must be opened rather than the small pilot gas supply valve. If a mistake is made, a relatively large amount of unburned gas enters the firebox, compared to the small amount from a pilot. This large amount of unburned gas is an unsafe condition and has the potential to form an explosive mixture within the firebox.

13.2.1.3 Corrective Action

Ensure that the fuel gas is flowing to the pilot. If the fuel piping has been pressure tested or purged with an inert gas, this gas must be completely displaced with the fuel before a successful light-off can occur. Also check for closed valves or blinds, plugging of the small pilot gas lines, or a plugged strainer or filter. Since the ports in the pilot are small, they are also susceptible to plugging. Check for a plugged orifice, and clean it if it appears plugged.

Pilots are lit using a spark igniter such as the one seen in Figure 13.1 or a handheld torch as seen in Figure 13.2. If the lighting torch flame is unstable or is not properly positioned, the torch flame might not contact

FIGURE 13.1
High-energy portable igniter.

Burner Troubleshooting

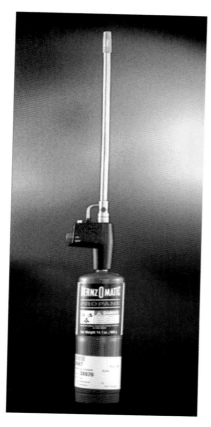

FIGURE 13.2
Portable torch.

the flammable mixture leaving the pilot tip. Adjust the torch flame to ensure stability, and take care in the positioning of the flame relative to the pilot tip. If spark ignition is used, clean the spark plug whenever possible, and replace insulators on the ignition rod.

Sometimes the pilot gas–air mixture leaving the tip is not in the flammable region, usually too lean to support combustion. Close the primary air door to enrich the mixture. The pilot gas may be of the wrong composition or pressure to induce adequate primary air flow. Check both the gas composition and the pilot fuel pressure against the manufacturer's specifications and correct either if necessary.

A natural-draft premix pilot will not stay lit if the firebox pressure at the pilot location is positive. Adjust dampers to obtain a negative pressure at the burners. There may be some cases where the use of compressed air should be considered. Consult with the burner manufacturer for the correct pilot in this service.

The pilot may be wet, from condensed steam after firebox purging, for example, and refuse to light. Allow the pilot to dry, or blow it out with compressed air. Light the pilot and observe the pilot flame for stability. If the pilot flame appears to be too high above the pilot tip, close the primary air door on the pilot mixer. If the flame appears to be burning with a yellow flame, open the primary air door on the mixer. If a flame rectification rod is being used to monitor the pilot, the operator needs to ensure this device is functioning properly.

The following information can be used to troubleshoot pilots with flame rods: Flame rods, as seen in Figures 13.3 through 13.5, are used on process burner pilots to confirm that a flame is present. Flame rods work due to the ionization/rectification process to complete a circuit. When the flame rod is energized, the current produces a positive charge that attracts the negative ions in the flame. The positive ions normally will be attracted to the grounding area of the pilot tip. The theory is to collect more positive ions, through the ground, than negative ions, from the flame rod, so the flow of electrons is "rectified" or flowing in one direction. This produces the direct current signal that is used to indicate the presence of flame.

FIGURE 13.3
Pilot with electric ignition and flame rod.

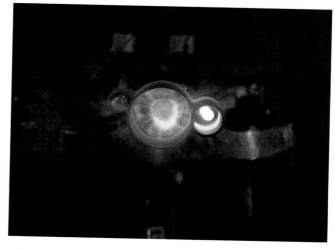

FIGURE 13.4
Pilot with shrouded flame rod.

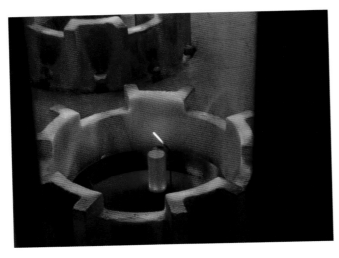

FIGURE 13.5
Pilot with external flame rod.

FIGURE 13.6
ST-1SE-FR pilot which includes a flame rod.

13.2.1.4 Flame Rod Testing

> **WARNING!**
>
> High voltages capable of causing death may be used with this equipment. Use extreme caution when servicing control cabinets and electrically actuated components. Testing should be performed by qualified personnel only. Use the required personal protective equipment (PPE) including safety glasses and leather gloves.

Testing was performed to establish a procedure to troubleshoot pilots with flame rods. The test pilot was an ST-1SE-FR (shown in Figure 13.6) which is a standard pilot for process burners. The pilot was tested on natural gas at various fuel pressures and air door settings.

A "test circuit" assembly was used that consisted of an 820,000 ohm resistor wired in series with a 1N4004 diode. The red, or positive clip, is the diode end, and the black, or negative clip, is the resistor end of the test circuit. This can be seen in Figure 13.7. The test meter was a "Fluke 87 V true RMS multimeter."

Pilot flame rods work on the ionization/rectification process to complete a circuit. There is always voltage to the flame rod when the power is on. A relay panel (see Figure 13.8) used a small transformer to increase the voltage from 120 VAC to approximately 300 VAC at the flame rod (see Figure 13.9). This voltage is with the pilot off and no presence of flame. This voltage was measured by touching the red probe (+), on the Fluke meter, to the flame rod extension rod and the black probe (−) to the pilot stand for grounding. It is important to note that this voltage is the result of alternating current (AC).

FIGURE 13.7
Test circuit.

FIGURE 13.8
Relay panel.

Burner Troubleshooting

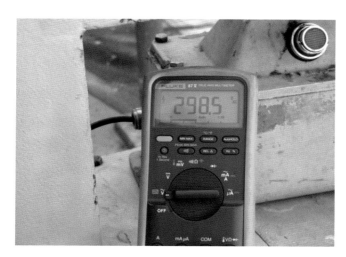

FIGURE 13.9
Flame rod voltage (VAC)/no flame/power on.

> **NOTE**
>
> The system used a FIREYE "Rectification Amplifier—Type 72DRT1." Different types of amplifiers may result in voltages and currents that are higher or lower from those noted in this procedure.

The test circuit was then connected with the red clip on the high-tension wire to the flame rod and the black clip to the pilot mounting plate. Figures 13.10 and 13.11 show these connections. The test circuit simulates the flame rectification circuit to energize the "Flame On" light on the relay panel, as seen in Figure 13.12.

The red clip was connected to the flame rod, near the pilot tip and the black clip to the pilot mounting plate, and the light came on.

FIGURE 13.10
Positive lead/test circuit.

FIGURE 13.11
Negative lead/test circuit.

FIGURE 13.12
Relay panel/"Flame ON" light energized.

The next step was to light the pilot at 5 psig (0.35 barg) fuel pressure with the air door 50% open. The light came on, and the measured voltage was 193 V direct current (VDC) (see Figures 13.13 and 13.14).

The current at this point was 44 µA. To measure the current, the Fluke meter has to be connected in series with the red lead to the flame rod wire and the black lead to the extension rod (see Figure 13.15).

The pilot pressure was increased to 10 psig (0.7 barg), and the voltage increased from 170 to 240 VDC. Increasing the pressure further to 14.5 psig (1 barg) increased the voltage to 250 VDC.

At this point, the fuel pressure was adjusted down to 5 psig (0.3 barg) and then the mixer air door was closed to approximately 25% (0.07–0.09 in., 1.8–2.3 mm). The voltage dropped from a constant 200 VDC to fluctuating between 90 and 150 VDC.

Opening the mixer air door to 100% increased the voltage to 240 VDC.

FIGURE 13.13
Connecting test probes for flame voltage.

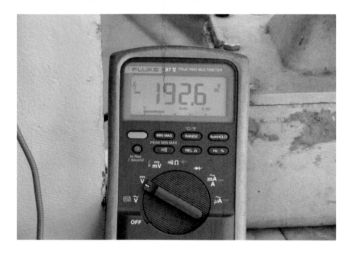

FIGURE 13.14
Flame rod voltage (VDC).

FIGURE 13.15
Flame rod current (μA).

13.2.1.5 Recommendations and Conclusions

The pilot fuel pressure did not have a significant impact on the voltage generated in the circuit. Running on natural gas, the recommended pressure would be 7–15 psig (0.5–1.0 barg). The light on the relay panel actually stayed on down to 1.5 psig (0.1 barg).

The position of the mixer air door does reduce the flame voltage significantly. The air door should be set between 50% and 100% open for the best signal.

During normal operation, the flame rod should generate a voltage of 220–250 VDC. The minimum voltage, for a reliable signal, should be 90 VDC. The current should be 20–45 μA. Current levels less than 5 μA may cause "loss of flame" trips.

13.2.1.6 Field Test Procedure

To prove the system wiring is functional, close the pilot fuel valve. Then connect the red clip, on the test circuit, to the high-tension wire at the pilot and black clip to the mounting plate. *Use extreme caution since the system will have power on.* The "Flame On" light on the relay or the panel should come on.

To prove the flame rod is working will require that the pilot be removed from the burner. Close the pilot fuel valve, and remove the pilot. Be sure the pilot is grounded. Connect the red clip, on the test circuit, to the flame rod and the black clip to the mounting plate. The "Flame On" light should come on. If the light does not come on, then the problem is with the flame rod, the insulators, or the extension rods. *Use extreme caution; do not touch the pilot or the flame rod when the power is on.*

Inspect the flame rod for any type of corrosion or buildup that could interfere with the ionization process. Moisture buildup on the insulators and connecting rods is a frequent problem. *Turn the power off before cleaning or replacing the pilot or pilot parts.*

13.2.2 Main Burner Fails to Light Off

13.2.2.1 Indications of the Problem

The operator follows the usual purge and ignition procedures for lighting a burner, but there is no indication of main burner ignition. After the heater is in operation, one or more burners flame out.

13.2.2.2 Effect on Operation

If a burner fails to light, the process unit outlet temperature may not be achieved, or the heater start-up may be delayed. Additionally, if the burner fails to ignite or flames out, the furnace may fill with a dangerous mixture of gas and air that can result in an explosion in the firebox.

13.2.2.3 Corrective Action

The most common cause of ignition failure is improper positioning of the pilot burner in relation to the main burner. If the pilot is not located so that its flame is directed into the fuel/air mixture leaving the main burner, the ignition temperature is not achieved, and the main flame is not initiated. The operator should check all components of the pilot and burner to ensure that they are positioned according to the specifications in the manufacturer's design drawings.

Check that fuel gas is being supplied to the burner. Before start-up, the fuel supply line may be purged or pressure tested with an inert gas. If this gas is not completely displaced with fuel, the operator will be attempting ignition of an inert material. Procedures should be in place for removing the inert gas and checking for the presence of a flammable material.

Check for closed or blocked valves, fuel gas line blinds, plugged strainers or filters, and plugged burner ports. Plugging is a frequent problem during start-up if the fuel piping has debris or scale left in the lines. Old piping may have internal scale that dislodges and enters the burners, thereby plugging the ports. Consideration should be made to change old piping to low-grade stainless steel. The burner ports and fuel lines may need to be cleaned.

Adjust the burner damper to 25%–50% open if the main flame will not ignite. For forced-draft units, reduce the fan to light-off conditions. Too much air can make it difficult to light the burner.

The draft could also be too high for light-off. Close the stack damper or induced-draft fan to reduce the draft at the burner.

Be sure that the gas tips are located per the burner drawing. The gas tip orientation is also critical for proper burner operation.

13.3 Flame/Flue Gas Patterns

13.3.1 Long Flames

13.3.1.1 Indications of the Problem

Visual observation inside the firebox reveals long flames, possibly reaching into the convection section of the heater (see Figure 13.16). Smoke may be observed exiting the stack. The flame, rather than being confined in the flame pattern space within the firebox, may be impinging on the process tubes in the convection section. The combustion zone may appear hazy, rather than bright and clear.

The heater operator may notice that the process outlet temperature cannot be achieved. The stack temperature may be above the design specifications.

FIGURE 13.16
Long flames.

13.3.1.2 Effect on Operation

Long flames are, theoretically, less efficient than short flames in transferring heat to the radiant tubes because of their lower average temperature. A lower radiant efficiency could lead to higher fuel consumption for the same radiant absorbed duty.

Smoky flames are the result of incomplete combustion and can consume far more fuel than necessary to achieve the desired heat transfer and outlet temperature. Soot (unburned carbon) will be deposited on tube surfaces in the convection section, reducing heat transfer. Coke formation may be observed in the burner tile. The required process outlet temperatures may not be achieved. The stack CO emission level may show elevated values.

Impingement of flames on the tubes may lead to hot spots and increased coking rates inside the tubes, resulting in short run lengths between decokes. Prolonged operation with flame impingement will reduce the life of the coil due to local carburization and embrittlement.

13.3.1.3 Corrective Action

Long flames can be caused due to

- Incorrect burner arrangement or firebox design
- Burner gas tip problems
- Combustion air problems

- Operation of the burner outside the design envelope
- Fuels that are heavier than the burner was designed to accommodate

13.3.1.3.1 Incorrect Burner Arrangement or Firebox Design

The firebox design or the burner arrangement can cause long flames in several different ways:

- Insufficient burner spacing
- Burner circles that are too small
- Adverse flue gas patterns caused by asymmetric firebox shapes or burner arrangements

In each of these cases, the flames of the burners can be forced to merge. This risk is higher for ultralow NOx burners that rely on large amounts of flue gas recirculation and entrainment. Once the heater has been constructed, it is often difficult to correct the root cause for the flame issues. The burner vendor can, however, try to mitigate the problem by shortening the flames or biasing the fuel injection. The use of computational fluid dynamics (CFD); see Volume 1, Chapter 13 to diagnose and resolve the issues is indispensible.[1]

13.3.1.3.2 Burner Gas Tip Problems

Whenever air and fuel are mixed in the incorrect proportions, elongated flames may occur. This is the case when fuel gas ports are too large, too small, or completely plugged.

Check the size of the ports in the burner tips. The ports may be enlarged due to repeated cleaning efforts. The wrong tips may have been installed during maintenance. In the latter case, the tips may even vary from burner to burner. Enlarged ports cause uneven fuel input for the available air. Replace any damaged or improper tips.

Tip fuel ports get plugged when liquid carryover in the gas fuel condenses and forms coke inside the gas tips. Plugged gas tips should be cleaned or replaced immediately. Figure 13.17 shows an example of a plugged gas tip.

13.3.1.3.3 Combustion Air Problems

Whenever air and fuel are mixed in incorrect proportions, elongated flames may occur. This is the case when one or several burners are starved for combustion air.

The corrective actions are aimed at ensuring that adequate air is provided to each burner and that this air is mixed quickly and completely with the fuel so as to achieve rapid combustion and a smaller flame volume. Air that is not mixed with fuel at the burner passes into the less turbulent firebox where mixing of the fuel and air is less likely and ignition temperatures may be too low for combustion.

FIGURE 13.17
Fouled gas tip.

The operator should check the draft and excess oxygen level at the top of the radiant section of the heater to determine if there is sufficient oxygen to burn the fuel. If the excess oxygen is lower than the target, the draft and excess oxygen should be adjusted to the correct levels.

13.3.1.3.4 Operation of the Burner outside the Original Design Envelope

If the burners still show long flames and most of the burners have the same flame length, the firing rate needs to be checked. Firing above the rated heat release, particularly with closely sized burners or with some burners out of service, will cause a shortage of oxygen at the burners and long flames that smoke. Place more burners in service, or reduce the fuel input. Contact the burner manufacturer if larger burners are required to increase the heat release.

13.3.1.3.5 Heavier Fuels

Another reason for long flames could be a change in the fuel gas composition. The substitution of methane or heavier hydrocarbons may increase the oxygen requirement. The air registers and possibly the stack damper must be opened to provide more air. The heavier fuel may also result in a reduced fuel gas pressure at the burner. This will lead to higher NOx emissions due to less flue gas entrainment and longer flames due to a reduced mixing efficiency between fuel and air. If the heavier fuel has become the new standard, contact the burner manufacturer to resize the fuel tips for this new composition.

13.3.2 Leaning Flames

13.3.2.1 Indications of the Problem

In floor- or ceiling-fired furnaces, the flames may lean to one side rather than burning in a vertical line. In

FIGURE 13.18
Leaning flames.

side- or end-fired furnaces, the flames may lean to the side rather than firing horizontally and curling upward. Figure 13.18 shows an example of leaning flames.

Observation of the flame pattern inside the firebox reveals that the centerline of the flame does not follow the designed path as specified by the burner manufacturer. The flame is commonly expected to propagate in the general direction of the centerline of the air orifice or refractory tile.

13.3.2.2 Effect on Operation

Flames that do not have the designed pattern and direction can create problems such as impingement on the process tubes which results in hot spots, increased fouling and coking rates, and may eventually cause tube rupture.

13.3.2.3 Causes and Corrective Action

Leaning flames can be caused due to

- Incorrect burner arrangement or firebox design
- Burner gas tip problems
- Firebox and burner maintenance issues
- Combustion air problems

13.3.2.3.1 Incorrect Burner Arrangement or Firebox Design

The firebox design or the burner arrangement can cause leaning flames in several different ways:

- Insufficient or uneven burner spacing
- Uneven spacing from burners to process tubes relative to the burner duty
- Adverse flue gas patterns caused by asymmetric firebox shapes or burner arrangements

In each of these cases, the flames of the burners can be forced to lean in certain directions, such as toward the center of the heater, toward the radiant coils, or toward the flue gas exit of the firebox. This risk is higher for ultralow NOx burners that rely on large amounts of flue gas entrainment. Once the heater has been constructed, it is often difficult to correct the root cause for the flame issues. The burner vendor can, however, try to mitigate the problem by shortening the flames or biasing the fuel injection. The use of computational fluid dynamics (see Volume 1, Chapter 13) to diagnose and resolve the issues is indispensable.[1] Figure 13.19 shows an example of this use for CFD.

13.3.2.3.2 Burner Gas Tip Problems

Incorrect positioning and orientation of the burner tips with respect to the refractory walls or floor of the firebox can cause leaning flames. This may be due to incorrect burner installation or to deformation of the heater steel. Both should be checked and corrected. The wrong burner tip, an improperly drilled tip, or a tip turned in the wrong direction can also cause flames to lean. All should be checked and the manufacturer's drawings referenced to ensure that orientation and locations are correct.

13.3.2.3.3 Firebox and Burner Maintenance Issues

Improperly installed or damaged refractory can affect the direction of flame propagation. When observed, such sections should be repaired. Damaged diffuser cones can cause leaning and uneven flame patterns and should be replaced. Damaged burner tiles should be replaced if they are contributing to flame leaning.

Air infiltration when operating with minimum excess oxygen can cause flames to lean toward the source of the leaking air. Air leakage points should be identified with smoke tests or draft gauges and sealed.

13.3.2.3.4 Combustion Air Problems

Flames can lean whenever a side of the burner becomes starved of oxygen. Sometimes the reason for this is easily identified, such as a blocked air inlet or (partially)

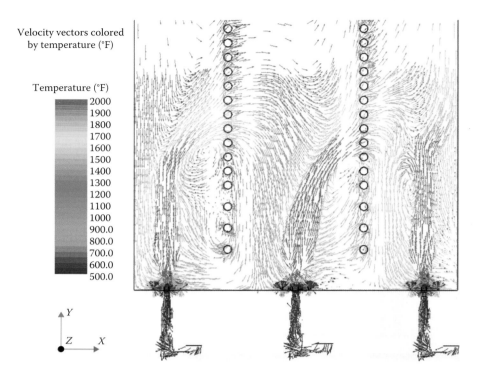

FIGURE 13.19
CFD modeling of leaning flames caused by irregular burner spacing.

closed air registers. Roof refractory can become loose and fall into the burner, blocking air flow.

If none of these apply, this can be a difficult case to diagnose, and CFD is a useful tool to do so. Oxygen starvation can be caused by an asymmetric burner inlet, forcing too much air to one side of the burner throat. This is more likely to happen to burners with low pressure loss (<0.3 in H_2O 7.6 mmH_2O). Contact the burner manufacturer to see if a different style air inlet could solve the issues.

Strong gusting wind can cause uneven air distribution as well. When this is the case, the leaning of the flames changes with the direction and speed of the wind. Installing a wind screen around the heater can help to mitigate any wind effects.

13.3.3 Irregular/Nonuniform Flames

13.3.3.1 Indications of the Problem

For multiple burners in a heater, an irregular flame pattern means that the flames from one burner to another burner are not of the same shape or volume for the same conditions. When a single burner has an irregular flame, it means that the flame pattern is nonsymmetrical. An irregular or nonuniform flame pattern implies that the flame varies in length or shape across the width of a burner.

Irregular flame patterns can appear in two ways. A single burner may have one side of the flame longer than the other side, or part of the flame may be emerging at a different angle from the main flame. In a multi-burner furnace, some burners may have a longer flame than other burners in the furnace when the same fuel pressure and same air register opening is being provided at each of the burners. Figures 13.20 and 13.21 show two examples of irregular flame patterns.

FIGURE 13.20
Irregular flame pattern.

13.3.3.2 Effect on Operation

In either the single burner or multi-burner case, the irregular pattern may cause the furnace to have hot

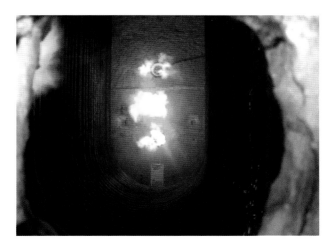

FIGURE 13.21
Nonuniform flame patterns.

FIGURE 13.23
Eroded gas tip.

spots on the tubes. In the case of multi-burner furnaces, the burners with the longer flames may create flame impingement on the process tubes, while the other burners have short, compact flames and cause no hot spots on the tubes. Irregular flame patterns can result in higher operating costs or afterburning due to incomplete combustion of the fuel being burned.

13.3.3.3 Cause and Corrective Action

In the case of an irregular flame from a single burner, the problem may be a dirty burner tip, an eroded tip, or a tip improperly oriented with respect to the refractory. In a dirty burner gas tip, as seen in Figure 13.22, some orifices may be partially plugged while others are operating normally.

If part of the flame is emerging at a different angle from the main flame, tip erosion, as shown in Figure 13.23, may have occurred. Improperly placed tips may be too close, too far, or turned at the wrong orientation with respect to the refractory tile ignition ledge. The irregular flame may also be caused by a foreign material, such as refractory or plenum insulation, in the burner throat or by a damaged tile. Figure 13.24 shows an example of a piece of arch brick that fell into the burner causing an irregular flame.

Either will affect the local mixing of fuel and air, delaying mixing in part of the throat, and lengthening the flame on that burner.

In the case of several burners in the same firebox, the problem may be more complicated. Some of the burners may be plugged, while others are operating normally. The air registers may be at different settings, causing uneven air flow across each of the burners. A poorly designed plenum may be supplying more combustion air to some of the burners and less to others or more to one side of the burners. Maintenance or installation personnel may have placed the burner tips in an incorrect

FIGURE 13.22
Plugged gas tip.

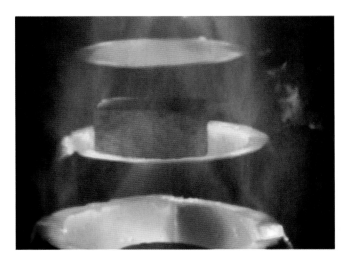

FIGURE 13.24
Arch brick in burner tile.

position with respect to the refractory ledge. Burner tips designed to be oriented in a specific direction with respect to other tips or the refractory ledge may be turned to the wrong position, causing flame-to-flame interference. The refractory may be damaged, causing poor or uneven air circulation around the burner tips.

Obviously, it is good operational practice to maintain the burner tips and ensure that they do not become dirty or plugged. Many irregular flame patterns are corrected by simply keeping the burner tips in good operating condition. While cleaning the tips, the orifices should be inspected to ensure that no erosion has occurred that would change the burning characteristics of the tip. If the fuel is oil, some of the orifices on the tip may become completely plugged with coke. Manually inserting a twist drill bit into the orifice and blowing steam through the orifice should clean the orifice. Do not use powered drills to clean orifices. The use of a power drill may change the size or shape of the orifice. A larger or different-sized orifice then flows fuel at a different rate thereby causing the tip to produce an irregularly-shaped flame.

In multi-burner furnaces, all of the burner tips should be inspected, cleaned, and have the same orifice design. All air registers should be set to the same approximate opening. All fuel pressures to the burners should be at the same level. If the flames are still irregular, check the manufacturer's drawings for the burner fuel tip position and orientation. Then check each tip on each burner, and ensure that the tips are at the correct height and orientation with respect to the ignition ledge. Inspect the burner refractory tile for damage, especially in the immediate area of the tips. Some tips are designed to operate in close proximity to the refractory, and a difference of ±0.25 in. (6.4 mm) can cause irregular flames.

Air plenum distribution problems are difficult to diagnose. Such problems usually occur when the plenum is not correctly sized. As a result, some burners get more combustion air, and other burners are starved for combustion air. The operator needs to ensure that all burner air registers have the same size opening. The designer should ensure that the air plenums are properly sized for good air distribution to all burners. In some cases, it may be beneficial to have the air plenums computer modeled so that they are correctly sized. In large plenums, benefit may be obtained by adding air inlets to improve the distribution of air.

13.3.4 Pulsating Flames/Burners Out of Air

13.3.4.1 Indications of the Problem

The pulsating flames phenomenon is sometimes called "woofing" or "breathing." The woofing or breathing noise is a very low-frequency noise that is different from the normal combustion noise around the heater. Instead of a steady consistent flame volume, the flame is changing in size and creating pressure surges as the fuel searches for air. This movement or pulsing in the flame front causes the huffing to occur. Huffing is that audible condition where the "sound" of the combustion is not continuous or constant, accompanied with visible inconsistent flame pattern and flame root position.

There have been many cases where this low-frequency noise can be heard at great distances from the heater. In one instance, the huffing from a large heater was heard upon immediately leaving the control room. The control room was over 600 ft (180 m) away from the heater!

13.3.4.2 Cause and Effect on Operation

The cause of the pulsating flames is lack of oxygen in the combustion reaction (see Figures 13.25 and 13.26). When the oxygen or air flow is inadequate, the flame will burn where there is air available. This alternates between the inside of the heater as the air flows across the burner and outside the heater as it becomes starved for air. As the flame moves into the heater, a pressure front is generated, again causing the air flow to cease, and the flame moves back to the burner for oxygen. The movement of the flame and pressure continue to increase in intensity and with such force as to cause damage to the heater. In extreme cases, the flame may oscillate so far from the burner that combustion is extinguished, and the flame is lost.

The condition of insufficient oxygen in the combustion zone of the furnace may exist even when the oxygen levels measured in the stack flue gas indicate that sufficient oxygen is available. Air can leak into the heater through the flanges between the convection section and the radiant section, cracks or corrosion in the heater casing, sight ports left open, and access doors. Air leakage into the heater will add oxygen to the flue gases, causing

FIGURE 13.25
Heater out of air.

FIGURE 13.26
Flames short of air.

inaccurate indications of oxygen available for combustion in the throat of the burner.

13.3.4.3 Corrective Action

As soon as a condition of pulsating flames or huffing is observed, the firing rate should be immediately reduced. *Do not shut the fuel off or give the heater more air*! Increasing the air before cutting back on the fuel may fill the furnace with a large combustible mixture of fuel and air, which might result in an explosion and damage to the equipment and loss of life. Cut the fuel rate by 10%–15% to establish sufficient oxygen for the combustion reaction to go to completion. If the first cut does not stabilize the heater within 3 minutes then cut the fuel again. The operator must reduce the firing rate until the pulsation has stopped and no woofing noise is heard. When no woofing noise is heard and the flames are not pulsating, the operator should observe good stable combustion. There should be a measurable excess of oxygen in the firebox flue gases. Then the operator can open the air registers on the burner and increase the stack damper's opening to adjust the excess oxygen and draft to the correct levels required for the firing rate desired. The firing rate can then be increased to the burner capacity or to the required heat release requested by the heater control system.

If the air registers or the stack damper are fully opened, then the heater has reached its maximum capacity. Any wind blowing under the heater or across the top of the stack may interrupt the air flow to the burners and start the pulsating flame problem. Sometimes, a windscreen or fence around the furnace is necessary to prevent the wind effects on heaters operating at maximum capacity in high-wind locations. If the flue gas analyzer is in the stack, then it should be relocated to the top of the radiant firebox. This will eliminate the air leaking into the convection section from giving a false indication if excess oxygen.

13.3.5 Flame Lift-Off

13.3.5.1 Indications of the Problem

The first indication of a flame lift-off problem is when the operator observes that the flame is detached from the burner as seen in Figure 13.27. The operator may also hear the sound of the flame front lifting and trying to reattach to the burner. The lifted flame can also cause the draft in the furnace to fluctuate rapidly. A single lifted flame can become so violent that it will influence the adjacent burners to go into the same type of unstable operation. If the flames go out, then raw fuel is being dumped into the heater which is an extremely dangerous situation. An explosive mixture of raw fuel and air can happen in a matter of seconds depending on the actual operating conditions.

The pulsating flame phenomenon is sometimes called "woofing" or "breathing." The woofing or breathing noise is a very low-frequency noise that is different from the normal combustion noise around the heater. A flame that has lifted from its stabilization point will normally try to reattach to the tip. This movement or pulsing in the flame front causes the huffing to occur. Huffing is that audible condition where the "sound" of the combustion is not continuous or constant, accompanied with visible inconsistent flame pattern and flame root position.

FIGURE 13.27
Flame lift-off.

The effect on operation during the woofing condition is normally just a passing condition if corrected immediately. However, if the woofing condition is not resolved and is allowed to continue, the intensity of the pressure surges within the furnace may cause the refractory insulation to begin to break up and fall onto the heater floor. The burner tile may even begin to deteriorate and fall apart. The heater vibration can break piping, tubing, and instruments. The incomplete combustion causes a drop in heat release, and the heater cannot fulfill its required duty. The heater may have to be shut down for major repairs.

13.3.5.2 Effect on Operation

Flame lift-off from the burner is a very significant safety hazard. If the lift-off is extreme, there may be a total loss of flame at the burner, and unburned fuel will be flowing into the firebox. If the refractory remains at a sufficiently high temperature or if the pilot remains lit, then reignition may occur. Reignition may also be initiated by adjacent burner flames. The reignition may cause a minor or major explosion, depending on the amount of fuel injected into the firebox, with the extent of damage dependent on the heater design and configuration. If the explosion within the firebox is minor, the explosion doors (if present) will open and relieve the internal pressure built up within the firebox. If there are no explosion doors on the heater to relieve the internal pressure buildup, the heater may be damaged by an explosion. If the explosion within the firebox is a major explosion, the complete heater may be torn apart, resulting in the heater and process being shut down. The loss of the heater will cause a loss of product and hence a loss of profits being generated from the process unit. In the most severe explosions within the firebox, there may be loss of life or injury to operating personnel.

13.3.5.3 Corrective Action

Flame speed and air/fuel delivery speed must be balanced to ensure that the flame is attached to the tip rather than rising above it.

When flame lift-off from a burner is first observed, immediately reduce the fuel pressure on that burner by partially closing the specific burner block valve. If the flame continues to lift-off, the operator should completely shut off the burner. In either case, take the following corrective actions. Flame lift-off produces unsafe conditions and should be corrected immediately before a more serious condition occurs.

The alignment and positioning of the gas burner tip and the oil burner tip should be checked to ensure they are correctly installed and positioned in accordance with the burner manufacturer's drawings and specifications.

The gas tips should be located in relation to the tile ledge or flame holder as shown on the drawings, and the oil tip should be located in relation to the inlet tile throat or diffuser as per the drawings. If the oil tip is too high in relation to the tile or flame holder, the oil flame may lift-off.

The fuel gas firing ports and ignition ports should be checked to ensure they are not plugged. If they are plugged, the ports should be cleaned by manually inserting a twist drill the same size as the port and twisting the drill to remove all foreign material in the port.

If the burner is a premix burner design, then the primary air door should be adjusted to the correct position. On a premix burner, the gas/air mixture exits the gas tip at a given velocity. The flame burning above the firing port has a flame velocity that is traveling in the opposite direction, that is, it is trying to get back to the source of the fuel and air mixture. If the gas/air velocity is much higher than the flame speed then the flame begins lift-off from the burner gas tip. To correct the exit velocity from the firing port, the primary air door is closed to reduce the amount of primary combustion air entering with the gas, hence a reduction in the gas/air velocity. With the reduced gas/air velocity, the flame reattaches to the burner gas tip or the flame holder.

If an oil-fired burner is experiencing flame lift-off, the operator will need to adjust the steam atomization pressure per the burner manufacturer's instructions. If the atomizing steam pressure is too high, the oil flame will tend to lift-off from the burner. Raw gas burners that utilize a stabilizing cone can experience lift-off if the cone is missing or damaged. Shut off any such burner where the cone is partially or completely missing or is improperly installed, and replace the diffuser cone or flame holder.

13.3.6 Flashback

13.3.6.1 Indications of the Problem

Flashback is the phenomenon that occurs in premix burners when the flame velocity is greater than the velocity of the flowing mixture. Under the right conditions, the flame front propagates back through the mixer and burns inside the mixer (see Figure 13.28). In some cases, the flame may flash back to the orifice spud where the fuel and primary air are being mixed. Flashback is most likely to occur in burners using fuels having a high ratio between the upper and lower explosive limits of the fuel. Table 13.1 reveals that the gases most susceptible to flashback include carbon disulfide, acetylene, ethylene oxide, hydrogen, hydrogen sulfide, and ethylene. Another reason for flashback may be that the gas tip design is not optimized for the fuel that is being burned. If the velocity of the gas and air exiting the gas tip is very low, because of a large-diameter firing

FIGURE 13.28
Flashback inside the burner mixer.

port, then the velocity of the flame front may be greater than the velocity of the fuel/air mixture exiting the tip and flashback will occur.

When flashback occurs within the mixer or venturi, a flame will be observed burning in the venturi or mixer. If flashback has occurred sometime in the past, the mixer or venturi will show signs of oxidation on the outside of the cast-iron venturi. When flashback occurs in a premix burner, there is little doubt in the operator's mind that flashback is occurring within the burner. A sharp barking noise in the mixer is continually emitted until corrected.

13.3.6.2 Effect on Operation

When flashback occurs and remains uncorrected, the burner mixer or venturi is damaged from the high temperatures generated within the mixer from the combustion reaction that is occurring. The damage to the burner parts, as seen in Figures 13.29 and 13.30, will result in higher maintenance costs.

When flashback occurs, the capacity of the burner is restricted; if flashback occurs on many burners within the heater, the outlet temperature of the process cannot be obtained. The burning inside the venturi tube and intermittent open flame constitute a safety hazard.

13.3.6.3 Corrective Action

The first step is to shut the burner off and allow the mixer to cool down. The solution to the problem will vary depending on what is available to the operator. The operator should immediately check the gas tip

TABLE 13.1

Ratio of the Upper and Lower Explosive Limits and Flashback Probability in Premix Burners for Various Fuels

Fuel	Ratio[a]	Probability[b]
Acetone	5.01	1.67
Acetylene	32.00	Infinite
Acrylonitrile	5.57	1.85
Ammonia	1.71	0.57
Aromatics (mean)	5.00	1.66
Butadiene	5.75	1.92
Butane	4.52	1.51
Butylene	4.88	1.63
Carbon disulfide	40.00	Infinite
Carbon monoxide	5.93	1.97
Cyanogen	6.45	2.15
Ethane	4.02	1.34
Ethyl alcohol	5.77	1.92
Ethyl chloride	3.70	1.23
Ethylene	10.04	3.33
Ethylene oxide	26.66	Infinite
Gasoline (mean)	5.06	1.68
Hydrocyanic acid	7.14	2.38
Hydrogen	18.55	Infinite
Hydrogen sulfide	10.60	3.52
Methane	3.00	1.00
Methyl alcohol	5.43	1.81
Methyl chloride	2.26	0.75
Naphtha	5.45	1.81
Oil gas	6.84	2.28
Propane	5.25	1.75
Propylene	5.55	1.85
Vinyl chloride	5.42	1.80

Source: Reed, R.D., *Petrol. Eng.*, C7, 1950.
[a] Ratio of the upper and lower explosive limits.
[b] Probability of flashback as compared to methane.

discharge port and the main gas orifice to ensure that both are clean. If the gas orifice is dirty, the fuel flow may be reduced to the point of creating a flashback condition. If the gas tip discharge port is dirty, the flow of the fuel/air is reduced on the exit discharge port to the point of creating the flashback condition. Hence, the operator must clean both the gas metering orifice and the gas tip discharge port to allow the fuel/air velocity to remain higher than the flame velocity so that no flashback will occur. If the fuel composition or process heat requirement changes, resulting in lower operating pressures, burners must be shut off to raise the fuel gas pressure, or new orifices may be required in the mixers or venturi to keep the fuel gas pressure at a level sufficient to prevent flashback.

If raising the fuel pressure does not resolve the flashback problem, one can look to the flame velocity of the fuel/air mixture for another solution. The flame velocity is related to the percentage of air in the air-fuel gas mixture. A change in this percentage, by adjusting the burner air door, can raise or lower the flame velocity. Try reducing the primary air flow to lower the mixture flame velocity. Adjust the secondary air register to maintain the target excess oxygen level.

13.3.7 Flame Impingement on Tubes

13.3.7.1 Indications of the Problem

The most direct indication of flame impingement is visual observation by the operator of the flames contacting the external tube surface inside the firebox (see Figure 13.31).

If flame impingement is suspected but cannot be directly observed, several infrared photographs of the tubes should be taken to determine if there are any high tube-skin temperatures as a result of direct or indirect flame impingement. Also, components of the flame radiating at frequencies not visible to the naked eye (see Volume 1, Chapter 7 on radiation) may be contacting the tubes. To establish if the flame is contacting the tubes, the operator can inject some baking soda or activated carbon particles into the combustion air. The flame temperature causes the baking soda or carbon to glow a bright yellow as it oxidizes. The bright yellow of the baking soda dissipates at 1800°F–2000°F (980°C–1100°C) which indicates the end of the hot combustion gases from the flame. The burning carbon will show the flame pattern being emitted from the burner for a short time span and will also indicate if there is any flame impinging on the tubes. Some of the carbon particles will also glow orange and be carried by the flue gas as it moves throughout the furnace.[7]

FIGURE 13.29
Flame burning inside gas tip.

FIGURE 13.30
Gas tips glowing due to flashback.

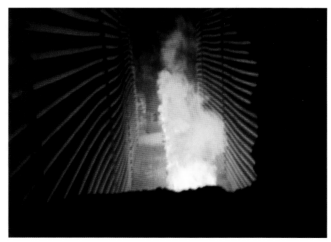

FIGURE 13.31
Flame impingement.

Burner Troubleshooting

13.3.7.2 Cause and Effect on Operation

During normal operation, the process fluid flowing through the tubes will provide sufficient cooling of the tube surface to cause the tube color to be essentially black in contrast to the tube hangers or brackets. The uncooled tube hangers or brackets that support the tubes will normally be a dull to bright red color (see Figure 13.32).

Flame impingement on the tubes can create hot spots, causing the tubes to appear red or orange in color. Since the majority of process heaters have a feed that is a hydrocarbon, flame impingement can cause serious problems. Visual observation of the burners may show that the flames are contacting the tubes. In some cases, a gradual increase in the tube metal temperature (TMT) could also indicate possible flame impingement. The operators should make it a point to look into each of their fired heaters at least once a shift to check for any problems with the flame patterns such as that seen in Figure 13.33.

The tube color indicates excessive TMTs that may result in localized coke formation. The layer of coke, on the inside of the tube, insulates the tube wall from the cooling effect of the process fluid. This insulating effect creates two undesirable conditions: (1) heat transfer to the process fluid is impeded, thereby reducing efficiency; and (2) the tube is inadequately cooled by the process fluid, resulting in hot spots, more coke deposition, and eventually tube rupture within the heater. Hot spots will normally develop in progressive stages.

When the flames contact the tube surface, there is a cooling effect on the flame. This results in ash being laid down on the tube. This buildup will lead to scale on the tubes as the outer layer of the tube starts to burn away (see Figure 13.34).

Here are the various stages during this process (see Figure 13.35):

1. Dark areas first start to appear from the carbon coating on the side of the tubes facing the burners.
2. Silver or light gray spots form within the dark areas. This is caused by the carbon being burned off.

FIGURE 13.32
Tube hangers glowing red.

FIGURE 13.33
Flames short of air and impinging on tubes.

FIGURE 13.34
Tube scale.

FIGURE 13.35
Stages of overheating on process tubes.

3. These light gray spots will enlarge and cover more area.
4. As the coking continues, red spots will begin to appear in the gray areas of the tubes. In some cases, the tube will take on a "mirror" finish that looks almost like a chromed piece of pipe.
5. The tube will eventually start to bulge and then develop "pinhole" leaks (see Figure 13.36). At this point, the tube is ready to rupture, as seen in Figure 13.37, and immediate action must be taken.

Some processed liquids do not coke when overheated, but form vapor. If not considered in the heater design, the vapor may significantly increase the resistance to flow and the pressure drop. The lowered flow rate combined with film boiling at the location of impingement will reduce the heat transfer coefficient and raise the local tube temperature.

A possible cause of flame impingement on the tubes may be a deficiency of combustion air in the combustion reaction zone, resulting in non-combusted fuel that reacts with oxygen in other locations than the flame zone. The deficiency of combustion air for the combustion reaction may be a result of overfiring or of air leaking into the firebox and not flowing across the burner air orifice. Air leakage through other openings on the heater does not mix well with the combustion air moving across the throat of the burner. Overfiring may result from instrument failure, installation of wrong burner tips, or a change in fuel gas composition, affecting the flame pattern. Another possible cause may be that the burner firing ports are eroded such that the fuel is being injected through the

FIGURE 13.36
Pinhole leak in tube.

FIGURE 13.37
Tube rupture due to coking.

combustion air flow directly toward the tube. Another possible cause may be that flue gas recirculation within the firebox is preventing the flame from forming an acceptable pattern in its allotted space. Flue gas circulation may be pushing some burner flames onto the tube surface.

13.3.7.3 Corrective Action

The most important thing is to keep the flames off of the tubes! If flame impingement is noticed, the first step should be to adjust the burner causing the impingement to get the flame off the tube.

- Check the burner air register to confirm that it is open. Then look at the gas and oil tips, and determine if there is any plugging that would cause the flame to impinge on the tube. If there is plugging, then remove the tips and clean them. Be sure the gas tips are properly oriented by looking at the burner drawing.
- Confirm that the excess oxygen and draft requirements are per the heater design specifications.

FIGURE 13.38
Reed walls between burners and tubes.

If the heater cannot be shut down, there are three options:

1. Take the burner out of service, or reduce the firing rate by manually closing the block valve.
2. Increase the excess air to help cool the firebox.
3. Increase the process flow to the overheated pass.

There are other options such as wrapping the tubes or clamping them with the heater in service. These options would have to be considered extreme risk situations and approved by the safety department.

If the flue gas circulation patterns within the firebox are pushing the flames into the tubes, then a Reed wall[8] or division wall[9] may need to be installed in the heater firebox. The Reed wall, which can be seen in Figure 13.38, redirects the flue gas circulation pattern within the heater, while the division wall interrupts the circulation. Each allows the burner flame pattern to develop in the space designed for the flame pattern. Note that emissions of NOx are likely to increase when flue gas recirculation flows are impeded or obstructed by internal walls. It is recommended to use CFD modeling to investigate the impact of reed or dividing walls on flame shape and flue gas flow patterns.

Damaged or missing burner tile sections, as seen in Figure 13.39, can cause unequal distribution of air

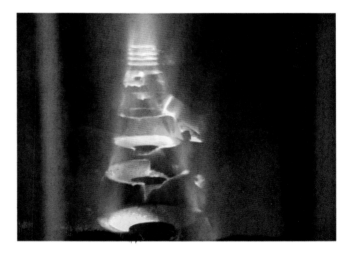

FIGURE 13.39
Damaged burner tile.

within the burner. This will lead to fuel-rich zones, locally longer flame segments, and the potential to lean toward and into the tubes. Check the tile condition and repair if necessary.

Impingement may be overcome by changing the flame shape. For example, if the firebox dimensions allow a longer flame, the burner tip port included angle can be reduced to obtain a more slender flame and move the flame envelope further from the tubes.

13.3.8 Burner Spacing/Flame Interaction

13.3.8.1 Indications of the Problem

Whenever burners are spaced too close together, the operator may observe the following symptoms:

- Noticeable interaction between the flames
- Increased emissions of NOx and CO
- Irregular flame patterns
- Elongated flames
- Flame "rollover," impingement into radiant tubes, hot spots
- Increased bridgewall temperature, lower thermal efficiency

13.3.8.2 Cause and Effect on Operations

Ultralow NOx burners achieve their low emissions by internally recirculating flue gas back into the burner. As such, they require sufficient space around each burner to allow this large volume of flue gas to reach the gas tips. If the burners are spaced too close together, this space will be inadequate, but the fuel jets will still try to conserve momentum. The result may be that flames from one burner are pulled into another. The resulting lack of dilution by flue gas will increase the local flame temperatures and increase NOx emissions. Entrainment of one flame into the other can also disturb the flame stoichiometry, resulting in uneven air/fuel distributions, elongation of flames, and irregular flame patterns.

Spacing burners too close together is most likely to happen in heater revamp situations where conventional burners are replaced by ultralow NOx burners without changing the burner layout. In conventional style raw gas burners, the fuel and air mix and combust near the throat of the tile without any staging of fuel or air. They produce a small flame envelope resulting in high flame temperatures and consequently high NOx emissions. As a result, process heaters designed to utilize these conventional burners typically have very tight burner spacing. Fitting ultralow NOx burners into the same arrangement can quickly lead to flame interactions as seen in Figure 13.40.

Ultralow NOx burners can also be interacting in new furnace designs when a firebox layout negatively impacts the flue gas recirculation patterns. Common examples are vertical cylindrical heaters with burner circles that are too small compared to the tube circle diameter. The inability of the inner fuel tips to recirculate the flue gas causes the flames to merge into one long flame.

Flame rollover is typically observed in box type heaters with flat flame burners when the burners are

FIGURE 13.40
Flame interaction.

interacting as seen in Figure 13.41. The local disturbance in stoichiometry causes unburned fuel to flow into the recirculation vortices. The fuel burns in the vortex as soon as it mixes with available oxygen, which is typically observed as flame detaching and rolling into the coils, hence the name "flame rollover." In addition to affecting emissions, flame rollover causes flame impingement and hot spots on the tube coils, which causes additional problems such as increased fouling rates, reduced operating cycles between decokes, potential for tube damage, and reduced coil life.

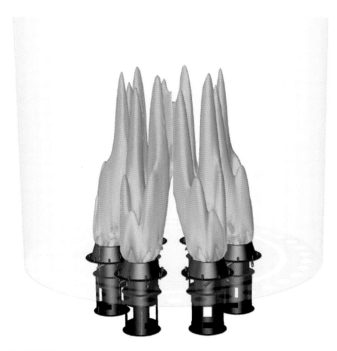

FIGURE 13.41
CFD model shows flames merging as a result of a tight spacing.

13.3.8.3 Corrective Action

Correcting burner spacing problems can be very challenging once the furnace is constructed. Burner cutouts can hardly ever be changed due to the associated cost and structural steel interference, so a solution often must be found within those constraints. The burner vendor may be able to change the burner design to better cope with the situation. Solutions may consist of biasing fuel away from areas where an excess appears to be. For example, in the earlier example of a tight burner circle, fuel may be moved away from the inner tips. This reduces the entrainment requirements for these inner tips and subsequently reduces the tendency of the flames to merge together.

In certain cases, Reed walls may change the recirculated flue gas flow patterns sufficiently to reduce flame interactions. CFD modeling (see Volume 1, Chapter 13) will be helpful to evaluate this option.

Other solutions exist to reduce the burner entrainment needs, for example, by moving fuel from the outside of the tile to the burner throat. Such solutions will also affect NOx production and must be very carefully evaluated before implementation. The solution to each situation depends highly on the type of burner and the specific flame interaction that is observed. Therefore, each individual case requires a tailored approach.

Changes in wall-fired burners to eliminate flame rollover have to be considered together with their impact on the process side heat transfer, as the changes could modify the incident radiant flux profile. An example of such a solution is to create high-output and low-output combustion zones that have lower tendencies to interact with each other, by varying burner liberation rates along the length of the heater. Choosing the location of high- and low-duty zones will have to be done keeping the specific flux requirements of the process coils in mind.

These are just a few examples of the possible solutions. As mentioned earlier, each flame interaction situation needs to be analyzed individually to determine what solution will work best. Due to the complexity of the problem, CFD is often the right approach for this. Figures 13.41 through 13.43 show examples of CFD modeling of this problem.

13.3.9 Low-Temperature Operation (<1300°F [700°C])/Flame Stability/CO Formation

13.3.9.1 Indications of the Problem

High levels of CO are present in the stack gas. Flames are irregular, pulsating, occasionally lifting off, or even completely extinguished. The problem is worse for fuels without hydrogen.

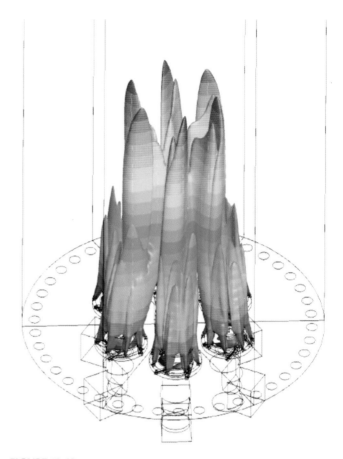

FIGURE 13.42
CFD model showing flame interaction.

FIGURE 13.43
CFD model showing flame interaction and recirculation zone.

13.3.9.2 Causes

Ultralow NOx burners employ fuel staging and internal flue gas recirculation in order to reduce peak flame temperatures and thermal NOx production (see Figures 13.44 and 13.45). All fuel tips are outside the tile to entrain flue gas into the primary and secondary zones.

FIGURE 13.44
Cold firebox operation (floor and walls are darker in color).

CO is produced in significant quantities early in the combustion process and then oxidized to CO_2 via

$$CO + O + M \Leftrightarrow CO_2 + M \quad (13.1)$$

$$CO + O_2 \Leftrightarrow CO_2 + O \quad (13.2)$$

$$CO + OH \Leftrightarrow CO_2 + H \quad (13.3)$$

$$CO + HO_2 \Leftrightarrow CO_2 + OH \quad (13.4)$$

Reactions 13.1 and 13.2 are the dominant CO oxidation steps in a hydrogen-free environment and are very slow compared to reactions 13.3 and 13.4 that predominantly occur when hydrogen is present. CO is the last component in the flame to be oxidized, and this process can be prematurely stopped if the flame is quenched by large amounts of flue gas. This means that oxidation of CO to CO_2 can be problematic for ultralow NOx burners operating at temperatures below 1300°F (700°C) without any hydrogen in the fuel.

CO can rise quickly in a narrow temperature range, as shown in Figure 13.46 where the CO emissions are presented for a hydrogen-free fuel gas as a function of bridgewall temperature.

13.3.9.3 Corrective Action

Temperature is the most dominant factor in the production of CO. The CO concentration can exponentially increase from nothing to 2000 ppm over a temperature range of just 100°C (180°F). Consequently, the most

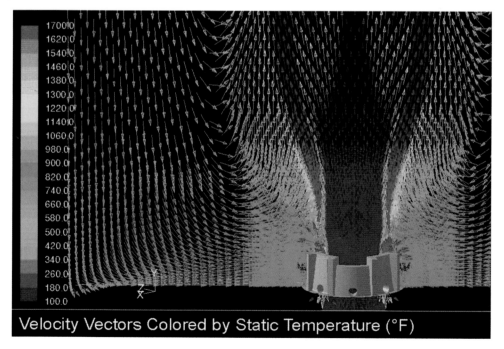

FIGURE 13.45
CFD model showing velocity vectors.

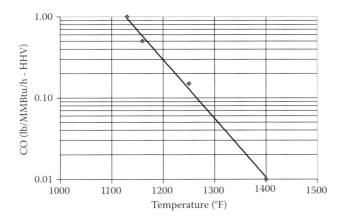

FIGURE 13.46
CO versus firebox temperature.

obvious solution to excessive CO production or flame instability is to look for ways to increase the firebox temperature. Running at higher duty increases the firebox temperature, but this is not always possible. The firebox floor temperature can be increased by the addition of a Reed wall. A Reed wall forms a radiative shield between the floor around the burners and the lower part of the radiant coils. This solution must be carefully analyzed, since the Reed wall can also create unwanted flue gas recirculations and increased NOx emissions.

Changing the fuel composition by adding some hydrogen would also be of great benefit, since CO oxidation will then preferentially take place according to reactions 13.3 and 13.4. The critical temperature for increased CO production is about 150°F (65°C) lower for hydrogen-containing fuels than for a natural gas.

Running at higher excess air may benefit some burners, as this increases the oxygen content of the recirculated flue gas and facilitates reactions 13.1 and 13.2. However, the higher excess air also cools down the flame envelope and can actually lead to increased CO emissions in other burner types. It will also lower the thermal efficiency and may not be a desired solution for heaters that run at these low temperatures for extended periods of time.

Certain burner types are not sensitive to low temperatures and may be a better solution for very low-temperature applications. Staged air burners such as the John Zink HAWAstar and Enviromix produce virtually no CO at temperatures in the range of 700°F–900°F (370°C–480°C). Compared to ultralow NOx burners, the NOx production may be higher, but that is offset by the reduced temperatures. If very low NOx numbers are required as well as low CO at low temperatures, the John Zink HALO burner provides the most optimal solution. Partial premix or lean premix burners should typically be avoided for low-temperature applications.

The burner vendor may be able to change some of the burner characteristics and lower the CO formation tendency at the expense of increased NOx. This balance can be optimized during burner testing. Besides a significant improvement in flame stability, an additional advantage of this approach is that flame lengths can be substantially reduced as well.[10]

13.3.10 High Stack Temperature

13.3.10.1 Indication of the Problem

Measured stack outlet temperature is elevated.

13.3.10.2 Effect on Operations

Under normal furnace operation, the flue gas temperature measured in the stack will be near the operating temperature predicted by the furnace designer at the design duty. High stack temperatures indicate decreased heater efficiency, higher fuel consumption, and increased operating cost. High stack temperatures can indicate excessive heat in the convective section of the heater. Long-term operation at higher than normal stack temperatures can damage the furnace, especially in the roof area of the convective section. High convection section temperatures can cause serious problems with tubes, tube supports, failed refractory, and overall heater operation.

High convection temperatures may indicate afterburning in the convection section. The afterburning may be caused by lack of excess oxygen within the burner throat and firebox. Air leaks into the convection section, and the temperature of the flue gases is sufficient to cause the completion of the combustion reaction to occur (afterburning) in the convection section of the heater. The afterburning will result in the destruction of the extended surface on the tubes in the convection section of the heater. The loss of the extended surfaces results in a loss of heat transfer in the convection section. The afterburning within the convection section may ultimately result in tube failures in the convection section of the heater.

13.3.10.3 Causes and Corrective Actions

These two problems can be the result of common causes, or they may be independent of each other. Some of the more common causes are the following:

- High excess air (high stack temperature)
- Insufficient air/positive pressure at the arch
- Internally or externally fouling on convection tubes
- Missing fins on the convection tubes
- Process leak in the convection section
- Fouled or partially plugged shock/radiant tubes
- Flue gas bypassing the convection tubes

13.3.10.3.1 High Excess Air

The most common cause of high stack temperatures is running with too much excess air. If the process and temperature in the convection section are within specified limits, the first item to check is the excess oxygen. If the excess oxygen is above the specified target, then steps should be taken to reduce the air coming through the burners.

High excess air levels result in high stack temperatures because the increased mass flow of flue gas through the heater reduces the heat transfer efficiency. The convection section is not sized to adequately cool the increased flue gas rate down to design stack temperatures. The higher air and flue gas flow rates typically also require an increased firing rate in order to maintain the firebox temperature and process duty of the heater. The higher nitrogen content of higher excess air combustion products also reduces the radiative properties of the flue gas, since nitrogen does not participate in the radiative heat transfer process.

13.3.10.3.2 Insufficient Air/Positive Pressure at the Arch

Positive pressure at the arch of the furnace can cause both high stack temperature and high convection section temperature. Closing the stack damper too far will lower the excess air and impede the flow of the flue gas. This results in a "stagnant" zone in the convection section allowing the hot flue gas to be in contact with the convection tubes longer than necessary. This can cause high TMTs, high process temperatures, failed tube supports, and failed refractory. This condition can also cause a reduction in the amount of air being supplied to the burners.

13.3.10.3.3 Internally or Externally Fouling on Convection Tubes

The condition of the tubes, in the convection section, is also very important. If the tubes are coking internally, fouled or developing scale buildup, or missing fins (see Figure 13.47), then the flue gas temperature will be higher than designed.

13.3.10.3.4 Process Leak in the Convection Section

A leak in one of the tubes will cause extremely high temperatures due to the oxidation of the fluid. These "pinhole" leaks take time to develop, but once the fissure is open, the hydrocarbon will spray into the firebox. Obviously, this situation should be corrected immediately by shutting down the heater and making the necessary repairs.

13.3.10.3.5 Flue Gas Bypassing the Convection Tubes

Sometimes the high stack gas temperature is caused by the heater construction. Hot flue gas will follow the path of least resistance to the stack entrance. This is usually at the ends of the convection section tubes, where the surface to absorb heat is minimal. Unless steps are taken to minimize this bypassing, which may occur in header boxes or at tube ends in the main flue gas passage, a significant amount of hot gas reaches the stack without transferring much heat to the convection tubes. The convection section may require more than one exit to provide a uniform flow profile across the tube bank. This will help reduce the chance of the flue gas exiting at higher than design temperatures.

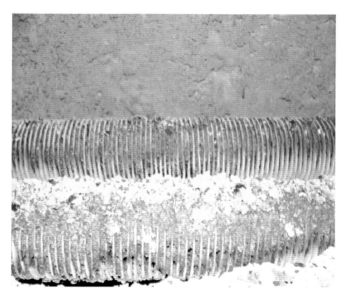

FIGURE 13.47
Plugged fins on convection section tubes.

13.3.11 Overheating of the Convection Section

13.3.11.1 Indications of the Problem

Upon visually inspecting the inside of the firebox, there is refractory lying on the floor and in the burners. The heater shell and structure on the convection section show signs of overheating. The shock tubes and shock tube hangers are failing. The draft at the top of the radiant section is at a positive pressure. When the sight ports are opened, hot flue gases are forced out, thereby causing a safety hazard to the operator.

13.3.11.2 Effect on Operation

The convection section is designed to remove heat from the hot flue gases exiting the radiant section at the bridgewall temperature of 1200°F–2000°F (650°C–1100°C) and transfer the heat from the flue gases to the process liquid primarily by convection. If the hot flue gases leak out of the cracks in the convection section, then the structural steel and heater convection section shell are overheated, resulting in damage to the heater.

The hot flue gases leaking out of the cracks in the heater shell result in overheating the carbon steel heater shell and structural steel supporting the convection section of the heater above the radiant section. With the overheating of the heater shell, the refractory anchors are damaged and leave nothing to support the refractory in the arch section of the heater, and the refractory falls to the heater floor. As a result, more heat reaches the structural steel, and finally, the convection section falls into the radiant section of the heater, and the unit is shut down for repair.

13.3.11.3 Corrective Action

The operator must first obtain the draft reading at the top of the radiant section and compare this to the targeted draft reading. If the draft reading indicates positive pressure, then the hot flue gases exiting all cracks must be stopped before the carbon steel shell is overheated and the anchors holding the refractory are damaged. Upon observing the positive pressure in the heater, the operator must immediately open the stack damper on a natural-draft heater or the induced-draft fan damper on a balanced-draft system if there is a measured excess of oxygen at the firebox exit. If there is no excess of oxygen, reduce the firing rate until an excess is attained; then increase fuel input while maintaining the draft and excess oxygen at the targets. A negative pressure must be established to eliminate the overheating of the convection section.

If the stack damper or induced-draft fan damper is completely opened, then the operator must reduce the firing rate on the heater such that there is a slightly negative pressure at the top of the radiant section of the heater.

13.4 Fuel Gas Problems

Problems associated with fuel gas are the most common burner problem encountered in refineries and petrochemical plants. No other single issue causes more problems relating to the operation of the burner.

13.4.1 Burner Fuel Pressure/Impact on Operation

One of the most common problems associated with burners is the difference between actual fuel pressure and the theoretical pressure shown on the burner capacity curve (see Figure 13.48).

There are at least six possible reasons why the fuel pressure would be different from the capacity curve:

- Burner fuel gas tips problems.
- A change in the fuel composition.
- The wrong gas tips have been installed on the burner.
- Actual fired duty is different than specified.
- The fuel flow measurement is incorrect.
- The fuel pressure is incorrect or measured at the wrong location.

13.4.1.1 Fuel Gas Tip Problems

The holes in the fuel gas tips may be plugged. This is commonly caused by dirty fuel, but can also be caused by coking inside the tip as a result of exposure to elevated temperatures without adequate gas flow. The ports could also be too large caused by cleaning them many times or by cleaning them with a power motor.

13.4.1.2 Fuel Composition

Different fuel compositions result in changes to the heating value and molecular weight of the fuel. Both of these factors will affect the operating pressure of the burners. If the fuel contains more hydrogen than originally specified, the heating value, on a volume basis, will decrease. This decrease in heating value may result in higher fuel pressures to maintain a given firing rate. Pull a fuel sample, and compare it to the fuel shown

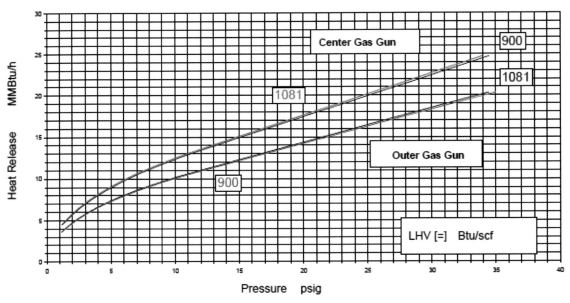

FIGURE 13.48
Burner capacity curve.

on the burner data sheet. If the compositions have changed, then consult the burner manufacturer for updated capacity curves.

If you know the specific gravity and higher heating value of the current fuel and the original fuel, you can calculate the Wobbe indices for comparison. The Wobbe index is used to evaluate the interchangeability of fuels. The formula is shown as this:

$$\text{Wobbe index} = \frac{\text{HHV}}{\sqrt{\text{SG}}} \quad (13.5)$$

The HHV of the fuel is divided by the square root of the specific gravity. For example, the Wobbe index for methane is

$$\text{WI} = \frac{1010\,\text{Btu/ft}^3}{\sqrt{0.55}} = 1361.88$$

Any volumetric unit of higher heating value can be used as long as the units are consistent.

Replacing the gas tips should be considered if the Wobbe indices vary by more than 5%.

13.4.1.3 Wrong Gas Tips

This can and does happen. Obviously, if the wrong gas tips are on the burner (see Figure 13.49), then the fuel pressure will not match the capacity curve.

FIGURE 13.49
Identical gas tips with different drillings.

13.4.1.4 Fired Duty Has Changed

Many factors can impact the amount of duty required from the burners. Using the heater for another service may result in reduced firing rates. Adding more feed or deterioration of the tubes can result in higher fired duties being required.

When the fired duty has changed, then the original gas tips may require too low or too high of fuel pressure for the new conditions. This results in fuel pressure alarm points being invalid causing a safety concern. Operating at too high of fuel pressures can cause instability due to the flame "lifting" off and trying to reattach to the gas tip. Too low of fuel pressure can result in poor mixing of the fuel and air. This can lead to flame stability problems if the mixture becomes too fuel lean.

13.4.1.5 Fuel Flow Measurement Is Incorrect

This is a very common issue associated with flowmeters. The meters are normally very accurate. The problem comes from changes in the fuel composition, pressure, or temperature (see Chapter 7 and Volume 1, Chapter 9). When the density of the fuel changes, the correction factors used in the flow calculations have to change. The meter documentation should be reviewed, and a calibration check should be considered if there are any doubts about the accuracy of the measurements.

13.4.1.6 Fuel Pressure Is Incorrect

Pressure gauges and transmitters (see Chapter 7) are not fool proof. Pressure gauges should be checked and calibrated at least once a year. Pressure transmitters should be checked and calibrated on a regular basis since they provide the input into the safety shutdown system.

Both the pressure gauges and transmitters should be located as close to the burners as possible. They should be downstream of orifice runs, block valves, and pressure regulators.

The proper range or scale should also be considered when selecting pressure gauges. The majority of burners operate between 5 and 25 psig (0.3–1.7 barg). A pressure gauge with a range from 0 to 30 psig (0 to 2 barg), as shown in Figure 13.50, would be a good choice for this application.

13.4.1.7 Operation outside Design Fuel Pressure Range

Operation outside the burner design fuel pressure range can cause the following to happen:

- Flame instability due to low fuel pressure
- Flame lift-off at higher fuel pressure
- Running the heater out of air at higher fuel pressure

13.4.1.7.1 Flame Quality/Instability at Low Fuel Pressure

Operating below the recommended minimum pressure can cause instability due to improper mixing of the fuel and air. When this happens, the mixture can locally become too lean or too rich for stable combustion to occur (see Figure 13.51). Too much air, in the combustion zone, can cool the reaction below the ignition temperature of the fuel. Low fuel pressure may not generate enough energy to assure proper mixing of the fuel with the air. This would result in a fuel-rich condition that may be above the upper flammability limit of the fuel. Both cases can cause instability in the flame or even extinguish the flame.

13.4.1.7.2 Flame Lift-Off at High Fuel Pressure

Operating a burner at too high of fuel pressure may result in flame "lift-off" (see Section 13.3.5) as seen in

FIGURE 13.50
Fuel pressure gauge.

FIGURE 13.51
Poor mixing due to low fuel pressure.

FIGURE 13.52
Flame lift-off.

Figure 13.52. This phenomenon is caused when the flame front moves away from the stabilization point. If the flame front does not return to the stabilization point, a portion of the flame may blow out. A flame that has lifted from its stabilization point will normally try to reattach to the gas tip. This movement or pulsing in the flame front causes "huffing" to occur. The huffing is actually pressure waves created by the energy of the oscillating flame front. If this movement becomes too severe, the flame can be extinguished.

13.4.1.8 Running the Heater Out of Air/Flame Impingement

Running the heater out of air can be the most dangerous of the three scenarios. Because complete combustion requires enough air to burn the fuel, the ultimate safe fired duty of the burner is normally limited by the amount of combustion air available. Burners are virtually unlimited in the amount of fuel that can be introduced into the firebox. Simply increasing the fuel pressure will result in more fuel to the burners.

The ability to pull in combustion air is a function of the heater draft, the forced-draft fan, and the burner tile. Since most heaters are natural draft, the available draft and burner tile are often the controlling criteria. Another area for caution is that since the combustion air can be reduced by closing the stack damper or burner air registers, it is possible to run out of air at heat releases below the maximum rate. This is why it is so critical to have an accurate oxygen analyzer located in the radiant section of the furnace. High fuel pressure alarms alone cannot be relied upon to ensure that a heater is running with a safe level of excess air!

If more fuel is introduced into the burner than there is air available to burn the fuel, oxygen deficiency will occur. Anytime that a fired heater is in an oxygen-deficient condition resulting in unburned fuel, the chance of an explosion is present. If something changes that would suddenly allow air into the firebox, an explosion could occur. Afterburning in the convection section, due to air leakage, is also a very real concern.

Flame impingement, as seen in Figure 13.53, on the process tubes can lead to external scaling of the pipe and internal coking. The external pipe scale leads to reduced heat transfer to the process, while the internal coking can result in a tube rupture. Table 13.2 shows some simple troubleshooting guidelines to determine what could be the cause of a particular problem with a gas burner and the solution required for that problem.

FIGURE 13.53
Flame impingement.

Burner Troubleshooting

TABLE 13.2

Troubleshooting Gas Burners

Trouble	Causes	Solutions
Burners go out	Gas/air mixture too lean or too much draft	Reduce total air. Reduce primary air
Flame flashback	Low gas pressure	Shut off burners to raise the fuel gas pressure to the operating burners; reducing the burner orifice size can also be helpful
	High hydrogen concentration in fuel gas	Reduce primary air; tape the primary air shut if flashback continues; a new burner or tip drilling may be required
Insufficient heat release	Low gas flow	Increase gas flow; increase burner tip orifice size; make sure that sufficient air will be available for the increased fuel rate
	Desired heat release exceeds design capacity	Larger burner tips or new burners may be required
Pulsating fire or breathing (flame alternately ignites and goes out)	Lack of oxygen/draft	Reduce firing rate immediately; establish complete combustion at lower rate; open stack damper and/or air registers to increase air and draft; reduce fuel before increasing air
Erratic flame	Lack of combustion air	Adjust air register and/or stack damper
	Incorrect position of burner tip	Locate tips per manufacturer's drawings
	Damaged burner block	Repair burner block to manufacturer's tolerances
Gas flame too long	Excessive firing	Reduce firing rates
	Too little primary air	Increase primary air; decrease secondary air
	Worn burner tip	Replace tip
	Tip drilling angle too narrow	Change to wide drilling angle tip
Gas flame too short	Too much primary air	Increase secondary air; decrease primary air
	Tip drilling angle too wide	Change to narrow drilling angle tip

13.5 Oil Firing Problems

Firing oil properly, as seen in Figure 13.54, can be very difficult or relatively easy depending on the condition of the oil and steam. Basically, good oil firing with clean flames comes down to three key items:

1. Has the burner/oil gun been designed for the type of oil to be fired? (See Figures 13.55 and 13.56.)
2. The oil must be at the right pressure and temperature at the burner.
3. The steam must be at the correct pressure and should be free of condensate.

13.5.1 Effect on Operations

The impact on operation, without the proper design and operating conditions, can result in serious and dangerous situations. Improper oil firing can result in many hazardous conditions including

1. Oil spillage and dripping from the burners to grade (see Figure 13.57)
2. Incomplete combustion resulting in CO and UHCs in the flue gas
3. Smoke coming out of the stack
4. Fouled and plugged oil guns (see Figure 13.58)
5. Unstable flames with possible lift-off or going out

FIGURE 13.54
DEEPstar oil burner.

FIGURE 13.55
Combination burner with Regen (oil) tile.

FIGURE 13.56
Oil gun in center of Regen (oil) tile (burner partially assembled).

FIGURE 13.57
Oil spillage on burners.

FIGURE 13.58
Fouled oil guns.

 6. Flame impingement on the tubes (see Figure 13.59)
 7. Coke buildup on the oil guns and burner tile (see Figure 13.60)

13.5.2 Oil Combustion

13.5.2.1 How Does Oil Burn?

A liquid fuel does not readily burn like a gas fuel. To burn a liquid requires that the liquid be changed into a vapor. This is accomplished by what is known as

Burner Troubleshooting

FIGURE 13.59
Flame impingement.

FIGURE 13.61
Typical oil gun components.

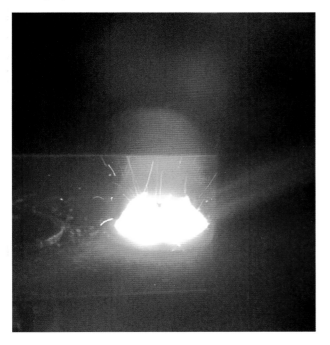

FIGURE 13.60
Coke mound on burner.

"atomization." When the proper level of atomization exists, the oil will be converted to a fine mist, which increases the liquid surface area by several orders of magnitude. This large surface area allows for a fast and efficient vaporization of the liquid that can then be oxidized. This is done through specially designed components in the oil guns such as those seen in Figure 13.61.

13.5.2.2 Viscosity and Temperature

The single most important item when firing heavy oil is the temperature of the oil at the burner. The temperature controls the viscosity of the oil. Having the right viscosity is required for proper atomization. The viscosity of a liquid is an indicator of the fluid's resistance to internal shear. A higher viscosity makes it more difficult to atomize the oil properly. It is recommended that the viscosity of the oil be no more than 200 SSU (43 cSt) at the burner.

13.5.2.3 Steam Atomization

Steam is the best method to atomize heavy oil. Light oils such as No. 2 diesel and naphtha can be atomized with air in some cases. For the majority of oils, steam is the best alternative for clean oil fires and proper atomization. The condition of the steam is very important to good atomization. This means that the steam must be dry (i.e., no condensate) and delivered to the burner at a higher temperature than the oil. Normally, the standard 150 psig (10 barg) steam, available in most plants, will work for oil firing applications. In some cases, such as a heavy pitch, the use of 400 psig (28 barg) steam may be required to ensure the steam is hotter than the oil temperature. Oil is generally labeled as "pitch" if the API gravity is less than 10.

13.5.2.4 Contaminants in the Oil

Heavy oil contains components that can be harmful to the burner and the environment. The most common of these problem contaminants are the following:

- Chemically bound nitrogen
- Sulfur
- Vanadium
- Sodium

FIGURE 13.62
Vanadium attack on burner tile.

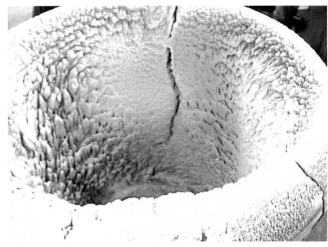

FIGURE 13.63
Catalyst buildup on oil tile.

- Nickel
- Catalyst fines from a fluidized catalytic cracking unit (FFC)
- Ash and sediment

Elemental nitrogen that is chemically bound in a component of the oil will increase the amount of NOx during the combustion reaction (see Volume 1, Chapter 15). This is due to the formation of "fuel NOx" which can exceed the amount of "thermal NOx."

Sulfur will oxidize to sulfur dioxide (SO_2) and sulfur trioxide (SO_3) which increase pollution and corrosion. Sulfur dioxide is the principal component of acid rain. The sulfur trioxide can form sulfuric acid (H_2SO_4) under the right conditions. Sulfuric acid attacks the carbon steel stack and casing of the heater. Sulfur in the oil will also attack the metal parts of the oil gun causing corrosion.

Vanadium (V) forms vanadium pentoxide (V_2O_5) during combustion. This compound can attack refractory, as that seen in Figure 13.62, resulting in failure. The compound has a characteristic that leaches the binder from the refractory causing it to fall apart. Vanadium oxide also helps convert SO_2 to SO_3, promoting the increase of H_2SO_4.

Sodium and nickel are heavy metals that form heavy particulate oxides. These will show up in stack testing as PM_{10} particulate matter.

The catalyst fines from a FCC unit will destroy oil guns and atomizers due to the abrasive nature of the fines. Since many catalysts are platinum/nickel base, they are extremely erosive at high velocities. The fines can also collect on the oil tile as shown in Figure 13.63.

Ash and sediment form particulate matter and contribute to atomizer and tip plugging.

13.5.3 Oil System

Unlike a typical fuel gas system, the oil system requires more attention and maintenance. A typical heavy oil system operates at 200°F–300°F (90°C–150°C) with 150–200 psig (10–14 barg) internal pressure. The heat tracing and insulation must be inspected and kept in working order. The pumps, strainers, heat exchangers, flowmeters, and pressure gauges are all subject to the effects of hot oil. A schematic for a heavy oil system can be seen in Figure 13.64.

13.5.3.1 Heating and Storage

Fuel oils that are heavier than diesel normally need to be heated for proper combustion. The oil must be heated to achieve the proper viscosity at the burners. The most common method of heating the oil is to use a steam heat exchanger. Electric units can be used at a higher operating cost. The outlet temperature has to be high enough to offset any losses in the system so that the burners receive oil at the correct temperature for atomization. A problem could occur if the oil temperature gets too high causing the oil to polymerize or flash.

The storage tank should be located inside a retaining wall to prevent the release of oil in case there is a leak in the tank or piping. The tank may require a heating element and an agitator to mix the oil so that it stays at the correct temperature for pumping. This temperature is normally 80°F–150°F (27°C–66°C).

13.5.3.2 Recirculation System

The pump used to pull the oil from the tank should be a positive displacement type unit. The pump should be sized to provide a flow rate that is 150%–200% of the

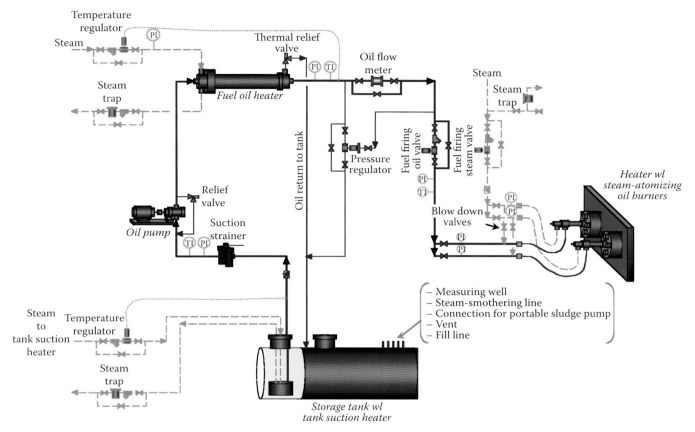

FIGURE 13.64
Heavy oil system.

required burner capacity. This will allow for a good recirculation flow that will help to keep the temperature uniform throughout the system. The oil delivery system should have a Coriolis-type flowmeter (see Chapter 7) to measure the oil flow to the burners. The meter must be downstream of the recirculation takeoff.

13.5.3.3 Heat Tracing and Insulation

The insulation used on the oil piping should be well maintained and covered with a stainless jacket as shown in Figure 13.65. The jacket will help to minimize heat losses which can be as high as 150 Btu/ft² h (475 W/m²) with exposed insulation. Some climates may require that the oil lines be heat traced under the insulation. A common problem is to leave the last 5–6 ft (1.5–1.8 m) of the oil piping uninsulated. This is a mistake that may allow the oil to cool to a point where it is difficult to atomize.

13.5.3.4 Pressure Gauges

A calibrated pressure gauge should be located as close to the burner as possible. The gauge should be liquid filled and have a range that covers the maximum and minimum operating pressure of the burner. For example, a burner that uses an oil pressure of 40–100 psig (2.7–6.8 barg) should have a gauge with a range from 0 to 150 psig (0–10 barg). The gauge should have the standard shutoff valve so that it can be blocked in and removed for service with the burner in operation.

13.5.4 Steam System

Two of the most common problems, when using steam atomization, are the lack of insulation on the steam piping and faulty steam traps that allow condensate into the burner. The steam piping should be fully insulated all the way to the oil gun on the burner. Leaving 3–4 ft (0.9–1.2 m) of pipe uninsulated can cool the steam enough to produce condensate. The piping should be laid out to avoid any unnecessary low points. All low points should be provided with steam traps to expel condensate.

13.5.4.1 Insulation

The steam insulation is important to prevent cooling and condensation in the steam piping. As noted earlier,

FIGURE 13.65
Insulated fuel skid.

the atomizing steam should be slightly superheated to be sure that it is moisture-free. Missing or deteriorated insulation will allow the steam to cool. Wet steam does not atomize with the same efficiency and satisfactory results as dry steam. The insulation should be run all the way to the connection at the burner to avoid any cooling effect.

13.5.4.2 Checking Steam Traps

When a steam trap is working properly, there will be a substantial difference between the temperature of the inlet and outlet piping to the trap. A difference of 50°F–70°F (10°C–21°C) usually indicates normal operation. A downstream temperature within a few degrees of the upstream temperature indicates that the trap may not be working as it should. The downstream temperature should always be lower than the inlet temperature.

13.5.4.3 Superheated Steam

Another "best operation" recommendation is to provide steam with a slight amount of superheat such as 40°F–50°F (4°C–10°C). This will keep the steam dry and free of condensate.

13.5.4.4 Pressure Indication and Control

The steam has to be delivered to the burner at the right pressure for the oil gun to work as designed. Most of John Zink's oil guns are designed to operate with a constant differential pressure of 20–30 psig (1.4–2 barg). This means the steam pressure will be 20–30 psig (1.4–2 barg) higher than the oil pressure. In some cases, a higher differential may be required for proper atomization. Steam consumption varies with the type of oil gun. Normal rates are between 0.15 and 0.3 lb steam/lb oil (0.15 and 0.3 kg steam/kg oil).

The steam pressure should be controlled with a differential pressure regulator. The sensing line should be tied into the oil line so that the steam pressure maintains a constant differential pressure.

John Zink provides a very detailed "Troubleshooting" guide in the "Installation and Operating Manual" provided with all of its oil-fired burners. The operators and maintenance people can use this to resolve many of the common problems associated with oil firing. A summary is shown in Table 13.3.

13.5.5 Smoke Emission from the Stack

13.5.5.1 Indication of the Problem

Smoke appears at the top of the stack.

13.5.5.2 Effect on Operation and Equipment

Smoke indicates either incomplete combustion or a process tube rupture if the process fluid is a hydrocarbon.

TABLE 13.3

Troubleshooting for Oil Burners

Trouble	Causes	Solutions
Burners dripping; coke deposits on burner blocks; coking of burner tip when firing fuel oil only	Improper atomization due to high oil viscosity, clogging of burner tip, insufficient atomizing steam, improper location of burner tip	Check fuel oil type; increase fuel temperature to lower viscosity to proper level; check composition for heavier fractions; clean or replace burner tip; confirm burner tip is in proper location; increase atomizing steam; place tip in location as per burner drawings
Failure to maintain ignition	Too much atomizing steam	Reduce atomizing steam until ignition is stabilized; during start-up, have atomizing steam on low side until ignition is well established
	Too much primary air at firing rates	Reduce primary air
	Too much moisture in atomizing steam	Ensure appropriate insulation is on steam lines; confirm steam traps are functioning; adjust quality of atomizing steam to appropriate levels.
Coking of oil tip when firing oil in combination	High rate of gas with a low rate of oil; high heat radiation to the fuel oil tip	Increase atomizing steam to produce sufficient cooling effect to avoid coking; reduce gas firing rate; dedicate individual burners to either fuel
	Incorrect oil gun position	Place tip in locations as per burner drawings. If this fails, readjust burner tip ±0.5 in. until coking ceases
Erratic flame	Lack of combustion air	Adjust air damper or register
	Plugged burner gun	Clean burner gun
	Worn burner gun	Replace burner gun
	High rate of gas firing while firing a low rate of oil	Reduce gas rate; dedicate burners to either fuel
	Damaged burner block	Repair burner block
Excess smoke at stack	Insufficient atomizing steam	Increase atomizing steam
	Moisture in atomizing steam	Requires knockout drum or increase in superheat, check steam line insulation
	Low excess air	Increase excess air

As a regulated emission, continued smoking could lead to sanctions, fines, and termination of operations.

13.5.5.3 Corrective Action

If smoke appears while burning oil, suspect poor atomization. Check the fuel oil temperature, the atomizing medium conditions, oil and atomizing medium pressures, and the condition of the oil gun. Consider a switch to a gaseous fuel until the problem is resolved.

Check the excess oxygen in the firebox to identify any deficiency. Determine if there has been any change in fuel composition. When operating at low excess air conditions, a change in fuel gas composition—adding some higher molecular weight gases—can initiate smoking because of a lack of increase in air flow.

A process tube leak (see Figure 13.36), spilling hydrocarbon contents into the firebox, will create an incomplete combustion condition and smoke at the stack. This situation requires a heater shutdown and a steam purge into the firebox to cool any conflagration and to reduce the amount of air and the intensity of combustion.

13.6 Emissions

13.6.1 Nitrogen Oxides

The formation of NOx during the combustion reaction is unavoidable (see Volume 1, Chapter 15). This is due to the fact that atmospheric air contains nitrogen and oxygen, the two components that form NOx. The two main species of NOx are nitrogen monoxide (also called nitric oxide, NO) and nitrogen dioxide (NO_2). Inside the fired heater, 90%–95% of the total NOx is usually nitric oxide with the remainder being nitrogen dioxide. The problem is that once the nitric oxide exits the stack, it will convert to nitrogen dioxide in the atmosphere. Nitrogen dioxide is a toxic gas that contributes to acid rain, smog, and respiratory problems in people.

The amount of NOx exiting the heater stack is a function of several items:

- Burner type
- Firebox temperature
- Fuel composition

- Excess air
- Combustion air temperature
- Relative humidity

13.6.1.1 Burner Type

Older process heaters may have conventional type premix or raw gas burners such as the one shown in Figure 13.66. These older style burners do not offer any type of NOx reduction mechanism. The NOx levels from these burners will normally range from 100 to 300 ppm depending on the fuel and the operating conditions.

Early versions of low NOx burners used staged air to reduce NOx. Staging the air helps to limit NOx in the primary combustion zone. This sub-stoichiometric combustion results in the formation of hydrogen and carbon monoxide. These two gases then combine with oxygen in the outer staged air zone to limit the oxygen available for NO formation.

The staged air is normally an area around the outside of the burner tile which can be seen in Figure 13.67. This zone is often called the "tertiary" air since many early low NOx burners had three air zones: primary, secondary, and tertiary. The maximum NOx reduction is normally achieved with the tertiary air full open. The primary and secondary registers are used to control the excess oxygen in the combustion reaction.

Staged fuel burners, as seen in Figure 13.68, were the next generation of low NOx burners. These burners staged the fuel gas instead of the combustion air. This type of staging resulted in a very lean primary combustion zone and a fuel-rich secondary or "staged" zone inside the heater.

The staged fuel-type burners offered three main advantages over staged air burners:

- Lower NOx emissions
- Better flame patterns
- A single air control device

The latest versions of low NOx burners are designated as "ultralow NOx" burners or simply "ULNB" burners. These burners use the principle of staged fuel combustion in combination with internal flue gas recirculation. This type of burner can be seen in Figure 13.69.

The energy of the fuel exiting the ports on the gas tips is used to entrain flue gas into the fuel gas. This mixing of flue gas and fuel gas lowers the peak combustion temperature. This is due to the inert content of the flue gas. Flue gas is basically made up of nitrogen, carbon dioxide, and water vapor. These three components help to suppress the formation of NOx by reducing the flame temperature at strategic zones in the burner.

More information on NOx reduction techniques and various process burner types can be found in Volume 3, Chapter 1.

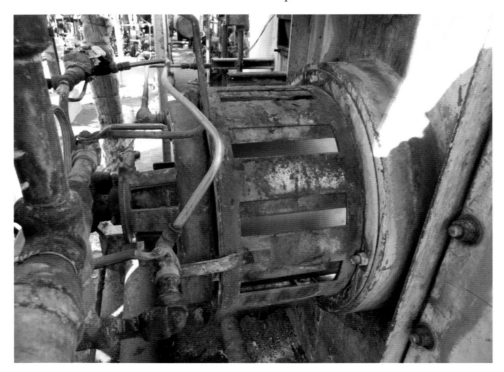

FIGURE 13.66
Conventional DBA-style burner.

FIGURE 13.67
Staged air low NOx burner.

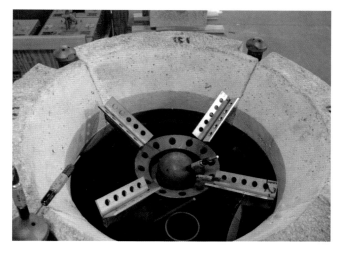

FIGURE 13.68
Staged fuel low NOx burner.

FIGURE 13.69
COOLstar™ ultralow NOx burner.

13.6.1.2 Firebox Temperature

There are three basic types of NOx formed during combustion:

- Thermal NOx
- Fuel NOx
- Prompt NOx

Thermal NOx is the predominant mechanism in gas-fired burners. This means that the hotter the firebox, the more NOx will be generated. A linear NOx formation interpolation formula which is valid at higher firebox temperatures looks like this:

$$NOx_2 = NOx_1 \left(\frac{T_2 - 400}{T_1 - 400} \right) \tag{13.6}$$

FIGURE 13.70
Air leakage around convection tube.

For example, a firebox at 1600°F will generate 1.2 times more NOx than a firebox at 1400°F.

> **NOTE**
>
> Since NOx formation is not linearly dependent on temperature, this linear interpolation is only valid within a limited temperature range.

13.6.1.3 Fuel Composition

Changes in the fuel can have a great impact on the NOx formation. Adding hydrogen or heavy hydrocarbons will increase the flame temperature. This will increase the amount of Thermal NOx. Chemically bound nitrogen contained in ammonia or amine compounds can produce high levels of fuel NOx. Fuels that contain molecular nitrogen (N_2) or carbon dioxide (CO_2) will lower the flame temperature resulting in lower NOx levels.

13.6.1.4 Excess Air

In raw gas or diffusion burners (in other words, when fuel and air enter the burner separate from each other), the lower the excess air, the lower the NOx levels. In contrast, the NOx emission from 100% premix burners decreases with increased excess air.

Low excess air not only reduces NOx, it is also good for improving the thermal efficiency (see Volume 1, Chapter 12) of the heater. The issue will be sealing the heater so that the air is coming through the burners and not through cracks, openings, or around the process tubes. An example of air leaking around the process tubes can be seen in Figure 13.70. Using flexible tube seals as a solution to this problem is shown in Figure 13.71.

For most process burners and heaters, target excess air levels should be 10%–15% for optimization of NOx and efficiency. This amount of excess air would be equivalent to 2%–3% excess oxygen (vol, dry) in the flue gas.

FIGURE 13.71
Flexible tube seals. (Courtesy of Thorpe Corp, Houston, TX.)

13.6.1.5 Combustion Air Temperature

Adding an air preheater to a unit will improve the efficiency. However, the thermal NOx will also increase as the air temperature increases. As a general rule, combustion air at 600°F (320°C) will roughly double the NOx formation compared to ambient air.

13.6.1.6 Relative Humidity

Dry climates will generate more NOx than humid areas. The water vapor in the air lowers the adiabatic flame temperature of the combustion zone.

13.6.2 Sulfur Oxides

Sulfur dioxide (SO_2) and sulfur trioxide (SO_3) will be formed during combustion if sulfur is present in the fuel. The normal source of sulfur in the fuel is hydrogen sulfide (H_2S). Some off gases contain sulfur dioxide in their composition. Heavy oil may contain elemental sulfur or other sulfur compounds.

Acid rain is formed by the reaction of SO_2 and SO_3 with moisture in the atmosphere. Acid rain destroys vegetation and changes the pH level of lakes and streams making them uninhabitable to plants and wildlife.

During combustion, approximately 95% of the sulfur will oxidize to SO_2. The remaining 5% will exit the stack as SO_3 unless it combines with water to form sulfuric acid (H_2SO_4). This is why it is important to be sure that the flue gas temperature is always above the dew point. Sulfuric acid can also condense in the impingers during particulate testing which will increase the amount of total particulate matter. As the amount of SO_3 increases, there is also an increase in the dew point temperature of the flue gas.

The only way to control SO_2 and SO_3 formation is to reduce or eliminate it from the fuel source.

13.6.3 Carbon Monoxide

Carbon monoxide is a deadly gas that is odorless and colorless. Carbon monoxide in the flue gas is a product of incomplete combustion (see Volume 1, Chapter 4). This could be a result of poor mixing, of the fuel and air, or simply not enough oxygen to burn all of the fuel. Firebox temperatures below 1200°F (650°C) can also contribute to CO formation especially with ultralow NOx-type burners.

One of the most common causes of high CO levels is a false excess O_2 reading in the flue gas. The main issue is leakage or "tramp" air that does not come through the burners. The tramp air shows up as excess oxygen causing the burners to operate at lower than design air

FIGURE 13.72
Open inspection port.

levels. One possible cause of tramp air is leaving an inspection port open as shown in Figure 13.72.

Plugged or fouled gas and oil tips can also contribute to CO formation during the combustion reaction. Check the tips to be sure they are clean, have the proper size ports, and are orientated as shown on the burner drawing.

13.6.4 Combustibles/Volatile Organic Compounds/Unburned Hydrocarbons

Combustibles and VOCs are two of the most misunderstood terms in the industry. Combustibles can be any unburned fuel such as hydrogen, methane, propane, or carbon monoxide. Flue gas analyzers that measure "combustibles" generally only measure hydrogen and carbon monoxide.

Volatile organic compounds are actually hexanes and heavier organic compounds with high vapor pressures due to low boiling points between 50°C and 250°C (120°F and 480°F). At standard atmospheric conditions, the majority of VOCs are in either the liquid or solid state. The combustion reaction in normal process heaters is not a significant contributor to atmospheric VOC levels.

If the local air quality district requires sampling for unburned hydrocarbons (UHCs), a special analyzer will usually be required.

13.6.5 Particulate Matter

This is another potentially confusing emission definition. The air quality district is concerned about the total particulate coming out of the stack. The burner manufacturer can only guarantee the particulate emmisions from the combustion in the burners. Generally, this would be in the form of solid carbon matter generated from incomplete combustion. The total particulate will include insulation fibers, rust particles, pipe scale, refractory dust, and contaminants in the air.

Process heaters firing a gas fuel normally emit very low amounts of particulate matter into the atmosphere. A normal amount would be less than 0.003 lb/MMBtu based on HHV of the fuel. Heavy oils and fuels with high amounts of sulfur will produce higher amounts of particulates.

13.7 Combustion Air Issues

13.7.1 How Ambient Weather Conditions Can Affect Burner Performance

The two main concerns with ambient weather conditions are the temperature and humidity. A burner is somewhat like a fan in that it "pulls" in a volumetric flow rate of air for the combustion process. As the ambient temperature and humidity change so does the oxygen content in each cubic foot of air. If the oxygen content becomes too low, then the overall burner performance may be less than acceptable. Ambient weather conditions change during normal operation. These effects are considered next.

13.7.1.1 Effect on Operations

As ambient conditions change, the amount of excess oxygen in the flue gas will decrease or increase, as shown in Figures 13.73 and 13.74. This change in oxygen can lead to flame impingement on the tubes, instability in the flames, and higher or lower emissions levels and also affect the heater efficiency.

Heater draft is a function of the temperature inside the firebox and the ambient temperature outside the firebox. If the ambient temperature changes, then the draft inside the heater will change.

The barometric pressure will also change due to the temperature and humidity. This, of course, impacts the density of the air and the oxygen content in the air stream.

Atmospheric air at standard conditions of 60°F (16°C), 0% humidity, and 14.7 psia (1 bara) contains approximately 21% oxygen on a volume basis. If the air coming to the burners is 80°F (27°C) with 80% humidity, then the oxygen content is approximately 20%. This change would result in the excess oxygen, in the flue gas, dropping to 2.2% instead of 3%. This translates roughly to running with 10% excess air instead of 15% excess air.

Humidity will also have an impact on the NOx levels coming out of the heater. During cooler weather with dry air, the NOx may increase noticeably because of less water in the air which cools the flame.

Rain and low air temperatures can cool the stack which may reduce the amount of draft.

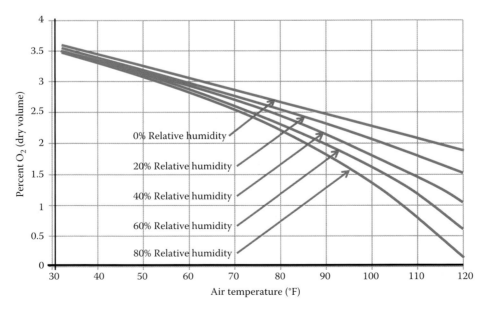

FIGURE 13.73
Oxygen versus relative humidity%.

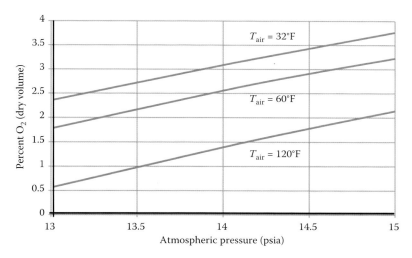

FIGURE 13.74
Oxygen versus atmospheric pressure.

The jobsite elevation also has a major impact on the amount of oxygen in the combustion reaction. A heater designed to operate in Houston, Texas, would require 20% more draft in Cody, Wyoming, to achieve the same fired duty and excess air level. The atmospheric pressure in Cody is 12.2 psia (0.83 bara) compared to 14.7 psia (1 bara) in Houston.

13.7.1.2 Corrective/Preventive Actions

The main corrective action is to be sure that you have an accurate oxygen analyzer and draft sampling system are being used. The oxygen probe and draft tube should be located at the top of the radiant section as shown in Figure 13.75. This is the control point for both excess oxygen and heater draft.

Plan ahead if possible for significant changes in the weather. During the summer months, the heater may need additional air and more draft to achieve the same process duty.

Adjust the stack damper to provide more draft, and use the burner air registers to increase the excess oxygen before the ambient conditions change.

If a heater is normally operated with 2% excess oxygen and −0.1 in. (2.5 mm) of draft at the arch, then increase these to 3% excess oxygen and −0.2 in. (5 mm) of draft.

The heater can be adjusted for improved efficiency after the ambient conditions have stabilized.

13.7.2 Burner Sizing

A process burner is "sized" based on the amount of air it will flow at a given draft loss. The required air flow is a function of the fired duty and excess air required by the process. For example, say the specification calls for 7.5×10^6 Btu/h (2.2 MW) with 15% excess air. There are various sizes of burners that can be designed to achieve the desired duty on the fuel side. When the available draft is added to the equation, then the selection becomes limited. The fact is that there is normally only one burner size, that will have the air capacity at the given draft to provide optimal performance.

For this example, let's outline the burner requirements for capacity are as follows:

Maximum fired duty: 7.5×10^6 Btu/h

Fuel: natural gas

Excess air: 15%

Draft available: 0.35 in. of water column

The burner designer selects a burner with a large enough air throat to meet the specification. Now, what if the heater designer and end user decide to add some additional capacity for future cases? The new design is now 9.75 MMBtu/h with 15% excess air at 0.35 in. of draft. The burner designer has to select a larger burner to handle the increase in air flow.

This larger burner will now require less draft at the original design conditions:

$$\Delta P_2 = \left(\frac{Q_1}{Q_2}\right)^2 \Delta P_1 = \left(\frac{7.5}{9.75}\right)^2 0.35 \text{ in.} = 0.21 \text{ in.}$$

This means that the burner air damper will have to be closed to achieve the 15% excess air at the lower duty. Taking a large pressure loss at the burner damper, instead of the tile throat, could impact the flame quality.

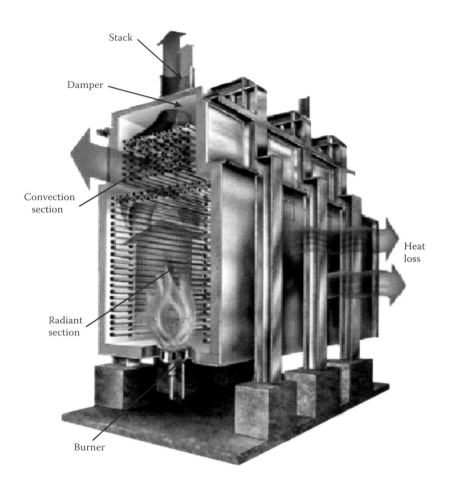

FIGURE 13.75
Typical fired heater.

The point of selecting the proper burner size is to avoid operating at less than optimum conditions. In the earlier example, what if the "future" case never happens? Then the burners and heater may be operated for years with continuing problems associated with poor mixing and controlling excess air because the burner was over-sized.

13.7.3 Forced-Draft System

The addition of a fan and air preheater can improve the efficiency and operation of a fired heater. They can also add many problems if the system is not designed correctly.

13.7.3.1 Selecting the Correct Size Burner

Selecting the correct size burner is often easier than it sounds. The difficulty is deciding how to handle the natural-draft firing mode if the forced-draft fan goes down.

A typical design specification will include the fired duty for both natural-draft and forced-draft operation. In almost all cases, the natural-draft fired duty is higher than the forced-draft mode. The reason for the higher duty is because of the loss of the sensible heat contained in the preheated air under forced-draft operation. This is not the correct way to design the system as the following example will show:

The burner specification for forced-draft operation is

Maximum fired duty: 7.5×10^6 Btu/h

Fuel: natural gas

Excess air: 10%

Pressure drop available: 1.0 in. of water column

Combustion air temperature: 500°F

The burner specification for natural-draft operation is

Maximum fired duty: 8.6×10^6 Btu/h

Fuel: natural gas

Excess air: 15%

Draft available: 0.4 in. of water column

Combustion air temperature: 60°F

The burner designer has to select a burner for the natural-draft case. Based on this, the pressure drop under the forced-draft conditions can be calculated:

$$\Delta P_2 = \Delta P_1 \left(\frac{Q_2}{Q_1}\right)^2 \left(\frac{XA_2}{XA_1}\right)^2 \left(\frac{460+T_2}{460+T_1}\right)$$

$$\Delta P_2 = 0.40 \left(\frac{7.5}{8.6}\right)^2 \left(\frac{1.1}{1.15}\right)^2 \left(\frac{460+500}{460+60}\right)$$

$$= 0.51 \text{ in. of water column}$$

The burner pressure drop on forced draft is 0.51 in. of water. The draft in the heater is 0.4 in. of water which means the positive pressure in the duct system is only 0.11 in. of water column! This is not enough pressure to promote uniform air distribution in the system! The correct method would be to size the burner for the forced-draft case and accept the reduced firing under natural-draft operation.

13.7.3.2 Designing for Uniform Air Distribution

Nothing is more important than being sure that a proposed forced-draft system will provide uniform air flow to all the burners in the system. If the air flow is not right, then there could be many burner-related issues including

- High NOx levels
- Leaning or long flames
- Flame impingement on the tubes
- Flame instability

Trying to fix an air flow problem after the heater is online is time consuming and in many cases simply not possible.

An established method to ensure uniform air flow is the constant velocity approach. This requires that the duct sizing be continuously reduced in cross-sectional flow area. This reduction in area will maintain the constant velocity at lower air flows due to a reduced number of burners that need air. For ambient air systems, the velocity should be no more than 15 ft/s (4.5 m/s) and for preheat systems no more than 30 ft/s (9 m/s).

Modern technology offers a more detailed analysis of the system by running a CFD model (Volume 1, Chapter 13). These computer-generated profiles are based on the actual system using the detailed layout drawings of the duct work. An example model is shown in Figure 13.76.

Based on the velocities in the system, the model shows nonuniform air flow patterns in the system.

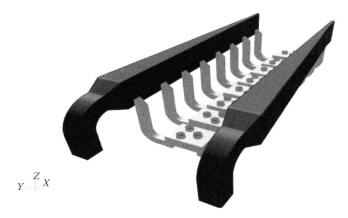

FIGURE 13.76
Typical CFD model of a duct system.

13.8 Summary

In summary, troubleshooting burners in a furnace involves (1) observing the problem, (2) identifying the problem, (3) determining the effect on the operation of the furnace, and (4) determining the solution and the corrective action that should be taken to correct the problem. If there is a problem that cannot be identified and resolved, one should consult with the burner manufacturer to obtain advice on how to proceed.

References

1. N.P. Lieberman, *Troubleshooting Process Operations*, 3rd edn., Penn Well Publishing, Tulsa, OK, 1991.
2. *Burners for Fired Heaters in General Refinery Services*, 1st edn., API Publication 535, Washington, DC, July 1995.
3. R.A. Meyers, *Handbook of Petroleum Refining Processes*, 2nd edn., McGraw-Hill, New York, 1997.
4. W. Bartok and A.F. Sarofim, Eds., *Fossil Fuel Combustion: A Source Book*, John Wiley & Sons, New York, 1991.
5. John Zink Co. LLC, John Zink Burner school notes, Tulsa, OK, September 16–18, 1998.
6. E.A. Barrington, Fired process heaters, Course Notes, 1999.
7. R.D. Reed, *Furnace Operations*, 3rd edn., Gulf Publishing, Houston, TX, 1981.
8. R.D. Reed, A new approach to design for radiant heat transfer in process work, *Petroleum Engineer*, August, C7–C10, 1950.
9. *Fired Heaters for General Refinery Services*, 2nd edn., API Standard 560, September, 1995.
10. T.M. Korb, I.P. Chung, and X. Chen, CO emissions challenges for process heaters operating with ultra low NOx burners, *AFRC Meeting*, Maui, HI, 2010.

14

Flare Operations, Maintenance, and Troubleshooting

Robert E. Schwartz and Zachary L. Kodesh

CONTENTS

- 14.1 Flare Operations 378
  - 14.1.1 Safety in Flare Operations 378
    - 14.1.1.1 Purging 378
    - 14.1.1.2 Suitable Purge Gases 378
    - 14.1.1.3 Admission Point for the Purge Gas 379
    - 14.1.1.4 Alarm for Purge Failure 379
    - 14.1.1.5 Minimum Purge Gas Rate 379
    - 14.1.1.6 Staged Post-Purge Concept 379
  - 14.1.2 Pre-Start-Up 379
    - 14.1.2.1 Pre-Start-Up Checks for All Flares 379
    - 14.1.2.2 Pre-Start-Up Checks for Steam-Assisted Flares 380
    - 14.1.2.3 Pre-Start-Up Checks for Air-Assisted Flares 380
    - 14.1.2.4 Pre-Start-Up Checks for Staged Flare Systems 381
    - 14.1.2.5 Pre-Start-Up Checks for Knockout Drums and Liquid Seals 383
    - 14.1.2.6 Pre-Start-Up Checks for Molecular Seal 384
  - 14.1.3 Start-Up and Shutdown 384
    - 14.1.3.1 Unassisted Flares 384
    - 14.1.3.2 Steam-Assisted Flares 384
    - 14.1.3.3 Air-Assisted Flares 385
    - 14.1.3.4 Staged Flare Systems 385
  - 14.1.4 Pilot Lighting Procedure 386
    - 14.1.4.1 Flame Front Generator (FFG) 386
    - 14.1.4.2 Visual Pilot Verification with FFG 387
    - 14.1.4.3 Other Pilot Ignition Systems 387
- 14.2 Maintenance 387
  - 14.2.1 Purge System 388
    - 14.2.1.1 Shutdown Items 388
    - 14.2.1.2 Periodic On-Line Items 388
  - 14.2.2 Pilots 388
    - 14.2.2.1 Shutdown Items: 388
    - 14.2.2.2 Periodic On-Line Items 388
  - 14.2.3 Flame Front Generator 389
    - 14.2.3.1 Periodic On-Line Items 389
  - 14.2.4 Flare Tips 389
    - 14.2.4.1 Shutdown Items for All Flare Tip Types 389
    - 14.2.4.2 Shutdown Items for Steam-Assisted Flare Tips 389
    - 14.2.4.3 Shutdown Items for Air-Assisted Flare Tips 389
    - 14.2.4.4 Shutdown Items for Staged Flare Systems 389
    - 14.2.4.5 Periodic On-Line Items for Air-Assisted Flare Tips 389

14.2.5 Liquid Seals and Knockout Drums...390
    14.2.5.1 Shutdown Items..390
    14.2.5.2 Periodic On-Line Items for Knockout Drums..390
    14.2.5.3 Periodic On-Line Items for Liquid Seals...390
14.2.6 Molecular Seals 390
    14.2.6.1 Shutdown Items..390
    14.2.6.2 Periodic On-Line Items...390
14.3 Flare Troubleshooting 390

14.1 Flare Operations

> **WARNING**
>
> The information presented in this section is not intended to replace the manufacturer's or owner's operation instructions. All safety precautions and operation instructions developed for specific equipment shall be followed.

This section contains a generic discussion of flare start-up, operation, and shutdown. Any action taken or withheld shall follow the flare owner's operating instructions. The flare is a part of a pressure relief or vent system. Such systems can become large, extending back into the process area. The flare owner/operator shall be responsible for integrating the flare start-up, operation, shutdown, and operator training into the overall plant operations. This section does not directly address operator training.

14.1.1 Safety in Flare Operations

14.1.1.1 Purging

Flares can pose one of the most serious safety hazards to a facility if improperly operated. The potential exists for an explosion or detonation to occur in a flare system.

Three elements are required before such an explosion or detonation will occur. There must be fuel, a source of ignition, and oxygen. Two of these elements, fuel (waste gas or a combustible purge gas) and a source of ignition (pilots), are always present in a flare system. The safety of the system is highly dependent upon the prevention of air infiltration into the flare. An effective method for the prevention of air infiltration is the introduction of a purge gas. As energy costs have risen, more attention has been given to the cost of purge gas. At the same time, efforts have increased to reduce or eliminate relief valve leakage. The net result of these efforts has been a significant reduction in the normal flow to the flare, which in turn has increased the importance of maintaining the proper purge gas flow.

If air (or more specifically, oxygen in air) is present in a flare system, there is a danger of an explosion when the flare pilot(s) are ignited. To prevent this, the flare system is purged from the beginning of the flare gas collection system all the way to the flare tip. An initial purge volume of noncondensable gas equal to 10 or more times the volume of the flare system and collection system should be used to ensure low or zero oxygen levels unless an oxygen analyzer is used to measure oxygen levels at the flare stack. The flare system volume includes all the piping and vessels from the relief valves/pressure control valves to the flare stack and up to the exit of the flare tip. The initial purge volume should be injected at a rate substantially higher than the normal continuous purge rate for the system to ensure complete removal of air pockets.

In the case of a "hot gas" relief (above 200°F = 93°C), the gas remaining in the system after the relief can cool rapidly. As the gas cools, it contracts and can draw air into the flare system through piping leaks or through the flare tip. To protect the system, it is necessary to provide a flow rate of purge gas equal to at least twice the anticipated shrinkage rate.

14.1.1.2 Suitable Purge Gases

A suitable purge gas for the flare system is any gas or mixture of gases, which (a) cannot reach a dew point at any condition of ambient temperature normal for the location, (b) which cannot self-detonate, and (c) which does not contain free oxygen. If the average molecular weight of the flared gases exceeds 30, there should be a calculation of the dew point potential for the flared gases. The volume of the flared gas that can go to dew point should be added to the volume required for a calculated flare tip velocity.

Since a dew point can occur readily during the cold season, it is suggested that the extra purge gas volume be on temperature control for atmospheric temperature and also for admission only when a low temperature can cause a dew point situation to exist. This will save purge gas volume, which can be quite expensive.

Suitable purge gases include natural gas, propane, nitrogen, or carbon dioxide. Steam, as a purge gas, is not recommended for two reasons. The first is that the steam is initially at an elevated temperature and the volume occupied by the steam will shrink as the steam cools and condenses. This will allow air to be drawn back into the flare system. The second is that as the steam condenses, water will be left in the flare system to partially block the system, create a freezing hazard at low temperatures, and, by its "wetting" action, encourage accelerated corrosion (see Chapter 4).

14.1.1.3 Admission Point for the Purge Gas

In all cases, the purge gas must enter the flare system immediately downstream of the first relief valve/pressure control valve so that the purge gas will "sweep" the entire system. If there is more than one header feeding the flare, each header must be purged. That is, there must be an entry of gas to each header that enters the system. Minimum purge gas flow should be distributed proportionally among all flare headers.

The pilots (see Volume 3, Chapter 12) should be ignited only after the system has been purged as recommended and while the purge gas is flowing. If the purge gas is combustible, the burning of the purge gas at the flare will be an indication of pilot ignition.

14.1.1.4 Alarm for Purge Failure

It is recommended that a flow switch(es) or flow transmitter(s) be utilized to verify adequate purge flow into the flare gas collection system. As a minimum there should be a pressure switch immediately upstream of the orifice, which regulates the purge flow so that an alarm will sound if the purge gas pressure at that point falls below a set level. The pressure setting should correspond to the minimum purge rate required by the flare system. Note that a pressure switch will not detect a purge gas regulating orifice that has plugged, hence the recommendation to use a flow switch or flow transmitter. Upstream of the regulating orifice and pressure switch, if utilized, it is further recommended that the purge gas pass through a strainer, in which the mesh openings are not greater than one-quarter of the diameter of the limiting orifice.

WARNING

Failure to provide a purge flow adequate to prevent air ingress or failure to limit free oxygen entering the flare system can create an explosion potential. Confirm that the flare system is essentially oxygen-free prior to pilot ignition.

14.1.1.5 Minimum Purge Gas Rate

Flare systems require a minimum continuous purge gas flow. For the purge rate of a specific flare, see the manufacturer's recommendation. Initial purging of the flare system may require a substantially higher flow rate to establish the oxygen-free condition.

14.1.1.6 Staged Post-Purge Concept

Staged flare systems utilize multiple burner heads mounted to manifolds. A group of these burners that share a common manifold with no valves in between are called "stages." The flow of waste gas to each stage of burners is commonly controlled via an automated on/off valve, referred to as a "staging" valve. As the fuel flow to the flare increases, stages are opened to allow fuel to flow to more burners thus limiting the pressure in the gas collection system to an acceptable range. As the fuel flow to the flare decreases, stages are closed to reduce the number of active burners and thus maintain the gas collection system at a suitable pressure for good flare performance.

For staged flare systems, it is desirable to execute a short-duration, high-volume purge of a stage that is no longer active. Once the purge of a stage is complete, no additional purge gas is required until the stage once again becomes active. The purge is required to be injected downstream of the staging control device(s). Generally, an inert purge is used for this purpose. This purge is injected for a short time at a high rate to clear the system downstream of the staging control device(s). The purpose is to minimize corrosion and/or to prevent a flashback. For the post-purge rate of a specific flare, see the manufacturer's recommendation. This purge rate must be maintained until at least 10 system volumes of inert gas have been injected.

14.1.2 Pre-Start-Up

14.1.2.1 Pre-Start-Up Checks for All Flares

Before starting up a flare, checks should be made of the following items:

1. Verify that all flare components are installed in accordance with applicable drawings.
2. Verify that all nuts and bolts are tightened to the appropriate torque.
3. Verify that all flare components are properly grounded.
4. Verify that all electrical devices are connected to the proper power sources.

5. Verify that all control wiring is connected to the proper terminals.
6. If thermocouples are utilized, verify that the thermocouple extension wire used is adequate for the radiation load and is proper for the thermocouple used (i.e., KX wire for type-K thermocouples using a high-temperature insulation).

> **NOTE**
>
> Cross connecting the positive and negative wires anywhere in the circuit will adversely affect the thermocouple signal.

7. All system lines should be dry and free of dirt and foreign material. This should be verified prior to connecting the pilot gas lines at the pilots. Pilot gas lines should be confirmed as free and dry by removing the mixer spud and strainer screen and blowing them with clean dry air. Reinstall the spud and screen only after clear flow is confirmed.
8. Verify that all drain and vent valves are closed and that all drain and vent plugs are tightly secured.
9. Verify that the manual valves in the pilot gas and waste gas systems are closed.
10. Verify that all instruments are properly calibrated. Check that all setpoints are properly adjusted.
11. For systems that use a flame front generator (FFG) for pilot ignition, verify that the pilot ignition lines are dry and unobstructed so that the flame front is not quenched or blocked. In freezing climates, heat tracing or other methods that prevent condensation are recommended. A drain plug or valve is required at each low point in each pilot ignition line.
12. For systems that use an electronic or other proprietary ignition system, refer to the suppliers' documentation for pre-start-up system checks.
13. Verify that all loop seals are filled with an appropriate fluid. In freezing climates, heat tracing is recommended.
14. Verify that all purge gas injection points are connected and ready for use.
15. Verify that all shutdown maintenance items described in Section 14.1.5 have been covered.
16. Verify sufficient spare parts are on hand for start-up.

14.1.2.2 Pre-Start-Up Checks for Steam-Assisted Flares

Before starting up a steam-assisted flare, checks should be made of the following items:

1. Visually inspect the steam equipment for damage during shipment or installation.
2. Verify that steam traps at the base of the stack are working. It is recommended that steam traps and condensate return systems be located downstream of the steam control valves and be designed to accommodate very-low-pressure steam (near atmospheric pressure) as well as the design pressure of the steam line.
3. Verify the steam piping is properly insulated.
4. Verify proper operation of the steam control valves.
5. Verify the cooling steam bypass lines are insulated around steam control valves. Verify sizing of the cooling steam orifices.

> **NOTE**
>
> A cooling steam bypass typically consists of a small bypass line around a steam control valve. An orifice is located in the bypass to allow adequate steam flow to the flare when the steam control valve is closed to protect the flare tip from mechanical damage due to heat.

> **CAUTION**
>
> Evacuate personnel in the vicinity of the flare before proceeding with steam equipment checkout. Steam equipment can produce very high noise levels. Initial steam injection may result in hot water spray at the tip.

6. Initiate cooling steam flows to the steam injectors. Maintain the cooling steam flows to establish steady piping temperatures.
7. Apply sufficient steam pressure (approximately 20% of the maximum operating pressure) to the steam piping and check for leaks.
8. Reduce the steam flow to cooling rates on all lines and maintain cooling steam flow through start-up.

14.1.2.3 Pre-Start-Up Checks for Air-Assisted Flares

1. Documentation supplied by the blower vendor takes precedence over these instructions. Carefully review the blower vendor literature

and the flare manufacturer's instructions before proceeding.
2. Visually inspect the blower for damage during shipment or installation.
3. Verify the blower blades are free to rotate.

> **CAUTION**
>
> Ensure all power sources to the blower are disabled when verifying free blade rotation.

> **NOTE**
>
> Before proceeding with the checkout procedure, advise personnel in the area of the flare that the blower will be operated.

4. Insure that the blower is installed and properly lubricated per the manufacturer's recommendations.
5. Place blower in manual operation and operate at a nominal rotation. Verify the blower rotation is in the proper direction. (Air should be pulled into the blower inlet.)

> **WARNING**
>
> Operation of the flare with improper rotation of the blower can cause flammable gases or fire to discharge from the blower inlet.

> **CAUTION**
>
> Operating the flare with the blower turned off, but not blocked in, can also cause a hazard when flaring gases heavier than 29 MW or colder than ambient temperature.

6. With the blower operating at full speed, verify the motor amperage is similar to amperage listed on the blower vendor data sheet. Amperage may vary due to ambient temperature.
7. Insure all process instrumentation is properly calibrated per the manufacturer's supplied documentation. Instruments typically utilized to control airflow include pressure switches, pressure transmitters, flow switches, and flow transmitters.
8. Place the blower in automatic operation.
9. Simulate input(s) into the control system. Verify the blower accelerates and decelerates as expected per the control algorithm. (The blower control algorithm can be developed by the flare manufacturer or by the flare operator.)
10. Remove the simulation equipment and return the process sensor(s) to normal operation.

> **NOTE**
>
> If blower operation does not perform as expected, troubleshoot and correct before proceeding.

> **NOTE**
>
> Regarding the air control instrumentation and control logic, some field adjustment of setpoints may be required during start-up for best operation. If field adjustment is required, make note of the optimum setpoints for future reference.

14.1.2.4 Pre-Start-Up Checks for Staged Flare Systems

1. Visually check that all staging valves are closed. Check that all individual stage manual block valves are closed. Verify that all flare headers are blocked in.
2. Check that instrument air is connected to the staging valve solenoids. Regulate instrument air pressure to the level required for the staging valve actuators.
3. Open all manual block valves that deliver instrument air to the staging valves.

> **CAUTION**
>
> Commanding a staging valve to OPEN or to operate in AUTO mode is accomplished via the control system. When switched to the AUTO mode, the staging valves should close. Advise personnel in the area of the staging valves prior to switching the valves to AUTO mode.

4. Power ON staging control system if separate from the DCS.

5. Switch all staging valve selector switches (located in DCS or local control panel depending on design) to AUTO. (With no pressure on the header, all staging valves should close.)
6. All the staging valve open indicators should indicate "NOT OPEN." All the staging valve closed indicators should indicate "CLOSED." All bypass device closed indicators should indicate "CLOSED." All pilot-proved indicators should indicate "PILOT NOT PROVED," while all the pilot failure indicators should indicate "PILOT FAILURE."

> **CAUTION**
>
> If any of these indicators are NOT in the proper condition, trace and correct the cause before proceeding.

7. Bypass devices can be various technologies including rupture disks, buckling pin relief valves, and possibly other fail-safe technologies. For each bypass, verify the proper status feedback to the control system via the method described in the flare manufacturer's supplied O&M manual.

> **CAUTION**
>
> If any of the bypass status indicators fail to operate as described in the O&M manual, trace and correct the problem before proceeding.

8. Staging valve stroke test
    This test requires communication between observers at the staging valve area and at the staging control operator interface.
    For each staging valve, perform the following operations:
    a. Command valve to OPEN mode.
    b. Verify that the associated stage open indicator is ON and the associated stage closed indicator is OFF.
    c. Verify that the associated valve opens within the time interval specified by the flare manufacturer.
    d. Command valve to AUTO mode.
    e. Verify that the associated stage open indicator is OFF and the associated stage closed indicator is ON.
    f. Verify that the associated valve closes within the time interval specified by the flare manufacturer.

> **CAUTION**
>
> If any of the component actions do not match or if any of the indicators fail to operate as described, trace and correct the problem before proceeding.

If the closing stroke time of a value is too long, check for anything that restricts flow into the actuator. If the opening stroke time of a value is too long, check to anything that restricts the exhaust flow out of the actuator. Remove any obstructions and repeat the stroke test.

9. Logic sequencing test
    Pressure transmitters control the operation of the staged flare. It is very important that these instruments be properly calibrated. Refer to manufacturer's O&M manual for proper calibration spans and procedures.
    A test should be performed to verify the proper operation of the pressure transmitters and logic system. This test checks the staging setpoints and sequencing in the logic system as well as the performance of the pressure transmitter(s). All staging valves must be in AUTO mode.

> **CAUTION**
>
> Before performing this test, verify that all individual stage manual block valves are closed.

Staging pressure setpoints and post-purge times are provided by the flare manufacturer.

> **NOTE**
>
> Capacity curves that relate these setpoints to flow rates are provided by the flare manufacturer. Improper adjustment of setpoints may affect bypass device operation, staging control hysteresis, and smokeless performance.

14.1.2.4.1 Logic Sequencing Test

1. Close the hand valve that isolates the pressure transmitter(s).

> **NOTE**
>
> Some staged flare systems use multiple pressure transmitters. Such transmitters can be configured logically for 1 out of 2, 2 out of 2, 2 out of 3, etc. Multiple transmitters may need to be manifolded together in order to get the staging logic to operate properly. See flare manufacturer's O&M manual for the details of the transmitter logic used for a specific flare system.

2. Connect the transmitter(s) to a gas supply. This can be a small pressurizing hand bulb, a bottled air or gas supply, a compressor, etc. The gas supply should be equipped with a calibrated pressure gauge appropriate for reading the pressures.
3. Raise and lower the pressure to verify the setpoints and staging sequence order described in the flare manufacturer's O&M manual. First, raise the pressure to check each of the upstage setpoints. Then lower the pressure to check each of the destage set points. While the staging valve sequencing can vary from flare to flare, the basic concept of staging is constant. As the relief gas flow increases, more of the flare's burners are brought into service. As the relief gas flow decreases, burners are taken out of service. The following is a verbal description of the basic concept:
   a. When the flare header pressure reaches the "upstage" level, a certain number of burners are brought into service by opening a particular staging valve or valves.
   b. When the flare header pressure drops to the "destage" level, a certain number of burners are removed from service by closing a particular staging valve or valves.
   c. When a staging valve is commanded to close, a post-purge valve will open to sweep relief gas out of the piping downstream of the staging valve. Once the staging valve and bypass device are both proved closed, a purge timer will start. Once the purge timer times out, the post-purge valve will close. See the flare manufacturer's O&M manual for purge times.

> **NOTE**
>
> If multiple pressure transmitters are used in a system, inputting pressure into a single transmitter will cause a discrepancy between it and the other transmitters. This should cause a discrepancy alarm to occur.

> **CAUTION**
>
> The correct staging sequence is essential to proper flare operation. If any valve actions other than those described in the O&M manual are observed, trace and correct the problem before proceeding.

4. Disconnect the gas supply that was connected in step 2.
5. Reconnect the pressure transmitter(s) to the header.
6. Open the hand valve to allow the pressure transmitter(s) to sense flare header pressure.

> **CAUTION**
>
> The pressure transmitter signal is essential to safe and proper flare operation. The isolating valve should be locked open or tagged to prevent inadvertent closure.

7. For bypass devices that use a valve, manually open the bypass on each stage. If rupture disks are used, open the burst disk indicator circuit on each stage. The post-purge valve for that stage should open and remain open until the bypass proves closed. Once the bypass is closed, the post-purge timer is started in the logic system.

14.1.2.5 Pre-Start-Up Checks for Knockout Drums and Liquid Seals

1. Check that any hydrotest plugs are tightly secured. Visually inspect all pressure-containing surfaces for proper sealing.
2. For a liquid seal, fill drum to the normal liquid level (NLL). For ease of checking during start-up, mark the NLL on the level indicator.
3. Verify setpoints for level controls. See drum drawing for NLL and alarm levels.

14.1.2.6 Pre-Start-Up Checks for Molecular Seal

The optimum position of the molecular seal is immediately upstream of the flare tip. See the supplier's documentation for the recommended minimum purge rate.

> **WARNING**
>
> If the molecular seal is not purged, the molecular seal will not function properly and air can enter through the flare tip. This could possibly cause an internal explosion in the flare system.

In order to maintain the optimum effectiveness of the molecular seal, the condensate drain line from the molecular seal will require a loop seal leg filled to a depth of 1.75 times the pressure drop at the maximum rate of flow. The pressure drop is that which is caused by the flare tip and the molecular seal together with any interconnecting piping.

> **WARNING**
>
> Failure to maintain the liquid level in the molecular seal condensate drain loop seal may allow waste gas to flow to the drainage system. This could possibly cause an internal explosion in the drainage system.

In colder climates, freeze protection of the drain lines may be required to prevent ice plugging.

> **WARNING**
>
> Blockage of the molecular seal condensate drain line by ice or other foreign matter can cause total blockage of the molecular seal and prevent adequate relief of the waste gas. This line should be regularly checked to assure open flow.

14.1.3 Start-Up and Shutdown

14.1.3.1 Unassisted Flares

14.1.3.1.1 Start-Up and Operation

1. Prepare the system for start-up per Section 14.1.2.
2. Purge the system in accordance with Section 14.1.1.
3. Verify the oxygen concentration in the flare stack is appropriate for the expected waste gases to prevent a flashback or explosion.
4. Light the pilots per Section 14.1.4.
5. The flare is now ready to receive waste gas.

14.1.3.1.2 Shutdown

1. Shut down the waste gas flow.
2. Extinguish the pilots by closing the pilot fuel gas block valve.
3. Shut down the purge gas flow.
4. Close all hand valves.
5. Close all utility gas supply systems.

14.1.3.2 Steam-Assisted Flares

14.1.3.2.1 Start-Up and Operation

1. Prepare the system for start-up per Section 14.1.2.
2. Introduce the cooling steam flow to the steam injection equipment.

> **CAUTION**
>
> Operation without cooling steam flow can cause a reduction in steam equipment life.

> **WARNING**
>
> The following warning is for tips that use center steam, steam that is injected into the body of the flare and does not inspirate air. In cold climates, center steam condensation can cause ice blockage of the flare tip under certain conditions. During extremely cold weather, the center steam should be blocked in.

3. Purge the system in accordance with Section 14.1.1.
4. Verify the oxygen concentration in the flare stack is appropriate for the expected waste gases to prevent a flashback or explosion.
5. Light the pilots per Section 14.1.4.
6. The flare is now ready to receive waste gas.
7. Start the waste gas flow at a standard base load rate.
8. Center steam adjustment: The purpose of the center steam is to minimize burning inside the flare tip, which can reduce tip life. Internal burning is affected by several factors including the base load flow rate, waste composition, and wind speed. Since some of these factors can be highly variable, it is recommended that nightly observations take place when the waste gas flow is low to ensure the flare operation has not transitioned

into a state where internal burning occurs. Look for a glow (excessive heat) below the flare tip exit. If such a glow is observed, increase center steam flow to minimize internal burning.

9. During a flaring event, the steam flow to the flare should be adjusted to the point where smoke is not visible and the flame is a yellow-orange color. Excessive steam injection will cause high noise and can cause reduced destruction efficiency.

   For flare tip designs that use multiple steam supply inlets, careful coordination of steam supplies is needed to prevent "capping" of the flare. Capping will force waste gas down into the body of the tip potentially reducing tip life. See the flare manufacturer's recommendations for steam coordination of such designs.

10. After cessation of a flaring event, steam rates should be returned to their cooling rates or a higher rate as necessary to produce a smokeless yellow-orange flame.

14.1.3.2.2 Shutdown

1. Shut down the waste gas flow.
2. Close the steam supply valves to the flare tip.
3. Extinguish the pilots by closing the pilot fuel gas block valve.
4. Shut down the purge gas flow.
5. Close all hand valves.
6. Close all utility gas supply systems

14.1.3.3 Air-Assisted Flares

14.1.3.3.1 Start-Up and Operation

1. Prepare the system for start-up per Section 14.1.2.
2. Purge the system in accordance with Section 14.1.1.
3. Verify the oxygen concentration in the flare stack is appropriate for the expected waste gases to prevent a flashback or explosion.
4. Light the pilot(s) per Section 14.1.4.
5. The flare is now ready to accept relief gas flows.
6. The blower(s) will now operate per the control algorithm.

> **WARNING**
>
> Operating the flare with no assist-air is dangerous to personnel and can cause damage to the flare equipment.

14.1.3.3.2 Shutdown

1. Shut down the waste gas flow.
2. Extinguish the pilots by closing the pilot fuel gas block valve.
3. Shut down the purge gas flow.
4. Turn off the blower control system.
5. Close all hand valves.
6. Close all utility gas supply systems.

14.1.3.4 Staged Flare Systems

14.1.3.4.1 Start-Up and Operation

1. Prepare the system for start-up per Section 14.1.2.
2. In order to purge the entire flare header system, including the staging manifold, it is necessary to open the staging valve furthest from the staging manifold inlet along with its corresponding manual block valve and close off all other stages. If the staging manifold inlet occurs in the middle of the manifold, both ends of the manifold will need to be purged. This should be done sequentially with one end being purged till clear of oxygen, then the other. The intent is for the gas used to purge the flare header to flow through the entire staging manifold.
3. Purge the system in accordance with Section 14.1.1. The purge gas is to be injected at the upstream ends of the flare header. If there are multiple branches at the upstream end of the flare header, distribute the purge gas evenly to all major branches.
4. An O_2 analyzer can be used to confirm removal of air from the system. Sample points should be included at low points and high points in header piping and vessels and should include a sampling nozzle just downstream of the staging valves.
5. When the O_2 level at all sample points is zero, close the staging valve(s). If the first stage does not use an automatic staging valve, open the stage 1 manual block valve. Reduce the purge rate to the continuous rate. If there are multiple branches at the upstream end of this flare header, the continuous purge gas flow should be distributed evenly to all major branches.
6. If steam is used as assist media for any of the burners, introduce the cooling steam flow.

> **CAUTION**
>
> For those burners that utilize steam, operation without cooling steam flow will cause a reduction in steam equipment life.

7. Light the pilots per Section 14.1.4.
8. Confirm all staging valves are in "AUTO" mode.
9. Open all manual block valves to the various stages of the flare.
10. The staged flare is now ready to receive waste gas.

> **CAUTION**
>
> For steam-assisted burners, operation without steam flow may cause a smoke plume. Such a smoke plume may become a breathing or a visibility hazard.

11. Staging is controlled automatically to ensure smokeless operation. Waste gas pressure in the staging manifold is used as a control signal to the control system. Refer to the flare manufacturer's documentation for staging sequence and setpoints.

14.1.3.4.2 Shutdown

1. Shut down the waste gas flow.
2. If steam is used, close the steam supply valves.
3. Extinguish the pilots by closing the pilot fuel gas block valves.
4. Shut down the purge gas flow. Close all the individual stage manual block valves.
5. Close all utility gas supply systems.

14.1.4 Pilot Lighting Procedure

14.1.4.1 Flame Front Generator (FFG)

1. Blow down the air and gas lines to the FFG in order to remove condensate and/or foreign material. This blowdown should occur through blowdown valves to avoid plugging the FFG flow orifices. Open all ignition line drains. Blow down each ignition line, one at a time.
2. Close all drains. Close the ignition gas block valve. Check each ignition line for blockage using the following procedure:
   a. Open an ignition selector valve.
   b. Open the ignition air block valve allowing air to flow through the ignition selector valve and ignition line.
   c. Close the air block valve. Pressure should fall off rapidly on the pressure gauge. If the pressure does not fall off rapidly, investigate and correct the blockage in the line.
3. With the ignition gas supply valve closed and the ignition chamber purged with air, energize the FFG control panel. Depress the "ignition" pushbutton. Verify a spark is present through the sight port on the ignition chamber.
4. At initial start-up, blow down pilot gas lines to pilots at a blowdown valve to remove condensate and/or foreign material. Periodically, when the flare is out of service, repeat the pilot gas blowdown procedure to assure dry and clean piping. A strainer at grade in the pilot gas line shall be used. In addition, a strainer located just upstream of the pilot gas orifice shall be used.
5. Flow pilot gas to the pilots and set the pilot gas regulator to the flare manufacturer's recommended pressure. Allow time for the fuel gas to completely purge the line to the pilots. The purge time will depend on the distance from the pilot gas valve to the pilots.
6. Open the FFG air and ignition gas block valves along with an ignition selector valve. Set the air and ignition gas regulators to the manufacturer's recommended pressures.
7. Purge the ignition line to the pilot for sufficient time to completely fill the ignition line with a combustible gas/air mixture. The purge time will depend on the distance from the FFG to the pilot(s).
8. Quickly press and release the "ignition" push button.

> **CAUTION**
>
> Do not hold the "ignition" pushbutton down. This could create a steady flame inside the ignition line, which will damage the ignition selector valves and the paint on the ignition piping as well as posing a personnel hazard.

9. The pilot verification system supplied with the pilots should indicate if the pilot has ignited. If thermocouples are used for pilot verification, some time is required to allow the pilot to heat up and verify ignition. If the accuracy of the pilot verification system is in doubt, use the visual pilot verification procedure in the following text to confirm pilot ignition. If the pilot does not light, repeat steps #7 and #8.

10. If repeating steps #7 and #8 two or three times does not light the pilot, reduce the ignition gas pressure slightly. Then again repeat steps #7 and #8. If necessary, continue to adjust fuel and air pressures and repeat the ignition attempts until the pilot is lit.

> **NOTE**
>
> It is recommended that the operator make a record of the air and ignition gas pressure that worked best in order to simplify future start-up attempts.

11. Repeat steps #7 and #8 for the other pilots one at a time.

12. After all the pilots are lit and ignition is verified, close the ignition gas valve at the FFG. Open all the ignition selector valves. Allow the air to flush the ignition lines clear for a few minutes. This will remove combustion products left over from the ignition process, which can be corrosive. Close the air valve. Close the ignition selector valves.

14.1.4.2 Visual Pilot Verification with FFG

1. Open the ignition selector valve for the pilot to be verified. Make sure all other ignition selector valves are closed.
2. Close the air block valve on the FFG.
3. Open the ignition gas block valve on the FFG for about 2 min. This fills a section of ignition line with fuel gas.
4. Open the air block valve on the FFG. This blows the slug of fuel gas through the ignition hood at a fairly high flow rate.
5. If the selected pilot is lit, a prominent yellow flame will appear at the pilot. The yellow flame will last for about 10 s.

6. Close the ignition gas block valve on the FFG. Allow airflow to clear the ignition line for about 1 min.
7. Close the air block valve on the FFG. Close the ignition selector valve.

14.1.4.3 Other Pilot Ignition Systems

1. At initial start-up, blow down the pilot gas lines to the pilots at a blowdown valve to remove condensate and/or foreign material. Periodically, when the flare is out of service, repeat the pilot gas blowdown procedure to assure dry and clean piping. A strainer shall be located at grade in the pilot gas line.
2. Flow pilot gas to the pilots and set the pilot gas regulator to the flare manufacturer's recommended pressure. Allow time for the fuel gas to completely purge the line to the pilots. The purge time will depend on the distance from the pilot gas valve to the pilots.
3. Refer to the supplier's documentation for operational details of a proprietary ignition system.

14.2 Maintenance

Periodic maintenance is essential to keep the flare system operating properly. The following recommendations are provided as minimum guidelines for maintaining a safe and effective flare system. These recommendations are in addition to the basic maintenance requirements of individual components. Refer to supplier documentation for system and component maintenance recommendations.

The following sections are divided into shutdown items and periodic on-line items. Shutdown items are those items that can only be accomplished safely during a flare shutdown. Periodic on-line items are those items that can be accomplished safely with the flare on-line. Periodic on-line items should also be checked during a flare shutdown.

> **WARNING**
>
> Access to flare tips and pilots while the flare is on-line is not safe. Excessive radiation, flame lick, and/or toxic gas exposure may cause injury or death.

14.2.1 Purge System

14.2.1.1 Shutdown Items

1. Pull, clean, and confirm size of purge gas restriction orifice(s).
2. Test purge verification system for purge failure alarm. Check instrument(s) calibration and logic set point(s) per supplier documentation.
3. For staged purge systems, check out purge staging controls per Section 14.1.2.4, logic sequencing test.

14.2.1.2 Periodic On-Line Items

1. Check the purge gas regulator. Confirm correct outlet pressure. Replace diaphragm on a regular schedule.
2. Check filters and strainers. Clean if necessary.

14.2.2 Pilots

14.2.2.1 Shutdown Items

1. Inspect pilot tip for visible deformations and weld cracks. Replace in kind if necessary.
2. Inspect pilot mixer for cracks. Replace in kind if necessary.
3. Inspect the orifice spud. Ensure spud is straight and plumb. Clean the orifice and confirm the drilling. Replace in kind if necessary.
4. Inspect the pilot fuel gas strainer at the pilot mixer. Clean the screen and replace in kind if necessary.
5. If thermocouples are used for pilot verification, replace the pilot thermocouple. Ensure that the thermocouple tip seats fully in the thermowell.

> **NOTE**
> If a thermocouple has been in service more than 12 months, replace it regardless of apparent condition.

14.2.2.2 Periodic On-Line Items

Access to pilots while the flare is on-line is not safe. Inspection can be accomplished using binoculars when necessary. From a safe distance, inspect the pilot at night for any sign of a red glow.

For some manufacturers pilots designed to accommodate FFG ignition:

1. The back of the ignition hood can glow a dull red during normal operation. This is caused by a small pilot stability fire, which burns continuously inside the hood.
2. The normal ignition hood fire may cause a dull glow, slightly farther down the ignition line of the downwind pilot. Reinspect after the wind direction changes.
3. A bright red glow on the ignition hood and/or smoke at the pilot indicates that raw gas is being fed into the ignition line. Close the ignition gas valve at the FFG. Open the air valve at the FFG for several minutes to cool the pilot hood. Close the air valve and reinspect the pilot.
4. Paint damage and/or a red glow at the FFG also indicates an open or leaky ignition gas valve on the FFG. Close the gas valve immediately.
5. Inspect ignition lines for low points and sagging. If possible, correct low points by adjusting the slope of the line. Any low points that cannot be eliminated must be equipped with drain plugs. Drain all low points.
6. Break the ignition line at the base of the stack (and/or any other accessible low point). Vigorously rattle the ignition line to remove scale.

For all pilots designed with a venturi mixer:

7. A glow near the pilot mixer indicates that the pilot has flashed back to the mixer. This usually indicates that the pilot tip has deteriorated significantly. Replace the pilot tip as soon as possible. The following procedure should be followed to try to eliminate the burning at the mixer.
   a. Release enough combustible gas to the flare to establish a steady flame on the flare tip. Maintain this gas flow to the flare throughout the procedure.
   b. Extinguish the pilots by shutting off the pilot gas supply valve.
   c. Allow time for the mixers to cool.
   d. Open the pilot gas supply valve and relight the pilots.
8. When pilot ignition is confirmed, resume normal purge rates to the flare.
9. Inspect the pilot fuel gas strainer at grade. Clean the screen and replace in kind if necessary.
10. Check the pilot gas regulator. Replace the diaphragm on a regular schedule.

14.2.3 Flame Front Generator

Since the FFG is located at an accessible location, maintenance can be performed at any time. All periodic on-line items should be checked during any shutdown.

14.2.3.1 Periodic On-Line Items

1. Check the regulators on air and ignition gas. Replace the diaphragms on a regular schedule.
2. Clean the air and ignition gas orifices.
3. Clean and gap the spark plug. Refer to the supplier's documentation for correct gap width. Replace the spark plug if the ceramic insulator is damaged.
4. Clean the sight glass on ignition chamber. Replace if cracked or opaque.
5. Check the FFG panel alarm and indicating lamps on a regular schedule. Replace burnt-out lamps.
6. Drain the ignition lines at the low point drains on a regular schedule, at least weekly.

14.2.4 Flare Tips

14.2.4.1 Shutdown Items for All Flare Tip Types

1. Check the purge system. (See Section 14.1.5.1.)
2. Check the pilots. (See Section 14.1.5.2.)
3. Check all welds for cracking and repair or replace as necessary.
4. Check all brackets for integrity and repair or replace as necessary.
5. If a casting is used as part for the tip design, repair or replace if the casting is cracked.
6. Check for any visible deformation and repair or replace as necessary.
7. Check the flame stabilization mechanism such as flame retention segments. Repair or replace as necessary.
8. Check all flange gaskets and replace if necessary.
9. Check all flange bolts for proper torque.
10. If insulation is part of the tip construction, repair as needed.
11. Verify all drains are clear.
12. Clean any coke or debris that may have accumulated in the tip or, if applicable, the molecular seal.

14.2.4.2 Shutdown Items for Steam-Assisted Flare Tips

1. Check and clean the steam ports. Replace steam injectors if ports show any indication of erosion.
2. Check the alignment of the steam injectors.
3. If steam tubes are used in the design, check tubes for deformation. Replace any tubes that have deformed to the point of reduced exit area.
4. Check strainers.

14.2.4.3 Shutdown Items for Air-Assisted Flare Tips

1. Check blower controls.
    a. Confirm the control system contains the latest version of the system logic.
    b. Check all process instrumentation is properly calibrated.
    c. Check all process instrumentation is properly installed.
    d. Verify blower control logic is working properly.

14.2.4.4 Shutdown Items for Staged Flare Systems

1. Pressure test the staging valve and bypass assemblies.
    a. Close the manual block valve and staging valve for a stage.
    b. Pressurize piping between the block valve and the staging valve.
    c. Leave this piping pressurized for 24 hours.
    d. Check to see how much pressure remains. Note that it may be necessary to adjust the expected pressure at the end of the test for large ambient temperature changes. A significant drop in pressure indicates either a leaky staging valve, leaky bypass valve, leaky manual block valve, or leaking connections in the piping. Investigate the cause and correct.

14.2.4.5 Periodic On-Line Items for Air-Assisted Flare Tips

1. Check blower.
    a. Lubricate on manufacturer's recommended cycle.
    b. Check the motor amperage seasonally. A blower motor will draw more amps during cold weather.

c. Check for vibration.
d. Check and clean the blower inlet screen.
e. Confirm damper open and closed position (if applicable).

14.2.5 Liquid Seals and Knockout Drums

14.2.5.1 Shutdown Items

1. Drain the drum completely. Inspect inside for accumulated solids. Remove any such solids.

> **CAUTION**
> Follow safe vessel entry procedures as required by plant standards.

2. For liquid seals, inspect the liquid seal internals for accumulated solids, cracks, plugging, or other obvious damage. Clean or repair as required.
3. Check the level instruments per vendor information.
4. Check metal thickness.

14.2.5.2 Periodic On-Line Items for Knockout Drums

1. Check the physical liquid level of the level indicator on a regular schedule. Use level indicator level to check level control instruments.
2. Check that all isolating valves on all level sensing equipment are open to allow the level sensing equipment to monitor levels.
3. Drain the drum to the minimum level whenever possible.

14.2.5.3 Periodic On-Line Items for Liquid Seals

1. Check the physical liquid level of the level indicator on a regular schedule. Use level indicator to check level control instruments.
2. Check that all isolating valves on all level sensing equipment are open to allow the level sensing equipment to monitor levels.
3. Skim any accumulated hydrocarbon liquids from the surface of the liquid on a regular schedule.
4. If the waste gas contains any acidic or alkaline compounds, monitor the pH of the seal fluid. Treat or replace the seal fluid before the pH becomes corrosive to vessel metallurgy.

14.2.6 Molecular Seals

14.2.6.1 Shutdown Items

1. Remove any foreign material that may have accumulated in the bottom of the molecular seal.
2. Use ultrasonic examination or other suitable methods to check material thickness on bottom of unit. This is especially important in corrosive gas service or if any obstruction of the drain line has allowed water to remain in the unit for an extended period of time.

14.2.6.2 Periodic On-Line Items

1. Flush the loop seal piping to remove any foreign material that may have washed down from the molecular seal.
2. Ensure that the siphon break vent line is open and clear of obstructions.

14.3 Flare Troubleshooting

The symptoms, possible causes, and recommended responses for various problems pertaining to flare operation, pilots, front generators, electronic ignition at or near pilot tips, and pilot verification systems are listed in Tables 14.1 through 14.5.

TABLE 14.1
Flare Operation

Symptom	Possible Cause	Response
Flame puffs or surges	Liquid seal level too high	Check liquid seal level and adjust to manufacturer's recommended level
	Damage to liquid seal internals	Inspect liquid seal internals at first opportunity, and repair/replace as required
	Liquid in molecular seal	Verify molecular seal is properly drained
	Capping of flare with assist-steam	Check steam flow rate. Adjust rate to achieve a yellow-orange nonsmoking flame
	Knockout drum level too high	Check knockout drum liquid level and drain to avoid excessive level
Burning liquids around the flare	Carryover of liquid from knockout drum	Control liquid level in knockout drum. Search for sources of liquid within the process area
	Liquid entrained in vent gas	Remove liquid from vent gas
	Liquid accumulating in piping	Verify flare lines have no low point and slope to a knockout drum
	Hydrocarbon liquid accumulating in liquid seal	Verify continuous overflow or periodic skimming of hydrocarbon liquid from liquid seal is adequate for liquid load
Unusual noise from flare. Possibly periodic, occurring at regular intervals	Flashback into flare stack	Check purge gas flow rate and verify it is equal or greater than the manufacturer's recommended minimum purge rate
		At first opportunity, inspect molecular seal device for damage and replace/repair as needed
	Flashback into staged flare runner	Leaky staging valve or bypass device, maintenance as required
Flame present in flare tip muffler	Capping of flare with assist-steam	Reduce the steam flow to the upper steam ring. Increase steam flow to the lower steam ring. The upper steam ring flow rate should not be significant until the lower steam ring flow rate is sufficient to avoid capping
Smoke	Loss of steam flow or incorrect amount of steam flow	Verify steam is flowing to the flare. Verify the steam control system is operating properly. Increase steam flow until flame is a yellow-orange color and nonsmoking
	Damaged flare tip	Visually inspect tip for damage. At first opportunity, repair/replace as needed
	Vent flow exceeds design capabilities	Check vent flow rate and composition with flare design basis. Possible solutions include upgrading flare capability or diverting certain feeds to a different flare
	Loss of airflow or incorrect amount of airflow	Verify air is flowing to the flare. Verify the air control system is operating properly. Verify blower rotation is correct and not surging. Increase airflow until flame is a yellow-orange color and nonsmoking
	Destage pressure too low for given gas composition	Consult with manufacturer to determine if staging pressure adjustments can be made
Steam leak observed at tip	Damaged steam manifold	Increase minimum cooling steam rate to compensate for steam lost from leak. At first opportunity, repair/replace as needed
Ice buildup on flare tip and/or structure	Steam condensing and freezing in cold weather	Increase hydrocarbon flow to the flare. Restrict access to flare area to prevent injury from falling ice
Flame pulldown on outer shell	Low vent flow in high wind	For a steam-assisted flare with an upper steam ring, increase the steam flow to the upper ring. If tip design does not include an upper steam ring, increase steam flow

(*continued*)

TABLE 14.1 (continued)
Flare Operation

Symptom	Possible Cause	Response
Flare makes too much noise	High assist (air or steam) flow causing unstable combustion	Control assist flow to each injection point. The result should be a yellow-orange, nonsmoking flame
	Damaged muffler	Repair/replace at first opportunity
High back pressure on flare system	High liquid level in knockout drum	Drain liquid from knockout drum
	High liquid level in liquid seal	Control liquid seal level to manufacturer's recommendation
	Liquid in molecular seal	Drain liquid from molecular seal
	Damaged equipment	At first opportunity, repair any damage that causes obstruction in any part of the flare system including flare tip, molecular seal, liquid seal, and knockout drum
	If steam flare, buildup of ice internal of flare tip	Inject an environmentally acceptable ice-melting substance into center steam injector
	Coking inside of flare tip due to low vent flow rates burning inside of tip	Increase center steam flow For a steam-assisted tip with a lower steam ring, increase steam flow to lower ring
	Coking inside of flare tip due to air leaking into body of flare tip	Repair any air leaks in body of tip or internal tubes
	Formation of hydrate in flare system.	Insulate and/or heat trace affected piping Add heat or environmentally acceptable substance to melt hydrate
Loss of flame	Pilot failure	See pilot troubleshooting Section 14.2.3
	Unstable combustion at high flow rates due to damaged flame stability device(s)	Repair/replace flame retention ring or segments
	Heating value of waste gas too low	For traditional unassisted, steam-assisted, and air-assisted flare tips, add supplemental fuel gas to increase the heating value of the vent gas to enable stable combustion. For other flare designs, consult with manufacturer
	Process condition exceeds flare capability	Verify maximum operating capability of flare
Hot spot(s) on flare tip	For tips with internal refractory, damaged refractory	Repair internal refractory
	Combustion inside the flare tip	Increase center steam flow Increase lower steam ring flow Decrease upper steam ring flow Increase assist-air flow Increase vent gas flow or replace
Flame present on a row of burners of a staged flare system, which should be off line. Note that a stage of burners will have a brief period of flame after the staging valve closes	Bypass device open	Reset or replace bypass device
	Leaky bypass device or staging valve	Perform maintenance on bypass device and/or staging valve as required
	Staging valve open due to insufficient air supply. The air supply can be deficient in either pressure or flow capacity. Lack of capacity can allow quick exhaust valves to flutter between open and close, venting air and never allowing sufficient pressure to build to close the staging valve	Adjust instrument air pressure to design requirement. If air supply capacity is choked at some point, remove choke point or add a volume bottle to prevent flutter

Flare Operations, Maintenance, and Troubleshooting

TABLE 14.2

Pilots

Symptom	Possible Cause	Response
Pilot fails to ignite or weak pilot flame	Improper pilot gas composition or flow rate	Confirm pilot gas composition and flow rate meets pilot requirements. Verify pilot strainers in pilot gas line are clear
	Obstruction in pilot	Verify the following components are clear of obstructions: pilot mixer, pilot gas orifice, pilot tip, and all piping between tip and mixer
	Pilot ignition system failure	Troubleshoot ignition system
	Pilot verification system failure	Observe pilot(s) at night to confirm status. See Section 14.2.5
	Pilot damage	Replace at first opportunity
Glow or flame at pilot mixer	Pilot flame has flashed back to mixer due to improper pilot gas composition	Confirm the gas compositions the pilot is designed to accept. See Section 14.1.5.2 item 7 for a procedure to correct a pilot that has flashed back
	Pilot flame has flashed back to mixer due to pilot damage	Replace pilot at first opportunity. See Section 14.1.5.2 item 7 for a procedure to correct a pilot that has flashed back

TABLE 14.3

Flame Front Generator

Symptom	Possible Cause	Response
No spark in ignition chamber.	Power problem	Confirm primary power is reaching the ignition transformer. Confirm the transformer is producing the manufacturer's specified output voltage and it is reaching the spark plug
	Failed transformer	Replace
	Moisture shorting circuit	Confirm ignition chamber is moisture-free. Confirm conduit and wiring to/from spark plug is moisture-free
	High pressure in ignition chamber (>5 psig)	
	No selector valve open	Open a valve
	Liquid in ignition line	Drain line
	Long ignition line	Reduce ignition gas and air flow rates
	Improper ignition line design	Reduce ignition gas and air flow rates
	Spark plug gap too wide or shorted	Confirm spark gap is correct
Spark present, but no fireball generated	Incorrect fuel air mixture	Confirm fuel and airflow orifices are the correct size for the ignition gas utilized and that they are unobstructed
		Confirm air pressure is set at the manufacturer's recommended pressure. Adjust fuel pressure until ignition is achieved
	Faulty solenoid valve (automatic systems) causing pulsating flow or no flow	Confirm via pressure gauges that both fuel and air flows are steady. Repair/replace solenoid as required
Fireball is generated, but pilot does not ignite	Condensation in ignition line(s)	Confirm ignition lines are clear of any liquid. Drain ignition lines at all low points. Be aware that sagging lines can produce low spots that do not have drains
	Ignition occurring before fuel/air mixture reaches pilot	Allow sufficient time for the fuel/air mixture to travel to the pilot
	Detonation in ignition line. A loud bang during ignition indicates a detonation is occurring in the ignition line. This is typically caused by short FFG lines	Reduce the gas pressure to around 2 psig (14 kPa). Reduce the air pressure to below 2 psig (14 kPa). Press the ignition push button and verify if ignition is occurring in the sight glass. If not, increase the air pressure slightly. Repeat incrementing the air pressure and pressing the ignition push button until ignition occurs. Once ignition is achieved, allow sufficient time for the ignition line to fill with the fuel/air mixture
	Pilot failure	See Section 14.3.2
	Failure of pilot verification system	See Section 14.3.5
	Open drain(s) in ignition line	Confirm all drains are closed

TABLE 14.4

Electronic Ignition at or near Pilot Tip

Symptom	Possible Cause	Response
Pilot will not ignite	Pilot failure	See Section 14.3.2
	Incorrect installation	Verify installation and wiring is per manufacturer's documentation.
	Short or discontinuity in wiring	A short or break in the wiring can prevent a spark from being generated at the tip. Correct/replace wiring as required
	Power problem	Confirm primary power is reaching the ignition components. Confirm the ignition component output is per the manufacturer's specifications
	Damage to spark-generating components	Replace damaged components at first opportunity
	Failure of pilot verification system	See Section 14.3.5

TABLE 14.5

Pilot Verification System

Symptom	Possible Cause	Response
Pilot is ignited, but will not prove	Broken thermocouple	Disconnect thermocouples from terminals in panel and check continuity. If an open circuit is detected, replace thermocouple at first opportunity
	Pilot-proved set point for thermocouple too high	Use a volt-ohm meter to read the millivolt signal from the type-K thermocouple. If the signal is greater than approximately 10 mV and not falling, the pilot is lit. Lower the set point until the pilot proves
	Incorrect thermocouple wiring	Confirm that the type-K thermocouples are correctly wired, yellow to (+), red to (−). Confirm that the ignition lines are paired up with the correct thermocouples
	Flame-ionization rod or wiring failure	Repair/replace as required
	Poorly aimed optical system	Adjust sighting of system
	Optical system view obscured	Clean lens. If view is obscured by process equipment plumes, consider relocating optical device to more advantageous location
	Ignition line obstructed limiting sound to acoustic flame monitor	Drain ignition lines and clear any other obstructions

Note: For additional troubleshooting guidance, refer to API 537.

15

Thermal Oxidizer Installation and Maintenance

Dale Campbell

CONTENTS

- 15.1 Introduction 396
- 15.2 Burner/Pilot 397
  - 15.2.1 Installation 397
  - 15.2.2 Maintenance 398
    - 15.2.2.1 Shutdown Items 398
    - 15.2.2.2 Periodic On-Line Items 399
- 15.3 Thermal Oxidizer 399
  - 15.3.1 Installation 399
  - 15.3.2 Maintenance 400
    - 15.3.2.1 Shutdown Items 400
    - 15.3.2.2 Periodic On-Line Items 401
- 15.4 Boilers 401
  - 15.4.1 Installation 403
  - 15.4.2 Maintenance 404
    - 15.4.2.1 Shutdown Items 404
    - 15.4.2.2 Periodic On-Line Items 404
- 15.5 Miscellaneous Control Items 405
  - 15.5.1 Installation 405
  - 15.5.2 Maintenance 406
    - 15.5.2.1 Shutdown Items 406
    - 15.5.2.2 Periodic On-Line Items 406
- 15.6 Liquid, Air/Steam Atomizing Guns (Fuel/Waste Oil, Quench Water, Aqueous Wastes) 406
  - 15.6.1 Installation 407
  - 15.6.2 Maintenance 407
    - 15.6.2.1 Shutdown Items 407
    - 15.6.2.2 Inspect the Following Individual Items 407
    - 15.6.2.3 Periodic On-Line Items 408
- 15.7 Water Weir, Spray Quench Contactor, Quench Tank 408
  - 15.7.1 Installation 408
  - 15.7.2 Maintenance 408
    - 15.7.2.1 Shutdown Items 408
    - 15.7.2.2 Periodic On-Line Items 409
- 15.8 Heat Exchangers 409
  - 15.8.1 Installation 410
  - 15.8.2 Maintenance 410
    - 15.8.2.1 Shutdown Items 410
    - 15.8.2.2 Periodic On-Line Items 411
- 15.9 Liquid Seals 411
  - 15.9.1 Installation 411
  - 15.9.2 Maintenance 411
    - 15.9.2.1 Shutdown Items 411
    - 15.9.2.2 Periodic On-Line Items 412

15.10 Absorbers and Scrubbers 412
    15.10.1 Installation 412
    15.10.2 Maintenance 413
        15.10.2.1 Shutdown Items 413
        15.10.2.2 Periodic On-Line Items 413
15.11 Conclusion 414
Reference 414

15.1 Introduction

Correct installation and maintenance of a thermal oxidizer (TO) system (see Figure 15.1) requires experience, training, and a proactive maintenance and spare parts program to maximize the uptime of the total system. A TO system is not only the TO, but also the control system and the other associated equipment that attaches to the TO such as a boiler, quench system, absorbers, scrubbing equipment, blowers, heat exchangers, and pumps. Almost every TO system will be configured differently in their supply of equipment in order to match the exact needs at each site and for different waste streams. Since TO systems can vary so much in their scope of supply, this chapter addresses some of the more basic items of installation and maintenance of the major pieces of this equipment. These following installation instructions are intended for general guidance to be used only by erection or installation contractors having experience in this type of equipment.

Do not attempt to install or erect the TO system without first being thoroughly familiar with the original installation and operation instructions that came with the equipment. Loss of life and damage to equipment may result if the instructions are not strictly followed.

Verify the weight of each piece of equipment to assure that the machinery used for moving and lifting is properly rated, capable of safely handling the weight of the equipment.

Consult reference drawings listed in the original manual and other included drawings before attempting to erect or install any equipment.

FIGURE 15.1
Drawing of a typical TO system.

15.2 Burner/Pilot

15.2.1 Installation

Refer to Chapter 11 for additional information on burner installation.

Always refer to the original job drawings and operating manual as a reference before proceeding with the installation. The exact order of the installation may vary slightly from job to job. The majority of TOs have a single burner (see Figure 15.2), but sometimes the TO system may also have multiple burners. Prior to installing the connecting lines to the pilot and burner, blow down any of these upstream lines with an inert gas to a safe location to remove any scale, rust, and any other particulate matter prior to the first introduction of the fuel gases to prevent plugging the pilot and the burner:

- Bolt the burner to the TO with the required nuts, bolts, and gasket.

  Confirm that the burner has the pilot and sight glasses installed. If they are not installed, then proceed with their installation per the job drawings. Do not over-torque the sight glasses to prevent breaking the glass. A flanged 2 in. (5 cm) sight port with a 4 in. (10 cm) glass should have bolts tightened by fingers first, and then torque them in a crisscross star pattern to 8 ft-lb (11 N-m). A flanged 4 in. (10 cm) sight port with a 6 in. (15 cm) glass should have bolts tightened by fingers first, and then torque them in a back and forth star pattern to 18 ft-lb (24 N-m).
- Install the flame scanners.
- Install the burner tile if it has not already been installed using the proper refractory mortar.
- Install the piping from the necessary utilities (i.e., fuel gas, combustion air, waste streams, etc.) to the connections on the burner per the job drawings. Flex hoses and/or expansion joints should be used to be able to withstand the necessary thermal movement from the TO.
- Install the flame scanners.
- Install the purge air tubing, and purge air rotameters to the required purge connections such as sight glass and flame scanner.
- Install the ignition transformer and install the proper ignition wire to the igniter on the pilot.
- Complete the flame scanner wiring.
- Inspect the burner tips and cone to ensure that there are no blockages or plugged ports, and make sure flameholder(s) and gas tip(s) are located in the correct position. Ensure gas tips are spaced/aligned correctly to the burner tile per the job drawing.
- Look for proper placement of the pilot tip in the guide tube.
- Check the pilot to confirm that the ceramic insulators are not broken and the igniter tip is in the proper position.

FIGURE 15.2
Typical TO burner.

15.2.2 Maintenance

Refer to Chapter 11 for additional information on burner maintenance.

The maintenance of the burner and pilot is divided into shutdown items and periodic on-line items. Shutdown items are those items that can only be accomplished safely during a TO shutdown. Periodic on-line items are those items that can be accomplished safely with the TO on-line. Periodic on-line items should also be checked during a TO shutdown.

15.2.2.1 Shutdown Items

1. Inspect the gas tip(s) for visible deformations and weld cracks. Replace in kind if necessary. Pull the gas gun out and inspect the gas tip for blockage, coking (see Figure 15.3), and/or damage (see Figure 15.4). This should be done approximately every 6 months or during a planned outage.

2. Inspect the strainers on the fuel gas line. Clean screen and replace in kind if necessary.

3. Inspect the burner tile for deep cracks, with evidence of pulling away from the shell (see Figure 15.5). This should be done every 6 months or during a planned outage.

4. Clean the sight port(s) on the burner, as needed. Confirm the proper purge flow rate that should be about 5–7 SCFM (0.14–0.20 Nm^3/h) for a 2 in. (5 cm) sight port and 20–25 SCFM (0.57–0.71 Nm^3/h) for 4 in. (10 cm) sight port.

5. Inspect the burner stabilization cone for damage (see Figure 15.6) approximately every 6 months or during a planned outage.

FIGURE 15.4
Damaged burner tip.

FIGURE 15.5
Damaged burner tile.

6. Inspect the pilot assembly approximately every 6 months or during a planned outage for the following:

   a. Inspect the pilot orifice spud. Ensure that the spud is straight and plumb. Check for plugged orifice, clean orifice, and confirm drilling. Replace in kind if necessary.

   b. Inspect the pilot tip for damage.

   c. Inspect the ignition rod tip for damage.

FIGURE 15.3
Coked-up burner tip.

FIGURE 15.6
Damaged burner cone.

d. Inspect the ignition rod insulators to see if they are cracked, broken, or dirty (see Figure 15.7).

e. Inspect to see if the ignition rod is grounding out along the side of the pilot.

f. Check/inspect high-voltage ignition wire and replace as needed.

g. Inspect the ignition transformer for proper spark every 6 months or during a planned shutdown.

h. Inspect the pilot mixer for cracks. Replace in kind if necessary.

7. Clean the flame scanner lens as needed.

8. Check for overall corrosion on the burner housing approximately every 6 months or during a planned outage.

9. Check the main gas/pilot gas regulators. Replace the diaphragms on a regular schedule.

FIGURE 15.7
Dirty pilot ceramic insulator.

15.2.2.2 Periodic On-Line Items

Access to the burner(s) while the TO is on-line is not safe. However, inspection can be accomplished using the sight ports when necessary:

- Monitor the flame pattern on a regular basis and if the flame pattern suddenly changes, becomes unstable, becomes smoky, nonsymmetrical, or has sparklers in the flame. Be sure to inspect the items mentioned earlier at the next shutdown.

15.3 Thermal Oxidizer

A TO is a refractory lined vessel, which is fitted with one or more burners (see Volume 3, Chapter 8). It is designed to provide adequate residence time, at an elevated temperature for controlled completion of the required oxidation of combustible materials (see Figure 15.8). TOs are designed, selected, and manufactured to meet the specific customer requirements.

15.3.1 Installation

Always refer to the original job drawings and operating manual as a reference before proceeding with the installation. The exact order of the installation may vary slightly from job to job:

- Set and secure the TO to the foundation—Note: If supplied, install the Fluorogold® slide plates; they typically will need to be welded to both of the saddle base plates and both of the foundation base plates to handle the thermal expansion. If there are slotted holes in the base plate for a sliding saddle, be sure that the anchor bolt nut is loose and not tight or binding that would prevent the free movement of the saddle (see Figure 15.9).

- Make sure there are no ridged structures that could be damaged during expansion and contraction of the TO vessel.

- Install the sight glasses with the nuts, bolts, and gaskets. Do not overtorque the sight glasses to prevent breaking the glass. A flanged 2 in. (5 cm) sight port with a 4 in. (10 cm) glass should have bolts tightened by fingers first, and then torque them in a crisscross star pattern to 8 ft-lb (11 N-m). A flanged 4 in. (10 cm) sight port with a 6 in. (15 cm) glass should have bolts tightened by fingers first, and then torque them in a back and forth star pattern to 18 ft-lb (24 N-m).

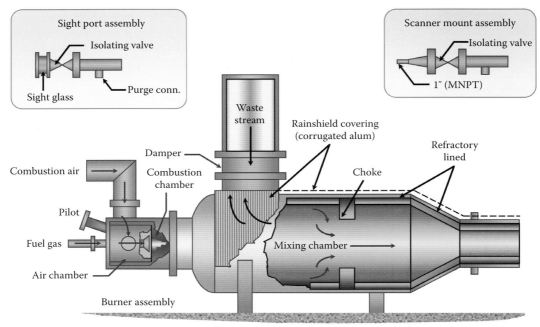

FIGURE 15.8
Typical TO.

FIGURE 15.9
Sliding anchor bolt.

- Install the sight glass and flame scanner purge air tubing, and purge air rotameters.
- Install the thermowells/thermocouples.
- Install the rain shield—if supplied.
- Install/inspect the refractory. Look for refractory defects, missing refractory, and broken or missing burner tile. Ensure all designated openings in the refractory are in place. Measure the refractory for correct thickness per drawing. Make sure hangers/anchors are in place, if required.
- Look to make sure all openings (i.e., flame scanners, combustion/quench air, quench water, waste, etc.) are not blocked by refractory.

15.3.2 Maintenance

The maintenance of the TO is divided into shutdown items and periodic on-line items. Shutdown items are those items that can only be accomplished safely during a TO shutdown. Periodic on-line items are those items that can be accomplished safely with the TO on-line. Periodic on-line items should also be checked during a TO shutdown.

15.3.2.1 Shutdown Items

Inspect the refractory for deep cracks in the castable/brick section with evidence of spalling (see Figure 15.10) or pulling away from the shell and for broken keepers on ceramic blanket linings. (Note: Small hairline expansion cracks are normal once the vessel has cooled down from normal operating temperatures. These cracks will close upon heating back to normal operating temperatures). Note: Also refer to Chapter 5 for additional information on refractory:

- Take pictures of the refractory in each inspection in order to provide a historical reference. This should be done approximately every 6 months or when there is a planned outage. As

Thermal Oxidizer Installation and Maintenance

FIGURE 15.10
Spalled refractory.

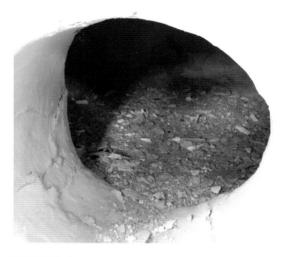

FIGURE 15.12
Example of solids in a TO.

extended operating experience is gained, the frequency of inspection should be reevaluated. If a wooden lead pencil cannot be placed into the crack or pinch spall for a depth of at least 1 in. past the sharpened point, then nothing should be done with the crack. A qualified refractory repair vendor should perform routine maintenance on the refractory system.

- Clean sight ports on the TO as needed. Confirm the proper purge flow rate, which should be about 5–7 SCFM (0.14–0.20 Nm^3/h) for a 2 in. (5 cm) sight port and 20–25 SCFM (0.57–0.71 Nm^3/h) for 4 in. (10 cm) sight port (see Figure 15.11).
- Clean out any solid buildup inside of the TO (see Figure 15.12).

FIGURE 15.13
Example of corrosion in the shell.

15.3.2.2 Periodic On-Line Items

- Perform yearly external infrared thermal imaging survey of the external shell in order to detect hot spots from missing refractory.
- Check for external corrosion on shell and add touch-up paint as necessary. Conduct yearly thickness checks of the steel shells and maintain in maintenance records (see Figure 15.13).

15.4 Boilers

Boilers are typically either a fire-tube boiler (see Figure 15.14) (hot flue gases are inside of the tubes) or a watertube boiler (see Figure 15.15) (water/steam is inside of the tubes). The maintenance of the boiler is divided

FIGURE 15.11
Dirty sight port.

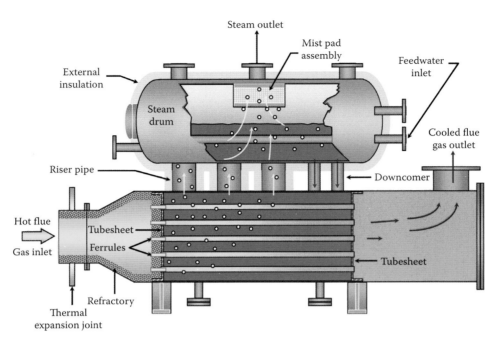

FIGURE 15.14
Typical fire-tube boiler.

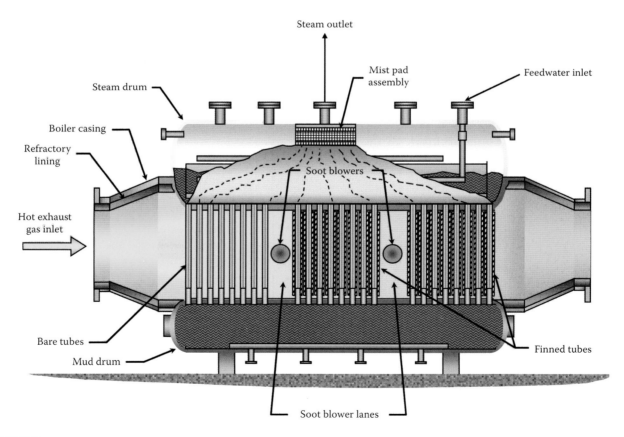

FIGURE 15.15
Typical watertube boiler.

into shutdown items and periodic on-line items. Shutdown items are those items that can only be accomplished safely during a TO shutdown. Periodic on-line items are those items that can be accomplished safely with the TO on-line. Periodic on-line items should also be checked during a TO shutdown.

15.4.1 Installation

- Set and secure the boiler to the foundation—Note: If supplied, the Fluorogold slide plates will need to be welded to boiler sliding saddle base plate and both of the foundation base plates. If there are slotted holes in the base plate for a sliding saddle, be sure that the anchor bolt nut is loose and not tight or binding that would prevent the free movement of the saddle.
- Make sure there are no ridged structures that could be damaged during expansion and contraction of the boiler vessel.
- Bolt or weld the boiler to the TO as shown by the drawings.
- Install the boiler steam drum and downcomer piping if supplied loose on a fire-tube boiler.
- Note: A final field hydrostatic test will need to be performed before the boiler can receive a final ASME stamp/approval and be put into service. Contact the local Authorized Inspector (AI) for information.
- Install all of the boiler controls, valves, level gauges, relief valves, pressure gauges, water column (see Figure 15.16), etc., with the proper nuts, bolts, and gaskets, and install the necessary electrical interconnecting wiring.
- Install the steam outlet piping, chemical feed, blowdown, and feed water piping with the proper nuts, bolts, and gaskets.
- Install the tube sheet ferrules and refractory to the inlet tube sheet if the boiler is a fire-tube boiler.
- Boiler and piping is then typically externally insulated and cladding applied by the insulating contractor as is directed by the job drawings.
- Before the first heat up of the boiler, the boiler is filled with cleaning chemicals to initially "boil out" or clean the boiler of the remaining oils, rust, and loose weld material left over from the fabrication of the boiler. This cleanout of the boiler insures a clean boiler that maximizes the heat transfer in the boiler. Consult with both the local chemical treatment company and the boiler manufacturer on the exact chemicals and heat up schedule that are to be used for the boiler.
- Take pictures of the ASME boiler stamped nameplate and maintain the records along with the ASME code paperwork supplied by the fabricator.

FIGURE 15.16
Typical boiler water column and gauge glass.

15.4.2 Maintenance

The instructions and procedures outlined in this section are intended to allow general maintenance to be carried out by competent operators who are familiar with boiler operations. All maintenance and operating problems are not presumed covered since every installation is different to some degree. It is therefore important that operators be familiar with the general control logic of not only the boiler but all associated equipment. The maintenance of the boiler is divided into shutdown items and periodic on-line items. Shutdown items are those items that can only be accomplished safely during a TO shutdown. Periodic on-line items are those items that can be accomplished safely with the TO on-line. Periodic on-line items should also be checked during a TO shutdown.

15.4.2.1 Shutdown Items

- Check the tube side and the shell side for signs of plugged tubes or passages or signs of corrosion and/or erosion.
- Remove any solid buildup.
- Inspect the refractory and ceramic ferrules for damage, signs of cracks, overheating, or erosion (see Figures 15.17 and 15.18). Inspect the refractory for deep cracks in the castable/brick section with evidence of pulling away from the shell and for broken keepers on ceramic blanket linings. (Note: Small hairline expansion cracks are normal once the vessel has cooled down from normal operating temperatures. These cracks will close up upon heating back up to normal operating temperatures.) Take pictures of the refractory at each inspection in order to provide a historical reference. This should be done approximately every 6 months or when there is a planned outage. As extended operating experience is gained, the frequency of inspection should be reevaluated. If a wooden lead pencil cannot be inserted into the crack or pinch spall for a depth of at least 1 in. (2.5 cm) past the sharpened point, then nothing should be done with the crack. A qualified refractory repair vendor should perform routine maintenance on the refractory system.
- Inspect all level gauge glasses for cracks, leakage, or damage and replace and/or clean if necessary.
- Inspect, calibrate, and confirm proper operation of all major interlocked shutdowns at least once a year.
- Calibrate all pressure switches and gauges at least once a year.
- Test all valves for proper operation at least once a year.
- Check all strainers and steam traps for fouling or damage at least every 6 months.
- Calibrate and test all transmitters at least once a year.
- Test and calibrate the safety relief vent valves at least once a year.

15.4.2.2 Periodic On-Line Items

- At least once per week, an operator should ascertain whether the pipe, fittings, and valves between the boiler and the water glass are free and open by blowing down the water columns and water glasses and noting the promptness of the return of the water to the glasses. This

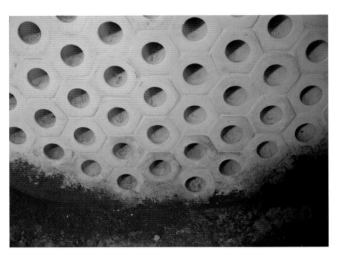

FIGURE 15.18
Boiler ferrules—after repair.

FIGURE 15.17
Damaged boiler ferrules—before repair.

should be done at least once a week and more frequently if trouble is experienced with boiler compounds, foaming, priming, and other feed water troubles that are apt to cause choking of the connections. It should also be done after replacing the water glasses. Typically, there is a manually held temporary bypass push button that allows this to be performed without shutting down the TO system.

- Keep the water column well illuminated, and keep the glass clean. A dirty mark in a gauge glass can be mistaken all too easily for the water level. Do not allow steam or water to leak from the water-level glass. Do not allow steam or water to leak from the water column or its connections, as this will cause the water glass to show a false water level. Keep the outlet end of the drain pipes from the water column, water glass, and gauge cocks free from obstructions.

- When the level of the water is not visible in the water glass, follow the procedure and blow down the water column or gauge glass to determine whether the existing level of the water is above or below the water glass. If the water level is below the water glass, shut down the flue gas unless certain that feed water is being supplied, in which case reduce the flue gas rate by adjusting the process.

- Perform regular boiler feed water treatments and analyze the boiler feed water on a routine basis as established by the plant's water treating consultant.

- Blow down the boiler regularly as determined by an analysis of the boiler water. The object of blowing down is to maintain the concentration of dissolved and suspended solids within safe limits to avoid priming and carry over. Blow down the boiler at a period during the day when the steam demand is lowest. When blowing down the boiler, open fully the valve next to the boiler first, then open the second valve. When closing, close fully the second (outer) valve first, and then the valve next to the boiler.

15.5 Miscellaneous Control Items

15.5.1 Installation

For the installation of miscellaneous control items, i.e., control valves, regulators, pressure/temperature/flow switches, etc., it is best to follow each individual vendor's instructions per their own installation and operating manual.

Carefully install the necessary thermowells/thermocouples (see Figure 15.19) to avoid breakage, and verify

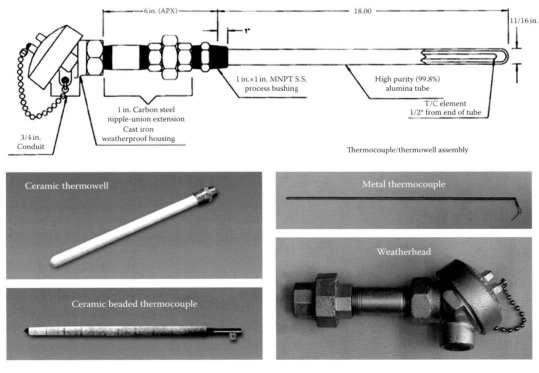

FIGURE 15.19
Thermocouples and thermowell.

FIGURE 15.20
Cracked damaged expansion joint.

if the correct thermocouples are installed (type K, J, R, etc.) and are hooked up properly to the correct extension wire. Carefully observe the special polarity of the extension wire to the thermal couple (e.g., type-K wire is usually colored red and yellow and the red wire is actually the negative wire).

15.5.2 Maintenance

This section on the maintenance of the miscellaneous control items is divided into shutdown items and periodic on-line items. Shutdown items are those items that can only be accomplished safely during a TO shutdown. Periodic on-line items are those items that can be accomplished safely with the TO on-line. Periodic on-line items should also be checked during a TO shutdown.

15.5.2.1 Shutdown Items

- Inspect, calibrate, and confirm proper operation of all major interlocked shutdowns at least once a year.
- Calibrate pressure switches and gauges at least once a year.
- Test all valves for proper operation at least once a year.
- Check all strainers for fouling or damage at least every 6 months.
- Calibrate and test all transmitters at least once a year.
- Inspect the blower(s) for damage and for proper operation at least once a year (clean and check inlet screen, lubricate bearings, inspect couplings and/or belts). Note: Also see Chapter 3 for more information on fans and blowers.
- Inspect all thermocouples for damage or breakage every 6 months and replace as necessary.
- Check front of control panel for proper operation (test light bulbs, meters, etc.), at least once a year.

15.5.2.2 Periodic On-Line Items

- Inspect all flexible hoses and expansion joints for cracking, leakage, and/or failure at least once a year (see Figure 15.20).

15.6 Liquid, Air/Steam Atomizing Guns (Fuel/Waste Oil, Quench Water, Aqueous Wastes)

> **CAUTION**
>
> The alignment pins in the John Zink liquid body receiver assembly should align with holes in the body assembly to allow it to be inserted in only one way. If the liquid and atomizing steam/air connections are reversed, the operation of the liquid gun will be severely affected.

> **CAUTION**
>
> New guns are sometimes shipped with a protective cap over the exit ports. This should be removed prior to light-off.

15.6.1 Installation

Confirm that the liquid atomizing gun is installed per the original drawings supplied with the equipment. Prior to installing the connecting lines to the liquid gun, blow down any upstream lines to a safe location to remove any scale, rust, and any other particulate prior to the first introduction of the liquid and atomizing medium to prevent plugging the liquid gun. Install the liquid line and atomizing steam and/or airline to each gun and insure that it is connected properly and is not installed with the connections reversed. Use flexible hoses to all connections to allow for thermal movements and to allow for adjusting the gun insertion length (see Volume 1, Chapter 10 on oil atomization for more information).

15.6.2 Maintenance

The maintenance of the liquid atomizing gun is divided into shutdown items and periodic on-line items. Shutdown items are those items that can only be accomplished safely during a TO shutdown. Periodic on-line items are those items that can be accomplished safely with the TO on-line. Periodic on-line items should also be checked during a TO shutdown. Refer to the job drawings for additional details on the detailed parts of the liquid gun.

15.6.2.1 Shutdown Items

General points to remember when servicing any liquid gun:

- Correct parts: Check the liquid tip and atomizer to make sure it is the correct tip for that gun. The tip's part number should be stamped on the tip and should be checked against the burner drawing.
- Correct orientation: Check the gun for correct length, orientation, and installation according to the burner drawings. This will ensure it is correctly positioned inside the burner or vessel, thus providing the correct capacity and flame pattern. Be sure that oil, steam, air, and flex hoses are installed in the correct orientation and are connected to the correct connections on the oil body receiver.
- Proper seals and alignment: Inspect and replace all gaskets, seals, and o-rings to ensure proper interface seals. The correct operation of the liquid gun depends on several machined orifices, channels, and seals properly aligned in their place.
- Corrosion and erosion: Liquid gun components are subject to high velocities, abrasive flows, and corrosive action. The resulting wear on these components is compounded by the contaminants found in many liquids. Coke or carbon particles, catalyst fines, and silica particles have a highly erosive action on metal parts when subjected to high-pressure, high-velocity metering. Sulfur chloride compounds and, in some cases, anhydrous acids will severely attack, through corrosion, the materials of the atomizer and dispersion nozzle.

15.6.2.2 Inspect the Following Individual Items

- Liquid tip: External inspection of the port region will reveal much about the condition of the liquid gun. Visual carbon or liquid buildup around the exit ports indicate three possible problems: tip inserted too far, low atomizing medium pressure/flow, or erosive/corrosive action on tip. Clean all foreign materials from the face of the liquid tip and inspect the exit ports for wear. Visual erosion, egg-shaping, of the exit ports will cause disruption of the flame pattern and is an indication that the tip needs replacement.
- Atomizer: External inspection of the steam ports should show them free of any foreign material. The exit port should be concentric and not eroded or corroded. The labyrinth seal should be clean and unscored in the longitudinal direction.
- Liquid spud: Liquid spud should be free of foreign materials and not eroded or corroded.
- Tip/atomizer assembly: Atomizer labyrinth seal should insert into the liquid tip with a minimum amount of clearance. New atomizers often require a twisting motion to achieve insertion. If atomizer/tip fit is not "firm," steam bypassing of the atomizing chamber is possible. Check the atomizer in a new tip to determine if it is the tip or the atomizer that is worn.

- Receiver and body assembly: Inspect the body and receiver for wear, corrosion, misalignment, failed gaskets, and missing pins and/or sleeves. Check to make sure the liquid and atomizing steam/air connections are not reversed.
- Replace parts that show the following signs of deterioration:
  - Enlargement of the liquid orifice: Caused by high liquid flow, low atomizing medium ratio, poor atomization, or burner overfiring
  - Enlargement of the atomizer exit: Caused by lowered mixing chamber pressure, reduced atomization quality, and burner overfiring
  - Deterioration of the atomizer labyrinth seal: Steam bypassing the atomizing chamber, poor atomization, instability, and nonsymmetrical flame patterns
  - Enlargement of the exit ports: Will increase the diameter of the exit port causing poor atomization, instability, and nonsymmetrical flame patterns

15.6.2.3 Periodic On-Line Items

- Check for leakage from gaskets located between the liquid body and the receiver. Replace as necessary. Monitor the flame pattern or spray pattern on a regular basis. If it suddenly changes or if the flame becomes smoky, perform maintenance as necessary if the system allows safe removal of the gun. Otherwise, be sure to inspect the items mentioned earlier at the next shutdown.
- Check the liquid pressure and atomizing medium pressure and the liquid flow rate if a flowmeter is available. Check these numbers against the capacity curve of the original liquid gun.
- Note: The viscosity of the liquid should always be maintained below 200 SSU (34 cP) at the normal operating temperature of the liquid.

15.7 Water Weir, Spray Quench Contactor, Quench Tank

Water quenching is used in TO systems either with no heat recovery or downstream of heat recovery equipment. Typically, the hot flue gases pass through an annular weir assembly, through a contactor tube fitted with spray guns, and through a downcomer into a quench tank (see Figure 15.21). The spray guns must provide efficient atomization across the full cross section of the contactor tube, if quenching is to be effective and complete saturation of flue gases to be achieved.

15.7.1 Installation

Review the job drawings showing the installation detail. First confirm that the concrete support base for the flat-bottomed quench pot vessel will provide full and uniform support over the entire tank/vessel bottom area and is level. The support base must be properly designed to prevent settling or deflection under maximum design loads. The support base surface shall be nonporous and free of all cracks, depressions, vertical projections, and steep changes in plane. Reinforced concrete, towel finished to American Concrete Institute Specifications (ACI-301-72, Section 11.7.3, "Trowel Finish"), is most often used as a support base.[1] If the quench pot and spray quench contactor are made out of fiberglass materials, be sure to observe the special lifting requirements from the fiberglass supplier.

Install the quench tank on the foundation anchor bolts. Bolt the contactor and water weir using the proper nuts, bolts, and gaskets. Typically, the contactor is installed first and then the refractory brick is installed later. Typically, multiple spray guns will need to be installed in the contactor; refer to installation instructions included in the original operating manual. Use Teflon®-lined flexible hoses to connect up to the spray nozzle guns. Insure that the contactor spray tips clear the inside refractory by ½ in. (1.3 cm). It is very important to observe any requirement of each nozzle orientation as some quench nozzles are flat sprays that must be installed in the horizontal position. Care must be taken that the water weir is level after installation is complete. The water weir depends upon gravity for even flow of liquid over a baffle wall, and proper operation of this device is predicated on proper orientation.

15.7.2 Maintenance

The maintenance of the water weir, spray quench contactor, and quench pot is divided into shutdown items and periodic on-line items. Shutdown items are those items that can only be accomplished safely during a TO shutdown. Periodic on-line items are those items that can be accomplished safely with the TO on-line. Periodic on-line items should also be checked during a TO shutdown.

15.7.2.1 Shutdown Items

- Inspect, calibrate, and confirm proper operation of all major interlocked shutdowns at least once a year.

Thermal Oxidizer Installation and Maintenance

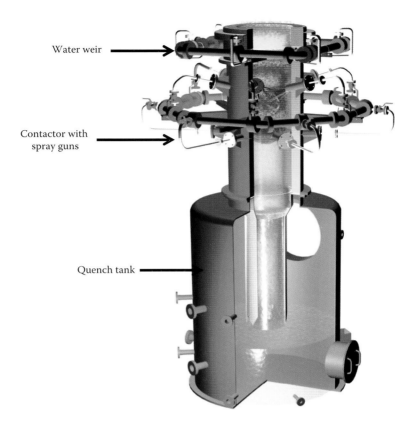

FIGURE 15.21
Typical quench system.

- Confirm that the water weir is level and that water flow out of the weir opening covers the entire weir and that there is an even coverage over the contactor brick below the weir. Check for signs of corrosion and that no passages or holes are plugged.
- Check for signs of plugged passages and/or holes and corrosion and erosion on the contactor spray tips at least once a year, and replace as needed.
- Check for signs of corrosion, erosion, and hot spots on the quench pot. Remove any solid buildup in the bottom of the quench pot.
- Check and calibrate the level instruments at least once a year.
- Inspect and clean any level gauge glasses.

15.7.2.2 Periodic On-Line Items

- Perform a yearly external infrared thermal imaging survey of the external shell in order to detect any potential hot spots.
- Check for any leaking flanges that need to have the gaskets replaced.
- Check the liquid pressure and atomizing medium pressure and the liquid flow rate if a flowmeter is available. Check these numbers against the capacity curve of the original liquid gun.

15.8 Heat Exchangers

The instructions and procedures outlined in this section are intended to allow general maintenance to be carried out by competent operators who are familiar with heat exchanger operations. All maintenance and operating problems are not presumed covered since every installation is different to some degree. It is therefore important that operators be familiar with the general control logic of not only the heat exchanger but all associated equipment. See individual vendor literature in the operating manual for additional details. The heat exchanger typically takes the higher-temperature flue gases and transfers heat and preheats the cooler combustion air or waste gas to recover the heat that would be lost if the heat exchanger was not installed (see Figure 15.22).

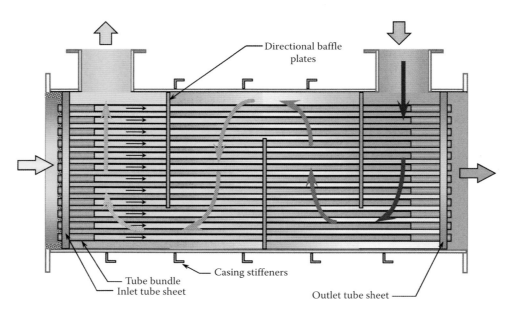

FIGURE 15.22
Typical heat exchanger.

15.8.1 Installation

- Set and secure the heat exchanger to the ductwork and/or support structure, if used.
- Make sure there are no ridged structures that could be damaged during expansion and contraction of the heat exchanger.
- Bolt the heat exchanger to all of the necessary ductwork as shown by the drawings using the correct nuts, bolts, and gaskets.
- Install all of the heat exchanger control items, i.e., valves, flowmeters, thermowells/thermocouples, temperature switches, pressure gauges, etc.
- If not already applied, install insulation and cladding to ductwork (usually applied by the insulating contractor). Check the installation drawings to confirm what is necessary.

15.8.2 Maintenance

The maintenance of the heat exchanger is divided into shutdown items and periodic on-line items. Shutdown items are those items that can only be accomplished safely during a TO shutdown. Periodic on-line items are those items that can be accomplished safely with the TO on-line. Periodic on-line items should also be checked during a TO shutdown.

15.8.2.1 Shutdown Items

- Inspect, calibrate, and confirm proper operation of all major interlocked shutdowns at least once a year.
- Check the tube side and the shell side of the heat exchanger to see if the heat exchanger tube or passageways are plugged or show signs of corrosion, and/or erosion, and any damage or cracking in the tubes.
- Remove any solid buildup.
- Inspect refractory for signs of cracks, overheating, or erosion. Inspect refractory for deep cracks in the castable/brick section, if used, with evidence of pulling away from the shell and for broken keepers on ceramic blanket linings. (Note: Small hairline expansion cracks are normal once the vessel has cooled down from normal operating temperatures. These cracks will close upon heating back to normal operating temperatures.) Take pictures of refractory in each inspection in order to provide a historical reference. This should be done approximately every 6 months or when there is a planned outage. As extended operating experience is gained, the frequency of inspection should be reevaluated. If a wooden lead pencil cannot be inserted into the crack or pinch spall for a

depth of at least 1 in. (2.5 cm) past the sharpened point, then nothing should be done with the crack. A qualified refractory repair vendor should perform routine maintenance on the refractory system.

15.8.2.2 Periodic On-Line Items

- Perform a yearly external infrared thermal imaging survey of external shell in order to detect any potential hot spots.
- Keep track of the temperature and pressure differentials across both sides of the heat exchanger as this can indicate if there are any internal problems occurring with the heat exchanger.

15.9 Liquid Seals

Sometimes liquid seals are used on waste vapor lines to minimize the possibility of a flashback (see Figure 15.23).

15.9.1 Installation

- Inspect, calibrate, and confirm proper operation of all major interlocked shutdowns at least once a year.
- Set and secure the liquid seal to the concrete foundation and connect to the anchor bolts.
- Bolt the liquid seal to all of the necessary piping as shown by the drawings using the correct nuts, bolts, and gaskets.
- Install all of the liquid seal control items, i.e., valves, level switches, thermowells/thermocouples, temperature switches, pressure gauges, etc.
- Heat trace and insulate as necessary for the site atmospheric conditions.

15.9.2 Maintenance

The maintenance of the liquid seal is divided into shutdown items and periodic on-line items. Shutdown items are those items that can only be accomplished safely during a TO shutdown. Periodic on-line items are those items that can be accomplished safely with the TO on-line. Periodic on-line items should also be checked during a TO shutdown.

15.9.2.1 Shutdown Items

- Drain the drum completely and replace the liquid at least annually. Inspect inside the bottom head for accumulated solids and corrosion. Remove any such solids.
- Inspect, calibrate, and confirm proper operation of all major interlocked shutdowns at least once a year.

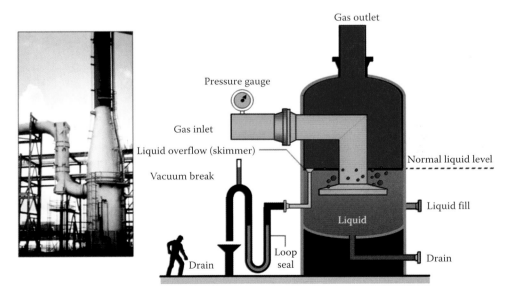

FIGURE 15.23
Typical liquid seal.

> **CAUTION**
>
> Follow safe vessel entry procedures as required by plant standards.

- Inspect liquid seal internals for accumulated solids, cracks, plugged holes, or other obvious damage. Clean or repair as required.
- Check level instruments per vendor information.

15.9.2.2 Periodic On-Line Items

- Check physical liquid level in the gauge glass on a regular schedule. Use a gauge glass level to check level control instruments.
- Skim any accumulated hydrocarbon liquids from the surface of the liquid on a regular schedule.
- If the waste gas contains any acidic or alkaline compounds, monitor the pH of the seal fluid. Treat or replace the seal fluid before the pH becomes corrosive to the vessel metallurgy.

15.10 Absorbers and Scrubbers

Review the job drawings showing the installation detail (see Figures 15.24 and 15.25). First, confirm that the concrete support base for the flat-bottomed absorber/scrubber will provide full and uniform support over the entire tank/vessel bottom area. The support base must be properly designed to prevent settling or deflection under maximum design loads. The support base surface shall be nonporous and free of all cracks, depressions, vertical projections, and steep changes in plane. Reinforced concrete, trowel finished to specifications for structural concrete (ACI-301-10, Section 11, "Trowel Finish"), is most often used as a support base. If the absorber/scrubber is made out of fiberglass materials, be sure to observe the special lifting requirements from the fiberglass supplier.

15.10.1 Installation

- Install the absorber/scrubber on the foundation anchor bolts. Bolt the absorber/scrubber using the proper nuts, bolts, and gaskets. Confirm that the absorber/scrubber is level.
- Confirm that any necessary lateral support from a tower structure is installed.
- If required, install any necessary refractory/brick lining.
- Install the necessary internals into the tower as indicated by the job drawings. These typically can include grid bar supports, support plates, tower packing, liquid distributors,

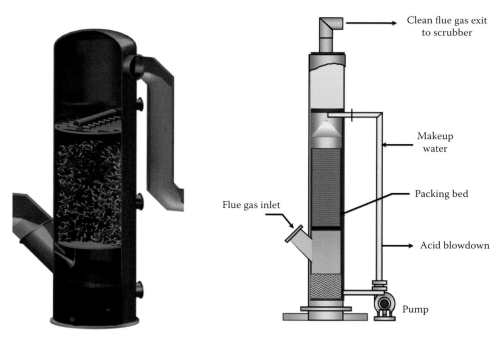

FIGURE 15.24
Typical packed quench absorber tower.

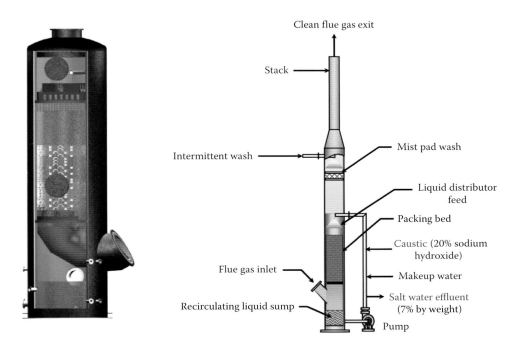

FIGURE 15.25
Typical scrubber.

mist eliminator pads, and wash down spray nozzles. Follow the vendor's installation instructions.

- Install any necessary inlet and outlet ductwork, expansion joints, and connecting piping, with the correct nuts, bolts, and gaskets.
- If all of the tower internals have been installed and inspected, install all manway covers with the correct nuts, bolts, and gaskets.
- Install all necessary control items, i.e., level gauges, level switches, thermowells/thermocouples, etc.

15.10.2 Maintenance

The maintenance of the absorber/scrubber is divided into shutdown items and periodic on-line items. Shutdown items are those items that can only be accomplished safely during a TO shutdown. Periodic on-line items are those items that can be accomplished safely with the TO on-line. Periodic on-line items should also be checked during a TO shutdown.

15.10.2.1 Shutdown Items

- Inspect, calibrate, and confirm proper operation of all major interlocked shutdowns at least once a year.

- Check for signs of plugged passages and corrosion and erosion on the absorber/scrubber packing at least once a year, and replace as needed.
- Check for signs of corrosion, erosion, and hot spots on the absorber/scrubber. Remove any solid buildup in the bottom of the absorber/scrubber.
- Check and calibrate level instruments at least once a year.
- Inspect and clean any level gauge glasses.

15.10.2.2 Periodic On-Line Items

- Perform yearly external infrared thermal imaging survey of external shell in order to detect any potential hot spots.
- Check for any leaking flanges that need to have the gaskets changed out.
- If the absorber/scrubber tower is displaying higher than normal pressure drop, there are companies available that can run gamma-ray scans into the packed column and can confirm if certain areas of the packing are damaged, plugged, or fouled.

15.11 Conclusion

As discussed in the introduction, the correct installation and maintenance of a TO system requires experience, training, and a proactive maintenance and spare parts program to maximize the uptime of the total system. It is highly recommended to keep good maintenance records along with pictures for an ongoing historical database for the TO system.

Reference

1. Specifications for Structural Concrete. ACI-301-10. American Concrete Institute Committee 301.

16
Thermal Oxidizer Operations and Troubleshooting

Dale Campbell

CONTENTS

16.1 Introduction 416
    16.1.1 Safety Warnings 416
        16.1.1.1 Fire and Explosion Hazards 417
        16.1.1.2 Elevated Temperatures 417
        16.1.1.3 Electrical Hazards 417
        16.1.1.4 Rotating and Mechanical Equipment 417
16.2 Training 418
16.3 Burner/Pilot 418
    16.3.1 Operations 418
    16.3.2 Troubleshooting 419
16.4 Thermal Oxidizer 420
    16.4.1 Operations 420
    16.4.2 Troubleshooting 422
16.5 Boiler 423
    16.5.1 Operations 423
        16.5.1.1 Boiler Start-Up 423
        16.5.1.2 Start-Up after Temporary Shutdown 423
        16.5.1.3 Boiler Shutdown Procedure 425
        16.5.1.4 Normal Boiler Operation 426
        16.5.1.5 Placing the Boiler into Service 426
        16.5.1.6 Care When Out of Service 426
    16.5.2 Troubleshooting 427
        16.5.2.1 Foaming and Priming 427
        16.5.2.2 Scale in Boiler 427
        16.5.2.3 Corrosion or Pitting 428
16.6 Miscellaneous Control Items 428
    16.6.1 Operations 428
    16.6.2 Troubleshooting 428
16.7 Liquid and Air/Steam Atomizing Guns (Fuel/Waste Oil, Quench Water, Aqueous Wastes) 428
    16.7.1 Operations 428
    16.7.2 Troubleshooting 428
16.8 Water Weir, Spray Quench Contactor, Quench Tank 428
    16.8.1 Operations 428
        16.8.1.1 Pre-Start-Up Checklist 429
        16.8.1.2 Start-Up Philosophy and Overview 430
        16.8.1.3 Normal Operation 431
    16.8.2 Troubleshooting 431
16.9 Heat Exchangers 431
    16.9.1 Operations 431
    16.9.2 Troubleshooting 433
16.10 Liquid Seals 433
    16.10.1 Operations 433
    16.10.2 Troubleshooting 433

16.11 Absorbers and Scrubbers 433
       16.11.1 Operations 433
       16.11.2 Troubleshooting 435
16.12 Conclusion.. 435
References.. 435

16.1 Introduction

Correct operations and troubleshooting of a thermal oxidizer (TO) system requires experience, training (see Volume 1, Chapter 17), and a proactive maintenance (see Chapter 15) and spare parts program in order to maximize the uptime of the total system. A TO system is not only the TO vessel, but also the control system and other downstream equipment that attaches to the TO such as a boiler, quench system, absorbers, scrubbing equipment, blowers, heat exchangers, and pumps. Almost every TO system will be configured differently, in order to match the needs at each site and for different waste streams. Since TO systems can vary so much in their scope of supply, this chapter addresses the basic items of operations and troubleshooting. It is important to follow the operating manual for the originally supplied TO system that was supplied with the equipment. Refer to Chapter 15 for additional information on TO installation and maintenance.

16.1.1 Safety Warnings

The following are general typical safety precautions for the operation and maintenance of a TO system. Instructions contained in this section are in addition to, and do not replace, the operating company's and owner's existing operating procedures and policies with regard to standard safety precautions for the TO, burner, and fuel system operation.

> **WARNING**
>
> High voltages capable of causing death are used with this equipment. Use extreme caution when servicing control cabinets and electrically actuated components.

> **WARNING**
>
> Do not enter the TO vessel(s) until an adequate cool-off period has been observed and your company's vessel entry procedures have been completed. Enter the vessels only in the presence of someone who is capable of rendering aid.

> **WARNING**
>
> Use or operation of the TO system at other than the specified conditions can result in hazardous operation, unsatisfactory performance, and/or deterioration or destruction of the equipment.

> **WARNING**
>
> Any attempt to defeat or compromise the performance of the purge function could result in severe damage to the unit, personal injury, and/or death.

> **WARNING**
>
> All fuels are flammable and potentially explosive. Familiarize yourself with the specific welding, hot work guidelines, torquing, draining, venting, purging, bleed-down procedures, leak checks, and line-entry instructions for the components serviced before starting work on the fuel system.

> **WARNING**
>
> Failure to use proper respiratory equipment may result in exposure to hazardous materials. Dust generated by refractory may contain crystalline silica, which is known to cause cancer. In addition to the use of proper respiratory equipment, please refer to the refractory company's latest MSDS to avoid exposure.

> **CAUTION**
>
> Failure to adhere to the refractory heat-up/cool-down procedures as listed in the operating instructions for your TO system may result in damage to your equipment.

Thermal Oxidizer Operations and Troubleshooting 417

> **NOTE**
>
> At least the following personal protection equipment will be necessary when operating this equipment: fire retardant/resistant clothing, ear protection, gloves, and eye protection. When installing the equipment, additional safety equipment, such as steel-toed shoes, hard hat, and respiratory protective equipment, will be necessary. Refer to the personal protection equipment requirements of the operator and owner and the regulatory authorities (such as OSHA) to determine the personnel protection equipment appropriate for the work being performed.

> **WARNING**
>
> Safeguard interlocks shall not be bypassed at any time during system start-up or operating modes.

See Chapter 1 for more general safety information.

16.1.1.1 Fire and Explosion Hazards

The TO burner utilizes fuel, which is flammable and potentially explosive. Extreme care should be exercised when making fuel piping connections. Use the correct gaskets, bolts, thread lubricants, and tightening torques to prevent leaks. It is recommended that drain and/or vent piping be channeled to safe locations. Valve packings should be periodically tightened and a rigorous leak check program be implemented as part of your preventive maintenance. Before providing fuel train component maintenance, make sure the component is isolated from the fuel source and the applicable equipment purge and bleed-down requirements are followed. Consult your company's safety department for line entry and other safety procedures.

Flame management sequencing and interlocks should never be bypassed. Consult your company's safety department and your insurance underwriter's guidelines for support to determine safety-related requirements.

Never attempt welding on any pipe train or in its vicinity without consulting your safety department for "hot work" guidelines.

16.1.1.2 Elevated Temperatures

The TO burner and associated equipment generate very high internal temperatures. Refractory and insulation are typically provided to maintain acceptable external surface temperatures. Personnel protection shields are often used, but care must be exercised to prevent burns and other thermal hazards when in near proximity to the burner/TO system. Never enter a fired vessel until an adequate cool-off period has been observed and the company's vessel entry procedures have been completed.

16.1.1.3 Electrical Hazards

Potentially hazardous voltages exist in control cabinets and electrically actuated control components. These components should only be serviced when system power is removed and only by qualified electrical/instrument service personnel.

16.1.1.4 Rotating and Mechanical Equipment

This TO system may contain rotating equipment (force draft fans, pumps, etc.), mechanically automated devices (louvers or register with linkages and actuators), or electrically and pneumatically operated control components (control valves, block valves, etc.). Never operate this equipment unless guides, shields, or covers are in place. Be cognizant of moving components and never place oneself in a position where mechanical actuation could provide "pinch points" or other dangers.

First and foremost, only qualified and properly trained personnel should be allowed to operate and maintain any part of the burner and TO.

This safety section contains general safety guidelines for a typical TO. If there is a safety procedure in place at the job site, it should be the ruling document. It should also be noted that this section does not cover all areas of safety that may be of concern, but is merely a guideline to be used if no procedures are in place.

The following safety equipment may be necessary when operating a TO system: face shield, gloves, fire retardant/resistant clothing, and earplugs/protection.

When installing the TO system, additional safety equipment may be necessary such as safety glasses and a hard hat.

When working inside the TO vessel(s) during installation and maintenance, at least the following safety procedures should be put in place before entering the equipment:

- Disconnect and blind all fuel supply lines.
- Close and blind all fuel sources including waste vent gas dampers prior to entry into the vessel.
- Lock out/tag out (LOTO) all rotating equipment (force draft fans, pumps, etc.), mechanically automated devices (louvers or register with linkages and actuators), or electrically and

pneumatically operated control components (control valves, block valves, etc.).

- Purge TO vessel(s) for not less than five system air volume changes.
- If the TO vessel(s) has been in operation, make sure that the refractory has cooled and that the TO internal temperature is below 100°F (37°C).
- Place "Man in Vessel" signs at all manways, fuel control panels, supply valves, and fuel line blind locations.
- Provide radio contact between the inside and outside of the vessel or stand a watch at the TO manway.

There are some additional rules based on actual operating experiences, which are essential for personnel protection:

- Never look directly into a furnace or firebox without proper eye protection. When viewing flames, always wear tinted goggles or a face shield.
- Do not stand directly in front of open manways, viewing ports, or furnace doors. Variations in the firing rate of the equipment could cause pulsation and blast hot flue gases or slag out of these openings.
- Do not use open-ended pipes or uninsulated equipment to clear debris or slag from observation ports. The hot gases can be channeled and discharged through the open end of the pipe, or the pipe or uninsulated instrument can get extremely hot.
- Do not enter any confined space or any vessel until it is completely cooled and purged from combustibles or dangerous fumes and it is well ventilated. Always station another person at the entrance, and use a safety harness.
- Use grounded or low-voltage extension cords and explosion-proof lightbulbs and flashlights.
- Never open or enter any rotating equipment until it has come to a complete stop and it is secured with a braking or locking mechanism.
- Always secure the driving mechanism for dampers, gates, and doors before going through them.
- Always be prepared for the presence of hot water and its leakage when opening any manways or ports for inspection.

Important note: "LOTO" refers to specific practices and procedures to safeguard employees from the unexpected energization or start-up of machinery and equipment, or the release of hazardous energy during service or maintenance activities.

Approximately 3 million workers service equipment and face the greatest risk of injury if LOTO is not properly implemented. Compliance with the LOTO standard (29 CFR 1910.147)[1] prevents an estimated 120 fatalities and 50,000 injuries each year. Workers injured on the job from exposure to hazardous energy lose an average of 24 workdays for recuperation. In a study conducted by the United Auto Workers (UAW), 20% of the fatalities (83 of 414) that occurred among their members between 1973 and 1995 were attributed to inadequate hazardous energy control procedures, specifically LOTO procedures. Be sure to follow OSHA, the local state, and the specific jobsite requirements for LOTO.

16.2 Training

Training of the personnel that will have the responsibility of operating and maintaining any TO system must be done to insure the safety and long-term reliability of the TO system. Not only is the initial operator training important, but also the ongoing refresher training of the operators of the TO equipment. In fact, the National Fire Protection Association (NFPA), in their 2011 edition of their NFPA 86 Code, Standard for Ovens and Furnaces[2], states that personnel who operate, maintain, or supervise the TO system are to be thoroughly instructed and trained in their job function by qualified personnel and to also receive regularly scheduled refresher training and demonstrate their understanding in how to operate the equipment. Refer to Volume 1, Chapter 17 for additional information on combustion training. Reference [3] discusses TO training designed for operators in a plant.

16.3 Burner/Pilot

16.3.1 Operations

Refer to Chapter 12 for additional information on burner/pilot operation.

Gas-fired pilots will typically have a gas orifice that is drilled for a specific heat release at a given operating gas pressure. Refer to the documentation on the specific pilot to confirm the proper operating pressure and capacity for the pilot.

TO burners, much like other types of burners, are sized for a specific maximum heat release on gas and/or fuel

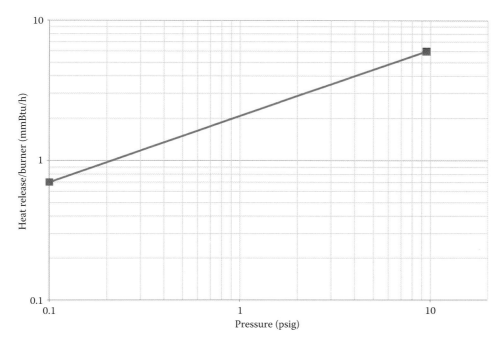

FIGURE 16.1
Typical gas capacity curve.

oil at a given maximum pressure. Every burner also has a minimum turndown point, below which the burner flame can go out and give you a flame failure. This is typically shown on the capacity curve that comes with the burner (see Figure 16.1). As an example, if the turndown range on a gas burner is 10 to 1, therefore if the maximum heat release is 20×10^6 Btu/h (5.9 MW), then the minimum turndown is 2×10^6 Btu/h (0.6 MW). This heat release versus pressure capacity curve details the stable operating range from minimum fire to maximum fire for that particular burner. Refer to the documentation on the specific burner to confirm the proper firing range of that burner.

Confirm that the proper amount of combustion air for the heat release of the burner is always maintained. Consult the operating manual for the specific burner and confirm by combustion airflow measurement or by excess oxygen in the flue gas.

Confirm proper purge flow rate to the burner in order to keep the sight ports and flame scanner ports clean and free from moisture, which should be 5–7 SCFM (0.14–0.20 Nm³/h) for a 2 in. (5 cm) sight port and 20–25 SCFM (0.57–0.71 Nm³/h) for a 4 in. (10 cm) sight port.

16.3.2 Troubleshooting

Refer to Chapter 13 for additional information on burner troubleshooting.

Failure of a pilot to ignite and burn steadily within a 10 s timed trial for ignition interval indicates a problem. Only after locating and correcting the problem should relighting of the pilot be attempted.

Visually observe the pilot through the sight port during the ignition period.

If no sparks are seen at the ignition spark gap at the pilot tip, check for improper spark gap or shorting in the high-tension spark system.

If sparks can be seen, but the pilot will not light, check for air in the natural gas supply line, insufficient natural gas pressure, plugged pilot orifice, or incorrect combustion air damper setting.

If the pilot lights, but the flame scanner does not activate the "Flame On" light, misalignment between the pilot flame and the scanner sight tube is the most probable cause.

Slight misalignment can often be overcome by simply turning up the pilot gas pressure to produce a larger flame body, which the scanner can "see" without difficulty.

If the preceeding step fails, *carefully* remove flame scanner from scanner tube and look into scanner tube. When properly aligned, the pilot flame body will completely cover the viewing area of the scanner tube.

If the flame cannot be seen, or it is off center of the sight tube, realign the pilot by carefully bending or shimming.

If alignment is correct, but scanner does not activate "Flame On" light, check the flame scanner by holding a match or other flame (flashlight will not work) in front of the scanner eye.

If scanner does not respond to flame, see the scanner troubleshooting guide.

If there is a flowmeter on the fuel that is being fired in the burner, compare this to the burner capacity curve to confirm if the burner is doing one of the following three things:

1. The burner is operating in the correct range of the curve and is performing properly.
2. The burner pressure is too high or the capacity is too low; then this is an indication that the ports in the tip(s) have been plugged with debris or has coked up. Inspect the burner and correct as necessary.
3. The burner pressure is too low or the capacity is too high; then this is an indication that the ports have been enlarged or a tip has fallen off due to damage. Inspect the burner and correct as necessary.

If the burner runs with too little combustion air, the burner can exhibit a woofing or pulsing. If this should happen, reduce the heat input into the burner first to remedy this condition. Do not try to increase the combustion airflow as this can actually make the pulsations worse.

In some cases, high-intensity spin-type burners can create high noise/vibrations in the downstream TO. See Volume 1, Chapter 16 for additional information on noise.

This is a phenomenon that cannot be predicted or modeled by computational fluid dynamics (CFD—see Volume 1, Chapter 13). It is influenced by the downstream dimensions of the TO system. Acoustic analysis of the system may be necessary of the equipment in the field to diagnose the cause of the noise. If the equipment frequency noise produced matches the resonant frequency of the equipment, this can greatly contribute to the production of noise. Some of the solutions in the past to resolve this type of noise issue include adding downstream random stacking or a choke of brick, repositioning the gas tip, reducing the operating pressure of the gas (especially if hydrogen is a large percentage of the fuel gas), attaching quarter-wave tubes to the large vessel, and changing the air to fuel ratio.

16.4 Thermal Oxidizer

16.4.1 Operations

A TO is a refractory-lined vessel, which is fitted with one or more burners (see Figure 16.2). It is designed to provide adequate residence time, at an elevated temperature for controlled completion of the required oxidation of combustible materials. TOs are designed,

FIGURE 16.2
Typical horizontal TO.

selected, and manufactured to meet the specific customer requirements. Each TO is designed to operate at a specific temperature and with a specific volume (sufficient residence time) to meet the customer's emission requirements. Refer to the operating manual for the specific TO; this manual should specify the operating temperature and the specific waste streams at which it was designed to operate. The operating temperature of the TO is typically measured by either a thermocouple or an optical pyrometer that sends a signal to a temperature controller. The temperature controller sends a signal to the fuel control valve. The fuel control valve then modulates to control the setpoint of the temperature controller. The temperature controller must be tuned to match the fluctuations from the various flow rate and composition changes in the waste stream. See Chapter 2 for additional information on combustion controls.

The initial refractory cure-out is the most crucial part of starting any unit; proper cure-out allows physical and chemical changes to occur within the refractory, which give the refractory its resistance to high temperatures (see Figure 16.3).

Another reason for the slow rate at which the refractory is cured-out is the removal of water that may exist in the refractory. It is necessary to heat the refractory at a slow rate so that the entrapped water will have an opportunity to escape (via evaporation). Any water that becomes trapped and is cut off from escape will expand as it is heated and may lead to damage to the refractory in the form of sudden spalling.

> **CAUTION!**
>
> If the TO should need to be turned off at any time during cure-out, the cure-out should be resumed at the temperature at which the unit is restarted.

> **CAUTION!**
>
> Sudden temperature changes may cause damage to the refractory installed in the unit.

The common types of refractory used in TOs are brick, castable, blanket, and plastic. Brick and blanket refractories require no cure-out, while the castables follow the cure-out procedures as recommended by each refractory manufacturer. The length of time necessary to cure-out plastic refractory is dependent upon the thickness but is generally longer than the castable. When using plastic refractory, the manufacturer should be consulted in order to determine the proper cure-out schedule.

When more than one type of refractory is used, such as brick and castable or blanket and castable, the cure-out schedule must allow for proper cure-out of all refractory (i.e., the longer cure-out schedule must

FIGURE 16.3
Typical TO refractory.

be used). See Chapter 5 for additional information on refractory in TOs.

The actual operating procedures for a TO vary depending upon the specific application. The controls that are being used on the system vary greatly from one application to another, depending upon the customer's specifications and the nature of the application. Always review the specific operating manual for the detailed operation of the TO.

16.4.2 Troubleshooting

Table 16.1 shows a general TO troubleshooting guide. Each specific TO may have different requirements than this guide.

Improper placement of thermocouples in the TO will increase the response time to changes in the operating system. Effective control of a system depends upon receiving accurate readings from control instrumentation.

Thermocouples are an important part of such control instrumentation, and their placement is crucial. The thermocouples must be placed directly into the flow stream. Dead flow zones, such as corners, do not respond to temperature changes as quickly as areas in which there is flow.

The length of the thermocouples also needs to be verified. Remember to take into account the thickness of the refractory and shell, as well as the nozzle projection. As a rule of thumb, thermocouples should extend from the refractory at least 6 in. (15 cm) out into the flow stream.

On horizontal units, it is preferred to mount the thermocouples on the top of the unit and have the thermocouple hanging vertically into the unit. This helps to keep the thermocouples (especially ceramic) from sagging at high temperatures.

You should also be aware of the thermal expansion that is present when a TO is being brought up to operating temperature. Depending on the design and orientation of the unit, different methods of accounting for such expansion may be utilized. Assure that movements due to thermal expansion have been given proper allowances, including connecting piping, ductwork, supports for conduit, and cable trays.

TABLE 16.1

General TO Troubleshooting Guide

Symptom	Possible Cause	Corrective Action
High TO temperature shutdown	Operating temperature of unit too close to high-temp shutdown	Reduce operating temperature of TO
	Faulty control thermocouple or temperature switch input that causes temperature switch to misread TO temperature	Troubleshoot temperature loop and correct any faulty components
	Combustion air blower or damper not allowing enough air into the TO	Troubleshoot combustion air control and correct any faulty components
	Thermocouple failure	Replace TC
	Loose thermocouple wire connection	Check TC connections
High fuel pressure shutdown	Faulty regulator	Verify if fuel gas regulator is set at the correct pressure and operating properly; confirm that plug is allowed to fully close on seat
	Faulty pressure switch	Verify if pressure switch is calibrated and operating properly
Low fuel pressure shutdown	Faulty regulator	Verify if fuel gas regulator is set at the correct pressure and operating properly
	Fuel gas supply low pressure	Confirm fuel gas supply pressure is adequate and not experiencing "dips" in pressure
Low combustion airflow shutdown	Blower problem	Confirm blower and VFD is operating properly
	Air damper problem	Confirm damper is modulating as controlled by the DCS
	Airflow or pressure switch problem	Confirm airflow or pressure switch is operating properly
Flame failure	Unstable combustion	Visually verify if flame is unstable or pulsating. Troubleshoot air control to stabilize flame
	Lack of combustible fuel or waste	Confirm fuel gas pressure to burner and ensure inert gas has not entered fuel gas piping
	Faulty block valve	Confirm proper operation of fuel gas block valves and waste gas block valve
	Damaged burner	Confirm burner internals are in good condition and repair or replace damaged components
	Dirty scanner lens blocking UV signal	Clean scanner lens

16.5 Boiler

16.5.1 Operations

The instructions and procedures outlined in this manual are intended to allow start-up, operating, and general maintenance to be carried out by competent operators who are familiar with boiler operations. All maintenance and operating problems are not presumed covered since every installation is different to some degree. It is therefore important that operators be familiar with the general control logic of not only the boiler, but all associated equipment (see Figure 16.4).

Consult the boiler vendor's operating manual first in order to see their recommended operating procedures.

As a general guideline, please also see Table 16.2: U.S. Department of Energy—Federal Energy Management Program, Operations and Maintenance Best Practices—A Guide to Achieving Operational Efficiency, Release 3.0, August, 2010.[4]

16.5.1.1 Boiler Start-Up

If the boiler is new, or has been out of service for an extended period, or has been opened for cleaning or repairs, before closing the waterside examine both the gas side and the waterside to make sure that each is free from tools and foreign matter. Loose material, e.g., dirt, trash, mill scale, or deposits, should be removed by washing and draining. Drum internals should be inspected for tightness and satisfactory condition.

NOTE: A final field hydrostatic test will need to be completed before the boiler can be final ASME stamped/approved and put into service. Contact the local Authorized Inspector (AI) for information.

NOTE: Before the first heat-up of the boiler, the boiler is filled with cleaning chemicals to initially "boil out" or clean the boiler of the remaining oils, rust, and loose weld material left over from the fabrication of the boiler. This cleanout of the boiler insures a clean boiler that maximizes the heat transfer in the boiler. Consult with both the local chemical treatment company and the boiler manufacturer on the exact chemicals and heat-up schedule that are to be used for the boiler.

16.5.1.2 Start-Up after Temporary Shutdown

This procedure applies if the boiler has been out of service for less than 1 month.

Close all manholes and refill the boiler with feedwater (at a minimum temperature of 20°C [68°F] and within 25°C [77°F] of the drum temperature) up to the low-level alarm. Leave the drum vent valve open.

1. Light the burner and gradually increase the burner capacity so that the boiler water temperature increases by no more than 55°C (130°F) per hour.

FIGURE 16.4
Typical watertube boiler.

TABLE 16.2

U. S. Department of Energy—Federal Energy Management Program, Operations and Maintenance Best Practices—A Guide to Achieving Operational Efficiency

Description	Comments	Maintenance			
		Daily	Weekly	Monthly	Annually
Boiler use/sequencing	Turn off/sequence unnecessary boilers	X			
Overall visual inspection	Complete overall visual inspection to be sure all equipment is operating and safety systems are in place	X			
Follow manufacturer's recommended procedures in lubricating all components	Compare temperatures with tests performed after annual cleaning	X			
Check stream pressure	Is variation in stream pressure as expected under different loads? Wet stream may be produced if the pressure drops too fast	X			
Check unstable water level	Unstable levels can be sign of contaminants in feedwater, overloading	X			
Check burner	Check for proper control and cleanliness	X			
Check motor condition	Check for proper function temperatures	X			
Check air temperature in boiler rule	Temperatures should not exceed or drop below design limits	X			
Boiler blowdown	Verify if the bottom, surface, and water column blowdowns are occurring and/or effective	X			
Boiler logs	Keep daily logs on • Type and amount of fuel used • Flue gas temperature • Makeup water volume • Stream pressure temperature and amount generated	X			
Check oil filter assemblies	Check and clean/replace oil filters and strainers	X			
Inspect oil heaters	Check to ensure that oil is at proper temperature prior to burning	X			
Check boiler water treatments	Confirm water treatment system is functioning properly	X			
Check flue gas temperatures and composition	Measure flue gas composition and temperatures at selected firing position—recommended $O_2\%$ and $CO_2\%$		X		
	Fuel $O_2\%$ $CO_2\%$ Natural gas 1.5 10 No.2 fuel oil 2.0 11.5 No.6 fuel oil 2.0 12.5				
	Note: Percentages may vary due to fuel composition variations				
Check all relief valves	Check for leaks		X		
Check water level control	Stop feedwater pump and allow control to stop fuel flow to burner. Do not allow water level to drop below recommended level		X		
Check pilot and burner assemblies	Clean pilot and burner following manufacturer's guidelines. Examine for mineral and corrosion buildup		X		
Check boiler operating characteristics	Stop fuel flow and observe flame failure. Start boiler and observe characteristics of flames		X		
Inspect system for water/stream leaks and leakage opportunities	Look for leaks, defective valves and tapes, corroded piping, condition of insulation		X		
Inspect all linkages on combustion air dampers and fuel valves	Check for proper setting and tightness		X		
Check blowdown and water treatment procedures	Determine if blowdown is adequate to prevent solid buildup			X	
Flue gases	Measure and compare last month's readings of flue gas composition over entire firing range			X	
Inspect boiler for air leaks	Check damper seals			X	

TABLE 16.2 (continued)

U. S. Department of Energy—Federal Energy Management Program, Operations and Maintenance Best Practices—A Guide to Achieving Operational Efficiency

Description	Comments	Maintenance			
		Daily	Weekly	Monthly	Annually
Combustion air supply	Check combustion air inlet to boiler room and boiler to make sure openings are adequate and clean			X	
Check fuel system	Check pressure gauge, pumps, filters, and transfer lines. Clean filters as required			X	
Check belts and packing glands	Check belts for proper tension. Check packing for compression leakage			X	
Check for air leaks	Check for air leaks around access openings and flame scanner assembly			X	
Check all blower belts	Check for tightness and minimum slippage			X	
Check all gaskets	Check gaskets for tight sealing; replace if do not provide tight seal			X	
Inspect boiler insulation	Inspect all boiler insulation and casings for hot spots			X	
Steam control valves	Calibrate steam control valves as specified by manufacturers			X	
Pressure reducing/regulating	Check for proper operation valves			X	
Perform water quality test	Check water quality for proper chemical balance			X	
Clean waterside surfaces	Follow manufacturers recommendation on cleaning and preparing waterside surfaces				X
Clean fireside	Follow manufacturers recommendation on cleaning and preparing fireside surfaces				X
Inspect and repair refectories on fireside	Use recommended material and procedures				X
Relief value	Remove and recondition or replace				X
Feedwater system	Clean and recondition feedwater pumps. Clean condensate receivers and deaeration system				X
Fuel system	Clean and recondition system pumps, filters pilots, oil preheaters, oil storage tanks, etc.				X
Electrical systems	Clean all electrical terminals. Check electronic controls and replace any defective parts				X
Hydraulic and pneumatic valves	Check operation and repair as necessary				X
Flue gases	Make adjustments to give optimal flue gas composition. Record composition, firing position, and temperature				X
Eddy-current test	As required, conduct eddy-current test to assess tube wall thickness				X

2. Close the steam drum vent as soon as it indicates that steam is being generated and all air is purged.
3. Recheck the level control settings as the drum build up pressure. Also examine the boiler for leaks at this time.
4. Establish normal flows and controls once the operating pressure is attained in the steam drum.

16.5.1.3 Boiler Shutdown Procedure

1. Before shutting down the boiler, use both the continuous and intermittent valves to blow down the boiler. This will insure a low solid concentration in the boiler water and should minimize the solids that could collect on the tubes (scaling) during shutdown.
2. Decrease the drum water temperature at a rate of 55°C (130°F) per hour until 65°C ± 15°C (149°F ± 59°F) is reached. This is accomplished by lowering the operating pressure and reducing the burner firing rate. Maintain normal water levels in the boiler for as long as possible during the shutdown period.

During an emergency situation, decrease the operating temperature as slowly as possible (i.e., from the operating condition to 65°C, 150°F, drum water temperature in 1 h).

16.5.1.4 Normal Boiler Operation

Normal operating instructions for the boiler follow. Keep in mind that the two most important rules in the safe operation of a boiler are to maintain the proper water level and to maintain recommended solids concentration in the boiler:

1. At least once per shift an operator should ascertain whether the pipe, fittings, and valves between the boiler and the water glass are free and open by blowing down the water columns and water glasses and noting the promptness of the return of the water to the glasses. This should be done at the beginning of each shift and preferably before the relieved shift has gone off duty and more frequently if trouble is experienced with boiler compounds, foaming, priming, and other feedwater troubles that are apt to cause choking of the connections. It should also be done after replacing the water glasses.

2. Keep the water column well illuminated, and keep the glass clean. A dirty mark in a gauge glass can be mistaken all too easily for the water level. Do not allow steam or water to leak from the water level. Do not allow steam or water to leak from the water column or its connections as this will cause the water glass to show a false water level. Keep the outlet end of the drain pipes from the water column, water glass, and gauge cocks free from obstructions.

   When the level of the water is not visible in the water glass, blow down the water column or gauge glass to determine whether the existing level of the water is above or below the water glass. If the water level is below the water glass, shut down the flue gas unless certain feedwater is being supplied, in which case reduce the flue gas rate by adjusting the process.

   Check the feedwater lines. If low water is caused by operating conditions, remedy it immediately before resuming normal steaming rates. If any uncertainty exists, do not change the feedwater supply, do not open the safety valves or change the steam outlet valves, or do not make any adjustment that will cause a sudden change in the stresses acting on the boiler. Determine the cause of low water and remedy it.

   To determine if scale buildup in the feedwater lines is causing low feedwater flow, install a pressure gauge at the discharge of the feedwater pump. If an unusually high pressure difference between the steam drum and the feedwater pump discharge is observed, then there may be a scale buildup problem and the line should be cleaned.

   In case of high water level, a quick check of the controlling instrument should be made. If no problem can be found, blow down until the low water level is reached, then observe the instrument and control valve operation. If no problems can be found and the conditions persist, either manually control feedwater rate or blow down or shut unit down until corrective steps can be completed.

3. Perform regular boiler feedwater treatments and analyze the boiler feedwater on a routine basis as established by the plant's water treating consultant.

4. Blow down the boiler regularly as determined by analysis of the boiler water. The object of blowing down is to maintain the concentration of dissolved and suspended solids within safe limits to avoid priming and carryover. Blow down the boiler at a period during the day when the steam demand is lowest. When blowing down the boiler, open fully the valve next to the boiler first, then open the second valve. When closing, close fully the second (outer) valve first, and then the valve next to the boiler.

16.5.1.5 Placing the Boiler into Service

"Cutting-in" of a boiler to a stem header already in service requires that the new boiler steam piping be heated prior to opening the stop-check valve. This reduces thermal shock.

When the pressures in the boiler and the header are approximately equal, slowly open the steam stop valve to full open while slowly opening the header valve to full open.

16.5.1.6 Care When Out of Service

A boiler, which is to be out of service for more than 24 h, should be kept either absolutely dry or completely filled with water in order to minimize corrosion. If the boiler is not exposed to freezing temperatures while out of service, it is simpler to hold it completely filled with water.

For a period of up to 1 month, build up alkalinity concentration in the boiler water to 300 ppm and fill the boiler completely. In addition, maintain a sodium sulfite concentration of 50–100 ppm for a protection against oxygen in the water.

If the boiler is to be out of service for more than 1 month, drain and wash it carefully, then refill with fresh water, adding 1 kg of caustic soda per 1000 kg (2200 lb) of cold water required to completely fill the boiler. Dissolve, in a separate container, caustic soda in water and pour into the drum. Add sodium sulfite, 3/1000 kg (6.6/2200 lb) of water, before closing the boiler.

For drained out of service periods, it is essential that the internal surfaces be completely dry. Place a shallow container of quicklime in the drums to absorb the moisture in the air, and then close the boiler tightly. If desired, silica gel can be used as a desiccator and is more satisfactory than quicklime.

16.5.2 Troubleshooting

The best way to approach the problem is by forming an outline of the problem and the possible causes. Once the outline is fully developed, systematically go through the outline eliminating possible causes until the problem is resolved.

Two of the most common problems are (1) not being able to generate the required amount of steam or (2) not being able to maintain the required steam pressure. While these are the most common problems, other examples of possible problems include safety valves actuating at all times, water levels not being maintained, water carryover, corrosion, erosion, and fouling.

The problems that boilers develop are typically found in the following three areas: (1) fireside, (2) waterside, and (3) controls. Depending on the nature of the problem, there are many questions that should be asked. Their answers can be valuable clues to solving the problem:

1. Some of the questions for the fireside are as follows:
   a. Are the flue gases coming in at the right temperature?
   b. Is the flow of gases at the designed flow rate or are they too low?
   c. Is the pressure drop excessive or not enough?
   d. Is the flue gas composition the same as the design case?
   e. Are there any obstructions in the flue gas duct?
   f. Is there any gas bypass around the boiler?
   g. What about cold air leaks in the duct? Especially in an induced draft system operating at negative pressures.
2. Waterside questions include the following:
   a. Is the feedwater temperature too low?
   b. Is the flow rate correct? If not, are there any restrictions on the lines?
   c. Are the valves fully open?
   d. Are the glass gauges clean and indicating the correct water level?
   e. Are the drain valves closed? Are the drain lines free and clear of debris?
   f. Is the feedwater pump operating?
3. From the controls and instrumentation side, ask questions such as the following:
   a. Are the pressure gauges correct? Have they been calibrated? Are the gauges' isolating valves open?
   b. Are the pressure switches correctly set? Are they operating?
   c. Are the level controllers and water columns properly set? Are they calibrated?
   d. Are the safety valves set at the proper pressure? Have they been tested? Are there any obstructions in the discharge line?
   e. Are the control circuits properly grounded?

In dealing with miscellaneous problems having to do with corrosion, erosion, fouling, and the like, some questions that can be asked are as follows:

- Is the deaerator working?
- Is the water treatment adequate? Does it need to be upgraded or modified?
- Are the chemicals being injected at the recommended flow?
- Is the blowdown of the boiler done on a regular basis?

16.5.2.1 Foaming and Priming

When foaming and priming occur in a steam boiler, it can cause large quantities of water to pass into the steam main. It can be detected by wide water-level fluctuations, as shown in the gauge glass.

Foaming and priming can be attributed to (1) dirt or oil in the boiler water, (2) too high a water level, or (3) overloading the boiler. To eliminate this situation, alternately blow down and feed in fresh water to the boiler several times. If foaming or priming does not stop, cool the boiler completely, empty and wash out the boiler, and refill with fresh feedwater.

16.5.2.2 Scale in Boiler

When operating the boiler, it is good practice to return as much condensate to the boiler as is practical. This practice keeps the amount of raw makeup water to an absolute minimum. Raw waters contain minerals and salts in solution; when heated, these impurities precipitate or separate as solids. These solids then are either carried in suspension or settle out to form mud or hard scale.

In a steam boiler, water is distilled and discharged as steam through the steam outlet, while the solids remain inside the unit. There is, therefore, a gradual accumulation of solids inside the boiler when raw feedwater is used. Salts that are carried in suspension can cause overheating and damage.

Any amount of raw water fed to the boiler increases the mineral concentration, and unless corrective measures are taken, there will be unsatisfactory results and possible damage to the boiler.

In order to minimize the impact the solids have, it is important that the instructions given for the treatment of feedwater by an experienced boiler water chemist are followed.

16.5.2.3 Corrosion or Pitting

Corrosion of the metal inside the boiler can be of serious consequence. Examine the boiler regularly for signs of pitting or corrosion. This problem can be controlled by proper water treatment.

16.6 Miscellaneous Control Items

16.6.1 Operations

For the operation of miscellaneous control items, i.e., control valves, regulators, pressure/temperature/flow switches, etc., it is best to follow each individual vendor's instructions per their own installation and operating manual.

16.6.2 Troubleshooting

For the troubleshooting of miscellaneous control items, i.e., control valves, regulators, pressure/temperature/flow switches, etc., it is best to follow each individual vendor's instructions per their own installation and operating manual.

16.7 Liquid and Air/Steam Atomizing Guns (Fuel/Waste Oil, Quench Water, Aqueous Wastes)

16.7.1 Operations

See Volume 1, Chapter 10 on oil atomization for more information. Check the operating manual on your specific liquid atomizing gun to confirm the proper liquid pressures, air/steam pressure ranges, and flow rate ranges that each gun is capable of properly operating to maintain good atomization. The air/steam pressure will typically be operated either at a constant pressure or at a differential pressure as compared to the liquid pressure. Please note, in general, that air atomization is typically better for water-based, aqueous, waste streams and, in general, that steam atomization is typically better for organic-based waste streams (see Figure 16.5).

16.7.2 Troubleshooting

The viscosity of the liquid must always be kept below 200 SSU (35 cP) in order to maintain good atomization quality. Typical problems are either plugging or coking of the ports or erosion/corrosion of the gun. This can be confirmed by monitoring the flow rates versus the pressure of the liquid over time. If the liquid pressure is gradually increasing versus the same flow points, then the tip is either plugging or coking and the liquid gun needs to be removed for cleaning. If the liquid pressure is gradually decreasing versus the same flow points, then the tip is either eroded or corroded and the liquid gun needs to be removed and the tip/atomizer needs to be inspected and replaced.

16.8 Water Weir, Spray Quench Contactor, Quench Tank

16.8.1 Operations

Check the operating manual on the specific quench system to confirm the proper operating conditions as the following statements cover a generic range of quench systems (see Figure 16.6). The high temperature of the flue gases from the TO enters into the quench section. The purpose of the quench section is to reduce the gas temperature and presaturate the gas before it enters the downstream absorber and/or scrubber (180°F–200°F = 85°C–95°C). In addition, the gas scrubbing conditions are enhanced by lowering the temperature and raising the moisture content of the gas. The wetted-wall water inlet of the water weir is the first point of liquid contact for the incoming hot flue gas. The weir is supplied with a constant pressure stream of water and/or recycled liquid, and this provides continuous wetted-wall coverage of the contactor quench section refractory walls for proper cooling. This wetted wall provides water that helps to cool the flue gases and protects the contactor tube from excessive temperature. It is important that the weir be level so that the water flow over it will be uniform around its circumference. Makeup water for the quench system is usually supplied at the weir nozzles. The typical materials of construction for the

FIGURE 16.5
Spray gun examples using water with air atomization.

weir are zirconium or Hastelloy®-type alloys. The next point of liquid contact for the incoming hot gas is a multiple-spray nozzle contactor. The contactor shell is typically made of FRP or of an alloy material and is brick lined. The contactor is typically supplied with water at 60 psig (410 kPa) assuring complete spray coverage. The material of construction of the spray guns and tips is typically zirconium, Hastelloy alloys, or Teflon®. The quench tank provides the sump for the collection and recycling of liquor, which can then be pumped to the sprays in the contactor. An integral part of the system is the downcomer tube in the quench tank, which allows for the contraction of the flue gases and provides a further saturation area at the base, where the wetted-wall liquor meets the quench tank liquid level. FRP is typically used for tank construction.

16.8.1.1 Pre-Start-Up Checklist

1. Check equipment installation with design drawings.
2. Ensure all piping installation is completed and all control components, valves, pumps, thermocouples, etc., have been commissioned in line with individual control philosophy requirements.
3. Ensure all utilities are commissioned and available for start-up.
4. Ensure that all mechanical installation work has been completed and that all personnel in the unit area have been notified of start-up operations.

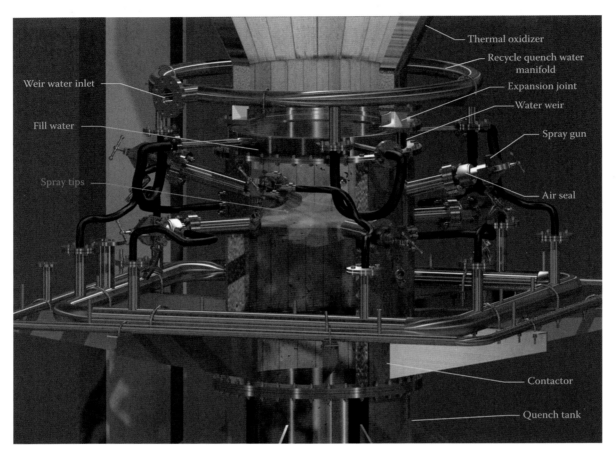

FIGURE 16.6
Typical quench system.

16.8.1.2 Start-Up Philosophy and Overview

> **WARNING**
>
> The quench system has been designed for operation within specific parameters. This document has been generated as a basic guide to quench operations. The operating philosophy may differ for individual cases. This document is not intended for use as the details of day-to-day operations.

The purpose of the water quench system is to reduce incinerator flue gas temperature to the flue gas adiabatic saturation temperature (AST) (this includes saturating the gases with water). This is achieved through a train of three pieces of equipment, a water weir, a contactor tube, and a quench tank.

Before starting up, the quench system should be visually inspected for the following items:

Verify that the quench pot, contactor, and weir are level. Failure to level the equipment may cause dry spots on the weir and could cause damage to the contactor refractory.

Check expansion joints to see that they are properly installed.

Ensure that contactor spray tips clear the refractory by ½ in. (13 mm). If fan sprays are used, be sure that fan is oriented horizontally.

Ensure that tips are not plugged with debris.

Visually inspect the water flow over the weir to insure that there are no dry spots and that flow is even.

Rinse FRP equipment with fresh water before start-up to prevent any loose glass fibers from plugging contactor nozzles.

The water to the weir and contactor should be turned on before the oxidizer is started up. The flow rate to the weir must be equal to or greater than the minimum flow rates listed in the operating manual for proper operation of the weir. The drains in the quench pot are closed to allow a water level to build up in the quench.

Once the normal water level has been reached in the quench pot, the recirculation pump can be started and water can be recycled to the contactor. The normal water level is usually set at 6 in. (15 cm) below the bottom of the downcomer for best operation.

Setting the NWL too low could cause the flue gases to be unsaturated, because they do not come in contact with the quench water. Setting the NWL too close to the downcomer could cause a bubbling effect and make level control difficult.

> **CAUTION**
>
> Starting pumps before sufficient water level has been established could cause cavitation that can damage the pumps.
>
> The contactor sprays should be checked for plugging regularly during start-up as debris from construction and shipping accumulate at the spray tip. Refer to troubleshooting guide in Section 16.8.2 for details on this.

16.8.1.3 Normal Operation

A thermocouple and a temperature switch are always provided to monitor the quench temperature. The high set point of this switch should be set 15°F (8°C) above normal saturation temperature, but should not exceed the maximum service temperature of the FRP. During operation, a high temperature in the quench section will shut down the entire system, excluding pumps. A loss of flow to contactor or weir will also cause a system shutdown. The troubleshooting portion of this section offers solutions to frequent high-temperature shutdowns.

16.8.2 Troubleshooting

Usually, the biggest problem with the quench system is contactor sprays becoming plugged. It is convenient that an isolation valve be supplied for each individual spray, so that the guns can be cleaned easily. If plugging is suspected, each hose should be checked for flow. If the hose is cool, the gun is probably clogged.

Pulling the contactor guns out while the system is running is not advised; however, if this is necessary, a temporary blind should be placed over the connection to prevent hot gases from escaping and causing injury to personnel. Once the tip has been cleaned, the hose should be flushed to clean out any sludge that collected in the line while it was plugged. The gun can then be put back into service.

A level control is always supplied in the quench tank. The normal water level, as discussed previously, is set at 6 in. (15 cm) below the bottom of the downcomer for best operation. Low-level alarms and low-level shutdown set points also need to be provided. These set points will depend on the net positive suction head (NPSH) needed for the pump. The water level should never be allowed to drop below the NPSH level required for the pump.

In a case where the makeup water to the weir is used in the level control loop, the minimum stop on the control valve should be set for the minimum weir flow required.

The maximum service temperature of the FRP resin for a particular job must be confirmed by the operator. Operating the equipment above the maximum service temperature may cause significant damage to the equipment.

Another area of caution is the water level of the tank. In cases where the particulate loading is high, solids may settle in the bottom of the tank and displace the water. The level indicator will then give a false reading. This is also true in down-fired systems where refractory degradation can contribute to insoluble solids in the quench tank. Table 16.3 shows some basic troubleshooting steps for a quench system.

16.9 Heat Exchangers

16.9.1 Operations

The instructions and procedures outlined in this section are intended to be carried out by competent operators who are familiar with heat exchanger operations (see Figure 16.7). All operating problems are not presumed covered since every installation is different to some degree. It is therefore important that operators be familiar with the general control logic of not only the heat exchanger, but all associated equipment (see individual vendor literature in the operating manual for additional details). The heat exchanger typically takes the higher-temperature flue gases and transfers heat and preheats the cooler combustion air or waste gas to recover the heat that would be lost if the heat exchanger was not installed. Consult the manufacturer's operating manual for details of start-up procedures as typically there is

TABLE 16.3

Quench System Troubleshooting

Problem	Possible Causes	Solutions
1. Quench temperature too high	1a. Check the contactor sprays for solid pluggage	1a. Check the hoses to all contactor sprays. The hoses cool to the touch are probably plugged. Clean plugged tips and place back on line
	1b. Contactor spray orifices too small for design flow	1b. Check vendor literature for spray tip capacity. Replace with larger spray tips
	1c. Pump not delivering required pressure for spray	1c. Check pump capacity curve for design pressure. Replace spray tips to handle flow at lower pressure
	1d. Thermocouple not working properly	1d. Check thermocouple, connections, and wiring
2. Hot spots on contactor shell	2a. Weir is not level	2a. Level weir
	2b. Water flow over weir annulus not even around inside circumference, causing dry spots	2b. If leveling is not possible, grind lip of the weir down to even out flow
3. Pressure to contactor guns increasing	3. Spray tips are plugged	3. Check the hoses to the contactor for flow. The hoses should be hot; if cool, the tip is plugged. Close valve and isolate plugged gun. Remove and clean tip. Flush hose before placing gun in service
4. Pressure to contactor guns decreasing	4a. Spray tips are eroded, are corroded, or have completely come off of the gun	4a. Check condition of spray tips, if spray tips have come off of the guns. If so, also check the refractory brick directly across from this gun for erosion damage
	4b. Pump is not putting out sufficient pressure	4b. Switch to spare pump. Take damaged pump out of service for repair

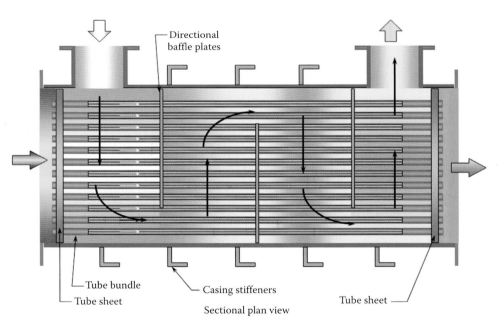

FIGURE 16.7
Gas to gas exchanger.

always a maximum heat-up and cooldown rate to keep from damaging the heat exchanger. There is also a maximum-use temperature that must not be exceeded to keep from damaging the heat exchanger. Always monitor the temperature drop and pressure drop across the heat exchanger to see if the heat exchanger is plugging or corroding.

Prior to the start-up of any heat exchanger, a visual inspection is recommended in order to verify that the equipment is properly installed including all expansion joints. Additionally, the following checklist should be consulted:

1. Important—Ascertain that the sliding saddle(s) will be capable of sliding.
2. Check all flanged connections to insure that they are properly bolted, gasketed, and tight.
3. All manways, handholes, and inspection holes should be closed.
4. All instruments shall be installed and be in working order.

In order to prevent thermal shock, the airflow or the cool gas flow through the exchanger must be established prior to allowing hot flue gases to flow through the unit.

Follow the refractory curing instructions or the start-up instructions for the units upstream of the gas exchanger, i.e., TO, until the design entering flue gas operating temperature is reached.

At the time of shutdown, it is recommended that the burner and exchanger cold stream be shut off simultaneously. This is commonly referred as the "soak" state.

Upon restarting the unit, follow the refractory start-up procedure that has the lowest flue gas temperature increase per hour.

If none is available or has been prescribed, then increase the flue gas temperature through the unit at a rate of 200°F (93°C) per hour.

On the rare occasions where HCl, SO_3, HBr, etc., are present, the start-ups as indicated earlier still apply, but the shutdowns require a special procedure to be outlined and covered in the operations and maintenance (O&M) manual for the specific job.

16.9.2 Troubleshooting

Typically, the main problems with heat exchangers are corrosion and fouling. Always monitor the temperature drop and pressure drop across the heat exchanger to see if the heat exchanger is plugging, fouling, or corroded. By knowing the operating history of the heat exchanger, one can determine when one is starting to have a problem.

If fouling is a problem, this greatly reduces the heat transfer rate of the heat exchanger. Do an analysis of the solids that are building up on the heat exchanger. This will determine where the material is originating (i.e., waste stream, corrosion, etc.) and will determine the best method of cleaning (i.e., compressed air, water blast, steam, etc.).

If corrosion is a problem, determine the composition of waste stream and determine the compound that is causing the corrosion (i.e., chlorine, bromine, etc.). The metallurgy of the heat exchanger must be compatible with all components in the waste stream.

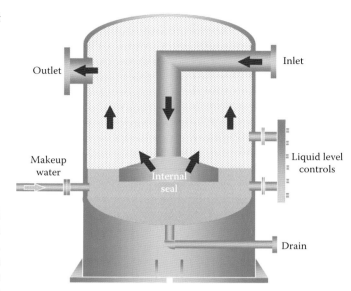

FIGURE 16.8
Typical liquid seal.

16.10 Liquid Seals

16.10.1 Operations

A liquid seal functions as a flame arrestor on a waste vapor line to stop flame propagation (see Figure 16.8). The waste vapor stream is distributed and bubbles through the water in the liquid seal, and the water provides the seal to prevent flame propagation. The main operating point on a liquid seal is to always be sure the water level is maintained in the liquid seal.

16.10.2 Troubleshooting

Make sure that the liquid level is monitored and kept at the proper height. The internals of the liquid seal can be subject to fouling due to solid buildup and corrosion. At least once yearly, completely drain, inspect, flush out, and replace the water in the liquid seal.

16.11 Absorbers and Scrubbers

16.11.1 Operations

The instructions and procedures outlined later are intended to allow start-up, operating, and general maintenance to be carried out by competent operators who are familiar with absorber/scrubber operations (see Figure 16.9). All maintenance and operating problems are not presumed covered since every installation is different to some degree. It

FIGURE 16.9
Typical scrubber.

is therefore important that operators be familiar with the general control logic of not only the absorber/scrubber, but all associated equipment. Check the operating manual on the specific absorber/scrubber system to confirm the proper operating conditions as the following statements cover a generic range of absorber/scrubber systems.

The following sequence should be used for the start-up of a gas absorber and/or scrubber. All these steps should be done before the TO is lit.

Visually check the equipment for flaws and to familiarize oneself with the piping and P&ID.

Open the hand valves in the water recirculation line so that makeup water can enter and to make sure the water recirculation pump will have an open suction line and also will not be deadheaded.

With the makeup water controller (water-level control) in MANUAL, admit water into the absorber and/or scrubber until the set point water level is reached. Place the controller in AUTO.

Bump the water recirculation pump to make sure the pump shaft is rotating in the correct direction. Start the water recirculation pump.

Open the blowdown valve to the expected blowdown flow rate. Check the water level to ensure the level controller is turned properly.

Calibrate the pH probe and place it in position. The probe should have been kept wet until this time. If it had been installed during construction, it spent a lengthy amount of time dry (the time prior to pump start-up) and probably will no longer work.

Start the caustic injection pump and place the pH controller in AUTO at the proper set point.

At this point, the incinerator may be lit and brought through its heat-up schedule. Check often during this time period to make sure that the flue gas inlet temperature is below 200°F (93°C) and that the absorber water level is stable. If there are strainers in the water recirculation line, check them every hour for buildup, as particulate matter from construction and start-up will be flowing in the lines.

> **WARNING**
>
> When cleaning the water recirculation strainer and when calibrating the pH probe, always wear rubber gloves and safety glasses. The water is hot and may be acidic. Eye or skin burns may result.

TABLE 16.4

Absorber Scrubber System Troubleshooting

Problem	Possible Causes	Solutions
1. Recycle flow rate is too low	1a. The strainers and or liquid distributors have solid pluggage	1a. Check the strainers and liquid distributors for pluggage
	1b. Pump not delivering required pressure flow	1b. Check pump capacity curve for design pressure and flow
2. Pressure drop too high across the packed beds	2. Packing is plugged with solids	2. Check packing for pluggage. Run gamma-ray scans into your packed column to confirm pluggage. Confirm if you have enough blowdown to keep the solids content/conductivity low enough to prevent salting up the packing. Clean/replace as necessary. Check internals for damage
3. Pressure drop too low across packed beds	3. Support trays have collapsed	3. Check the internals of the absorber/scrubber for damage

16.11.2 Troubleshooting

Always monitor the pressure drop across the packed beds of the absorber and/or scrubber. Table 16.4 outlines some absorber scrubber system troubleshooting.

16.12 Conclusion

As discussed in the introduction, correct operations and troubleshooting of a TO system requires experience, training, and a proactive maintenance and spare parts program in order to maximize the uptime of the total system. As exemplified throughout the chapter, a deep understanding of the operation of all of the equipment provides the groundwork of which successful operation and troubleshooting of the TO system can occur.

References

1. *The Control of Hazardous Energy (Lockout/Tagout)*, United States Department of Labor, Occupational Health and Safety Administration (OSHA), 29 CFR 1910.147.
2. NFPA 86, 2011 edn., *Standard for Ovens and Furnaces*.
3. T. Gilder, D. Campbell, T. Robertson, and C. Baukal, Customize operator training for your thermal oxidizers, *Hydrocarbon Processing*, 89(11), 55–59, 2010.
4. United States Department of Energy-Federal Energy Management Program, *Operations & Maintenance Best Practices—A Guide to Achieving Operational Efficiency*, Release 3.0, August, 2010.

Appendix A: Units and Conversions

TABLE A.1
Prefixes

Multiplier	Prefix	Symbol
10^{18}	Exa	E
10^{15}	Peta	P
10^{12}	Tera	T
10^{9}	Giga	G
10^{6}	Mega	M
10^{3}	Kilo	k
10^{2}	Hecta	h
10	Deca	da
10^{-1}	Deci	d
10^{-2}	Centi	c
10^{-3}	Milli	m
10^{-6}	Micro	m
10^{-9}	Nano	n
10^{-12}	Pico	p
10^{-15}	Femto	f
10^{-18}	Atto	a

Source: Annamalai, K. and Puri, I.K., *Combustion Science and Engineering*, CRC Press, Table A.1A, CRC Press, Boca Raton, FL, p. 981, 2007.

TABLE A.2
Basic Units, Conversions, and Molecular Properties

Area
1 acre = 4046.9 m²
1 m² = 10^{-6} km² = 10^{4} cm² = 10^{6} mm²
1 m² = 10.764 ft² = 1550 in.²
1 ft² = 144 in.² = 0.0929 m²
1 hectare = 10,000 m² = 2.5 acres = 108,000 ft²

Density
1 g/cm³ = 1 kg/L = 1000 kg/m³ = 62.43 lb_m/ft³ = 0.03613 lb_m/in³
1 kg/m³ = 0.06243 lb_m/ft³, 1 lb_m/ft³ = 16.018 kg/m³
Specific gravity = density/reference density
For liquids, reference density of water at 15.74°C (60°F) = 999 kg/m³, 62.4 lb/ft³
For gases, reference density of air at 15.74°C (60°F) = 1.206 kg/m³

Energy
1 eV ≈ 1.602 × 10^{-19} J.
1 mBtu = 1 kBtu = 1000 Btu, 1 mmBtu = 1000 kBtu = 10^{6} Btu
1 TBtu = 10^{9} Btu or 1 GBtu
1 quad = 10^{15} Btu or 1.05 × 10^{15} kJ or 2.93 × 10^{11} kWh = 172.4 million barrels of crude oil
1 kWh = 0.0036 GJ = 3.6 MJ = 3412 Btu, 1 hp h = 0.00268 GJ = 2.68 MJ = Btu
1 Btu = 778.14 ft lb_f = 1.0551 kJ, 1 kJ = 0.94782 Btu = 25,037 lb_m ft/s²
1 cal = 4.1868 J, 1 (food) cal = 1000 cal or 1 kcal

TABLE A.2 (continued)
Basic Units, Conversions, and Molecular Properties

1 kJ/kg = 0.43 Btu/lb, 1 Btu/lb = 2.326 kJ/kg, 1 kg/GJ = 1 g/MJ = 2.326 lb_m/mmBtu
1 Btu/SCF = 37 kJ/m³, 1 m³/GJ = 37.3 ft³/mmBtu, 1 lb_m/mmBtu = 0.430 kg/GJ = 0.430 g/MJ
1 Therm = 10^{5} Btu = 1.055 × 10^{5} kJ
1 hp = 0.7064 Btu/s = 0.7457 kW = 745.7 W = 550 lb_f·ft/s = 42.41 Btu/min
1 boiler hp = 33,475 Btu/h, 1 Btu/h = 1.0551 kJ/h
1 barrel (42 gallons) of crude oil = 5,800,000 Btu = 6120 MJ
1 gallon of gasoline = 124,000 Btu = 131 MJ
1 gallon of heating oil = 139,000 Btu = 146.7 MJ
1 gallon of diesel fuel = 139,000 Btu = 146.7 MJ
1 barrel of residual fuel oil = 6,287,000 Btu = 6633 MJ
1 cubic foot of natural gas = 1,026 Btu = 1.082 MJ
1 gallon of propane = 91,000 Btu = 96 MJ
1 short ton of coal = 20,681,000 Btu = 21,821 MJ

Force
1 lb_f = 4.4482 N = 32.174 lb_m · ft/s² or g_c = 32.174 lb_m ft/s² lb_f

Ideal Gas Law
$PV = RT; PV = mRT; PV = n\bar{R}T, P\bar{v} = \bar{R}T,$

$\bar{R} = 8.314$ kPa m³/k mol k = 0.08314 bar m³/k mol K

$= 1.986$ Btu/lb mol °R = 1545 ft lb_f/lb mol °R

$= 0.7299$ atm ft³/lb mol °R

Length/Velocity
1 in. = 0.0254 m
1 ft = 12 in. = 0.3048 m
1 mile = 5280 ft = 1609.3 m
1 statute mile = 0.87 nmi = 1.609 km
1 nautical mile = 1.15 smi = 1.85 km
1 mi/h = 1.46667 ft/s = 0.447 m/s = 1.609 km/h
1 m/s = 3.2808 ft/s = 2.237 mi/h = 1.96 kt = 1.15 smi/h = 3.63 km/h
Speed of light in vacuum, c = 2.998 × 10^{8} m/s
Speed of sound = $\sqrt{\gamma RT}$

Mass
1 teragram (Tg) = 1 million metric tonnes
Mass of an electron = 0.5 MeV (1 MeV = 10^{6} eV; for mass, use $E = mc^2$) = 9.109 × 10^{31} kg
Mass of proton = 940 MeV = 1.67 × 10^{-27} kg, Mass of neutron = 1.675 × 10^{-27} kg
1 lb_m = 0.45359 kg = 7000 grains
1 short ton = 2000 lb = 907.2 kg
1 long ton = 2240 lb or 1016.1 kg
1 metric ton = 1000 kg
1 ounce = 28.3495 g
1 kg = 2.2046 lb

(continued)

TABLE A.2 (continued)

Basic Units, Conversions, and Molecular Properties

Molecular Properties
1 Angstrom = 1.0×10^{-10} m
N_{Avog} = 6.023×10^{26} molecules/kmol for a molecular substance (e.g., oxygen)
= 6.023×10^{26} atoms/atom mole for an atomic substance (e.g., He)
Boltzmann constant, k_B = 1.38×10^{-26} kJ/molecule K
Planck's constant, h_P = 6.626×10^{-37} kJ s/molecule
Stefan–Boltzmann constant, σ = 5.66961×10^{-11} kW/m² K⁴
Charge of an electron = 1.602×10^{-19} coulombs, orbit radius (nm) = $0.0529n^2$, n: orbit number
Energy level of an orbit (eV) = $13.56/n^2$

Numbers
ln x = $2.303 \log_{10} x$
$\log_{10} x$ = 0.4343 ln x
e = 2.718
π = 3.142
1 deg = 0.0175 radians

Pressure
1 bar = 10^5 Pa, 1 mm Hg = 133.3 Pa
1 in Hg = 3.387 kPa = 0.491 psi
1 in water (4°C) = 0.03613 psi
1 atm = 14.696 lb$_f$/in.² = 1.0133 bar = 10.3323 mm of H₂O (4°C) = 760 mm of Hg(0°C)
1 psi = 1 lb$_f$/in.² = 144 lb$_f$/ft² = 6.894 kPa = 6894 Pa = 27.653 in water (4°C)

Specific Heat
1 Btu/lb °F = 4.1868 kJ/kg °C
1 kJ/kg °C = 0.23885 Btu/lb °F

Temperature
T(°C) = (T(°F) − 32) * (5/9)
T(°F) = T(°C) * 1.8 + 32
T(K) = T(°C) + 273.15
T(°R) = T(°F) + 459.67
1°R = 0.556 K, 1 K = 1.8°R
To convert electron volts into the corresponding temperature in Kelvin, multiply by 11,604.

Volume
1 m³ = 1000 L
1 fluid ounce = 29.5735 cm³ = 0.0295735 L
1 m³/kg = 1000 L/kg = 16.02 ft³/lb, 1 m³/GJ = 37.26 ft³/mmBtu
1 ft³/lb$_m$ = 0.062428 m³/kg
1 U.S. gallon = 128 fluid ounce = 3.786 L
1 barrel = 42 U.S. gallons = 35 imperial gallons = 158.98 L = 5.615 ft³ = 231 in.³ = 0.1337 ft³

TABLE A.2 (continued)

Basic Units, Conversions, and Molecular Properties

Volume of 1 kmol (SI) and 1 lb mol (English) of an ideal gas at STP conditions as defined below:

Scientific or SATP	U.S. Standard (1976) or ISA	Chemists' Standard or CSA	NTP (Gas Industry)
25°C (77°F), 101.3 kPa (14.7 psi, 29.92 in. of Hg)	15°C (60°F), 101.33 kPa (1 atm, 14.696 psi, 29.92 in. of Hg)	0°C (32°F), 101.33 kPa (1 atm, 14.7 psi, 29.92 in. of Hg)	20°C (65°F), 101.33 kPa (1 atm)
24.5 m³/kmol (392 ft³/lb mol)	23.7 m³/kmol (375.6 ft³/lb mol)	22.4 m³/kmol (359.2 ft³/lb mol)	23.89 m³/kmol (382.7 ft³/lb mol)

SATP, standard ambient temperature and pressure; ISA, International Standard Atmosphere; NTP, normal temperature and pressure.

Air Composition

Species	Mole %	Mass %	Molecular Weight
Ar	0.934	1.288287	39.948
CO_2	0.0314	0.047715	44.01
N_2	78.084	75.51721	28.01
O_2	20.9476	23.14489	32
Ne	0.001818	0.001267	20.18
He	0.000524	7.24E−05	4.0026
Krypton	0.000114	0.00033	83.8
Xe	8.70E−06	3.94E−05	131.3
H_2	0.00005	3.48E−06	2.016
CH_4	0.0002	0.000111	16.043
N_2O	0.00005	7.6E−05	44.013
SO_2, NO_2, CO, I_2	0.000235	—	—

Source: Annamalai, K. and Puri, I.K., *Combustion Science and Engineering*, Table A.1A, CRC Press, Boca Raton, FL, p. 981, 2007.

Note: Molecular weight (mass) of air = 28.96 kg/kmol.

TABLE A.3
Atomic Weights for Common Elements

Name	Symbol	Atomic Number	Atomic Weight
Aluminum	Al	13	26.98
Antimony	Sb	51	121.76
Argon	Ar	18	39.95
Arsenic	As	33	74.92
Barium	Ba	56	137.32
Beryllium	Be	4	9.01
Bismuth	Bi	83	208.98
Boron	B	5	10.811
Bromine	Br	35	79.90
Cadmium	Cd	48	112.41
Calcium	Ca	20	40.08
Carbon	C	6	12.01
Cesium	Cs	55	132.91
Chlorine	Cl	17	35.45
Chromium	Cr	24	52.00
Cobalt	Co	27	58.93
Copper	Cu	29	63.55
Fluorine	F	9	19.00
Germanium	Ge	32	72.61
Gold	Au	79	196.97
Helium	He	2	4.00
Hydrogen	H	1	1.01
Indium	In	49	114.82
Iodine	I	53	126.90
Iridium	If	77	192.22
Iron	Fe	26	55.85
Krypton	Kr	36	83.80
Lead	Pb	82	207.20
Lithium	Li	3	6.94
Magnesium	Mg	12	24.31
Manganese	Mn	25	54.94
Mercury	Hg	80	200.59
Molybdenum	Mo	42	95.94
Neon	Ne	10	20.18
Nickel	Ni	28	58.69
Nitrogen	N	7	14.01
Oxygen	O	8	16.00
Palladium	Pd	46	106.42
Phosphorus	P	15	30.97
Platinum	Pt	78	195.08
Plutonium	Pu	94	244.00
Potassium	K	19	39.10
Radium	Ra	88	226.00
Radon	Rn	86	222.00
Rhodium	Rh	45	102.91
Selenium	Se	34	78.96
Silicon	Si	14	28.09

TABLE A.3 (continued)
Atomic Weights for Common Elements

Name	Symbol	Atomic Number	Atomic Weight
Silver	Ag	47	107.87
Sodium	Na	11	22.99
Strontium	Sr	38	87.62
Sulfur	S	16	32.07
Tantalum	Ta	73	180.95
Thallium	Tl	81	204.38
Tin	Sn	50	118.71
Titanium	Ti	22	47.87
Tungsten	W	74	183.84
Uranium	U	92	238.03
Vanadium	V	23	50.94
Xenon	Xe	54	131.29
Zinc	Zn	30	65.39
Zirconium	Zr	40	92.22

Source: Annamalai, K. and Puri, I.K., *Combustion Science and Engineering*, CRC Press, Table A.1B, CRC Press, Boca Raton, FL, p. 985, 2007.

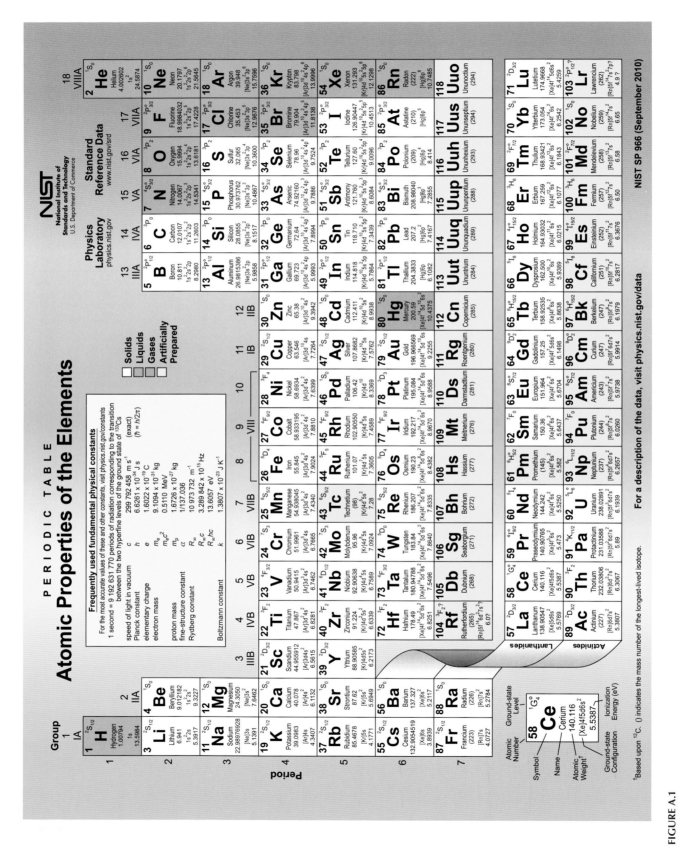

FIGURE A.1
Periodic table of the elements. (Courtesy of National Institute of Standards and Technology, Washington, DC.)

TABLE A.4

Conversion Table for Length

Length	mm	cm	in.	ft	yd	m	km	mile
mm	1[a]	0.1[a]	0.03937	3.2808 × 10^{-3}	1.093 × 10^{-3}	10^{-3}[a]		
cm	10[a]	1[a]	0.393701	0.032808	0.010936	0.01[a]		
in.	25.4[a]	2.54[a]	1[a]	0.83333	0.02778	0.0254[a]		
ft	304.8[a]	30.48[a]	12[a]	1[a]	0.333333	0.3048[a]	3.048 × 10^{-4}[a]	1.894 × 10^{-4}
yd	914.4[a]	91.44[a]	36[a]	3[a]	1[a]	0.9144[a]	9.144 × 10^{-4}[a]	5.682 × 10^{-4}
m	1000[a]	100[a]	39.3701	3.28084	1.09361	1[a]	10^{-3}[a]	6.214 × 10^{-4}
km	10^{6}[a]	100,000	30,370.1	3280.84	1093.61	1000[a]	1[a]	0.621371
mile	1.60934 × 10^{6}	160,934	63,360[a]	5280[a]	1760[a]	1609.34	1.60934	1[a]

Source: Adapted from Cornforth, J.R. (ed.), *Combustion Engineering and Gas Utilisation*, 3rd edn., E&F Spon, London, U.K., 1992, Table I, p. 861.
Note: The tables are read from row to column (i.e., 1 unit in the row = # units of the column.
[a] Denotes exact conversions.

TABLE A.5

Conversion Table for Area

Area	mm^2	cm^2	in.^{2}	ft^2	yd^2	m^2	acre	km^2	mile^2
mm^2	1	0.01	1.550 × 10^{-3}	1.076 × 10^{-5}	1.196 × 10^{-6}	10^{-6}			
cm^2	100	1	0.155	1.076 × 10^{-3}	1.196 × 10^{-4}	10^{-4}			
in.^{2}	645.16	6.4516	1	6.944 × 10^{-3}	7.716 × 10^{-4}	6.452 × 10^{-4}			
ft^2	92,903	929	144	1	0.1111	0.0929	2.30 × 10^{-5}	9.29 × 10^{-8}	3.587 × 10^{-8}
yd^2	836,127	8361	1296	9	1	0.8361	2.066 × 10^{-4}	8.361 × 10^{-7}	3.228 × 10^{-7}
m^2	10^6	10,000	1550	10.764	1.196	1	2.471 × 10^{-4}	10^{-6}	3.861 × 10^{-7}
acre				43,560	4840	4047	1	4.047 × 10^{-3}	1.562 × 10^{-3}
km^2				1.0764 × 10^7	1.196 × 10^6	10^6	247.1	1	0.3861
mile^2				2.7878 × 10^7	3.0976 × 10^6	2.590 × 10^6	640	2.590	1

Source: Adapted from Cornforth, J.R. (ed.), *Combustion Engineering and Gas Utilisation*, 3rd edn., E&F Spon, London, U.K., 1992, Table II, p. 862.
Note: The tables are read from row to column (i.e., 1 unit in the row = # units of the column.

TABLE A.6

Conversion Table for Volume

Volume	mm^3	mL	in.^{3}	L	gal	ft^3	yd^3	m^3
mm^3	1	10^{-3}	6.1024 × 10^{-5}	10^{-6}	2.642 × 10^{-7}	3.531 × 10^{-8}	1.308 × 10^{-9}	10^{-9}
mL	1000	1	0.061026	10^{-3}	2.642 × 10^{-4}	3.532 × 10^{-5}	1.308 × 10^{-6}	10^{-6}
in.^{3}	16,387	16.39	1	0.01639	4.329 × 10^{-3}	5.787 × 10^{-4}	2.143 × 10^{-5}	1.639 × 10^{-5}
L	10^6	1000	61.026	1	0.2642	0.03532	1.308 × 10^{-3}	10^{-3}
gal	3.785 × 10^6	3785	231.0	3.785	1	0.1337	4.951 × 10^{-3}	3.785 × 10^{-3}
ft^3	2.832 × 10^7	2.832 × 10^4	1728	28.32	7.4805	1	0.03704	0.02832
yd^3	7.6456 × 10^8	7.6453 × 10^5	46,656	764.53	202.0	27	1	0.76456
m^3	10^9	10^6	61,024	1000	264.2	35.31	1.308	1

Source: Adapted from Cornforth, J.R. (ed.), *Combustion Engineering and Gas Utilisation*, 3rd edn., E&F Spon, London, U.K., 1992, Table III, p. 863.
Note: The tables are read from row to column (i.e., 1 unit in the row = # units of the column.

TABLE A.7

Conversion Table for Mass

Mass	g	oz	lb	kg	cwt	ton	t (tonne)
g	1	0.03527	2.2046×10^{-3}	10^{-3}			
oz	28.3495	1	0.0625	0.028350			
lb	453.592	16	1	0.453592	8.9286×10^{-3}	5.00×10^{-4}	4.5359×10^{-4}
kg	10^3	35.2740	2.20462	1	0.019684	1.1023×10^{-3}	10^{-3}
cwt	50,802.3	1792	112	50.8023	1	0.056	0.05080
ton	907,185	32,000	2000	907.185	17.8571	1	0.907185
t (tonne)	10^6	35,273.9	2204.62	1000	19.6841	1.10231	1

Source: Adapted from Cornforth, J.R. (ed.), *Combustion Engineering and Gas Utilisation*, 3rd edn., E&F Spon, London, U.K., 1992, Table IV, p. 864.

Note: The tables are read from row to column (i.e., 1 unit in the row = # units of the column.

TABLE A.8

Conversion Table for Density

Density	kg/m³	lb/ft³	g/cm³	lb/in³
kg/m³	1	0.062428	10^{-3}	3.6046×10^{-5}
lb/ft³	16.0185	1	0.0160185	5.7870×10^{-4}
g/cm³	1000	62.4280	1	0.036127
lb/in³	27,679.9	1728	27.6799	1

Source: Adapted from Cornforth, J.R. (ed.), *Combustion Engineering and Gas Utilisation*, 3rd edn., E&F Spon, London, U.K., 1992, Table V, p. 865.

Note: The tables are read from row to column (i.e., 1 unit in the row = # units of the column.

TABLE A.9

Conversion Table for Velocity

Velocity	mm/s	ft/min	km/h	ft/s	mile/h	m/s
mm/s	1	0.19685	3.6×10^{-3}	3.281×10^{-3}	2.237×10^{-3}	10^{-3}
ft/min	5.08	1	0.018288	0.016667	0.01136	5.08×10^{-3}
km/h	277.778	54.6086	1	0.911344	0.621371	0.277778
ft/s	304.8	60	1.09728	1	0.681818	0.3048
mile/h	447.04	88	1.609344	1.46667	1	0.44704
m/s	1000	196.840	3.6	3.28084	2.23694	1

Source: Adapted from Cornforth, J.R. (ed.), *Combustion Engineering and Gas Utilisation*, 3rd edn., E&F Spon, London, U.K., 1992, Table VI, p. 866.

Note: The tables are read from row to column (i.e., 1 unit in the row = # units of the column.

TABLE A.10
Conversion Table for Mass Rate of Flow

Mass Rate of Flow	lb/h	kg/h	g/s	lb/min	t/h	lb/s	kg/s
lb/h	1	0.4536	0.126	0.01667	4.536×10^{-4}	2.778×10^{-4}	1.260×10^{-4}
kg/h	2.205	1	0.2778	0.03674	1×10^{-3}	6.124×10^{-4}	2.778×10^{-4}
g/s	7.937	3.6	1	0.1323	3.6×10^{-3}	2.205×10^{-3}	10^{-3}
lb/min	60	27.216	7.56	1	2.722×10^{-2}	1.667×10^{-2}	7.56×10^{-3}
t/h	2205	1000	277.8	36.74	1	0.6124	0.2778
lb/s	3600	1633	453.6	60	1.633	1	0.4536
kg/s	7937	3600	1000	132.3	3.6	2.205	1

Source: Adapted from Cornforth, J.R. (ed.), *Combustion Engineering and Gas Utilisation*, 3rd edn., E&F Spon, London, U.K., 1992, Table VII, p. 867.
Note: The tables are read from row to column (i.e., 1 unit in the row = # units of the column.

TABLE A.11
Conversion Table for Volume Rate of Flow

Volume Rate of Flow	L/h	mL/s	gal/h	L/min	gal/min	m³/h	ft³/min	L/s	ft³/s	m³/s	U.K. gal/min
L/h	1	0.2778	0.2642	0.01667	4.403×10^{-3}	10^{-3}	5.886×10^{-4}	2.778×10^{-4}	0.810×10^{-6}	2.778×10^{-7}	3.666×10^{-3}
mL/s	3.6	1	0.9510	0.0600	0.0158	3.6×10^{-3}	2.119×10^{-3}	10^{-3}	3.532×10^{-5}	10^{-6}	0.01320
Gal/h	3.785	1.052	1	0.0631	0.01667	3.785×10^{-3}	2.228×10^{-3}	1.052×10^{-3}	3.714×10^{-5}	1.052×10^{-6}	0.01388
L/min	60	16.67	15.85	1	0.2642	0.0600	0.03531	0.01667	5.886×10^{-4}	1.667×10^{-5}	0.2200
gal/min	227.2	63.09	60	3.785	1	0.2272	0.1336	0.0631	2.228×10^{-3}	6.309×10^{-5}	0.8326
m³/h	1000	277.8	264.2	16.67	4.403	1	0.5886	0.2778	0.810×10^{-3}	2.778×10^{-4}	3.666
ft³/min	1699	471.9	448.8	28.31	7.481	1.699	1	0.4719	0.01667	4.719×10^{-4}	6.229
L/s	3600	1000	951.2	60	15.84	3.6	2.119	1	0.03531	10^{-3}	13.20
ft³/s	1.019×10^5	2.832×10^4	2.693×10^4	1699	448.8	101.9	60	28.32	1	0.02832	373.7
m³/s	3.6×10^6	10^6	9.510×10^5	6×10^4	1.585×10^4	3600	2119	1000	35.31	1	1.320×10^4
U.K. gal/min	272.8	75.77	72.06	4.546	1.201	0.2728	0.1605	0.07577	2.677×10^{-3}	7.577×10^{-5}	1

Source: Adapted from Cornforth, J.R. (ed.), *Combustion Engineering and Gas Utilisation*, 3rd edn., E&F Spon, London, U.K., 1992, Table VIII, p. 868.
Note: The tables are read from row to column (i.e., 1 unit in the row = # units of the column.
gal = U.S. gal.
1 U.K. or Imperial gallon = 1.201 U.S. gal.

TABLE A.12
Conversion Table for Pressure

Pressure	Pa	mbar	mm Hg	in H₂O	kPa	in Hg	lb_f/in.²	kgf/cm²	bar	atm
Pa	1	0.0100	7.501×10^{-3}	4.015×10^{-3}	10^{-3}	2.953×10^{-4}	1.450×10^{-4}	1.020×10^{-5}	10^{-5}	9.869×10^{-6}
mbar	100	1	0.7501	0.4015	0.1000	0.02953	0.01450	1.020×10^{-3}	10^{-3}	9.869×10^{-4}
mm Hg	133.3	1.333	1	0.5352	0.1333	0.03937	0.01934	1.360×10^{-3}	1.333×10^{-3}	1.316×10^{-3}
in H₂O	249.1	2.491	1.868	1	0.2491	0.07356	0.03613	2.540×10^{-3}	2.491×10^{-3}	2.458×10^{-3}
kPa	1000	10	7.501	4.015	1	0.2953	0.1450	0.01020	0.0100	9.869×10^{-3}
in Hg	3386	33.86	25.40	13.60	3.386	1	0.4912	0.03453	0.03386	0.03342
lb_f/in.²	6895	68.95	51.71	27.68	6.895	2.036	1	0.07031	0.06895	0.06805
kgf/cm²	9.807×10^4	980.7	735.6	393.7	98.07	28.96	14.22	1	0.9807	0.9678
bar	10^5	1000	750.1	401.5	100	29.53	14.50	1.020	1	0.9869
atm	1.013×10^5	1013	760.0	406.8	101.3	29.92	14.70	1.033	1.013	1

Source: Adapted from Cornforth, J.R. (ed.), *Combustion Engineering and Gas Utilisation*, 3rd edn., E&F Spon, London, U.K., 1992, Table IX, p. 869.
Note: The tables are read from row to column (i.e., 1 unit in the row = # units of the column.
1 kgf/cm² = 1 kp/cm² = 1 technical atmosphere = 14.22 lb_f/in.².
1 torr = 1 mm Hg (to within 1 part in 7 million).

TABLE A.13

Conversion Table for Energy, Work, Heat

Energy, Work, Heat	J	kJ	Btu	kcal	MJ	hph	kWh	therm	GJ
J	1	10^{-3}	9.478×10^{-4}	2.388×10^{-4}	10^{-6}	3.725×10^{-7}	2.778×10^{-7}	0.478×10^{-9}	10^{-9}
kJ	1000	1	0.9478	0.2388	10^{-3}	3.725×10^{-4}	2.778×10^{-4}	0.478×10^{-6}	10^{-6}
Btu	1055.1	1.0551	1	0.2520	1.055×10^{-3}	3.930×10^{-4}	2.931×10^{-4}	10^{-5}	1.055×10^{-6}
kcal	4186.8	4.1868	3.9683	1	4.187×10^{-3}	1.560×10^{-3}	1.163×10^{-3}	3.968×10^{-5}	4.187×10^{-6}
MJ	10^6	1000	947.82	238.85	1	0.3725	0.2778	9.478×10^{-3}	10^{-3}
hph	2.6845×10^6	2684.5	2544.4	641.19	2.6845	1	0.7457	0.02544	2.6845×10^{-3}
kWh	3.6000×10^6	3600	3412.1	859.84	3.600	1.3410	1	0.03412	3.6×10^{-3}
therm	1.0551×10^8	1.0551×10^5	1×10^5	25,200	105.51	39.301	29.307	1	0.10551
GJ	10^9	10^6	9.4782×10^5	2.3885×10^5	10^3	372.5	277.8	9.478	1

Source: Adapted from Cornforth, J.R. (ed.), *Combustion Engineering and Gas Utilisation*, 3rd edn., E&F Spon, London, U.K., 1992, Table X, p. 870.
Note: The tables are read from row to column (i.e., 1 unit in the row = # units of the column).
1 thermie = 1.163 kWh = 4.186 MJ = 999.7 kcal = 3,967.1 Btu.

TABLE A.14

Conversion Table for Power, Heat Flow Rate

Power, Heat Flow Rate	Btu/h	W	kcal/h	hp	kW	MW
Btu/h	1	0.2931	0.2520	3.930×10^{-4}	2.931×10^{-4}	2.93×10^{-7}
W	3.4121	1	0.8598	1.341×10^{-3}	10^{-3}	10^{-6}
kcal/h	3.9683	1.163	1	1.560×10^{-3}	1.163×10^{-3}	1.16×10^{-6}
hp	2544	745.70	641.19	1	0.7457	7.457×10^{-4}
kW	3412.1	1000	859.8	1.3410	1	10^{-3}
MW	3.4121×10^6	10^6	8.598×10^6	1341	1000	1

Source: Adapted from Cornforth, J.R. (ed.), *Combustion Engineering and Gas Utilisation*, 3rd edn., E&F Spon, London, U.K., 1992, Table XI, p. 871.
Note: The tables are read from row to column (i.e., 1 unit in the row = # units of the column).
1 W = 1 J/s.
1 cal/s = 3.6 kcal/h.
1 ton of refrigeration = 3,517 W = 12,000 Btu/h.

TABLE A.15

Conversion Table for Specific Energy

A. Mass Basis	kJ/kg	Btu/lb	kcal/kg	MJ/kg
kJ/kg	1	0.4299	0.2388	10^{-3}
Btu/lb	2.326	1	0.5556	2.326×10^{-3}
kcal/kg	4.187	1.8	1	4.187×10^{-3}
MJ/kg	1000	429.9	238.8	1
B. Volume Basis	kJ/m³	kcal/m³	Btu/ft³	MJ/m³
kJ/m³	1	0.2388	0.02684	10^{-3}
kcal/m³	4.187	1	0.1124	4.187×10^{-3}
Btu/ft³	37.26	8.899	1	0.03726
MJ/m³	1000	238.8	26.84	1

Source: Adapted from Cornforth, J.R. (ed.), *Combustion Engineering and Gas Utilisation*, 3rd edn., E&F Spon, London, U.K., 1992, Table XII, p. 872.
Note: The tables are read from row to column (i.e., 1 unit in the row = # units of the column).
1 therm (10^5 Btu)/U.K. gal = 23,208 MJ/m³.
1 therm/L = 4185 MJ/m³.

TABLE A.16

Conversion Table for Thermal Conductance

Thermal Conductance, U Value	W/(m² K)	kcal/(m²h K)	Btu/(ft²h °F)
W/(m² K)	1	0.8598	0.1761
kcal/(m²h K)	1.163	1	0.2048
Btu/(ft²h °F)	5.678	4.882	1

Source: Adapted from Cornforth, J.R. (ed.), *Combustion Engineering and Gas Utilisation*, 3rd edn., E&F Spon, London, U.K., 1992, Table XIII, p. 874.

Note: The tables are read from row to column (i.e., 1 unit in the row = # units of the column.

TABLE A.17

Conversion Table for Thermal Conductivity

Thermal Conductivity	Btu in./(ft²h °F)	W/(m K)	kcal/(m h K)
Btu in./(ft²h °F)	1	0.1442	0.124
W/(m K)	6.933	1	0.8598
kcal/(m h K)	8.064	1.163	1

Source: Adapted from Cornforth, J.R. (ed.), *Combustion Engineering and Gas Utilisation*, 3rd edn., E&F Spon, London, U.K., 1992, Table XIV, p. 874.

Note: The tables are read from row to column (i.e., 1 unit in the row = # units of the column.

TABLE A.18

Conversion Table for Density of Heat Flow Rate

Density of Heat Flow Rate	W/m²	kcal/(m²h)	Btu/(ft²h)
W/m²	1	0.8598	0.317
kcal/(m²h)	1.163	1	0.3687
Btu/(ft²h)	3.155	2.712	1

Source: Adapted from Cornforth, J.R. (ed.), *Combustion Engineering and Gas Utilisation*, 3rd edn., E&F Spon, London, U.K., 1992, Table XV, p. 875.

Note: The tables are read from row to column (i.e., 1 unit in the row = # units of the column.

TABLE A.19

Conversion Table for Specific Heat Capacity

Specific Heat Capacity	kJ/(kg K)	Btu/(lb °F)	kcal/(kg K)
kJ/(kg K)	1	0.2388	0.2388
Btu/(lb °F)	4.1868	1	1
kcal/(kg K)	4.1868	1	1

Source: Adapted from Cornforth, J.R. (ed.), *Combustion Engineering and Gas Utilisation*, 3rd edn., E&F Spon, London, U.K., 1992, Table XVI, p. 875.

Note: The tables are read from row to column (i.e., 1 unit in the row = # units of the column.

References

1. Annamalai, K. and Puri, I.K., *Combustion Science and Engineering*, CRC Press, Boca Raton, FL, 2007.
2. Cornforth, J.R. (ed.), *Combustion Engineering and Gas Utilisation*, 3rd edn., E&F Spon, London, U.K., 1992.

Appendix B: Physical Properties of Materials

TABLE B.1

Areas and Circumferences of Circles and Drill Sizes

Drill Size	Diameter (in.)	Circumference (in.)	Area (in.)	Area (ft)
80	0.0135	0.04241	0.000143	0.0000009
79	0.0145	0.04555	0.000165	0.0000011
1/64 in.	0.0156	0.04909	0.000191	0.0000013
78	0.0160	0.05027	0.000201	0.0000014
77	0.0180	0.05655	0.000254	0.0000018
76	0.0200	0.06283	0.000314	0.0000022
75	0.0210	0.06597	0.000346	0.0000024
74	0.0225	0.07069	0.000398	0.0000028
73	0.0240	0.07540	0.000452	0.0000031
72	0.0250	0.07854	0.000491	0.0000034
71	0.0260	0.08168	0.000531	0.0000037
70	0.0280	0.08796	0.000616	0.0000043
69	0.0292	0.09173	0.000670	0.0000047
68	0.0310	0.09739	0.000755	0.0000052
1/32 in.	0.0313	0.09818	0.000765	0.0000053
67	0.0320	0.10053	0.000804	0.0000056
66	0.0330	0.10367	0.000855	0.0000059
65	0.0350	0.10996	0.000962	0.0000067
64	0.0360	0.11310	0.001018	0.0000071
63	0.0370	0.11624	0.001075	0.0000075
62	0.0380	0.11938	0.001134	0.0000079
61	0.0390	0.12252	0.001195	0.0000083
60	0.0400	0.12566	0.001257	0.0000087
59	0.0410	0.12881	0.001320	0.0000092
58	0.0420	0.13195	0.001385	0.0000096
57	0.0430	0.13509	0.001452	0.0000101
56	0.0465	0.14608	0.001698	0.0000118
3/64 in.	0.0469	0.14726	0.00173	0.0000120
55	0.0520	0.16336	0.00212	0.0000147
54	0.0550	0.17279	0.00238	0.0000165
53	0.0595	0.18693	0.00278	0.0000193
1/16 in.	0.0625	0.19635	0.00307	0.0000213
52	0.0635	0.19949	0.00317	0.0000220
51	0.0670	0.21049	0.00353	0.0000245
50	0.0700	0.21991	0.00385	0.0000267
49	0.0730	0.22934	0.00419	0.0000291
48	0.0760	0.23876	0.00454	0.0000315
5/64 in.	0.0781	0.24544	0.00479	0.0000333
47	0.0785	0.24662	0.00484	0.0000336
46	0.0810	0.25447	0.00515	0.0000358
45	0.0820	0.25761	0.00528	0.0000367
44	0.0860	0.27018	0.00581	0.0000403
43	0.0890	0.27960	0.00622	0.0000432
42	0.0935	0.29374	0.00687	0.0000477
3/32 in.	0.0937	0.29452	0.00690	0.0000479

TABLE B.1 (continued)

Areas and Circumferences of Circles and Drill Sizes

Drill Size	Diameter (in.)	Circumference (in.)	Area (in.)	Area (ft)
41	0.0960	0.30159	0.00724	0.0000503
40	0.0980	0.30788	0.00754	0.0000524
39	0.0995	0.31259	0.00778	0.0000540
38	0.1015	0.31887	0.00809	0.0000562
37	0.1040	0.32673	0.00849	0.0000590
36	0.1065	0.33458	0.00891	0.0000619
7/64 in.	0.1094	0.34361	0.00940	0.0000652
35	0.1100	0.34558	0.00950	0.0000660
34	0.1110	0.34872	0.00968	0.0000672
33	0.1130	0.35500	0.01003	0.0000696
32	0.1160	0.36443	0.01057	0.0000734
31	0.1200	0.37699	0.01131	0.0000785
1/8 in.	0.1250	0.39270	0.01227	0.0000852
30	0.1285	0.40370	0.01296	0.0000901
29	0.1360	0.42726	0.01453	0.0001009
28	0.1405	0.44139	0.01549	0.0001077
9/64 in.	0.1406	0.44179	0.01553	0.0001079
27	0.1440	0.44239	0.01629	0.0001131
26	0.1470	0.46182	0.01697	0.0001179
25	0.1495	0.46967	0.01755	0.0001219
24	0.1520	0.47752	0.01815	0.0001260
23	0.1540	0.48381	0.01863	0.0001294
5/32 in.	0.1562	0.49087	0.01917	0.0001331
22	0.1570	0.49323	0.01936	0.0001344
21	0.1590	0.49951	0.01986	0.0001379
20	0.1610	0.50580	0.02036	0.0001414
19	0.1660	0.52151	0.02164	0.0001503
18	0.1695	0.53250	0.02256	0.0001567
11/64 in.	0.1719	0.53996	0.02320	0.0001611
17	0.1730	0.54350	0.02351	0.0001632
16	0.1770	0.55606	0.02461	0.0001709
15	0.1800	0.56549	0.02545	0.0001767
14	0.1820	0.57177	0.02602	0.0001807
13	0.1850	0.58120	0.02688	0.0001867
3/16 in.	0.1875	0.58905	0.02761	0.0001917
12	0.1890	0.59376	0.02806	0.0001948
11	0.1910	0.60005	0.02865	0.0001990
10	0.1930	0.60633	0.02940	0.0002032
9	0.1960	0.61575	0.03017	0.0002095
8	0.1990	0.62518	0.03110	0.0002160
7	0.2010	0.63146	0.03173	0.0002204
13/64 in.	0.2031	0.63814	0.03241	0.0002248
6	0.2040	0.64089	0.03269	0.0002270
5	0.2055	0.64560	0.03317	0.0002303
4	0.2090	0.65659	0.03431	0.0002382

(continued)

TABLE B.1 (continued)

Areas and Circumferences of Circles and Drill Sizes

Drill Size	Diameter (in.)	Circumference (in.)	Area (in.)	Area (ft)
3	0.2130	0.66916	0.03563	0.0002475
7/32 in.	0.2187	0.68722	0.03758	0.0002610
2	0.2210	0.69429	0.03836	0.0002664
1	0.2280	0.71628	0.04083	0.0002835
A	0.2340	0.73513	0.04301	0.0002987
15/64 in.	0.2344	0.73631	0.04314	0.0002996
B	0.2380	0.74770	0.04449	0.0003089
C	0.2420	0.76027	0.04600	0.0003194
D	0.2460	0.77283	0.04753	0.0003301
E = 1/4 in.	0.2500	0.78540	0.04909	0.0003409
F	0.2570	0.80739	0.05187	0.0003602
G	0.2610	0.81996	0.05350	0.0003715
17/64 in.	0.2656	0.83441	0.05542	0.0003849
H	0.2660	0.83567	0.05557	0.0003859
I	0.2720	0.85452	0.05811	0.0004035
J	0.2770	0.87022	0.06026	0.0004185
K	0.2810	0.88279	0.06202	0.0004307
9/32 in.	0.2812	0.88357	0.06213	0.0004315
L	0.2900	0.91106	0.06605	0.0004587
M	0.2950	0.92677	0.06835	0.0004747
19/64 in.	0.2969	0.93266	0.06922	0.0004807
N	0.3030	0.95190	0.07163	0.0005007
5/16 in.	0.3125	0.98175	0.07670	0.0005326
O	0.3160	0.99275	0.07843	0.0005446
P	0.3230	1.01474	0.08194	0.0005690
21/64 in.	0.3281	1.0308	0.08456	0.0005872
Q	0.3320	1.0430	0.08657	0.0006012
R	0.3390	1.0650	0.09026	0.0006268
11/32 in.	0.3437	1.0798	0.09281	0.0006445
S	0.3480	1.0933	0.09511	0.0006605
T	0.3580	1.1247	0.1006	0.0006990
23/64 in.	0.3594	1.1290	0.1014	0.0007044
U	0.3680	1.1561	0.1064	0.0007386
3/8 in.	0.3750	1.1781	0.1105	0.0007670
V	0.3770	1.1844	0.1116	0.0007752
W	0.3860	1.2127	0.1170	0.0008127
25/64 in.	0.3906	1.2272	0.1198	0.0008322
X	0.3970	1.2472	0.1238	0.0008596
Y	0.4040	1.2692	0.1282	0.0008902
13/32 in.	0.4062	1.2763	0.1296	0.0009001
Z	0.4130	1.2975	0.1340	0.0009303
27/64 in.	0.4219	1.3254	0.1398	0.0009708
7/16 in.	0.4375	1.3745	0.1503	0.001044
29/64 in.	0.4531	1.4235	0.1613	0.001120
15/32 in.	0.4687	1.4726	0.1726	0.001198
31/64 in.	0.4844	1.5217	0.1843	0.001280
1/2 in.	0.5000	1.5708	0.1964	0.001364
33/64 in.	0.5156	1.6199	0.2088	0.001450
17/32 in.	0.5313	1.6690	0.2217	0.001539
35/64 in.	0.5469	1.7181	0.2349	0.001631

TABLE B.1 (continued)

Areas and Circumferences of Circles and Drill Sizes

Drill Size	Diameter (in.)	Circumference (in.)	Area (in.)	Area (ft)
9/16 in.	0.5625	1.7672	0.2485	0.001726
37/64 in.	0.5781	1.8162	0.2625	0.001823
19/32 in.	0.5938	1.8653	0.2769	0.001923
39/64 in.	0.6094	1.9144	0.2917	0.002025
5/8 in.	0.6250	1.9635	0.3068	0.002131
41/64 in.	0.6406	2.0126	0.3223	0.002238
21/32 in.	0.6562	2.0617	0.3382	0.002350
43/64 in.	0.6719	2.1108	0.3545	0.002462
11/16 in.	0.6875	2.1598	0.3712	0.002578
23/32 in.	0.7188	2.2580	0.4057	0.002818
3/4 in.	0.7500	2.3562	0.4418	0.003068
25/32 in.	0.7812	2.4544	0.4794	0.003329
13/16 in.	0.8125	2.5525	0.5185	0.003601
27/32 in.	0.8438	2.6507	0.5591	0.003883
7/8 in.	0.8750	2.7489	0.6013	0.004176
29/32 in.	0.9062	2.8471	0.6450	0.004479
15/16 in.	0.9375	2.9452	0.6903	0.004794
31/32 in.	0.9688	3.0434	0.7371	0.005119
1 in.	1.0000	3.1416	0.7854	0.005454
1 1/16 in.	1.0625	3.3379	0.8866	0.006157
1 1/8 in.	1.1250	3.5343	0.9940	0.006903
1 3/16 in.	1.1875	3.7306	1.1075	0.007691
1 1/4 in.	1.2500	3.9270	1.2272	0.008522
1 5/16 in.	1.3125	4.1233	1.3530	0.009396
1 3/8 in.	1.3750	4.3170	1.4849	0.01031
1 7/16 in.	1.4375	4.5160	1.6230	0.01127
1 1/2 in.	1.5000	4.7124	1.7671	0.01227
1 9/16 in.	1.5625	4.9087	1.9175	0.01332
1 5/8 in.	1.6250	5.1051	2.0739	0.01440
1 11/16 in.	1.6875	5.3014	2.2365	0.01553
1 3/4 in.	1.7500	5.4978	2.4053	0.01670
1 13/16 in.	1.8125	5.6941	2.5802	0.01792
1 7/8 in.	1.8750	5.8905	2.7612	0.01918
1 15/16 in.	1.9375	6.0868	2.9483	0.02047
2 in.	2.0000	6.2832	3.1416	0.02182
2 1/16 in.	2.0625	6.4795	3.3410	0.02320
2 1/8 in.	2.1250	6.6759	3.5466	0.02463
2 3/16 in.	2.1875	6.8722	3.7583	0.02610
2 1/4 in.	2.2500	7.0686	3.9761	0.02761
2 5/16 in.	2.3125	7.2649	4.2000	0.02917
2 3/8 in.	2.3750	7.4613	4.4301	0.03076
2 7/16 in.	2.4375	7.6576	4.6664	0.03241
2 1/2 in.	2.5000	7.8540	4.9087	0.03409
2 9/16 in.	2.5625	8.0503	5.1572	0.03581
2 5/8 in.	2.6250	8.2467	5.4119	0.03758
2 11/16 in.	2.6875	8.4430	5.6727	0.03939
2 3/4 in.	2.7500	8.6394	5.9396	0.04125
2 13/16 in.	2.8125	8.8357	6.2126	0.04314
2 7/8 in.	2.8750	9.0323	6.4918	0.04508
2 15/16 in.	2.9375	9.2284	6.7771	0.04706

TABLE B.1 (continued)
Areas and Circumferences of Circles and Drill Sizes

Drill Size	Diameter (in.)	Circumference (in.)	Area (in.)	Area (ft)
3 in.	3.0000	9.4248	7.0686	0.04909
3 1/16 in.	3.0625	9.6211	7.3662	0.05115
3 1/8 in.	3.1250	9.8175	7.6699	0.05326
3 3/16 in.	3.1875	10.014	7.9798	0.05542
3 1/4 in.	3.2500	10.210	8.2958	0.05736
3 5/16 in.	3.3125	10.407	8.6179	0.05985
3 3/8 in.	3.3750	10.603	8.9462	0.06213
3 7/16 in.	3.4375	10.799	9.2806	0.06445
3 1/2 in.	3.5000	10.996	9.6211	0.06681
3 9/16 in.	3.5625	11.192	9.9678	0.06922
3 5/8 in.	3.6250	11.388	10.321	0.07167
3 11/16 in.	3.6875	11.585	10.680	0.07417
3 3/4 in.	3.7500	11.781	11.045	0.07670
3 13/16 in.	3.8125	11.977	11.416	0.07928
3 7/8 in.	3.8750	12.174	11.793	0.08190
3 15/16 in.	3.9375	12.370	12.177	0.08456
4 in.	4.0000	12.566	12.566	0.08726
4 1/16 in.	4.0625	12.763	12.962	0.09002
4 1/8 in.	4.1250	12.959	13.364	0.09281
4 3/16 in.	4.1875	13.155	13.772	0.09564
4 1/4 in.	4.2500	13.352	14.186	0.09852
4 5/16 in.	4.3125	13.548	14.607	0.1014
4 3/8 in.	4.3750	13.745	15.033	0.1043
4 7/16 in.	4.4375	13.941	15.466	0.1074
4 1/2 in.	4.5000	14.137	15.904	0.1104
4 9/16 in.	4.5625	14.334	16.349	0.1135
4 5/8 in.	4.6250	14.530	16.800	0.1167
4 11/16 in.	4.6875	14.726	17.257	0.1198
4 3/4 in.	4.7500	14.923	17.721	0.1231
4 13/16 in.	4.8125	15.119	18.190	0.1263
4 7/8 in.	4.8750	15.315	18.665	0.1296
4 15/16 in.	4.9375	15.512	19.147	0.1330
5 in.	5.0000	15.708	19.635	0.1364
5 1/16 in.	5.0625	15.904	20.129	0.1398
5 1/8 in.	5.1250	16.101	20.629	0.1433
5 3/16 in.	5.1875	16.297	21.135	0.1468
5 1/4 in.	5.2500	16.493	21.648	0.1503
5 5/16 in.	5.3125	16.690	22.166	0.1539
5 3/8 in.	5.3750	16.886	22.691	0.1576
5 7/16 in.	5.4375	17.082	23.221	0.1613
5 1/2 in.	5.5000	17.279	23.758	0.1650
5 9/16 in.	5.5625	17.475	24.301	0.1688
5 5/8 in.	5.6250	17.671	24.851	0.1726
5 11/16 in.	5.6875	17.868	25.406	0.1764
5 3/4 in.	5.7500	18.064	25.967	0.1803
5 13/16 in.	5.8125	18.261	26.535	0.1843
5 7/8 in.	5.8750	18.457	27.109	0.1883
5 15/16 in.	5.9375	18.653	27.688	0.1923
6 in.	6.0000	18.850	28.274	0.1963
6 1/8 in.	6.1250	19.242	29.465	0.2046
6 1/4 in.	6.2500	19.649	30.680	0.2131
6 3/8 in.	6.3750	20.028	31.919	0.2217
6 1/2 in.	6.5000	20.420	33.183	0.2304
6 5/8 in.	6.6250	20.813	34.472	0.2394
6 3/4 in.	6.7500	21.206	35.785	0.2485
6 7/8 in.	6.8750	21.598	37.122	0.2578
7 in.	7.0000	21.991	38.485	0.2673
7 1/8 in.	7.1250	22.384	39.871	0.2769
7 1/4 in.	7.2500	22.777	41.283	0.2867
7 3/8 in.	7.3750	23.169	42.718	0.2967
7 1/2 in.	7.5000	23.562	44.179	0.3068
7 5/8 in.	7.6250	23.955	45.664	0.3171
7 3/4 in.	7.7500	24.347	47.173	0.3276
7 7/8 in.	7.8750	24.740	48.707	0.3382
8 in.	8.0000	25.133	50.266	0.3491
8 1/8 in.	8.1250	25.525	51.849	0.3601
8 1/4 in.	8.2500	25.918	53.456	0.3712
8 3/8 in.	8.3750	26.301	55.088	0.3826
8 1/2 in.	8.5000	26.704	56.745	0.3941
8 5/8 in.	8.6250	27.096	58.426	0.4057
8 3/4 in.	8.7500	27.489	60.132	0.4176
8 7/8 in.	8.8750	27.882	61.862	0.4296
9 in.	9.0000	28.274	63.617	0.4418
9 1/8 in.	9.1250	28.667	65.397	0.4541
9 1/4 in.	9.2500	29.060	67.201	0.4667
9 3/8 in.	9.3750	29.452	69.029	0.4794
9 1/2 in.	9.5000	29.845	70.882	0.4922
9 5/8 in.	9.6250	30.238	72.760	0.5053
9 3/4 in.	9.7500	30.631	74.662	0.5185
9 7/8 in.	9.8750	31.023	76.589	0.5319
10 in.	10.0000	31.416	78.540	0.5454
10 1/8 in.	10.1250	31.809	80.516	0.5591
10 1/4 in.	10.2500	32.201	82.516	0.5730
10 3/8 in.	10.3750	32.594	84.541	0.5871
10 1/2 in.	10.5000	32.987	86.590	0.6013
10 5/8 in.	10.6250	33.379	88.664	0.6157
10 3/4 in.	10.7500	33.772	90.763	0.6303
10 7/8 in.	10.8750	34.165	92.886	0.6450
11 in.	11.0000	34.558	95.033	0.6600
11 1/8 in.	11.1250	34.950	97.205	0.6750
11 1/4 in.	11.2500	35.343	99.402	0.6903
11 3/8 in.	11.3750	35.736	101.6	0.7056
11 1/2 in.	11.5000	36.128	103.9	0.7213
11 5/8 in.	11.6250	36.521	106.1	0.7371
11 3/4 in.	11.7500	36.914	108.4	0.7530
11 7/8 in.	11.8750	37.306	110.8	0.7691
12 in.	12.0000	37.699	113.1	0.7854
12 1/4 in.	12.2500	38.485	117.9	0.819
12 1/2 in.	12.5000	39.269	122.7	0.851
12 3/4 in.	12.7500	40.055	127.7	0.886

(continued)

TABLE B.1 (continued)
Areas and Circumferences of Circles and Drill Sizes

Drill Size	Diameter (in.)	Circumference (in.)	Area (in.)	Area (ft)
13 in.	13.0000	40.841	132.7	0.921
13 1/4 in.	13.2500	41.626	137.9	0.957
13 1/2 in.	13.5000	42.412	143.1	0.995
13 3/4 in.	13.7500	43.197	148.5	1.031
14 in.	14.0000	43.982	153.9	1.069
14 1/4 in.	14.2500	44.768	159.5	1.109
14 1/2 in.	14.5000	45.553	165.1	1.149
14 3/4 in.	14.7500	46.339	170.9	1.185
15 in.	15.0000	47.124	176.7	1.228
15 1/4 in.	15.2500	47.909	182.7	1.269
15 1/2 in.	15.5000	48.695	188.7	1.309
15 3/4 in.	15.7500	49.480	194.8	1.352
16 in.	16.0000	50.266	201.1	1.398
16 1/4 in.	16.2500	51.051	207.4	1.440
16 1/2 in.	16.5000	51.836	213.8	1.485
16 3/4 in.	16.7500	52.622	220.4	1.531
17 in.	17.0000	53.407	227.0	1.578
17 1/4 in.	17.2500	54.193	233.7	1.619
17 1/2 in.	17.5000	54.978	240.5	1.673
17 3/4 in.	17.7500	55.763	247.5	1.719
18 in.	18.0000	56.548	254.5	1.769
18 1/4 in.	18.2500	57.334	261.6	1.816
18 1/2 in.	18.5000	58.120	268.8	1.869
18 3/4 in.	18.7500	58.905	276.1	1.920
19 in.	19.0000	59.690	283.5	1.969
19 1/4 in.	19.2500	60.476	291.0	2.022
19 1/2 in.	19.5000	61.261	298.7	2.075
19 3/4 in.	19.7500	62.047	306.4	2.125
20 in.	20.0000	62.832	314.2	2.182
20 1/4 in.	20.2500	63.617	322.1	2.237
20 1/2 in.	20.5000	64.403	330.1	2.292
20 3/4 in.	20.7500	65.188	338.2	2.348
21 in.	21.0000	65.974	346.4	2.405
21 1/4 in.	21.2500	66.759	354.7	2.463
21 1/2 in.	21.5000	67.544	363.1	2.521
21 3/4 in.	21.7500	68.330	371.5	2.580
22 in.	22.0000	69.115	380.1	2.640
22 1/4 in.	22.2500	69.901	388.8	2.700
22 1/2 in.	22.5000	70.686	397.6	2.761
22 3/4 in.	22.7500	71.471	406.5	2.823
23 in.	23.0000	72.257	415.5	2.885
23 1/4 in.	23.2500	73.042	424.6	2.948
23 1/2 in.	23.5000	73.828	433.7	3.012
23 3/4 in.	23.7500	74.613	443.0	3.076
24 in.	24.0000	75.398	452.4	3.142
24 1/4 in.	24.2500	76.184	461.9	3.207
24 1/2 in.	24.5000	76.969	471.4	3.274
24 3/4 in.	24.7500	77.755	481.1	3.341
25 in.	25.0000	78.540	490.9	3.409
25 1/4 in.	25.2500	79.325	500.7	3.477
25 1/2 in.	25.5000	80.111	510.7	3.547
25 3/4 in.	25.7500	80.896	520.8	3.616
26 in.	26.0000	81.682	530.9	3.687
26 1/4 in.	26.2500	82.467	541.2	3.758
26 1/2 in.	26.5000	83.252	551.6	3.830
26 3/4 in.	26.7500	84.038	562.0	3.903
27 in.	27.0000	84.823	572.6	3.976
27 1/4 in.	27.2500	85.609	583.2	4.050
27 1/2 in.	27.5000	86.394	594.0	4.125
27 3/4 in.	27.7500	87.179	604.8	4.200
28 in.	28.0000	87.965	615.8	4.276
28 1/4 in.	28.2500	88.750	626.8	4.353
28 1/2 in.	28.5000	89.536	637.9	4.430
28 3/4 in.	28.7500	90.321	649.2	4.508
29 in.	29.0000	91.106	660.5	4.587
29 1/4 in.	29.2500	91.892	672.0	4.666
29 1/2 in.	29.5000	92.677	683.5	4.746
29 3/4 in.	29.7500	93.463	695.1	4.827
30 in.	30.0000	94.248	706.9	4.909
31 in.	31.0000	97.390	754.8	5.241
32 in.	32.0000	100.53	804.3	5.585
33 in.	33.0000	103.67	855.3	5.940
34 in.	34.0000	106.81	907.9	6.305
35 in.	35.0000	109.96	962.1	6.681
36 in.	36.0000	113.10	1017.9	7.069
37 in.	37.0000	116.24	1075.2	7.467
38 in.	38.0000	119.38	1134.1	7.876
39 in.	39.0000	122.52	1194.6	8.296
40 in.	40.0000	125.66	1256.6	8.727
41 in.	41.0000	128.81	1320.3	9.168
42 in.	42.0000	131.95	1385.4	9.621
43 in.	43.0000	135.09	1452.2	10.08
44 in.	44.0000	138.23	1520.5	10.56
45 in.	45.0000	141.37	1590.4	11.04
46 in.	46.0000	144.51	1661.9	11.54
47 in.	47.0000	147.66	1734.9	12.04
48 in.	48.0000	150.80	1809.6	12.57
49 in.	49.0000	153.94	1885.7	13.10
50 in.	50.0000	157.08	1963.5	13.64

Appendix B: Physical Properties of Materials

TABLE B.2
Physical Properties of Pipe

Nominal Pipe Size, OD (in.)	Schedule Number a	Schedule Number b	Schedule Number c	Wall Thickness (in.)	ID (in.)	Inside Area (in.²)	Metal Area (in.²)	Sq. Ft. Outside Surface (per ft)	Sq. Ft. Inside Surface (per ft)	Weight per ft (lb)	Weight of Water per ft (lb)	Moment of Inertia (in.⁴)	Section Modulus (in.³)	Radius Gyration (in.)
1/8 0.405	—	—	10S	0.049	0.307	0.0740	0.0548	0.106	0.0804	0.186	0.0321	0.00088	0.00437	0.1271
	40	Std	40S	0.068	0.269	0.0568	0.0720	0.106	0.0705	0.245	0.0246	0.00106	0.00525	0.1215
	80	XS	80S	0.095	0.215	0.0364	0.0925	0.106	0.0563	0.315	0.0157	0.00122	0.00600	0.1146
1/4 0.540	—	—	10S	0.065	0.410	0.1320	0.0970	0.141	0.1073	0.330	0.0572	0.00279	0.01032	0.1694
	40	Std	40S	0.088	0.364	0.1041	0.1250	0.141	0.0955	0.425	0.0451	0.00331	0.01230	0.1628
	80	XS	80S	0.119	0.302	0.0716	0.1574	0.141	0.0794	0.535	0.0310	0.00378	0.01395	0.1547
3/8 0.675	—	—	10S	0.065	0.545	0.2333	0.1246	0.177	0.1427	0.423	0.1011	0.00586	0.01737	0.2169
	40	Std	40S	0.091	0.493	0.1910	0.1670	0.177	0.1295	0.568	0.0827	0.00730	0.02160	0.2090
	80	XS	80S	0.126	0.423	0.1405	0.2173	0.177	0.1106	0.739	0.0609	0.00862	0.02554	0.1991
1/2 0.840	—	—	10S	0.083	0.674	0.3570	0.1974	0.220	0.1765	0.671	0.1547	0.01431	0.0341	0.2692
	40	Std	40S	0.109	0.622	0.3040	0.2503	0.220	0.1628	0.851	0.1316	0.01710	0.0407	0.2613
	80	XS	80S	0.147	0.546	0.2340	0.3200	0.220	0.1433	1.088	0.1013	0.02010	0.0478	0.2505
	160	—	—	0.187	0.466	0.1706	0.3830	0.220	0.1220	1.304	0.0740	0.02213	0.0527	0.2402
	—	XXS	—	0.294	0.252	0.0499	0.5040	0.220	0.0660	1.714	0.0216	0.02425	0.0577	0.2192
3/4 1.050	—	—	5S	0.065	0.920	0.6650	0.2011	0.275	0.2409	0.684	0.2882	0.02451	0.0467	0.349
	—	—	10S	0.083	0.884	0.6140	0.2521	0.275	0.2314	0.857	0.2661	0.02970	0.0566	0.343
	40	Std	40S	0.113	0.824	0.5330	0.3330	0.275	0.2157	1.131	0.2301	0.0370	0.0706	0.334
	80	XS	80S	0.154	0.742	0.4320	0.4350	0.275	0.1943	1.474	0.1875	0.0448	0.0853	0.321
	160	—	—	0.218	0.614	0.2961	0.5700	0.275	0.1607	1.937	0.1284	0.0527	0.1004	0.304
	—	XXS	—	0.308	0.434	0.1479	0.7180	0.275	0.1137	2.441	0.0641	0.0579	0.1104	0.284
1 1.315	—	—	5S	0.065	1.185	1.1030	0.2553	0.344	0.3100	0.868	0.478	0.0500	0.0760	0.443
	—	—	10S	0.109	1.097	0.9450	0.4130	0.344	0.2872	1.404	0.409	0.0757	0.1151	0.428
	40	Std	40S	0.133	1.049	0.8640	0.4940	0.344	0.2746	1.679	0.374	0.0874	0.1329	0.421
	80	XS	80S	0.179	0.957	0.7190	0.6390	0.344	0.2520	2.172	0.311	0.1056	0.1606	0.407
	160	—	—	0.250	0.815	0.5220	0.8360	0.344	0.2134	2.844	0.2261	0.1252	0.1903	0.387
	—	XXS	—	0.358	0.599	0.2818	1.0760	0.344	0.1570	3.659	0.1221	0.1405	0.2137	0.361
1 1/4 1.660	—	—	5S	0.065	1.530	1.839	0.326	0.434	0.401	1.107	0.797	0.1038	0.1250	0.564
	—	—	10S	0.109	1.442	1.633	0.531	0.434	0.378	1.805	0.707	0.1605	0.1934	0.550
	40	Std	40S	0.140	1.380	1.496	0.669	0.434	0.361	2.273	0.648	0.1948	0.2346	0.540
	80	XS	80S	0.191	1.278	1.283	0.881	0.434	0.335	2.997	0.555	0.2418	0.2913	0.524
	160	—	—	0.250	1.160	1.057	1.107	0.434	0.304	3.765	0.458	0.2839	0.342	0.506
	—	XXS	—	0.382	0.896	0.631	1.534	0.434	0.2346	5.214	0.2732	0.341	0.411	0.472
	—	—	5S	0.065	1.770	2.461	0.375	0.497	0.463	1.274	1.067	0.1580	0.1663	0.649
	—	—	10S	0.109	1.682	2.222	0.613	0.497	0.440	2.085	0.962	0.2469	0.2599	0.634

(continued)

TABLE B.2 (continued)

Physical Properties of Pipe

Nominal Pipe Size, OD (in.)	Schedule Number a	Schedule Number b	Schedule Number c	Wall Thickness (in.)	ID (in.)	Inside Area (in.²)	Metal Area (in.²)	Sq. Ft. Outside Surface (per ft)	Sq. Ft. Inside Surface (per ft)	Weight per ft (lb)	Weight of Water per ft (lb)	Moment of Inertia (in.⁴)	Section Modulus (in.³)	Radius Gyration (in.)
1 1/2 1.900	40	Std	40S	0.145	1.610	2.036	0.799	0.497	0.421	2.718	0.882	0.310	0.326	0.623
	80	XS	80S	0.200	1.500	1.767	1.068	0.497	0.393	3.631	0.765	0.391	0.412	0.605
	160	—	—	0.281	1.338	1.406	1.429	0.497	0.350	4.859	0.608	0.483	0.508	0.581
	—	XXS	—	0.400	1.100	0.950	1.885	0.497	0.288	6.408	0.412	0.568	0.598	0.549
	—	—	5S	0.065	2.245	3.960	0.472	0.622	0.588	1.604	1.716	0.315	0.2652	0.817
	—	—	10S	0.109	2.157	3.650	0.776	0.622	0.565	2.638	1.582	0.499	0.420	0.802
2 2.375	40	Std	40S	0.154	2.067	3.360	1.075	0.622	0.541	3.653	1.455	0.666	0.561	0.787
	80	XS	80S	0.218	1.939	2.953	1.477	0.622	0.508	5.022	1.280	0.868	0.731	0.766
	160	—	—	0.343	1.689	2.240	2.190	0.622	0.442	7.444	0.971	1.163	0.979	0.729
	—	XXS	—	0.436	1.503	1.774	2.656	0.622	0.393	9.029	0.769	1.312	1.104	0.703
	—	—	5S	0.083	2.709	5.76	0.728	0.753	0.709	2.475	2.499	0.710	0.494	0.988
	—	—	10S	0.120	2.635	5.45	1.039	0.753	0.690	3.531	2.361	0.988	0.687	0.975
2 1/2	40	Std	40S	0.203	2.469	4.79	1.704	0.753	0.646	5.793	2.076	1.530	1.064	0.947
2.875	80	XS	80S	0.276	2.323	4.24	2.254	0.753	0.608	7.661	1.837	0.193	1.339	0.924
	160	—	—	0.375	2.125	3.55	2.945	0.753	0.556	10.01	1.535	2.353	1.637	0.894
	—	XXS	—	0.552	1.771	2.46	4.030	0.753	0.464	13.70	1.067	2.872	1.998	0.844
	—	—	5S	0.083	3.334	8.73	0.891	0.916	0.873	3.03	3.78	1.301	0.744	1.208
	—	—	10S	0.120	3.260	8.35	1.274	0.916	0.853	4.33	3.61	1.822	1.041	1.196
3	40	Std	40S	0.216	3.068	7.39	2.228	0.916	0.803	7.58	3.20	3.02	1.724	1.164
3.500	80	XS	80S	0.300	2.900	6.61	3.020	0.916	0.759	10.25	2.864	3.90	2.226	1.136
	160	—	—	0.437	2.626	5.42	4.210	0.916	0.687	14.32	2.348	5.03	2.876	1.094
	—	XXS	—	0.600	2.300	4.15	5.470	0.916	0.602	18.58	1.801	5.99	3.43	1.047
	—	—	5S	0.083	3.834	11.55	1.021	1.047	1.004	3.47	5.01	1.960	0.980	1.385
3 1/2	—	—	10S	0.120	3.760	11.10	1.463	1.047	0.984	4.97	4.81	2.756	1.378	1.372
4.000	40	Std	40S	0.226	3.548	9.89	2.68	1.047	0.929	9.11	4.28	4.79	2.394	1.337
	80	XS	80S	0.318	3.364	8.89	3.68	1.047	0.881	12.51	3.85	6.28	3.14	1.307
	—	—	5S	0.083	4.334	14.75	1.152	1.178	1.135	3.92	6.40	2.811	1.249	1.562
	—	—	10S	0.120	4.260	14.25	1.651	1.178	1.115	5.61	6.17	3.96	1.762	1.549
4	40	Std	40S	0.237	4.026	12.73	3.17	1.178	1.054	10.79	5.51	7.23	3.21	1.510
4.500	80	XS	80S	0.337	3.826	11.50	4.41	1.178	1.002	14.98	4.98	9.61	4.27	1.477
	120	—	—	0.437	3.626	10.33	5.58	1.178	0.949	18.96	4.48	11.65	5.18	1.445
	160	—	—	0.531	3.438	9.28	6.62	1.178	0.900	22.51	4.02	13.27	5.90	1.416
	—	XXS	—	0.674	3.152	7.80	8.10	1.178	0.825	27.54	3.38	15.29	6.79	1.374
	—	—	5S	0.109	5.345	22.44	1.868	1.456	1.399	6.35	9.73	6.95	2.498	1.929
	—	—	10S	0.134	5.295	22.02	2.285	1.456	1.386	7.77	9.53	8.43	3.03	1.920

Appendix B: Physical Properties of Materials

Nominal	OD	Sch No.	Sch	Sch S	Wall	ID									
5	5.563	40	Std	40S	0.258	5.047	20.01	4.30	1.456	1.321	14.62	8.66	15.17	5.45	1.878
		80	XS	80S	0.375	4.813	18.19	6.11	1.456	1.260	20.78	7.89	20.68	7.43	1.839
		120	—	—	0.500	4.563	16.35	7.95	1.456	1.195	27.04	7.09	25.74	9.25	1.799
		160	—	—	0.625	4.313	14.61	9.70	1.456	1.129	32.96	6.33	30	10.8	1.760
		—	XXS	—	0.750	4.063	12.97	11.34	1.456	1.064	38.55	5.62	33.6	12.1	1.722
		—	—	5S	0.109	6.407	32.20	2.231	1.734	1.677	5.37	13.98	11.85	3.58	2.304
		—	—	10S	0.134	6.357	31.70	2.733	1.734	1.664	9.29	13.74	14.4	4.35	2.295
6	6.625	40	Std	40S	0.280	6.065	28.89	5.58	1.734	1.588	18.97	12.51	28.14	8.5	2.245
		80	XS	80S	0.432	5.761	26.07	8.40	1.734	1.508	28.57	11.29	40.5	12.23	2.195
		120	—	—	0.562	5.501	23.77	10.70	1.734	1.440	36.39	10.30	49.6	14.98	2.153
		160	—	—	0.718	5.189	21.15	13.33	1.734	1.358	45.30	9.16	59	17.81	2.104
		—	XXS	—	0.864	4.897	18.83	15.64	1.734	1.282	53.16	8.17	66.3	20.03	2.060
		—	—	5S	0.109	8.407	55.5	2.916	2.258	2.201	9.91	24.07	26.45	6.13	3.01
		—	—	10S	0.148	8.329	54.5	3.94	2.258	2.180	13.40	23.59	35.4	8.21	3.00
		20	—	—	0.250	8.125	51.8	6.58	2.258	2.127	22.36	22.48	57.7	13.39	2.962
		30	—	—	0.277	8.071	51.2	7.26	2.258	2.113	24.70	22.18	63.4	14.69	2.953
		40	Std	40S	0.322	7.981	50.0	8.40	2.258	2.089	28.55	21.69	72.5	16.81	2.938
8	8.625	60	—	—	0.406	7.813	47.9	10.48	2.258	2.045	35.64	20.79	88.8	20.58	2.909
		80	XS	80S	0.500	7.625	45.7	12.76	2.258	1.996	43.39	19.80	105.7	24.52	2.878
		100	—	—	0.593	7.439	43.5	14.96	2.258	1.948	50.87	18.84	121.4	28.14	2.847
		120	—	—	0.718	7.189	40.6	17.84	2.258	1.882	60.63	17.60	140.6	32.6	2.807
		140	—	—	0.812	7.001	38.5	19.93	2.258	1.833	67.76	16.69	153.8	35.7	2.777
		—	XXS	—	0.875	6.875	37.1	21.30	2.258	1.800	72.42	16.09	162	37.6	2.757
		160	—	—	0.906	6.813	36.5	21.97	2.258	1.784	74.69	15.80	165.9	38.5	2.748
		—	—	5S	0.134	10.482	86.3	4.52	2.815	2.744	15.15	37.4	63.7	11.85	3.75
		—	—	10S	0.165	10.420	85.3	5.49	2.815	2.728	18.70	36.9	76.9	14.3	3.74
		20	—	—	0.250	10.250	82.5	8.26	2.815	2.683	28.04	35.8	113.7	21.16	3.71
		—	—	—	0.279	10.192	81.6	9.18	2.815	2.668	31.20	35.3	125.9	23.42	3.70
		30	—	—	0.307	10.136	80.7	10.07	2.815	2.654	34.24	35.0	137.5	25.57	3.69
10	10.750	40	Std	40S	0.365	10.020	78.9	11.91	2.815	2.623	40.48	34.1	160.8	29.9	3.67
		60	XS	80S	0.500	9.750	74.7	16.10	2.815	2.553	54.74	32.3	212	39.4	3.63
		80	—	—	0.593	9.564	71.8	18.92	2.815	2.504	64.33	31.1	244.9	45.6	3.60
		100	—	—	0.718	9.314	68.1	22.63	2.815	2.438	76.93	29.5	286.2	53.2	3.56
		120	—	—	0.843	9.064	64.5	26.24	2.815	2.373	89.20	28.0	324	60.3	3.52
		140	—	—	1.000	8.750	60.1	30.6	2.815	2.291	104.13	26.1	368	68.4	3.47
		160	—	—	1.125	8.500	56.7	34.0	2.815	2.225	115.65	24.6	399	74.3	3.43
		—	—	5S	0.165	12.420	121.2	6.52	3.34	3.25	19.56	52.5	129.2	20.27	4.45
		—	—	10S	0.180	12.390	120.6	7.11	3.34	3.24	24.20	52.2	140.5	22.03	4.44
		20	—	—	0.250	12.250	117.9	9.84	3.34	3.21	33.38	51.1	191.9	30.1	4.42
		30	—	—	0.330	12.090	114.8	12.88	3.34	3.17	43.77	49.7	248.5	39.0	4.39
		—	Std	40S	0.375	12.000	113.1	14.58	3.34	3.14	49.56	49.0	279.3	43.8	4.38

(continued)

TABLE B.2 (continued)
Physical Properties of Pipe

Nominal Pipe Size, OD (in.)	Schedule Number a	Schedule Number b	Schedule Number c	Wall Thickness (in.)	ID (in.)	Inside Area (in.²)	Metal Area (in.²)	Sq. Ft. Outside Surface (per ft)	Sq. Ft. Inside Surface (per ft)	Weight per ft (lb)	Weight of Water per ft (lb)	Moment of Inertia (in.⁴)	Section Modulus (in.³)	Radius Gyration (in.)
12	40	—	—	0.406	11.938	111.9	15.74	3.34	3.13	53.53	48.5	300	47.1	4.37
12.750	—	XS	80S	0.500	11.750	108.4	19.24	3.34	3.08	65.42	47.0	362	56.7	4.33
	60	—	—	0.562	11.626	106.2	21.52	3.34	3.04	73.16	46.0	401	62.8	4.31
	80	—	—	0.687	11.376	101.6	26.04	3.34	2.978	88.51	44.0	475	74.5	4.27
	100	—	—	0.843	11.064	96.1	31.5	3.34	2.897	107.20	41.6	562	88.1	4.22
	120	—	—	1.000	10.750	90.8	36.9	3.34	2.814	125.49	39.3	642	100.7	4.17
	140	—	—	1.125	10.500	86.6	41.1	3.34	2.749	139.68	37.5	701	109.9	4.13
	160	—	—	1.312	10.126	80.5	47.1	3.34	2.651	160.27	34.9	781	122.6	4.07
	10	—	—	0.250	13.500	143.1	10.80	3.67	3.53	36.71	62.1	255.4	36.5	4.86
	20	—	—	0.312	13.376	140.5	13.42	3.67	3.5	45.68	60.9	314	44.9	4.84
	30	Std	—	0.375	13.250	137.9	16.05	3.67	3.47	54.57	59.7	373	53.3	4.82
	40	—	—	0.437	13.126	135.3	18.62	3.67	3.44	63.37	58.7	429	61.2	4.80
	—	XS	—	0.500	13.000	132.7	21.21	3.67	3.4	72.09	57.5	484	69.1	4.78
	—	—	—	0.562	12.876	130.2	23.73	3.67	3.37	80.66	56.5	537	76.7	4.76
14	60	—	—	0.593	12.814	129.0	24.98	3.67	3.35	84.91	55.9	562	80.3	4.74
14.000	—	—	—	0.625	12.750	127.7	26.26	3.67	3.34	89.28	55.3	589	84.1	4.73
	—	—	—	0.687	12.626	125.2	28.73	3.67	3.31	97.68	54.3	638	91.2	4.71
	80	—	—	0.750	12.500	122.7	31.2	3.67	3.27	106.13	53.2	687	98.2	4.69
	—	—	—	0.875	12.250	117.9	36.1	3.67	3.21	122.66	51.1	781	111.5	4.65
	100	—	—	0.937	12.126	115.5	38.5	3.67	3.17	130.73	50.0	825	117.8	4.63
	120	—	—	1.093	11.814	109.6	44.3	3.67	3.09	150.67	47.5	930	132.8	4.58
	140	—	—	1.250	11.500	103.9	50.1	3.67	3.01	170.22	45.0	1127	146.8	4.53
	160	—	—	1.406	11.188	98.3	55.6	3.67	2.929	189.12	42.6	1017	159.6	4.48
	10	—	—	0.250	15.500	188.7	12.37	4.19	4.06	42.05	81.8	384	48	5.57
	20	—	—	0.312	15.376	185.7	15.38	4.19	4.03	52.36	80.5	473	59.2	5.55
	30	Std	—	0.375	15.250	182.6	18.41	4.19	3.99	62.58	79.1	562	70.3	5.53
	—	—	—	0.437	15.126	179.7	21.37	4.19	3.96	72.64	77.9	648	80.9	5.50
	40	XS	—	0.500	15.000	176.7	24.35	4.19	3.93	82.77	76.5	732	91.5	5.48
	—	—	—	0.562	14.876	173.8	27.26	4.19	3.89	92.66	75.4	813	106.6	5.46
	—	—	—	0.625	14.750	170.9	30.2	4.19	3.86	102.63	74.1	894	112.2	5.44
16	60	—	—	0.656	14.688	169.4	31.6	4.19	3.85	107.50	73.4	933	116.6	5.43
16.000	—	—	—	0.687	14.626	168.0	33.0	4.19	3.83	112.36	72.7	971	121.4	5.42
	—	—	—	0.750	14.500	165.1	35.9	4.19	3.8	122.15	71.5	1047	130.9	5.40
	80	—	—	0.842	14.314	160.9	40.1	4.19	3.75	136.46	69.7	1157	144.6	5.37
	—	—	—	0.875	14.250	159.5	41.6	4.19	3.73	141.35	69.1	1193	154.1	5.36
	100	—	—	1.031	13.938	152.6	48.5	4.19	3.65	164.83	66.1	1365	170.6	5.30

Appendix B: Physical Properties of Materials

	120	—	—	1.218	13.564	144.5	56.6	4.19	3.55	192.29	62.6	1556	194.5	5.24
	140	—	—	1.437	13.126	135.3	65.7	4.19	3.44	223.50	58.6	1760	220.0	5.17
	160	—	—	1.593	12.814	129.0	72.1	4.19	3.35	245.11	55.9	1894	236.7	5.12
18	10	—	—	0.250	17.500	240.5	13.94	4.71	4.58	47.39	104.3	549	61.0	6.28
18.000	20	—	—	0.312	17.376	237.1	17.34	4.71	4.55	59.03	102.8	678	75.5	6.25
	—	Std	—	0.375	17.250	233.7	20.76	4.71	4.52	70.59	101.2	807	89.6	6.23
	30	—	—	0.437	17.126	230.4	24.11	4.71	4.48	82.06	99.9	931	103.4	6.21
	—	XS	—	0.500	17.000	227.0	27.49	4.71	4.45	93.45	98.4	1053	117.0	6.19
	40	—	—	0.562	16.876	223.7	30.8	4.71	4.42	104.75	97.0	1172	130.2	6.17
	—	—	—	0.625	16.750	220.5	34.1	4.71	4.39	115.98	95.5	1289	143.3	6.15
	—	—	—	0.687	16.626	217.1	37.4	4.71	4.35	127.03	94.1	1403	156.3	6.13
	60	—	—	0.750	16.500	213.8	40.6	4.71	4.32	138.17	92.7	1515	168.3	6.10
	—	—	—	0.875	16.250	207.4	47.1	4.71	4.25	160.04	89.9	1731	192.8	6.06
	80	—	—	0.937	16.126	204.2	50.2	4.71	4.22	170.75	88.5	1834	203.8	6.04
	100	—	—	1.156	15.688	193.3	61.2	4.71	4.11	207.96	83.7	2180	242.2	5.97
	120	—	—	1.375	15.250	182.6	71.8	4.71	3.99	244.14	79.2	2499	277.6	5.90
	140	—	—	1.562	14.876	173.8	80.7	4.71	3.89	274.23	75.3	2750	306	5.84
	160	—	—	1.781	14.438	163.7	90.7	4.71	3.78	308.51	71.0	3020	336	5.77
	10	—	—	0.250	19.500	298.6	15.51	5.24	5.11	52.73	129.5	757	75.7	6.98
	—	—	—	0.312	19.376	294.9	19.30	5.24	5.07	65.40	128.1	935	93.5	6.96
20	20	Std	—	0.375	19.250	291.0	23.12	5.24	5.04	78.60	126.0	1114	111.4	6.94
20.000	—	—	—	0.437	19.126	287.3	26.86	5.24	5.01	91.31	124.6	1286	128.6	6.92
	30	XS	—	0.500	19.000	283.5	30.6	5.24	4.97	104.13	122.8	1457	145.7	6.90
	—	—	—	0.562	18.876	279.8	34.3	5.24	4.94	116.67	121.3	1624	162.4	6.88
	40	—	—	0.593	18.814	278.0	36.2	5.24	4.93	122.91	120.4	1704	170.4	6.86
	—	—	—	0.625	18.750	276.1	38.0	5.24	4.91	129.33	119.7	1787	178.7	6.85
	—	—	—	0.687	18.626	272.5	41.7	5.24	4.88	141.71	118.1	1946	194.6	6.83
	60	—	—	0.750	18.500	268.8	45.4	5.24	4.84	154.20	116.5	2105	210.5	6.81
	—	—	—	0.812	18.376	265.2	48.9	5.24	4.81	166.40	115.0	2257	225.7	6.79
	80	—	—	0.875	18.250	261.6	52.6	5.24	4.78	178.73	113.4	2409	240.9	6.77
	100	—	—	1.031	17.938	252.7	61.4	5.24	4.70	208.87	109.4	2772	277.2	6.72
	120	—	—	1.281	17.438	238.8	75.3	5.24	4.57	256.10	103.4	3320	332	6.63
	140	—	—	1.500	17.000	227.0	87.2	5.24	4.45	296.37	98.3	3760	376	6.56
	160	—	—	1.750	16.500	213.8	100.3	5.24	4.32	341.10	92.6	4220	422	6.48
	—	—	—	1.968	16.064	202.7	111.5	5.24	4.21	379.01	87.9	4590	459	6.41
	10	—	—	0.250	23.500	434	18.65	6.28	6.15	63.41	188.0	1316	109.6	8.40
	—	—	—	0.312	23.376	430	23.20	6.28	6.12	78.93	186.1	1629	135.8	8.38
	20	Std	—	0.375	23.250	425	27.83	6.28	6.09	94.62	183.8	1943	161.9	8.35
	—	—	—	0.437	23.126	420	32.4	6.28	6.05	109.97	182.1	2246	187.4	8.33
	—	XS	—	0.500	23.000	415	36.9	6.28	6.02	125.49	180.1	2550	212.5	8.31

(continued)

TABLE B.2 (continued)
Physical Properties of Pipe

Nominal Pipe Size, OD (in.)	Schedule Number a	Schedule Number b	Schedule Number c	Wall Thickness (in.)	ID (in.)	Inside Area (in.²)	Metal Area (in.²)	Sq. Ft. Outside Surface (per ft)	Sq. Ft. Inside Surface (per ft)	Weight per ft (lb)	Weight of Water per ft (lb)	Moment of Inertia (in.⁴)	Section Modulus (in.³)	Radius Gyration (in.)
24	30	—	—	0.562	22.876	411	41.4	6.28	5.99	140.80	178.1	2840	237.0	8.29
24.000	—	—	—	0.625	22.750	406	45.9	6.28	5.96	156.03	176.2	3140	261.4	8.27
	40	—	—	0.687	22.626	402	50.3	6.28	5.92	171.17	174.3	3420	285.2	8.25
	—	—	—	0.750	22.500	398	54.8	6.28	5.89	186.24	172.4	3710	309	8.22
	60	—	—	0.968	22.064	382	70.0	6.28	5.78	238.11	165.8	4650	388	8.15
	80	—	—	1.218	21.564	365	87.2	6.28	5.65	296.36	158.3	5670	473	8.07
	100	—	—	1.531	20.938	344	108.1	6.28	5.48	367.40	149.3	6850	571	7.96
	120	—	—	1.812	20.376	326	126.3	6.28	5.33	429.39	141.4	7830	652	7.87
	140	—	—	2.062	19.876	310	142.1	6.28	5.20	483.13	134.5	8630	719	7.79
	160	—	—	2.343	19.314	293	159.4	6.28	5.06	541.94	127.0	9460	788	7.70
	10	—	—	0.312	29.376	678	29.1	7.85	7.69	98.93	293.8	3210	214	10.50
30	20	—	—	0.500	29.000	661	46.3	7.85	7.59	157.53	286.3	5040	336	10.43
30.000	30	—	—	0.625	28.750	649	57.6	7.85	7.53	196.08	281.5	6220	415	10.39

[a] ASA B36.10 Steel-pipe schedule numbers.
[b] ASA B36.10 Steel-pipe nominal wall-thickness designations.
[c] ASA B36.19 Stainless-steel-pipe schedule numbers.

TABLE B.3
Commercial Copper Tubing[a]

Size, OD		Wall Thickness			Flow Area		Metal Area (in.²)	Surface Area		Weight (lb/ft)
in.	mm	in.	mm	gage	in.²	mm²		Inside (ft²/ft)	Outside (ft²/ft)	
1/8	3.2	0.030	0.76	A	0.003	1.9	0.012	0.017	0.033	0.035
3/16	4.76	0.030	0.76	A	0.013	8.4	0.017	0.034	0.049	0.058
1/4	6.4	0.030	0.76	A	0.028	18.1	0.021	0.050	0.066	0.080
1/4	6.4	0.049	1.24	18	0.018	11.6	0.031	0.038	0.066	0.120
5/16	7.94	0.032	0.81	21A	0.048	31.0	0.028	0.065	0.082	0.109
3/8	9.53	0.032	0.81	21A	0.076	49.0	0.033	0.081	0.098	0.134
3/8	9.53	0.049	1.24	18	0.060	38.7	0.050	0.072	0.098	0.195
1/2	12.7	0.032	0.81	21A	0.149	96.1	0.047	0.114	0.131	0.182
1/2	12.7	0.035	0.89	20L	0.145	93.6	0.051	0.113	0.131	0.198
1/2	12.7	0.049	1.24	18K	0.127	81.9	0.069	0.105	0.131	0.269
1/2	12.7	0.065	1.65	16	0.108	69.7	0.089	0.97	0.131	0.344
5/8	15.9	0.035	0.89	20A	0.242	156	0.065	0.145	0.164	0.251
5/8	15.9	0.040	1.02	L	0.233	150	0.074	0.143	0.164	0.285
5/8	15.9	0.049	1.24	18K	0.215	139	0.089	0.138	0.164	0.344
3/4	19.1	0.035	0.89	20A	0.363	234	0.079	0.178	0.196	0.305
3/4	19.1	0.042	1.07	L	0.348	224	0.103	0.174	0.196	0.362
3/4	19.1	0.049	1.24	18K	0.334	215	0.108	0.171	0.196	0.418
3/4	19.1	0.065	1.65	16	0.302	195	0.140	0.162	0.196	0.542
3/4	19.1	0.083	2.11	14	0.268	173	0.174	0.151	0.196	0.674
7/8	22.2	0.045	1.14	L	0.484	312	0.117	0.206	0.229	0.455
7/8	22.2	0.065	1.65	16K	0.436	281	0.165	0.195	0.229	0.641
7/8	22.2	0.083	2.11	14	0.395	255	0.206	0.186	0.229	0.800
1	25.4	0.065	1.65	16	0.594	383	0.181	0.228	0.262	0.740
1	25.4	0.083	2.11	14	0.546	352	0.239	0.218	0.262	0.927
1 1/8	28.6	0.050	1.27	L	0.825	532	0.176	0.268	0.294	0.655
1 1/8	28.6	0.065	1.65	16K	0.778	502	0.216	0.261	0.294	0.839
1 1/4	31.8	0.065	1.65	16	0.985	636	0.242	0.293	0.327	0.938
1 1/4	31.8	0.083	2.11	14	0.923	596	0.304	0.284	0.327	1.18
1 3/8	34.9	0.055	1.40	L	1.257	811	0.228	0.331	0.360	0.884
1 3/8	34.9	0.065	1.65	16K	1.217	785	0.267	0.326	0.360	1.04
1 1/2	38.1	0.065	1.65	16	1.474	951	0.294	0.359	0.393	1.14
1 1/2	38.7	0.083	2.11	14	1.398	902	0.370	0.349	0.393	1.43
1 5/8	41.3	0.060	1.52	L	1.779	1148	0.295	0.394	0.425	1.14
1 5/8	41.3	0.072	1.83	K	1.722	1111	0.351	0.388	0.425	1.36
2	50.8	0.083	2.11	14	2.642	1705	0.500	0.480	0.628	1.94
2	50.8	0.109	2.76	12	2.494	1609	0.620	0.466	0.628	2.51
2 1/8	54.0	0.070	1.78	L	3.095	1997	0.449	0.520	0.556	1.75
2 1/8	54.0	0.083	2.11	14K	3.016	1946	0.529	0.513	0.556	2.06
2 5/8	66.7	0.080	2.03	L	4.77	3078	0.645	0.645	0.687	2.48
2 5/8	66.7	0.095	2.41	13K	4.66	3007	0.760	0.637	0.687	2.93
3 1/8	79.4	0.090	2.29	L	6.81	4394	0.950	0.771	0.818	3.33
3 1/8	79.4	0.109	2.77	12K	6.64	4284	1.034	0.761	0.818	4.00

(continued)

TABLE B.3 (continued)
Commercial Copper Tubing[a]

Size, OD		Wall Thickness			Flow Area		Metal Area (in.^{2})	Surface Area		Weight (lb/ft)
in.	mm	in.	mm	gage	in.^{2}	mm^2		Inside (ft^2/ft)	Outside (ft^2/ft)	
3 5/8	92.1	0.100	2.54	L	9.21	5942	1.154	0.897	0.949	4.29
3 5/8	92.1	0.120	3.05	11K	9.00	5807	1.341	0.886	0.949	5.12
4 1/8	104.8	0.110	2.79	L	11.92	7691	1.387	1.022	1.080	5.38
4 1/8	104.8	0.134	3.40	10K	11.61	7491	1.682	1.009	1.080	6.51

Source: The CRC Handbook of Mechanical Engineering, CRC Press, Boca Raton, FL, 1998.

Notes: The table above gives dimensional data and weights of copper tubing used for automotive, plumbing, refrigeration, and heat exchanger services. For additional data see the standards handbooks of the Copper Development Association, Inc., the ASTM standards, and the "SAE Handbook."

Dimensions in this table are actual specified measurements, subject to accepted tolerances. Trade size designations are usually by actual OD, except for water and drainage tube (plumbing), which measures 1/8 in. larger OD. A 1/2 in. plumbing tube, for example, measures 5/8 in. OD, and a 2 in. plumbing tube measures 2 1/8 in. OD.

Key to Gage Sizes

Standard-gage wall thicknesses are listed by numerical designation (14–21), BWG or Stubs gage. These gage sizes are standard for tubular heat exchangers. The letter A designates SAE tubing sizes for automotive service. Letter designations *K* and *L* are the common sizes for plumbing services, soft or hard temper.

Other Materials

These same dimensional sizes are also common for much of the commercial tubing available in aluminum, mild steel, brass, bronze, and other alloys. Tube weights in this table are based on copper at 0.323 lb/in^3. For other materials the weights should be multiplied by the following approximate factors:

Aluminum	0.30
Mild steel	0.87
Brass	0.95
Monel	0.96
Stainless steel	0.89

[a] Compiled and computed.

TABLE B.4

Standard Grades of Bolts—SAE Grades for Steel Bolts

SAE Grade No.	Size Range Incl.	Proof Strength (kpsi)[a]	Tensile Strength (kpsi)[a]	Material	Head Marking
1	$\frac{1}{4}$ to $1\frac{1}{2}$			Low- or medium-carbon steel	
2	$\frac{1}{4}$ to $\frac{3}{4}$	55	74		
	$\frac{7}{8}$ to $1\frac{1}{2}$	33	60		
5	$\frac{1}{4}$ to 1	85	120	Medium-carbon steel, Q & T	
	$1\frac{1}{8}$ to $1\frac{1}{2}$	74	105		
5.2	$\frac{1}{4}$ to 1	85	120	Low-carbon martensite steel, Q & T	
7	$\frac{1}{4}$ to $1\frac{1}{2}$	105	133	Medium-carbon alloy steel, Q & T[b]	
8	$\frac{1}{4}$ to $1\frac{1}{2}$	120	150	Medium-carbon alloy steel, Q & T	
8.2	$\frac{1}{4}$ to 1	120	150	Low-carbon martensite steel, Q & T	

Sources: *Helpful Hints*, Russell, Burdsall & Ward Corp., Mentor, OH, and Chapter 23; *The CRC Press Handbook of Mechanical Engineering*, CRC Press, Boca Raton, FL, 1998.

[a] Minimum values.
[b] Roll threaded after heat treatment.

TABLE B.5

Standard Grades of Bolts—ASTM Grades for Steel Bolts

ASTM Designation	Size Range Incl.	Proof Strength (kpsi)[a]	Tensile Strength (kpsi)[a]	Material	Head Marking
A307	$\frac{1}{4}$ to 4			Low-carbon steel	(plain hex)
A325 type 1	$\frac{1}{2}$ to 1	85	120	Medium-carbon steel, Q & T	A325
	$1\frac{1}{8}$ to $1\frac{1}{2}$	74	105		
A325 type 2	$\frac{1}{2}$ to 1	85	120	Low-carbon steel, Q & T	A325
	$1\frac{1}{8}$ to $1\frac{1}{2}$	74	105		
A325 type 3	$\frac{1}{2}$ to 1	85	120	Weathering steel, Q & T	A325
	$1\frac{1}{8}$ to $1\frac{1}{2}$	74	105		
A354 grade BC				Alloy steel, Q & T	BC
A354 grade BD	$\frac{1}{4}$ to 4	120	150	Alloy steel, Q & T	(6 radial marks)
A449	$\frac{1}{4}$ to	85	120	Medium-carbon steel. Q & T	(3 radial marks)
	$1\frac{1}{8}$ to $1\frac{1}{2}$	74	105		
	$1\frac{3}{4}$ to 3	55	90		
A490 type	$\frac{1}{2}$ to $1\frac{1}{2}$	120	150	Alloy steel, Q & T	A490
A490 type 3				Weathering steel, Q & T	A490

Source: *Helpful Hints*, Russell, Burdsall & Ward Corp., Mentor, OH, and Chapter 23; *The CRC Press Handbook of Mechanical Engineering*, CRC Press, Boca Raton, FL, 1998.

[a] Minimum values.

Appendix B: Physical Properties of Materials

TABLE B.6

Standard Grades of Bolts—Metric Mechanical Property Classes for Steel Bolts, Screws, and Studs

Property Class	Size Range Incl.	Proof Strength (MPa)	Tensile Strength (MPa)	Material	Head Marking
4.6	M5–M36	225	400	Low- or medium-carbon steel	4.6
4.8	M1.6–M16	310	420	Low- or medium-carbon steel	4.8
5.8	M5–M24	380	520	Low- or medium-carbon steel	5.8
8.8	M16–M36	600	830	Medium-carbon steel, Q & T	8.8
9.8	M1.6–M16	650	900	Medium-carbon steel Q & T	9.8
10.9	M5–M36	830	1040	Low-carbon martensite steel, Q & T	10.9
12.9	M1.6–M36	970	1220	Alloy steel, Q & T	12.9

Source: *Helpful Hints*, Russell, Burdsall & Ward Corp., Mentor, OH, and Chapter 23; SAE standard J1199, and ASTM standard F568; *The CRC Press Handbook of Mechanical Engineering*, CRC Press, Boca Raton, FL, 1998.

TABLE B.7

Flange Size Data

Nominal Pipe Size, NPS (in.)	Class 150			
	Diameter of Flange (in.)	No. of Bolts	Diameter of Bolts (in.)	Bolt Circle (in.)
1/4	3 3/8	4	1/2	2 1/4
1/2	3 1/2	4	1/2	2 3/8
3/4	3 7/8	4	1/2	2 3/4
1	4 1/4	4	1/2	3 1/8
1 1/4	4 5/8	4	1/2	3 1/2
1 1/2	5	4	1/2	3 7/8
2	6	4	5/8	4 3/4
2 1/2	7	4	5/8	5 1/2
3	7 1/2	4	5/8	6
3 1/2	8 1/2	8	5/8	7
4	9	8	5/8	7 1/2
5	10	8	3/4	8 1/2
6	11	8	3/4	9 1/2
8	13 1/2	8	3/4	11 3/4
10	16	12	7/8	14 1/4
12	19	12	7/8	17
14	21	12	1	18 3/4
16	23 1/2	16	1	21 1/4
18	25	16	1 1/8	22 3/4
20	27 1/2	20	1 1/8	25
24	32	20	1 1/4	29 1/2
	Class 300			
1/4	3 3/8	4	1/2	2 1/4
1/2	3 3/4	4	1/2	2 5/8
3/4	4 5/8	4	5/8	3 1/4
1	4 7/8	4	5/8	3 1/2
1 1/4	5 1/4	4	5/8	3 7/8
1 1/2	6 1/8	4	3/4	4 1/2
2	6 1/2	8	5/8	5
2 1/2	7 1/2	8	3/4	5 7/8
3	8 1/4	8	3/4	6 5/8
3 1/2	9	8	3/4	7 1/4
4	10	8	3/4	7 7/8
5	11	8	3/4	9 1/4
6	12 1/2	12	3/4	10 5/8
8	15	12	7/8	13
10	17 1/2	16	1	15 1/4
12	20 1/2	16	1 1/8	17 3/4
14	23	20	1 1/8	20 1/4
16	25 1/2	20	1 1/4	22 1/2
18	28	24	1 1/4	24 3/4
20	30 1/2	24	1 1/4	27
24	36	24	1 1/2	32

Appendix C: Properties of Gases and Liquids

TABLE C.1
Ideal Gas Properties of Air (SI Units)

T (K)	h (kJ/kg)	p_r	u (kJ/kg)	v_r	s° (kJ/kg k)
200	199.97	0.3363	142.56	1707.0	1.29559
210	209.97	0.3987	149.69	1512.0	1.34444
220	219.97	0.4690	156.82	1346.0	1.39105
230	230.02	0.5477	164.00	1205.0	1.43557
240	240.02	0.6355	171.13	1084.0	1.47824
250	250.05	0.7329	178.28	979.0	1.51917
260	260.09	0.8405	185.45	887.8	1.55848
270	270.11	0.9590	192.60	808.0	1.59634
280	280.13	1.0889	199.75	738.0	1.63279
285	285.14	1.1584	203.33	760.1	1.65055
290	290.16	1.2311	206.91	676.1	1.66802
295	295.17	1.3068	210.49	647.9	1.68515
300	300.19	1.3860	214.07	621.2	1.70203
305	305.22	1.4686	217.67	596.0	1.71865
310	310.24	1.5546	221.25	572.3	1.73498
315	315.27	1.6442	224.85	549.8	1.75106
320	320.29	1.7375	228.42	528.6	1.76690
325	325.31	1.8345	232.02	508.4	1.78249
330	330.34	1.9352	235.61	489.4	1.79783
340	340.42	2.149	242.82	454.1	1.82790
350	350.49	2.379	250.02	422.2	1.85708
360	360.58	2.626	257.24	393.4	1.88543
370	370.67	2.892	264.46	367.2	1.91313
380	380.77	3.176	271.69	343.4	1.94001
390	390.88	3.481	278.93	321.5	1.96633
400	400.98	3.806	286.16	301.6	1.99194
410	411.12	4.153	293.43	283.3	2.01699
420	421.26	4.522	300.69	266.6	2.04142
430	431.43	4.915	307.99	251.1	2.06533
440	441.61	5.332	315.30	236.8	2.08870
450	451.80	5.775	322.62	223.6	2.11161
460	462.02	6.245	329.97	211.4	2.13407
470	472.24	6.742	337.32	200.1	2.15604
480	482.49	7.268	344.70	189.5	2.17760
490	492.74	7.824	352.08	179.7	2.19876
500	503.02	8.411	359.49	170.6	2.21952
510	513.32	9.031	366.92	162.1	2.23993
520	523.63	9.684	374.36	154.1	2.25997
530	533.98	10.37	381.84	146.7	2.27967
540	544.35	11.10	389.34	139.7	2.29906
550	554.74	11.68	396.86	133.1	2.31809
560	565.17	12.66	404.42	127.0	2.33685
570	575.59	13.50	411.97	121.2	2.35531
580	586.04	14.38	419.55	115.7	2.37348
590	596.52	15.31	427.15	110.6	2.39140
600	607.02	16.28	434.78	105.8	2.40902
610	617.53	17.30	442.42	101.2	2.42644
620	628.07	18.36	450.09	96.92	2.44356
630	638.63	19.84	457.78	92.84	2.46048
640	649.22	20.64	465.50	88.99	2.47716
650	659.84	21.86	473.25	85.34	2.49364
660	670.47	23.13	481.01	81.89	2.50985
670	681.14	24.46	488.81	78.61	2.52589
680	691.82	25.85	496.62	75.50	2.54175
690	702.52	27.29	504.45	72.56	2.55731
700	713.27	28.80	512.33	69.76	2.57277
710	724.04	30.38	520.23	67.07	2.58810
720	734.82	32.02	528.14	64.53	2.60319
730	745.62	33.72	536.07	62.13	2.61803
740	756.44	35.50	544.02	59.82	2.63280
750	767.29	37.35	551.99	57.63	2.64737
760	778.18	39.27	560.01	55.54	2.66176
770	789.11	41.31	568.07	53.39	2.67595
780	800.03	43.35	576.12	51.64	2.69013
790	810.99	45.55	584.21	49.86	2.70400
800	821.95	47.75	592.30	48.08	2.71787
820	843.98	52.59	608.59	44.84	2.74504
840	866.08	57.60	624.95	41.85	2.77170
860	888.27	63.09	641.40	39.12	2.79783
880	910.56	68.98	657.95	36.61	2.82344
900	932.93	75.29	674.58	34.31	2.84856
920	955.38	82.05	691.28	32.18	2.87324
940	977.92	89.28	708.08	30.22	2.89748
960	1000.55	97.00	725.02	28.40	2.92128
980	1023.25	105.2	741.98	26.73	2.94468
1000	1046.04	114.0	758.94	25.17	2.96770
1020	1068.89	123.4	776.10	23.72	2.99034
1040	1091.85	133.3	793.36	22.39	3.01260
1060	1114.86	143.9	810.62	21.14	3.03449
1080	1137.89	155.2	827.88	19.98	3.05608
1100	1161.07	167.1	845.33	18.896	3.07732
1120	1184.28	179.7	862.79	17.886	3.09825
1140	1207.57	193.1	880.35	16.946	3.11883
1160	1230.92	207.2	897.91	16.064	3.13916
1180	1254.34	222.2	915.57	15.241	3.15916
1200	1277.79	238.0	933.33	14.470	3.17888
1220	1301.31	254.7	951.09	13.747	3.19834
1240	1324.93	272.3	968.95	13.069	3.21751

(*continued*)

TABLE C.1 (continued)
Ideal Gas Properties of Air (SI Units)

T (K)	h (kJ/kg)	p_r	u (kJ/kg)	v_r	s° (kJ/kg k)
1260	1348.55	290.8	986.90	12.435	3.23638
1280	1372.24	310.4	1004.76	11.835	3.25510
1300	1395.97	330.9	1022.82	11.275	3.27345
1320	1419.76	352.5	1040.88	10.747	3.29160
1340	1443.60	375.3	1058.94	10.247	3.30959
1360	1467.49	399.1	1077.10	9.780	3.32724
1380	1491.44	424.2	1095.26	9.337	3.34474
1400	1515.42	450.5	1113.52	8.919	3.36200
1420	1539.44	478.0	1131.77	8.526	3.37901
1440	1563.51	506.9	1150.13	8.153	3.39586
1460	1587.63	537.1	1168.49	7.801	3.41247
1480	1611.79	568.8	1186.95	7.468	3.42892
1500	1635.97	601.9	1205.41	7.152	3.44516
1520	1660.23	636.5	1223.87	6.854	3.46120
1540	1684.51	672.8	1242.43	6.569	3.47712
1560	1708.82	710.5	1260.99	6.301	3.49276
1580	1733.17	750.0	1279.65	6.046	3.50829
1600	1757.57	791.2	1298.30	5.804	3.52364
1620	1782.00	834.1	1316.96	5.574	3.53879
1640	1806.46	878.9	1335.72	5.355	3.55381
1660	1830.96	925.6	1354.48	5.147	3.56867
1680	1855.50	974.2	1373.24	4.949	3.58335
1700	1880.1	1025	1392.7	4.761	3.5979
1750	1941.6	1161	1439.8	4.328	3.6336
1800	2003.3	1310	1487.2	3.944	3.6684
1850	2065.3	1475	1534.9	3.601	3.7023
1900	2127.4	1655	1582.6	3.295	3.7354
1950	2189.7	1852	1630.6	3.022	3.7677
2000	2252.1	2068	1678.7	2.776	3.7994
2050	2314.6	2303	1726.3	2.555	3.8303
2100	2377.4	2559	1775.3	2.356	3.8605
2150	2440.3	2837	1823.8	2.175	3.8901
2200	2503.2	3138	1872.4	2.012	3.9191
2250	2566.4	3464	1921.3	1.864	3.9474

Source: CRC Handbook of Thermal Engineering, Table A.8, Part a, p. A-30.

Note: The properties p_r (relative pressure) and v_r (relative specific volume) are dimensionless quantities used in the analysis of isentropic processes, and should not be confused with the properties pressure and specific volume.

h is enthalpy, u is internal energy, and $s°$ is entropy.

TABLE C.2
Ideal Gas Properties of Air (English Units)

T (°R)	h (Btu/lb)	p_r	u (Btu/lb)	v_r	s° (Btu/lb °R)
360	85.97	0.3363	61.29	396.6	0.50369
380	90.75	0.4061	64.70	346.6	0.51663
400	95.53	0.4858	68.11	305.0	0.52890
420	100.32	0.5760	71.52	270.1	0.54058
440	105.11	0.6776	74.93	240.6	0.55172
460	109.90	0.7913	78.36	215.33	0.56235
480	114.69	0.9182	81.77	193.65	0.57255
500	119.48	1.0590	85.20	174.90	0.58233
520	124.27	1.2147	88.62	158.58	0.59172
537	128.34	1.3593	91.53	146.34	0.59945
540	129.06	1.3860	92.04	144.32	0.60078
560	133.86	1.5742	95.47	131.78	0.60950
580	138.66	1.7800	98.90	120.70	0.61793
600	143.47	2.005	102.34	110.88	0.62607
620	148.28	2.249	105.78	102.12	0.63395
640	153.09	2.514	109.21	94.30	0.64159
660	157.92	2.801	112.67	87.27	0.64902
680	162.73	3.111	116.12	80.96	0.65621
700	167.56	3.446	119.58	75.25	0.66321
720	172.39	3.806	123.04	70.07	0.67002
740	177.23	4.193	126.51	65.38	0.67665
760	182.08	4.607	129.99	61.10	0.68312
780	186.94	5.051	133.47	57.20	0.68942
800	191.81	5.526	136.97	53.63	0.69558
820	196.69	6.033	140.47	50.35	0.70160
840	201.56	6.573	143.98	47.34	0.70747
860	206.46	7.149	147.50	44.57	0.71323
880	211.35	7.761	151.02	42.01	0.71886
900	216.26	8.411	154.57	39.64	0.72438
920	221.18	9.102	158.12	37.44	0.72979
940	226.11	9.834	161.68	35.41	0.73509
960	231.06	10.61	165.26	33.52	0.74030
980	236.02	11.43	168.83	31.76	0.74540
1000	240.98	12.30	172.43	30.12	0.75042
1040	250.95	14.18	179.66	27.17	0.76019
1080	260.97	16.28	186.93	24.58	0.76964
1120	271.03	18.60	194.25	22.30	0.77880
1160	281.14	21.18	201.63	20.29	0.78767
1200	291.30	24.01	209.05	18.51	0.79628
1240	301.52	27.13	216.53	16.93	0.80466
1280	311.79	30.55	224.05	15.52	0.81280
1320	322.11	34.31	231.63	14.25	0.82075
1360	332.48	38.41	239.25	13.12	0.82848
1400	342.90	42.88	246.93	12.10	0.83604
1440	353.37	47.75	254.66	11.17	0.84341
1480	363.89	53.04	262.44	10.34	0.85062
1520	374.47	58.78	270.26	9.578	0.85767
1560	385.08	65.00	278.13	8.890	0.86456
1600	395.74	71.73	286.06	8.263	0.87130

TABLE C.2 (continued)
Ideal Gas Properties of Air (English Units)

T (°R)	h (Btu/lb)	p_r	u (Btu/lb)	v_r	s° (Btu/lb °R)
1650	409.13	80.89	296.03	7.556	0.87954
1700	422.59	90.95	306.06	6.924	0.88758
1750	436.12	101.98	316.16	6.357	0.89542
1800	449.71	114.0	326.32	5.847	0.90308
1850	463.37	127.2	336.55	5.388	0.91056
1900	477.09	141.5	346.85	4.974	0.91788
1950	490.88	157.1	357.20	4.598	0.92504
2000	504.71	174.0	367.61	4.258	0.93205
2050	518.61	192.3	378.08	3.949	0.93891
2100	532.55	212.1	388.60	3.667	0.94564
2150	546.54	233.5	399.17	3.410	0.95222
2200	560.59	256.6	409.78	3.176	0.95868
2250	574.69	281.4	420.46	2.961	0.96501
2300	588.82	308.1	431.16	2.765	0.97123
2350	603.00	336.8	441.91	2.585	0.97732
2400	617.22	367.6	452.70	2.419	0.98331
2450	631.48	400.5	463.54	2.266	0.98919
2500	645.78	435.7	474.40	2.125	0.99497
2550	660.12	473.3	485.31	1.996	1.00064
2600	674.49	513.5	496.26	1.876	1.00623
2650	688.90	556.3	507.25	1.765	1.01172
2700	703.35	601.9	518.26	1.662	1.01712
2750	717.83	650.4	529.31	1.566	1.02244
2800	732.33	702.0	540.40	1.478	1.02767
2850	746.88	756.7	551.52	1.395	1.03282
2900	761.45	814.8	562.66	1.318	1.03788
2950	776.05	876.4	573.84	1.247	1.04288
3000	790.68	941.4	585.04	1.180	1.04779
3050	805.34	1011	596.28	1.118	1.05264
3100	820.03	1083	607.53	1.060	1.05741
3150	834.75	1161	618.82	1.006	1.06212
3200	849.48	1242	630.12	0.9546	1.06676
3250	864.24	1328	641.46	0.9069	1.07134
3300	879.02	1418	652.81	0.8621	1.07585
3350	893.83	1513	664.20	0.8202	1.08031
3400	908.66	1613	675.60	0.7807	1.08470
3450	923.52	1719	687.04	0.7436	1.08904
3500	938.40	1829	698.48	0.7087	1.09332
3550	953.30	1946	709.95	0.6759	1.09755
3600	968.21	2068	721.44	0.6449	1.10172
3650	983.15	2196	732.95	0.6157	1.10584
3700	998.11	2330	744.48	0.5882	1.10991
3750	1013.1	2471	756.04	0.5621	1.11393
3800	1028.1	2618	767.60	0.5376	1.11791
3850	1043.1	2773	779.19	0.5143	1.12183
3900	1058.1	2934	790.80	0.4923	1.12571
3950	1073.2	3103	802.43	0.4715	1.12955
4000	1088.3	3280	814.06	0.4518	1.13334
4050	1103.4	3464	825.72	0.4331	1.13709
4100	1118.5	3656	837.40	0.4154	1.14079
4150	1133.6	3858	849.09	0.3985	1.14446
4200	1148.7	4067	860.81	0.3826	1.14809
4300	1179.0	4513	884.28	0.3529	1.15522
4400	1209.4	4997	907.81	0.3262	1.16221
4500	1239.9	5521	931.39	0.3019	1.16905
4600	1270.4	6089	955.04	0.2799	1.17575
4700	1300.9	6701	978.73	0.2598	1.18232
4800	1331.5	7362	1002.5	0.2415	1.18876
4900	1362.2	8073	1026.3	0.2248	1.19508
5000	1392.9	8837	1050.1	0.2096	1.20129
5100	1423.6	9658	1074.0	0.1956	1.20738
5200	1454.4	10539	1098.0	0.1828	1.21336
5300	1485.3	11481	1122.0	0.1710	1.21923

Source: Adapted from Moran, M.J. and Shapiro, H.N., *Fundamentals of Engineering Thermodynamics*, 3rd edn., Wiley, New York, 1995, as based on Keenan, J.H. and Kaye, J., *Gas Tables*, Wiley, New York, 1945; *CRC Handbook of Thermal Engineering*, Table A.8, Part b, p, A-32.

Note: The properties p_r (relative pressure) and v_r (relative specific volume) are dimensionless quantities used in the analysis of isentropic processes, and should not be confused with the properties pressure and specific volume.

h is enthalpy, u is internal energy, and $s°$ is entropy.

TABLE C.3
Ideal Gas Properties of Nitrogen, N_2 (SI Units)

T (K)	$\bar{h}$ (kJ/kmol)	$\bar{u}$ (kJ/kmol)	$\bar{s}°$ (kJ/kmol·K)
0	0	0	0
220	6,391	4,562	182.639
230	6,683	4,770	183.938
240	6,975	4,979	185.180
250	7,266	5,188	186.370
260	7,558	5,396	187.514
270	7,849	5,604	188.614
280	8,141	5,813	189.673
290	8,432	6,021	190.695
298	8,669	6,190	191.502
300	8,723	6,229	191.682
310	9,014	6,437	192.638
320	9,306	6,645	193.562
330	9,597	6,853	194.459
340	9,888	7,061	195.328
350	10,180	7,270	196.173
360	10,471	7,478	196.995
370	10,763	7,687	197.794
380	11,055	7,895	198.572
390	11,347	8,104	199.331
400	11,640	8,314	200.071
410	11,932	8,523	200.794
420	12,225	8,733	201.499
430	12,518	8,943	202.189
440	12,811	9,153	202.863
450	13,105	9,363	203.523
460	13,399	9,574	204.170
470	13,693	9,786	204.803
480	13,988	9,997	205.424
490	14,285	10,210	206.033
500	14,581	10,423	206.630
510	14,876	10,635	207.216
520	15,172	10,848	207.792
530	15,469	11,062	208.358
540	15,766	11,277	208.914
550	16,064	11,492	209.461
560	16,363	11,707	209.999
570	16,662	11,923	210.528
580	16,962	12,139	211.049
590	17,262	12,356	211.562
600	17,563	12,574	212.066
610	17,864	12,792	212.564
620	18,166	13,011	213.055
630	18,468	13,230	213.541
640	18,772	13,450	214.018
650	19,075	13,671	214.489
660	19,380	13,892	214.954
670	19,685	14,114	215.413
680	19,991	14,337	215.866
690	20,297	14,560	216.314
700	20,604	14,784	216.756
710	20,912	15,008	217.192
720	21,220	15,234	217.624
730	21,529	15,460	218.059
740	21,839	15,686	218.472
750	22,149	15,913	218.889
760	22,460	16,141	219.301
770	22,772	16,370	219.709
780	23,085	16,599	220.113
790	23,398	16,830	220.512
800	23,714	17,061	220.907
810	24,027	17,292	221.298
820	24,342	17,524	221.684
830	24,658	17,757	222.067
840	24,974	17,990	222.447
850	25,292	18,224	222.822
860	25,610	18,459	223.194
870	25,928	18,695	223.562
880	26,248	18,931	223.927
890	26,568	19,168	224.288
900	26,890	19,407	224.647
910	27,210	19,644	225.002
920	27,532	19,883	225.353
930	27,854	20,122	225.701
940	28,178	20,362	226.047
950	28,501	20,603	226.389
960	28,826	20,844	226.728
970	29,151	21,086	227.064
980	29,476	21,328	227.398
990	29,803	21,571	227.728
1000	30,129	21,815	228.057
1020	30,784	22,304	228.706
1040	31,442	22,795	229.344
1060	32,101	23,288	229.973
1080	32,762	23,782	230.591
1100	33,426	24,280	231.199
1120	34,092	24,780	231.799
1140	34,760	25,282	232.391
1160	35,430	25,786	232.973
1180	36,104	26,291	233.549
1200	36,777	26,799	234.115
1220	37,452	27,308	234.673
1240	38,129	27,819	235.223
1260	38,807	28,331	235.766
1280	39,488	28,845	236.302
1300	40,170	29,361	236.831
1320	40,853	29,378	237.353
1340	41,539	30,398	237.867

TABLE C.3 (continued)
Ideal Gas Properties of Nitrogen, N_2 (SI Units)

T (K)	$\bar{h}$ (kJ/kmol)	$\bar{u}$ (kJ/kmol)	$\bar{s}°$ (kJ/kmol·K)
1360	42,227	30,919	238.376
1380	42,915	31,441	238.878
1400	43,605	31,964	239.375
1420	44,295	32,489	239.865
1440	44,988	33,014	240.350
1460	45,682	33,543	240.827
1480	46,377	34,071	241.301
1500	47,073	34,601	241.768
1520	47,771	35,133	242.228
1540	48,470	35,665	242.685
1560	49,168	36,197	243.137
1580	49,869	36,732	243.585
1600	50,571	37,268	244.028
1620	51,275	37,806	244.464
1640	51,980	38,344	244.896
1660	52,686	38,884	245.324
1680	53,393	39,424	245.747
1700	54,099	39,965	246.166
1720	54,807	40,507	246.580
1740	55,516	41,049	246.990
1760	56,227	41,594	247.396
1780	56,938	42,139	247.798
1800	57,651	42,685	248.195
1820	58,363	43,231	248.589
1840	59,075	43,777	248.979
1860	59,790	44,324	249.365
1880	60,504	44,873	249.748
1900	61,220	45,423	250.128
1920	61,936	45,973	250.502
1940	62,654	46,524	250.874
1960	63,381	47,075	251.242
1980	64,090	47,627	251.607
2000	64,810	48,181	251.969
2050	66,612	49,567	252.858
2100	68,417	50,957	253.726
2150	70,226	52,351	254.578
2200	72,040	53,749	255.412
2250	73,856	55,149	256.227
2300	75,676	56,553	257.027
2350	77,496	57,958	257.810
2400	79,320	59,366	258.580
2450	81,149	60,779	259.332
2500	82,981	62,195	260.073
2550	84,814	63,613	260.799
2600	86,650	65,033	261.512
2650	88,488	66,455	262.213
2700	90,328	67,880	262.902
2750	92,171	69,306	263.577
2800	94,014	70,734	264.241

TABLE C.3 (continued)
Ideal Gas Properties of Nitrogen, N_2 (SI Units)

T (K)	$\bar{h}$ (kJ/kmol)	$\bar{u}$ (kJ/kmol)	$\bar{s}°$ (kJ/kmol·K)
2850	95,859	72,163	264.895
2900	97,705	73,593	265.538
2950	99,556	75,028	266.170
3000	101,407	76,464	266.793
3050	103,260	77,902	267.404
3100	105,115	79,341	268.007
3150	106,972	80,782	268.601
3200	108,830	82,224	269.186
3250	110,690	83,668	269.763

Source: Adapted from Kenneth Wark, *Thermodynamics*, 4th edn., McGraw-Hill, New York, 1983, pp. 787–798. Originally published in JANAF, *Thermochemical Tables*, NSRDS-NBS-37, 1971; *Thermodynamics: An Engineering Approach*, 7th edn., McGraw Hill, New York, Table A-18, p. 936.

$\bar{h}$ is enthalpy, $\bar{u}$ is internal energy, and $\bar{s}°$ is entropy.

TABLE C.4
Ideal Gas Properties of Nitrogen, N_2 (English Units)

T (°R)	$\bar{h}$ (Btu/lbmol)	$\bar{u}$ (Btu/lbmol)	$\bar{s}°$ (Btu/lbmol·°R)
300	2,082.0	1,486.2	41.695
320	2,221.0	1,585.5	42.143
340	2,360.0	1,684.4	42.564
360	2,498.9	1,784.0	42.962
380	2,638.0	1,883.4	43.337
400	2,777.0	1,982.6	43.694
420	2,916.1	2,082.0	44.034
440	3,055.1	2,181.3	44.357
460	3,194.1	2,280.6	44.665
480	3,333.1	2,379.9	44.962
500	3,472.2	2,479.3	45.246
520	3,611.3	2,578.6	45.519
537	3,729.5	2,663.1	45.743
540	3,750.3	2,678.0	45.781
560	3,889.5	2,777.4	46.034
580	4,028.7	2,876.9	46.278
600	4,167.9	2,976.4	46.514
620	4,307.1	3,075.9	46.742
640	4,446.4	3,175.5	46.964
660	4,585.8	3,275.2	47.178
680	4,725.3	3,374.9	47.386
700	4,864.9	3,474.8	47.588
720	5,004.5	3,574.7	47.785
740	5,144.3	3,674.7	47.977
760	5,284.1	3,774.9	48.164
780	5,424.2	3,875.2	48.345
800	5,564.4	3,975.7	48.522
820	5,704.7	4,076.3	48.696
840	5,845.3	4,177.1	48.865
860	5,985.9	4,278.1	49.031

(continued)

TABLE C.4 (continued)

Ideal Gas Properties of Nitrogen, N_2 (English Units)

T (°R)	$\bar{h}$ (Btu/lbmol)	$\bar{u}$ (Btu/lbmol)	$\bar{s}°$ (Btu/lbmol·°R)
880	6,126.9	4,379.4	49.193
900	6,268.1	4,480.8	49.352
920	6,409.6	4,582.6	49.507
940	6,551.2	4,684.5	49.659
960	6,693.1	4,786.7	49.808
980	6,835.4	4,889.3	49.955
1000	6,977.9	4,992.0	50.099
1020	7,120.7	5,095.1	50.241
1040	7,263.8	5,198.5	50.380
1060	7,407.2	5,302.2	50.516
1080	7,551.0	5,406.2	50.651
1100	7,695.0	5,510.5	50.783
1120	7,839.3	5,615.2	50.912
1140	7,984.0	5,720.1	51.040
1160	8,129.0	5,825.4	51.167
1180	8,274.4	5,931.0	51.291
1200	8,420.0	6,037.0	51.143
1220	8,566.1	6,143.4	51.534
1240	8,712.6	6,250.1	51.653
1260	8,859.3	6,357.2	51.771
1280	9,006.4	6,464.5	51.887
1300	9,153.9	6,572.3	51.001
1320	9,301.8	6,680.4	52.114
1340	9,450.0	6,788.9	52.225
1360	9,598.6	6,897.8	52.335
1380	9,747.5	7,007.0	52.444
1400	9,896.9	7,116.7	52.551
1420	10,046.6	7,226.7	52.658
1440	10,196.6	7,337.0	52.763
1460	10,347.0	7,447.6	52.867
1480	10,497.8	7,558.7	52.969
1500	10,648.0	7,670.1	53.071
1520	10,800.4	7,781.9	53.171
1540	10,952.2	7,893.9	53.271
1560	11,104.3	8,006.4	53.369
1580	11,256.9	8,119.2	53.465
1600	11,409.7	8,232.3	53.561
1620	11,562.8	8,345.7	53.656
1640	11,716.4	8,459.6	53.751
1660	11,870.2	8,573.6	53.844
1680	12,024.3	8,688.1	53.936
1700	12,178.9	8,802.9	54.028
1720	12,333.7	8,918.0	54.118
1740	12,488.8	9,033.4	54.208
1760	12,644.3	9,149.2	54.297
1780	12,800.2	9,265.3	54.385
1800	12,956.3	9,381.7	54.472
1820	13,112.7	9,498.4	54.559
1840	13,269.5	9,615.5	54.645
1860	13,426.5	9,732.8	54.729
1900	13,742	9,968	54.896
1940	14,058	10,205	55.061
1980	14,375	10,443	55.223
2020	14,694	10,682	55.383
2060	15,013	10,923	55.540
2100	15,334	11,164	55.694
2140	15,656	11,406	55.846
2180	15,978	11,649	55.995
2220	16,302	11,893	56.141
2260	16,626	12,138	56.286
2300	16,951	12,384	56.429
2340	17,277	12,630	56.570
2380	17,604	12,878	56.708
2420	17,392	13,126	56.845
2460	18,260	13,375	56.980
2500	18,590	13,625	57.112
2540	18,919	13,875	57.243
2580	19,250	14,127	57.372
2620	19,582	14,379	57.499
2660	19,914	14,631	57.625
2700	20,246	14,885	57.750
2740	20,580	15,139	57.872
2780	20,914	15,393	57.993
2820	21,248	15,648	58.113
2860	21,584	15,905	58.231
2900	21,920	16,161	58.348
2940	22,256	16,417	58.463
2980	22,593	16,675	58.576
3020	22,930	16,933	58.688
3060	23,268	17,192	58.800
3100	23,607	17,451	58.910
3140	23,946	17,710	59.019
3180	24,285	17,970	59.126
3220	24,625	18,231	59.232
3260	24,965	18,491	59.338
3300	25,306	18,753	59.442
3340	25,647	19,014	59.544
3380	25,989	19,277	59.646
3420	26,331	19,539	59.747
3460	26,673	19,802	59.846
3500	27,016	20,065	59.944
3540	27,359	20,329.	60.041
3580	27,703	20,593	60.138
3620	28,046	20,858	60.234
3660	28,391	21,122	60.328
3700	28,735	21,387	60.422
3740	29,080	21,653	60.515
3780	29,425	21,919	60.607

TABLE C.4 (continued)
Ideal Gas Properties of Nitrogen, N_2 (English Units)

T (°R)	$\bar{h}$ (Btu/lbmol)	$\bar{u}$ (Btu/lbmol)	$\bar{s}°$ (Btu/lbmol·°R)
3820	29,771	22,185	60.698
3860	30,117	22,451	60.788
3900	30,463	22,718	60.877
3940	30,809	22,985	60.966
3980	31,156	23,252	61.053
4020	31,503	23,520	61.139
4060	31,850	23,788	61.225
4100	32,198	24,056	61.310
4140	32,546	24,324	61.395
4180	32,894	24,593	61.479
4220	33,242	24,862	61.562
4260	33,591	25,131	61.644
4300	33,940	25,401	61.726
4340	34,289	25,670	61.806
4380	34,638	25,940	61.887
4420	34,988	26,210	61.966
4460	35,338	26,481	62.045
4500	35,688	26,751	62.123
4540	36,038	27,022	62.201
4580	36,389	27,293	62.278
4620	36,739	27,565	62.354
4660	37,090	27,836	62.429
4700	37,441	28,108	62.504
4740	37,792	28,379	62.578
4780	38,144	28,651	62.652
4820	38,495	28,924	62.725
4860	38,847	29,196	62.798
4900	39,199	29,468	62.870
5000	40,080	30,151	63.049
5100	40,962	30,834	63.223
5200	41,844	31,518	63.395
5300	42,728	32,203	63.563

Source: Adapted from Kenneth Wark, *Thermodynamics*, 4th edn., McGraw-Hill, New York, 1983, pp. 834–844. Originally published in Keenan, J.H. and Kaye, J., *Gas Tables*, John Wiley & Sons, New York, 1945; *Thermodynamics: An Engineering Approach*, 7th edn., McGraw Hill, New York, Table A-18E, p. 984.

$\bar{h}$ is enthalpy, $\bar{u}$ is internal energy, and $\bar{s}°$ is entropy.

TABLE C.5
Ideal Gas Properties of Oxygen, O_2 (SI Units)

T (K)	$\bar{h}$ (kJ/kmol)	$\bar{u}$ (kJ/kmol)	$\bar{s}°$ (kJ/kmol·K)
0	0	0	0
220	6,404	4,575	196.171
230	6,694	4,782	197.461
240	6,984	4,989	198.696
250	7,275	5,197	199.885
260	7,566	5,405	201.027
270	7,858	5,613	202.128
280	8,150	5,822	203.191
290	8,443	6,032	204.218
298	8,682	6,203	205.033
300	8,736	6,242	205.213
310	9,030	6,453	206.177
320	9,325	6,664	207.112
330	9,620	6,877	208.020
340	9,916	7,090	208.904
350	10,213	7,303	209.765
360	10,511	7,518	210.604
370	10,809	7,733	211.423
380	11,109	7,949	212.222
390	11,409	8,166	213.002
400	11,711	8,384	213.765
410	12,012	8,603	214.510
420	12,314	8,822	215.241
430	12,618	9,043	215.955
440	12,923	9,264	216.656
450	13,228	9,487	217.342
460	13,525	9,710	218.016
470	13,842	9,935	218.676
480	14,151	10,160	219.326
490	14,460	10,386	219.963
500	14,770	10,614	220.589
510	15,082	10,842	221.206
520	15,395	11,071	221.812
530	15,708	11,301	222.409
540	16,022	11,533	222.997
550	16,338	11,765	223.576
560	16,654	11,998	224.146
570	16,971	12,232	224.708
580	17,290	12,467	225.262
590	17,609	12,703	225.808
600	17,929	12,940	226.346
610	18,250	13,178	226.877
620	18,572	13,417	227.400
630	18,895	13,657	227.918
640	19,219	13,898	228.429
650	19,544	14,140	228.932
660	19,870	14,383	229.430
670	20,197	14,626	229.920
680	20,524	14,871	230.405

(*continued*)

TABLE C.5 (continued)
Ideal Gas Properties of Oxygen, O_2 (SI Units)

T (K)	$\bar{h}$ (kJ/kmol)	$\bar{u}$ (kJ/kmol)	$\bar{s}°$ (kJ/kmol·K)
690	20,854	15,116	230.885
700	21,184	15,364	231.358
710	21,514	15,611	231.827
720	21,845	15,859	232.291
730	22,177	16,107	232.748
740	22,510	16,357	233.201
750	22,844	16,607	233.649
760	23,178	16,859	234.091
770	23,513	17,111	234.528
780	23,850	17,364	234.960
790	24,186	17,618	235.387
800	24,523	17,872	235.810
810	24,861	18,126	236.230
820	25,199	18,382	236.644
830	25,537	18,637	237.055
840	25,877	18,893	237.462
850	26,218	19,150	237.864
860	26,559	19,408	238.264
870	26,899	19,666	238.660
880	27,242	19,925	239.051
890	27,584	20,185	239.439
900	27,928	20,445	239.823
910	28,272	20,706	240.203
920	28,616	20,967	240.580
930	28,960	21,228	240.953
940	29,306	21,491	241.323
950	29,652	21,754	241.689
960	29,999	22,017	242.052
970	30,345	22,280	242.411
980	30,692	22,544	242.768
990	31,041	22,809	242.120
1000	31,389	23,075	243.471
1020	32,088	23,607	244.164
1040	32,789	24,142	244.844
1060	33,490	24,677	245.513
1080	34,194	25,214	246.171
1100	34,899	25,753	246.818
1120	35,606	26,294	247.454
1140	36,314	26,836	248.081
1160	37,023	27,379	248.698
1180	37,734	27,923	249.307
1200	38,447	28,469	249.906
1220	39,162	29,018	250.497
1240	39,877	29,568	251.079
1260	40,594	30,118	251.653
1280	41,312	30,670	252.219
1300	42,033	31,224	252.776
1320	42,753	31,778	253.325
1340	43,475	32,334	253.868
1360	44,198	32,891	254.404
1380	44,923	33,449	254.932
1400	45,648	34,008	255.454
1420	46,374	34,567	255.968
1440	47,102	35,129	256.475
1460	47,831	35,692	256.978
1480	48,561	36,256	257.474
1500	49,292	36,821	257.965
1520	50,024	37,387	258.450
1540	50,756	37,952	258.928
1560	51,490	38,520	259.402
1580	52,224	39,088	259.870
1600	52,961	39,658	260.333
1620	53,696	40,227	260.791
1640	54,434	40,799	261.242
1660	55,172	41,370	261.690
1680	55,912	41,944	262.132
1700	56,652	42,517	262.571
1720	57,394	43,093	263.005
1740	58,136	43,669	263.435
1760	58,880	44,247	263.861
1780	59,624	44,825	264.283
1800	60,371	45,405	264.701
1820	61,118	45,986	265.113
1840	61,866	46,568	265.521
1860	62,616	47,151	265.925
1880	63,365	47,734	266.326
1900	64,116	48,319	266.722
1920	64,868	48,904	267.115
1940	65,620	49,490	267.505
1960	66,374	50,078	267.891
1980	67,127	50,665	268.275
2000	67,881	51,253	268.655
2050	69,772	52,727	269.588
2100	71,668	54,208	270.504
2150	73,573	55,697	271.399
2200	75,484	57,192	272.278
2250	77,397	58,690	273.136
2300	79,316	60,193	273.891
2350	81,243	61,704	274.809
2400	83,174	63,219	275.625
2450	85,112	64,742	276.424
2500	87,057	66,271	277.207
2550	89,004	67,802	277.979
2600	90,956	69,339	278.738
2650	92,916	70,883	279.485
2700	94,881	72,433	280.219
2750	96,852	73,987	280.942
2800	98,826	75,546	281.654

TABLE C.5 (continued)
Ideal Gas Properties of Oxygen, O_2 (SI Units)

T (K)	$\bar{h}$ (kJ/kmol)	$\bar{u}$ (kJ/kmol)	$\bar{s}°$ (kJ/kmol·K)
2850	100,808	77,112	282.357
2900	102,793	78,682	283.048
2950	104,785	80,258	283.728
3000	106,780	81,837	284.399
3050	108,778	83,419	285.060
3100	110,784	85,009	285.713
3150	112,795	86,601	286.355
3200	114,809	88,203	286.989
3250	116,827	89,804	287.614

Source: Thermodynamics: An Engineering Approach, 7th edn., McGraw Hill, New York, Table A-19, p. 938.

$\bar{h}$ is enthalpy, $\bar{u}$ is internal energy, and $\bar{s}°$ is entropy.

TABLE C.6
Ideal Gas Properties of Oxygen, O_2 (English Units)

T (°R)	$\bar{h}$ (Btu/lbmol)	$\bar{u}$ (Btu/lbmol)	$\bar{s}°$ (Btu/lbmol·°R)
300	2,073.5	1,477.8	44.927
320	2,212.6	1,577.1	45.375
340	2,351.7	1,676.5	45.797
360	2,490.8	1,775.9	46.195
380	2,630.0	1,875.3	46.571
400	2,769.1	1,974.8	46.927
420	2,908.3	2,074.3	47.267
440	3,047.5	2,173.8	47.591
460	3,186.9	2,273.4	47.900
480	3,326.5	2,373.3	48.198
500	3,466.2	2,473.2	48.483
520	3,606.1	2,573.4	48.757
537	3,725.1	2,658.7	48.982
540	3,746.2	2,673.8	49.021
560	3,886.6	2,774.5	49.276
580	4,027.3	2,875.5	49.522
600	4,168.3	2,976.8	49.762
620	4,309.7	3,078.4	49.993
640	4,451.4	3,180.4	50.218
660	4,593.5	3,282.9	50.437
680	4,736.2	3,385.8	50.650
700	4,879.3	3,489.2	50.858
720	5,022.9	3,593.1	51.059
740	5,167.0	3,697.4	51.257
760	5,311.4	3,802.4	51.450
780	5,456.4	3,907.5	51.638
800	5,602.0	4,013.3	51.821
820	5,748.1	4,119.7	52.002
840	5,894.8	4,226.6	52.179
860	6,041.9	4,334.1	52.352
880	6,189.6	4,442.0	52.522
900	6,337.9	4,550.6	52.688
920	6,486.7	4,659.7	52.852
940	6,636.1	4,769.4	53.012

TABLE C.6 (continued)
Ideal Gas Properties of Oxygen, O_2 (English Units)

T (°R)	$\bar{h}$ (Btu/lbmol)	$\bar{u}$ (Btu/lbmol)	$\bar{s}°$ (Btu/lbmol·°R)
960	6,786.0	4,879.5	53.170
980	6,936.4	4,990.3	53.326
1000	7,087.5	5,101.6	53.477
1020	7,238.9	5,213.3	53.628
1040	7,391.0	5,325.7	53.775
1060	7,543.6	5,438.6	53.921
1080	7,696.8	5,552.1	54.064
1100	7,850.4	5,665.9	54.204
1120	8,004.5	5,780.3	54.343
1140	8,159.1	5,895.2	54.480
1160	8,314.2	6,010.6	54.614
1180	8,469.8	6,126.5	54.748
1200	8,625.8	6,242.8	54.879
1220	8,782.4	6,359.6	55.008
1240	8,939.4	6,476.9	55.136
1260	9,096.7	6,594.5	55.262
1280	9,254.6	6,712.7	55.386
1300	9,412.9	6,831.3	55.508
1320	9,571.9	6,950.2	55.630
1340	9,730.7	7,069.6	55.750
1360	9,890.2	7,189.4	55.867
1380	10,050.1	7,309.6	55.984
1400	10,210.4	7,430.1	56.099
1420	10,371.0	7,551.1	56.213
1440	10,532.0	7,672.4	56.326
1460	10,693.3	7,793.9	56.437
1480	10,855.1	7,916.0	56.547
1500	11,017.1	8,038.3	56.656
1520	11,179.6	8,161.1	56.763
1540	11,342.4	8,284.2	56.869
1560	11,505.4	8,407.4	56.975
1580	11,668.8	8,531.1	57.079
1600	11,832.5	8,655.1	57.182
1620	11,996.6	8,779.5	57.284
1640	12,160.9	8,904.1	57.385
1660	12,325.5	9,029.0	57.484
1680	12,490.4	9,154.1	57.582
1700	12,655.6	9,279.6	57.680
1720	12,821.1	9,405.4	57.777
1740	12,986.9	9,531.5	57.873
1760	13,153.0	9,657.9	57.968
1780	13,319.2	9,784.4	58.062
1800	13,485.8	9,911.2	58.155
1820	13,652.5	10,038.2	58.247
1840	13,819.6	10,165.6	58.339
1860	13,986.8	10,293.1	58.428
1900	14,322	10,549	58.607
1940	14,658	10,806	58.782
1980	14,995	11,063	58.954
2020	15,333	11,321	59.123

(continued)

TABLE C.6 (continued)

Ideal Gas Properties of Oxygen, O_2 (English Units)

T (°R)	$\bar{h}$ (Btu/lbmol)	$\bar{u}$ (Btu/lbmol)	$\bar{s}°$ (Btu/lbmol·°R)
2060	15,672	11,581	59.289
2100	16,011	11,841	59.451
2140	16,351	12,101	59.612
2180	16,692	12,363	59.770
2220	17,036	12,625	59.926
2260	17,376	12,888	60.077
2300	17,719	13,151	60.228
2340	18,062	13,416	60.376
2380	18,407	13,680	60.522
2420	18,572	13,946	60.666
2460	19,097	14,212	60.808
2500	19,443	14,479	60.946
2540	19,790	14,746	61.084
2580	20,138	15,014	61.220
2620	20,485	15,282	61.354
2660	20,834	15,551	61.486
2700	21,183	15,821	61.616
2740	21,533	16,091	61.744
2780	21,883	16,362	61.871
2820	22,232	16,633	61.996
2860	22,584	16,905	62.120
2900	22,936	17,177	62.242
2940	23,288	17,450	62.363
2980	23,641	17,723	62.483
3020	23,994	17,997	62.599
3060	24,348	18,271	62.716
3100	24,703	18,546	62.831
3140	25,057	18,822	62.945
3180	25,413	19,098	63.057
3220	25,769	19,374	63.169
3260	26,175	19,651	63.279
3300	26,412	19,928	63.386
3340	26,839	20,206	63.494
3380	27,197	20,485	63.601
3420	27,555	20,763	63.706
3460	27,914	21,043	63.811
3500	28,273	21,323	63.914
3540	28,633	21,603	64.016
3580	28,994	21,884	64.114
3620	29,354	22,165	64.217
3660	29,716	22,447	64.316
3700	30,078	22,730	64.415
3740	30,440	23,013	64.512
3780	30,803	23,296	64.609
3820	31,166	23,580	64.704
3860	31,529	23,864	64.800
3900	31,894	24,149	64.893
3940	32,258	24,434	64.986
3980	32,623	24,720	65.078
4020	32,989	25,006	65.169

TABLE C.6 (continued)

Ideal Gas Properties of Oxygen, O_2 (English Units)

T (°R)	$\bar{h}$ (Btu/lbmol)	$\bar{u}$ (Btu/lbmol)	$\bar{s}°$ (Btu/lbmol·°R)
4060	33,355	25,292	65.260
4100	33,722	25,580	65.350
4140	34,089	25,867	64.439
4180	34,456	26,155	65.527
4220	34,824	26,144	65.615
4260	35,192	26,733	65.702
4300	35,561	27,022	65.788
4340	35,930	27,312	65.873
4380	36,300	27,602	65.958
4420	36,670	27,823	66.042
4460	37,041	28,184	66.125
4500	37,412	28,475	66.208
4540	37,783	28,768	66.290
4580	38,155	29,060	66.372
4620	38,528	29,353	66.453
4660	38,900	29,646	66.533
4700	39,274	29,940	66.613
4740	39,647	30,234	66.691
4780	40,021	30,529	66.770
4820	40,396	30,824	66.848
4860	40,771	31,120	66.925
4900	41,146	31,415	67.003
5000	42,086	32,157	67.193
5100	43,021	32,901	67.380
5200	43,974	33,648	67.562
5300	44,922	34,397	67.743

Source: *Thermodynamics: An Engineering Approach*, 7th edn., McGraw Hill, New York, Table A-19E, p. 986.

$\bar{h}$ is enthalpy, $\bar{u}$ is internal energy, and $\bar{s}°$ is entropy.

TABLE C.7

Ideal Gas Properties of CO_2 (SI Units)

T (K)	$\bar{h}$ (kJ/kmol)	$\bar{u}$ (kJ/kmol)	$\bar{s}°$ (kJ/kmol·K)
0	0	0	0
220	6,601	4,772	202.966
230	6,938	5,026	204.464
240	7,280	5,285	205.920
250	7,627	5,548	207.337
260	7,979	5,817	208.717
270	8,335	6,091	210.062
280	8,697	6,369	211.376
290	9,063	6,651	212.660
298	9,364	6,885	213.685
300	9,431	6,939	213.915
310	9,807	7,230	215.146
320	10,186	7,526	216.351
330	10,570	7,826	217.534
340	10,959	8,131	218.694
350	11,351	8,439	219.831
360	11,748	8,752	220.948
370	12,148	9,068	222.044
380	12,552	9,392	223.122
390	12,960	9,718	224.182
400	13,372	10,046	225.225
410	13,787	10,378	226.250
420	14,206	10,714	227.258
430	14,628	11,053	228.252
440	15,054	11,393	229.230
450	15,483	11,742	230.194
460	15,916	12,091	231.144
470	16,351	12,444	232.080
480	16,791	12,800	233.004
490	17,232	13,158	233.916
500	17,678	13,521	234.814
510	18,126	13,885	235.700
520	18,576	14,253	236.575
530	19,029	14,622	237.439
540	19,485	14,996	238.292
550	19,945	15,372	239.135
560	20,407	15,751	239.962
570	20,870	16,131	240.789
580	21,337	16,515	241.602
590	21,807	16,902	242.405
600	22,280	17,291	243.199
610	22,754	17,683	243.983
620	23,231	18,076	244.758
630	23,709	18,471	245.524
640	24,190	18,869	246.282
650	24,674	19,270	247.032
660	25,160	19,672	247.773
670	25,648	20,078	248.507
680	26,138	20,484	249.233
690	26,631	20,894	249.952
700	27,125	21,305	250.663
710	27,622	21,719	251.368
720	28,121	22,134	252.065
730	28,622	22,522	252.755
740	29,124	22,972	253.439
750	29,629	23,393	254.117
760	30,135	23,817	254.787
770	30,644	24,242	255.452
780	31,154	24,669	256.110
790	31,665	25,097	256.762
800	32,179	25,527	257.408
810	32,694	25,959	258.048
820	33,212	26,394	258.682
830	33,730	26,829	259.311
840	34,251	27,267	259.934
850	34,773	27,706	260.551
860	35,296	28,125	261.164
870	35,821	28,588	261.770
880	36,347	29,031	262.371
890	36,876	29,476	262.968
900	37,405	29,922	263.559
910	37,935	30,369	264.146
920	38,467	30,818	264.728
930	39,000	31,268	265.304
940	39,535	31,719	265.877
950	40,070	32,171	266.444
960	40,607	32,625	267.007
970	41,145	33,081	267.566
980	41,685	33,537	268.119
990	42,226	33,995	268.670
1000	42,769	34,455	269.215
1020	43,859	35,378	270.293
1040	44,953	36,306	271.354
1060	46,051	37,238	272.400
1080	47,153	38,174	273.430
1100	48,258	39,112	274.445
1120	49,369	40,057	275.444
1140	50,484	41,006	276.430
1160	51,602	41,957	277.403
1180	52,724	42,913	278.361
1200	53,848	43,871	297.307
1220	54,977	44,834	280.238
1240	56,108	45,799	281.158
1260	57,244	46,768	282.066
1280	58,381	47,739	282.962
1300	59,522	48,713	283.847
1320	60,666	49,691	284.722
1340	61,813	50,672	285.586

(*continued*)

TABLE C.7 (continued)
Ideal Gas Properties of CO_2 (SI Units)

T (K)	$\bar{h}$ (kJ/kmol)	$\bar{u}$ (kJ/kmol)	$\bar{s}°$ (kJ/kmol·K)
1360	62,963	51,656	286.439
1380	64,116	52,643	287.283
1400	65,271	53,631	288.106
1420	66,427	54,621	288.934
1440	67,586	55,614	289.743
1460	68,748	56,609	290.542
1480	66,911	57,606	291.333
1500	71,078	58,606	292.114
1520	72,246	59,609	292.888
1540	73,417	60,613	292.654
1560	74,590	61,620	294.411
1580	76,767	62,630	295.161
1600	76,944	63,741	295.901
1620	78,123	64,653	296.632
1640	79,303	65,668	297.356
1660	80,486	66,592	298.072
1680	81,670	67,702	298.781
1700	82,856	68,721	299.482
1720	84,043	69,742	300.177
1740	85,231	70,764	300.863
1760	86,420	71,787	301.543
1780	87,612	72,812	302.217
1800	88,806	73,840	302.884
1820	90,000	74,868	303.544
1840	91,196	75,897	304.198
1860	92,394	76,929	304.845
1880	93,593	77,962	305.487
1900	94,793	78,996	306.122
1920	95,995	80,031	306.751
1940	97,197	81,067	307.374
1960	98,401	82,105	307.992
1980	99,606	83,144	308.604
2000	100,804	84,185	309.210
2050	103,835	86,791	310.701
2100	106,864	89,404	312.160
2150	109,898	92,023	313.589
2200	112,939	94,648	314.988
2250	115,984	97,277	316.356
2300	119,035	99,912	317.695
2350	122,091	102,552	319.011
2400	125,152	105,197	320.302
2450	128,219	107,849	321.566
2500	131,290	110,504	322.808
2550	134,368	113,166	324.026
2600	137,449	115,832	325.222
2650	140,533	118,500	326.396
2700	143,620	121,172	327.549
2750	146,713	123,849	328.684
2800	149,808	126,528	329.800

TABLE C.7 (continued)
Ideal Gas Properties of CO_2 (SI Units)

T (K)	$\bar{h}$ (kJ/kmol)	$\bar{u}$ (kJ/kmol)	$\bar{s}°$ (kJ/kmol·K)
2850	152,908	129,212	330.896
2900	156,009	131,898	331.975
2950	159,117	134,589	333.037
3000	162,226	137,283	334.084
3050	165,341	139,982	335.114
3100	168,456	142,681	336.126
3150	171,576	145,385	337.124
3200	174,695	148,089	338.109
3250	177,822	150,801	339.069

Source: *Thermodynamics: An Engineering Approach*, 7th edn., McGraw Hill, New York, Table A-20, p. 940. $\bar{h}$ is enthalpy, $\bar{u}$ is internal energy, and $\bar{s}°$ is entropy.

TABLE C.8
Ideal Gas Properties of CO_2 (English Units)

T (°R)	$\bar{h}$ (Btu/lbmol)	$\bar{u}$ (Btu/lbmol)	$\bar{s}°$ (Btu/lbmol·°R)
300	2,108.2	1,512.4	46.353
320	2,256.6	1,621.1	46.832
340	2,407.3	1,732.1	47.289
360	2,560.5	1,845.6	47.728
380	2,716.4	1,961.8	48.148
400	2,874.7	2,080.4	48.555
420	3,035.7	2,201.7	48.947
440	3,199.4	2,325.6	49.329
460	3,365.7	2,452.2	49.698
480	3,534.7	2,581.5	50.058
500	3,706.2	2,713.3	50.408
520	3,880.3	2,847.7	50.750
537	4,027.5	2,963.8	51.032
540	4,056.8	2,984.4	51.082
560	4,235.8	3,123.7	51.408
580	4,417.2	3,265.4	51.726
600	4,600.9	3,409.4	52.038
620	4,786.6	3,555.6	52.343
640	4,974.9	3,704.0	52.641
660	5,165.2	3,854.6	52.934
680	5,357.6	4,007.2	53.225
700	5,552.0	4,161.9	53.503
720	5,748.4	4,318.6	53.780
740	5,946.8	4,477.3	54.051
760	6,147.0	4,637.9	54.319
780	6,349.1	4,800.1	54.582
800	6,552.9	4,964.2	54.839
820	6,758.3	5,129.9	55.093
840	6,965.7	5,297.6	55.343
860	7,174.7	5,466.9	55.589
880	7,385.3	5,637.7	55.831
900	7,597.6	5,810.3	56.070
920	7,811.4	5,984.4	56.305
940	8,026.8	6,160.1	56.536

TABLE C.8 (continued)
Ideal Gas Properties of CO_2 (English Units)

T (°R)	$\bar{h}$ (Btu/lbmol)	$\bar{u}$ (Btu/lbmol)	$\bar{s}°$ (Btu/lbmol·°R)
960	8,243.8	6,337.4	56.765
980	8,462.2	6,516.1	56.990
1000	8,682.1	6,696.2	57.212
1020	8,903.4	6,877.8	57.432
1040	9,126.2	7,060.9	57.647
1060	9,350.3	7,245.3	57.861
1080	9,575.8	7,431.1	58.072
1100	9,802.6	7,618.1	58.281
1120	10,030.6	7,806.4	58.485
1140	10,260.1	7,996.2	58.689
1160	10,490.6	8,187.0	58.889
1180	10,722.3	8,379.0	59.088
1200	10,955.3	8,572.3	59.283
1220	11,189.4	8,766.6	59.477
1240	11,424.6	8,962.1	59.668
1260	11,661.0	9,158.8	59.858
1280	11,898.4	9,356.5	60.044
1300	12,136.9	9,555.3	60.229
1320	12,376.4	9,755.0	60.412
1340	12,617.0	9,955.9	60.593
1360	12,858.5	10,157.7	60.772
1380	13,101.0	10,360.5	60.949
1400	13,344.7	10,564.5	61.124
1420	13,589.1	10,769.2	61.298
1440	13,834.5	10,974.8	61.469
1460	14,080.8	11,181.4	61.639
1480	14,328.0	11,388.9	61.800
1500	14,576.0	11,597.2	61.974
1520	14,824.9	11,806.4	62.138
1540	15,074.7	12,016.5	62.302
1560	15,325.3	12,227.3	62.464
1580	15,576.7	12,439.0	62.624
1600	15,829.0	12,651.6	62.783
1620	16,081.9	12,864.8	62.939
1640	16,335.7	13,078.9	63.095
1660	16,590.2	13,293.7	63.250
1680	16,845.5	13,509.2	63.403
1700	17,101.4	13,725.4	63.555
1720	17,358.1	13,942.4	63.704
1740	17,615.5	14,160.1	63.853
1760	17,873.5	14,378.4	64.001
1780	18,132.2	14,597.4	64.147
1800	18,391.5	14,816.9	64.292
1820	18,651.5	15,037.2	64.435
1840	18,912.2	15,258.2	64.578
1860	19,173.4	15,479.7	64.719
1900	19,698	15,925	64.999
1940	20,224	16,372	65.272
1980	20,753	16,821	65.543
2020	21,284	17,273	65.809
2060	21,818	17,727	66.069
2100	22,353	18,182	66.327
2140	22,890	18,640	66.581
2180	23,429	19,101	66.830
2220	23,970	19,561	67.076
2260	24,512	20,024	67.319
2300	25,056	20,489	67.557
2340	25,602	20,955	67.792
2380	26,150	21,423	68.025
2420	26,699	21,893	68.253
2460	27,249	22,364	68.479
2500	27,801	22,837	68.702
2540	28,355	23,310	68.921
2580	28,910	23,786	69.138
2620	29,465	24,262	69.352
2660	30,023	24,740	69.563
2700	30,581	25,220	69.771
2740	31,141	25,701	69.977
2780	31,702	26,181	70.181
2820	32,264	26,664	70.382
2860	32,827	27,148	70.580
2900	33,392	27,633	70.776
2940	33,957	28,118	70.970
2980	34,523	28,605	71.160
3020	35,090	29,093	71.350
3060	35,659	29,582	71.537
3100	36,228	30,072	71.722
3140	36,798	30,562	71.904
3180	37,369	31,054	72.085
3220	37,941	31,546	72.264
3260	38,513	32,039	72.441
3300	39,087	32,533	72.616
3340	39,661	33,028	72.788
3380	40,236	33,524	72.960
3420	40,812	34,020	73.129
3460	41,388	34,517	73.297
3500	41,965	35,015	73.462
3540	42,543	35,513	73.627
3580	43,121	36,012	73.789
3620	43,701	36,512	73.951
3660	44,280	37,012	74.110
3700	44,861	37,513	74.267
3740	45,442	38,014	74.423
3780	46,023	38,517	74.578
3820	46,605	39,019	74.732
3860	47,188	39,522	74.884
3900	47,771	40,026	75.033
3940	48,355	40,531	75.182

(*continued*)

TABLE C.8 (continued)

Ideal Gas Properties of CO_2 (English Units)

T (°R)	$\bar{h}$ (Btu/lbmol)	$\bar{u}$ (Btu/lbmol)	$\bar{s}°$ (Btu/lbmol · °R)
3980	48,939	41,035	75.330
4020	49,524	41,541	75.477
4060	50,109	42,047	75.622
4100	50,695	42,553	75.765
4140	51,282	43,060	75.907
4180	51,868	43,568	76.048
4220	52,456	44,075	76.188
4260	53,044	44,584	76.327
4300	53,632	45,093	76.464
4340	54,221	45,602	76.601
4380	54,810	46,112	76.736
4420	55,400	46,622	76.870
4460	55,990	47,133	77.003
4500	56,581	47,645	77.135
4540	57,172	48,156	77.266
4580	57,764	48,668	77.395
4620	58,356	49,181	77.581
4660	58,948	49,694	77.652
4700	59,541	50,208	77.779
4740	60,134	50,721	77.905
4780	60,728	51,236	78.029
4820	61,322	51,750	78.153
4860	61,916	52,265	78.276
4900	62,511	52,781	78.398
5000	64,000	54,071	78.698
5100	65,491	55,363	78.994
5200	66,984	56,658	79.284
5300	68,471	57,954	79.569

Source: *Thermodynamics: An Engineering Approach*, 7th edn., McGraw Hill, New York, Table A-20E, p. 988.

$\bar{h}$ is enthalpy, $\bar{u}$ is internal energy, and $\bar{s}°$ is entropy.

Appendix C: Properties of Gases and Liquids

TABLE C.9
Properties of Gases at Atmospheric Pressure (101.3 kPa = 14.7 psia): Air (Gas Constant = 286.8 J/(kg K) = 53.3 ft lbf/lbm °R; $\gamma = c_p/c_v = 1.4$)

Temp, T		Density, ρ		Specific Heat, c_p		Kinematic Viscosity, ν		Thermal Conductivity, k		Thermal Diffusivity, α		Prandtl Number, P_r
K	°R	kg/m³	lbm/ft³	J/kg·K	Btu/lbm·°R	m²/s	ft²/s	W/m·K	Btu/h·ft·°R	m²/s	ft²/h	
100	180	3.601	0.225	1026.6	0.245	1.923×10^{-6}	2.070×10^{-5}	0.009246	0.005342	0.02501×10^{-6}	0.0869	0.770
150	270	2.368	0.148	1009.9	0.241	4.343	4.674	0.013735	0.007936	0.05745	0.223	0.753
200	360	1.768	0.110	1006.1	0.240	7.490	8.062	0.01809	0.01045	0.10165	0.394	0.739
250	450	1.413	0.0882	1005.3	0.240	9.49	10.2	0.02227	0.02287	0.13161	0.510	0.722
300	540	1.177	0.0735	1005.7	0.240	15.68	16.88	0.02624	0.01516	0.22160	0.859	0.708
350	630	0.998	0.0623	1009.0	0.241	20.76	22.35	0.03003	0.01735	0.2983	1.156	0.697
400	720	0.883	0.0551	1014.0	0.242	25.90	27.88	0.03365	0.01944	0.3760	1.457	0.689
450	810	0.783	0.489	1020.7	0.244	28.86	31.06	0.037.7	0.02142	0.4222	1.636	0.683
500	900	0.705	0.0440	1029.5	0.245	37.90	40.80	0.04038	0.02333	0.5564	2.356	0.680
550	990	0.642	0.0401	1039.2	0.248	44.34	47.73	0.04360	0.02519	0.6532	2.531	0.680
600	1000	0.589	0.0367	1055.1	0.252	51.34	55.26	0.04659	0.02682	0.7512	2.911	0.680
650	1170	0.543	0.0339	1063.5	0.254	58.51	62.98	0.00953	0.02862	0.8578	3.324	0.682
700	1260	0.503	0.0314	1075.2	0.257	66.25	7131	0.05230	0.030023	0.9672	3.748	0.684
750	1350	0.471	0.0594	1085.6	0.259	73.91	79.56	0.05509	0.03183	1.0774	4.175	0.686
800	1440	0.441	0.0275	1097.8	0.262	8229	88.58	0.05779	0.03339	1.1951	4.631	0.689
850	1530	0.415	0.0259	1109.5	0.265	90.75	97.68	0.06028	0.03483	1.3097	5.075	0.692
900	1620	0.393	0.0245	1121.2	0.268	99.3	107	0.06279	0.03628	1.4278	5.530	0.696
950	1710	0.372	0.0232	1132.1	0.270	108.2	116.5	0.06525	0.03770	1.5510	6.010	0.699
1000	1800	0.352	0.0220	1141.7	0.273	117.8	126.8	0.06752	0.03901	1.6779	6502	0.702
1100	1980	0.320	0.0120	1160	0.277	138.6	149.2	0.0732	0.0423	1.969	7.630	0.704
1200	2160	0.295	0.0184	1179	0.282	159.1	171.3	0.0782	0.0423	1.969	7.630	0.707
1300	2340	0.271	0.0189	1197	0.286	182.1	196.0	0.0837	0.0434	2.583	10.01	0.705
1400	2520	0.252	0.0157	1214	0.290	205.5	221.2	0.0891	0.0515	2.920	11.32	0.705
1500	2700	0.236	0.0147	1230	0.294	229.1	246.6	0.0946	0.0547	3.262	1264	0.705
1600	2880	0.221	0.0138	1248	0.298	254.5	273.9	0.100	0.0578	3.609	13.98	0.705
1700	36060	0.208	0.0130	1267	0.303	280.5	301.9	0.105	0.0607	3.977	15.41	0.705
1800	3240	0.197	0.0123	1287	0.307	308.1	331.6	0.111	0.0641	4.379	16.97	0.704
1900	3420	0.186	0.0115	1309	0.383	338.5	364.4	0.117	0.0676	4.811	18.64	0.704
2000	3600	0.176	0.0110	1338	0.320	369.0	397.2	0.124	0.0716	5.260	20.38	0.702
2100	3780	0.168	0.0105	1372	0.328	399.6	430.1	0.131	0.0757	5.715	22.15	0.700
2200	3960	0.160	0.0100	1419	0.339	432.6	465.6	0.139	0.0803	6120	2372	0.707
2300	4140	0.154	0.00955	1482	0.354	464.0	499.4	0.149	0.0861	6.540	25.34	0.710
2400	4320	0.146	0.00905	1574	0376	504.0	542.5	0.161	0.0930	7.020	27.20	0.718
2500	4500	0.139	0.00868	1688	0.403	543.5	585.0	0.175	0.101	7.441	28.83	0.730

Source: Engineering Heat Transfer, 2nd edn., CRC Press, Table D.1, p. 654.

TABLE C.10

Properties of Gases at Atmospheric Pressure (101.3 kPa = 14.7 psia): Nitrogen (Gas Constant = 296.8 J/(kg K) = 55.16 ft lbf/lbm °R; $\gamma = c_p/c_v = 1.40$)

Temp, T		Density, ρ		Specific Heat, c_p		Kinematic Viscosity, v		Thermal Conductivity, K		Thermal Diffusivity, α		Prandtl Number, P_r
K	°R	kg/m³	lbm/ft³	J/kg·K	Btu/lbm·°R	m²/s	ft²/s	W/m·K	Btu/h·ft·°R	m²/s	ft²/h	
100	180	3.4808	0.2173	1072.2	0.2561	1.971×10^{-6}	2.122×10^{-5}	0.009450	0.005460	0.025319×10^{-4}	0.09811	0.786
200	360	1.7108	0.1068	1042.9	0.2491	7.568	8.146	0.01824	0.01054	0.10224	0.3962	0.747
300	540	1.1421	0.0713	1040.8	0.2486	15.63	16.82	0.02620	0.01514	0.22044	0.8542	0.713
400	720	0.8538	0.0533	1045.9	0.2498	25.74	27.71	0.03335	0.01927	0.3734	1.447	0.691
500	900	0.6824	0.0426	1055.5	0.2521	37.66	40.54	0.03984	0.02302	0.5530	2.143	0.684
600	1080	0.5687	0.0355	1075.6	0.2569	51.19	55.10	0.4580	0.02646	0.7486	2.901	0.686
700	1260	0.4934	0.0308	1096.9	0.2620	65.13	7010	0.05123	0.02960	0.9466	3.668	0.691
800	1440	0.4277	0.0267	1122.5	0.2681	81.46	87.68	0.05609	0.03241	1.1685	4.528	0.700
900	1620	0.3796	0.0237	1146.4	0.2738	91.06	98.02	0.06070	0.03507	1.3946	5.404	0.711
1000	1800	0.3412	0.0213	1167.7	0.2789	117.2	126.2	0.06475	0.03741	1.6250	6.297	0.724
1100	1980	0.3108	0.0194	1185.7	0.2382	136.0	146.4	0.06850	0.03958	1.8591	7.204	0.736
1200	2160	0.2851	0.0178	1203.7	0.2875	156.1	168.0	0.07184	0.04151	2.0932	8.111	0.748

Source: *Engineering Heat Transfer*, 2nd edn., CRC Press, Table D.5, p. 657.

Appendix C: Properties of Gases and Liquids

TABLE C.11

Properties of Gases at Atmospheric Pressure (101.3 kPa = 14.7 psia): Oxygen (Gas Constant = 260 J/(kg K) = 48.3 ft lbf/lbm °R; $\gamma = c_p/c_v = 1.40$)

Temp, T		Density, ρ		Specific Heat, c_p		Kinematic Viscosity, ν		Thermal Conductivity, k		Thermal Diffusivity, α		Prandtl Number, P_r
K	°R	kg/m³	lbm/ft³	J/kg·K	Btu/lbm·°R	m²/s	ft²/h	W/m·K	Btu/h·ft·°R	m²/s	ft²/h	
100	180	3.9118	0.2492	947.9	0.2264	1.946×10^{-6}	2.095×10^{-5}	0.00903	0.00522	0.023876×10^{-4}	0.09252	0.815
150	270	26190	0.1635	917.8	0.2192	4.387	4.722	0.01367	0.00790	0.05688	0.2204	0.773
200	360	1.9559	0.1221	913.1	0.2181	7.593	8.173	0.01824	0.01054	0.10214	0.3958	0.745
250	450	1.5618	0.0975	915.7	0.2187	11.45	12.32	0.02259	0.01305	0.15794	0.6120	0.725
300	540	1.3007	0.0812	920.3	0.2198	15.86	17.07	0.02676	0.01546	0.22353	0.8662	0.709
350	630	1.1133	0.0695	929.1	0.2219	20.80	22.39	0.03070	0.01774	0.2968	1.150	0.702
400	720	0.9755	0.0609	942.0	0.2250	26.18	2818	0.03461	0.02000	0.3768	1.460	0.695
450	810	0.8682	0.0542	956.7	0.2285	31.99	34.43	0.03828	0.02212	0.4609	1.786	0.694
500	900	0.7801	0.0487	972.2	0.2322	38.37	41.27	0.04173	0.02411	0.5502	2.132	0.697
550	990	0.7096	0.0443	988.1	0.2360	45.05	48.49	0.04517	0.02610	06441	2.496	0.700
600	1080	0.6508	0.0406	1004.4	0.2399	52.15	56.13	0.04882	0.02792	0.7399	2.867	0.704

Source: *Engineering Heat Transfer*, 2nd edn., CRC Press, Table D.6, p. 658.

TABLE C.12

Properties of Gases at Atmospheric Pressure (101.3 kPa = 14.7 psia): Carbon Dioxide (Gas Constant = 188.9 J/(kg K) = 35.11 ft lbf/lbm °R; $\gamma = c_p/c_v = 1.30$)

Temp, T		Density, ρ		Specific Heat, c_p		Kinematic Viscosity, v		Thermal Conductivity, k		Thermal Diffusivity, α		Prandtl Number, P_r
K	°R	kg/m³	lbm/ft³	J/kg·K	Btu/lbm·°R	m²/s	ft²/s	W/m·K	Btu/h·ft·°R	m²/s	ft²/h	
220	396	2.4733	0.1544	783	0.187	4.490×10^{-6}	4.833×10^{-5}	0.010805	0.006243	0.05920×10^{-4}	0.2294	0.818
250	450	2.1657	0.1352	804	0.192	5.813	6.257	0.012884	0.007444		0.2868	0.793
300	540	1.7973	0.1122	871	0.208	8.321	8.957	0.016572	0.009575	0.10588	0.4103	0.770
350	630	1.5362	0.0959	900	0.215	11.19	12.05	0.02047	0.01183	0.14808	0.5738	0.755
400	720	1.3424	0.0838	942	0.225	14.39	15.49	0.02461	0.01422	0.19463	0.7542	0.738
450	810	1.1918	0.0744	980	0.234	17.90	19.27	0.02897	0.01674	0.24813	0.9615	0.721
500	900	1.0732	0.0670	1013	0.242	21.67	23.33	0.03352	0.01937	0.3084	1.195	0.702
550	990	0.9739	0.0608	1047	0.250	25.74	27.71	0.03821	0.02208	0.3750	1.453	0.685
600	1080	0.8938	0.0558	1076	0.257	30.02	32.31	0.04313	0.02491	0.4483	1.737	0.668

Source: Engineering Heat Transfer, 2nd edn., CRC Press, Table D.2, p. 655.

Appendix C: Properties of Gases and Liquids

TABLE C.13

Properties of Gases at Atmospheric Pressure (101.3 kPa = 14.7 psia): Water Vapor or Steam (Gas Constant = 461.5 J/(kg K) = 85.78 ft lbf/lbm °R; $\gamma = c_p/c_v = 1.33$)

Temp, T		Density, ρ		Specific Heat, c_p		Kinematic Viscosity, v		Thermal Conductivity, k		Thermal Diffusivity, α		Prandtl Number, P_r
K	°R	kg/m³	lbm/ft³	J/kg·K	Btu/lbm·°R	m²/s	ft²/s	W/m·K	Btu/h·ft·°R	m²/s	ft²/h	
380	684	0.5863	0.0366	2060	0.492	2.16 × 10⁻⁶	2.33 × 10⁻⁵	0.0246	0.0142	0.2036 × 10⁻⁴	0.789	1.060
400	720	0.5542	0.0346	2014	0.481	2.42	2.61	0.0261	0.0151	0.2338	0.906	1.040
450	810	0.4902	0.0306	1980	0.473	3.11	3.35	0.0299	0.0173	0.307	1.19	1.010
500	900	0.4005	0.0275	1985	0.474	3.86	4.16	0.0339	0.0196	0.387	1.50	0.996
550	990	0.4005	0.0250	1997	0.477	4.70	5.06	0.0379	0.0219	0.475	1.84	0.991
600	1080	0.3652	0.0228	2026	0.484	5.66	6.09	0.0422	0.0244	0.573	2.22	0.986
650	1170	0.3380	0.0211	2056	0.491	6.64	7.15	0.0464	0.0268	0.666	2.58	0.995
700	1260	13140	0.0196	2085	0.498	7.75	8.31	0.0505	0.0292	0.772	2.99	1.000
750	1350	0.2931	0.0183	2119	0.506	8.88	9.56	0.0549	0.0317	0.883	3.42	0.005
800	1440	0.2739	0.0171	2152	0.514	10.20	10.98	0.0592	0.0342	1.001	3.88	1.010
850	1530	0.2579	0.0161	2186	0.522	11.52	12.40	0.0637	0.0368	1.130	4.38	1.019

Source: Engineering Heat Transfer, 2nd edn., CRC Press, Table D.7, p. 659.

TABLE C.14

Thermodynamic Properties of Steam: Temperature Table (SI Units)

T_{sat} (°C)	P_{sat} (kPa)	Specific Volume (m³/kg)			Internal Energy (kJ/kg)			Enthalpy (kJ/kg)			Entropy (kJ/kg K)		
		v_f	v_{fg}	v_g	u_f	u_{fg}	u_g	h_f	h_{fg}	h_g	s_f	s_{fg}	s_g
0	0.61	0.001000	206.13	206.13	0.00	2373.9	2373.9	0.0	2500.0	2500.0	−0.0012	9.1590	9.1578
5	0.87	0.001000	147.20	147.20	21.04	2361.1	2382.1	21.0	2489.6	2510.6	0.0757	8.9510	9.0267
10	1.23	0.001000	106.36	106.36	42.02	2347.8	2389.8	42.0	2478.4	2520.4	0.1509	8.7511	8.9020
15	1.71	0.001001	78.036	78.037	62.95	2333.7	2396.7	63.0	2466.8	2529.7	0.2244	8.5582	8.7827
20	2.34	0.001002	57.801	57.802	83.86	2319.9	2403.7	83.9	2455.0	2538.9	0.2965	8.3718	8.6684
25	3.17	0.001003	43.446	43.447	104.75	2305.5	2410.3	104.8	2443.1	2547.9	0.3672	8,1919	8.5591
30	4.24	0.001004	32.907	32.908	125.63	2291.6	2417.2	125.6	2431.2	2556.8	0.4367	8.0180	8.4546
35	5.62	0.001006	25.250	25.251	146.50	2277.3	2423.8	146.5	2419.2	2565.7	0.5049	7.8496	8.3545
40	7.37	0.001008	19.536	19.537	167.37	2263.2	2430.6	167.4	2407.3	2574.6	0.5720	7.6864	8.2584
45	9.58	0.001010	15.262	15.263	188.24	2249.1	2437.3	188.3	2395.3	2583.5	0.6381	7.5281	8.1662
50	12.33	0.001012	12.046	12.047	209.12	2234.7	2443.8	209.1	2383.2	2592.3	0.7031	7.3745	8.0776
55	15.74	0.001014	9.5771	9.5781	230.01	2220.4	2450.4	230.0	2371.1	2601.1	0.7672	7.2253	7.9925
60	19.92	0.001017	7.6776	7.6786	250.91	2206.0	2456.9	250.9	2358.9	2609.8	0.8303	7.0804	7.9107
65	25.00	0.001020	6.1996	6.2006	271.83	2191.6	2463.4	271.9	2346.6	2618.4	0.8926	6.9394	7.8320
70	31.15	0.001023	5.0452	5.0462	292.76	2177.0	2469.7	292.8	2334.2	2626.9	0.9540	6.8023	7.7563
75	38.54	0.001026	4.1328	4.1338	313.70	2162.3	2476.0	313.7	2321.6	2635.4	1.0146	6.6687	7.6834
80	47.35	0.001029	3.4074	3.4085	334.67	2147.6	2482.3	334.7	2309.8	2643.7	1.0744	6.5387	7.6131
85	57.80	0.001032	2.8276	2.8286	355.65	2132.8	2488.4	355.7	2296.2	2651.9	1.1335	6.4118	7.5453
90	70.10	0.001036	2.3604	2.3614	376.66	2117.8	2494.5	376.7	2283.3	2660.0	1.1917	6.2881	7.4798
95	84.52	0.001039	1.9806	1.9817	397.69	2102.8	2500.5	397.8	2270.2	2668.0	1.2493	6.1673	7.4166
100	101.32	0.001043	1.6689	1.6699	418.75	2087.9	2506.6	418.9	2257.0	2675.8	1.3062	6.0492	7.3554
105	120.80	0.001047	1.4142	1.4152	439.83	2072.8	2512.6	440.0	2243.6	2683.6	1.3624	5.9338	7.2962
110	143.27	0.001051	1.2063	1.2074	460.95	2057.2	2518.2	461.1	2230.0	2691.1	1.4179	5.8209	7.2388
115	169.07	0.001056	1.0350	1.0361	482.10	2041.3	2523.4	482.3	2216.3	2698.6	1.4728	5.7105	7.1833
120	198.55	0.001060	0.89100	0.8921	503.28	2025.4	2528.7	503.5	2202.3	2705.8	1.5271	5.6023	7.1293
125	232.11	0.001065	0.76938	0.7704	524.51	2009.6	2534.1	524.8	2188.2	2712.9	1.5807	5.4962	7.0770
130	270.15	0.001070	0.66702	0.6681	545.78	1993.6	2539.4	546.1	2173.8	2719.9	1.6338	5.3922	7.0261
135	313.09	0.001075	0.58074	0.5818	567.09	1977.3	2544.4	567.4	2159.2	2726.6	1.6864	5.2902	6.9766
140	361.39	0.001080	0.50739	0.5085	588.46	1960.9	2549.3	588.8	2144.3	2733.1	1.7384	5.1900	6.9284
145	415.53	0.001085	0.44462	0.4457	609.88	1944.3	2554.2	610.3	2129.1	2739.4	1.7899	5.0916	6.8815
150	475.99	0.001091	0.39100	0.3921	631.35	1927.5	2558.8	631.9	2113.6	2745.5	1.8409	4.9948	6.8358
155	543.30	0.001096	0.34514	0.3462	652.89	1910.3	2563.2	653.5	2097.8	2751.3	1.8915	4.8996	6.7911
160	618.00	0.001102	0.30566	0.3068	674.50	1892.8	2567.3	675.2	2081.7	2756.9	1.9416	4.8059	6.7475
165	700.68	0.001108	0.27131	0.2724	696.18	1875.1	2571.3	697.0	2065.2	2762.2	1.9912	4.7135	6.7048
170	791.86	0.001114	0.24141	0.2425	717.93	1857.2	2575.2	718.8	2048.4	2767.2	2.0405	4.6224	6.6630
175	892.20	0.001121	0.21538	0.2165	739.77	1839.0	2578.8	740.8	2031.2	2772.0	2.0894	4.5325	6.6220
180	1,002.3	0.001127	0.19266	0.1938	761.69	1820.5	2582.1	762.8	2013.6	2776.4	2.1380	4.4437	6.5817
185	1,122.9	0.001134	0.17272	0.1739	783.70	1803.6	2585.3	785.0	1995.5	2780.5	2.1862	4.3559	6.5421
190	1,254.5	0.001141	0.15513	0.1563	805.80	1782.4	2588.3	807.2	1977.1	2784.3	2.2341	4.2691	6.5032
195	1,398.0	0.001148	0.13964	0.1408	828.01	1762.9	2590.9	829.6	1958.1	2787.8	2.2817	4.1834	6.4651
200	1,553.9	0.001156	0.12597	0.1271	850.32	1743.0	2593.3	852.1	1938.8	2790.9	2.3290	4.0986	6.4276
205	1,723.1	0.001164	0.11386	0.1150	872.74	1722.7	2595.4	874.7	1918.9	2793.6	2.3761	4.0147	6.3908
210	1,906.3	0.001172	0.10307	0.1042	895.28	1702.0	2597.3	897.5	1898.5	2796.0	2.4230	3.9314	6.3544
215	2,104.3	0.001180	0.09345	0.0946	917.94	1681.0	2598.9	920.4	1877.6	2798.0	2.4696	3.8485	6.3181
220	2,317.8	0.001189	0.08486	0.0860	940.73	1659.5	2600.2	943.5	1856.2	2799.7	2.5161	3.7661	6.2821
225	2,547.8	0.001198	0.07716	0.0784	963.66	1637.6	2601.3	966.7	1834.2	2800.9	2.5623	3.6841	6.2464
230	2,795.0	0.001208	0.07022	0.0714	986.73	1615.4	2602.2	990.1	1811.7	2801.8	2.6084	3.6025	6.2109
235	3,060.3	0.001218	0.06400	0.0652	1010.0	1592.7	2602.7	1033.7	1788.6	2802.3	2.6544	3.5213	6.1757

TABLE C.14 (continued)
Thermodynamic Properties of Steam: Temperature Table (SI Units)

T_{sat} (°C)	P_{sat} (kPa)	Specific Volume (m³/kg)			Internal Energy (kJ/kg)			Enthalpy (kJ/kg)			Entropy (kJ/kg K)		
		v_f	v_{fg}	v_g	u_f	u_{fg}	u_g	h_f	h_{fg}	h_g	s_f	s_{fg}	s_g
240	3,344.7	0.001228	0.05851	0.0597	1033.6	1569.1	2602.5	1037.5	1764.8	2802.3	2.7002	3.4404	6.1406
245	3,649.0	0.001239	0.05353	0.0548	1056.9	1545.2	2602.1	1061.4	1740.5	2801.9	2.7460	3.3597	6.1057
250	3,974.2	0.001250	0.04893	0.0502	1080.7	1521.0	2601.7	1085.6	1715.5	2801.2	2.7917	3.2792	6.0708
255	4,321.3	0.001262	0.04471	0.0460	1104.6	1496.7	2601.3	1110.1	1689.9	2800.0	2,8373	3.1986	6.0359
260	4,691.2	0.001275	0.04086	0.0421	1128.8	1471.9	2600.7	1134.8	1663.5	2798.3	2.8829	3.1180	6.0009
265	5,085.0	0.001288	0.03738	0.0387	1153.2	1446.4	2599.6	1159.8	1636.5	2796.3	2.9286	3.0372	5.9657
270	5,503.8	0.001302	0.03424	0.0355	1177.9	1420.3	2598.1	1185.1	1608.7	2793.7	2.9743	2.9560	5.9303
275	5,948.6	0.001317	0.03139	0.0327	1202.8	1393.4	2596.3	1210.7	1580.1	2790.8	3.0200	2.8745	5.8945
280	6,420.5	0.001333	0.02878	0.0301	1228.1	1366.0	2594.0	1236.6	1550.8	2787.4	3.0660	2.7924	5.8584
285	6,920.8	0.001349	0.02639	0.0277	1253.7	1337.9	2591.6	1263.0	1520.6	2783.6	3.1121	2.7097	5.8218
290	7,450.6	0.001366	0.02418	0.0255	1279.6	1297.7	2577.3	1289.8	1477.9	2767.7	3.1585	2.6262	5.7847
295	8,011.1	0.001385	0.02214	0.0235	1306.0	1265.5	2571.5	1317.1	1442.8	2759.9	3.2052	2.5417	5.7469
300	8,603.7	0.001404	0.02025	0.0217	1332.8	1232.0	2564.8	1344.9	1406.2	2751.1	3.2523	2.4560	5.7083
305	9,214.4	0.001425	0.01850	0.0199	1360.2	1197.5	2557.6	1373.3	1367.9	2741.2	3.3000	2.3688	5.6687
310	9,869.4	0.001447	0.01688	0.0183	1388.0	1161.1	2549.2	1402.3	1327.8	2730.1	3.3483	2.2797	5.6279
315	10,561.0	0.001470	0.01538	0.0169	1416.5	1123.0	2539.6	1432.1	1285.5	2717.6	3.3973	2.1884	5.5858
320	11,289.0	0.001499	0.01398	0.0155	1445.7	1083.0	2528.7	1462.6	1240.9	2703.5	3.4473	2.0947	5.5420
325	12,056.0	0.001528	0.01267	0.0142	1475.5	1040.9	2516.4	1494.0	1193.6	2687.3	3.4984	1.9979	5.4962
330	12,862.0	0.001561	0.01143	0.0130	1506.2	996.3	2502.5	1526.3	1143.3	2669.6	3.5507	1.8973	5.4480
335	13,712.0	0.001598	0.01026	0.0119	1537.8	949.0	2486.8	1559.7	1089.6	2649.3	3.6045	1.7922	5.3967
340	14,605.0	0.001639	0.00914	0.0108	1570.4	898.7	2469.0	1594.3	1032.2	2626.5	3.6601	1.6820	5.3420
345	15,545.0	0.001686	0.00808	0.0098	1606.3	842.1	2448.4	1632.5	967.7	2600.2	3.7176	1.5658	5.2834
350	16,535.0	0.001741	0.00706	0.0088	1643.0	780.5	2423.5	1671.8	897.2	2569.0	3.7775	1.4416	5.2191
355	17,577.0	0.001808	0.00605	0.0079	1682.1	710.9	2393.0	1713.9	817.3	2531.2	3.8400	1.3054	5.1454
360	18,675.0	0.001896	0.00504	0.0069	1726.2	629.5	2355.7	1761.6	723.7	2485.3	3.9056	1.1531	5.0587
365	19,833.0	0.002016	0.00400	0.0060	1777.9	531.0	2308.9	1817.8	610.3	2428.1	3.9746	0.9822	4.9569
370	21,054.0	0.002225	0.00274	0.0050	1843.3	394.1	2237.3	1890.1	451.9	2342.0	4.0476	0.7555	4.8030
374.4	22,090.0	0.00315	0.0	0.00315	2029.6	0.0	2029.6	2099.3	0.0	2099.3	4.4298	0.0	4.4298

Source: Properties obtained from software, *STEAMCALC*, John Wiley & Sons, New York, 1983; *Introduction to Thermal and Fluid Engineering*, CRC Press, Table A.3, p. 901.

TABLE C.15

Thermodynamic Properties of Steam: Pressure Table (SI Units)

P_{sat} (kPa)	T_{sat} (°C)	Specific Volume (m³/kg)			Internal Energy (kJ/kg)			Enthalpy (kJ/kg)			Entropy (kJ/kg K)		
		v_f	v_{fg}	v_g	u_f	u_{fg}	u_g	h_f	h_{fg}	h_g	s_f	s_{fg}	s_g
1.00	7.0	0.001000	129.08	129.08	29.40	2356.1	2385.5	29.4	2485.2	2514.6	0.1058	8.8704	8.9763
1.50	13.0	0.001001	88.067	88.068	54.68	2339.3	2394.0	54.7	2471.4	2526.3	0.1956	8.6337	8.8292
2.00	17.5	0.001001	67.073	67.074	73.41	2326.8	2400.2	73.4	2460.9	2534.3	0.2607	8.4642	8.7249
2.50	21.1	0.001002	54.290	54.291	88.41	2316.7	2405.1	88.4	2452.4	2540.8	0.3120	8.3322	8.6442
3.00	24.1	0.001003	45.751	45.752	100.96	2308.0	2409.0	101.0	2445.3	2546.2	0.3545	8.2240	8.5786
3.50	26.7	0.001003	39.483	39.484	111.81	2300.9	2412.7	111.8	2439.1	2550.9	0.3908	8.1324	8.5233
4.00	29.0	0.001004	34.779	34.780	121.37	2294.5	2415.9	121.4	2433.6	2555.0	0.4226	8.0529	8.4756
4.50	31.0	0.001005	31.128	31.129	129.95	2288.6	2418.6	130.0	2428.7	2558.7	0.4509	7.9827	8.4336
5.00	32.9	0.001005	28.194	28.195	137.73	2283.3	2421.0	137.7	2424.3	2562.0	0.4764	7.9197	8.3961
5.50	34.6	0.001006	25.773	25.774	144.86	2278.4	2423.3	144.9	2420.2	2565.0	0.4996	7.8626	8.3622
6.00	36.2	0.001006	23.742	23.743	151.45	2274.0	2425.4	151.5	2416.4	2567.9	0.5209	7.8104	8.3313
6.50	37.7	0.001007	22.013	22.014	157.58	2269.8	2427.4	157.6	2412.9	2570.5	0.5407	7.7623	8.3030
7.00	39.0	0.001007	20.522	20.523	163.31	2266.0	2429.3	163.3	2409.6	2572.9	0.5590	7.7177	8.2768
7.50	40.3	0.001008	19.225	19.226	168.70	2262.3	2431.0	168.7	2406.5	2575.2	0.5763	7.6762	8.2524
8.00	41.5	0.001008	18.086	18.087	173.79	2258.9	2432.7	173.8	2403.6	2577.4	0.5924	7.6372	8.2296
8.50	42.7	0.001009	17.080	17.081	178.61	2255.6	2434.2	178.6	2400.8	2579.4	0.6077	7.6006	8.2083
9.00	43.8	0.001009	16.185	16.186	183.19	2252.5	2435.7	183.2	2398.2	2581.4	0.6222	7.5660	8.1881
9.50	44.8	0.001010	15.383	15.384	187.56	2249.5	2437.1	187.6	2395.7	2583.2	0.6359	7.5332	8.1691
10.00	45.8	0.001010	14.660	14.661	191.74	2246.7	2438.4	191.7	2393.3	2585.0	0.6490	7.5021	8.1511
15.00	54.0	0.001014	10.020	10.021	225.83	2223.2	2449.0	225.8	2373.5	2599.4	0.7544	7.2548	8.0092
20.00	60.1	0.001017	7.6483	7.6493	251.28	2205.7	2457.0	251.3	2358.7	2610.0	0.8314	7.0779	7.9093
25.00	65.0	0.001020	6.2015	6.2025	271.80	2191.6	2463.3	271.8	2346.6	2618.4	0.8925	6.9396	7.8321
30.00	69.1	0.001022	5.2277	5.2287	289.09	2379.5	2468.6	289.1	2336.3	2625.5	0.9433	6.8260	7.7693
35.00	72.7	0.001024	4.5249	4.5259	304.11	2169.0	2473.1	304.1	2327.4	2631.5	0.9869	6.7295	7.7164
40.00	75.9	0.001026	3.9918	3.9929	317.42	2159.7	2477.1	317.5	2319.4	2636.8	1.0253	6.6455	7.6707
45.00	78.7	0.001028	3.5744	3.5755	329.40	2151.3	2480.7	329.4	2312.2	2641.6	1.0595	6.5710	7.6305
50.00	81.3	0.001030	3.2389	3.2398	340.31	2143.6	2483.9	340.4	2305.5	2645.9	1.0904	6.5042	7.5946
60.00	86.0	0.001033	2.7305	2.7316	359.66	2129.9	2489.6	359.7	2293.7	2653.5	1.1446	6.3880	7.5326
70.00	90.0	0.001036	2.3638	2.3648	376.49	2117.9	2494.4	376.6	2283.4	2659.9	1.1913	6.2891	7.4804
80.00	93.5	0.001038	2.0859	2.0869	391.43	2107.2	2498.7	391.5	2274.1	2665.6	1.2323	6.2029	7.4352
90.00	96.7	0.001041	1.8667	1.8678	404.90	2097.7	2502.6	405.0	2265.7	2670.7	1.2689	6.1265	7.3954
100.00	99.6	0.001043	1.6898	1.6908	417.20	2089.0	2506.2	417.3	2258.0	2675.3	1.3020	6.0578	7.3598
101.32	100.0	0.001043	1.66895	1.6700	418.74	2087.9	2506.6	418.8	2257.0	2675.8	1.3062	6.0493	7.3554
125.00	106.0	0.001048	1.36965	1.3707	444.01	2069.7	2513.7	444.1	2240.9	2685.1	1.3734	5.9113	7.2847
150.00	111.4	0.001053	1.15612	1.1572	466.74	2052.9	2519.6	466.9	2226.3	2693.2	1.4330	5.7904	7.2234
175.00	116.1	0.001057	1.00248	1.0035	486.58	2037.9	2524.5	486.8	2213.4	2700.1	1.4844	5.6873	7.1717
200.00	120.2	0.001060	0.88498	0.8860	504.25	2024.7	2529.0	504.5	2201.7	2706.2	1.5295	5.5974	7.1269
225.00	124.0	0.001064	0.79229	0.7934	520.22	2012.8	2533.0	520.5	2191.1	2711.5	1.5700	5.5175	7.0874
250.00	127.4	0.001067	0.71751	0.7186	534.82	2001.9	2536.7	535.1	2181.2	2716.3	1.6066	5.4455	7.0521
275.00	130.6	0.001070	0.65602	0.6571	548.30	1991.7	2540.0	548.6	2172.1	2720.7	1.6401	5.3800	7.0201
300.00	133.5	0.001073	0.60457	0.6056	560.83	1982.1	2542.9	561.2	2163.5	2724.6	1.6710	5.3199	6.9910
325.00	136.3	0.001076	0.56082	0.5619	572.57	1973.1	2545.7	572.9	2155.4	2728.3	1.6998	5.2643	6.9641
350.00	138.9	0.001079	0.52305	0.5241	583.60	1964.6	2548.2	584.0	2147.7	2731.6	1.7266	5.2126	6.9392
375.00	141.3	0.001081	0.49007	0.4911	594.03	1956.6	2550.6	594.4	2140.3	2734.8	1.7519	5.1642	6.9161
400.00	143.6	0.001084	0.46105	0.4621	603.93	1948.9	2552.8	604.4	2133.3	2737.7	1.7757	5.1187	6.8944
425.00	145.8	0.001086	0.43534	0.4364	613.36	1941.6	2554.9	613.8	2126.6	2740.4	1.7982	5.0758	6.8740
450.00	147.9	0.001088	0.41242	0.4135	622.35	1934.5	2556.9	622.8	2120.1	2743.0	1.8196	5.0352	6.8548
475.00	149.9	0.001091	0.39188	0.3930	630.97	1927.7	2558.7	631.5	2113.9	2745.4	1.8400	4.9966	6.8366
500.00	151.8	0.001093	0.37336	0.3745	639.24	1921.2	2560.4	639.8	2107.9	2747.6	1.8595	4.9598	6.8193

TABLE C.15 (continued)

Thermodynamic Properties of Steam: Pressure Table (SI Units)

P_{sat} (kPa)	T_{sat} (°C)	Specific Volume (m³/kg)			Internal Energy (kJ/kg)			Enthalpy (kJ/kg)			Entropy (kJ/kg K)		
		v_f	v_{fg}	v_g	u_f	u_{fg}	u_g	h_f	h_{fg}	h_g	s_f	s_{fg}	s_g
550.00	155.5	0.001097	0.34129	0.3424	654.86	1908.7	2563.5	655.5	2096.4	2751.8	1.8960	4.8910	6.7871
600.00	158.8	0.001101	0.31443	0.3155	669.42	1896.9	2566.3	670.1	2085.5	2755.6	1.9298	4.8278	6.7576
650.00	162.0	0.001104	0.29151	0.2926	683.06	1885.8	2568.8	683.8	2075.2	2759.0	1.9613	4.7692	6.7305
700.00	164.9	0.001108	0.27168	0.2728	695.93	1875.3	2571.2	696.7	2065.4	2762.1	1.9907	4.7146	6.7053
750.00	167.7	0.001111	0.25439	0.2555	708.11	1865.3	2573.4	708.9	2056.0	2765.0	2.0183	4.6634	6.6817
800.00	170.4	0.001115	0.23919	0.2403	719.68	1855.7	2575.4	720.6	2047.0	2767.6	2.0445	4.6152	6.6597
850.00	172.9	0.001118	0.22572	0.2268	730.71	1846.5	2577.2	731.7	2038.4	2770.0	2.0692	4.5696	6.6388
900.00	175.3	0.001121	0.21372	0.2148	741.27	1837.7	2578.9	742.3	20300	2.772.3	2.0928	4.5264	6.6192
950.00	177.7	0.001124	0.20295	0.2041	751.39	1829.1	2580.5	752.5	2021.9	2774.3	2.1152	4.4853	6.6005
1,000	179.9	0.001127	0.19322	0.1943	761.11	1820.8	2581.9	762.2	2014.0	2776.3	2.1367	4.4460	6.5827
1,100	184.1	0.001133	0.17631	0.1774	779.52	1805.1	2584.6	780.8	1999.0	2779.8	2.1771	4.3725	6.5495
1,200	187.9	0.001138	0.16209	0.1632	796.71	1790.2	2586.9	798.1	1984.7	2782.8	2.2145	4.3047	6.5191
1,300	191.6	0.001143	0.14998	0.1511	812.87	1776.1	2589.0	814.4	1971.1	2785.4	2.2493	4.2417	6.4910
1,400	195.0	0.001148	0.13956	0.1407	828.13	1762.7	2590.8	829.7	1958.0	2787.8	2.2820	4.1829	6.4649
1,500	198.3	0.001153	0.13050	0.1317	842.61	1749.8	2592.4	844.3	1945.5	2789.8	2.3127	4.1277	6.4405
1,600	201.4	0.001158	0.12254	0.1237	856.39	1737.3	2593.7	858.2	1933.4	2791.7	2.3418	4.0757	6.4176
1,700	204.3	0.001163	0.11549	0.1167	869.56	1725.4	2595.0	871.5	1921.7	2793.3	2.3695	4.0265	6.3960
1,800	207.1	0.001167	0.10918	0.1104	882.18	1713.9	2596.1	884.3	1910.4	2794.7	2.3958	3.9797	6.3755
1,900	209.8	0.001172	0.10351	0.1047	894.30	1702.7	2597.0	896.5	1899.4	2795.9	2.4210	3.9350	6.3559
2,000	212.4	0.001176	0.09839	0.0996	905.97	1691.9	2597.9	908.3	1888.7	2797.0	2.4450	3.8922	6.3372
2,250	218.4	0.001186	0.08751	0.0887	933.41	1666.2	2599.6	936.1	1863.1	2799.2	2.5012	3.7925	6.2936
2,500	223.9	0.001196	0.07873	0.0799	958.75	1642.1	2600.9	961.7	1839.0	2800.7	2.5525	3.7015	6.2540
2,750	229.1	0.001206	0.07147	0.0727	982.37	1619.4	2601.8	985.7	1816.0	28017	2.5997	3.6178	6.2176
3,000	233.8	0.001215	0.06538	0.0666	1004.5	1597.9	2602.4	1008.2	1794.0	2802.2	2.6437	3.5402	6.1839
3,250	238.3	0.001225	0.06028	0.0615	1025.4	1577.0	2602.5	1029.4	1772.9	2802.3	2.6848	3.4676	6.1524
3,500	242.5	0.001234	0.05594	0.0572	1045.3	1556.8	2602.1	1049.6	1752.6	2802.2	2.7234	3.3995	6.1229
3,750	246.5	0.001242	0.05208	0.0533	1064.2	1537.6	2601.8	1068.8	1732.9	2801.7	2.7600	3.3350	6.0950
4,000	250.3	0.001251	0.04864	0.0499	1082.2	1519.3	2601.5	1087.2	1713.9	2801.1	2.7947	3.2739	6.0686
5,000	263.9	0.001285	0.03811	0.0394	1147.9	1451.9	2599.8	1154.3	1642.4	2796.7	2.9186	3.0548	5.9734
6,000	275.5	0.001319	0.03109	0.0324	1205.6	1390.4	2596.0	1213.5	1576.9	2790.5	3.0251	2.8655	5.8906
7,000	285.8	0.001352	0.02603	0.0274	1257.8	1333.5	2591.3	1267.2	1515.7	2782.9	3.1194	2.6965	5.8159
8,000	295.0	0.001385	0.02214	0.0235	1305.9	1265.8	2571.7	1317.0	1442.9	2759.9	3.2050	2.5421	5.7471
9,000	303.3	0.001418	0.01907	0.0205	1351.0	1209.3	2560.3	1363.7	1380.9	2744.7	3.2840	2.3981	5.6821
10,000	311.0	0.001452	0.01658	0.0180	1393.7	1153.8	2547.5	1408.2	1319.5	2727.7	3.3580	2.2616	5.6196
11,000	318.1	0.001489	0.01450	0.0160	1434.6	1098.6	2533.1	1451.0	1258.0	2709.0	3.4283	2.1305	5.5588
12,000	324.7	0.001527	0.01273	0.0143	1474.0	1043.3	2517.3	1492.0	1196.0	2688.4	3.4958	2.0028	5.4986
13,000	331.0	0.001568	0.01119	0.0128	1512.4	987.6	2499.9	1532.8	1133.1	2665.8	3.5611	1.8771	5.4382
14,000	336.9	0.001612	0.00984	0.0114	1549.8	931.1	2480.9	1572.4	1068.8	2641.2	3.6249	1.7520	5.3769
15,000	342.4	0.001661	0.00862	0.0103	1586.6	873.4	2460.0	1611.5	1002.8	2614.3	3.6877	1.6265	5.3142
16,000	347.7	0.001715	0.00752	0.0092	1626.1	810.1	2436.2	1653.6	930.4	2584.0	3.7498	1.4996	5.2494
17,000	352.3	0.001769	0.00660	0.0084	1660.2	750.3	2410.5	1690.3	862.5	2552.8	3.8054	1.3819	5.1872
18,000	357.0	0.001839	0.00566	0.0075	1698.6	680.8	2379.4	1731.7	782.6	2514.3	3.8652	1.2481	5.1134
19,000	361.4	0.001926	0.00475	0.0067	1740.1	603.6	2343.6	1776.7	693.9	2470.5	3.9249	1.1063	5.0312
20,000	365.7	0.002037	0.00384	0.0059	1785.8	515.1	2300.9	1826.6	591.9	2418.5	3.9846	0.9568	4.9414
21,000	369.8	0.002208	0.00281	0.0050	1839.7	401.7	2241.4	1886.0	460.8	2346.8	4.0443	0.7681	4.8124
22,000	373.7	0.002623	0.00114	0.0038	1944.6	174.0	2118.6	2002.3	199.0	2201.3	4.1042	0.4563	4.5605
22,090	374.4	0.00315	0.0	0.00315	2029.6	0.0	2029.6	2099.3	0.0	2099.3	4.4298	0.0	4.4298

Source: Properties obtained from software, STEAMCALC, John Wiley & Sons, New York, 1983; *Introduction to Thermal and Fluid Engineering*, CRC Press, Table A.4, p. 904.

TABLE C.16

Thermodynamic Properties of Steam: Superheated Vapor Table (SI Units)

P (kPa)	T (°C)	v (m³/kg)	u (kJ/kg)	h (kJ/kg)	s (kJ/kg k)
10 (T_{sat} = 45.8°C)					
	100	17.196	2516.2	2688.1	8.4498
	150	19.513	2588.2	2783.3	8.6893
	200	21.826	2661.2	2879.5	8.9040
	250	24.136	2735.5	2976.8	9.0996
	300	26.446	2811.2	3075.6	9.2799
	350	28.754	2888.3	3175.9	9.4476
	400	31.063	2967.1	3277.7	9.6048
	450	33.371	3047.4	3381.1	9.7530
	500	35.679	3129.4	3486.2	9.8935
	550	37.987	3213.1	3593.0	10.027
	600	40.295	3298.5	3701.5	10.155
	650	42.603	3385.7	3811.7	10.278
	700	44.911	3474.5	3923.7	10.396
	750	47.219	3565.2	4037.4	10.510
	800	49.526	3657.6	4152.9	10.620
	850	51.834	3751.8	4270.2	10.727
50 (T_{sat} = 81.3°C)					
	100	3.4182	2512.0	2682.9	7.6959
	150	3.8894	2586.0	2780.5	7.9413
	200	4.3561	2659.8	2877.6	8.1583
	250	4.8206	2734.5	2975.6	8.3551
	300	5.2840	2810.5	3074.7	8.5360
	350	5.7468	2887.8	3175.1	8.7040
	400	6.2092	2966.6	3277.1	8.8614
	450	6.6715	3047.1	3380.6	9.0098
	500	7.1336	3129.1	3485.8	9.1504
	550	7.5956	3212.9	3592.6	9.2843
	600	8.0575	3298.3	3701.2	9.4123
	650	8.5193	3385.5	3811.4	9.5351
	700	8.9811	3474.4	3923.4	9.6532
	750	9.4428	3565.0	4037.2	9.7672
	800	9.9045	3657.5	4152.7	9.8775
	850	10.366	3751.7	4270.0	9.9843
100 (T_{sat} = 99.6°C)					
	100	1.6956	2506.4	2676.0	7.3610
	150	1.9363	2583.1	2776.8	7.6146
	200	2.1724	2658.1	2875.3	7.8347
	250	2.4062	2733.3	2974.0	8.0329
	300	2.6388	2809.6	3073.5	8.2146
	350	2.8708	2887.1	3174.2	8.3831
	400	3.1025	2966.1	3276.4	8.5407
	450	3.3340	3046.6	3380.0	8.6893
	500	3.5654	3128.8	3485.3	8.8300
	550	3.7966	3212.5	3592.2	8.9640
	600	4.0277	3298.0	3700.8	9.0921
	650	4.2588	3385.2	3811.1	9.2149
	700	4.4898	3474.1	3923.1	9.3331

TABLE C.16 (continued)

Thermodynamic Properties of Steam: Superheated Vapor Table (SI Units)

P (kPa)	T (°C)	v (m³/kg)	u (kJ/kg)	h (kJ/kg)	s (kJ/kg k)
	750	4.7208	3564.8	4036.9	9.4471
	800	4.9518	3657.3	4152.4	9.5574
	850	5.1827	3751.5	4269.8	9.6643
101.32 (T_{sat} = 100.0°C)					
	150	1.9109	2583.1	2776.7	7.6084
	200	2.1439	2658.0	2875.3	7.8286
	250	2.3747	2733.3	2973.9	8.0268
	300	2.6043	2809.6	3073.4	8.2085
	350	2.8334	2887.1	3174.2	8.3770
	400	3.0621	2966.1	3276.3	8.5347
	450	3.2905	3046.6	3380.0	8.6832
	500	3.5189	3128.7	3485.3	8.8240
	550	3.7471	3212.5	3592.2	8.9580
	600	3.9752	3298.0	3700.8	9.0860
	650	4.2033	3385.2	3811.1	9.2089
	700	4.4313	3474.1	3923.1	9.3271
	750	4.6593	3564.8	4036.9	9.4411
	800	4.8872	3657.3	4152.4	9.5513
	850	5.1152	3751.5	4269.8	9.6582
200 (T_{sat} = 120.2°C)					
	150	0.9596	2577.2	2769.1	7.2804
	200	1.0804	2654.5	2870.6	7.5072
	250	1.1989	2730.9	2970.7	7.7084
	300	1.3162	2807.8	3071.1	7.8916
	350	1.4329	2885.8	3172.4	8.0610
	400	1.5492	2965.0	3274.9	8.2192
	450	1.6653	3045.7	3378.8	8.3682
	500	1.7813	3128.0	3484.3	8.5092
	550	1.8971	3211.9	3591.3	8.6434
	600	2.0129	3297.4	3700.0	8.7716
	650	2.1286	3384.7	3810.4	8.8945
	700	2.2442	3473.7	3922.5	9.0128
	750	2.3598	3564.4	4036.4	9.1268
	800	2.4754	3656.9	4152.0	9.2372
	850	2.5909	3751.1	4269.3	9.3441
300 (T_{sat} = 133.5°C)					
	150	0.6338	2570.8	2760.9	7.0779
	200	0.7164	2650.8	2865.7	7.3122
	250	0.7965	2728.5	2967.4	7.5165
	300	0.8753	2806.1	3068.7	7.7014
	350	0.9535	2884.4	3170.5	7.8717
	400	1.0314	2963.9	3273.4	8.0305
	450	1.1091	3044.8	3377.6	8.1798
	500	1.1866	3127.2	3483.2	8.3211
	550	1.2639	3211.2	3590.4	8.4554
	600	1.3413	3296.9	3699.2	8.5838

TABLE C.16 (continued)
Thermodynamic Properties of Steam: Superheated Vapor Table (SI Units)

P (kPa)	T (°C)	v (m³/kg)	u (kJ/kg)	h (kJ/kg)	s (kJ/kg k)
	650	1.4185	3384.2	3809.7	8.7068
	700	1.4957	3473.2	3921.9	8.8252
	750	1.5728	3564.0	4035.8	8.9393
	800	1.6499	3656.5	4151.5	9.0497
	850	1.7270	3750.8	4268.9	9.1566
400 (T_{sat} = 143.6°C)					
	150	0.4707	2563.9	2752.2	6.9287
	200	0.5343	2647.0	2860.7	7.1712
	250	0.5952	2726.0	2964.1	7.3789
	300	0.6549	2804.3	3066.2	7.5655
	350	0.7139	2883.1	3168.6	7.7367
	400	0.7725	2962.9	3271.9	7.8961
	450	0.8309	3044.0	3376.3	8.0458
	500	0.8892	3126.5	3482.2	8.1873
	550	0.9474	3210.6	3589.5	8.3219
	600	1.0054	3296.3	3698.5	8.4504
	650	1.0635	3383.7	3809.0	8.5735
	700	1.1214	3472.7	3921.3	8.6919
	750	1.1793	3563.5	4035.3	8.8062
	800	1.2372	3656.1	4151.0	8.9166
	850	1.2951	3750.4	4268.4	9.0236
600 (T_{sat} = 158.8°C)					
	200	0.3521	2639.0	2850.2	6.9669
	250	0.3939	2720.8	2957.2	7.1819
	300	0.4344	2800.6	3061.3	7.3719
	350	0.4742	2880.3	3164.8	7.5451
	400	0.5136	2960.7	3268.9	7.7057
	450	0.5528	3042.2	3373.8	7.8562
	500	0.5919	3125.0	3480.1	7.9982
	550	0.6308	3209.3	3587.7	8.1332
	600	0.6696	3295.1	3696.9	8.2619
	650	0.7084	3382.6	3807.7	8.3853
	700	0.7471	3471.8	3920.1	8.5039
	750	0.7858	3562.7	4034.2	8.6182
	800	0.8245	3655.3	4150.0	8.7287
	850	0.8631	3749.7	4267.6	8.8358
800 (T_{sat} = 170.4°C)					
	200	0.2608	2630.4	2839.1	6.8156
	250	0.2932	2715.5	2950.1	7.0388
	300	0.3241	2796.9	3056,2	7.2326
	350	0.3544	2877.5	3161.0	7.4079
	400	0.3842	2958.5	3265.8	7.5697
	450	0.4137	3040.4	3371.4	7.7209
	500	0.4432	3123.5	3478.0	7.8635
	550	0.4725	3208.0	3586.0	7.9988
	600	0.5017	3294.0	3695.4	8.1278
	650	0.5309	3381.6	3806.3	8.2514
	700	0.5600	3470.9	3918.9	8.3702

TABLE C.16 (continued)
Thermodynamic Properties of Steam: Superheated Vapor Table (SI Units)

P (kPa)	T (°C)	v (m³/kg)	u (kJ/kg)	h (kJ/kg)	s (kJ/kg k)
	750	0.5891	3561.9	4033.1	8.4846
	800	0.6181	3654.6	4149.1	8.5953
	850	0.6471	3749.0	4266.7	8.7024
1000 (T_{sat} = 179.9°C)					
	200	0.2059	2621.4	2827.3	6.6930
	250	0.2328	2710.0	2942.8	6.9251
	300	0.2580	2793.1	3051.1	7.1229
	350	0.2824	2874.7	3157.1	7.3003
	400	0.3065	2956.3	3262.7	7.4633
	450	0.3303	3038.5	3368.8	7.6154
	500	0.3540	3121.9	3475.9	7.7585
	550	0.3775	3206.6	3584.2	7.8942
	600	0.4010	3292.8	3693.8	8.0235
	650	0.4244	3380.6	3805.0	8.1473
	700	0.4477	3470.0	3917.7	8.2662
	750	0.4710	3561.0	4032.0	8.3808
	800	0.4943	3653.8	4148.1	8.4916
	850	0.5175	3748.3	4265.8	8.5988
1500 (T_{sat} = 198.3°C)					
	250	0.15199	2695.4	2923.4	6.7093
	300	0.16971	2783.3	3037.8	6.9183
	350	0.18654	2867.4	3147.2	7.1014
	400	0.20292	2950.6	3255.0	7.2677
	450	0.21906	3034.0	3362.5	7.4219
	500	0.23503	3118.1	3470.6	7.5664
	550	0.25089	3203.3	3579.7	7.7030
	600	0.26666	3289.9	3689.9	7.8331
	650	0.28237	3378.0	3801.5	7.9574
	700	0.29803	3467.6	3914.7	8.0767
	750	0.31364	3558.9	4029.4	8.1917
	800	0.32921	3651.8	4145.7	8.3027
	850	0.34475	3746.5	4263.6	8.4101
2000 (T_{sat} = 212.4°C)					
	250	0.11145	2679.5	2902.4	6.5451
	300	0.12550	2772.9	3023.9	6.7671
	350	0.13856	2860.0	3137.1	6.9565
	400	0.15113	2944.8	3247.1	7.1263
	450	0.16343	3029.3	3356.1	7.2826
	500	0.17556	3114.2	3465.3	7.4286
	550	0.18757	3200.0	3575.1	7.5662
	600	0.19950	3287.0	3686.0	7.6970
	650	0.21137	3375.4	3798.1	7.8218
	700	0.22318	3465.3	3911.6	7.9416
	750	0.23494	3556.8	4026.7	8.0569
	800	0.24667	3649.9	4143.2	8.1681
	850	0.25836	3744.7	4261.5	8.2758

(continued)

TABLE C.16 (continued)
Thermodynamic Properties of Steam: Superheated Vapor Table (SI Units)

P (kPa)	T (°C)	v (m³/kg)	u (kJ/kg)	h (kJ/kg)	s (kJ/kg k)
2500 (T_{sat} = 223.9°C)					
	250	0.08699	2662.2	2879.7	6.4076
	300	0.09893	2762.0	3009.3	6.6446
	350	0.10975	2852.2	3126.6	6.8409
	400	0.12004	2938.9	3239.1	7.0145
	450	0.13005	3024.6	3349.7	7.1730
	500	0.13987	3110.3	3459.9	7.3205
	550	0.14958	3196.6	3570.6	7.4592
	600	0.15921	3284.1	3682.1	7.5906
	650	0.16876	3372.8	3794.7	7.7161
	700	0.17827	3462.9	3908.6	7.8362
	750	0.18772	3554.6	4024.0	7.9518
	800	0.19714	3648.0	4140.8	8.0633
	850	0.20653	3742.9	4259.3	8.1712
3000 (T_{sat} = 233.8°C)					
	250	0.07055	2643.2	2854.9	6.2855
	300	0.08116	2750.6	2994.1	6.5399
	350	0.09053	2844.3	3115.9	6.7437
	400	0.09931	2932.9	3230.9	6.9213
	450	0.10779	3019.8	3343.1	7.0822
	500	0.11608	3106.3	3454.5	7.2312
	550	0.12426	3193.2	3566.0	7.3709
	600	0.13234	3281.1	3678.1	7.5031
	650	0.14036	3370.2	3791.3	7.6291
	700	0.14832	3460.6	3905.6	7.7497
	750	0.15624	3552.5	4021.2	7.8656
	800	0.16412	3646.0	4138.4	7.9774
	850	0.17197	3741.1	4257.1	8.0855
4000 (T_{sat} = 250.3°C)					
	300	0.058835	2725.8	2961.2	6.3622
	350	0.066448	2827.6	3093.4	6.5835
	400	0.073377	2920.6	3214.1	6.7699
	450	0.079959	3010.0	3329.8	6.9358
	500	0.086343	3098.2	3443.6	7.0879
	550	0.092599	3186.4	3556.8	7.2298
	600	0.098764	3275.2	3670.2	7.3636
	650	0.10486	3364.9	3784.3	7.4907
	700	0.11090	3455.9	3899.5	7.6121
	750	0.11690	3548.2	4015.8	7.7287
	800	0.12285	3642.1	4133.5	7.8411
	300	0.045302	2698.2	2924.7	6.2085
	350	0.051943	2809.9	3069.6	6.4512
	400	0.057792	2907.7	3196.6	6.6474
	850	0.12878	3737.6	4252.7	7.9496

P (kPa)	T (°C)	v (m³/kg)	u (kJ/kg)	h (kJ/kg)	s (kJ/kg k)
5000 (T_{sat} = 263.9°C)					
	300	0.045302	2698.2	2924.7	6.2085
	350	0.051943	2809.9	3069.6	6.4512
	400	0.057792	2907.7	3196.6	6.6474
	450	0.063252	3000.0	3316.2	6.8188
	500	0.068495	3090.0	3432.5	6.9743
	550	0.073603	3179.5	3547.5	7.1184
	600	0.078617	3269.2	3662.3	7.2538
	650	0.083560	3359.6	3777.4	7.3820
	700	0.088447	3451.1	3893.4	7.5044
	750	0.093289	3543.9	4010.4	7.6216
	800	0.098094	3638.2	4128.7	7.7345
	850	0.102867	3734.0	4248.3	7.8435
6000 (T_{sat} = 275.6°C)					
	300	0.036146	2667.1	2884.0	6.0669
	350	0.042223	2790.9	3044.2	6.3354
	400	0.047380	2894.3	3178.6	6.5429
	450	0.052104	2989.7	3302.3	6.7202
	500	0.056592	3081.7	3421.3	6.8793
	550	0.060937	3172.5	3538.1	7.0258
	600	0.065185	3263.1	3654.2	7.1627
	650	0.069360	3354.3	3770.4	7.2922
	700	0.073479	3446.4	3887.2	7.4154
	750	0.077552	3539.6	4004.9	7.5333
	800	0.081588	3634.3	4123.8	7.6467
	850	0.085592	3730.4	4243.9	7.7562
7000 (T_{sat} = 285.8°C)					
	300	0.029459	2631.9	2838.1	5.9299
	350	0.035234	2770.6	3017.2	6.2301
	400	0.039922	2880.4	3159.8	6.4504
	450	0.044132	2979.1	3288.1	6.6343
	500	0.048087	3073.2	3409.8	6.7971
	550	0.051890	3165.4	3528.6	6.9460
	600	0.055591	3257.0	3646.1	7.0846
	650	0.059218	3348.9	3763.4	7.2153
	700	0.062788	3441.6	3881.1	7.3394
	750	0.066312	3535.3	3999.5	7.4580
	800	0.069798	3630.3	4118.9	7.5720
	850	0.073253	3726.7	4239.5	7.6818

Source: Introduction to Thermal and Fluid Engineering, CRC Press, Table A.5, p. 907.

TABLE C.17

Thermodynamic Properties of Steam: Temperature Table (English Units)

		Specific Volume (ft³/lbm)			Internal Energy (Btu/lbm)			Enthalpy (Btu/lbm)			Entropy (Btu/lbm · °R)		
Temp., T (°F)	Sat. Press., P_{sat} (psia)	Sat. Liquid, v_f	Evap., v_{fg}	Sat. Vapor, v_g	Sat. Liquid, u_f	Evap., u_{fg}	Sat. Vapor, u_g	Sat. Liquid, h_f	Evap., h_{fg}	Sat. Vapor, h_g	Sat. Liquid, s_f	Evap., s_{fg}	Sat. Vapor, s_g
32.018	0.08871	0.01602	3299.88	3299.9	0.000	1021.0	1021.0	0.000	1075.2	1075.2	0.00000	2.18672	2.1867
35	0.09998	0.01602	2945.68	2945.7	3.004	1019.0	1022.0	3.004	1073.5	1076.5	0.00609	2.17011	2.1762
40	0.12173	0.01602	2443.58	2443.6	8.032	1015.6	1023.7	8.032	1070.7	1078.7	0.01620	2.14271	2.1589
45	0.14756	0.01602	2035.78	2035.8	13.05	1012.2	1025.3	13.05	1067.8	1080.9	0.02620	2.11587	2.1421
50	0.17812	0.01602	1703.08	1703.1	18.07	1008.9	1026.9	18.07	1065.0	1083.1	0.03609	2.08956	2.1256
55	0.21413	0.01603	1430.38	1430.4	23.07	1005.5	1028.6	23.07	1062.2	1085.3	0.04586	2.06377	2.1096
60	0.25638	0.01604	1206.08	1206.1	28.08	1002.1	1030.2	28.08	1059.4	1087.4	0.05554	2.03847	2.0940
65	0.30578	0.01604	1020.78	1020.8	33.08	998.76	1031.8	33.08	1056.5	1089.6	0.06511	2.01366	2.0788
70	0.36334	0.01605	867.16	867.18	38.08	995.39	1033.5	38.08	1053.7	1091.8	0.07459	1.98931	2.0639
75	0.43016	0.01606	739.25	739.27	43.07	992.02	1035.1	43.07	1050.9	1093.9	0.08398	1.96541	2.0494
80	0.50745	0.01607	632.39	632.41	48.06	988.65	1036.7	48.07	1048.0	1096.1	0.09328	1.94196	2.0352
85	0.59659	0.01609	542.78	542.80	53.06	985.28	1038.3	53.06	1045.2	1098.3	0.10248	1.91892	2.0214
90	0.69904	0.01610	467.38	467.40	58.05	981.90	1040.0	58.05	1042.4	1100.4	0.11161	1.89630	2.0079
95	0.81643	0.01612	403.72	403.74	63.04	978.52	1041.6	63.04	1039.5	1102.6	0.12065	1.87408	1.9947
100	0.95052	0.01613	349.81	349.83	68.03	975.14	1043.2	68.03	1036.7	1104.7	0.12961	1.85225	1.9819
110	1.2767	0.01617	264.94	264.96	78.01	968.36	1046.4	78.02	1031.0	1109.0	0.14728	1.80970	1.9570
120	1.6951	0.01620	202.92	202.94	88.00	961.56	1049.6	88.00	1025.2	1113.2	0.16466	1.76856	1.9332
130	2.2260	0.01625	157.07	157.09	97.99	954.73	1052.7	97.99	1019.4	1117.4	0.18174	1.72877	1.9105
140	2.8931	0.01629	122.79	122.81	107.98	947.87	1055.9	107.99	1013.6	1121.6	0.19855	1.69024	1.8888
150	3.7234	0.01634	96.91	96.929	117.98	940.98	1059.0	117.99	1007.8	1125.7	0.21508	1.65291	1.8680
160	4.7474	0.01639	77.17	77.185	127.98	934.05	1062.0	128.00	1001.8	1129.8	0.23136	1.61670	1.8481
170	5.9999	0.01645	61.97	61.982	138.00	927.08	1065.1	138.02	995.88	1133.9	0.24739	1.58155	1.8289
180	7.5197	0.01651	50.16	50.172	148.02	920.06	1068.1	148.04	989.85	1137.9	0.26318	1.54741	1.8106
190	9.3497	0.01657	40.90	40.920	158.05	912.99	1071.0	158.08	983.76	1141.8	0.27874	1.51421	1.7930
200	11.538	0.01663	33.60	33.613	168.10	905.87	1074.0	168.13	977.60	1145.7	0.29409	1.48191	1.7760
210	14.136	0.01670	27.78	27.798	178.15	898.68	1076.8	178.20	971.35	1149.5	0.30922	1.45046	1.7597
212	14.709	0.01671	26.77	26.782	180.16	897.24	1077.4	180.21	970.09	1150.3	0.31222	1.44427	1.7565
220	17.201	0.01677	23.12	23.136	188.22	891.43	1079.6	188.28	965.02	1153.3	0.32414	1.41980	1.7439
230	20.795	0.01684	19.36	19.374	198.31	884.10	1082.4	198.37	958.59	1157.0	0.33887	1.38989	1.7288
240	24.985	0.01692	16.30	16.316	208.41	876.70	1085.1	208.49	952.06	1160.5	0.35342	1.36069	1.7141
250	29.844	0.01700	13.80	13.816	218.54	869.21	1087.7	218.63	945.41	1164.0	0.36779	1.33216	1.6999
260	35.447	0.01708	11.74	11.760	228.68	861.62	1090.3	228.79	938.65	1167.4	0.38198	1.30425	1.6862
270	41.877	0.01717	10.04	10.059	238.85	853.94	1092.8	238.98	931.76	1170.7	0.39601	1.27694	1.6730
280	49.222	0.01726	8.63	8.6439	249.04	846.16	1095.2	249.20	924.74	1173.9	0.40989	1.25018	1.6601
290	57.573	0.01735	7.44	7.4607	259.26	838.27	1097.5	259.45	917.57	1177.0	0.42361	1.22393	1.6475
300	67.028	0.01745	6.45	6.4663	269.51	830.25	1099.8	269.73	910.24	1180.0	0.43720	1.19818	1.6354
310	77.691	0.01755	5.61	5.6266	279.79	822.11	1101.9	280.05	902.75	1182.8	0.45065	1.17289	1.6235
320	89.667	0.01765	4.90	4.9144	290.11	813.84	1104.0	290.40	895.09	1185.5	0.46396	1.14802	1.6120
330	103.07	0.01776	4.29	4.3076	300.46	805.43	1105.9	300.80	887.25	1188.1	0.47716	1.12355	1.6007
340	118.02	0.01787	3.77	3.7885	310.85	796.87	1107.7	311.24	879.22	1190.5	0.49024	1.09945	1.5897
350	134.63	0.01799	3.32	3.3425	321.29	788.16	1109.4	321.73	870.98	1192.7	0.50321	1.07570	1.5789
360	153.03	0.01811	2.94	2.9580	331.76	779.28	1111.0	332.28	862.53	1194.8	0.51607	1.05227	1.5683
370	173.36	0.01823	2.61	2.6252	342.29	770.23	1112.5	342.88	853.86	1196.7	0.52884	1.02914	1.5580
380	195.74	0.01836	2.32	2.3361	352.87	761.00	1113.9	353.53	844.96	1198.5	0.54152	1.00628	1.5478
390	220.33	0.01850	2.07	2.0842	363.50	751.58	1115.1	364.25	835.81	1200.1	0.55411	0.98366	1.5378
400	247.26	0.01864	1.85	1.8639	374.19	741.97	1116.2	375.04	826.39	1201.4	0.56663	0.96127	1.5279

(continued)

TABLE C.17 (continued)

Thermodynamic Properties of Steam: Temperature Table (English Units)

Temp., T (°F)	Sat. Press., P_{sat} (psia)	Specific Volume (ft³/lbm)			Internal Energy (Btu/lbm)			Enthalpy (Btu/lbm)			Entropy (Btu/lbm·°R)		
		Sat. Liquid, v_f	Evap., v_{fg}	Sat. Vapor, v_g	Sat. Liquid, u_f	Evap., u_{fg}	Sat. Vapor, u_g	Sat. Liquid, h_f	Evap., h_{fg}	Sat. Vapor, h_g	Sat. Liquid, s_f	Evap., s_{fg}	Sat. Vapor, s_g
410	276.69	0.01878	1.65	1.6706	384.94	732.14	1117.1	385.90	816.71	1202.6	0.57907	0.93908	1.5182
420	308.76	0.01894	1.48	1.5006	395.76	722.08	1117.8	396.84	806.74	1203.6	0.59145	0.91707	1.5085
430	343.64	0.01910	1.33	1.3505	406.65	711.80	1118.4	407.86	796.46	1204.3	0.60377	0.89522	1.4990
440	381.49	0.01926	1.20	1.2178	417.61	701.26	1118.9	418.97	785.87	1204.8	0.61603	0.87349	1.4895
450	422.47	0.01944	1.08	1.0999	428.66	690.47	1119.1	430.18	774.94	1205.1	0.62826	0.85187	1.4801
460	466.75	0.01962	0.98	0.99510	439.79	679.39	1119.2	441.48	763.65	1205.1	0.64044	0.83033	1.4708
470	514.52	0.01981	0.88	0.90158	451.01	668.02	1119.0	452.90	751.98	1204.9	0.65260	0.80885	1.4615
480	565.96	0.02001	0.80	0.81794	462.34	656.34	1118.7	464.43	739.91	1204.3	0.66474	0.78739	1.4521
490	621.24	0.02022	0.72	0.74296	473.77	644.32	1118.1	476.09	727.40	1203.5	0.67686	0.76594	1.4428
500	680.56	0.02044	0.66	0.67558	485.32	631.94	1117.3	487.89	714.44	1202.3	0.68899	0.74445	1.4334
510	744.11	0.02067	0.59	0.61489	496.99	619.17	1116.2	499.84	700.99	1200.8	0.70112	0.72290	1.4240
520	812.11	0.02092	0.54	0.56009	508.80	605.99	1114.8	511.94	687.01	1199.0	0.71327	0.70126	1.4145
530	884.74	0.02118	0.49	0.51051	520.76	592.35	1113.1	524.23	672.47	1196.7	0.72546	0.67947	1.4049
540	962.24	0.02146	0.44	0.46553	532.88	578.23	1111.1	536.70	657.31	1194.0	0.73770	0.65751	1.3952
550	1044.8	0.02176	0.40	0.42465	545.18	563.58	1108.8	549.39	641.47	1190.9	0.75000	0.63532	1.3853
560	1132.7	0.02207	0.37	0.38740	557.68	548.33	1106.0	562.31	624.91	1187.2	0.76238	0.61284	1.3752
570	1226.2	0.02242	0.33	0.35339	570.40	532.45	1102.8	575.49	607.55	1183.0	0.77486	0.59003	1.3649
580	1325.5	0.02279	0.30	0.32225	583.37	515.84	1099.2	588.95	589.29	1178.2	0.78748	0.56679	1.3543
590	1430.8	0.02319	0.27	0.29367	596.61	498.43	1095.0	602.75	570.04	1172.8	0.80026	0.54306	1.3433
600	1542.5	0.02362	0.24	0.26737	610.18	480.10	1090.3	616.92	549.67	1166.6	0.81323	0.51871	1.3319
610	1660.9	0.02411	0.22	0.24309	624.11	460.73	1084.8	631.52	528.03	1159.5	0.82645	0.49363	1.3201
620	1786.2	0.02464	0.20	0.22061	638.47	440.14	1078.6	646.62	504.92	1151.5	0.83998	0.46765	1.3076
630	1918.9	0.02524	0.17	0.19972	653.35	418.12	1071.5	662.32	480.07	1142.4	0.85389	0.44056	1.2944
640	2059.3	0.02593	0.15	0.18019	668.86	394.36	1063.2	678.74	453.14	1131.9	0.86828	0.41206	1.2803
650	2207.8	0.02673	0.14	0.16184	685.16	368.44	1053.6	696.08	423.65	1119.7	0.88332	0.38177	1.2651
660	2364.9	0.02767	0.12	0.14444	702.48	339.74	1042.2	714.59	390.84	1105.4	0.89922	0.34906	1.2483
670	2531.2	0.02884	0.10	0.12774	721.23	307.22	1028.5	734.74	353.54	1088.3	0.91636	0.31296	1.2293
680	2707.3	0.03035	0.08	0.11134	742.11	269.00	1011.1	757.32	309.57	1066.9	0.93541	0.27163	1.2070
690	2894.1	0.03255	0.06	0.09451	766.81	220.77	987.6	784.24	253.96	1038.2	0.95797	0.22089	1.1789
700	3093.0	0.03670	0.04	0.07482	801.75	146.50	948.3	822.76	168.32	991.1	0.99023	0.14514	1.1354
705.10	3200.1	0.04975	0	0.04975	866.61	0	866.6	896.07	0	896.1	1.05257	0	1.0526

Source: Thermodynamics: An Engineering Approach, 7th edn., McGraw Hill, New York, Table A-4E, p. 964.

Appendix C: Properties of Gases and Liquids

TABLE C.18

Thermodynamic Properties of Steam: Pressure Table (English Units)

Press., P (psia)	Sat. Temp., T_{sat} (°F)	Specific Volume (ft³/lbm)			Internal Energy (Btu/lbm)			Enthalpy (Btu/lbm)			Entropy (Btu/lbm·°R)		
		Sat. Liquid, v_f	Evap., v_{fg}	Sat. Vapor, v_g	Sat. Liquid, u_f	Evap., u_{fg}	Sat. Vapor, u_g	Sat. Liquid, h_f	Evap., h_{fg}	Sat. Vapor, h_g	Sat. Liquid, s_f	Evap., s_{fg}	Sat. Vapor, s_g
1	101.69	0.01614	333.47	333.49	69.72	973.99	1043.7	69.72	1035.7	1105.4	0.13262	1.84495	1.9776
2	126.02	0.01623	173.69	173.71	94.02	957.45	1051.5	94.02	1021.7	1115.8	0.17499	1.74444	1.9194
3	141.41	0.01630	118.68	118.70	109.39	946.90	1056.3	109.40	1012.8	1122.2	0.20090	1.68489	1.8858
4	152.91	0.01636	90.61	90.629	120.89	938.97	1059.9	120.90	1006.0	1126.9	0.21985	1.64225	1.8621
5	162.18	0.01641	73.51	73.525	130.17	932.53	1062.7	130.18	1000.5	1130.7	0.23488	1.60894	1.8438
6	170.00	0.01645	61.97	61.982	138.00	927.08	1065.1	138.02	995.88	1133.9	0.24739	1.58155	1.8289
8	182.81	0.01652	47.33	47.347	150.83	918.08	1068.9	150.86	988.15	1139.0	0.26757	1.53800	1.8056
10	193.16	0.01659	38.41	38.425	161.22	910.75	1072.0	161.25	981.82	1143.1	0.28362	1.50391	1.7875
14.696	211.95	0.01671	26.79	26.805	180.12	897.27	1077.4	180.16	970.12	1150.3	0.31215	1.44441	1.7566
15	212.99	0.01672	26.28	26.297	181.16	896.52	1077.7	181.21	969.47	1150.7	0.31370	1.44441	1.7549
20	227.92	0.01683	20.08	20.093	196.21	885.63	1081.8	196.27	959.93	1156.2	0.33582	1.39606	1.7319
25	240.03	0.01692	16.29	16.307	208.45	876.67	1085.1	208.52	952.03	1160.6	0.35347	1.36060	1.7141
30	250.30	0.01700	13.73	13.749	218.84	868.98	1087.8	218.93	945.21	1164.1	0.36821	1.33132	1.6995
35	259.25	0.01708	11.88	11.901	227.92	862.19	1090.1	228.03	939.16	1167.2	0.38093	1.30632	1.6872
40	267.22	0.01715	10.48	10.501	236.02	856.09	1092.1	236.14	933.69	1169.8	0.39213	1.28448	1.6766
45	274.41	0.01721	9.39	9.4028	243.34	850.52	1093.9	243.49	928.68	1172.2	0.40216	1.26506	1.6672
50	280.99	0.01727	8.50	8.5175	250.05	845.39	1095.4	250.21	924.03	1174.2	0.41125	1.24756	1.6588
55	287.05	0.01732	7.77	7.7882	256.25	840.61	1096.9	256.42	919.70	1176.1	0.41958	1.23162	1.6512
60	292.69	0.01738	7.16	7.1766	262.01	836.13	1098.1	262.20	915.61	1177.8	0.42728	1.21697	1.6442
65	297.95	0.01743	6.64	6.6560	267.41	831.90	1099.3	267.62	911.75	1179.4	0.43443	1.20341	1.6378
70	302.91	0.01748	6.19	6.2075	272.50	827.90	1100.4	272.72	908.08	1180.8	0.44112	1.19078	1.6319
75	307.59	0.01752	5.80	5.8167	277.31	824.09	1101.4	277.55	904.58	1182.1	0.44741	1.17895	1.6264
80	312.02	0.01757	5.46	5.4733	281.87	820.45	1102.3	282.13	901.22	1183.4	0.45335	1.16783	1.6212
85	316.24	0.01761	5.15	5.1689	286.22	816.97	1103.2	286.50	898.00	1184.5	0.45897	1.15732	1.6163
90	320.26	0.01765	4.88	4.8972	290.38	813.62	1104.0	290.67	894.89	1185.6	0.46431	1.14737	1.6117
95	324.11	0.01770	4.64	4.6532	294.36	810.40	1104.8	294.67	891.89	1186.6	0.46941	1.13791	1.6073
100	327.81	0.01774	4.41	4.4327	298.19	807.29	1105.5	298.51	888.99	1187.5	0.47427	1.12888	1.6032
110	334.77	0.01781	4.02	4.0410	305.41	801.37	1106.8	305.78	883.44	1189.2	0.48341	1.11201	1.5954
120	341.25	0.01789	3.71	3.7289	312.16	795.79	1107.9	312.55	878.20	1190.8	0.49187	1.09646	1.5883
130	347.32	0.01796	3.44	3.4557	318.48	790.51	1109.0	318.92	873.21	1192.1	0.49974	1.08204	1.5818
140	353.03	0.01802	3.20	3.2202	324.45	785.49	1109.9	324.92	868.45	1193.4	0.50711	1.06858	1.5757
150	358.42	0.01809	3.00	3.0150	330.11	780.69	1110.8	330.61	863.88	1194.5	0.51405	1.05595	1.5700
160	363.54	0.01815	2.82	2.8347	335.49	776.10	1111.6	336.02	859.49	1195.5	0.52061	1.04405	1.5647
170	368.41	0.01821	2.66	2.6749	340.62	771.68	1112.3	341.19	855.25	1196.4	0.52682	1.03279	1.5596
180	373.07	0.01827	2.51	2.5322	345.53	767.42	1113.0	346.14	851.16	1197.3	0.53274	1.02210	1.5548
190	377.52	0.01833	2.39	2.4040	350.24	763.31	1113.6	350.89	847.19	1198.1	0.53839	1.01191	1.5503
200	381.80	0.01839	2.27	2.2882	354.78	759.32	1114.1	355.46	843.33	1198.8	0.54379	1.00219	1.5460
250	400.97	0.01865	1.83	1.8440	375.23	741.02	1116.3	376.09	825.47	1201.6	0.56784	0.95912	1.5270
300	417.35	0.01890	1.52	1.5435	392.89	724.77	1117.7	393.94	809.41	1203.3	0.58818	0.92289	1.5111
350	431.74	0.01912	1.31	1.3263	408.55	709.98	1118.5	409.79	794.65	1204.4	0.60590	0.89143	1.4973
400	444.62	0.01934	1.14	1.1617	422.70	696.31	1119.0	424.13	780.87	1205.0	0.62168	0.86350	1.4852

(continued)

TABLE C.18 (continued)

Thermodynamic Properties of Steam: Pressure Table (English Units)

Press., P (psia)	Sat. Temp., T_{sat} (°F)	Specific Volume (ft³/lbm)			Internal Energy (Btu/lbm)			Enthalpy (Btu/lbm)			Entropy (Btu/lbm·°R)		
		Sat. Liquid, v_f	Evap., v_{fg}	Sat. Vapor, v_g	Sat. Liquid, u_f	Evap., u_{fg}	Sat. Vapor, u_g	Sat. Liquid, h_f	Evap., h_{fg}	Sat. Vapor, h_g	Sat. Liquid, s_f	Evap., s_{fg}	Sat. Vapor, s_g
450	456.31	0.01955	1.01	1.0324	435.67	683.52	1119.2	437.30	767.86	1205.2	0.63595	0.83828	1.4742
500	467.04	0.01975	0.91	0.92819	447.68	671.42	1119.1	449.51	755.48	1205.0	0.64900	0.81521	1.4642
550	476.97	0.01995	0.82	0.84228	458.90	659.91	1118.8	460.93	743.60	1204.5	0.66107	0.79388	1.4550
600	486.24	0.02014	0.75	0.77020	469.46	648.88	1118.3	471.70	732.15	1203.9	0.67231	0.77400	1.4463
700	503.13	0.02051	0.64	0.65589	488.96	627.98	1116.9	491.62	710.29	1201.9	0.69279	0.73771	1.4305
800	518.27	0.02087	0.55	0.56920	506.74	608.30	1115.0	509.83	689.48	1199.3	0.71117	0.70502	1.4162
900	532.02	0.02124	0.48	0.50107	523.19	589.54	1112.7	526.73	669.46	1196.2	0.72793	0.67505	1.4030
1000	544.65	0.02159	0.42	0.44604	538.58	571.49	1110.1	542.57	650.03	1192.6	0.74341	0.64722	1.3906
1200	567.26	0.02232	0.34	0.36241	566.89	536.87	1103.8	571.85	612.39	1184.2	0.77143	0.59632	1.3677
1400	587.14	0.02307	0.28	0.30161	592.79	503.50	1096.3	598.76	575.66	1174.4	0.79658	0.54991	1.3465
1600	604.93	0.02386	0.23	0.25516	616.99	470.69	1087.7	624.06	539.18	1163.2	0.81972	0.50645	1.3262
1800	621.07	0.02470	0.19	0.21831	640.03	437.86	1077.9	648.26	502.35	1150.6	0.84144	0.46482	1.3063
2000	635.85	0.02563	0.16	0.18815	662.33	404.46	1066.8	671.82	464.60	1136.4	0.86224	0.42409	1.2863
2500	668.17	0.02860	0.10	0.13076	717.67	313.53	1031.2	730.90	360.79	1091.7	0.91311	0.31988	1.2330
3000	695.41	0.03433	0.05	0.08460	783.39	186.41	969.8	802.45	214.32	1016.8	0.97321	0.18554	1.1587
3200.1	705.10	0.04975	0	0.04975	866.61	0	866.6	896.07	0	896.1	1.05257	0	1.0526

Source: Thermodynamics: An Engineering Approach, 7th edn., McGraw Hill, New York, Table A-5E, p. 966.

TABLE C.19
Thermodynamic Properties of Steam: Superheated Vapor Table (English Units)

T (°F)	v (ft³/lbm)	u (Btu/lbm)	h (Btu/lbm)	s (Btu/lbm °R)	v (ft³/lbm)	u (Btu/lbm)	h (Btu/lbm)	s (Btu/lbm °R)	v (ft³/lbm)	u (Btu/lbm)	h (Btu/lbm)	s (Btu/lbm °R)
	$P = 1.0$ psia (101.69°F)[a]				$P = 5.0$ psia (162.18°F)				$P = 10$ psia (193.16°F)			
Sat.[b]	333.49	1043.7	1105.4	1.9776	73.525	1062.7	1130.7	1.8438	38.425	1072.0	1143.1	1.7875
200	392.53	1077.5	1150.1	2.0509	78.153	1076.2	1148.5	1.8716	38.849	1074.5	1146.4	1.7926
240	416.44	1091.2	1168.3	2.0777	83.009	1090.3	1167.1	1.8989	41.326	1089.1	1165.5	1.8207
280	440.33	1105.0	1186.5	2.1030	87.838	1104.3	1185.6	1.9246	43.774	1103.4	1184.4	1.8469
320	464.20	1118.9	1204.8	2.1271	92.650	1118.4	1204.1	1.9490	46.205	1117.6	1203.1	1.8716
360	488.07	1132.9	1223.3	2.1502	97.452	1132.5	1222.6	1.9722	48.624	1131.9	1221.8	1.8950
400	511.92	1147.1	1241.8	2.1722	102.25	1146.7	1241.3	1.9944	51.035	1146.2	1240.6	1.9174
440	535.77	1161.3	1260.4	2.1934	107.03	1160.9	1260.0	2.0156	53.441	1160.5	1259.4	1.9388
500	571.54	1182.8	1288.6	2.2237	114.21	1182.6	1288.2	2.0461	57.041	1182.2	1287.8	1.9693
600	631.14	1219.4	1336.2	2.2709	126.15	1219.2	1335.9	2.0933	63.029	1219.0	1335.6	2.0167
700	690.73	1256.8	1384.6	2.3146	138.09	1256.7	1384.4	2.1371	69.007	1256.5	1384.2	2.0605
800	750.31	1295.1	1433.9	2.3553	150.02	1294.9	1433.7	2.1778	74.980	1294.8	1433.5	2.1013
1000	869.47	1374.2	1535.1	2.4299	173.86	1374.2	1535.0	2.2524	86.913	1374.1	1534.9	2.1760
1200	988.62	1457.1	1640.0	2.4972	197.70	1457.0	1640.0	2.3198	98.840	1457.0	1639.9	2.2433
1400	1107.8	1543.7	1748.7	2.5590	221.54	1543.7	1748.7	2.3816	110.762	1543.6	1748.6	2.3052
	$P = 15$ psia (212.99°F)				$P = 20$ psia (227.92°F)				$P = 40$ psia (267.22°F)			
Sat.	26.297	1077.7	1150.7	1.7549	20.093	1081.8	1156.2	1.7319	10.501	1092.1	1169.8	1.6766
240	27.429	1087.8	1163.9	1.7742	20.478	1086.5	1162.3	1.7406				
280	29.085	1102.4	1183.2	1.8010	21.739	1101.4	1181.9	1.7679	10.713	1097.3	1176.6	1.6858
320	30.722	1116.9	1202.2	1.8260	22.980	1116.1	1201.2	1.7933	11.363	1112.9	1197.1	1.7128
360	32.348	1131.3	1221.1	1.8496	24.209	1130.7	1220.2	1.8171	11.999	1128.1	1216.9	1.7376
400	33.965	1145.7	1239.9	1.8721	25.429	1145.1	1239.3	1.8398	12.625	1143.1	1236.5	1.7610
440	35.576	1160.1	1258.8	1.8936	26.644	1159.7	1258.3	1.8614	13.244	1157.9	1256.0	1.7831
500	37.986	1181.9	1287.3	1.9243	28.458	1181.6	1286.9	1.8922	14.165	1180.2	1285.0	1.8143
600	41.988	1218.7	1335.3	1.9718	31.467	1218.5	1334.9	1.9398	15.686	1217.5	1333.6	1.8625
700	45.981	1256.3	1383.9	2.0156	34.467	1256.1	1383.7	1.9837	17.197	1255.3	1382.6	1.9067
800	49.967	1294.6	1433.3	2.0565	37.461	1294.5	1433.1	2.0247	18.702	1293.9	1432.3	1.9478
1000	57.930	1374.0	1534.8	2.1312	43.438	1373.8	1534.6	2.0994	21.700	1373.4	1534.1	2.0227
1200	65.885	1456.9	1639.8	2.1986	49.407	1456.8	1639.7	2.1668	24.691	1456.5	1639.3	2.0902
1400	73.836	1543.6	1748.5	2.2604	55.373	1543.5	1748.4	2.2287	27.678	1543.3	1748.1	2.1522
1600	81.784	1634.0	1861.0	2.3178	61.335	1633.9	1860.9	2.2861	30.662	1633.7	1860.7	2.2096

(continued)

TABLE C.19 (continued)

Thermodynamic Properties of Steam: Superheated Vapor Table (English Units)

T (°F)	v (ft³/lbm)	u (Btu/lbm)	h (Btu/lbm)	s (Btu/lbm °R)	v (ft³/lbm)	u (Btu/lbm)	h (Btu/lbm)	s (Btu/lbm °R)	v (ft³/lbm)	u (Btu/lbm)	h (Btu/lbm)	s (Btu/lbm °R)
	P = 60 psia (292.69°F)				P = 80 psia (312.02°F)				P = 100 psia (327.81°F)			
Sat.	7.1766	1098.1	1177.8	1.6442	5.4733	1102.3	1183.4	1.6212	4.4327	1105.5	1187.5	1.6032
320	7.4863	1109.6	1192.7	1.6636	5.5440	1105.9	1187.9	1.6271				
360	7.9259	1125.5	1213.5	1.6897	5.8876	1122.7	1209.9	1.6545	4.6628	1119.8	1206.1	1.6263
400	8.3548	1140.9	1233.7	1.7138	6.2187	1138.7	1230.8	1.6794	4.9359	1136.4	1227.8	1.6521
440	8.7766	1156.1	1253.6	1.7364	6.5420	1154.3	1251.2	1.7026	5.2006	1152.4	1248.7	1.6759
500	9.4005	1178.8	1283.1	1.7682	7.0177	1177.3	1281.2	1.7350	5.5876	1175.9	1279.3	1.7088
600	10.4256	1216.5	1332.2	1.8168	7.7951	1215.4	1330.8	1.7841	6.2167	1214.4	1329.4	1.7586
700	11.4401	1254.5	1381.6	1.8613	8.5616	1253.8	1380.5	1.8289	6.8344	1253.0	1379.5	1.8037
800	12.4484	1293.3	1431.5	1.9026	9.3218	1292.6	1430.6	1.8704	7.4457	1292.0	1429.8	1.8453
1000	14.4543	1373.0	1533.5	1.9777	10.8313	1372.6	1532.9	1.9457	8.6575	1372.2	1532.4	1.9208
1200	16.4525	1456.2	1638.9	2.0454	12.3331	1455.9	1638.5	2.0135	9.8615	1455.6	1638.1	1.9887
1400	18.4464	1543.0	1747.8	2.1073	13.8306	1542.8	1747.5	2.0755	11.0612	1542.6	1747.2	2.0508
1600	20.438	1633.5	1860.5	2.1648	15.3257	1633.3	1860.2	2.1330	12.2584	1633.2	1860.0	2.1083
1800	22.428	1727.6	1976.6	2.2187	16.8192	1727.5	1976.5	2.1869	13.4541	1727.3	1976.3	2.1622
2000	24.417	1825.2	2096.3	2.2694	18.3117	1825.0	2096.1	2.2376	14.6487	1824.9	2096.0	2.2130
	P = 120 psia (341.25°F)				P = 140 psia (353.03°F)				P = 160 psia (363.54°F)			
Sat.	3.7289	1107.9	1190.8	1.5883	3.2202	1109.9	1193.4	1.5757	2.8347	1111.6	1195.5	1.5647
360	3.8446	1116.7	1202.1	1.6023	3.2584	1113.4	1197.8	1.5811				
400	4.0799	1134.0	1224.6	1.6292	3.4676	1131.5	1221.4	1.6092	3.0076	1129.0	1218.0	1.5914
450	4.3613	1154.5	1251.4	1.6594	3.7147	1152.6	1248.9	1.6403	3.2293	1150.7	1246.3	1.6234
500	4.6340	1174.4	1277.3	1.6872	3.9525	1172.9	1275.3	1.6686	3.4412	1171.4	1273.2	1.6522
550	4.9010	1193.9	1302.8	1.7131	4.1845	1192.7	1301.1	1.6948	3.6469	1191.4	1299.4	1.6788
600	5.1642	1213.4	1328.0	1.7375	4.4124	1212.3	1326.6	1.7195	3.8484	1211.3	1325.2	1.7037
700	5.6829	1252.2	1378.4	1.7829	4.8604	1251.4	1377.3	1.7652	4.2434	1250.6	1376.3	1.7498
800	6.1950	1291.4	1429.0	1.8247	5.3017	1290.8	1428.1	1.8072	4.6316	1290.2	1427.3	1.7920
1000	7.2083	1371.7	1531.8	1.9005	6.1732	1371.3	1531.3	1.8832	5.3968	1370.9	1530.7	1.8682
1200	8.2137	1455.3	1637.7	1.9684	7.0367	1455.0	1637.3	1.9512	6.1540	1454.7	1636.9	1.9363
1400	9.2149	1542.3	1746.9	2.0305	7.8961	1542.1	1746.6	2.0134	6.9070	1541.8	1746.3	1.9986
1600	10.2135	1633.0	1859.8	2.0881	8.7529	1632.8	1859.5	2.0711	7.6574	1632.6	1859.3	2.0563
1800	11.2106	1727.2	1976.1	2.1420	9.6082	1727.0	1975.9	2.1250	8.4063	1726.9	1975.7	2.1102
2000	12.2067	1824.8	2095.8	2.1928	10.4624	1824.6	2095.7	2.1758	9.1542	1824.5	2095.5	2.1610

Appendix C: Properties of Gases and Liquids

T (°F)	P = 180 psia (373.07°F)			P = 200 psia (381.80°F)			P = 225 psia (391.80°F)		
	v	h	s	v	h	s	v	h	s
Sat.	2.5322	1197.3	1.5548	2.2882	1198.8	1.5460	2.0423	1200.3	1.5360
400	2.6490	1214.5	1.5752	2.3615	1210.9	1.5602	2.0728	1206.0	1.5427
450	2.8514	1243.7	1.6082	2.5488	1241.0	1.5943	2.2457	1237.6	1.5783
500	3.0433	1271.2	1.6376	2.7247	1269.0	1.6243	2.4059	1266.3	1.6091
550	3.2286	1297.7	1.6646	2.8939	1296.0	1.6516	2.5590	1293.8	1.6370
600	3.4097	1323.8	1.6897	3.0586	1322.3	1.6771	2.7075	1320.5	1.6628
700	3.7635	1375.2	1.7361	3.3796	1374.1	1.7238	2.9956	1372.7	1.7099
800	4.1104	1426.5	1.7785	3.6934	1425.6	1.7664	3.2765	1424.5	1.7528
900	4.4531	1478.0	1.8179	4.0031	1477.3	1.8059	3.5530	1476.5	1.7925
1000	4.7929	1530.1	1.8549	4.3099	1529.6	1.8430	3.8268	1528.9	1.8296
1200	5.4674	1636.5	1.9231	4.9182	1636.1	1.9113	4.3689	1635.6	1.8981
1400	6.1377	1746.0	1.9855	5.5222	1745.7	1.9737	4.9068	1745.4	1.9606
1600	6.8054	1859.1	2.0432	6.1238	1858.8	2.0315	5.4422	1858.6	2.0184
1800	7.4716	1975.6	2.0971	6.7238	1975.4	2.0855	5.9760	1975.2	2.0724
2000	8.1367	2095.4	2.1479	7.3227	2095.3	2.1363	6.5087	2095.1	2.1232

T (°F)	P = 250 psia (400.97°F)			P = 275 psia (409.45°F)			P = 300 psia (417.35°F)		
	v	h	s	v	h	s	v	h	s
Sat.	1.8440	1201.6	1.5270	1.6806	1202.6	1.5187	1.5435	1203.3	1.5111
450	2.0027	1234.0	1.5636	1.8034	1230.3	1.5499	1.6369	1226.4	1.5369
500	2.1506	1263.6	1.5953	1.9415	1260.8	1.5825	1.7670	1257.9	1.5706
550	2.2910	1291.5	1.6237	2.0715	1289.3	1.6115	1.8885	1287.0	1.6001
600	2.4264	1318.6	1.6499	2.1964	1316.7	1.6380	2.0046	1314.8	1.6270
650	2.5586	1345.1	1.6743	2.3179	1343.5	1.6627	2.1172	1341.9	1.6520
700	2.6883	1371.4	1.6974	2.4369	1370.0	1.6860	2.2273	1368.6	1.6755
800	2.9429	1423.5	1.7406	2.6699	1422.4	1.7294	2.4424	1421.3	1.7192
900	3.1930	1475.6	1.7804	2.8984	1474.8	1.7694	2.6529	1473.9	1.7593
1000	3.4403	1528.2	1.8177	3.1241	1527.4	1.8068	2.8605	1526.7	1.7968
1200	3.9295	1635.0	1.8863	3.5700	1634.5	1.8755	3.2704	1634.0	1.8657
1400	4.4144	1745.0	1.9488	4.0116	1744.6	1.9381	3.6759	1744.2	1.9284
1600	4.8969	1858.3	2.0066	4.4507	1858.0	1.9960	4.0789	1857.7	1.9863
1800	5.3777	1974.9	2.0607	4.8882	1974.7	2.0501	4.4803	1974.5	2.0404
2000	5.8575	2094.9	2.1116	5.3247	2094.7	2.1010	4.8807	2094.6	2.0913

Source: Thermodynamics: An Engineering Approach, 7th edn., McGraw Hill, New York, Table A-6E, p. 968.
[a] The temperature in parentheses is the saturation temperature at the specified pressure.
[b] Properties of saturated vapor at the specified pressure.

TABLE C.20

Enthalpy of Formation, Gibbs of Formation, and Absolute Entropy of Various Substances at 25°C (77°F) and 1 atm

Substance	Formula	$\bar{h}_f^o$ kJ/kmol	$\bar{h}_f^o$ Btu/lbmol	$\bar{g}_f^o$ kJ/kmol	$\bar{g}_f^o$ Btu/lbmol	$\bar{s}^o$ kJ/kmol K	$\bar{s}^o$ Btu/lbmol °R
Carbon	C(s)	0	0	0	0	5.74	1.36
Hydrogen	H_2(g)	0	0	0	0	130.57	31.21
Nitrogen	N_2(g)	0	0	0	0	191.5	45.77
Oxygen	O_2(g)	0	0	0	0	205.03	49
Carbon monoxide	CO(g)	−110,530	−47,540	−137,150	−59,010	197.54	47.21
Carbon dioxide	CO_2(g)	−393,520	−169,300	−394,380	−169,680	213.69	51.07
Water	H_2O(g)	−241,820	−104,040	−228,590	−98,350	188.72	45.11
Water	H_2O(l)	−285,830	−122,970	−237,180	−102,040	69.95	16.71
Hydrogen peroxide	H_2O_2(g)	−136,310	−58,640	−105,600	−45,430	232.63	55.6
Ammonia	NH_3(g)	−46,190	−19,750	−16,590	−7,140	192.33	45.97
Oxygen	O(g)	249,170	107,210	231,770	99,710	160.95	38.47
Hydrogen	H(g)	218,000	93,780	203,290	87,460	114.61	27.39
Nitrogen	N(g)	472,680	203,340	455,510	195,970	153.19	36.61
Hydroxyl	OH(g)	39,460	16,790	34,280	14,750	183.75	43.92
Methane	CH_4(g)	−74,850	−32,210	−50,790	−21,860	186.16	44.49
Acetylene	C_2H_2(g)	226,730	97,540	209,170	87,990	200.85	48
Ethylene	C_2H_4(g)	52,280	22,490	68,120	29,306	219.83	52.54
Ethane	C_2H_6(g)	−84,680	−36,420	−32,890	−14150	229.49	54.85
Propylene	C_3H_6(g)	20,410	8,790	62,720	26,980	266.94	63.8
Propane	C_3H_8(g)	−103,850	−44,680	−23,490	−10,105	269.91	64.51
Butane	C_4H_{10}(g)	−126,150	−54,270	−15,710	−6,760	310.03	74.11
Octane	C_8H_{18}(g)	−208,450	−89,680	17,320	7,110	463.67	111.55
Octane	C_8H_{18}(l)	−249,910	−107,530	6,610	2,840	360.79	86.23
Dodecane	$C_{12}H_{26}$(g)	−291,010	−125,190	50,150	21,570	622.83	148.86
Benzene	C_6H_6(g)	82,930	35,680	129,660	55,780	269.2	64.34
Methyl alcohol	CH_3OH(g)	−200,890	−86,540	−162,140	−69,700	239.7	57.29
Methyl alcohol	CH_3OH(l)	−238,810	−102,670	−166,290	−71,570	126.8	30.3
Ethyl alcohol	C_2H_5OH(g)	−235,310	−101,230	−168,570	−72,520	282.59	67.54
Ethyl alcohol	C_2H_5OH(l)	−277,690	−119,470	174,890	−75,240	160.7	38.4

Sources: Baukal, C.E. (ed.), *The John Zink Combustion Handbook*, CRC Press, Boca Raton, FL, 2001, Table B.2, p. 716; Cengel, Y. and Boles, M., *Thermodynamics: An Engineering Approach*, 7th edn., McGraw-Hill, New York, 2011, Table A-26E, p. 997.

Appendix C: Properties of Gases and Liquids

TABLE C.21
Combustion Data for Hydrocarbons (Metric and English Units)

Hydrocarbon	Formula	Higher Heating Value (Vapor) kJ/kg	Higher Heating Value (Vapor) Btu/lbm	Theor. Air/Fuel Ratio (by Mass)	Max Flame Speed m/s	Max Flame Speed ft/s	Adiabatic Flame Temp (in Air) °C	Adiabatic Flame Temp (in Air) °F	Ignition Temp (in Air) °C	Ignition Temp (in Air) °F	Flash Point °C	Flash Point °F	Flammability Limits (in Air) (% by Volume) LFL	Flammability Limits (in Air) (% by Volume) UFL
Paraffins or alkanes														
Methane	CH_4	55,533	23,875	17.195	0.34	1.1	1,918	3,484	705	1,301	Gas	Gas	5.0	15.0
Ethane	C_2H_6	51,923	22,323	15.899	0.40	1.3	1,949	3,540	520–630	968–1,166	Gas	Gas	3.0	12.5
Propane	C_3H_8	50,402	21,669	15.246	0.40	1.3	1,967	3,573	466	871	Gas	Gas	2.1	10.1
n-Butane	C_4H_{10}	49,593	21,321	14.984	0.37	1.2	1,973	3,583	405	761	−60	−76	1.86	8.41
iso-Butane	C_4H_{10}	49,476	21,271	14.984	0.37	1.2	1,973	3,583	462	864	−83	−117	1.80	8.44
n-Pentane	C_5H_{12}	49,067	21,095	15.323	0.40	1.3	2,232	4,050	309	588	<−40	<−40	1.40	7.80
iso-Pentane	C_5H_{12}	48,955	21,047	15.323	0.37	1.2	2,235	4,055	420	788	<−51	<−60	1.32	9.16
Neopentane	C_5H_{12}	48,795	20,978	15.323	0.34	1.1	2,238	4,060	450	842	Gas	Gas	1.38	7.22
n-Hexane	C_6H_{14}	48,767	20,966	15.238	0.40	1.3	2,221	4,030	248	478	−22	−7	1.25	7.00
Neohexane	C_6H_{14}	48,686	20,931	15.238	0.37	1.2	2,235	4,055	425	797	−48	−54	1.19	7.58
n-Heptane	C_7H_{16}	48,506	20,854	15.141	0.40	1.3	2,196	3,985	223	433	−4	25	1.00	6.00
Triptane	C_7H_{16}	48,437	20,824	15.141	0.37	1.2	2,224	4,035	454	849	—	—	1.08	6.69
n-Octane	C_8H_{18}	48,371	20,796	15.093	—	—	—	—	220	428	13	56	0.95	3.20
iso-Octane	C_8H_{18}	48,311	20,770	15.093	0.34	1.1	—	—	447	837	−12	10	0.76	5.94
Olefins or alkenes														
Ethylene	C_2H_4	50,325	21,636	14.807	0.67	2.2	2,343	4,250	490	914	Gas	Gas	2.75	28.6
Propylene	C_3H_6	48,958	21,048	14.807	0.43	1.4	2,254	4,090	458	856	gas	gas	2.00	11.1
Butylene	C_4H_8	48,506	20,854	14.807	0.43	1.4	2,221	4,030	443	829	Gas	Gas	1.98	9.65
iso-Butene	C_4H_8	48,234	20,737	14.807	0.37	1.2	—	—	465	869	Gas	Gas	1.80	9.00
n-Pentene	C_5H_{10}	48,195	20,720	14.807	0.43	1.4	2,296	4,165	298	569	—	—	1.65	7.70
Aromatics														
Benzene	C_6H_6	42,296	18,184	13.297	0.40	1.3	2,266	4,110	562	1,044	−11	12	1.35	6.65
Toluene	C_7H_8	43,033	18,501	13.503	0.37	1.2	2,232	4,050	536	997	4	40	1.27	6.75
p-Xylene	C_8H_{10}	43,410	18,663	13.663	—	—	2,210	4,010	464	867	17	63	1.00	6.00
Other hydrocarbons														
Acetylene	C_2H_2	50,014	21,502	13.297	1.40	4.6	2,632	4,770	406–440	763–824	Gas	Gas	2.50	81.0
Naphthalene	$C_{10}H_8$	40,247	17,303	12.932	—	—	2,260	4,100	515	959	79	174	0.90	5.9

Source: Adapted from Baukal, C.E. (ed.), *The John Zink Combustion Handbook*, CRC Press, Boca Raton, FL, 2001, Table B.1, p. 715.

TABLE C.22
Chemical, Physical, and Thermal Properties of Gases: Gases and Vapors, Including Fuels and Refrigerants, English and Metric Units

Common Name(s)	Acetylene (Ethyne)	Butadiene	n-Butane	Isobutane (2-Methyl Propane)
Chemical Formula	C_2H_2	C_4H_6	C_4H_{10}	C_4H_{10}
Refrigerant Number	—	—	600	600a
Chemical and physical properties				
Molecular weight	26.04	54.09	58.12	58.12
Specific gravity, air = 1	0.90	1.87	2.07	2.07
Specific volume, ft^3/lb	14.9	7.1	6.5	6.5
Specific volume, m^3/kg	0.93	0.44	0.405	0.418
Density of liquid (at atm bp), lb/ft^3	43.0		37.5	37.2
Density of liquid (at atm bp), kg/m^3	693.0		604.0	599.0
Vapor pressure at 25°C, psia			35.4	50.4
Vapor pressure at 25°C, MN/m^2			0.0244	0.347
Viscosity (abs), lbm/ft · s	6.72×10^{-6}		4.8×10^{-6}	
Viscosity (abs), centipoises[a]	0.01		0.007	
Sound velocity in gas, m/s	343	226	216	216
Thermal and thermodynamic properties				
Specific heat, c_p, Btu/lb · °F or cal/g · °C	0.40	0.341	0.39	0.39
Specific heat, c_p, J/kg · K	1,674.0	1427.0	1,675.0	1630.0
Specific heat ratio, c_p/c_v	1.25	1.12	1.096	1.10
Gas constant R, ft lb/lb · °R	59.3	28.55	26.56	26.56
Gas constant R, J/kg · °C	319	154.0	143.0	143.0
Thermal conductivity, Btu/h · ft · °F	0.014		0.01	0.01
Thermal conductivity, W/m °C	0.024		0.017	0.017
Boiling point (sat 14.7 psia), °F	−103	24.1	31.2	10.8
Boiling point (sat 760 mm), °C	−75	−4.5	−0.4	−11.8
Latent heat of evap. (at bp), Btu/lb	264		165.6	157.5
Latent heat of evap. (at bp), J/kg	614,000		386,000	366,000
Freezing (melting) point, °F (1 atm)	−116	−164.0	−217.0	−229
Freezing (melting) point, °C (1 atm)	−82.2	−109.0	−138	−145
Latent heat of fusion, Btu/lb	23.0		19.2	
Latent heat of fusion, J/kg	53,500		44,700	
Critical temperature, °F	97.1		306	273.0
Critical temperature, °C	36.2	171.0	152.0	134.0
Critical pressure, psia	907.0	652.0	550.0	537.0
Critical pressure, MN/m^2	6.25		3.8	3.7
Critical volume, ft^3/lb			0.070	
Critical volume, m^3/kg			0.0043	
Flammable (yes or no)	Yes	Yes	Yes	Yes
Heat of combustion, Btu/ft^3	1,450	2,950	3,300	3,300
Heat of combustion, Btu/lb	21,600	20,900	21,400	21,400
Heat of combustion, kJ/kg	50,200	48,600	49,700	49,700

TABLE C.22 (continued)

Chemical, Physical, and Thermal Properties of Gases: Gases and Vapors, Including Fuels and Refrigerants, English and Metric Units

Common Name(s)	1-Butene (Butylene)	cis-2-Butene	trans-2-Butene	Isobutene
Chemical Formula	C_4H_8	C_4H_8	C_4H_8	C_4H_8
Refrigerant Number	—	—	—	—
Chemical and physical properties				
Molecular weight	56.108	56.108	56.108	56.108
Specific gravity, air = 1	1.94	1.94	1.94	1.94
Specific volume, ft³/lb	6.7	6.7	6.7	6.7
Specific volume, m³/kg	0.42	0.42	0.42	0.42
Density of liquid (at atm bp), lb/ft³				
Density of liquid (at atm bp), kg/m³				
Vapor pressure at 25°C, psia				
Vapor pressure at 25°C, MN/m²				
Viscosity (abs), lbm/ft · s				
Viscosity (abs), centipoises[a]				
Sound velocity in gas, m/s	222	223.0	221.0	221.0
Thermal and thermodynamic properties				
Specific heat, c_p, Btu/lb · °F or cal/g °C	0.36	0.327	0.365	0.37
Specific heat, c_p, J/kg K	1,505.0	1368.0	1,527.0	15,48.0
Specific heat ratio, c_p/c_v	1.112	1.121	1.107	1.10
Gas constant R, ft lb/lb · °F	27.52			
Gas constant R, J/kg · °C	148.0			
Thermal conductivity, Btu/h ft °F				
Thermal conductivity, W/m °C				
Boiling point (sat 14.7 psia), °F	20.6	38.6	33.6	19.2
Boiling point (sat 760 mm), °C	−6.3	3.7	0.9	−7.1
Latent heat of evap. (at bp), Btu/lb	167.9	178.9	174.4	169.0
Latent heat of evap. (at bp), J/kg	391,000	416,000.0	406,000.0	393,000.0
Freezing (melting) point, °F (1 atm)	−301.6	−218.0	−158.0	
Freezing (melting) point, °C (1 atm)	−185.3	−138.9	−105.5	
Latent heat of fusion, Btu/lb	16.4	31.2	41.6	25.3
Latent heat of fusion, J/kg	38,100	72,600.0	96,800.0	58,800.0
Critical temperature, °F	291.0			
Critical temperature, °C	144.0	160.	155.0	
Critical pressure, psia	621.0	595.0	610.0	
Critical pressure, MN/m²	4.28	4.10	4.20	
Critical volume, ft³/lb	0.068			
Critical volume, m³/kg	0.0042			
Flammable (yes or no)	Yes	Yes	Yes	Yes
Heat of combustion, Btu/ft³	3,150	3,150.0	3,150.0	3,150.0
Heat of combustion, Btu/lb	21,000	21,000.0	21,000.0	21,000.0
Heat of combustion, kJ/kg	48,800	48,800.0	48,800.0	48,800.0

(continued)

TABLE C.22 (continued)

Chemical, Physical, and Thermal Properties of Gases: Gases and Vapors, Including Fuels and Refrigerants, English and Metric Units

Common Name(s)	Carbon Dioxide	Carbon Monoxide	Ethane	Ethylene (Ethene)
Chemical Formula	CO_2	CO	C_2H_6	C_2H_4
Refrigerant Number	744	—	170	1150
Chemical and physical properties				
Molecular weight	44.01	28.011	30.070	28.054
Specific gravity, air = 1	1.52	0.967	1.04	0.969
Specific volume, ft^3/lb	8.8	14.0	13.025	13.9
Specific volume, m^3/kg	0.55	0.874	0.815	0.87
Density of liquid (at atm bp), lb/ft^3	—		28.0	35.5
Density of liquid (at atm bp), kg/m^3	—		449.0	569.0
Vapor pressure at 25°C, psia	931.0			
Vapor pressure at 25°C, MN/m^2	6.42			
Viscosity (abs), $lbm/ft \cdot s$	9.4×10^{-6}	12.1×10^{-6}	64.0×10^{-6}	6.72×10^{-6}
Viscosity (abs), centipoises[a]	0.014	0.018	0.095	0.010
Sound velocity in gas, m/s	270.0	352.0	316.0	331.0
Thermal and thermodynamic properties				
Specific heat, c_p, Btu/lb °F or cal/g °C	0.205	0.25	0.41	0.37
Specific heat, c_p, J/kg K	876.0	1,046.0	1,715.0	1,548.0
Specific heat ratio, c_p/c_v	1.30	1.40	1.20	1.24
Gas constant R, ft lb/lb·°F	35.1	55.2	51.4	55.1
Gas constant R, J/kg·°C	189.0	297.0	276.0	296.0
Thermal conductivity, Btu/h ft °F	0.01	0.014	0.010	0.010
Thermal conductivity, W/m °C	0.017	0.024	0.017	0.017
Boiling point (sat 14.7 psia), °F	−109.4[b]	−312.7	−127.0	−155.0
Boiling point (sat 760 mm), °C	−78.5	−191.5	−88.3	−103.8
Latent heat of evap. (at bp), Btu/lb	246.0	92.8	210.0	208.0
Latent heat of evap. (at bp), J/kg	572,000.0	216,000.0	488,000.0	484,000.0
Freezing (melting) point, °F (1 atm)		−337.0	−278.0	−272.0
Freezing (melting) point, °C (1 atm)		−205.0	−172.2	−169.0
Latent heat of fusion, Btu/lb	—	12.8	41.0	51.5
Latent heat of fusion, J/kg	—		95,300.0	120,000.0
Critical temperature, °F	88.0	−220.0	90.1	49.0
Critical temperature, °C	31.0	−140.0	32.2	9.5
Critical pressure, psia	1,072.0	507.0	709.0	741.0
Critical pressure, MN/m^2	7.4	3.49	4.89	5.11
Critical volume, ft^3/lb		0.053	0.076	0.073
Critical volume, m^3/kg		0.0033	0.0047	0.0046
Flammable (yes or no)	No	Yes	Yes	Yes
Heat of combustion, Btu/ft^3	—	3,10.0		1,480.0
Heat of combustion, Btu/lb	—	4,340.0	22,300.0	20,600.0
Heat of combustion, kJ/kg	—	10,100.0	51,800.0	47,800.0

TABLE C.22 (continued)

Chemical, Physical, and Thermal Properties of Gases: Gases and Vapors, Including Fuels and Refrigerants, English and Metric Units

Common Name(s)	Hydrogen	Methane	Nitric Oxide	Nitrogen
Chemical Formula	H_2	CH_4	NO	N_2
Refrigerant Number	702	50	—	728
Chemical and physical properties				
Molecular weight	2.016	16.044	30.006	28.0134
Specific gravity, air = 1	0.070	0.554	1.04	0.967
Specific volume, ft³/lb	194.0	24.2	13.05	13.98
Specific volume, m³/kg	12.1	1.51	0.814	0.872
Density of liquid (at atm bp), lb/ft³	4.43	26.3		50.46
Density of liquid (at atm bp), kg/m³	71.0	421.0		808.4
Vapor pressure at 25°C, psia				
Vapor pressure at 25°C, MN/m²				
Viscosity (abs), lbm/ft·s	6.05×10^{-6}	7.39×10^{-6}	12.8×10^{-6}	12.1×10^{-6}
Viscosity (abs), centipoises[a]	0.009	0.011	0.019	0.018
Sound velocity in gas, m/s	1 315.0	446.0	341.0	353.0
Thermal and thermodynamic properties				
Specific heat, c_p, Btu/lb·°F or cal/g·°C	3.42	0.54	0.235	0.249
Specific heat, c_p, J/kg·K	14,310.0	2,260.0	983.0	1,040.0
Specific heat ratio, c_p/c_v	1.405	1.31	1.40	1.40
Gas constant R, ft lb/lb·°F	767.0	96.0	51.5	55.2
Gas constant R, J/kg·°C	4,126.0	518.0	277.0	297.0
Thermal conductivity, Btu/h·ft·°F	0.105	0.02	0.015	0.015
Thermal conductivity, W/m·°C	0.0182	0.035	0.026	0.026
Boiling point (sat 14.7 psia), °F	−423.0	−259.0	−240.0	−320.4
Boiling point (sat 760 mm), °C	20.4 K	−434.2	−151.5	−195.8
Latent heat of evap. (at bp), Btu/lb	192.0	219.2		85.5
Latent heat of evap. (at bp), J/kg	447,000.0	510,000.0		199,000.0
Freezing (melting) point, °F (1 atm)	−434.6	−296.6	−258.0	−346.0
Freezing (melting) point, °C (1 atm)	−259.1	−182.6	−161.0	−210.0
Latent heat of fusion, Btu/lb	25.0	14.0	32.9	11.1
Latent heat of fusion, J/kg	58,000.0	32,600.0	76,500.0	25,800.0
Critical temperature, °F	−399.8	−116.0	−136.0	−232.6
Critical temperature, °C	−240.0	−82.3	−93.3	−147.0
Critical pressure, psia	189.0	673.0	945.0	493.0
Critical pressure, MN/m²	1.30	4.64	6.52	3.40
Critical volume, ft³/lb	0.53	0.099	0.0332	0.051
Critical volume, m³/kg	0.033	0.0062	0.00207	0.00318
Flammable (yes or no)	Yes	Yes	No	No
Heat of combustion, Btu/ft³	320.0	985.0	—	—
Heat of combustion, Btu/lb	62,050.0	22,900.0	—	—
Heat of combustion, kJ/kg	144,000.0		—	

(*continued*)

TABLE C.22 (continued)
Chemical, Physical, and Thermal Properties of Gases: Gases and Vapors, Including Fuels and Refrigerants, English and Metric Units

Common Name(s)	Nitrous Oxide	Oxygen	Propane	Propylene (Propene)
Chemical Formula	N_2O	O_2		C_3H_6
Refrigerant Number	744A	732	290	1270
Chemical and physical properties				
Molecular weight	44.012	31.9988	44.097	42.08
Specific gravity, air = 1	1.52	1.105	1.52	1.45
Specific volume, ft³/lb	8.90	12.24	8.84	9.3
Specific volume, m³/kg	0.555	0.764	0.552	0.58
Density of liquid (at atm bp), lb/ft³	76.6	71.27	36.2	37.5
Density of liquid (at atm bp), kg/m³	1,227.0	1,142.0	580.0	601.0
Vapor pressure at 25°C, psia			135.7	166.4
Vapor pressure at 25°C, MN/m²			0.936	1.147
Viscosity (abs), lbm/ft·s	10.1×10^{-6}	13.4×10^{-6}	53.8×10^{-6}	57.1×10^{-6}
Viscosity (abs), centipoises[a]	0.015	0.020	0.080	0.085
Sound velocity in gas, m/s	268.0	329.0	253.0	261.0
Thermal and thermodynamic properties				
Specific heat, c_p, Btu/lb·°F or cal/g·°C	0.21	0.220	0.39	0.36
Specific heat, c_p, J/kg·K	879.0	920.0	1,630.0	1,506.0
Specific heat ratio, c_p/c_v	1.31	1.40	1.2	1.16
Gas constant R, ft lb/lb·°F	35.1	48.3	35.0	36.7
Gas constant R, J/kg·°C	189.0	260.0	188.0	197.0
Thermal conductivity, Btu/h·ft·°F	0.010	0.015	0.010	0.010
Thermal conductivity, W/m °C	0.017	0.026	0.017	0.017
Boiling point (sat 14.7 psia), °F	−127.3	−297.3	−44.0	−54.0
Boiling point (sat 760 mm), °C	−88.5	−182.97	−42.2	−48.3
Latent heat of evap. (at bp), Btu/lb	161.8	91.7	184.0	188.2
Latent heat of evap. (at bp), J/kg	376,000.0	213,000.0	428,000.0	438,000.0
Freezing (melting) point, °F (1 atm)	−131.5	−361.1	−309.8	−301.0
Freezing (melting) point, °C (1 atm)	−90.8	−218.4	−189.9	−185.0
Latent heat of fusion, Btu/lb	63.9	5.9	19.1	
Latent heat of fusion, J/kg	149,000.0	13,700.0	44,400.0	
Critical temperature, °F	97.7	−181.5	205.0	197.0
Critical temperature, °C	36.5	−118.6	96.0	91.7
Critical pressure, psia	1,052.0	726.0	618.0	668.0
Critical pressure, MN/m²	7.25	5.01	4.26	4.61
Critical volume, ft³/lb	0.036	0.040	0.073	0.069
Critical volume, m³/kg	0.0022	0.0025	0.0045	0.0043
Flammable (yes or no)	No	No	Yes	Yes
Heat of combustion, Btu/ft³	—	—	2,450.0	2,310.0
Heat of combustion, Btu/lb	—	—	21,660.0	21,500.0
Heat of combustion, kJ/kg	—	—	50,340.0	50,000.0

Source: The CRC Press Handbook of Thermal Engineering, CRC Press, Boca Raton, FL, 2000.
Note: The properties of pure gases are given at 25°C (77°F, 298 K) and atmospheric pressure (expect as stated).
[a] For N·s/m² divide by 1000.

TABLE C.23
Burning Velocities of Various Fuels

	φ = 0.7	φ = 0.8	φ = 0.9	φ = 1.0	φ = 1.1	φ = 1.2	φ = 1.3	φ = 1.4	S_{max}	φ at S_{max}
Saturated hydrocarbons										
Ethane	30.6	36.0	40.6	44.5	47.3	47.3	44.4	37.4	47.6	1.14
Propane			42.3	45.6	46.2	42.4	34.3		46.4	1.06
n-Butane		38.0	42.6	44.8	44.2	41.2	34.4	25.0	44.9	1.03
Methane		30.0	38.3	43.4	44.7	39.8	31.2		44.8	1.08
n-Pentane		35.0	40.5	42.7	42.7	39.3	33.9		43.0	1.05
n-Heptane		37.0	39.8	42.2	42.0	35.5	29.4		42.8	1.05
2,2,4-Trimethylpentane		37.5	40.2	41.0	37.2	31.0	23.5		41.0	0.98
2,2,3-Trimethylpentane		37.8	39.5	40.1	39.5	36.2			40.1	1.00
2,2-Dimethylbutane		33.5	38.3	39.9	37.0	33.5			40.0	0.98
Isopentane		33.0	37.6	39.8	38.4	33.4	24.8		39.9	1.01
2,2-Dimethylpropane			31.0	34.8	36.0	35.2	33.5	31.2	36.0	1.10
Unsaturated hydrocarbons										
Acetylene		107	130	144	151	154	154	152	155	1.25
Ethylene	37.0	50.0	60.0	68.0	73.0	72.0	66.5	60.0	73.5	1.13
Propyne		62.0	66.6	70.2	72.2	71.2	61.0		72.5	1.14
1,3-Butadiene			42.6	49.6	55.0	57.0	56.9	55.4	57.2	1.23
n-1-Heptyne		46.8	50.7	52.3	50.9	47.4	41.6		52.3	1.00
Propylene			48.4	51.2	49.9	46.4	40.8		51.2	1.00
n-2-Pentene		35.1	42.6	47.8	46.9	42.6	34.9		48.0	1.03
2,2,4-Trimethyl-3-pentene		34.6	41.3	42.2	37.4	33.0			42.5	0.98
Substituted alkyls										
Methanol		34.5	42.0	48.0	50.2	47.5	44.4	42.2	50.4	1.08
Isopropyl alcohol		34.4	39.2	41.3	40.6	38.2	36.0	34.2	41.4	1.04
Triethylamine		32.5	36.7	38.5	38.7	36.2	28.6		38.8	1 06
n-Butyl chloride	24.0	30.7	33.8	34.5	32.5	26.9	20.0		34.5	1.00
Allyl chloride	30.6	33.0	33.7	32.4	29.6				33.8	0.89
Isopropyl mercaptan		30.0	33.5	33.0	26.6				33.8	0.44
Ethylamine		28.7	31.4	32.4	31.8	29.4	25.3		32.4	1.00
Isopropylamine		27.0	29.5	30.6	29.8	27.7			30.6	1 01
n-Propyl chloride		24.7	28.3	27.5	24.1				28.5	0.93
Isopropyl chloride		24.8	27.0	27.4	25.3				27.6	0.97
n-Propyl bromide	No ignition									
Silanes										
Tetramethylsilane	39.5	49.5	57.3	58.2	57.7	54.5	47.5		58.2	1.01
Trimethylethoxysilane	34.7	41.0	47.4	50.3	46.5	41.0	35.0		50.3	1.00
Aldehydes										
Acrolein	47.0	58.0	66.6	65.9	56.5				67.2	0.95
Propionaldehyde		37.5	44.3	49.0	49.5	46.0	41.6	37.2	50.0	1.06
Acetaldehyde		26.6	35.0	41.4	41.4	36.0	30.0		42.2	1.05
Ketones										
Acetone		40.4	44.2	42.6	38.2				44.4	0.93
Methyl ethyl ketone		36.0	42.0	43.3	41.5	37.7	33.2		43.4	0.99
Esters										
Vinyl acetate	29.0	36.6	39.8	41.4	42.1	41.6	35.2		42.2	1.13
Ethyl acetate		30.7	35.2	37.0	35.6	30.0			37.0	1.00
Ethers										
Dimethyl ether		44.8	47.6	48.4	47.5	45.4	42.6		48.6	0.99
Diethyl ether	30.6	37.0	43.4	48.0	47.6	40.4	32.0		48.2	1.05
Dimethoxymethane	32.5	38.2	43.2	46.6	48.0	46.6	43.3		48.0	1.10
Diisopropyl ether		30.7	35.5	38.3	38.6	36.0	31.2		38.9	1.06

(*continued*)

TABLE C.23 (continued)
Burning Velocities of Various Fuels

	φ = 0.7	φ = 0.8	φ = 0.9	φ = 1.0	φ = 1.1	φ = 1.2	φ = 1.3	φ = 1.4	S_{max}	φ at S_{max}
Thio ethers										
Dimethyl sulfide		29.9	31.9	33.0	30.1	24.8			33.0	1.00
Peroxides										
Di-t-butyl peroxide		41.0	46.8	50.0	49.6	46.5	42.0	35.5	50.4	1.04
Aromatic compounds										
Furan	48.0	55.0	60.0	62.5	62.4	60.0			62.9	1.05
Benzene		39.4	45.6	47.6	44.8	40.2	35.6		47.6	1.00
Thiophane	33.8	37.4	40.6	43.0	42.2	37.2	24.6		43.2	1.03
Cyclic compounds										
Ethylene oxide	57.2	70.7	83.0	88.8	89.5	87.2	81.0	73.0	89.5	1.07
Butadiene monoxide		36.6	47.4	57.8	64.0	66.9	66.8	64.5	67.1	1.24
Propylene oxide	41.6	53.3	62.6	66.5	66.4	62.5	53.8		67.0	1.05
Dihydropyran	39.0	45.7	51.0	54.5	55.6	52.6	44.3	32.0	55.7	1.08
Cyclopropane		40.6	49.0	54.2	55.6	53.5	44.0		55.6	1.10
Tetrahydropyran	44.8	51.0	53.6	51.5	42.3				53.7	0.93
Cyclic compounds										
Tetrahydrofuran			43.2	48.0	50.8	51.6	49.2	44.0	51.6	1.19
Cyclopentadiene	36.0	41.8	45.7	47.2	45.5	40.6	32.0		47.2	1.00
Ethylenimine		37.6	43.4	46.0	45.8	43.4	38.9		46.4	1.04
Cyclopentane	31.0	38.4	43.2	45.3	44.6	41.0	34.0		45.4	1.03
Cyclohexane		41.3	43.5	43.9	38.0				44.0	1.08
Inorganic compounds										
Hydrogen	102	120	145	170	204	245	213	290	325	1.80
Carbon disulfide	50.6	58.0	59.4	58.8	57.0	55.0	52.8	51.6	59.4	0.91
Carbon monoxide				28.5	32.0	34.8	38.0		52.0	2.05
Hydrogen sulfide	34.8	39.2	40.9	39.1	32.3				40.9	0.90
Propylene oxide	74.0	86.2	93.0	96.6	97.8	94.0	84.0	71.5	97.9	1.09
Hydrazine	87.3	90.5	93.2	94.3	93.0	90.7	87.4	83.7	94.4	0.98
Furfural	62.0	73.0	83.3	87.0	87.0	84.0	77.0	65.5	87.3	1.05
Ethyl nitrate	70.2	77.3	84.0	86.4	83.0	72.3			86.4	1.00
Butadiene monoxide	51.4	57.0	64.5	73.0	79.3	81.0	80.4	76.7	81.1	1.23
Carbon disulfide	64.0	72.5	76.8	78.4	75.5	71.0	66.0	62.2	78.4	1.00
n-Butyl ether		67.0	72.6	70.3	65.0				72.7	0.91
Methanol	50.0	58.5	66.9	71.2	72.0	66.4	58.0	48.8	72.2	1.08
Diethyl cellosolve	49.5	56.0	63.0	69.0	69.7	65.2			70.4	1.05
Cyclohexene										
Monoxide	54.5	59.0	63.5	67.7	70.0	64.0			70.0	1.10
Epichlorohydrin	53.0	59.5	65.0	68.6	70.0	66.0	58.2		70.0	1.10
n-Pentane		50.0	55.0	61.0	62.0	57.0	49.3	42.4	62.9	1.05
n-Propyl alcohol	49.0	56.6	62.0	64.6	63.0	50.0	37.4		64.8	1.03
n-Heptane	41.5	50.0	58.5	63.8	59.5	53.8	46.2	38.8	63.8	1.00
Ethyl nitrite	54.0	58.8	62.6	63.5	59.0	49.5	42.0	36.7	63.5	1.00
Pinene	48.5	58.3	62.5	62.1	56.6	50.0			63.0	0.95
Nitroethane	51.5	57.8	61.4	57.2	46.0	28.0			61.4	0.92
Isooctane		50.2	56.8	57.8	53.3	50.5			58.2	0.98
Pyrrole		52.0	55.6	56.6	56.1	52.8	48.0	43.1	56.7	1.00
Aniline		41.5	45.4	46.6	42.9	37.7	32.0		46.8	0.98
Dimethyl formamide		40.0	43.6	45.8	45.5	40.7	36.7		46.1	1.04

Source: Compilation of laminar flame speed data from Gibbs and Calcote, *J. Chem. Eng. Data*, 4, 2226, 1959; *Combustion Science and Engineering*, CRC Press, Table A.39D, p. 1057.

Note: $T = 25°C$ (air–fuel temperature); $P = 1$ atm (0.31 mol % H_2O in air); burning velocity S as a function of equivalence ratio φ in cm/s. The data are for premixed fuel–air mixtures at 100°C and 1 atm pressure; 0.31 mol % H_2O in air; burning velocity S as a function of φ in cm/s.

Appendix C: Properties of Gases and Liquids

TABLE C.23

Burning Velocities of Various Fuels

	$\phi = 0.7$	$\phi = 0.8$	$\phi = 0.9$	$\phi = 1.0$	$\phi = 1.1$	$\phi = 1.2$	$\phi = 1.3$	$\phi = 1.4$	S_{max}	ϕ at S_{max}
Saturated hydrocarbons										
Ethane	30.6	36.0	40.6	44.5	47.3	47.3	44.4	37.4	47.6	1.14
Propane			42.3	45.6	46.2	42.4	34.3		46.4	1.06
n-Butane		38.0	42.6	44.8	44.2	41.2	34.4	25.0	44.9	1.03
Methane		30.0	38.3	43.4	44.7	39.8	31.2		44.8	1.08
n-Pentane		35.0	40.5	42.7	42.7	39.3	33.9		43.0	1.05
n-Heptane		37.0	39.8	42.2	42.0	35.5	29.4		42.8	1.05
2,2,4-Trimethylpentane		37.5	40.2	41.0	37.2	31.0	23.5		41.0	0.98
2,2,3-Trimethylpentane		37.8	39.5	40.1	39.5	36.2			40.1	1.00
2,2-Dimethylbutane		33.5	38.3	39.9	37.0	33.5			40.0	0.98
Isopentane		33.0	37.6	39.8	38.4	33.4	24.8		39.9	1.01
2,2-Dimethylpropane			31.0	34.8	36.0	35.2	33.5	31.2	36.0	1.10
Unsaturated hydrocarbons										
Acetylene		107	130	144	151	154	154	152	155	1.25
Ethylene	37.0	50.0	60.0	68.0	73.0	72.0	66.5	60.0	73.5	1.13
Propyne		62.0	66.6	70.2	72.2	71.2	61.0		72.5	1.14
1,3-Butadiene			42.6	49.6	55.0	57.0	56.9	55.4	57.2	1.23
n-1-Heptyne		46.8	50.7	52.3	50.9	47.4	41.6		52.3	1.00
Propylene			48.4	51.2	49.9	46.4	40.8		51.2	1.00
n-2-Pentene		35.1	42.6	47.8	46.9	42.6	34.9		48.0	1.03
2,2,4-Trimethyl-3-pentene		34.6	41.3	42.2	37.4	33.0			42.5	0.98
Substituted alkyls										
Methanol		34.5	42.0	48.0	50.2	47.5	44.4	42.2	50.4	1.08
Isopropyl alcohol		34.4	39.2	41.3	40.6	38.2	36.0	34.2	41.4	1.04
Triethylamine		32.5	36.7	38.5	38.7	36.2	28.6		38.8	1 06
n-Butyl chloride	24.0	30.7	33.8	34.5	32.5	26.9	20.0		34.5	1.00
Allyl chloride	30.6	33.0	33.7	32.4	29.6				33.8	0.89
Isopropyl mercaptan		30.0	33.5	33.0	26.6				33.8	0.44
Ethylamine		28.7	31.4	32.4	31.8	29.4	25.3		32.4	1.00
Isopropylamine		27.0	29.5	30.6	29.8	27.7			30.6	1 01
n-Propyl chloride		24.7	28.3	27.5	24.1				28.5	0.93
Isopropyl chloride		24.8	27.0	27.4	25.3				27.6	0.97
n-Propyl bromide	No ignition									
Silanes										
Tetramethylsilane	39.5	49.5	57.3	58.2	57.7	54.5	47.5		58.2	1.01
Trimethylethoxysilane	34.7	41.0	47.4	50.3	46.5	41.0	35.0		50.3	1.00
Aldehydes										
Acrolein	47.0	58.0	66.6	65.9	56.5				67.2	0.95
Propionaldehyde		37.5	44.3	49.0	49.5	46.0	41.6	37.2	50.0	1.06
Acetaldehyde		26.6	35.0	41.4	41.4	36.0	30.0		42.2	1.05
Ketones										
Acetone		40.4	44.2	42.6	38.2				44.4	0.93
Methyl ethyl ketone		36.0	42.0	43.3	41.5	37.7	33.2		43.4	0.99
Esters										
Vinyl acetate	29.0	36.6	39.8	41.4	42.1	41.6	35.2		42.2	1.13
Ethyl acetate		30.7	35.2	37.0	35.6	30.0			37.0	1.00
Ethers										
Dimethyl ether		44.8	47.6	48.4	47.5	45.4	42.6		48.6	0.99
Diethyl ether	30.6	37.0	43.4	48.0	47.6	40.4	32.0		48.2	1.05
Dimethoxymethane	32.5	38.2	43.2	46.6	48.0	46.6	43.3		48.0	1.10
Diisopropyl ether		30.7	35.5	38.3	38.6	36.0	31.2		38.9	1.06

(*continued*)

TABLE C.23 (continued)
Burning Velocities of Various Fuels

	φ = 0.7	φ = 0.8	φ = 0.9	φ = 1.0	φ = 1.1	φ = 1.2	φ = 1.3	φ = 1.4	S_{max}	φ at S_{max}
Thio ethers										
Dimethyl sulfide		29.9	31.9	33.0	30.1	24.8			33.0	1.00
Peroxides										
Di-*t*-butyl peroxide		41.0	46.8	50.0	49.6	46.5	42.0	35.5	50.4	1.04
Aromatic compounds										
Furan	48.0	55.0	60.0	62.5	62.4	60.0			62.9	1.05
Benzene		39.4	45.6	47.6	44.8	40.2	35.6		47.6	1.00
Thiophane	33.8	37.4	40.6	43.0	42.2	37.2	24.6		43.2	1.03
Cyclic compounds										
Ethylene oxide	57.2	70.7	83.0	88.8	89.5	87.2	81.0	73.0	89.5	1.07
Butadiene monoxide		36.6	47.4	57.8	64.0	66.9	66.8	64.5	67.1	1.24
Propylene oxide	41.6	53.3	62.6	66.5	66.4	62.5	53.8		67.0	1.05
Dihydropyran	39.0	45.7	51.0	54.5	55.6	52.6	44.3	32.0	55.7	1.08
Cyclopropane		40.6	49.0	54.2	55.6	53.5	44.0		55.6	1.10
Tetrahydropyran	44.8	51.0	53.6	51.5	42.3				53.7	0.93
Cyclic compounds										
Tetrahydrofuran			43.2	48.0	50.8	51.6	49.2	44.0	51.6	1.19
Cyclopentadiene	36.0	41.8	45.7	47.2	45.5	40.6	32.0		47.2	1.00
Ethylenimine		37.6	43.4	46.0	45.8	43.4	38.9		46.4	1.04
Cyclopentane	31.0	38.4	43.2	45.3	44.6	41.0	34.0		45.4	1.03
Cyclohexane		41.3	43.5	43.9	38.0				44.0	1.08
Inorganic compounds										
Hydrogen	102	120	145	170	204	245	213	290	325	1.80
Carbon disulfide	50.6	58.0	59.4	58.8	57.0	55.0	52.8	51.6	59.4	0.91
Carbon monoxide					28.5	32.0	34.8	38.0	52.0	2.05
Hydrogen sulfide	34.8	39.2	40.9	39.1	32.3				40.9	0.90
Propylene oxide	74.0	86.2	93.0	96.6	97.8	94.0	84.0	71.5	97.9	1.09
Hydrazine	87.3	90.5	93.2	94.3	93.0	90.7	87.4	83.7	94.4	0.98
Furfural	62.0	73.0	83.3	87.0	87.0	84.0	77.0	65.5	87.3	1.05
Ethyl nitrate	70.2	77.3	84.0	86.4	83.0	72.3			86.4	1.00
Butadiene monoxide	51.4	57.0	64.5	73.0	79.3	81.0	80.4	76.7	81.1	1.23
Carbon disulfide	64.0	72.5	76.8	78.4	75.5	71.0	66.0	62.2	78.4	1.00
n-Butyl ether		67.0	72.6	70.3	65.0				72.7	0.91
Methanol	50.0	58.5	66.9	71.2	72.0	66.4	58.0	48.8	72.2	1.08
Diethyl cellosolve	49.5	56.0	63.0	69.0	69.7	65.2			70.4	1.05
Cyclohexene										
Monoxide	54.5	59.0	63.5	67.7	70.0	64.0			70.0	1.10
Epichlorohydrin	53.0	59.5	65.0	68.6	70.0	66.0	58.2		70.0	1.10
n-Pentane		50.0	55.0	61.0	62.0	57.0	49.3	42.4	62.9	1.05
n-Propyl alcohol	49.0	56.6	62.0	64.6	63.0	50.0	37.4		64.8	1.03
n-Heptane	41.5	50.0	58.5	63.8	59.5	53.8	46.2	38.8	63.8	1.00
Ethyl nitrite	54.0	58.8	62.6	63.5	59.0	49.5	42.0	36.7	63.5	1.00
Pinene	48.5	58.3	62.5	62.1	56.6	50.0			63.0	0.95
Nitroethane	51.5	57.8	61.4	57.2	46.0	28.0			61.4	0.92
Isooctane		50.2	56.8	57.8	53.3	50.5			58.2	0.98
Pyrrole		52.0	55.6	56.6	56.1	52.8	48.0	43.1	56.7	1.00
Aniline		41.5	45.4	46.6	42.9	37.7	32.0		46.8	0.98
Dimethyl formamide		40.0	43.6	45.8	45.5	40.7	36.7		46.1	1.04

Source: Compilation of laminar flame speed data from Gibbs and Calcote, *J. Chem. Eng. Data*, 4, 2226, 1959; *Combustion Science and Engineering*, CRC Press, Table A.39D, p. 1057.

Note: T = 25°C (air–fuel temperature); P = 1 atm (0.31 mol % H_2O in air); burning velocity S as a function of equivalence ratio φ in cm/s. The data are for premixed fuel–air mixtures at 100°C and 1 atm pressure; 0.31 mol % H_2O in air; burning velocity S as a function of φ in cm/s.

Appendix C: Properties of Gases and Liquids

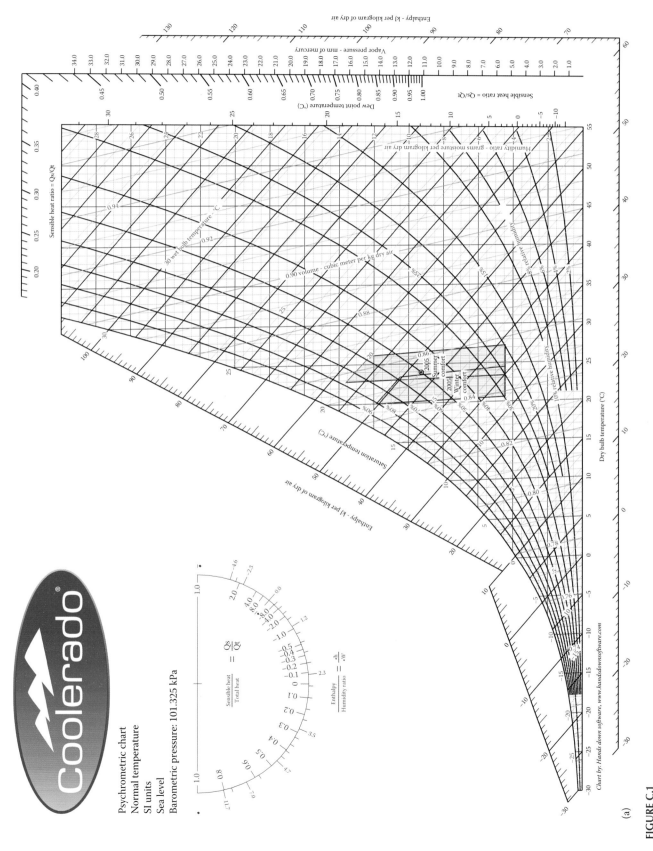

FIGURE C.1
Psychrometric chart. (a) SI units. (Courtesy of Coolerado.)

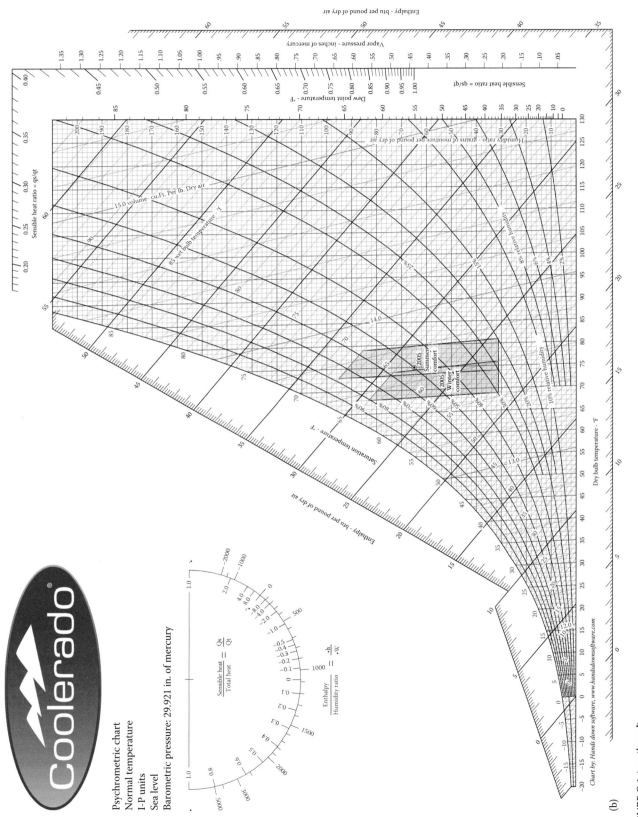

FIGURE C.1 (continued)
Psychrometric chart. (b) English units. (Courtesy of Coolerado.)

Appendix D: Properties of Solids

TABLE D.1

Thermal Properties of Selected Metallic Elements at 293 K (20°C) or 528°R (65°F)

Element	Specific Gravity	Specific Heat, c_p		Thermal Conductivity, k		Diffusivity, α		Melting Temperature	
		J/(kg K)	BTU/(lbm °R)	W/(m K)	BTU/(h ft °R)	m²/s × 10⁶	ft²/s × 10³	K	°R
Aluminum	2.702	896	0.214	236	136	97.5	1.05	933	1680
Beryllium	1.850	1750	0.418	205	118	63.3	0.681	1550	2790
Chromium	7.160	440	0.105	91.4	52.8	29.0	0.312	2118	3812
Copper	8.933	383	0.0915	399	231	116.6	1.26	1356	2441
Gold	19.300	129	0.0308	316	183	126.9	1.37	1336	2405
Iron	7.870	452	0.108	31.1	18.0	22.8	0.245	1810	3258
Lead	11.340	129	0.0308	35.3	20.4	24.1	0.259	601	1082
Magnesium	1.740	1017	0.243	156	90.1	88.2	0.949	923	1661
Manganese	7.290	486	0.116	7.78	4.50	2.2	0.0236	1517	2731
Molybdenum	10.240	251	0.0600	138	79.7	53.7	0.578	2883	5189
Nickel	8.900	446	0.107	91	52.6	22.9	0.246	1726	3107
Platinum	21.450	133	0.0318	71.4	41.2	25.0	0.269	2042	3676
Potassium	0.860	741	0.177	103	59.6	161.6	1.74	337	607
Silicon	2.330	703	0.168	153	88.4	93.4	1.01	1685	3033
Silver	10.500	234	0.0559	427	247	173.8	1.87	1234	2221
Tin	5.750	227	0.0542	67.0	38.7	51.3	0.552	505	909
Titanium	4.500	611	0.146	22.0	12.7	8.0	0.0861	1953	3515
Tungsten	19.300	134	0.0320	179	103	69.2	0.745	3653	6575
Uranium	19.070	113	0.0270	27.4	15.8	12.7	0.137	1407	2533
Vanadium	6.100	502	0.120	31.4	18.1	10.3	0.111	2192	3946
Zinc	7.140	385	0.0920	121	69.9	44.0	0.474	693	1247

Source: Data from *Engineering Heat Transfer*, 2nd edn., CRC Press, Table B.1, p. 643.

Notes: Density = ρ = specific gravity × 62.4 lbm/ft³ = specific gravity × 1000 kg/m³.
Diffusivity = α; for aluminum, α m²/s × 10⁶ = 97.5; so $\alpha = 97.5 \times 10^{-6}$ m²/s.
Also, $\alpha = k/\rho c_p$.

TABLE D.2

Thermal Properties of Selected Alloys

Alloy	Composition	Specific Gravity	Specific Heat, c_p J/(kg K)	Specific Heat, c_p BTU/(lbm °R)	Thermal Conductivity, k W/(m K)	Thermal Conductivity, k BTU/(h ft °R)	Diffusivity, α m²/s × 10⁵	Diffusivity, α ft²/s × 10⁴	Coeff. of Linear Expansion μm/m K	Coeff. of Linear Expansion μin./in. °F	Approximate Melting Point °C	Approximate Melting Point °F
Aluminum												
Aluminum alloy 3003, rolled	ASTM B221	2.73			155.7	90			23.2	12.9	649	1200
Aluminum alloy 2017, annealed	ASTM B221	2.8			164.4	95			22.9	12.7	641	1185
Aluminum alloy 380	ASTM SC84B	2.7			96.9	56			20.9	11.6	566	1050
Duralumin	95 Al, 5 Cu	2.787	833	0.199	164	94.7	6.676	7.187				
Silumin	87 Al, 13 Si	2.659	871	0.208	164	94.7	7.099	7.642				
Copper												
Copper	ASTM B152, B124, B133, B1, B2, B3	8.91			389.3	225			16.7	9.3	1082	1980
Red brass (cast)	ASTM B30, No. 4A	8.7			72.7	42			18.0	10.0	996	1825
Yellow brass (high brass)	ASTM B36, B134, B135	8.47			119.4	69			18.9	10.5	932	1710
Aluminum bronze	ASTM B169, Alloy A; ASTM B124, B150	7.8			70.9	41			16.6	9.2	1038	1900
Beryllium copper 25	ASTM B194	8.25			12.1	7			16.7	9.3	927	1700
A-bronze	95 Cu, 5 Al	8.666	410	0.0979	83	47.9	2.330	2.508				
Bronze	75 Cu, 25 Sn	8.666	343	0.0819	26	15.0	0.859	0.925				
Red brass	85 Cu, 9 Sn, 6 Zn	8.714	385	0.0920	61	35.2	1.804	1.942				
Brass	70 Cu, 30 Zn	8.522	385	0.0920	111	64.1	3.412	3.673				
German silver	62 Cu, 15 Ni, 22 Zn	8.618	394	0.0941	24.9	14.4	0.733	0.789				
Constantan	60 Cu, 40 Ni	8.922	410	0.0979	22.7	13.1	0.612	0.659				
Cupronickel	30%	8.95			29.4	17			15.3	8.5	1227	2240
Cupronickel	55-45 (Constantan)	8.9			22.5	13			14.6	8.1	1260	2300

Appendix D: Properties of Solids

Material	Type	Specification	Density		Specific heat									
Iron														
	Ingot iron		7.86			72.7	42			12.2	6.8	1538	2800
	Cast gray iron	ASTM A48-48, Class 25	7.2			45.0	26			12.1	6.7	1177	2150
	Malleable iron	ASTM A47	7.32							11.9	6.6	1232	2250
	Ductile cast iron	ASTM A339, A395	7.2			32.9	19			13.5	7.5	1149	2100
	Ni-resist cast iron	type 2	7.3			39.8	23			17.3	9.6	1232	2250
	Cast iron	4 C	7.272	420	0.100	52	30.0	1.702	1.832				
	Wrought iron	0.5 CH	7.849	460	0.110	59	34.1	1.626	1.750				
Steel														
	Plain carbon steel	AISI-SAE 1020	7.86			51.9	30			12.1	6.7	1516	2760
	Carbon steel	1 C	7.801	473	0.113	43	24.8	1.172	1.262				
		1.5 C	7.753	486	0.113	36	20.8	0.970	1.040				
	Chrome steel	1 Cr	7.865	460	0.110	61	35.2	1.665	1.792				
		5 Cr	7.833	460	0.110	40	23.1	1.110	1.195				
		10 Cr	7.785	460	0.110	31	17.9	0.867	0.933				
	Chrome-nickel steel	15 Cr, 10 Ni	7.865	460	0.110	19	11.0	0.526	0.577				
		20 Cr, 15 Ni	7.833	460	0.110	15.1	8.72	0.415	0.447				
	Nickel steel	10 Ni	7.945	460	0.110	26	15.0	0.720	0.775				
		20 Ni	7.993	460	0.110	19	11.0	0.526	0.566				
		40 Ni	8.169	460	0.110	10	5.78	0.279	0.300				
		60 Ni	8.378	460	0.110	19	11.0	0.493	0.531				
	Nickel-chrome steel	80 Ni, 15 C	8.522	460	0.110	17	9.82	0.444	0.478				
		40 Ni, 15 C	8.073	460	0.110	11.6	6.70	0.305	0.328				
	Manganese steel	1 Mn	7.865	460	0.110	50	28.9	1.388	1.494				
		5 Mn	7.849	460	0.110	22	12.7	0.637	0.686				
	Silicon steel	1 Si	7.769	460	0.110	42	24.3	1.164	1.164				
		5 Si	7.417	460	0.110	19	11.0	0.555	0.597				
	Stainless steel	Type 304	8.02	461	0.110	14.4	8.32	0.387	0.417	17.3	9.6	1427	2600
		Type 347	7.97	461	0.110	14.3	8.26	0.387	0.417				
	Tungsten steel	1 W	7.913	448	0.107	66	31.1	1.858	2.000				
		5 W	8.073	435	0.104	54	31.2	1.525	1.642				

(continued)

TABLE D.2 (continued)
Thermal Properties of Selected Alloys

Alloy	Composition	Specific Gravity	Specific Heat, c_p		Thermal Conductivity, k		Diffusivity, α		Coeff. of Linear Expansion		Approximate Melting Point	
			J/(kg K)	BTU/(lbm °R)	W/(m K)	BTU/(h ft °R)	m²/s × 10⁵	ft²/s × 10⁴	µ m/m K	µ in./in. °F	°C	°F
Other												
Chemical lead		11.35			34.6	20			29.5	16.4	327	621
Antimonial lead (hard lead)		10.9			29.4	17			27.2	15.1	290	554
Magnesium alloy	AZ31B	1.77			77.9	45			26.1	14.5	627	1160
Nickel	ASTM B160, B161, B162	8.89			60.6	35			11.9	6.6	1441	2625
Nickel silver 18% alloy A (wrought)	ASTM B122, No. 2	8.8			32.9	19			16.2	9.0	1110	2030
Commercial titanium		5			17.3	10			8.8	4.9	1816	3300
Zinc	ASTM B69	7.14			107.3	62			32.4	18.0	418	785
Zirconium, commercial		6.5			17.3	10			5.2	2.9	1843	3350
Cast 28-7 alloy (HD)	ASTM A297-63T	7.6			2.6	1.5			16.6	9.2	1482	2700
Hastelloy	C	3.94			8.7	5			11.3	6.3	1288	2350
Inconel	X, annealed	8.25			15.6	9			12.1	6.7	1399	2550
Haynes Stellite	alloy 25 (L605)	9.15			9.5	5.5			13.7	7.6	1371	2500
K Monel		8.47			19.0	11			13.3	7.4	1349	2460
Solder	50-50	8.89			45.0	26			23.6	13.1	216	420

Sources: Engineering Heat Transfer, 2nd edn., CRC Press, Table B.2, p. 644; *The CRC Handbook of Mechanical Engineering*, CRC Press, Table C.6.

TABLE D.3

Thermal Properties of Selected Building Materials and Insulations at 293 K (20°C) or 528°R (65°F)

Material	Specific Gravity	Specific Heat, c_p		Thermal Conductivity, k		Diffusivity, α	
		J/(kg K)	BTU/(lbm °R)	W/(m K)	BTU/(h ft °R)	m²/s × 10⁵	ft²/s × 10⁶
Asbestos	0.383	816	0.195	0.113	0.0653	0.036	3.88
Asphalt	2.120			0.698	0.403		
Bakelite	1.270			0.233	0.135		
Brick							
Carborundum (50% SiC)	2.200			5.82	3.36		
Common	1.800	840	0.201	0.38–0.52	0.22–0.30	0.028–0.034	3.0–3.66
Magnesite (50% MgO)	2.000			2.68	1.55		
Masonry	1.700	837	0.200	0.658	0.38	0.046	5.0
Silica (95% SiO₂)	1.900			1.07	0.618		
Cardboard				0.14–0.35	0.08–0.2		
Cement (hard)				1.047	0.605		
Clay (48.7% moist)	1.545	880	0.210	1.26	0.728	0.101	10.9
Coal (anthracite)	1.370	1260	0.301	0.238	0.137	0.013–0.015	1.4–1.6
Concrete (dry)	0.500	837	0.200	0.128	0.074	0.049	5.3
Cork board	0.150	1880	0.449	0.042	0.0243	0.015–0.044	1.6–4.7
Cork (expanded)	0.120			0.036	0.0208		
Earth (diatomaceous)	0.466	879	0.210	0.126	0.072	0.031	3.3
Earth (clay with 28% moist)	1.500			1.51	0.872		
Earth (sandy with 8% moist)	1.500			1.05	0.607		
Glass fiber	0.220			0.035	0.02		
Glass (window pane)	2.800	800	0.191	0.81	0.47	0.034	3.66
Glass (wool)	0.200	670	0.160	0.040	0.023	0.028	3.0
Granite	2.750			3.0	1.73		
Ice at 0°C	0.913	1830	0.437	2.22	1.28	0.124	13.3
Kapok	0.025			0.035	0.02		
Linoleum	0.535			0.081	0.047		
Mica	2.900			0.523	0.302		
Pine bark	0.342			0.080	0.046		
Plaster	1.800			0.814	0.47		

Source: Engineering Heat Transfer, 2nd edn., CRC Press, Table B.3, p. 645.

Notes: Density = ρ = specify gravity × 62.4 lbm/ft³ = specific gravity × 1000 kg/m³.

Diffusivity = α; for asbestos, $\alpha \times 10^3 = 0.036$ m²/s; so $\alpha = 0.036 \times 10^{-3}$ m²/s also, $\alpha = k/\rho\, c_p$.

Index

A

Absolute entropy, substances, 494
Absorbers and scrubbers
    installation, 412–413
    maintenance
        periodic on-line items, 413
        shutdown items, 413
    operations
        caustic injection pump, 434
        description, 433–434
        recirculation pump, 434
        start-up, 433
    packed quench absorber tower, 412
    troubleshooting, 435
Accident mitigation
    design engineering, *see* Design engineering
    emergency pull ring, shut down entire facility, 32, 33
    emergency shower, 32, 33
    examples, safety and medical kits, 32, 33
    fire alarm system, combustion test facility, 31, 32
    large portable fire extinguisher, 31, 32
    shutdowns, severity of incident, 31
    wind sock, 32, 34
Accident prevention
    cabinet example, store flammable, 30
    combustion testing, 27
    containment system designed, 30, 31
    description, 27
    emergency stop pushbuttons, control room, 28, 29
    furnace camera, 28, 29
    ignition control
        description, 27
        static electricity, 28
    liquid fuel containment, diesel storage tanks, 28, 30
    NO, CO, O_2 and combustible analyzers, 30, 31
    rubber mat over tripping hazard, 28
    safety signs, outside of building, 30, 31
    safety tape around test apparatus, 28, 29
    test furnace, emergency stop pushbutton, 28, 30
AIChEs, *see* American Institute of Chemical Engineers (AIChEs)
Air-assisted flares
    blower failure test, 219, 221
    blower goes, off to full on, 217, 220
    description, 118
    high flow rate, propane, 219
    internal burning and failure, stress corrosion cracking, 119, 120
    pre-start-up, 380–381
    start-up and shutdown, 385
Air delivery systems, burner test
    forced draft systems, 194–195
    mobile air preheater, 195, 196
    natural draft, 194
Airflow rate vs. air valve position, 65
Air leaks
    description, 327
    idled burners, 327
    open sight door, 327
    at sight and access door, 327, 328
    thermal image, open explosion door, 327
Air metering
    description, 156
    forced draft, 157
    natural draft, 156–157
    typical throat, raw gas burner, 156, 157
Air Movement and Control Association International (AMCA) standard, 73, 76
Air registers and dampers
    burner damper, 280, 281
    burners, air handles, 280, 281
    description
        dual-blade air dampers, 254, 262
        jackshaft system, 255, 263
        rotary-type air register, 254, 263
    dual-bladed opposed motion design, 280, 281
    E-Z Roll bearings, 280
    locking air control, 280, 281
    rotating-type air registers, 280
    smooth damper operation, 281, 282
Air valve characterizer
    data, 65
    shape, 65
    typical butterfly-type valve calculation, 65
Ambient atmospheric air, 152
American Institute of Chemical Engineers (AIChEs)
    PHA evaluation techniques, 39
    safety documentation feedback flowchart, 38, 39
    safety training program, 41
    SOPs, 40
American Petroleum Institute (API), 27
Analog control systems
    diagnostics and flexibility, 48
    feedback, 50
    feedforward loop, 50
    sensors and elements, 48
    setpoint (SP), 49
    simple analog loop, 49, 50
Analog devices
    carbon monoxide (CO) analyzer, 62
    conductivity analyzer, 61
    control valves, *see* Control valves
    description, 61
    flow meters, *see* Flow meters
    nitrogen oxide (NO_2) analyzer, 62
    oxygen (O_2) analyzer, 62
    pH analyzer, 61
    pressure transmitters, 60
    RTDs, 60
    thermocouples, 59
    velocity thermocouples, 59–60
Anchoring systems, refractory
    ceramic (refractory), 146
    curl, 147
    distance below refractory surface, 145
    examples, 142, 143
    Flexmesh, 147
    Hexmesh, 146–147
    K-Bar, 147
    primary function, 142
    S-Bar, 147
    spacing, 145
    steel fiber reinforcing, *see* Steel fiber reinforcing
    Tacko, 147
    temperature, stainless steel grades, 142
    typical anchor spacing, 142, 143
    V-Anchors, *see* V-Anchors
    wire anchors and components, 142
    Y-anchors, 146
Annunciators, 56
API, *see* American Petroleum Institute (API)
API-936 installation
    curing, drying and firing, 149
    description, 147
    installer certification, 148–149
    laboratory testing, 150
    materials requirement, 147–148
    refractory failure, 147, 148

refractory lining systems
  inspection, 149, 150
  repair, 149
shipping refractory equipment, 150
surface preparation, 148
Autoignition temperature, 4
Auto-oxidation, 4

B

Bearings and lubrication, blowers
  description, 86
  emergency shutdown, 86
  maintenance, arrangement 8
    bearings, 87
  oil-lubricated with reservoir, 86
  sleeve and oil, 86
  vibration, 86–87
Bench-scale testing, 192
Blowers, combustion systems
  AMCA standards, 73
  applications, 73
  backward curved blade operating curve, 79, 80
  balanced draft system, 77
  basic centrifugal fan curve, 78, 79
  basic vane axial fan curve, 78, 79
  bearings and lubrication, 86–87
  centrifugal fans, 74, 75
  construction materials, 84
  couplings and belts, see Couplings and belts, blowers
  density vs. horsepower, 78
  description, 73
  experience, usage and careful evaluation, 78
  fan arrangements, see Fan arrangements, blowers
  fan control, see Fan control, blowers
  fan wheel designs, 74
  filtration, 89
  forward-tip blade operating curve, 79, 80
  friction factor calculation, 78
  HP, basic centrifugal fan curve, 79
  inspection and testing, 90, 92
  maintenance and troubleshooting, 92
  motors and drives, see Motors and drives, blowers
  multistage high-speed centrifugal blower, landfill application, 75
  noise considerations, 89
  operational costs, see Operational costs, blowers
  pressure drop, 78
  primary and backup fan, field with ducting, 79, 81
  purge air blower, combustion chamber, 75
  relative characteristics, centrifugal, 75
  shaft seals, 89
  six-blade vane axial fan, 79, 81
  speed change fan laws, 79–82
  temperature and pressure vs. volume and horsepower, 77–78
  vane axial fan, 74–75
  vibration and installation, see Vibration and installation, blowers
BMS, see Burner management systems (BMS)
Boilers
  installation, 403
  maintenance
    periodic on-line items, 404–405
    shutdown items, 404
  operation
    description, 423
    normal operating instructions, 426
    out of service, 426–427
    into service, 426
    shutdown procedure, 425
    start-up after temporary shutdown, 423, 425
    U. S. Department of Energy-Federal Energy Management Program, 423–425
    watertube boiler, 423
  troubleshooting
    corrosion/pitting, 427–428
    fireside, 427
    foaming and priming, 427
    instrumentation side, 427
    scale, 427–428
    waterside, 427
Bourdon tube gauge
  calibration, pressure
    oil-type deadweight tester, 176
    thermal expansion, piston, 176
  common failure mechanisms
    harsh environments, 175
    pressure snubbers, 176
    vibrating environments, 175
  definition, 174
  design
    components, 174, 175
    gear and linkage system, 174, 175
    indicating needle, 174–175
    pressure gauge and its internal view, 174, 175
    sockets, 175
    tube, 175
  installation
    design considerations, static pressure taps, 177
    static pressure measurement, 176–177
  selection, 176
Bourdon tube pressure gauge, 243

Breakaway corrosion
  chromium, 100
  cyclic oxidation, 100
  defined, 100
  oxide scale removal, thermal cycling, 100
  service temperatures, 300-series SSs, 100, 101
Brick refractory products
  abrasion and corrosion resistance, 136
  characteristics
    challenges, 137
    common firebricks, 137
    firebricks cut and installed, circular pattern, 137, 139
    general duty, high-alumina and insulating firebricks, 137
    labor-intensive brick installation, 137, 139
    shapes and sizes, standard bricks, 137, 138
  description, 136
  expensive combustion chamber shell, 137
  installation
    hacking, 137
    pogo sticks, 137, 169
    problems, 137
Burner and combustion, 4
Burner controller
  advantages and disadvantages, 46–47
  sequence and safety parameters, 46
Burner design
  air control
    examples, 157–158
    fallacy, burner specification, 158
    industry groups, 157
    limitations, 157
    theoretical opening, 158, 159
  "batching" method, 152
  combustion, see Combustion, burner design
  definition, 152
  design functions, 152
  firing configuration, mounting and direction
    conventional burner, round flame, 168
    description, 168
    downfired, 170
    flat flame burner, see Flat flame burner
    radiant wall, 169–170
  fuel/air mixing
    co-flow, 160
    cross-flow, 160
    description, 158
    developmental mechanisms, 159

dual-reactant flame, 159
emission control, 159
entrainment, 159–160
flow stream disruption, 160
limitations, 159
ignition maintenance
   flame stabilizer/holder, 160
   fuel/air mixture, 161
   ledge in burner tile, 160
   near-stoichiometric proportion, 160
   swirler, 160, 161
materials selection, 170–171
metering
   air (combustion O_2), see Air metering
   fuel, see Fuel metering
minimizing, pollutants, 162
mold, patterned and controlled flame shape
   description, 161
   flat-shaped, 161, 162
   fuel injectors, 161
   furnace chamber and flames, 161
   round-shaped, 161
operations, 5 Ms, 153
specifications, operating conditions, 153
types
   conventional process heaters, 167
   description, 162
   high intensity, Keu combustor, see Keu combustor
   oil/liquid firing, see Oil/liquid firing
   premix and partial premix gas, 162–163
   raw gas/nozzle mix, 163
Burner gas, 247
Burner/heater operations
air leakage, oxygen analyzer, 300
description, 300
external inspection, see External inspection, heater
failure to purge heater before light-off, 300, 301
flame impingement, process tubes, 300
flames outside heater, 300
heater combustion control, see Heater combustion control
heater monitoring
   combustion air measurements, 312–314
   excess air and oxygen, see Excess air
   flue gas temperatures, 314
   fuel measurements, see Fuel measurements, burner/heater operations
   heater draft, 301–303
   operators, 317
   process fluid parameters, 316–317
   process tube temperature, 315–316
   typical control limits, 317, 318
heater safety, see Heater safety
leak in process tube, 300, 301
refining and petrochemical industries, 300
strategy and goals, 317
turndown operation, 324
visual inspection, see Visual inspection, heater
Burner installation
air registers and dampers, 254–255
burner inspection, 254, 261
electrical connections, see Electrical connections
fuel piping, see Fuel piping
gas tip/diffuser cone position, 254, 260
gas tip location, 254, 261
mounting, see Mounting, burner
radiant wall tip location, 254, 261
tile, see Tile installation
Burner maintenance
burner tile, 262
description, 262
fuel piping and gas tips, 262
gas tip cleaning, see Gas tip cleaning
recording operating information, 262, 266
visual inspection, 262, 265
Burner management systems (BMS)
critical input checking, 52
watchdog timer, 52
Burner/pilot
installation, 397
maintenance
   periodic on-line items, 399
   shutdown items, 398–399
operations, 418–419
troubleshooting, see Troubleshooting
Burner spacing/flame interaction
CFD models, 352, 353
cutouts, 353
description, 352
"flame rollover", 352
high-and low-duty zones, 353
indications, problem, 352
Reed walls, 353
ultralow NOx burners, 352
Burner testing
aerial view, industrial combustion testing facility, 193, 194
air delivery systems, 194–195
benefit and drawbacks, 192
vs. CFD, 192–193
commercial, 192
CO probe, 198
description, 191
flue gas, see Flue gas
fuel flow and composition
   advantage and drawbacks, 196
   Coriolis meter, 196
   measurement, 195–196
fuel selection
   LHV and RFG, 199
   process flow diagram, 199, 200
heat flux
   description, 197–198
   fired wall via access ports and critical parameters, 198
   hostile environment, full-scale furnace, 197
   plug-shaped thermopile element and thermopile-type sensor, 197
instrumentation and control, 195
noise, 198
parameter and measurements, 193
research and development purposes, 192
resolution, existing applications issues, 192
test furnaces
   cooling and temperature control, 194
   ethylene applications, 193–195
   insulation and cooling tubes, 194
   simulating terrace wall-fired heaters, 194, 195
   terrace wall firing and real-world applications, 193
   vertical cylindrical furnace, freestanding, upfired, 194, 196
test procedure, 199, 201
UHCs and oxides of sulfur, 198
Burner tile
air and fuel mixing, 246
combustion air metering, 246
emissions, 247
flame molding, 247
gas tip cleaning, see Gas tip cleaning
stability maintenance, 246–247
Burner troubleshooting
combustion air issues, see Combustion air issues
description, 333
effective and safe steps, 333
emissions, see Emissions
failure, light burners, see Light burners
flame/flue gas patterns, see Flame/flue gas patterns
fuel gas problems, see Fuel gas problems
oil firing problems, see Oil firing problems

operating problems, 333
symptoms, 333
Burning velocities, fuel, 501–502

C

Carbon monoxide (CO) analyzer, 62
Carbon steel
  compound
    classification, iron-carbon alloys, 96
    composition ranges, 96, 97
    iron, 96
    low-carbon and mild-carbon, 96
  iron oxide scale
    electron microscope image, corroded carbon steel, 97
    reddish-brown rust, 97
  welding process
    binary iron-carbon phase diagram, 120, 122
    martensite, 122
    maximum hardness vs. % carbon, weight, 122, 124
    medium-carbon alloy steel AISI 4340, 122, 123
    microstructural changes, 120, 122
    qualified welders, 122
    tensile and bend specimens, 124
    typical TTT diagram, medium-carbon steel, 122, 123
Carburization corrosion
  carbonaceous environments, 102
  internal carbide formation and metal dusting, 102
  metal dusting, 103
  pack hardening, 102
  petrochemical industry, 102
CARS, see Coherent anti-Stokes-Raman scattering (CARS)
Ceramic (refractory) anchors, 146
CFD, see Computational fluid dynamics (CFD)
Charpy impact test
  apparatus, 103, 104
  high-carbon, low-carbon and austenitic SS, 103, 104
  specimens, carbon steel at -50°F(-10°C), 103, 104
Chemically (phos) bonded castables, 135
Chemically bonded plastics
  anchoring system, 136
  installation, 136
  pieces, 136
Chemical, physical and thermal properties, gases, 496–500
Chemical species analysis
  analytical methods and instrumentation
    commercial test rack, 238, 239
    factors, 240
    FTIR gas analyzer, 238, 239
    gas chromatography (GC), 239–240
    in situ oxygen analyzer, 240
    mass and energy balances, 238
    portable combustion gas analyzer, 238, 239
  sample collection
    electrochemical sensors, 240
    extractive sample probe, 240
    factors, 240–241
Chlorination attack, metal failures, 102
Coen iScan® flame detector, 24
Coherent anti-Stokes-Raman scattering (CARS), 186
Cold flow visualization, 217
Combustible liquid, 4
Combustion air issues
  ambient weather conditions
    burner sizing, 373–374
    corrective/preventive actions, 373, 374
    description, 372
    humidity, rain and low air temperatures, 372
    jobsite elevation, 373
    oxygen vs. atmospheric pressure, 372, 373
    oxygen vs. relative humidity, 372
  burner sizing, 373–374
  forced-draft system
    designing, uniform air distribution, 375
    selection, correct size burner, 374–375
Combustion air measurements
  flow, 314
  pressure, 314
  temperature, 312, 314
Combustion, burner design
  "black art", 152
  definition, 152
  graph, methane sustainable combustion, 153
  O_2 sources, 152
  temperature vs. $\%O_2$ (gross) relationship, 152–153
  3 Ts, 153
  variance, fuel properties, 153
Combustion control
  agency approvals and safety
    combustion process feeds, 54
    double block and bleed, fuel supply, 51
    electrical safety and reliability, 51
    inside, large control panel, 54, 55
    large and small control panels, 54, 55
    MFT, see Master fuel trip (MFT)
    NFPA and NEC standards, 51
    pipe racks, 54
    PLC operation, see Programmable logic controller (PLC)
    system shutdown, see System shutdown
  analog control systems, 48–50
  controllers
    analog and digital, 67, 69
    automatic reset, 70
    CV and MV, 69
    defined, proportional band (PB), 69–70
    with manual reset, 69
    operation modes, 70
    reverse/direct acting, 70
    SP signals, 69
  control platforms, see Control platforms
  control schemes
    air SP, O_2 trim, 67, 68
    definition, controller output mode, 66
    flow meters and controllers, 65, 66
    high and low signal selectors, 66–67
    "lead-lag" control, 67
    meter output signal scaling, 65–66
    multiple fuels and O_2 sources, 67, 68
    multiplication function (X), 67
    O_2 trim, airflow rate, 67
    parallel positioning, see Parallel positioning, combustion
  DCSs, see Distributed control systems (DCSs)
  failure modes, 50
  physical and logical components, 46
  primary measurement
    analog devices, see Analog devices
    description, 54
    discrete devices, see Discrete devices
  safety parameters, 46
  tuning
    feedforward system, 71
    graphic recorder, 71
    parameters, 70
Combustion data, hydrocarbons, 495
Combustion diagnostics
  advanced
    FTIR, see Fourier transform infrared (FTIR) spectroscopy
    invasive techniques, 185–186
    LPLIF, see Liquid planar laser-induced fluorescence (LPLIF)
    NOx analysis and CO probing, 185
    optics/laser instruments, 186

Index

PDPA, *see* Phase Doppler
  particulate anemometer
  (PDPA)
flow measurement, *see* Flow
  measurements
pressure management
  Bourdon tube gauge, *see* Bourdon
    tube gauge
  manometer, 173–174
Combustion tetrahedron
  dry chemical/halogenated
    hydrocarbon, 6
  elements, 4
  fire, 4, 5
  oxidizers, 5
  solid fuels, 5
  standard temperature and pressure
    flammability limits, 5
    minimum ignition
      temperatures, 5–6
Computational fluid dynamics (CFD),
  192–193, 420
Conductivity analyzer, 61
Confined explosion and deflagration, 4
Control platforms
  burner controller, 46–47
  DCS, 47
  description, 46
  hybrid systems, 48
  loop controller, 47
  PLC, *see* Programmable logic
    controller (PLC)
  profibus/fieldbus, 48
  relay system
    advantages and disadvantages, 46
    definition, 46
    LCD faceplate, 46
    PLC functions, 46
  touchscreen, 48
Control valves
  actuator, 58
  body, 58
  characteristics, 58
  current-to-pressure transducer, 58
  description, 58
  mechanical stops, 59
  pneumatic, 58
  positioner, 58–59
  three-way (3-way) solenoid
    valve, 59
Conventional DBA-style burner,
  367, 368
COOLstar™ ultralow NOx burner,
  368, 369
Coriolis flow meters, 61
Coriolis meter, 196
Couplings and belts, blowers
  belt-driven centrifugal, 85, 86
  flexible coupling, 85
  protective guards, 86

Cryogenic service, metal failures
  Charpy impact test, 103–104
  description, 103
  normalized and not normalized
    carbon steel, 105
  steel transitions, 105
Curl anchors, 147
Current-to-pressure transducer, 58

D

Data acquisition system
  flare conditions, 215
  flare test control center, 211
  high-speed noise signal processor
    and front end, 212
  HMI, 211–212
  noise, 214–215
  performance parameters, 211
  thermal radiation, 212–214
  upwind and crosswind, flame, 212
DCSs, *see* Distributed control systems
  (DCSs)
Design engineering
  fire extinguishment, 38
  flammability characteristics
    and ignition, liquids and gases,
      33, 35
    liquids, 33–34, 36
    vapors, 34–36
  ignition control
    adiabatic compression,
      combustible/flammable
      materials, 37
    autoignition temperatures, 35, 37
    auto-oxidation, 37
    description, 36
    ethylene oxide plant explosion,
      37, 38
    flare burners, 36
    MIEs, 36, 37
    sources, major fires, 36, 37
    static electricity, 37
    thermodynamic relationships, 37
  industrial combustion applications,
    32–33
Diesel engine exhaust, 152
Digital signal processing (DSP), 215
Discrete devices
  annunciators, 56
  flame scanners, 57
  flow switches, 56–57
  ignition transformers, 57
  position switches, 56
  pressure switches, 56
  run indicators, 57
  solenoid valves, 57
  temperature switches, 56
Distributed control systems (DCSs)
  analog and discrete (on/off) control, 47

description, 47
devices, 47
safety monitoring and
  sequencing, 48
sensors, 48
simplified flow diagram, standard
  burner light-off sequence,
  48, 49
types, discrete devices, 48
vendor, 48
Double-block-and-bleed system, 51
Double-hooked V-anchors, 145
Downfired burners, 170
DSP, *see* Digital signal
  processing (DSP)

E

Electrical connections
  burners, flame scanners, 260, 265
  description, 259–260
  installation and troubleshooting,
    equipment, 262
  operating and installation
    manual, 260
  pilot conduit boxes, 260, 265
  "swivel"-type, 260
Emissions
  carbon monoxide
    combustibles/VOCs/UHCs, 371
    description, 371
    open inspection port, 371
  nitrogen oxides
    burner type, 368–369
    combustion air temperature, 371
    description, 367–368
    excess air, 370
    firebox temperature, 369–370
    fuel composition, 370
    relative humidity, 371
  particulate matter, 371–372
  sulfur oxides
    acid rain, 371
    description, 371
    SO_2 and SO_3 formation, 371
Endothermic waste gas streams
  description, 236
  H_2S/benzene, 236
  incineratability, hazardous air
    pollutants, 237
  John Zink Test facility, 237
  methane and carbon monoxide, 237
  physical properties, 236
  simulation, hazardous vent
    stream, 237
Enthalpy formation, 494
Equipment testing, TO
  burners
    configurations, 232
    cooled wall chambers, 231

cutaway view, fuel and air staging, 231
description, 231
geometries and pressure drops, 231
instrumentation, 232
modifications, 232
NOx emissions, 231
quantities, excess air, 231
chambers, 232
test equipment sizing, see Test equipment sizing
waste gas injection methods and configurations, 232
Excess air
alarm and trip settings, 309
combustion airflow measurement, 304
defined, 303
dry vs. wet oxygen measurement, 308
excess oxygen sampling, see Excess oxygen sampling
excess oxygen vs. CO measurement, 308–309
heater operation, 304
indication, oxygen content, 304
oxygen analyzers, see Oxygen analyzers
tunable diode lasers, 308
Excess oxygen sampling
adjustments, 305–306
air leakage around process tube, 304, 305
defined, "afterburning", 306
heater efficiency calculation, 306
large balanced draft cabin heater, 304, 305
multiple locations, 304, 305
open inspection port, tramp air into heater, 305, 306
operating cost, higher excess oxygen levels, 306, 307
oxygen probe extending into heater, 306
Exothermic waste streams
description, 237
properties, 237
simulator, 237
Explosion, fire and flame, 4
Explosions
combustibles testing, 19
description, 19
furnaces, 21–22
plans, 19–20
stacks, 19, 21
tanks and piping, 20–21
External inspection, heater
burner block valves, 328, 329
burner condition, 329

burner damper position, 329
heater shell/casing condition, 329
pressure gauges, 328
stack damper, 328

F

Factory mutual (FM), 27
Failure modes, 50
Fan arrangements, blowers
centrifugal fans AMCA standard, 76
location and maintenance, bearings, 76–77
Fan control, blowers
centrifugal fan, inlet and outlet dampers, 83
description, 83
inlet/outlet damper effects, 83
speed change effects, 83–84
and speed motors, 84
variable and controlled pitch change effects, 84
variable pitch blades, vane axial fan, 84
FFG, see Flame front generator (FFG)
Field testing, flares, 205
Field test procedure, 338
Firebricks, see Brick refractory products
Fires
heat damage
adequate instrumentation, 17, 18
automatic shutdown equipment and alarms, 19
combustible heat-transfer fluid, 18
flame impingement, tubes, 18
flow recorder, 18
furnace operators, 16–17
furnace wall temperatures, 17
heater tubes had carbon buildup, 18
natural gas-fired heater, 18
piping system, 18
process liquid flow, 18
safety instrumentation, 19
trapped steam, freeze and pipe failure, 17
tube rupture, fired heater, 16, 17
smoke generation
CO detector, 19, 20
description, 19
Flame/flue gas patterns
burner spacing/flame interaction, 352–353
CO formation, 353–355
flame impingement, tubes, see Flame impingement, tubes
flame lift-off, see Flame lift-off
flame stability, 353–355
flashback
corrective action, 347–348
effect on operation, 347
problem indications, 346–347

high stack temperature, see High stack temperature
irregular/nonuniform flames, see Irregular/nonuniform flames
leaning flames, see Leaning flames
long flames, see Long flames
low-temperature operation, 353–355
overheating, convection section, 357
pulsating flames, 344–345
Flame front generator (FFG)
flare troubleshooting, 392
flar maintenance, 389
pilot lighting procedure, 386–387
visual pilot verification, 387
Flame holders/stabilizers, 106
Flame impingement, tubes
burner adjustment, 351
burner firing ports, 350–351
burner tile damage, 351
burning carbon particles, 348
deficiency, combustion air, 350
description, 348
firebox dimensions, 351
flue gas recirculation, 351
hangers glowing red, 349
hot spots, 349
overfiring, 350
overheating stages, 349–350
Reed walls, burners and tubes, 351
short of air, 349
shut down, heater, 351
TMTs, 349
tube scale, 349
Flame instability
burner damaged, flashback, 23
burner design features, 24
Coen iScan® flame detector, 24
combustion chemistry, 22
Halon fire extinguishers, 22
"huffing", 24
laminar flame speeds, 23
premix burner lifting off, 23
prevention, design features, 23
stronger and higher temperature materials, 23
Flame lift-off
alignment and positioning, gas burner tip, 346
description, 345
effect, operation, 346
fuel gas firing ports and ignition ports, 346
speed and air/fuel delivery speed, 346
steam atomization pressure, 346
woofing/breathing, 345–346
Flame rod testing
current (µA), 337, 338
negative lead/test circuit, 337
positive lead/test circuit, 337
relay panel, 336

Index   519

relay panel/"Flame ON" light energization, 337
ST-1SE-FR pilot, 336
test circuit, 336
test probes connection, flame voltage, 337, 338
voltage (VDC), 337, 338
voltage (VAC)/no flame/power on, 336, 337
warning, 336
Flame scanners, 57
Flame stabilizer
  burner tile ledge, 278
  damaged diffuser cone, 279
  deflector, 278, 279
  diffuser cone, 278–280
  swirler, 278, 279
Flammable liquid, 4
Flare maintenance
  FFG, see Flame front generator (FFG)
  flare tips
    air-assisted flare tips, shutdown items, 389
    periodic on-line items, air-assisted flare tips, 389–390
    shutdown items, 389
    staged flare systems, shutdown items, 389
    steam-assisted flare tips, shutdown items, 389
  liquid seals and knockout drums, 390
  molecular seals, 390
  pilots
    periodic on-line items, 388
    shutdown items, 388
  purge system
    periodic on-line items, 388
    shutdown items, 388
Flare operations
  admission point, purge gas, 379
  alarm, purge failure, 379
  minimum purge gas rate, 379
  pilot lighting procedure, see Pilot lighting procedure, flare operations
  pre-start-up, see Pre-start-up, flare operations
  purging, 378
  staged post-purge concept, 379
  start-up and shutdown, see Start-up and shutdown, flare operations
  suitable purge gases, 378–379
  troubleshooting, 390–392
Flare pilot test facility, 215–216
Flare testing
  air-assisted, 217–219
  API 537, 203, 204
  classification, 203–204
  cold flow visualization, 217
  field, 205
  ground flare burner interactions, 217
  high-wind and high-rain conditions, 204–205
  hydrostatic, 216
  large-scale, see Large-scale flare test facility
  pilot test facility, see Flare pilot test facility
  steam-assisted, 219
  study variables, performance, 204
  unassisted, 217
  water-assisted, 219–221
Flare troubleshooting
  electronic ignition/near pilot tip, 393
  FFG, see Flame front generator (FFG)
  operations, see Flare operations
  pilots, 392
  pilot verification system, 393
Flashback, burner troubleshooting
  corrective action, 347–348
  description, 346
  flame burning inside gas tip, 347, 348
  gas tips glowing, 347, 348
  inside burner mixer, 346, 347
  mixer/venturi, 347
  upper and lower explosive limits, 346, 347
Flash point (FP), 4
Flat flame burner
  freestanding, 169
  NOx emission, 168
  typical staged-fuel, 168
  wall fired, 169
Flexmesh, 147
Flow control system
  control room and PHA, 209
  custom-designed hardware and software systems, 209
  flare test control screen, 209–210
  flare test facility and fuels rates, 207
  impact-resistant glass, 209
  manual and automatic operation graph, 210–211
  recording data and video images, 209
  safety devices and instrumentation, 209
  sophisticated computer control algorithm, 210
Flow measurements
  magnetic flow meter, 181–182
  orifice meter, see Orifice meter
  PD meter, see Positive displacement (PD) meter
  pilot tube, see Pilot tube
  thermal mass meter, 182–183
  TO
    combustion and quench air sources, 241
    description, 241
    orifice plate flowmeters, 241
    selection, considerations, 242
    thermal mass flowmeter, 241
    types, 241
  turbine flow meter, 180
  ultrasonic flow meter, 182
  venturi meter, see Venturi meter
  vortex flow meter, see Vortex flow meter
Flow meters
  Coriolis, 61
  description, 60
  magnetic, 61
  Orifice, 61
  positive displacement, 61
  turbine, 61
  ultrasonic, 61
  vortex shedder, 60–61
Flow switches, 56–57
Flue gas analysis
  CO and emission, 196
  electrochemical sensors and paramagnetic cells, 196
  NDIR technique, 196
  NOx, 196–197
Flue gas temperatures
  BWT, 314
  description, 314
  stack gas, 314
  standard thermocouples, 314
  suction pyrometer, 314
FM, see Factory mutual (FM)
Footed wavy V-anchor
  description, 144
  extensive use, 144
  refractory lining, steel surface, 145
  welding, 144
Fourier transform infrared (FTIR) spectroscopy
  definition, 186
  DRE, 186
  interferometry technique, 186
  organic compounds at concentrations, ppm range, 186
  simple FTIR spectrometer layout, 186
FP, see Flash point (FP)
FTIR, see Fourier transform infrared (FTIR) spectroscopy
Fuel delivery system
  gas manifold, 106
  gas/pilot tips, 106, 107
  gas risers, 106
  oil guns, 106
Fuel flow rate vs. control signal
  gas valve data, 65

predictable and repeatable
  calculations, 64
Fuel gas problems
  burner capacity curve, 357, 358
  description, 357
  fired duty change, 358–359
  flame lift-off, high fuel pressure, 359–360
  flame quality/instability, 359
  fuel composition, 357–358
  fuel gas tip problems, 357
  heater out of air/flame impingement, 360–361
  incorrect fuel flow measurement, 359
  incorrect fuel pressure, 359
  wrong gas tips, 358
Fuel measurements, burner/heater operations
  fired duty, 311–312
  fuel flow
    board operator, 309
    control valve, 309
    process outlet temperature, 309
    Wobbe index meter, 309
  fuel pressure
    alarm and trip points determination, 310
    defined, turndown, 309
    gauge, 309
    guidelines, 310
    maximum and minimum operating pressure, 310
    high-fuel and low-fuel pressure heater trip points, 311
    nuisance shutdowns, 311
  temperature, 312, 313
  typical burner fuel capacity curves, 311
Fuel metering
  description, 153–154
  gas
    "burner throat", 154
    calculations, mass flow, 154
    capacity curves, 154, 155
    heat release vs. pressure, 154
    typical premix metering orifice spud and air mixer assembly, 154
    typical raw gas burner tips, 154
  liquid
    description, 154
    EA, internal mix twin fluid atomizer, 154, 155
    MEA, internal mix twin fluid atomizer, 154, 155
    oil and atomization system, 154–155
    port mix twin fluid atomizer, PM, 154, 156

system of ports, 155
typical liquid fuel capacity curve, 155, 156
Fuel piping
  "blowdown", 259
  description, 255
  flexible metal hoses, 257
  gas tip fouling reduction, 259, 264
  oil firing units, 259
  and steam traced, 259, 264

G

Gases at atmospheric pressure
  air, 475
  carbon dioxide, 478
  nitrogen, 476
  oxygen, 477
  water vapor/steam, 479
Gas tip cleaning
  air registers and dampers, *see* Air registers and dampers
  burner tile
    catalyst buildup, 275, 277
    checking, 277, 278
    description, 273, 275
    honey/syrup, 275
    inspection, burner tiles, 275, 277
    large broken pieces, 275
    large crack, 275, 276
    mortar application, 275, 277
    multiple section, 277, 278
    radiant wall burner tile, 275, 276
    repaired cracks, 275, 277
    tile crumbling, 275, 276
    tile installation, 277, 278
    vanadium attack, 275, 276
  coke buildup, 265, 268
  corrective/preventive actions, 270–271
  description, 264–265
  flame stabilizer, *see* Flame stabilizer
  floor-mounted burners, 267, 269
  gas riser mounting flange, 267, 270
  gas riser/tip assembly, 268, 270
  gas tip drilling information, 264, 268
  maintenance recommended tools, 266–267, 269
  oil burner maintenance, *see* Oil burner maintenance
  overheated, 265, 268
  plugged, 265, 268
  premix gas burners, *see* Premix gas burners
  raw and premix, 264, 266, 267
  thread lubricant, 270
Gibbs formation, 494
Ground flare burner interactions
  array, 217
  fence type and design tools, 217

flame heights and cross lighting distances, 217, 218
single and dual burner test measurement, 217, 218

H

Handheld battery-operated digital manometers, 243
Hazards
  environmental, 24–25
  excessive temperature
    description, 7
    equipment, 8
    instrumentation, 8–9
    insulated temporary ductwork, 8
    metal mesh shielding personnel, hot exhaust stack, 8
    OSHA guidelines, 7–8
  explosions, *see* Explosions
  fires, *see* Fires
  flame instability, *see* Flame instability
  high pressure, 16, 17
  noise
    burners, 11
    combustion instability, 11
    combustion testing, 13
    convenience and comfort, 16
    cylindrical muffler, suction pyrometer, 14, 15
    defined, 10
    ear muffs designed, with hard hats, 16
    enclosed flares, 13, 14
    human hearing, 11
    industrial applications, 10
    large mufflers, two natural draft burners, 12, 13
    natural draft burner, air inlet damper, 11–12
    natural draft burner, no air inlet muffler, 11, 12
    radiant wall-fired natural draft burner, typical mufflers, 12
    reduction, pipe/tube, 11
    refractory linings, 13
    silencers, 11, 12
    sound pressure vs. frequency with and without muffler, 14, 15
    sound-proofed enclosure, 11, 16
    sound transmission mitigation, 12–13
    thermocouples, 14
    two radiant wall burners, large mufflers, 12, 13
    typical ear plugs and muffs, 15–16
    typical mufflers, natural draft burner, 14

Index

thermal radiation
  description, 9
  equipment damage and injure personnel, 9
  heat resistant suit, 10
  measurement, radiometer, 9, 10
  skin damage, 9
  source reduction, 10
  stainless steel fence shielding flow control equipment, 9
  from viewport, 9
  viewport with shutter, 10
Heater combustion control
  target draft level
    air supply and negative pressure, furnace, 322
    description, 321
    efficient furnace operation, 321–322
    firebox pressure, 322
  target excess air level
    balanced draft furnace adjustments, 322–324
    description, 322
    forced-/balanced-draft, 322
    heat and mass flow generation, 324
    natural-draft furnace adjustments, 322, 323
    typical values, gas burners, 322
Heater draft
  combustion air, 302
  definition, 301
  draft loss, 301
  FD and ID fan, 301
  profile, 301, 302
  sampling
    burners and draft measurement, 302
    damage/excessive fouling occurrence, 302
    inclined manometer, 303
    magnehelic draft gauge, 303
    negative pressure, 302
    and O_2 location, 302
    radiant firebox, 303
    transmitter, 303
  trip and alarm settings, 303, 304
Heater safety
  description, 317
  emergency procedures
    defined, 320
    mitigating actions, 321
    smothering steam, 321
    unburned combustibles collection, 321
  shutdown procedure, 320
  start-up procedures
    description, 317
    issues, 318–320

Heater turndown operation, 324
Heat exchangers
  description, 409
  installation, 410
  maintenance
    periodic on-line items, 411
    shutdown items, 410–411
  operations
    checklist, 432–433
    description, 431
    start-up instructions, 433
    temperature and pressure drop, 431–432
  troubleshooting, 433
Hexmesh, 146–147
High stack temperature
  afterburning, 355
  causes, 355
  flue gas bypassing, 356
  flue gas temperature measurement, 355
  fouling on convection tubes, 356
  high excess air, 356
  indications, problem, 355
  insufficient air/positive pressure at arch, 356
  process leak, convection section, 356
High-viscosity liquid fuels, 164
HMI, see Human-machine interface (HMI)
Hot corrosion, metal failures
  chlorination attack, 101
  definition, 101
  elements, salt water, 101
Human-machine interface (HMI), 211–212
Hydraulically bonded castables, 135
Hydrostatic testing, 216

I

Ideal gas properties
  air, 461–463
  CO_2, 471–474
  nitrogen (N_2), 464–467
  oxygen (O_2), 468–470
Ignition transformers, 57
Industrial insurance carriers
  description, 27
  IRI an FM, 27
Industrial risk insurers (IRI), 27
Input streams simulation, TO
  burner fuel, 234–235
  combustion air, 234
  pure components, 234
  quench medium, 234
  test facility metering skid, 234, 235
  waste, see Waste streams simulations
  waste gas properties, 234
In situ oxygen analyzer, 240

Inspection and testing, blowers, 90, 92
Instrumentation, TO testing
  chemical species analysis, 238–241
  description, 238
  durability and maintenance requirements, 238
  flow measurements, 241–242
  pressure measurement, 243–244
  temperature measurement, 242
Intergranular corrosion
  carbide precipitation, 99
  cracks formation, 99
  description, 99
  grain boundary, rich chromium content, 99
IRI, see Industrial risk insurers (IRI)
Irregular/nonuniform flames
  air plenum distribution problems, 344
  air-registers, 344
  arch brick, burner tile, 343
  effect, operation, 342–343
  eroded gas tip, 343
  indications, problem, 342, 343
  maintenance/installation personnel, 343–344
  orifices, 344
  plugged gas tip, 343

K

K-Bar anchors, 147
Keu combustor
  atomizing spray pattern, 166
  axial velocity within combustion chamber, 165, 167
  BAT, organic fine chemicals, 166
  centrifugal forces, 164–165
  defined, 164
  flame, 165, 167
  maximum pressure and temperature, 166
  with organ set, 164, 166
  schematic, 164, 165
  standard combustor normal-FD combustor short, 166, 167
Kiln and dryer off-gas, 152
Knockout drums and liquid seals
  flare maintenance
    periodic on-line items, 390
    shutdown items, 390
  pre-start-up, 383

L

Large-scale flare test facility
  air-assisted, 205, 206
  data acquisition system, 211–215
  description, 205
  design flaw, 206

flow control system, 207–211
pressure-assisted, Coanda principle, 206, 207
system description
    block diagram, 206, 208
    fuel processing system, 207, 208
    inert gases, 206
    LPG vaporizer and steam supply, 207
world-class, 205, 206
Laser Doppler anemometry (LDA), 186
Laser induced incandescence (LII), 186
LDA, see Laser Doppler anemometry (LDA)
Leaning flames
    burner gas tip problems, 341
    causes, 341
    combustion air problems, 341–342
    effect, operation, 341
    firebox and burner maintenance issues, 341
    incorrect burner arrangement/firebox design, 341, 342
    indications, problem, 340–341
LFL, see Lower flammability limit (LFL)
LHV, see Lower heating value (LHV)
Light burners
    fails to light off
        corrective action, 339
        effect, operation, 338
        indications, 338
    heater refractory work, 334
    installation and operating manual, 334
    pilot fails to ignite
        effect, operation, 334
        external flame rod, 335, 336
        field test procedure, 338
        flame rod testing, see Flame rod testing
        fuel gas flow through pilot, 334
        high-energy portable igniter, 334
        indications, 334
        portable torch, 334, 335
        recommendations and conclusions, 338
        shrouded flame rod, 335
        typical CFD model, duct system, 335
    start-up, 334
LII, see Laser induced incandescence (LII)
Liquid and air/steam atomizing guns
    cautions, 406–407
    installation, 407
    maintenance
        individual items inspection, 407–408
        periodic on-line items, 408
        shutdown items, 407
    operations, 428
    troubleshooting, 428
Liquid penetrant testing (PT)
    connection welds, flare, 125, 126
    dye penetrant indication example, 125, 126
    limitations and drawbacks, 126
    manufacturing flaws, 126
    water-/solvent-removable penetrant materials, 125
Liquid planar laser-induced fluorescence (LPLIF)
    cold spray studies, 188
    liquid mass distribution, 188
    spray images and scattered light, 188, 189
    typical multiline Ar-ion laser, 188
Liquid seals
    installation, 411
    maintenance
        description, 411
        periodic on-line items, 412
        shutdown items, 411–412
    operations, 433
    troubleshooting, 433
Lock out/tag out (LOTO)
    jobsite requirements, 419
    procedures, 419
    rotating equipment, 417
Logic sequencing test, 383
Long flames
    burner gas tip problems, 340
    burner operation outside design envelope, 340
    causes, 339–340
    combustion air problems, 340
    effect, operation, 339
    heavier fuels, 340
    incorrect burner arrangement/firebox design, 340
    indications, problem, 339
LOTO, see Lock out/tag out (LOTO)
Lower flammability limit (LFL), 4
Lower heating value (LHV), 199
Low-viscosity liquids, 164
LPLIF, see Liquid planar laser-induced fluorescence (LPLIF)

M

Magnetic flow meters
    advantages and disadvantages, 182
    description, 181
    principles, magnetic induction, 181
Magnetic particle testing (MT)
    AC yoke and DC prod, 127
    description, 127
    gas piping, AC yoke dry method, 127
    limitations and drawbacks, 127
    metallic particles, 127
Maintenance and troubleshooting, blowers, 92
Manometer, pressure management
    description, 173
    example, 174
    U-tube and inclined, 173, 174
Master fuel trip (MFT)
    BMS logics, 53
    circuit, 53
    RBS, 54
    "self-sealing" relay, 54
    shutoff valves, 53
Mechanical stops, 59
Metal dusting
    definition, 103
    inlet tube, heat exchanger unit, 103
    iron and nickel-based alloys, 103
Metal failures
    breakaway corrosion, see Breakaway corrosion
    carburization corrosion, see Carburization corrosion
    chlorination attack, 102
    cryogenic service, see Cryogenic service, metal failures
    hot corrosion, 101
    intergranular corrosion, see Intergranular corrosion
    stress corrosion cracking, 100
    sulfidation attack, 101
Metallographic replication
    austenitic SS alloys, 131
    description, 130
    equipment conditions, 130
    flare tip, fabrication, 131
    limitations, 131
    sensitization, welding process, 131
    service providers, 131
Metallurgy
    carbon steel, 96–97
    destruction factors, 96
    equipment failure, 96
    metal failure, combustion industry, see Metal failures
    NDT, see Nondestructive testing (NDT)
    process burners, see Process burners
    process flares, see Process flares
    stainless steel, 97–99
    welding, see Welding process
MFT, see Master fuel trip (MFT)
MIE, see Minimum ignition energies (MIEs)
Minimum ignition energies (MIEs), 4, 36
Molecular seal
    flare maintenance
        periodic on-line items, 390
        shutdown items, 390
    pre-start-up, 384

Moleseals and flare riser, 108
Monolithic refractory products
    chemically (phos) bonded castables, 135
    chemically bonded plastics, 136
    description, 134
    forms, 134
    hydraulically bonded castables, 135
Motors and drives, blowers
    area classification, 85
    centrifugal fans, 85
    combustion applications, 85
    electrical induction, 85
    service factor, 85
    steam turbines, 85
Mounting, burner
    burner stand and forklift, 250, 252
    cables and hoist, 253, 255
    crane-lifting burner, 250, 252
    description, 250
    down-fired burners, 254, 258–259
    and floor joint, 253, 257
    gas tips, 254, 257
    heater floor, 253, 254
    heater steel attachments, 253
    horizontal burners, 254, 258
    lifting device, 253
    lifting techniques, 253, 254
    open area, tip, 254, 257
    P-box-type burner, 253, 256
    ultralow-NOx radiant, 254, 260

N

National Electrical Code (NEC) standard, 51
National Fire Protection Association (NFPA)
    API, 27
    codes and standards, 26
    CSA International, 27
    European CEN, 27
    industrial insurance carriers, *see* Industrial insurance carriers
    NFPA 30, 26
    NFPA 54, 26
    NFPA 58, 26
    NFPA 70, 26
    NFPA 86
        furnace heating system, 26
        location and construction, 26
        safety equipment and application, 26
    NFPA 497, 26
    NFPA 921, 26
    testing laboratories, 27
NFPA, *see* National Fire Protection Association (NFPA)
Nitrogen oxide (NO_2) analyzer, 62
Noise, data acquisition system
    average sound profile, 215
    data logging system and DSP, 215
    flare test, sound measurement system, 214
    post-processing and control platform, 215
    pre-amplification and anti-aliasing filtration, 215
    TEDS and sound pressure record, 215
Nondestructive testing (NDT)
    description, 125
    liquid PT, *see* Liquid penetrant testing (PT)
    metallographic replication, 130–131
    methods, 125
    MT, *see* Magnetic particle testing (MT)
    PMI/AV, 130
    procedural requirements, 125
    RT, *see* Radiographic testing (RT)
    UT, *see* Ultrasonic testing (UT)
Nondispersive infrared (NDIR) technique, 196

O

Oil burner maintenance
    atomizer, 290
    disassembly, 289
    graphite/silicone-type lubricant, 286
    integral plenum box, 283, 285
    oil and gas LoNOx burner, 281, 282
    oil gun, *see* Oil gun
    oil gun insert removal, 289
    oil spud, 290
    oil tip, 289–290
    regen oil tile, 283, 284
    regen tile, 282, 283
    secondary and primary tile, 282
    and steam spray, 283
    tip/atomizer, 290
    typical rotary-type air registers, 283, 284
    vane-type air register, 283, 285
    Z-56 quick change oil gun, 289
Oil combustion
    catalyst buildup on oil tile, 364
    contaminants, 363–364
    defined, "atomization", 362–363
    steam atomization, 363
    sulfuric acid, 364
    "thermal NOx", 364
    typical oil gun components, 363
    vanadium attack on burner tile, 364
    viscosity and temperature, 363
Oil firing problems
    cleaning flames, 361
    coke mound on burner, 362, 363
    combination burner with Regen (oil) tile, 361, 362
    DEEPstar oil burner, 361, 362
    effect, operation, 361
    flame impingement, 362, 363
    fouled oil guns, 361, 362
    oil combustion, *see* Oil combustion
    oil gun center, Regen (oil) tile, 361, 362
    smoke emission from stack, 366–368
    spillage on burners, 361, 362
    steam, *see* Steam system
    system, *see* Oil system
Oil gun
    assembly, 290
    copper gaskets, sealing, 286
    EA-/SA-type oil tip, 286
    hand tools, 288
    HERO oil tip, 286–288
    high-temperature anti-seize, 286, 289
    insert removal, 289
    MEA oil gun parts and tips, 286, 287
    and oil body receiver, 286
    oil tip, 864, 286, 287
    pilots, *see* Pilots
    Z-56 quick change, 289
Oil/liquid firing
    definition, 164
    droplets, 164
    high-viscosity liquid fuels, 164, 165
    internal mix twin fluid atomizers, 164
    low-viscosity liquids, 164
    port mix twin fluid atomizers, 164
Oil system
    heating and storage, 364
    heat tracing and insulation, 365, 366
    heavy, 364, 365
    pressure gauges, 365
    recirculation system, 364–365
Operational costs, blowers
    blower efficiency, 90
    control options relative to design rate, 89
    description, 89
    inlet damper fan curve with HP, 90, 91
    outlet damper fan curve with HP, 89, 90
    potential controls cost savings, 89
    speed control fan curve with HP, 90, 91
Operator training and safety documentation
    AIChE guidance, 38, 41
    communication and transfer, knowledge and skills, 40
    design information, 38–39
    employees, 40
    feedback flowchart, 38, 39

PHA reports, *see* Process hazard analysis (PHA)
process knowledge and program, 38, 39
sessions, topics, 41
SOPs, *see* Standard operating procedures (SOPs)
Orifice flow meters, 61
Orifice meter
  accuracy, flow measurements, 179
  common pressure-tap arrangements, 177, 178
  description, 177
  mass flow rate calculation, 178–179
  plate, 177
  pressure taps, 177
  static pressure drop, 177
  upstream flow conditioners, 178
Oxygen (O_2) analyzer, 62
Oxygen analyzers
  calibration gas, 307–308
  close-coupled extractive analyzer, 306, 307
  conventional oxygen measurement, 306
  flue gas analyzer data panel, 307, 308
  installation, flame arrestors, 306–307
Oxygen-enriched streams, 152

P

Parallel positioning, combustion characterizer calculations
    airflow rate vs. air valve position, 65
    air valve characterizer, 65
    definition, 64
    fuel flow rate vs. control signal, 64–65
  description, 62
  electronic linkage
    convention aids system analysis, 63
    predictable and repeatable valve positions, 63
    signal inversion, 63
    TIC output, 64
    variation, parallel positioning, 63, 64
  mechanical linkage
    analytical feedback, 63
    description, 62
    TIC, 63
Particle image velocimetry (PIV), 186
PD, *see* Positive displacement (PD) meter
PDPA, *see* Phase Doppler particulate anemometer (PDPA)
PHA, *see* Process hazard analysis (PHA)

Phase Doppler particulate anemometer (PDPA)
  Doppler signal, 187
  drawbacks, optical instrument, 187
  droplet size and velocity measurements, 186–187
  limitations and measurement uncertainties, 187
  mass accumulation measurements, different mass ratios, 188
  oil gun and spray chamber, droplet size measurements, 187, 188
  spray combustion applications, 186
  typical PDPA droplet size measurements, 187, 188
Physical properties, materials
  areas and circumferences, circles and drill sizes, 445–448
  commercial copper tubing, 455–456
  flange size data, 460
  pipe, 449–454
  standard grades, bolts, 457–459
Pilot lighting procedure, flare operations
  FFG, *see* Flame front generator (FFG)
  pilot ignition systems, 387
Pilots
  ceramic insulator and ignition rod, 291, 295
  checking pilot orifice, 291, 292
  electrical connections, 294, 297–298
  end view, internal high-energy exciter, 291, 296
  exposed flame rod, 291, 294
  external igniter and ceramic insulators, 291, 295
  flame rod, 291, 294
  green light illumination, 291, 294
  internal high-energy igniter system, 291, 295
  new ST-1SE-FR pilot, 291, 295
  new-style pilot, 291, 295
  pilot ignition rod assembly, 291, 293
  pilot parts, 291, 292
  pilot shield, 290, 291
  ST-1SE-FR pilot, 294, 296, 297
  ST-1S high-stability burner pilot, 290, 291
  typical replay panel, 291, 293
Pilot tube
  averaging, 185
  description, 184
  inside a large duct, 185
  locations, traverse in round/rectangular duct, 184
  total volume and mass flow rate measurement, 184
PIV, *see* Particle image velocimetry (PIV)

PLC, *see* Programmable logic controller (PLC)
Position switches, 56
Positive displacement flow meters, 61
Positive displacement (PD) meter
  description, 183
  flow rate meters, 183–184
  mechanical wear, 183
  types, 183
Positive material identification/alloy verification (PMI/AV)
  description, 130
  handheld analyzer, process burner riser pipe, 130
  limitations and drawbacks, 130
Preheated atmospheric air, 152
Preinstallation work
  existing heater, 247–248
  new heater, 247
  receiving, handling and storage, 247
  safety, 248
Premix and partial premix gas
  air control and volumes of flame, 162–163
  defined, 162
  fuel metering, 163
  momentum conservation, 162
  "rate of reaction" and "flame speed", 163
Premix gas burners
  damaged, flashback, 273
  description, 271
  dirty mixer/primary air door, 273
  HEVD burner, muffer, 273–274
  HEVD spider, 273, 274
  JZV premix gas tip, 272
  LPM radiant wall, 272
  plugged QD orifice, 271, 272
  primary air door assembly, 271
  QD orifice spud, 271, 272
  single-port orifice spud, 271, 272
  spider, oxidation, 273
  worn gasket and dirty insulation, 273, 275
Pressure measurement
  Bourdon tube pressure gauge, 243
  diaphragm pressure transmitters, 243
  and flow, 243
  handheld battery-operated digital manometers, 243
  inclined manometer, 243, 244
  pressure drops, 243
  U-tube manometer, 243, 244
Pressure switches, 56
Pressure transmitters
  handheld communicator, 60
  and pressure gage, 60

Index

Pre-start-up, flare operations
- air-assisted flares, 380–381
- all flares, 379–380
- knockout drums and liquid seals, 383
- molecular seal, 384
- staged flare systems, 381–383
- steam-assisted flares, 380

Process burners
- burner damage
  - pilot materials, 113
  - support system failure, coker reactor, 112, 114
- control valve damage, 113, 114
- corroded orifice spud
  - furnace flue gas, 113
  - in 10 years service, 114, 115
- damaged premixed burner tip
  - carburization/metal dusting, 117
  - corroded and new burner tip, 116
  - erosion-corrosion damage, PSA burner gas, 116
  - fuel composition, 117
  - at high-temperature oxidation, 114, 115
- material selection
  - air plenum, 105–106
  - components, 105
  - designs, heating, 105
  - flame holders/stabilizers, 106
  - fuel delivery system, *see* Fuel delivery system
- oil gun damage, 113, 114
- oxide scale formation, heater process tubes
  - ceramic tube coatings, 109, 110
  - on outer surface, 109
- ruptured process tubes
  - after long-term overheating, 109, 111
  - on burner top and heater radiant section, 109, 111
  - external corrosion, heater process tube, 110, 112
  - failure and ruptured tube, 110, 112
  - heater tubes, 109, 111
  - SMR furnace and cracked tube, 110, 112
  - two-phase flow regimes, 110, 113
  - vacuum heater, oxidation signs, 112, 113
- tube distortion
  - rupture strength, low-carbon steel and 300-series SS, 112, 114
  - sagging, high-temperature creep, 112, 113

Process flares
- air-assisted, *see* Air-assisted flares
- flame retention segments
  - corrosion damage, 117
  - flare tip perimeter, 117
  - waste gas flow rates, 117
- material selection
  - corrosion damage, 108
  - flare tips, 108
  - moleseals and flare riser, 108
  - pilots, 108
- pilot, elevated flare
  - description, 118
  - external burning and windshield damage, 118, 119
  - natural gas treatment plant, 118
- pilot, enclosed flare
  - oxidizing flame, 118
  - scale and loss, sulfidation attack, 117, 118
- steam-assisted, *see* Steam-assisted flares

Process fluid parameters, 316–317
Process hazard analysis (PHA), 39, 209
Process tube temperature
- description, 315
- flow, effectiveness and temperature, hot gases, 316
- fouling due to flame impingement, 315
- infrared thermography, 315
- and tube hangers, 315
- tube scale due to flame impingement, 315
- tube-skin temperature, 315
- tube-skin thermocouples, 315
- "windows" in spectrum, 316

Programmable logic controller (PLC)
- advantages and disadvantages, 47
- critical input checking
  - electronic programmable logic devices, 52
  - power, 53
  - RMFT, 52
  - solid-state (triac) type, 53
  - switches and interlocks, 52
  - unsafe conditions, 52
- input and output modules, 47
- redundant/fault-tolerant, 47
- safety functions, 47
- watchdog timer
  - BMS logic system, 52
  - "strobe" output, 52

Psychrometric chart
- English units, 504
- SI units, 503

Pulsating flames
- air leakage, 344
- corrective action, 345
- flames short of air, 344, 345
- heater out of air, 344
- huffing, 344
- low-frequency noise, 344
- "woofing"/"breathing", 344

R

Radiant wall burners
- applications, 170
- cracking furnaces and hydrogen reformers, 169–170

Radiographic testing (RT)
- description, 128
- double-wall technique, 128
- flare component welds, iridium 192 radioactive source, 128
- gamma and x-ray radiations, 128
- limitations and drawbacks, 128
- recording media, 128
- single-wall technique, 128

Refinery fuel gas (RFG), 199
Refractory, combustion systems
- API-936, *see* API-936 installation
- brick products, *see* Brick refractory products
- chemical and physical properties
  - anchoring systems, *see* Anchoring systems, refractory
  - insulating castable backup lining, 142
  - properties, 141
  - relationship, 141
  - single-and multiple-component lining, 141
- description, 134
- destructive and extreme service conditions, 133
- example of everyday, 133, 134
- glue phase, 134, 135
- hard and soft, 134
- incinerators, 134
- materials, 134
- monolithic products, *see* Monolithic refractory products
- raw materials, 134, 135
- raw material type, 134
- RCRA, 133–134
- soft products, *see* Soft refractory products

Resistance temperature detectors (RTDs), 60
RFG, *see* Refinery fuel gas (RFG)
RTDs, *see* Resistance temperature detectors (RTDs)
Run indicators, 57

S

Safety
- accident mitigation, *see* Accident mitigation
- accident prevention, *see* Accident prevention
- advanced sensors/controls, 6
- catastrophic accidents, 6

checklist, damage/injury, 6
codes and standards
    definition, 25
    NFPA, see National Fire
        Protection Association (NFPA)
combustion tetrahedron, see
    Combustion tetrahedron
definitions, 4
documentation, 7
documentation and operator
    training, see Operator training
    and safety documentation
environmental permitting, 7
explosion hazard, 3
factors, accident, 2
fires and explosions danger, 2–3
furnace design modification, 3
hazards, see Hazards
HAZOP and What-If reviews, 6
human error, 3
industrial combustion, 2
industrial combustion process, 41
NFPA, 3
overpressured test heater, 2
principles, fire and explosion
    protection, 3
qualitative and quantitative
    methods, 6
Safety warnings, TO
    electrical hazards, 417
    elevated temperatures, 417
    fire and explosion hazards, 417
    precautions, burner and fuel system,
        416–417
    rotating and mechanical equipment
        description, 417
        LOTO, see Lock out/tag out
            (LOTO)
        personnel protection rule, 418
        "pinch points", 417
        procedures, safety, 417–418
S-Bar anchors, 147
Shaft seals, blowers, 89
Shock wave and spontaneous
    combustion, 4
Soft refractory products
    anchors penetration, 140
    blanket, 139
    ceramic anchoring components, 140
    ceramic installation, 140
    damage, thermal shock, 140
    description, 139
    self-locking washers, 139–140
    types, ceramic refractory, 139
Solenoid valves, 57
SOPs, see Standard operating
    procedures (SOPs)
Staged air low NOx burner, 368, 369
Staged flare systems
    pre-start-up, 381–383
    start-up and shutdown, 385–386
Staged fuel low NOx burner, 368, 369
Stainless steel
    chromium oxide scale
        defined, passive film, 97–98
        electron microscope image, SS
            surface, 98
    description, 97
    elements, 97
    process burner and flare industries
        metals, 98, 99
        300-series SS's, 98
    types, 98
    welding process
        burner and flare equipment
            materials, 124
        construction codes, 125
        $M_{23}C_6$ carbide formation, 124
        solidification cracking
            susceptibility, 124, 125
        solidification mode, FN, 124, 125
Standard operating procedures (SOPs)
    description, 39
    Phillips 66 Incident in Pasadena, 40
Start-up and shutdown, flare
    operations
    air-assisted flares, 385
    staged flare systems, 385–386
    steam-assisted flares, 384–385
    unassisted flares, 384
Steam-assisted flares
    description, 119
    deterioration, lower steam eductor
        tube, 119, 121
    effectiveness, smoke suppression,
        219, 222
    over-steaming conditions, 219, 222
    pre-start-up, 380
    start-up and shutdown, 384–385
    steam leaking, 119, 121
    steam spider deterioration, 119, 122
Steam system
    description, 365
    insulation, 365–366
    pressure indication and control,
        366, 367
    steamtraps checking, 366
    superheated steam, 366
Steel fiber reinforcing
    chopped fiber, 146
    cold-drawn fiber, 146
    examples, 145, 146
    melt-extracted fiber, 146
Stress corrosion cracking, 100
Sulfidation attack, metal failures, 101
System shutdown
    local reset requirement, 52
    unsatisfactory parameter
        fail-safe input to PLC, 51
        local reset requirement, 52
        MFT circuit, 52
        relay fails, 51, 52
        string, 52

T

Tacko anchors, 147
TEDS, see Transducer electronic data
    sheet (TEDS)
TEG, see Turbine exhaust gas (TEG)
Temperature measurement, TO
    ceramic-sheathed thermocouple, 242
    combustion applications, 242
    handheld radiometer, 242
    suction pyrometers, 242
Temperature switches, 56
Test equipment sizing
    CFD modeling, 234
    combustion processes, 232
    distance and/or vessel volume, 233
    fabrication limitations, 234
    injection geometries, 233
    matching pressure drops, 233
    minor pollutant emissions, 233
    NOx emissions and destruction
        efficiencies, 233
    velocity matching, 233
    waste injection nozzles, 233
Thermal mass meter
    description, 182, 183
    downstream vs. upstream, 182–183
    energy balance type, 182, 183
Thermal oxidizer (TO)
    absorbers and scrubbers, see
        Absorbers and scrubbers
    boilers, see Boilers
    burner/pilot, see Burner/pilot
    control items
        installation, 405–406
        maintenance, 406
    description, 396
    heat exchangers, 409–411
    installation, 399–400
    instructions, installation, 396
    liquid, air/steam atomizing gun,
        see Liquid and air/steam
        atomizing guns
    liquid seals, 411–412
    maintenance
        periodic on-line items, 401
        shutdown items, 400, 401
    operations and troubleshooting,
        see TO operations and
        troubleshooting
    testing, see TO testing
    water weir, spray quench contactor,
        quench tank, 408–409
Thermal properties
    alloys, 506–508

Index

building materials and insulations, 509
metallic elements, 505
Thermal radiation
  isoflux lines, 213–214
  measurement system and small-scale experiments, 213
  thermogram, flare flame, 212–213
  triangulation and large-scale test facility, 213
Thermocouples
  K-type, 59
  seebeck effect, 59
  and thermowell, 59
Thermodynamic properties, steam
  pressure table, 482–483, 489–490
  superheated vapor table, 484–486, 491–493
  temperature table, 480–481, 487–488
Three-way (3-way) solenoid valve, 59
Tile installation
  air-setting high-temperature mortar, 249, 250
  burner tile sections, 248, 249
  ceramic blanket/gasket material, 249
  description, 248
  dry fit, 248, 249
  gas tips, 249, 250
  groove-clearing bolt heads, 249, 251
  handle burner, 250, 251
  plenum, 249, 251
  temperatures, 248
Time-of-flight ultrasonic flow meter, see Ultrasonic flow meters
Tined anchors, 145
TMTs, see Tube metal temperatures (TMTs)
TO, see Thermal oxidizer (TO)
TO operations and troubleshooting
  absorbers and scrubbers, see Absorbers and scrubbers
  boiler, see Boilers
  brick and blanket refractories, 421
  Burner/pilot, see Burner/pilot
  control items
    operations, 428
    troubleshooting, 428
  customer's specifications, 422
  description, 416
  heat exchangers, see Heat exchangers
  installation and maintenance, 416
  liquid and air/steam atomizing guns, 428
  liquid seals, 433
  operating temperature, 420
  refractory cure-out, 421
  refractory-lined vessel, 420
  safety warnings, see Safety warnings, TO
  training, 418
  troubleshooting, 422
  water weir, spray quench contactor, quench tank, 428–431
TO testing
  applications, 229
  commercial system problems, 229
  companies and regulatory agencies, 227
  equipment and facility design objectives
    configuration, horizontal, 229
    design considerations, 230
    John Zink Test facility, 230
    parameters, configurations, 230
    provisions, 229–230
    stack design, 230
    test facility control room, 229
    waste streams, 229
  equipment testing, see Equipment testing, TO
  facilities, 227, 228
  input streams, see Input streams simulation, TO
  instrumentation, see Instrumentation, TO testing
  test data accuracy
    commercial systems, 230
    fuel gas flowmeters, 230
    mass and energy balance calculations, 230
    measurement errors, 231
    steady-state conditions, 230
  time constraints and distractions, 229
Transducer electronic data sheet (TEDS), 215
Troubleshooting
  burner capacity curve, 420
  CFD, 420
  description, 419
  "Flame On" light, 419
  flowmeter, 420
  high-intensity spin-type burners, 420
  scanner tube, 419
  sparks, 419
Tube metal temperatures (TMTs), 349, 356
Tunable diode lasers, 308
Turbine exhaust gas (TEG), 152
Turbine flow meters, 61, 180

U

UFL, see Upper flammability limit (UFL)
UHCs, see Unburned hydrocarbons (UHCs)
UL, see Underwriters laboratory (UL)
Ultrasonic flow meters, 61, 182
Ultrasonic testing (UT)
  angle-beam apparatus, 129
  description, 129
  limitations and drawbacks, 129
  sound amplitude, 129
  straight and angle beam methods, 129
  volumetric examination, 129
Unassisted flares
  Coanda effect, 217
  full-scale testing, 217
  multiple Indair® flare test, 217, 219
  shutdown, 384
  start-up and operation, 384
Unburned hydrocarbons (UHCs), 198, 371
Underwriters laboratory (UL), 27
Units and conversions
  atomic weights, common elements, 439
  and molecular properties, 437–438
  periodic table, elements, 440
  prefixes, 437
  tables, 441–444
Upper flammability limit (UFL), 4
U.S. Occupational Safety and Health Administration (OSHA) guidelines, 7–8
U-tube manometer, 243, 244

V

V-Anchors
  description, 142, 144, 145
  double-hooked, 145
  footed wavy V-anchor, see Footed wavy V-anchor
  tined, 145
Vapor pressure, 4
Velocity thermocouples
  combustion temperature measurement, 60
  suction pyrometers, 59
  and thermowell, 59
Venturi meter
  cutaway view, 179, 180
  mass flow rate, 179
  schematic, 179, 180
Vibration and installation, blowers
  fan bearing vibration limits, 87, 88
  fan foundation, 87
  fan vibration diagnostic clues, 87, 88
  inlet and outlet expansion joints, 87, 88
Visual inspection, heater
  air leaks, see Air leaks
  burner tile and diffuser condition, 326–327
  description, 324–325

flame pattern and stability
  API RP 560 cites, 325
  burner shapes, 325
  description, 325
  uneven patterns, 326
  uniform and refractory color, 325
process tubes, 326
protective gear, 325
refractory and tube support color, 326
sight doors, 325
VOCs, see Volatile organic compounds (VOCs)
Volatile organic compounds (VOCs), 371
Vortex flow meter
  design, 181
  NASA satellite image, clouds off Chilean coast, 180
  shedding frequency, bluff body, 181
  Strouhal number, 181
Vortex shedder flow meter, 60–61

W

Waste gas injection methods and configurations, 232
Waste streams simulations
  aqueous wastes, 238
  autoignition temperature, 236
  bulk properties, 235
  combustion properties, 236
  components, 238
  description, 235
  endothermic, see Endothermic waste gas streams
  exothermic, see Exothermic waste streams
  flame speeds and flammability limits, 236
  oxygen concentration, 235–236
Water-assisted flare
  full-scale testing, water injection, 219, 223
  noise reduction, water injection, 219, 224
  offshore oil production platform, 219, 223
  radiation reduction, water injection, 219, 224
Water weir, spray quench contactor and quench tank
  installation, 408
  maintenance
    periodic on-line item, 409
    shutdown items, 408–409
  operations
    description, 428–429
    normal operation, 430
    pre-start-up checklist, 429
    start-up philosophy, 430–431
    troubleshooting, 431
Welding process
  carbon steel
    binary iron-carbon phase diagram, 120, 122
    martensite, 122
    maximum hardness vs. % carbon, weight, 122, 124
    medium-carbon alloy steel AISI 4340, 122, 123
    microstructural changes, 120, 122
    qualified welders, 122
    tensile and bend specimens, 124
    typical TTT diagram, medium-carbon steel, 122, 123
  stainless steel
    burner and flare equipment materials, 124
    construction codes, 125
    $M_{23}C_6$ carbide formation, 124
    solidification cracking susceptibility, 124, 125
    solidification mode, FN, 124, 125
  types, 119–120
WI, see Wobbe index (WI)
Wobbe index (WI), 309, 358

Y

Y-anchors, 146